Lecture Notes in Networks and Systems

Volume 873

The series "Lecture Notes in Networks and Systems" publishes the latest developments in Networks and Systems—quickly, informally and with high quality. Original research reported in proceedings and post-proceedings represents the core of LNNS.

Volumes published in LNNS embrace all aspects and subfields of, as well as new challenges in, Networks and Systems.

The series contains proceedings and edited volumes in systems and networks, spanning the areas of Cyber-Physical Systems, Autonomous Systems, Sensor Networks, Control Systems, Energy Systems, Automotive Systems, Biological Systems, Vehicular Networking and Connected Vehicles, Aerospace Systems, Automation, Manufacturing, Smart Grids, Nonlinear Systems, Power Systems, Robotics, Social Systems, Economic Systems and other. Of particular value to both the contributors and the readership are the short publication timeframe and the world-wide distribution and exposure which enable both a wide and rapid dissemination of research output.

The series covers the theory, applications, and perspectives on the state of the art and future developments relevant to systems and networks, decision making, control, complex processes and related areas, as embedded in the fields of interdisciplinary and applied sciences, engineering, computer science, physics, economics, social, and life sciences, as well as the paradigms and methodologies behind them.

Indexed by SCOPUS, INSPEC, WTI Frankfurt eG, zbMATH, SCImago.

All books published in the series are submitted for consideration in Web of Science.

For proposals from Asia please contact Aninda Bose (aninda.bose@springer.com).

Vinit Kumar Gunjan · Jacek M. Zurada
Editors

Proceedings of 4th International Conference on Recent Trends in Machine Learning, IoT, Smart Cities and Applications

ICMISC 2023

Springer

Editors
Vinit Kumar Gunjan
Department of Computer Science and Engineering
CMR Institute of Technology
Hyderabad, Telangana, India

Jacek M. Zurada
Department of Electrical and Computer Engineering
University of Louisville
Louisville, KY, USA

ISSN 2367-3370 ISSN 2367-3389 (electronic)
Lecture Notes in Networks and Systems
ISBN 978-981-99-9441-0 ISBN 978-981-99-9442-7 (eBook)
https://doi.org/10.1007/978-981-99-9442-7

This Springer imprint is published by the registered company Springer Nature Singapore Pte Ltd.
The registered company address is: 152 Beach Road, #21-01/04 Gateway East, Singapore 189721, Singapore

Paper in this product is recyclable.

Preface

It is with immense pleasure and pride that I welcome you all to the 4th International Conference on Recent Trends in Machine Learning, IoT, Smart Cities and Applications 2023. This gathering represents a significant milestone in our journey to explore, understand, and harness the transformative power of emerging technologies that are shaping the world around us.

In today's rapidly evolving digital landscape, the convergence of machine learning, Internet of things (IoT), and the development of smart cities stands as a testament to human innovation. This conference is a testament to our collective dedication to staying at the forefront of these advancements, to push the boundaries of what is possible, and to share our knowledge and insights with the global community.

Machine learning, with its ability to analyze vast datasets and make data-driven predictions, is revolutionizing industries across the board. From health care and finance to manufacturing and transportation, the impact of machine learning is felt everywhere. The insights we gain from this technology are not only enhancing efficiency but also paving the way for groundbreaking discoveries.

The Internet of things, on the other hand, has ushered in a new era of interconnectedness. Billions of devices, from household appliances to industrial machines, are communicating and collaborating in ways that were once unimaginable. IoT is driving innovation in areas such as smart homes, agriculture, and health care, making our lives more convenient, sustainable, and secure.

Smart cities are the embodiment of this progress on a grand scale. They leverage technology to optimize urban infrastructure, enhance services, and improve the quality of life for citizens. The smart city concept is not just about efficiency; it is about creating inclusive, sustainable, and vibrant communities for future generations.

This conference provides a platform for researchers, practitioners, and industry leaders to share their knowledge, experiences, and visions. It is a forum for interdisciplinary collaboration, where we can explore the intersections of these technologies and chart the course for their future development.

As we engage in spirited discussions, listen to expert presentations, and witness groundbreaking research, let us not forget the ultimate purpose of our endeavors: to make the world a better place. Machine learning, IoT, and smart cities are not just buzzwords; they are tools that can help us address some of the most pressing challenges of our time, from climate change to healthcare disparities.

I encourage all of you to embrace the spirit of curiosity, collaboration, and innovation throughout this conference. Let us forge new partnerships, build bridges between disciplines, and pave the way for a future where technology serves humanity in its truest sense.

I extend my deepest gratitude to the organizing committee, sponsors, volunteers, and participants for making this conference possible. Your dedication and hard work have brought us together today, and I am confident that the knowledge and insights we gain here will shape the future of machine learning, IoT, smart cities, and their applications.

Together, we are embarking on a journey of discovery and transformation. Welcome to the International Conference on Recent Trends in Machine Learning, IoT, Smart Cities and Applications. Let us make the most of this extraordinary opportunity to learn, collaborate, and innovate.

Thank you, and may our shared endeavors lead to a brighter and more connected future.

Hyderabad, India
Louisville, USA

Vinit Kumar Gunjan
Jacek M. Zurada

Contents

About the Editors

Vinit Kumar Gunjan is Associate Professor in the Department of Computer Science and Engineering and Dean of Academics at CMR Institute of Technology Hyderabad (affiliated with Jawaharlal Nehru Technological University, Hyderabad). An active researcher, he has published research papers at IEEE, Elsevier, and Springer conferences, authored several books, and edited volumes of Springer series. Awarded the prestigious Early Career Research Award in 2016 by the Science Engineering Research Board, Department of Science and Technology, Government of India He is a senior member of IEEE, Additional Treasurer, chair of the IEEE Computer Society chapter Hyderabad, and an active volunteer in the IEEE Hyderabad section. He volunteered in the capacities of Treasurer, Secretary, and Chairman of CIS, IEEE Young Professionals Affinity Group, and IEEE Computer Society. He was involved as organizer in many technical and non-technical workshops, seminars, and conferences of IEEE and Springer. He had the privilege of working with the top IEEE executives during his tenure, and the IEEE Hyderabad Section awarded him the Best IEEE Young Professional Award in 2017.

Jacek M. Zurada (SM'85, F'96, LF'14) is Professor of Electrical and Computer Engineering and Director of the Computational Intelligence Laboratory at the University of Louisville, Louisville Kentucky, USA, where he served as Department Chair and a Distinguished University Scholar. He was Professor at Princeton University, Northeastern University, Auburn University, the National University of Singapore, Nanyang Technological University in Singapore, Chinese University of Hong Kong, the University of Chile, Santiago, Toyohashi University of Technology, Japan, the University of Stellenbosch, South Africa, the University of Marie-Curie, Paris, France, and was Postdoctoral Fellow at the Swiss Federal Institute of Technology, Zurich, Switzerland.

Smart Glasses for Visually Impaired

Swati Shilaskar, Devika Warghane, and Chirag Bhandari

Abstract Smart glasses for visually impaired refer to the system which assists the people facing difficulties of vision loss, covering cases from partial sight to totally blind. The proposed system helps visually challenged people to recognize the text and detect the objects in the environment using computer vision. Instead of using sensors which only detect the obstacles, the proposed system operates on vision input which also identifies the object. The device recognizes any text present and reads the text for the user after identification of an object. This will serve as an identification and detecting system, allowing people to complete challenging tasks such as reading, writing, walking as well as many other day-to-day tasks. The cost-effectiveness of the device makes it more accessible on the common grounds. Blind people can live independently with the help of this integrated module.

Keywords Computer vision · Object detection · Smart glasses · Text recognition · Visually impaired · Wearable system

1 Introduction

According to the 2010 fact sheet, 285 million individuals worldwide have poor eyesight, of which 39 million are blind [1]. Many people have lived much of their lives unable to perform tasks without assistance, with 82% of the blind people being fifty years of age or older [2]. These persons with visual impairments find it extremely difficult to perceive and deal with their surroundings, especially new ones. There are

S. Shilaskar · D. Warghane (✉) · C. Bhandari
E&TC Department, VIT, Pune, India
e-mail: devika.warghane20@vit.edu

S. Shilaskar
e-mail: swati.shilaskar@vit.edu

C. Bhandari
e-mail: chirag.bhandari20@vit.edu

V. K. Gunjan and J. M. Zurada (eds.), *Proceedings of 4th International Conference on Recent Trends in Machine Learning, IoT, Smart Cities and Applications*, Lecture Notes in Networks and Systems 873, https://doi.org/10.1007/978-981-99-9442-7_1

a few tools or navigational systems available for people with disabilities who are visually challenged. Most of the individuals typically rely on using a white cane to navigate while continually swaying it for detecting obstacles. Braille is widely used for reading scripts, but limited content is available in this form. Due to a lack of assistive technologies and financial obstacles, many schools and workplaces are still unable to accommodate them. 90% of them continue to live in poverty as a result [3]. The proposed system offers a number of characteristics that enable secure, independent movement for those who are visually impaired, as well as a feature that enables constant communication. The processor-based system detects the object using MobileNet SSD and recognizes the text using pytesseract in the range of 62.2 degrees horizontally and 48.8 degrees vertically. Pyttsx is used to generate the output as audio.

2 Related Work

The technology that detects and avoids obstacles is under constant evolution by using various methods such as camera-based [4–6], ultrasonic sensors-based [7]. When using an ultrasonic sensor-based method, the obstacle's distance can be measured and compared to a predetermined distance threshold to determine whether or not to proceed. However, this approach has some drawbacks, including the inability to precisely determine which way to move ahead and the possibility of sensor interference when utilizing an ultrasonic sensor array or ultrasonic radar. Also, ultrasonic sensor cannot recognize the object in front of it so it's quite difficult for the user [8]. Despite being popular for mobile robots due to their high precision and resolution, laser scanner-based navigation systems are unsuited for wearable navigation systems due to their high cost, mass, and power needs [9]. Many techniques rely on different cameras, including the RGB-D, mono, and stereo cameras. Recently, more research is being carried out on text extraction and detection. There are already a few solutions that enable blind people to read text on a variety of items. For example, blind people can use a product barcode reader to obtain voice and Braille versions of the product information that is stored in a database [10]. However, it is quite challenging for a blind individual to determine the location of the barcode on the item. Text may be extracted from photos using optical character recognition [11–14]. These tools help the user in reading paperwork, mail, receipts, flyers, etc. It has trouble reading text on complicated backdrops or cylindrical items like prescription bottles, although it can see black letters on a white background easily. Additionally, most of the time a blind person must manually localize text sections. A method that outlines a combination of Gaussian-based background removal techniques to determine the region of interest was proposed by Yi and Tian [15]. The text localization method makes advantage of gradient data such as stroke orientation and edge pixel distribution. OCR software that is available off the shelf can recognize text characters. The automated document reader for blind people using a Raspberry Pi is described in paper [16]. This process makes use of image sensing tools and computer programming to identify the printed characters using optical character recognition (OCR) technology. The authors of study [17] provide many character extraction techniques. In this instance, smaller characters are retrieved by a mathematical procedure called morphology, and the sobel edge detection approach and methods that use RGB color data are used to extract bigger letters. ImageMagick software was used to convert pictures into scanned images to give input to Tesseract OCR [18]. The Google Tesseract and Pico (text to speech) were used to translate images into verbal communication which include many preprocessing stages [19]. Components such as camera, Light Dependent Resistor, Global Positioning System module, Infrared sensors were interfaced with Raspberry Pi and connected via GPIO: General Purpose Input Output pins [20]. The novel characteristics of our system are it detects the text in the object along with the object itself and conveys both in the form of acoustic output to the user. Additionally, the device is simple to use.

3 Methodology

The paper describes smart eyewear for people with vision impairments made with the help of an embedded system. Figure 1 offers a diagrammatic representation of the system. The camera is used for capturing videos in the environment. The attributes are gathered from the recorded video and utilized for either object detection or text recognition depending on the user's input. The object detected or the text recognized is converted in the form of audio and received by the user through headphones.

The embedded system used here is the Raspberry Pi model and includes a quad-core 1.5 GHz CPU and 4 GB RAM, in contrast to the original Pi's single-core 700 MHz CPU and 256 MB RAM. The Raspberry Pi camera includes a 5-megapixel

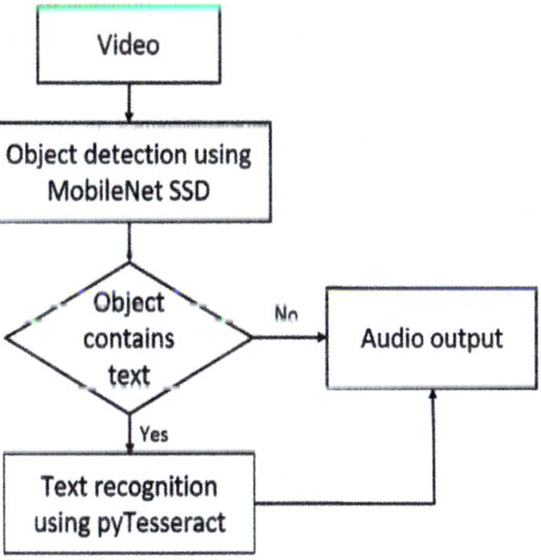

Fig. 1 Work flow of the system

sensor with a 2592 × 1944-pixel native resolution. It supports video resolutions of 640 × 480p60/90, 720p60, and 1080p30.

OpenCV, a significant image processing and computer vision open-source library. It currently contributes significantly to real-time functioning, which is essential in contemporary systems. OpenCv enables the quick process of videos and images to find objects or text in the file. To initiate text recognition, pytesseract is used. Pytesseract is an optical character recognition (OCR) program which is Python-based. It will identify and 'read' text contained in images as shown in Fig. 2. Google's Tesseract-OCR Engine is encased in Python-tesseract. This program can study any picture format that the Pillow and Leptonica imaging libraries support, including jpeg, png, gif, and bmp. As a result, it can be used as a stand-alone tesseract invocation script. To convert the object detected and the text recognized into audio output, pyttsx is used. A Python module named pyttsx supports text-to-speech engines. Platform-neutral text-to-speech software is available under the name pyttsx. The fact that this library can be used offline is a key benefit when utilizing it for text-to-speech conversion.

A Single-Shot Multibox Detection (SSD) network designed for object detection is the MobileNet model as shown in Fig. 3. The Caffe* framework is used in the implementation of this model. Two key ideas in SSD are (i) encoding information at different scales and (ii) collection of different filters that are specialized to predict boxes of a range of shapes and sizes. The VGG_16 network is used to get the feature map. SSD makes 8732 predictions for every single object using six layers. These

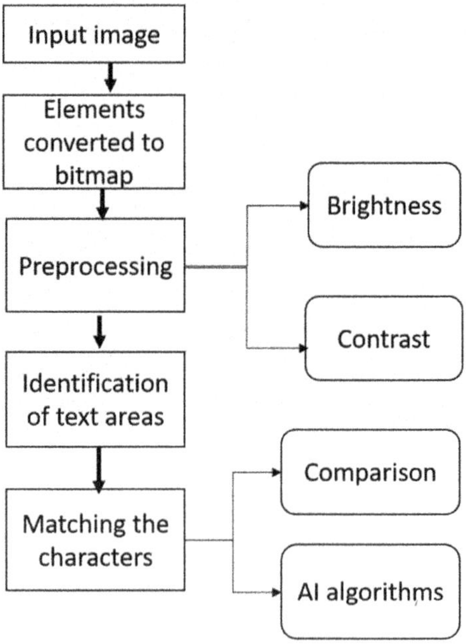

Fig. 2 Data flow for text recognition

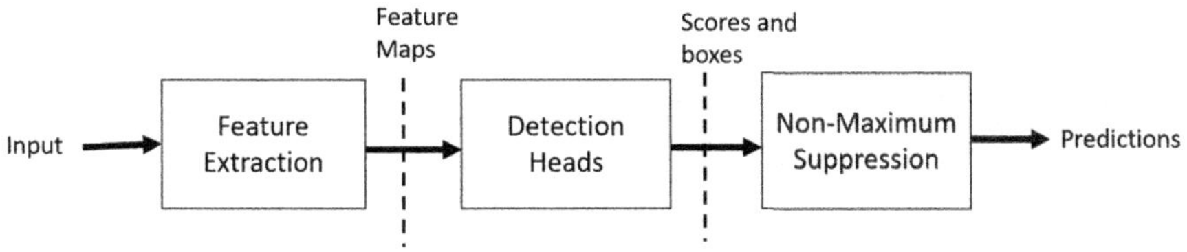

Fig. 3 Workflow for prediction of object using multibox detection

object detection algorithms' final phase, non-max suppression, is used to choose the object's ideal bounding box.

At the time of feature extraction, stack of convolutional networks generates feature maps to encode features at different scales. The feature maps are to encode nuanced semantic information that are localized in the different locations of the image which is used for making box proposals. The detection heads are used to generate boxes, box predictions, and class confidence of those boxes. The last stage (Non-Maximum suppression (NMS) is used to filter the predicted boxes to expel the repeated detections to get a detector with good performance. VGG is used to output a feature map, but the objective is to make detections on various scales. Many such layers are added and in each of these layers intermediate tensors are kept to generate a stack of feature maps with a variety of sizes and number of channels. For example, the 8 × 8 feature map as shown in Fig. 4a is being used to determine small-scale objects and the 4 × 4 feature map shown in Fig. 4b is often used to locate large-scale objects.

The loss function is utilized to assess how successfully the algorithm represents the dataset. The process of computing labels corresponds to identifying what is happening to every single of these default boxes inside the training image. The objective is to identify those default boxes that have large overlap with ground truth boxes. Overlap is the characterization of union measure, which is typically intersection over union measure of two boxes as in (1).

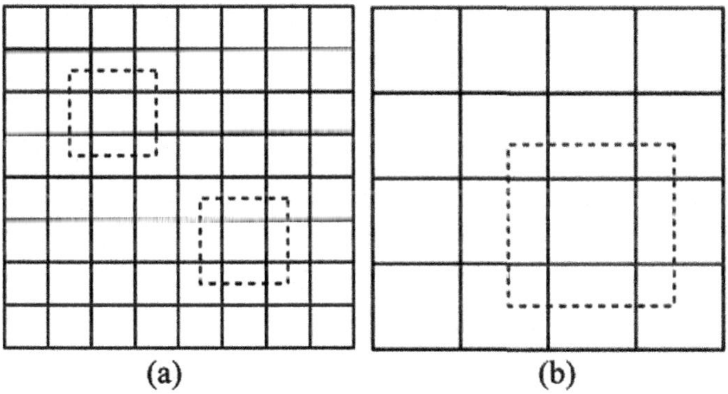

Fig. 4 **a** 8 × 8 Feature map to locate small-scale objects and **b** 4 × 4 feature maps to locate large-scale objects

$$0 < \frac{\text{Area of intersection}}{\text{Area of union}}(\text{threshold}) < 1 \tag{1}$$

$$\left.\begin{array}{l}\text{Pos} : (\hat{c}_{\text{d}}, \hat{\beta}_{\text{d}}) \sim (\text{class target, offsets target}) \text{ for default} \\ \text{Neg} : d \sim \text{dafault boxes containing no object}\end{array}\right\} \text{labels} \quad \text{box}$$

The threshold value for Pos (positive boxes) is greater than 0.45. This indicates that the ground truth boxes have an overlap of more than 45%. Pos (positive boxes) have additional data associated:

$\hat{c}_{\text{d}}$ class of the box
$\hat{\beta}_{\text{d}}$ offsets.

The localization loss in (2) is employed to educate the system to locate the target item by regressing a rectangular box, accounting for the discrepancy between the boundary box anticipated and the ground truth box. The loss is computed only over those boxes that have been labeled as positive boxes.

$$L_{\text{loc}} = \sum_{(\hat{c}_{\text{d}}, \hat{\beta}_{\text{d}}) \in \text{Pos}} L_{\text{loc}}\left(\beta_{\text{d}} - \hat{\beta}_{\text{d}}\right) \tag{2}$$

The inability to make a class prediction constitutes the confidence loss which is used to instruct the classifier head to identify the type of the target object. Confidence loss includes two different cases as shown in (3):

(i) Case of actual objects to be detected
(ii) Case of background.

First term deals with the classes to be predicted which includes the sum of all positive boxes, associated with ground truth class. Second term deals with the background class which includes negative boxes.

$$L_{\text{conf}} = -\sum_{(\hat{c}_{\text{d}} - \hat{\beta}_{\text{d}}) \in \text{Pos}} \log\left(\gamma_{\text{d}, \hat{c}_{\text{d}}}\right) - \sum_{\text{d} \in \text{Neg}} \log(\gamma_{\text{d,b}_{\text{g}}}) \tag{3}$$

where

γ softmax function, transforms a real number vector into a probability distribution of potential possibilities as depicted in Eq. (4).

$$\gamma_{d,c} = \frac{\exp(\sigma_{\text{c,d}})}{\sum_{\text{c}'} \exp(\sigma_{\text{c}',\text{d}})} \tag{4}$$

$\sigma_{\text{c,d}}$—confidence logits (round off the continuous values in the interval of [0,1])

$$L_{\text{final}} = \frac{1}{N}(L_{\text{conf}} + \alpha L_{\text{loc}}) \tag{5}$$

N number of successful matches.
α weight fraction of localization loss.

Combining these two losses together gives final loss as in (5) which is capable of forcing correct box classification to produce better localization. Only accurate forecasts are penalized by SSD. The forecasts resulting from the successful matches should be more accurate and unfavorable matches can be disregarded.

Object Detection: The MobileNet SSD (Single-Shot Detection) model is utilized for object detection. The input of images should be given in ground truth boxes. These boxes are made for better coverage of location and dnn: Deep Neural Networks are employed to categorize photos. The module supports deep learning interference on images and videos. One of the most often utilized large-scale tagged image collections that are open to the public is the coconames dataset used here. The dataset contains 1.5 million object instances and 80 object categories.

Text Recognition: The frames in the video are captured and read using OpenCv, while text recognition is done using pytesseract library. Pytesseract, image to string ('image name'), is used to convert to a string, and then, the result is saved in a variable. The function image_to_boxes () returns four values for each character while imagining boxes around each piece of text. The recognized text is then converted into speech using pyttsx.

```
Algorithm: Smart glasses for visually impaired
Input: Video captured using Raspberry Pi camera
Output: Object name and text in the form of audio
Frames(images) in video are captured, input given in ground
truth boxes
Passed through the dnn detection module
While(
  Image is read
   'detect 'method used to get confidence values, bounding boxes
and classId
  if (classId != 0)
    A rectangle displayed around the characters
    Text in terms of name
    Text in terms of confidence values
  if (threshold > 0.45)
    Object detected
Name of the object converted into speech
End While
For (Input image contains text)
 BGR image is converted to RGB
  Using pytesseract.image_to_data(), boxes are bound around each
word
 Words are accessed using counter as a key
 Text detected is converted into speech
End For
End of Algorithm
```

4 Results and Discussion

The processor-based system of smart glasses can make the life of a visually challenged person more convenient by detecting the object and recognizing text within the same module. Initially, the object is detected which is in front of the camera. If the object contains text, then the text detection is carried out. The pi camera starts taking real-time video of the surroundings and starts processing it for OCR. Thereafter, the text is extracted, and the user receives the output as audio using headphones that have been connected to the system. This algorithm was performed on the objects present in the surroundings. Results demonstrate that the suggested methodology performs well in both object detection and text recognition. This device will read the text from the objects detected such as newspapers, books, menus, and other screens and give the audio output. Real-time identification and recognition of objects was performed and tested with 71.8% accuracy.

5 Conclusion

A processor-based module for blind and visually challenged users is intended to assist blind and visually challenged persons to navigate their surroundings. The system is designed to guide a person for detecting objects in his surroundings and recognizing text from various sources. It is a user-friendly interface which provides assistance to the visually impaired. This system gives the blind individual a portable, dependable choice for living with ease. With this equipment, blind people can confidently and independently go from one location to another without the assistance of others. The embedded system used in this device is a bit heavy to carry along. Accuracy for object detection is just considerable which is taken into consideration and increased into future implementations. More beneficial applications for communicating to Wikipedia, Google, Voice Guidance, etc., using the smart glass system will be implemented in the future scope of this study. It may also recognize the monetary denominations and identify fake notes. To attain a very small form, the smart glass's general architecture and housing will be upgraded. It can be made interactive and advanced like, i.e., google assistance, Siri: in iOS systems, Cortana: in Windows, Bixby: for Galaxy and Alexa: Amazon cloud-based.

References

1. Pascolini D, Mariotti SP (2010) Global estimates of visual impairment: Br J Ophthalmol 96(5):614–618
2. He Y, Nie A, Pei J, Ji Z, Jia J, Liu H, Wan P, Ji M, Zhang C, Zhu Y, Wang X (2020) Prevalence and causes of visual impairment in populations more than 50 years old: The Shaanxi Eye Study. Medicine 99(20):e20109
3. WHO (2015) Visual impairment and blindness. WHO. http://www.who.int/mediacentre/factsheets/fs282/en/
4. Aladrén A, López-Nicolás G, Puig L, Guerrero JJ (2016) Navigation assistance for the visually impaired using RGB-D sensor with range expansion. IEEE Syst J 10(3):922–932
5. Fernandes H, Costa P, Filipe V, Hadjileontiadis L, Barroso J (2010) Stereo vision in blind navigation assistance. World Automation Congress, Kobe, pp 1–6
6. Lee YH, Medioni G (2014) Wearable RGBD indoor navigation system for the blind. In: ECCV workshops, vol 3, pp 493–508
7. Bousbia-Salah M, Bettayeb M, Larbi A (2011) A navigation aid for blind people. J. Intelligent Rob Syst 64(3):387–400
8. Miah MR, Hussain MS (2018) A unique smart eye glass for visually impaired people. In: 2018 international conference on advancement in electrical and electronic engineering (ICAEEE)
9. Khan S, Nazir S, Khan HU (2021) Analysis of navigation assistants for blind and visually impaired people: a systematic review. IEEE Access 9:26712–26734
10. ScanTalker (2006) Barcode scanning application to help blind identify over one million products
11. KReader Mobile User Guide (2008) knfb Reading Technology Inc
12. Shoval S, Borenstein J, Koren Y (1998) Auditory guidance with the Navbelt: a computerized travel for the blind. IEEE Trans Syst Man Cybern C Appl Rev 28(3):459–467
13. Yi C, Tian Y, Arditi A (2014) Portable camera-based assistive text and product label reading from hand-held objects for blind persons. IEEE/ASME Trans Mechatr 19(3)

14. Huang X, Shen T, Wang R, Gao C (2015) Text detection and recognition in natural scene images. In: 2015 international conference on estimation, detection and information fusion (ICEDIF 2015)
15. Epshtein B, Ofek E, Wexler Y (2010) Detecting text in natural scenes with stroke width transform. In: IEEE conference on computer vision and pattern recognition (CVPR), pp 2963–2970
16. Conaire CÓ (2005) Efficient Euler-number thresholding
17. Lin J-Y, Chiang C-L, Wu M-J, Yao C-C, Chen M-C (2020) Smart glasses application system for visually impaired people based on deep learning. In: 2020 Indo-Taiwan 2nd international conference on computing, analytics and networks (Indo-Taiwan ICAN)
18. Mainkar VV, Bagayatkar TU, Shetye SK, Tamhankar HR, Jadhav RG, Tendolkar RS (2020) Raspberry Pi based intelligent reader for visually impaired persons. In: 2020 2nd international conference on innovative mechanisms for industry applications (ICIMIA). IEEE, pp 323–326
19. Sonth S, Kallimani JS (2017) OCR based facilitator for the visually challenged. In: 2017 international conference on electrical, electronics, communication, computer, and optimization techniques (ICEECCOT). IEEE, pp 1–7
20. Paul IJL, Sasirekha S, Mohanavalli S, Jayashree C, Moohana Priya P, Monika K (2019) Smart eye for visually impaired—an aid to help the blind people. In: 2019 international conference on computational intelligence in data science (ICCIDS). IEEE, pp 1–5

Stock Market Prediction Using Machine Learning: A Review

Harish Kunder, B. C. Soundarya, K. Kishan Karyappa, M. Nithin, P. K. Mahesh, R. Sharath, and J. Uday Kiran

Abstract This review article explores the advancements in using machine learning algorithms to forecast the stock market. Various techniques are used to select features, evaluate models, and pre-process data. Examples of machine learning models discussed include support vector machines, random forests, and artificial neural networks (ANNs). The article also examines the challenges and limitations of using machine learning for stock market forecasting and identifies important components that improve the accuracy of these models. The study's objectives are to present a thorough evaluation of the state of the field today and to pinpoint potential areas for further investigation. Overall, while machine learning has great potential for market forecasting, more work is needed to improve the accuracy and reliability of these models.

Keywords Stock market prediction · Machine learning · Financial data · Model evaluation · Model comparison

1 Introduction

Academics and practitioners in the field of finance have long been intrigued by the idea of forecasting price and movement changes on the stock market. Stock market forecasting has become more sophisticated and challenging as methods and technology have advanced [1]. In this area, methods based on machine learning have shown to be useful tools, offering fresh perspectives on how the financial markets behave and facilitating wise investing choices. In this article, the management of risks, day trading, trading using algorithms, and time series financial projections in predicting the stock market are all examined [2]. It should be made clear that this

H. Kunder · B. C. Soundarya · K. K. Karyappa (✉) · M. Nithin · P. K. Mahesh · R. Sharath · J. U. Kiran
Department of Artificial Intelligence and Machine Learning, Alva's Institute of Engineering and Technology, Mangalore 574225, India
e-mail: kishankaryappak@gmail.com

V. K. Gunjan and J. M. Zurada (eds.), *Proceedings of 4th International Conference on Recent Trends in Machine Learning, IoT, Smart Cities and Applications*, Lecture Notes in Networks and Systems 873, https://doi.org/10.1007/978-981-99-9442-7_2

study is just based on a review of previously published research and does not draw any conclusions or real-world consequences. The major elements of the survey are a review of uses, a discussion of approaches for forecasting stock markets utilizing machine learning, an examination of obstacles and limits, and a conclusion. Each section's present state of the science is carefully examined, along with prospective future prospects for study and development [3].

The application of machine learning methods to stock market surveillance and manipulation allows for the making of educated judgments [4]. When used to solve particular issues, machine learning (ML) has been shown to be the most efficient way of processing time series data. Strong and well-liked ensemble approaches for prediction challenges include boosting and bagging [5]. Additionally, principal component analysis (PCA) has been employed to forecast the daily future returns of stock market indices, with deep neural networks surpassing artificial neural networks (ANNs) as classifiers based on PCA-represented data. Managing global investment portfolios often involves considering four layers of portfolio techniques [6].

2 Literature Review

We were able to obtain significant insights regarding pre-research, present circumstances, and prospective future advances in stock market forecasting using machine learning algorithms after reading ten relevant research publications. The conclusions of these publications are summarized in the reviews that follow:

The effectiveness of support vector machines (SVMs) in financial time series prediction has been shown [6] using machine learning alongside deep learning-inspired methodologies for stock market forecasting. Because they may be used to predict crises and hardship in the financial sector, the banking business, and various other areas, SVMs are an invaluable instrument. Combining modern deep learning and statistics machine learning methods yields the best results for financial projection predictive performance [3].

One such method is called Integrated Autoregressive Moving Average (ARIMA), which may be thought of as a time series forecasting problem. Despite the fact that nonlinear and non-stationary data are frequently found in the stock market, ARIMA struggles with them [7]. This issue can be partially solved with the use of ARIMA and SVM. The nonlinear component will be predicted using SVM, while the linear component will be predicted using ARIMA. A different approach uses a wavelet transformation to separate the stock data into smaller chunks before forecasting the stock data using SVM and a wavelet basis. Then, an ANN and ARIMA combination is employed to forecast the irregular component of the stock price input [8].

3 Methodology

The phases in the stock market forecast process are shown in Fig. 1.

In the first phase, real-time data are gathered from a variety of sources and previous datasets, such as NASDAQ, depending on their price index, as shown in Fig. 1. By comparing today's flat price to its previous valuation using the value index, which reflects a stock market section, investors may evaluate performance. To remove noise and other characteristics, the acquired data is first pre-processed. This processed data is crucial for precise stock market predictions [6]. Then, from a large amount of data, relevant characteristics are selected using choosing feature algorithms. Using data analysis tools or user-friendly software, the dataset is divided into two parts: present information and forecast information. These insights make it easier to make better stock market decisions [9]. After the publishing of a final judgment, the price indexes are announced in order to notify shareholders of the outcome, thereby giving crucial information for making investments. If the application generates a profit, investors may sell their stocks for a greater price; however, if the price gauge is low, they have to pay closer attention in order to make wise selections [7]. Figure 2 depicts the method of the supervising learning prediction of the stocks model.

Following is an expression of the stock prediction process according to the workflow shown in Fig. 2.

a. **Gathering and getting ready data**

The first stage is to gather past stock price data along with other important financial information, including news or indicators of the economy. After pre-processing, the data is ready for modeling.

b. **Design of features**

In this step, the pre-processed data is examined to find and extract useful properties. The important stage of feature selection can have a significant impact on how well the algorithm for machine learning performs.

c. **Choice of models**

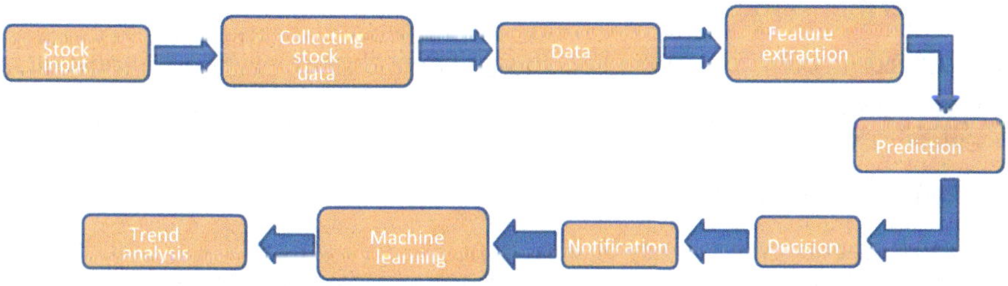

Fig. 1 Process used to forecast the stock market in steps

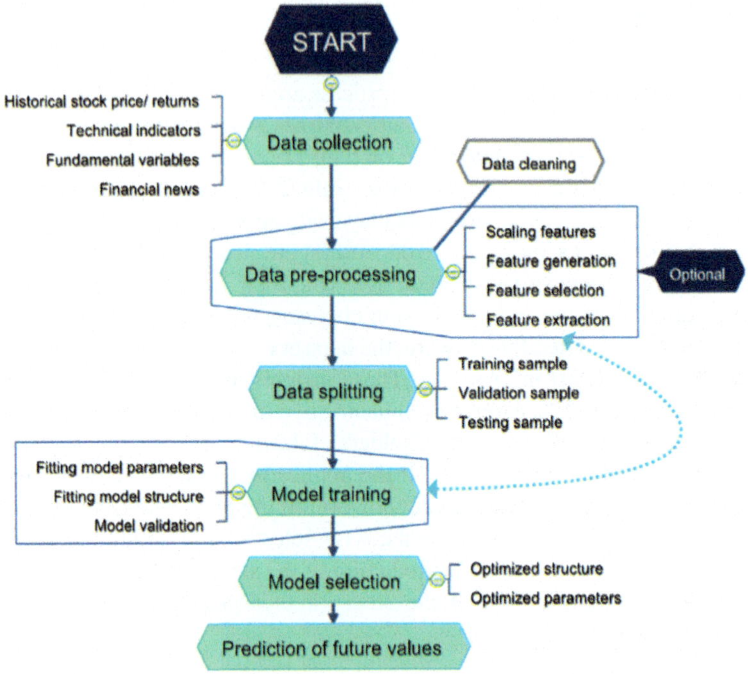

Fig. 2 Workflow for the stock market forecasting model utilizing supervised learning

A suitable machine learning strategy is chosen based on the level of accuracy that is required, the amount and nature of the data, and its forecasting goal (which might include classification or regression).

d. **Model building**

The pre-processed data is used as the training set for the machine learning model, with the extracted features serving as inputs. In order to reduce incorrect predictions, model parameters are adjusted.

e. **Model assessment**

The trained model is put to the test on a different dataset, such as a validation or test set, to gauge how well it performs. Assessment criteria like accuracy, precision, recall, or root mean squared error are used to evaluate the performance of the model.

f. **Applying the model**

If the model works well, it may be used to predict changes in the stock market. The model should be regularly updated with fresh data and retrained in order to retain its accuracy over time.

4 Conclusion

In conclusion, the use of prediction models using the fields of AI and machine learning is crucial for making accurate stock market forecasts as well as for being able to make well-informed decisions. The specific use application and the properties of the investigated data affect the language used for programming and model selection. However, the development and use of these models will contribute to further advancements in the fields of intelligence and machine learning.

The future of models of prediction in AI and machine learning offers significant potential, with ongoing innovations and improvements projected to yield ever more accurate and effective models. Application of innovative techniques like deep training and reinforcement training, which have had great success in sectors like artificial intelligence and gaming, is one potential direction for advancement. Furthermore, an increase in the use of cloud-based systems and tools may make it simpler to access powerful computing resources, enabling the deployment of larger and more complex models. The development of models that may give information about how AI makes judgments, increasing their transparency and trustworthiness, is also receiving increasing interest.

In conclusion, AI and machine learning prediction models have a promising future and provide a wide variety of fascinating prospects for more study and application. With the growing importance of knowledge in our daily lives comes an increased need for the ability to accurately predict future occurrences, which encourages more creativity in this field.

References

1. Kumbure MM, Lohrmann C, Luukka P, Porras J. Machine learning techniques and data for stock market forecasting: A literature review. School of Business and Management, LUT University, Yliopistonkatu 34, 53850 Lappeenranta, Finland; School of Engineering Science, LUT University
2. Mehtab S, Sen J, Dutta A. Stock price prediction using machine learning and LSTM-based deep learning models. Department of Data Science and Artificial Intelligence, Praxis Business School. Bakrahat Road, P.O. Rasapunja, Off Thakurpukur Road, Kolkata 700104, India
3. Bustos O, Pomares-Quimbayaa A. Stock market movement forecast: a systematic review. Facultad de Ingeniería. Departamento de Ingenieríade Sistemas, Pontificia Universidad Javeriana, Bogotá, Colombia
4. Kumar D, Sarangi PK, Verma R. A systematic review of stock market prediction using machine learning and statistical techniques
5. Zhong X, Enke D. Predicting the daily return direction of the stock market using hybrid machine learning algorithms
6. Jiang W. Applications of deep learning in stock market prediction: recent progress. Department of Electronic Engineering, Tsinghua University, Beijing 100084, China
7. Lee TK, Chob JH, Kwonb DS, Sohn SY. Global stock market investment strategies based on financial network indicators using machine learning techniques. Samsung Asset Management Co., Ltd., Seoul, Korea Department of Information and Industrial Engineering, Yonsei University, Seoul, Korea

8. Zhanga K, Zhonga G, Donga J, Wanga S, Wanga Y. Stock Market Prediction Based on Generative Adversarial Network. Department of Computer Science and Technology, Ocean University of China, Qingdao, 266100, China
9. Yadav K, Yadav M, Saini S. Stock values predictions using deep learning-based hybrid models

Implementation of Dual-Band Dielectric Resonator Antenna for 5G Applications

A. R. V. Manoj Kumar, Ch. Raghavendra, Shaik Ashraf, and B. Yaswanth

Abstract A suggested electromagnetic couple feeding method for dual-band operation uses a dual-band dielectric resonator MIMO antenna to improve the antenna's physical characteristics by utilising various slots. The operating frequencies of the antennas are 5.2–6.5 GHz and 4.96–5.04 GHz, respectively. HFSS is used to measure various element dimensions across a number of iterations. To obtain the necessary antenna properties, the fundamental mode is used. This antenna's benefit is that it offers many characteristics with various isolation strategies. WLAN, 5G, and sub-6 GHz communication applications are made possible by the performance of the recommended antenna. The antenna is ultimately built, examined in an anechoic laboratory, and modelled using HFSS software.

Keywords Electromagnetic coupling · Dielectric resonator antenna (DRA) · Dual-band polarisation · MIMO · And WLAN communication

1 Introduction

In our increasingly connected society, the need for high-speed data transmission has grown significantly. Our online interactions have changed significantly since the introduction of 5G technology. However, creating small, versatile antennas that can operate across a broad frequency range is a significant obstacle to achieving this ambition. Dielectric resonator antennas (DRAs), which provide a viable solution to this problem, have recently attracted interest because of their distinctive features. In this work, we examine a dual-band DRA designed specifically for 5G applications. Our research shows that our antenna has remarkable radiation efficiency, gain, and bandwidth, making it a top contender for the upcoming generation of wireless communication systems.

A. R. V. M. Kumar (✉) · Ch. Raghavendra · S. Ashraf · B. Yaswanth
V R Siddhartha Engineering College, Vijayawada, Andhra Pradesh, India
e-mail: saimanojkumar101@gmail.com

V. K. Gunjan and J. M. Zurada (eds.), *Proceedings of 4th International Conference on Recent Trends in Machine Learning, IoT, Smart Cities and Applications*, Lecture Notes in Networks and Systems 873, https://doi.org/10.1007/978-981-99-9442-7_3

Academics are thinking about novel antenna layouts that could handle the high rates of data and frequencies necessary for 5G networks because of the expanding need for rapid transmission of data in the context of the 5G revolution. A dielectric resonator antenna (DRA) is an electromagnetic resonator that capitalises on the resonant properties of a dielectric material.

DRAs offer numerous advantages over conventional antennas, such as a higher quality factor (Q-factor), a more compact form factor, and versatility in operating across various frequency bands. Dual-band DRAs are very useful for 5G applications because they can operate across a broad range of frequencies and manage a number of communication protocols, including frequencies.

A dual-band DRA's dimensions, dielectric material's characteristics, and feeding systems must all be taken into account while creating it. The antenna's geometry and size must be carefully tuned to attain the required resonance frequencies and bandwidths. The selection of the dielectric material can also have a big impact on how well the antenna performs. To achieve a robust Q-factor and minimal loss, it is essential for the dielectric material to possess both a high permittivity and a low loss tangent.

The antenna feeding system is a critical consideration, as it can influence both the antenna's radiation characteristics and impedance matching. Various feeding techniques are applicable to DRAs, including coaxial feeding, microstrip feeding, and aperture coupling.

This paper presents the design and analysis of a dual-band DRA tailored for 5G applications. We comprehensively discuss the antenna design procedure, encompassing the choice of feeding mechanism and dielectric material. Subsequently, we showcase the antenna's radiation characteristics, including gain, radiation efficiency, and bandwidth, as derived from simulation results. Our antenna's performance is then benchmarked against existing dual-band antennas documented in the literature, revealing its suitability as a superior solution for 5G communication systems.

2 Literature Survey

In recent times, DRAs have garnered significant attention as a promising choice for the upcoming generation of wireless communication systems. Their distinctive characteristics and capability to function across diverse frequency bands make them particularly noteworthy. Among these, dual-band DRAs have captured considerable interest due to their capacity to operate over a wide frequency spectrum and accommodate a variety of communication protocols and frequencies, making them well-suited for 5G applications (Table 1).

A quad-port dual-band dielectric resonator antenna has reportedly been developed that can be used for 5G and WLAN applications within the sub-6 GHz spectrum, according to [1].

Table 1 Performance and use of materials

References	Permittivity	Material used	Applications	Structure
[1]	4.40	FR-4	MIMO applications	Triangular
[2]	0.930	FR-4	Mobile-communication	Tri-band four-port
[3]	3.380	FR-4	MIMO application	Symmetrically
[4]	2.650	FR-4	MIMO application	Feed-structure
[5]	02.650	F4-B	5G applications	Feed-structure
[6]	03.380	Rogers	5G applications	air-substrate structure
[7]	4.40	FR-4	Wireless devices	Octagonal ring-shaped antenna
[8]	010.20	DR	Ultra-wideband	Shelf
[9]	02.940	Rogers	5G and IOT	Rectangular shaped
[10]	4.70	FR-4	MIMO application	Aperture feeding structure
[11]	4.40	FR4	Vehicle communication	Annular-ring

In his paper, Sharawi [2] proposed a dual-polarised UWB MIMO antenna with two T-shaped slots to achieve high isolation. The antenna's modest size and bandwidth of 2–12 GHz made it suitable for usage with mobile devices.

"Integrating Microscopic Printed Antennas into Miniature Diversity/MIMO Terminals" was offered to Karaboikis [3], but he declined. A planar UWB MIMO aerial was suggested by Karaboikis et al. The antenna was small, measuring 32×40 mm^2, and it had a 3.1–10.6 GHz bandwidth. The design used a meandering ground to improve isolation between the antennas.

Huang et al. introduced a proposal for a low-profile, dual-polarised patch antenna tailored for 5G MIMO applications [4]. This antenna featured a bandwidth spanning from 2.6 to 3.7 GHz, specifically engineered to operate seamlessly with 5G devices. The design incorporated a substrate-integrated waveguide (SIW) feeding system to achieve a low profile and enhance isolation.

In their study, Li et al. [5] proposed an antenna specifically designed for 5G sub-6 GHz base station applications. This antenna is characterised by differential feeding, dual-band operation, and dual polarisation, along with high selectivity. It includes a filtering element to reduce interference from neighbouring frequency bands within the ranges of 3.3–3.8 GHz and 4.8–5.3 GHz.

In their study, Le et al. proposed a small, dual-polarised, broadband antenna designed for 5G applications [6]. The antenna's bandwidth ranges from 3 to 5 GHz, and it was designed to be tiny and portable for use in portable devices. A tiny feeding structure was used in the design to achieve good isolation and a wide bandwidth. For S-band and applications involving human interface devices, an octagonal-shaped wideband MIMO antenna was proposed in [7]. The antenna is constructed from two octagonal-shaped radiators and is microstrip line-fed. Wideband performance and efficient radiator isolation are achieved.

An improved axial ratio small rack-shaped MIMO dielectric resonator antenna for UWB applications was suggested in [8]. The antenna consists of two rack-shaped radiators and a common dielectric resonator. The radiation pattern, isolation, and return loss of the suggested aerial are all good.

Rajat Girjashankar and Upadhyaya [9] proposed a millimetre-wave 5G wireless communication system featuring substrate-integrated waveguide feeding. The antenna achieves wide impedance bandwidth, high gain, and minimal cross-polarisation at 24 and 38 GHz frequencies.

In reference to [10], a MIMO dielectric resonator antenna was introduced for LTE applications, offering triple-band capabilities. This compact antenna features four rectangular radiators and a shared dielectric resonator, enabling both wideband and triple-band functionality.

A beam-tilting quad-port MIMO dielectric resonator antenna capabilities that is circularly polarised was suggested for use in automobile communication in [11]. The antenna enables circular polarisation and beam-tilting, and it has four ports. It is possible to attain high isolation with negligible cross-polarisation. Alibakhshikenari et al. [12] conducted a thorough review of the various decoupling techniques applicable to SAR and MIMO aerial systems. The primary subjects of the survey are the metamaterial and metasurface principles for decoupling processes.

3 Antenna Configuration

Figure 1a illustrates the completed MIMO dielectric resonator antenna, presenting both top and perspective views for clarity. At the outset, solely the rectangular patch resonator within the microstrip line was assembled. Traditional conductor-based antennas are significantly affected by metallic losses. In contrast, these antennas exhibit remarkable radiation efficiency, resulting in a high aerial gain. When an appropriate dielectric constant is utilised, these antennas offer a broader bandwidth compared to conventional patch antennas. The antenna was designed using an alumina substance having a loss tangent of 0.0001 and a dielectric constant of 9.9 due to its inexpensive cost and widespread availability. The ground is 30 mm long and 30 mm wide, respectively. The FR-4 substrate has the following measurements: 30 mm × 30 mm × 1.6 mm (Fig. 1b). Hence, the patch measures 19 mm by 24.5 mm in both length and width. Next we added a 15 mm long by 15 mm wide slot. Activating the dielectric material involves coupling an electromagnetic field generated by a conductive element using the aperture coupling technique. In the case of microstrip feed excitation, it stimulates the conductive elements. Notably, the microstrip feed method exhibits a simpler design compared to conventional patch antennas. The field associated with the dielectric resonator (DR) is further induced through the excited patch, facilitated using an aperture coupler. The widely used dielectric reflector (DR) aerial is built using a FR-4 substrate. On top of the FR4 dielectric resonator, standard copper sheets that are 70 μm thick were taken into account for modelling. The height, breadth, and length of the dielectric resonator are all 16 mm.

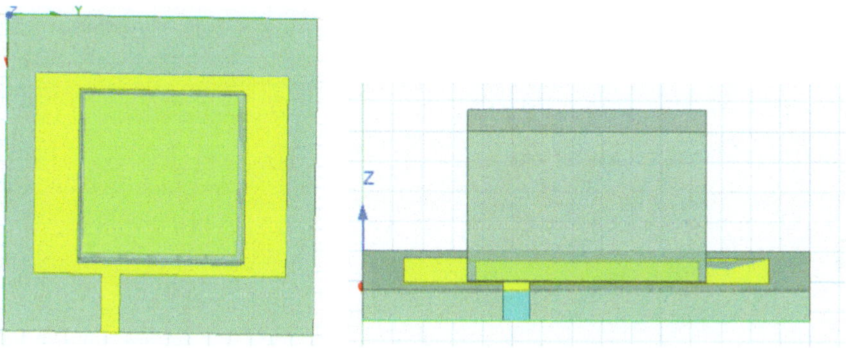

(a) (b)

Fig. 1 **a** Top view. **b** Side view

3.1 Antenna Feed

The component known as the antenna feed connects the transmitter/receiver to the antenna, enabling communication. It also includes other related equipment. A radio transmitter produces an alternating current with a radio frequency, which is fed to the antenna by the feed system, which converts the current's energy into radio waves. This current is transmitted through the feed system to the receiver, which analyses the signal. The incoming radio waves in a radio receiver start tiny alternating currents in the aerial (Fig. 2)

In the construction of microstrip antennas, one feeding technique utilised is rectangular microstrip feeding. This approach involves using a rectangular microstrip line to feed the antenna, with the ground plane situated atop the dielectric substrate, which, in turn, rests on the rectangular microstrip line—a slender strip of conductive material. To achieve optimal power transfer from the feedline to the antenna, the antenna's characteristic impedance is carefully designed to match that of the rectangular microstrip line. Accurate impedance matching relies on precise determination of the width and length of the rectangular microstrip line, making them critical design parameters.

Fig. 2 Microstrip feeding

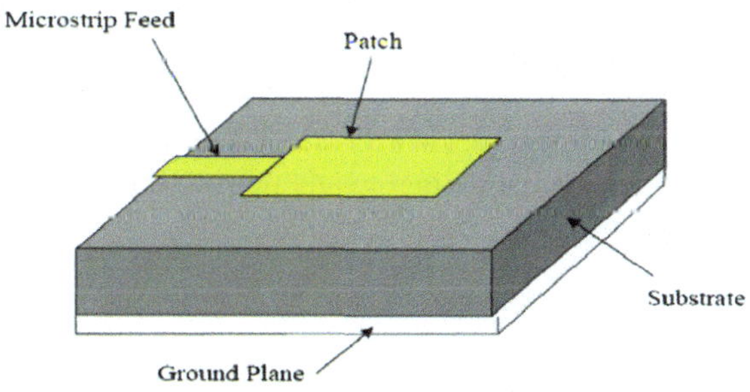

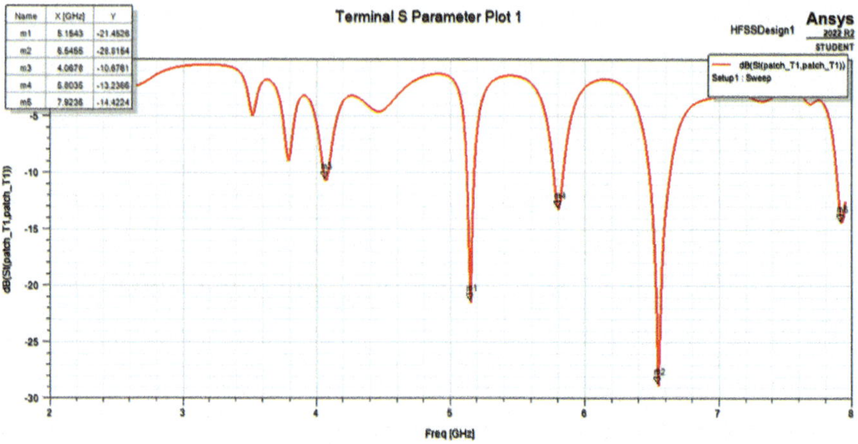

Fig. 3 S11 parameter plot

The advantage of rectangular microstrip feeding is that it makes it easier to integrate other components, including filters, amplifiers, and phase shifters. Multiple feeding locations can be employed as a feeding strategy to enable multi-frequency operation.

However, there are several disadvantages to rectangular microstrip feeding, such as the risk of radiation leakage from the feedline and the difficulty in attaining broadband impedance matching. To circumvent these limitations, methods include impedance matching networks and radiation pattern optimisation.

4 Simulation Results

S11-parameters, also known as scattering parameters, serve as a means to characterise the transmission of energy within an electrical network. S-parameters are particularly useful when describing the relationship between various ports, emphasising amplitude and phase across frequencies instead of voltages and currents. Frequencies of interest here encompass 5.14, 6.5, and 4 GHz (Figs. 3 and 4).

VSWR, or voltage standing wave ratio, quantifies the effectiveness of transferring radio-frequency power from a power source, through a transmission line, and into a load (Figs. 5, 6, and 7).

Gain of an antenna in a direction is the ratio of the intensity in that direction to the intensity of radiation that would be produced if the antenna's power were emitted isotropically.

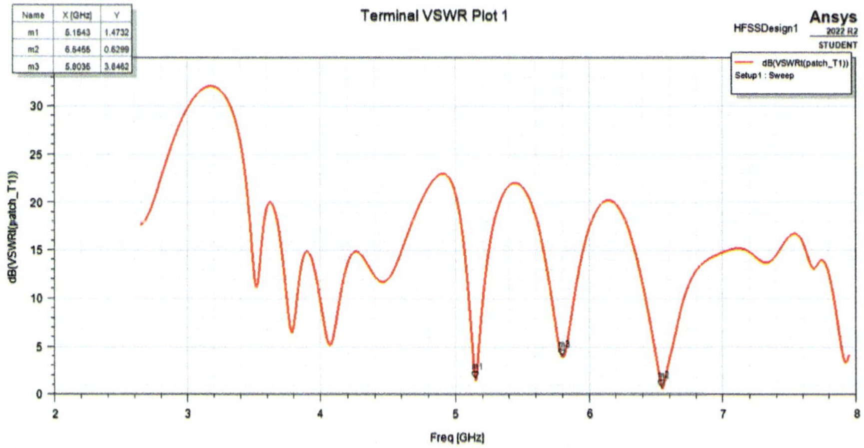

Fig. 4 Voltage standing wave ratio (VSWR) plot

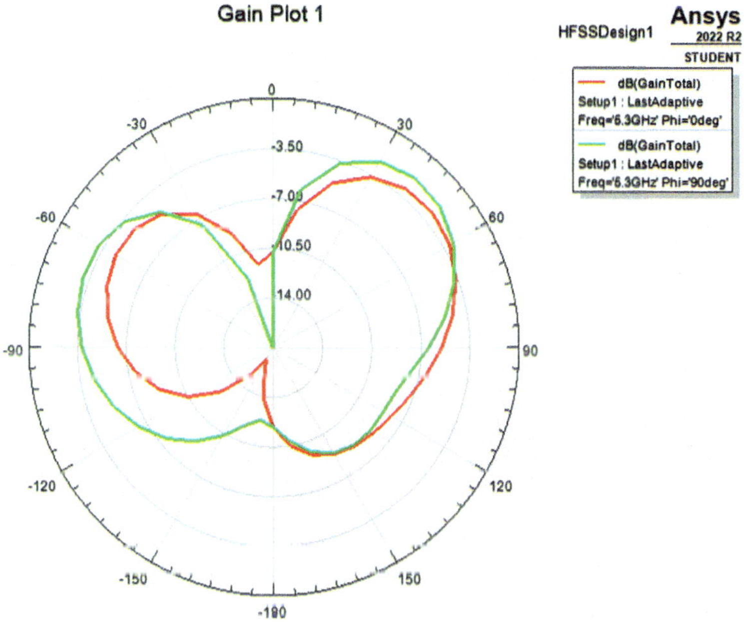

Fig. 5 Gain plot 1

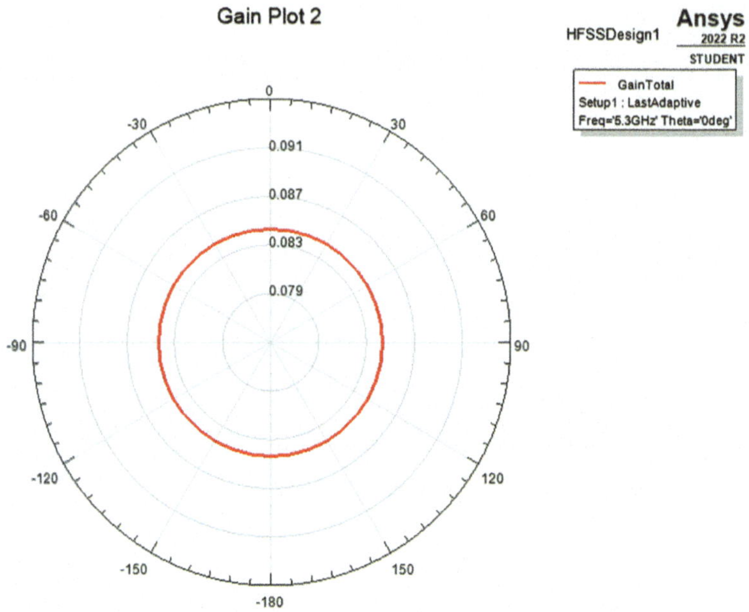

Fig. 6 Gain plot 2

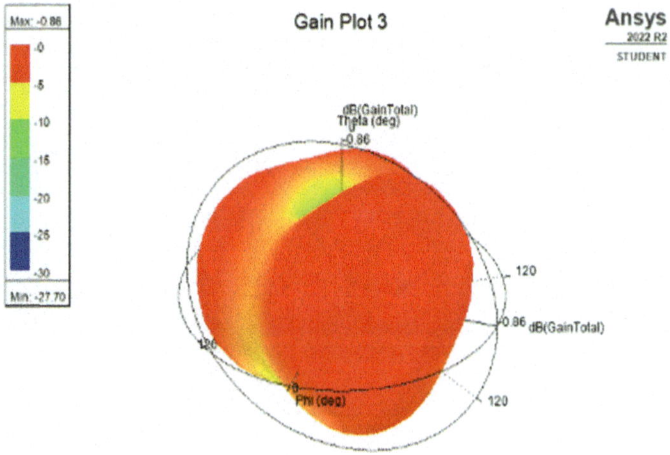

Fig. 7 Gain plot 3

5 Conclusion

The electromagnetic coupling technique was devised to enable dual-band operation by enhancing the antenna's physical characteristics through the incorporation of multiple slots. This study presents a dual-band dielectric resonator MIMO antenna that operates at two frequencies: 5.2–6.5 GHz and 4.96–5.04 GHz. Multiple rounds of the high-frequency structure simulator (HFSS) were used to evaluate various element sizes, eventually fine-tuning the antenna's attributes to suit the needed criteria. This antenna's particular benefit is its capacity to provide a variety of characteristics via various isolation approaches. Its performance paves the door for wireless LAN (WLAN), 5G, and communication applications within the sub-6 GHz spectrum. Finally, the antenna design was realised and confirmed further through simulations using HFSS.

References

1. Upa Dhyaya T, Park I, Pandey R, Patel U, Pandya K, Desai A, Pabari J, Byun G, Kosta Y (2022) Aperture-fed quad-port dual-band dielectric resonator-MIMO antenna for sub-6 GHz 5G and WLAN application. Int J Antennas Propag 2022:1–13
2. Sharawi MS (2013) Printed multi-band MIMO antenna systems and their performance metrics [wireless corner]. IEEE Antennas Propag Mag 55(5):218–232
3. Karaboikis MP, Papamichael VC, Tsachtsiris GF, Soras CF, Makios VT (2008) Integrating compact printed antennas onto small diversity/MIMO terminals. IEEE Trans Antennas Propag 56(7):2067–2078
4. Huang H, Li X, Liu Y (2019) A low-profile, dual-polarized patch antenna for 5G MIMO application. IEEE Trans Antennas Propag 67(2):1275–1279
5. Li Y, Zhao Z, Tang Z, Yin Y (2020) Differentially fed, dualband dual-polarized filtering antenna with high selectivity for 5G sub-6 GHz base station applications. IEEE Trans Antennas Propag 68(4):3231–3236
6. Le Thi CH, Ta SX, Nguyen XQ, Nguyen KK, Dao-Ngoc C (2021) Design of compact broadband dual-polarized antenna for 5G applications. Int J RF Microw Comput Aided Eng 31(5). Article ID e22615
7. Chouhan S, Panda DK, Kushwah VS, Mishra PK (2019) Octagonal-shaped wideband MIMO antenna for human interface device and S-band application. Int J Microw Wirel Technol 11(3):287–296
8. Yadav SK, Kaur A, Khanna R (2021) Compact rack shaped MIMO dielectric resonator antenna with improved axial ratio for UWB applications. Wirel Pers Commun 117(2):591–606
9. Rajat Girjashankar P, Upadhyaya T (2021) Substrate integrated waveguide fed dual band quad-elements rectangular dielectric resonator MIMO antenna for millimeter wave 5 G wireless communication systems. AEU Int J Electron Commun 137. Article ID 153821
10. Anuar SU, Jamaluddin MH, Din J, Kamardin K, Dahri MH, Idris IH (2020) Triple band MIMO dielectric resonator antenna for LTE applications. AEU Int J Electron Commun 118. Article ID 153172
11. Dwivedi AK, Sharma A, Singh AK, Singh V (2020) Circularly polarized quad-port MIMO dielectric resonator antenna with beam tilting feature for vehicular communication. In: IETE technical review, pp 1–13
12. Alibakhshikenari M, Babaeian F, Virdee BS et al (2020) A comprehensive survey on "various decoupling mechanisms with focus on metamaterial and metasurface principles applicable to SAR and MIMO antenna systems." IEEE Access 8:192965–193004

Defect Detection in Metal Surfaces Using Computer Vision

Krishna Kumar Singh and Manish Ghosh

Abstract Encoder models have shown remarkable success in various computer vision operations like object detection, image classification, and semantic segmentation. However, the results of one model showed that it was underfitting as it performed better on the validation set than on the training set. To gain insights into how the model was processing input images at different layers and which image features were being detected by different filters in the convolutional layers, an algorithm was used to visualize the activation maps for the output of convolutional layers in the model. To address the underfitting issue, the previous model was converted into a single feature-vector using global average pooling and used as a single layer in another deep learning model. As a result, the new model displayed a batch of images along with their corresponding segmentation masks and predicted segmentation masks generated by the encoder model. The output was a 3 × 1 image grid that included the original image, its ground truth segmentation mask, and the predicted segmentation mask. To optimize the model further, an algorithm was applied that trained an image encoder model on training data and evaluated it on validation data for a number of epochs. The algorithm then displayed a batch of samples with their ground truth and predicted segmentation masks. The results for this model were better than the previous model, as it was performing better with the train set as compared to the validation set. Overall, the usage of an encoder model and subsequent modifications allowed for improved performance in image segmentation tasks.

Keywords Computer vision · Defect detection · Metal surfaces · Encoder

K. K. Singh (✉) · M. Ghosh
Symbiosis Centre for Information Technology, Pune, India
e-mail: krishnakumar@scit.edu

M. Ghosh
e-mail: manish.ghosh@associates.scit.edu

V. K. Gunjan and J. M. Zurada (eds.), *Proceedings of 4th International Conference on Recent Trends in Machine Learning, IoT, Smart Cities and Applications*, Lecture Notes in Networks and Systems 873, https://doi.org/10.1007/978-981-99-9442-7_4

1 Introduction

Defect detection in metal materials is an essential task in the industrial and manufacturing sectors, as it ensures the quality and safety of metal products. In recent years, computer vision techniques have shown excellent defect detection potential due to the capacity to rapidly process large amounts of data, accurately. This research paper presents a detailed study of computer vision techniques to detect defects in metal materials. The designed system combines machine learning and image processing algorithms to analyse digital images of metal samples and identify defects such as cracks, voids, and inclusions. The image processing techniques applied include image enhancement, segmentation, and feature extraction methods to improve the visibility of defects and extract relevant information from the images. The machine learning algorithms used for classification include supervised learning techniques such as artificial neural networks (ANNs). In this research paper, we propose using jigsaw pretraining to enhance deep learning model efficiency for defect detection in metal materials. Jigsaw pretraining is a self-supervised learning method that trains a model to recognize the relationships between different parts of an image by solving jigsaw puzzles. By learning to identify patterns and relationships in the images, the model can better distinguish between defective and non-defective images. The proposed system has potential applications in the industrial and manufacturing sectors for ensuring the quality and safety of metal products, as well as in the discipline of non-destructive testing. Systems relying on computer vision have advantages over traditional methods, such as the ability to quickly process large amounts of data and detect defects in real time. Additionally, the proposed approach has the potential for automation and integration into existing manufacturing systems for more efficient and cost-effective defect detection. This research paper makes a valuable contribution to the field of defect identification based on computer vision in metal materials and has the potential to improve the quality and safety of metal products in industry and manufacturing. The findings of this research show that it is feasible. The effectiveness of using computer vision techniques for the detection of defects in metal materials, and further research is needed to optimize and improve the system.

2 Literature Review

Strict quality control procedures are frequently implemented to make sure that raw materials like plywood, metal, and plastics comply with industry requirements. These materials may be graded differently or even thrown away if defects are discovered. Machine learning is being utilized more and more to facilitate automation as Industrial IoT and Industry 4.0 take off. The detection of defects in complex goods like integrated circuit wafers is crucial to producing high-quality products in intelligent manufacturing systems [1]. However, current defect detection algorithms face difficulties when multiple types of flaws merge in a single wafer to create mixed-type defects. A knowledge-augmented wide learning system that consists of a broad selective sampling module and a multichannel selective sampling network was created to address this problem [2]. To overcome this, the authors proposed a system that can automatically classify three geometrically defined flaws using advanced sensor-based technologies. The system relies on image analysis methods such as the Hough transform, as opposed to conventional methods that require manual estimations, which are time-consuming and resource-intensive [3] which is accessible online and has a beneficial class imbalance. Through feature extraction and fine-tuning, the authors taught deep residual neural network (ResNet) classifiers to recognize external weaknesses [4]. In the textile industry, fabric defect detection has been a challenging problem, and manual labour was previously used for the detection of such defects, which had limitations such as human weariness, lack of focus, and time consumption [5]. In the past 20 years, many computer vision-based applications have been suggested to overcome these limitations. The goal of the paper is to provide a thorough analysis of various computer vision-based methods for detecting fabric flaws, including histograms, colour schemes, image segmentation, frequency domain operations, texture-based defect detection, sparse feature-based operations, image morphology operations, and deep learning developments [6, 7]. The support vector machine (SVM) technique and the artificial neural network (ANN)-based method are the two approaches the authors suggest for the detection of potato flaws. Based on the RGB values of each pixel, the SVM technique uses supervised colour segmentation based on multi-layer perceptron (MLP) and radial basis function (RBF) neural networks to identify potato pixels in the background. Importance of visual evaluation of civil infrastructure to guarantee its functionality and safety [8]. The paper provides a summary of the state of the art for assessing the visual health of civil infrastructure, both vertically and horizontally, including precast concrete tunnels, asphalt roads, underground concrete pipelines, and reinforced concrete bridges [9]. The authors propose a computer vision-based method to automatically identify aggregate segregation using Asymmetrical Appealing Repulsive Center-Symmetric Local Binary Pattern and Deep Convolutional Neural Network (DCNN) (ARCSLBP-XGBoost). The proposed approach shows excellent execution in terms of 95% of the time, the categorization was correctly identified using the classification accuracy rate, precision, recall, and $F1$-score [10]. One common sign of distress is the presence of cracks, which can indicate significant structural issues and pose risks to occupants and the structure itself. For uses in civil engineering, the precise and automated identification of cracks on structural surfaces poses a significant challenge. Recent advancements in hardware data-collecting systems have led to significant progress in automatic crack detection and measurement [11]. The study uses an active infrared thermography approach known as eTC-PCT, which is also used in biomedical and dental thermophotonic imaging, to detect manufacturing flaws such as cracks and holes in industrial materials. Three algorithm steps are used to process the eTC-PCT reconstructed image, which involve applying a filter for threshold selection, using the K-means algorithm to segment filtered images, and identifying discontinuity borders using the method for detecting canny edges [12]. However according to observations, vision-based examination methods perform poorly when shop floor images are used. To address this issue, the paper proposes using neural structure learning (NSL) adversarial training and histogram equalization, a powerful vision-based surface defect classification system is developed. A unique deep neural network architecture is developed to collect negative samples in the feature space rather than the original input image space [13]. The findings show that, in comparison with ENEU, the proposed method improves recognition accuracy by 92.4%, surpassing both competing deep learning approaches and conventional methods in terms of prediction accuracy. The paper concludes that the proposed technique's efficiency in boosting the generalization capabilities of machine learning models is demonstrated through qualitative and quantitative comparisons with results described in the literature.

3 Research Mythology

Detection of defect in metals using Jigsaw pretraining method in computer vision involves several stages, one of which is data collection. In this stage, the goal is to collect a dataset of images of metal surfaces with and without defects that will be used to train and test a deep learning model for defect detection. The following is a comprehensive description of the data collection stage for defect detection in metals using Jigsaw pretraining method in computer vision.

Defining the problem: The first step in data collection is to determine the problem at hand. In this case, the problem is to detect defects on metal surfaces using computer vision techniques. There must be a clear definition of the kinds of defects that must be found, their traits, and the desired degree of accuracy.

Data Sources: The next step is to identify the sources of data. These sources can be internal or external. Internal sources may include images of metal surfaces from previous inspections, while external sources may include images from industry-specific websites, open-source repositories, or even social media.

Image Capture: The images of metal surfaces can be captured in several ways. The most common methods include using a digital camera, a microscope, or even a drone. The images should be captured in a way that they show the defects clearly and are of high quality.

Image Annotation: Once the images are captured, the next step is to annotate them. Annotation involves labelling the images to indicate the presence or absence of defects. This can be carried out directly or automatically using tools, such as image segmentation algorithms. The annotations should be accurate and consistent.

Data Augmentation: The dataset can be augmented by generating new images from the existing ones. This can be done by applying transformations such as rotation, scaling, cropping, and adding noise to the images. Data augmentation can help expand the dataset, lessen overfitting, and increase the model's reliability.

Data Pre-processing: Next step is to pre-process the data. This involves converting the images into a format that can be used by the deep learning model. The images may need to be resized, normalized, and altered to a suitable data type, such as tensors.

Data Split: The dataset should be subjected to test-train-validation split. The training set on which the model was trained is examined using the validation set. The model's hyperparameters are adjusted, and the testing set is used to gauge the model's efficiency.

Data Balancing: To guarantee that the model does not favour one class over another, the dataset must be balanced. The minority class can be oversampled or the dominant class can be under-sampled to achieve this.

Data Storage: The final step is to store the data in a suitable format, such as a CSV file, a database, or a cloud-based storage solution. The data should be easily accessible and backed up regularly to avoid loss.

The data collection stage for defect detection in metals using Jigsaw pretraining method in computer vision involves defining the problem, identifying data sources, capturing images, annotating images, augmenting the dataset, pre-processing the data, splitting the data, balancing the dataset, and storing the data in a suitable format. The performance of the deep learning model can be greatly impacted by the quality and volume of the data gathered. Therefore, it is essential to guarantee that the data is accurate, varied, and indicative of the current issue. It's also essential to ensure that the data is representative of the real-world scenario and diverse enough to include different types of defects and variations in lighting and background. A balance between the number of defective and non-defective samples should also be maintained in order to train the model effectively. Data can be divided into a train-test set after it has been gathered. For the final performance of the model, the training set is used to train the model and assess the test set, respectively. For this research paper purpose, the data has been collected from several real-world and online sources. Real-world sources include the sample images from any forging and furnace companies which initially make test pieces for testing purposes (physical and chemical) and as the samples for their clients.

4 Methodology and Implementation

The above flowchart (in Fig. 1) represents the general methodology. Now, here in this case we are using the Jigsaw pretraining method for obtaining the result. Jigsaw pretraining is a method that can be used in computer vision to optimize the performance of a deep learning algorithm for categorizing image tasks. It is based on the idea that the model can learn better representations of the image by solving jigsaw puzzles. In jigsaw pretraining, the image is divided into smaller patches and then rearranged in a random order. The model is trained to reconstruct the original image from the rearranged patches. This process allows the model to learn to recognize and understand the relationships between different parts of the image. After the jigsaw pretraining, the model can be fine-tuned on the specific task of defect detection. By learning to recognize patterns and relationships in the images, the model is better able to distinguish between defective and non-defective images. This approach has demonstrated success in improving the performance of deep learning models for image classification tasks, including defect detection in metal materials. It can help the model to learn more robust representations of the images, which in turn can improve its ability to generalize to new images and detect defects more accurately. Now the methodology for defect detection in metal using computer vision can be divided into several steps. They are:

Data collection: The first step is to collect a large dataset of metal samples without defects, and a smaller dataset of metal samples with known defects. The images should be captured using high-resolution cameras or scanners and should be of sufficient quality to allow for the detection of defects.

Fig. 1 Flowchart showing the methodology in use

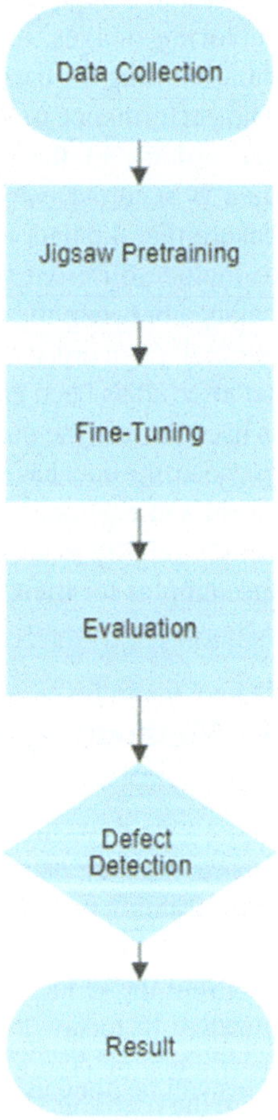

Image pre-processing: The collected images need to be pre-processed to remove noise and improve their quality. This can include techniques such as image enhancement, denoising, and histogram equalization.

Jigsaw pretraining: In this step, the images of the metal samples without defects are divided into smaller patches and then rearranged in a random order. The model is trained to reconstruct the original image from the rearranged patches, allowing the model to learn to recognize and understand the relationships between different parts of the image.

Fine-tuning: After the jigsaw pretraining, the model is fine-tuned on the dataset of metal samples with known defects. This step allows the model to learn to distinguish between defective and non-defective images.

Evaluation: A number of metrics, including recall, accuracy, $F1$-score, and precision, are used to assess the model's success. The model is also compared with more conventional identification techniques like X-ray imaging and ultrasonic testing.

Defect detection: The model can be used to categorize fresh images of metal samples as defective or non-defective after it has been trained and evaluated.

Optimization: The characteristics of the model, such as the number of neurons and the layers, learning rate, and batch size, can be optimized using approaches like grid search or random search to enhance the model's performance.

5 Data Analysis

After collecting the dataset, the next stage in defect detection in metals using Jigsaw pretraining method in computer vision is the data analysis stage. In this stage, the dataset is processed and analysed to extract valuable insights that can be used to build and evaluate the deep learning model for defect detection. The following is a comprehensive description of the data analysis stage for defect detection in metals using Jigsaw pretraining method in computer vision.

Data Cleaning: The first step in data analysis is to clean the dataset. This involves removing any duplicate or corrupted images, correcting any mislabeled annotations, and filling in any missing data. The data should be consistent and free of errors to ensure that the deep learning model learns from accurate data.

Exploratory Data Analysis (EDA): EDA involves exploring the dataset to gain insights into the data. This can be done by visualizing the data using techniques such as histograms, scatter plots, and box plots. EDA helps identify any trends, patterns, or outliers in the data that can inform the model building process.

Feature Extraction: The next step is to extract features from the images in the dataset. This can be done using techniques such as convolutional neural networks (CNNs) or handcrafted features such as texture, shape, and colour. Feature extraction helps reduce the dimensionality of the data and extract relevant information from the images.

Feature Selection: The next stage is to choose the most crucial features after feature extraction. Techniques like principal component analysis (PCA) or feature significance can be used for this. The model's accuracy is increased and its computational burden is decreased thanks to feature selection.

Model Building: Once the features have been extracted and selected, the next step is to build the deep learning model. This involves selecting an appropriate architecture, such as a CNN, and training the model using the training set. The model's hyperparameters, such as the epoch count, batch size, and learning rate, should be tuned using the validation set.

Model Evaluation: The trained model is assessed by the testing set. To analyse the model's performance, metrics like accuracy, precision, recall, and $F1$-score can be used. The evaluation's findings can be used to pinpoint any places where the model needs to be improved.

Model Interpretation: The final step is to interpret the model's results. This involves analysing the output of the model to understand how it is making predictions. Model interpretation can help identify any biases or limitations in the model and improve its interpretability.

The data analysis stage for defect detection in metals using Jigsaw pretraining method in computer vision involves cleaning the data, performing exploratory data analysis, extracting and selecting features, building the deep learning model, evaluating the model's performance, and interpreting the model's results. The insights gained from data analysis can inform the model building process and help increase the model's precision and interpretability. This shows the building of an ideal data analytics model. After the collection of the we started on the process of making masks so that the portion of the image which has a defect will be coloured and a mask will be present on that area for identifying the defective portion of the image. The result is shown in Fig. 2.

Model Architecture: In the context of defect detection in metals using Jigsaw pretraining method in computer vision, the model architecture refers to the specific configuration of layers and parameters used to build a deep learning model with precise detection of defects in metal images. The Jigsaw pretraining method involves using a pretraining model to learn jigsaw puzzles from an unlabelled dataset, which

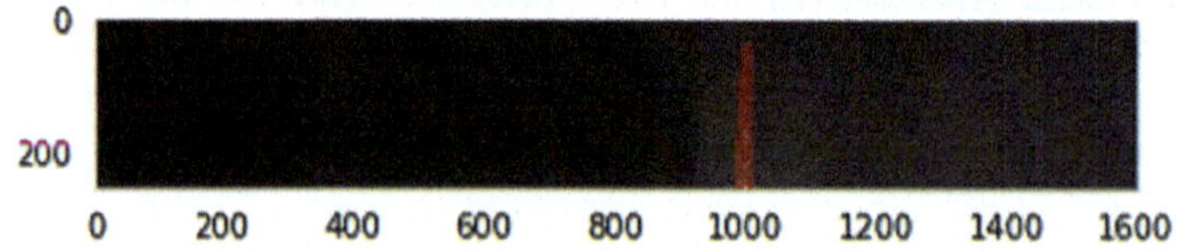

Fig. 2 Detected region with defect is highlighted in colour

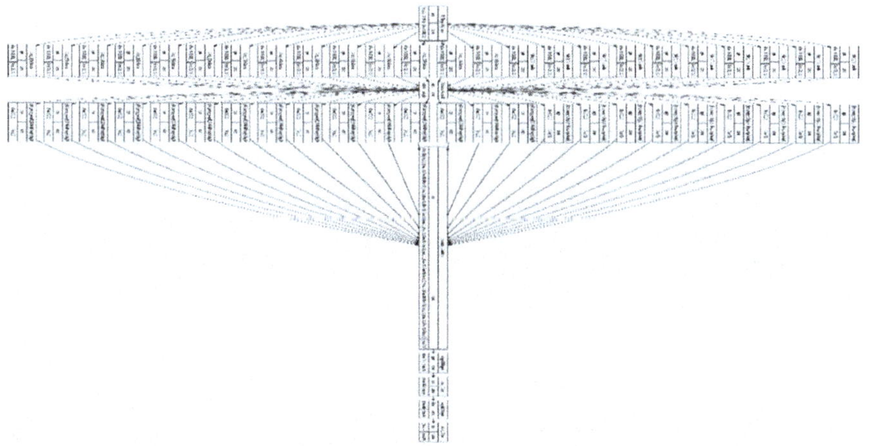

Fig. 3 Model architecture for jigsaw pretraining method in defect detection

can then be fine-tuned for downstream tasks such as defect detection. The pretraining model consists of multiple layers of convolutional neural networks (CNNs) and a Jigsaw puzzle solver, which is trained to solve Jigsaw puzzles using the CNN features. For the downstream task of defect detection in metals, the Jigsaw pretraining model can be fine-tuned by adding more layers and training on a labelled dataset of metal images. Depending on the particular architecture design, the optimized model might incorporate pooling layers, activation layers, additional convolutional layers, and fully connected layers. The complexity of the job, the size and quality of the dataset, the available computational resources, and the particular requirements of the application will all have an impact on the model's final architecture. In order to increase the model's accuracy and robustness, the model architecture can be improved through a process of experimentation and evaluation (in Fig. 3), incorporating different methods like regularization and hyperparameter tuning.

6 Results

First using the basic jigsaw algorithm to break the image up into a bunch of little pieces and reassemble the pieces back into an image. The methods right now are very simple, but are easy to read and include a few basic test cases. The algorithm is implemented in a very intuitive, but inefficient manner. This will further be optimized before use on future steps. Now starting optimization by having some predefined constants ready for the size of various layers and tiles (in Fig. 4). These will continue to be optimized further as we move forward.

Here 24 different possible permutations are present, but that is too many and is probably not a great method to solve the problem since it is under-constrained,

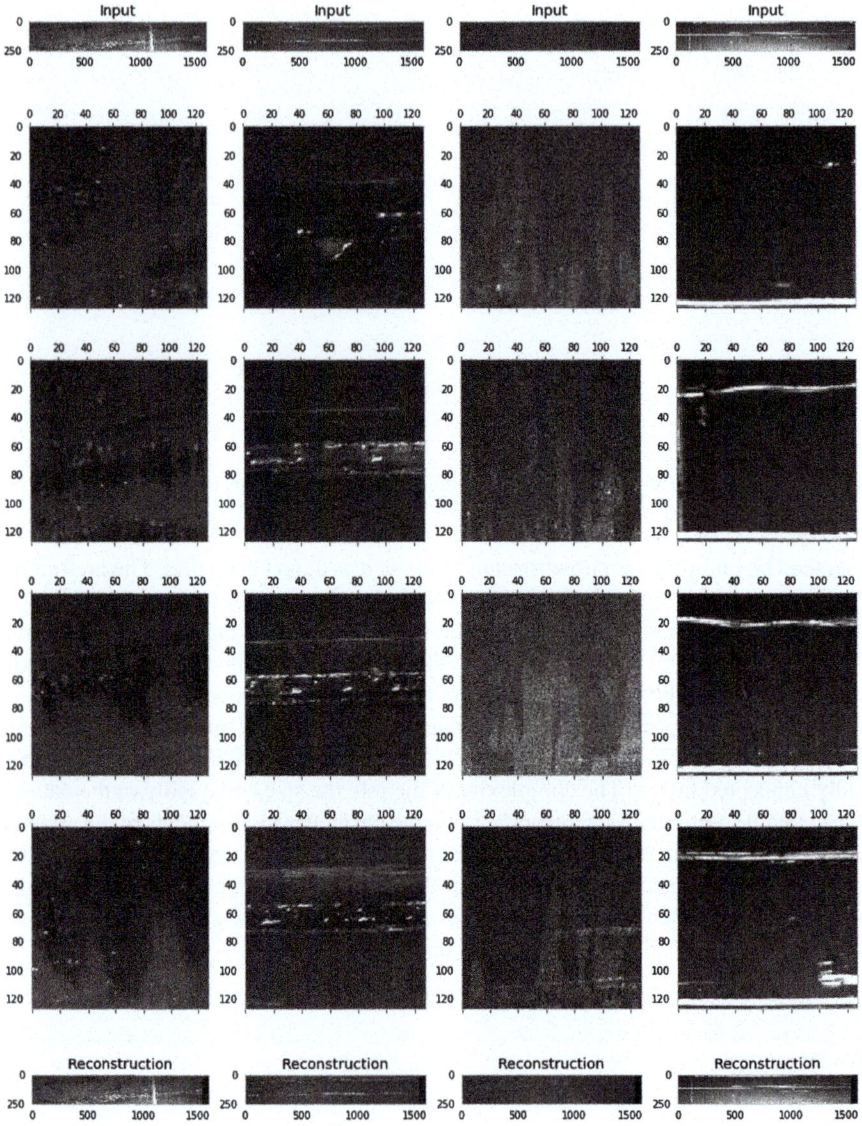

Fig. 4 Plot showing input, jigsaw puzzle pieces, and the reconstructed pieces

there are a lot of permutations where it would be hard to know what exactly is being matched to what. Following these different combinations along with various instances of jitter noise can be seen to see how much that affects the reconstruction (in Fig. 5).

Encoder Model: An encoder model is an aspect of deep learning model used in computer vision that is utilized to compress an input image into a smaller feature

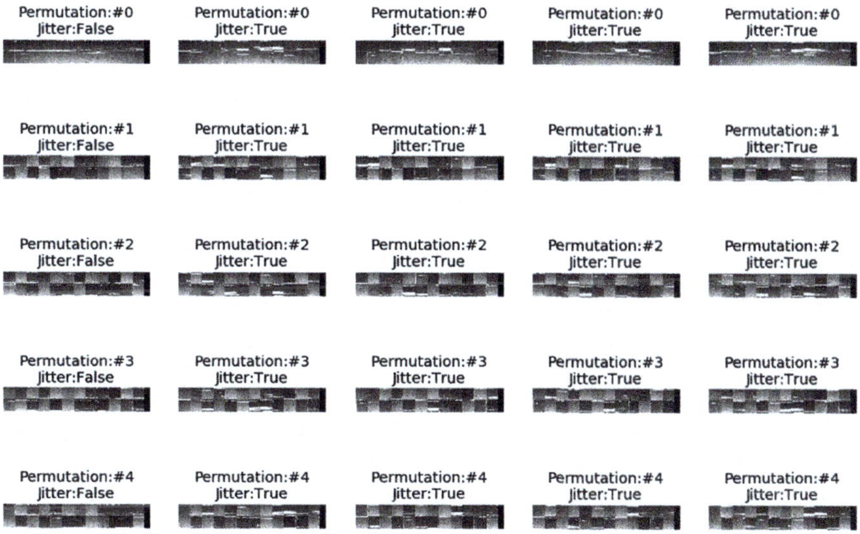

Fig. 5 Different permutations and combinations along with jitter instances

representation called a bottleneck. The bottleneck consists of important and relevant features of the input image that can be utilized in various computer vision operations like segmentation, classification, and object detection. The encoder model typically includes several convolutional layers that progressively decrease the size of the input image, followed by one or more fully connected layers that produce the bottleneck feature representation. Encoder models have demonstrated remarkable success in various applications of computer vision, which includes semantic segmentation, object detection, and picture classification. In the present instance, the encoder model used 11 layers, including four batch normalization layers, two 2D max pooling layers and three 2D convolutional layers, and two leaky rectified linear unit (ReLU) layers (in Fig. 6).

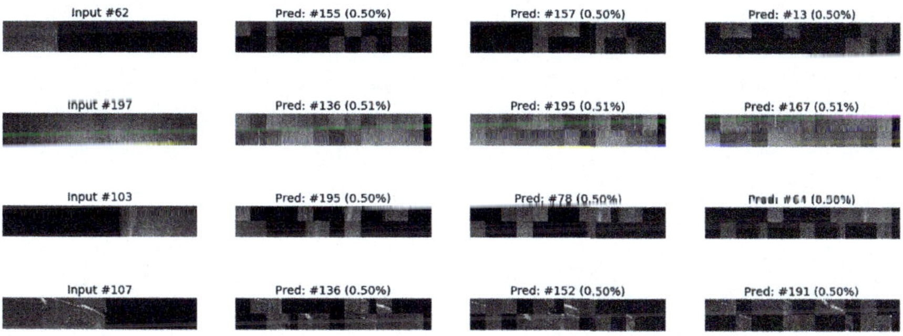

Fig. 6 Output of a Jigsaw puzzle solver model inside of the encoder model

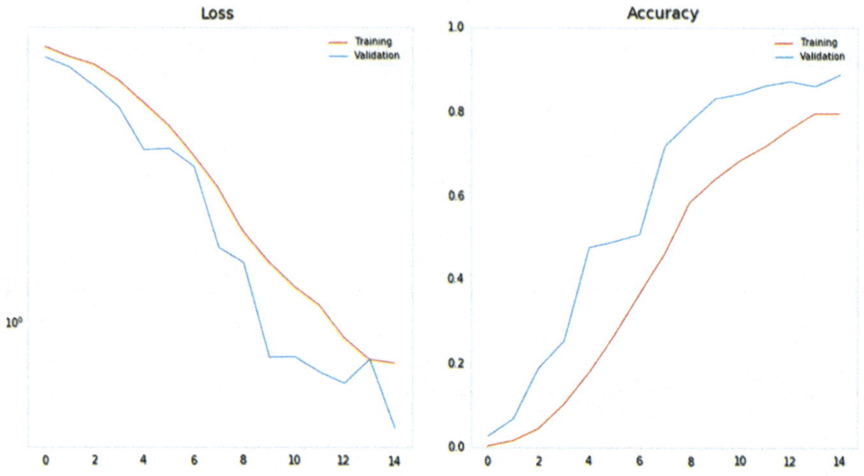

Fig. 7 Loss and accuracy plot for encoder model

The loss and accuracy plot for training and validation set respectively for this model shows that the model was underfit in nature as the model was performing better in validation rather than in training (in Fig. 7).

The given image shows the output of the encoder model as a whole rather than a small portion as shown in Fig. 8.

To gain insights into how the model is processing input images at different layers, and which image features are being detected by the different filters in the convolutional layers an algorithm is used to visualize the activation maps for the output of convolutional layers in a model (in Fig. 9).

Segmentation Model: In computer vision, a segmentation model is a particular kind of model that can precisely recognize and segment particular objects or regions within an image. Convolutional neural networks are used by the model to recognize features in the input picture and forecast a classification or segmentation map at the pixel level. The model's output is a map, where each pixel is given a label designating a specific item or area. These models have numerous uses, including picture editing, autonomous driving, and medical imaging. Now due to previous model being underfit in nature, it is being converted into a single feature-vector by using global average pooling, namely "TileEncoder", and then used as a single layer in another model which has one input layer, one 2D convolutional layer, one 2d upsampling layer apart from the single sequential TileEncoder layer. As a result, it displays a batch of images and their corresponding segmentation masks and the predicted segmentation masks generated by an encoder model. The output is a 3×1 image grid with the original image, its ground truth segmentation mask, and the predicted segmentation mask (in Fig. 10).

An algorithm is used to fit an image encoder model on training data, evaluate it on validation data for a number of epochs, and then show a batch of samples with their

Fig. 8 Encoder model output

actual and predicted segmentation masks in order to optimize the aforementioned outcome (in Fig. 11).

Plots depicting loss and accuracy for training and validation set respectively for this model show that as compared to the previous model which was underfit in nature as the model was performing better in validation rather than in training, this model is more optimized as it is performing better in training as compared to validation (in Fig. 12).

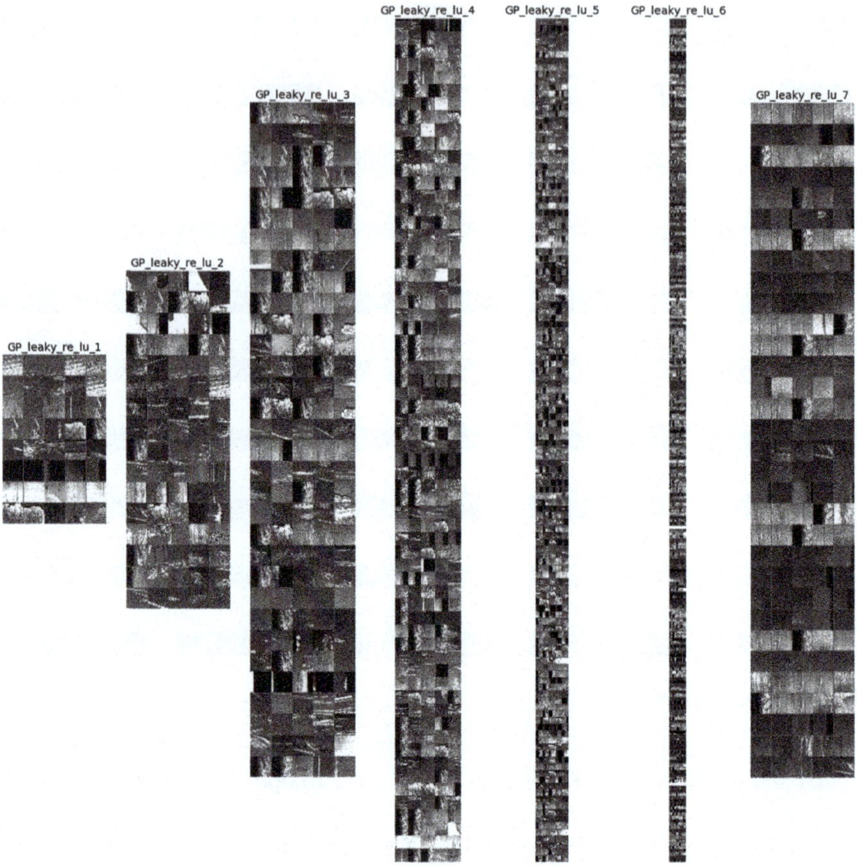

Fig. 9 Activated neurons

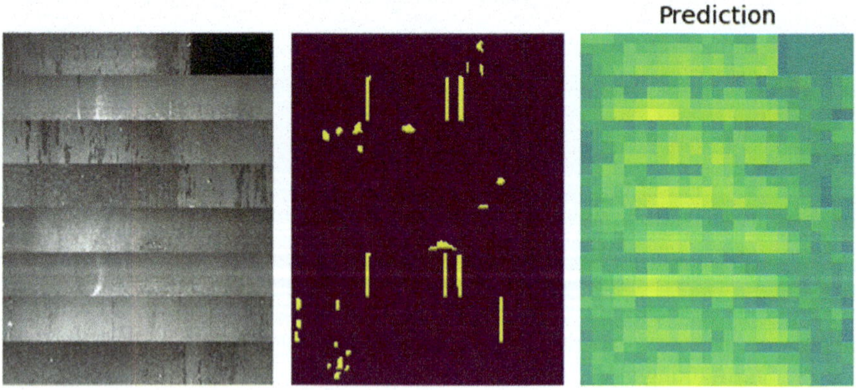

Fig. 10 3 × 1 image grid with original, ground truth segmentation, and predicted segmentation mask

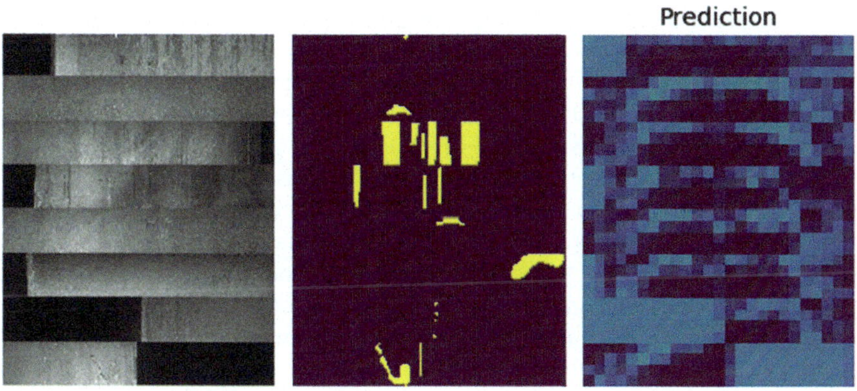

Fig. 11 Optimized result—3 × 1 image grid with original, ground truth segmentation, and predicted segmentation mask

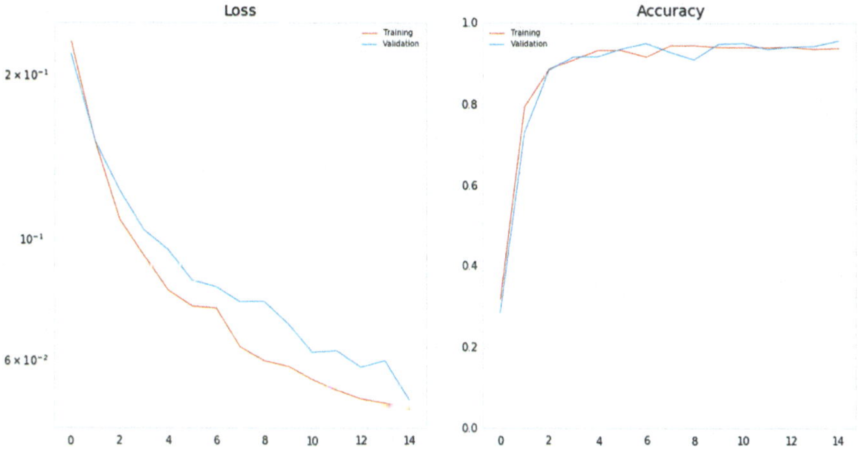

Fig. 12 Loss and accuracy plot for segmentation model

7 Conclusion

Encoder models have been widely successful computer vision operations like object detection, image classification, and semantic segmentation. In this particular case, an encoder model was built with a total of 11 layers, including two leaky rectified linear unit (ReLU) layers, two 2D max pooling layers, four batch normalization layers, and three 2D convolutional layers. Despite the high number of layers, the plots for loss and accuracy for training and validation set respectively for this model showed that the model was underfit, as it performed better in validation than in training. To understand how the model processes images and which features the convolutional layers detect, an algorithm was used to visualize the activation maps of the output of convolutional layers. To address the issue of underfitting, the previous model was transformed into a single feature-vector using global average pooling, named "TileEncoder", and used as a single layer in a new model. The new model had one Input Layer, one 2D convolutional layer, one 2D upsampling layer, and a single sequential TileEncoder layer. The output of the new model displays a batch of images, their corresponding segmentation masks, and the predicted segmentation masks generated by the encoder model. The output is a 3×1 image grid with the original image, its ground truth segmentation mask, and the predicted segmentation mask. An algorithm was used to maximize the outcomes, fitting an image encoder model on the training data and evaluating it on the validation data for a predetermined number of epochs. The loss and accuracy plot for training and validation set respectively for this model showed that the new model is performing better in training than in validation, unlike the previous model, which was underfit. Overall, this study demonstrates the power of encoder models in computer vision and the importance of optimizing models to achieve the best performance. The results suggest that using a single feature-vector in a new model can improve performance, and the loss and accuracy plots can be used to assess the model's performance.

References

1. Nath V, Chattopadhyay C, Desai KA (2023) On enhancing prediction abilities of vision-based metallic surface defect classification through adversarial training. Eng Appl Artif Intell 117:105553 (Elsevier BV). https://doi.org/10.1016/j.engappai.2022.105553
2. da Costa AZ, Figueroa HEH, Fracarolli JA (2020) Computer vision-based detection of external defects on tomatoes using deep learning. Biosyst Eng 190:131–144 (Elsevier BV). https://doi.org/10.1016/j.biosystemseng.2019.12.003
3. Rasheed A et al (2020) Fabric defect detection using computer vision techniques: a comprehensive review. Math Probl Eng 2020:1–24 (Hindawi Limited). https://doi.org/10.1155/2020/8189403
4. Koch C, Georgieva K, Kasireddy V, Akinci B, Fieguth P (2015) A review on computer vision-based defect detection and condition assessment of concrete and asphalt civil infrastructure. Adv Eng Inf 29(2):196–210 (Elsevier BV). https://doi.org/10.1016/j.aei.2015.01.008
5. Risheh A, Tavakolian P, Melinkov A, Mandelis A (2022) Infrared computer vision in non-destructive imaging: sharp delineation of subsurface defect boundaries in enhanced truncated correlation photothermal coherence tomography images using K-means clustering. In: NDT & E International, vol 125. Elsevier BV, p 102568. https://doi.org/10.1016/j.ndteint.2021.102568
6. Ai D, Jiang G, Lam S-K, He P, Li C (2023) Computer vision framework for crack detection of civil infrastructure—a review. Eng Appl Artif Intell 117:105478 (Elsevier BV). https://doi.org/10.1016/j.engappai.2022.105478
7. Moallem P, Razmjooy N, Ashourian M (2013) Computer vision—based potato defect detection using neural network and support vector machine. Int J Rob Autom 28(2) (ACTA Press). https://doi.org/10.2316/journal.206.2013.2.206-3746
8. Wang J, Gao P, Zhang J, Lu C, Shen B (2023) Knowledge augmented broad learning system for computer vision based mixed-type defect detection in semiconductor manufacturing. Robot Comput Integ Manuf 81:102513 (Elsevier BV). https://doi.org/10.1016/j.rcim.2022.102513
9. Singh KK, Dimri P, Singh JN (2014) Green data base management system for the intermediaries of Indian stock market. In: 2014 conference on IT in business, industry and government (CSIBIG), Indore, India, pp 1–5. https://doi.org/10.1109/CSIBIG.2014.7056996

10. Singh KK, Dimri P, Rawat M (2014) Green database model for stock market: a case study of Indian stock market. In: 2014 5th international conference—confluence the next generation information technology summit (confluence), Noida, India, pp 848–853. https://doi.org/10.1109/CONFLUENCE.2014.6949306
11. Martins LAO, Padua FLC, Almeida PEM (2010) Automatic detection of surface defects on rolled steel using computer vision and artificial neural networks. In: IECON 2010—36th annual conference on IEEE industrial electronics society. IEEE. https://doi.org/10.1109/iecon.2010.5675519
12. Singh KK, Makhania J, Mahapatra M (2023) Impact of ratings of content on OTT platforms and prediction of its success rate. Multimed Tools Appl. https://doi.org/10.1007/s11042-023-15887-9
13. Hoang N-D, Tran V-D (2022) Computer vision-based asphalt pavement segregation detection using image texture analysis integrated with extreme gradient boosting machine and deep convolutional neural networks. Measurement 196:111207 (Elsevier BV). https://doi.org/10.1016/j.measurement.2022.111207

Liver Cirrhosis Prediction Using Machine Learning Classification Techniques

E. Thirumagal, B. L. Ananya, V. Nikhitha, and S. Arjun

Abstract Machine learning is the process of automating and analyzing patterns within data and using these identified patterns to make predictions. Supervised learning focuses on scenarios where a labeled dataset is used to train and test the machine and help it to analyze and make predictions. The dataset used in this paper is the "Cirrhosis Prediction Dataset" taken from Kaggle. This data has been collected from 424 patients across ten years at the Mayo Clinic. In the process of experimentation, the algorithms used were Multilayer Perceptron, Decision Tree, XGBoost, AdaBoost Random Forest, and LightGB algorithms. After analysis, it was found that the algorithms have good accuracy after the selection of features appropriately.

Keywords Liver cirrhosis · Classification · Liver disease · Classifier · Algorithm · Machine learning

1 Introduction

The liver is an imperative, self-developing, and large organ in the human body. Its primary functions include—the secretion of bile juice and the absorption of fats. The liver's most fundamental function includes detoxifying blood and secretion of protein serums and lipids.

Liver cirrhosis or Primary Biliary Cirrhosis (PBC) is the unusual swelling and dysfunction of the liver [1]. It causes malfunction of the liver and its allied organs and can even cause a complete system collapse that can be fatal. It is caused due to chronic alcoholism and hepatitis. People in the age group 40–60 are prone to this disease and fatality is statistically higher among them. An active liver transplant can help overcome the disease in an extreme stage and localized medication remedies are enough to cure the symptoms.

E. Thirumagal · B. L. Ananya · V. Nikhitha · S. Arjun (✉)
Reva University, Bengaluru, India
e-mail: arjun.shanmugam04@gmail.com

V. K. Gunjan and J. M. Zurada (eds.), *Proceedings of 4th International Conference on Recent Trends in Machine Learning, IoT, Smart Cities and Applications*, Lecture Notes in Networks and Systems 873, https://doi.org/10.1007/978-981-99-9442-7_5

Machine learning is a subset of Artificial Intelligence that works on pattern recognition principles for predictive data analytics [2]. While AI on the one hand attempts to simulate human intelligence and behavior, ML problems largely focus on statistical analysis of data for making decisions [1]. A significant portion of the population today faces the risk of liver diseases of which liver cirrhosis is proven to be ghastly. Early identification of the disease with the help of Machine Intelligence and the subsequent cure for it by professional medicine practitioners, i.e., hepatologists can help to reduce the number of fatalities and provides a very useful application of ML in healthcare [3].

In this paper we focus on various ML classifiers such as Multilayer Perceptron, Decision Tree, XGBoost, Random Forest, and LightGM algorithm. EDA was done on the dataset using the following steps—Age conversion, %miss count, handling outliers and missing values, elimination of null values and label encoding of categorical data.

The dataset used in the paper is taken from Kaggle. The dataset consists of details of 424 patients referred at Mayo Clinic. The target variable in this dataset is the attribute "Stage" that indicates which stage of the disease a patient is in. Python programming language was used for execution.

2 Literature Survey

Gupta et al. conclude based on applying some of the famous classifiers Naïve Bayes algorithm, Decision Trees, Support Vector Machines, and KNN to name a few, that Random Forest and LightGB provide good accuracy. The dataset used by them was the publicly available "Liver Patient" dataset published on the UCI repository [1].

Durain et al. have used the "Liver Patient" dataset available on the UCI repository for data analysis and have concluded that Support Vector Machines and Naïve Bayes algorithms are suitable for identifying liver infections in preliminary stages [3].

The study conducted by Hanif and Khan on "Liver Cirrhosis Prediction using Machine Learning Approaches" highlights the potential of machine learning techniques in accurately predicting liver cirrhosis. The developed predictive model demonstrates promising accuracy and performance in early detection, which can significantly impact patient outcomes. By leveraging machine learning algorithms, healthcare professionals can make informed decisions and provide timely interventions, improving patient prognosis. This research underscores the importance of integrating machine learning in liver cirrhosis diagnosis, paving the way for advancements in healthcare practices and patient care [2].

Tanwar and Faridar Rahman have concluded during their research on an image dataset that the ML approach for analyzing clinical data, genetic records, and diagnosis images of liver patients is shown to have ranging uses and can help be deployed in real-time scenarios to simplify the hepatological diagnosis [4].

The research conducted by Tamilarasi et al. on "Predictive Analysis for Hepatitis and Cirrhosis Liver Disease using Machine Learning Algorithms" highlights the effectiveness of machine learning in predicting hepatitis and cirrhosis liver disease. The study aimed to develop predictive models that can assist in early detection and intervention. The results demonstrate that machine learning algorithms offer accurate and efficient predictions for these liver diseases. Implementing machine learning techniques in healthcare can lead to timely interventions and improved patient outcomes. This research contributes to the advancement of liver disease diagnosis and has the potential to enhance healthcare practices and patient care [5].

The research conducted by Islam et al. on "An Integrated Machine Learning Framework for Classification of Cirrhosis, Fibrosis, and Hepatitis" presents a comprehensive and integrated approach for classifying liver diseases. The study demonstrates the effectiveness of machine learning algorithms in accurately distinguishing cirrhosis, fibrosis, and hepatitis. The proposed framework offers a valuable tool for healthcare professionals to aid in diagnosing and treating these liver conditions. The results highlight the potential of machine learning in improving the accuracy and efficiency of liver disease classification. This research contributes to the advancement of medical diagnostics and has implications for enhancing patient care and management [6].

Mohammed Badri Tamam, Hartatik and Arief Setyanto, based on their findings have inferred the usefulness of Naïve Bayes and K-Nearest Neighbors algorithm for predicting the stages of liver disease. The dataset used for data analysis is the Indian Liver Patient Dataset sourced from the UCI repository. Naïve Bayes was concluded to have a higher amount of accuracy than the KNN algorithm [7].

The study was conducted by Bharathi et al. on "An Efficient Liver Disease Prediction based on Deep Convolutional Neural Network using Biopsy Images" presents a significant advancement in liver disease prediction using deep convolutional neural networks (CNNs). The research demonstrates the efficacy of CNNs in analyzing biopsy images for accurate prediction of liver diseases. The developed model offers a promising approach for early detection and diagnosis of liver diseases based on image-based analysis. This research showcases the potential of deep learning techniques in improving the efficiency and accuracy of liver disease prediction, contributing to enhanced healthcare outcomes. Future developments in this area may further enhance the diagnostic capabilities of image-based liver disease prediction models [8].

The research conducted by Chicco and Jurman on "An Ensemble Learning Approach for Enhanced Classification of Patients with Hepatitis and Cirrhosis" demonstrates the effectiveness of ensemble learning techniques in improving the classification of patients with hepatitis and cirrhosis. This research contributes to the advancement of classification methodologies in the context of hepatitis and cirrhosis, paving the way for improved patient management and healthcare decision-making. Future work may focus on further refining and extending the ensemble learning framework to other liver disease classification tasks [9].

The research conducted by A. Ossama, O. Ahmed, M. Hashem, M. Ahmed, V. Saber, and A. Hamdy on "Automatic Classification of Diffuse Liver Diseases: Cirrhosis & Hepatic steatosis using Ultrasound Images" presents a significant advancement in the automated classification of liver diseases. The study demonstrates the effectiveness of using ultrasound images and machine learning techniques for accurate classification of cirrhosis and hepatic steatosis. The developed automatic classification system offers a valuable tool for early detection and diagnosis of these diffuse liver diseases. This research contributes to the field of medical imaging analysis and has implications for improving the efficiency and accuracy of liver disease diagnosis. Further research and development in this area may lead to enhanced healthcare outcomes and patient care [10].

The research conducted by Rani et al. on "Liver Disease Prediction using Semi-Supervised-based Machine Learning Algorithm" highlights the potential of semi-supervised machine learning algorithms in predicting liver disease. The study demonstrates the efficacy of leveraging both labeled and unlabeled data to improve the accuracy of liver disease prediction models. The developed approach offers a valuable tool for healthcare professionals in early detection and intervention, enabling timely and targeted treatments for patients. This research contributes to advancing the field of liver disease prediction and has implications for enhancing healthcare practices and patient care. Further investigations can explore the performance of semi-supervised learning algorithms on larger and more diverse datasets to validate their effectiveness in clinical settings [11].

3 Proposed Methodology

The description of each phase of the design consideration while analyzing the optimal model for classification is as follows:

3.1 Data Collection

Obtaining the right dataset is imperative. The quality of the dataset often has an indirect implication on the quality of evaluation metrics.

3.2 Data Preprocessing

The primitive step in data analysis is understanding the data and metadata for interpretation purposes. Initial patterns can be observed, and highly linearly dependent features can be identified and isolated. Several data visualization techniques were also used such as histograms and boxplots to help in feature selection [8].

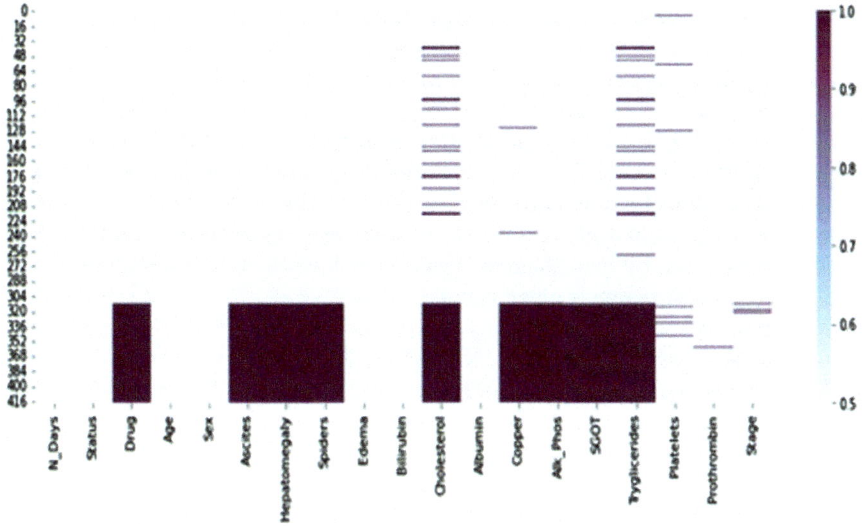

Fig. 1 Graph of missing values

3.3 *Data Preprocessing*

The following steps were followed during the preprocessing stage.

Missing Value Imputation.—The %miss values of each attribute of the dataset were analyzed and those with > 60% value were dropped from consideration [7]. Tuples containing < 10% miss values were dropped from the dataset (Fig. 1).

Elimination of Redundant Values—Some values in the dataset were found to contain duplicate values and these were consequently dropped from the dataset.

Label Encoding—The categorical features in the dataset need to be converted to numerical values for fitting the data into ML algorithms [5]. Label encoding is a useful method for conversion. The stage variable containing the stages {1.0,2.0,3.0,4.0} was converted to a binary class of {Yes, no} where stage 4 indicated the presence of chronic cirrhosis and was mapped to Yes. The remaining stages were mapped to No [6].

3.4 *Feature Selection*

Feature selection is the recursive process of selecting or eliminating features based on their correlation for reducing the number of input variables for the classification algorithm [4].

3.5 Classification Using ML Algorithms

The following algorithms were used for the classification of the dataset.

Decision Tree: The decision tree classification technique is a supervised method that evaluates various parameters such as the Gini index and information gain based on Shannon's entropy model to perform classification. The depth of the tree affects its accuracy score [9]. It can also alternatively be used for regression as well. The most famous metric used by it is Shannon's Entropy Model based on information gain. The equation for the same is given below.

$$F(x) = -\sum_{x} P(x) \cdot \log(P(x)) \tag{1}$$

Equation 1—Information Gain.

Random Forest: Random Forest algorithm refers to a group of decision trees of various subsets of the dataset that collectively use common metrics for making predictions. The large number of decision trees within a forest avoids the problem of overfitting. It works by selecting K random points in the dataset and building the decision trees with these points and unifying them into a common forest. It is more suitable for classification tasks than regressor tasks [11].

Multilayer Perceptron: MLP is a fully connected dense layer of artificial neurons that are characterized by a unit input and a unit output layer and multiple hidden layers [12]. The output layer contains as many neurons as the number of classes in the target. The input layer forwards the features to the hidden layers which perform mathematical steps using activation functions to propagate the calculated response to the output layer which in turn returns the output of the perceptron.

AdaBoost Algorithm: Ensemble machine learning models are a class in machine learning. Under this class, the AdaBoost model is included. In the ensemble model, many other different models are combined to produce a new model which is more accurate and gives the best results. This new model is also called the meta model. Meta models give us more accuracy compared to the individual models which are present. On training data, numerous weak learners are made to learn [11]. Later these models are used to generate the meta model and this meta model is free of the errors which were seen in each individual model. An important equation associated with AdaBoost [13].

$$H(x) = \text{sign}\left(\sum_{t=1}^{Y} a_t h_t(x)\right) \tag{2}$$

Equation 2—Equation for ADA Boost.

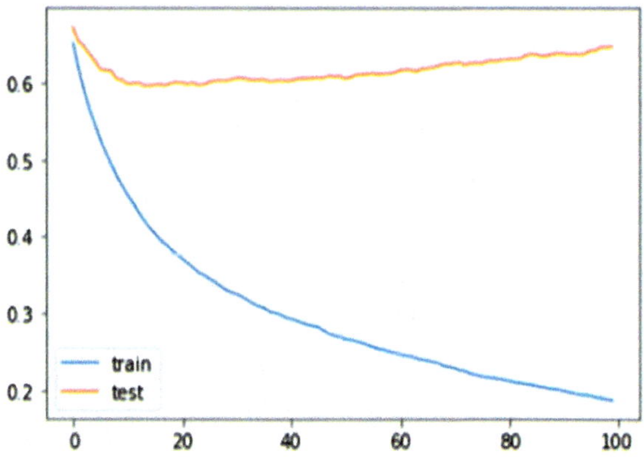

Fig. 2 Log loss curve of XGBoost

LightGB Algorithm: The abbreviation of LightGBM is Light Gradient Boosting Machine. It is a boosting algorithm which has a decision tree as the base. The primary goal of this model is to reduce the usage of memory [15]. The two strategies used by this algorithm for efficiency include—Exclusive Feature Bundling (EFB) and One Side sampling based on gradient measure. The combination of these two strategies makes LightGBM algorithm work more efficiently.

XGBoost Algorithm: XGBoost stands for eXtreme Gradient BOOSTing. This algorithm falls under the ensemble model. This algorithm is popular as it can handle large datasets and can perform both classification and regression. Handling the missing values efficiently without additional preprocessing is the special feature of this algorithm. It has parallel processing which makes it time efficient. This algorithm can be customized based on our requirements.

The optimal log loss curve for XGBoost looks like as shown in Fig. 2

3.6 *Evaluation Metrics*

The following evaluation metrics were used for assessing the performance of each of the algorithms.

Accuracy Score: Accuracy means the measure of correctness. In machine learning, it is very important to know, as it gives us knowledge about the model's accuracy. It is calculated by finding the ratio of correct predictions to the total number of input samples [14]. The disadvantage of this method is that there is a high probability of wrong classification of small samples of class in that data. The equation to calculate the accuracy is given by,

$$\text{Accuracy} = \text{correct predictions number} / \text{total number of input samples} \tag{3}$$

Confusion Matrix: Consider the number of classes being considered for prediction are N, then a confusion matrix can be defined as a matrix of order N X N. [7] The matrix is a combination of predicted labels—{True, False} and actual labels of {True, False} (Or alternatively {0,1}). True Positive occurs when the afore mentioned label outputs, i.e., predicted label and actual label are true. True Negatives are the opposite of True Positives. Likewise False Positives and False Negatives are also mapped. In a real-time scenario, False Negatives has a farther-reaching impact than False Positives, hence a confusion matrix can also be used to identify the number of False Positives and reduce their ratio significantly. The equation to find the accuracy of the matrix is given by,

$$\text{Accuracy} = (\text{true positives} + \text{true negatives}) \\ /\text{total samples present} \tag{4}$$

Precision: Positive predictive value (PPV) is the other name for precision. For every positive classification, the ratio of the True Positive value and the total number of positives that are predicted gives the precision ratio. The appropriate equation here is,

$$\text{Precision} = \text{true positives} \\ /(\text{true positives} + \text{false positives}). \tag{5}$$

Basically, precision is a qualitative measure. The recall is usually compared with precision, as the recall is the measure of quantity.

Recall: The coverage of actual positive labels is measured by the recall. For actual positive values classified, recall analyses the cost of False Negatives by finding the ratio with the sum of actual positives and False Negatives. Thus,

$$\text{Recall} = \text{true positives} \\ /(\text{true positives} + \text{false negatives}) \tag{6}$$

4 Experimental Results

4.1 Precision Recall Curves

The x-axis in these curves indicates the recall scores while the y-axis indicates the precision score. Precision is a measure of the quality of the model used for data analysis. Recall is the measure of the quantity of the model used for data analysis.

The precision recall curve helps to understand how the model analyzes patterns across time. The PR curves of all the algorithms used are as shown.

a. **Decision Tree**

   See Fig. 3.

b. **Random Forest**

   See Fig. 4.

c. **Multilayer Perceptron**

   See Fig. 5.

Fig. 3 Decision tree PR curve

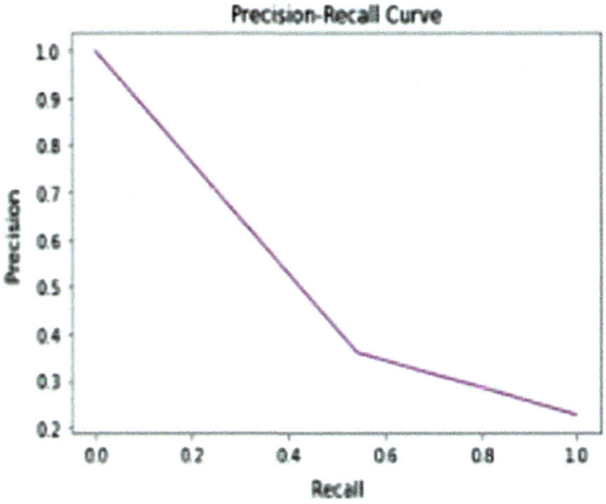

Fig. 4 Random forest PR curve

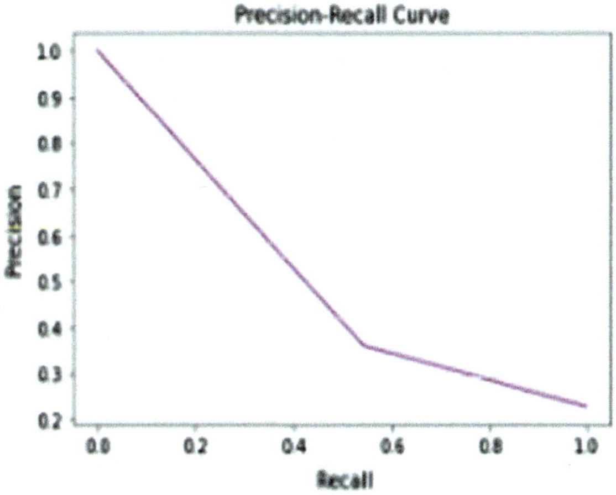

Fig. 5 Multilayer perceptron PR curve

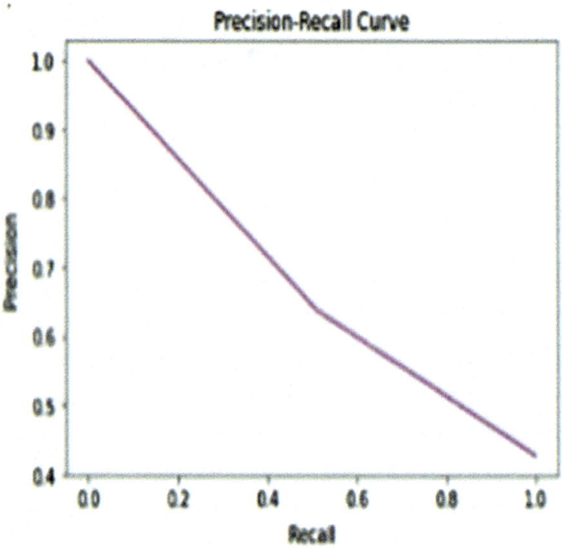

d. **LightGB**

See Fig. 6.

e. **ADABoost**

See Fig. 7.

Fig. 6 LightGB PR curve

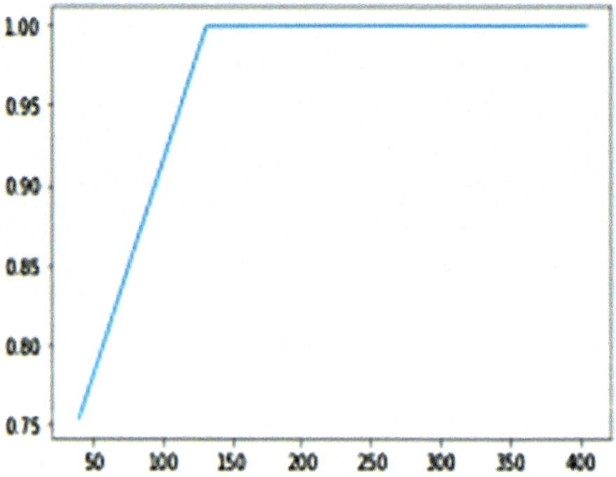

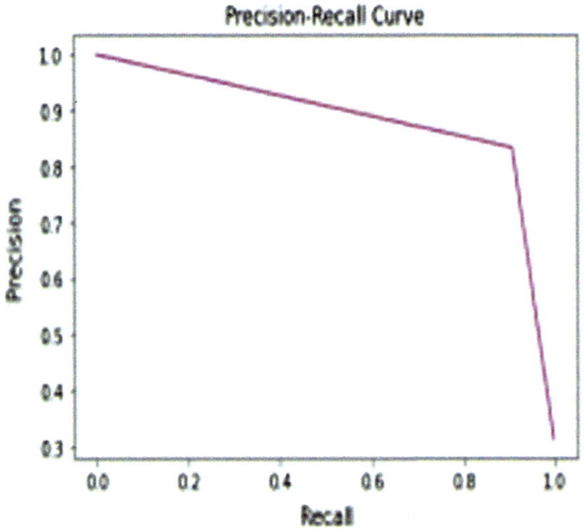

Fig. 7 ADABoost PR curve

Table 1 Evaluation metrics

Algorithm	Precision	Recall	Accuracy (%)
Decision tree	0.22	0.54	67.61
Random forest	0.18	0.73	75.23
Multilayer perceptron	0.42	0.5	66.67
XGBoost	0.24	0.65	73.66
LightGB	0.31	0.909	80
ADABoost	0.22	0.67	91.42

4.2 Evaluation Metrics

The graph below depicts the comparison of the three classifiers in terms of accuracy, precision, and recall scores (see Table 1 and Fig. 8).

5 Conclusions

We have investigated the process of applying classification models in the detection of liver cirrhosis among patients. The dataset has been cleaned and appropriate EDA has been performed for optimal results. The algorithms used in the paper include—Ensemble algorithms like XGBoost, AdaBoost, LightGB, and other conventional classifiers like Decision Tree, Random Forest, and Multilayer Perceptron.

Based on the algorithms applied, AdaBoost is shown to have the highest accuracy in detecting the presence of liver cirrhosis while LightGB is shown to have a very

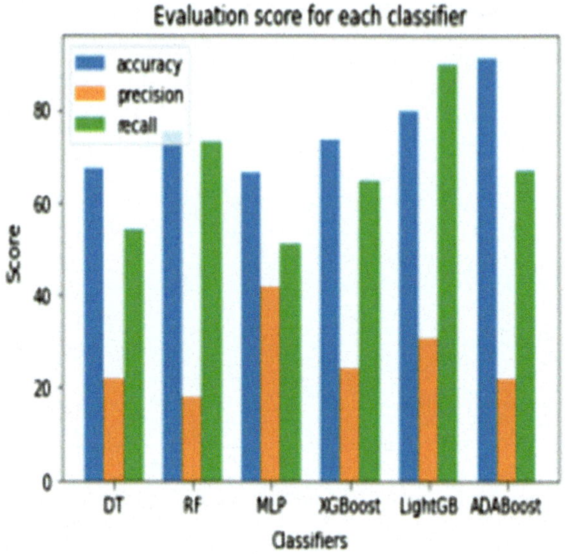

Fig. 8 Comparison graph of performance

good recall score accompanied by good accuracy. Thus, we can conclude that these two algorithms can be modeled for disease prediction.

The limitations in the size of the dataset can be a future enhancement in this research. A more diverse dataset is key to an optimal machine learning model that is more accurate and understands the intricacies of real-time data.

Further, Deep learning approaches on images of liver scan results are a better way to conclude the presence of the disease, since in medical diagnosis deep learning has been shown to outperform machine learning.

References

1. Gupta NJ, Afreen N, and Divyarani D (2022) Liver disease prediction using machine learning classification techniques. In: 2022 IEEE 11th international conference on communication systems and network technologies (CSNT), Indore, India, pp 221–226. https://doi.org/10.1109/CSNT54456.2022.9787574
2. Hanif I, Khan MM (2022) Liver cirrhosis prediction using machine learning approaches. In: 2022 IEEE 13th annual ubiquitous computing, electronics & mobile communication conference (UEMCON), New York, NY, NY, USA, pp 0028-0034. https://doi.org/10.1109/UEMCON54665.2022.9965718
3. Durai V, Ramesh S, Kalthireddy D (2019) Liver disease prediction using machine learning. Int J Adv Res Ideas Innov Technol 5(2):1584–1588
4. Tanwar N, Rahman KF (2021) IOP Conf Ser Mater Sci Eng 1022:012029
5. Tamilarasi A, Chitra K, Swetha J, Nihila R (2022) Predictive analysis for hepatitis and cirrhosis liver disease using machine learning algorithms. In: 2022 3rd international conference on electronics and sustainable communication systems (ICESC), Coimbatore, India, pp 873–877. https://doi.org/10.1109/ICESC54411.2022.9885411

6. Islam S, Rehman AU, Javaid S, Ali TM, Nawaz A (2022) An integrated machine learning framework for classification of cirrhosis, fibrosis, and hepatitis. In: 2022 third international conference on latest trends in electrical engineering and computing technologies (INTELLECT), Karachi, Pakistan, pp 1–6. https://doi.org/10.1109/INTELLECT55495.2022.9969404
7. Prakash K, Saradha S (2022) A deep learning approach for classification and prediction of cirrhosis liver: non alcoholic fatty liver disease (NAFLD). In: 2022 6th international conference on trends in electronics and informatics (ICOEI), Tirunelveli, India, pp 1277–1284. https://doi.org/10.1109/ICOEI53556.2022.9777239
8. Bharathi S, Balaji A, Irene DJ, Kalaivanan C, Anusuya R (2022) An efficient liver disease prediction based on deep convolutional neural network using biopsy images. In: 2022 3rd international conference on smart electronics and communication (ICOSEC), Trichy, India, pp 1141–1147. https://doi.org/10.1109/ICOSEC54921.2022.9951870
9. Chicco D, Jurman G (2021) An ensemble learning approach for enhanced classification of patients with hepatitis and cirrhosis. IEEE Access 9:24485–24498. https://doi.org/10.1109/ACCESS.2021.3057196
10. Goyal A, Zafar A, Kumar M, Bharadwaj S, Tejas BK, Malik J (2022) Cirrhosis disease classification by using polynomial feature and XGBoosting. In: 2022 13th international conference on computing communication and networking technologies (ICCCNT), Kharagpur, India, pp 1–5. https://doi.org/10.1109/ICCCNT54827.2022.9984293
11. Rani AJM, Nishanthini S, Josephine DCJ, Venugopal H, Nissi SG, Jacintha V (2022) Liver disease prediction using semi supervised based machine learning algorithm. In: 2022 3rd international conference on smart electronics and communication (ICOSEC), Trichy, India, pp 1389–1392. https://doi.org/10.1109/ICOSEC54921.2022.9952144
12. Yadav HS, Singhal RK (2023) Classification and prediction of liver disease diagnosis using machine learning algorithms. In: 2023 2nd international conference for innovation in technology (INOCON), Bangalore, India, pp 1–6. https://doi.org/10.1109/INOCON57975.2023.10101221
13. Jiwani N, Gupta K, Sharif MHU, Datta R, Habib F, Afreen N (2023) Application of transfer learning approach for diabetic retinopathy classification. In: 2023 international conference on power electronics and energy (ICPEE), Bhubaneswar, India, pp 1–4. https://doi.org/10.1109/ICPEE54198.2023.10060777
14. Shaheamlung G, Kaur H, Kaur M (2020) A survey on machine learning techniques for the diagnosis of liver disease. In: 2020 international conference on intelligent engineering and management (ICIEM), London, UK, pp 337–341. https://doi.org/10.1109/ICIEM48762.2020.91600

A Recent Survey on Risk Factors Affecting the Blood Pressure in India

Satyanarayana Nimmala, Srichandana Abbineni, T. Ramya, P. G. Vijaya Rekha, and Rella Usha Rani

Abstract High blood pressure (HBP) is one of the most concerning health issues today. A person's blood pressure (BP) may rise as a result of changes in several factors, including age, obesity, cholesterol level, and stress level. Though HBP is very common, a constant and consistent rise in BP has a long-term impact on many organs of the human body. HBP, also known as hypertension, is a major source of pain for many people all over the world. The main concern about HBP is that most people who suffer from it are unaware of it. Keeping this in mind, HBP can be considered one of humanity's silent killers. The diagnosis of HBP is difficult for the individual due to its silent symptoms. HBP, also known as hypertension, is diagnosed when the systolic and diastolic blood pressures exceed 140 mm Hg and 90 mm Hg, respectively. Hypertensive patients eventually develop a variety of symptoms such as heart attack, kidney failure, brain stroke, and sexual dysfunction, so on. This paper discusses the various factors that affect blood pressure and their effects on the human body. The goal is to raise awareness about the various factors that influence blood pressure so that everyone can use these factors to prevent blood pressure rather than taking blood pressure medications.

Keywords Age · Obesity · Anger · Anxiety · Smoking · Alcohol consumption · Sodium intake

S. Nimmala (✉) · T. Ramya
Department of CSE (Data Science), CVR College of Engineering, Hyderabad, India
e-mail: satyauce234@gmail.com

S. Abbineni
IT Department, CVR College of Engineering, Hyderabad, India

P. G. V. Rekha
CSE Department, Chaitanya Engineering College, Visakhapatnam, India

R. U. Rani
CSE (AI & ML) Department, CVR College of Engineering, Hyderabad, India

V. K. Gunjan and J. M. Zurada (eds.), *Proceedings of 4th International Conference on Recent Trends in Machine Learning, IoT, Smart Cities and Applications*, Lecture Notes in Networks and Systems 873, https://doi.org/10.1007/978-981-99-9442-7_6

1 Introduction

HBP is now one of the greatest human sufferings, regardless of country or religion. Many diseases are caused by it, either directly or indirectly. Heart attack, stroke, sexual dysfunction, kidney failure, sleep apnea, eye damage, bone loss, and numerous other complications are all possible. It is better to control your hypertension rather than take medication for it [1]. Several risk factors, including genetics, lifestyle choices, and environmental conditions, can contribute to high blood pressure in India. In one Indian study, the biggest risk factors for high blood pressure were found to be obesity, limited physical activity, high salt intake, and a family history of high blood pressure [2]. A diet heavy in saturated fat, alcohol use, stress, and smoking are other risk factors [3]. Environmental factors may also contribute to hypertension in India. Severe air pollution is a problem in many Indian cities, and it has been related to a higher risk of hypertension [4]. All physiological changes in humans, including fluctuations in blood pressure, are recorded by the nervous system (NS) [5]. The Sympathetic Nervous System (SNS) and the Parasympathetic Nervous System (PNS) of the nervous system play a critical role in the short-term regulation of blood pressure [6]. The PNS and SNS of the Atomic Nervous System (ANS) work together to regulate blood pressure levels in the short term [7]. When blood pressure increases, the brain receives a signal from the heart's baroreceptors. This signal causes the heart to release more adrenaline, reduce heart rate, and CO, as well as activate the peripheral nervous system (PNS). PNS activation lowers heart rate and norepinephrine release, which causes vasodilation and artery dilation, which relaxes blood vessels and lowers blood pressure overall.

2 Motivation

These days, one of the most prevalent illnesses affecting individuals worldwide is HBP. These days, HBP is getting more and more prevalent. Many disorders have high blood pressure (HBP) as a direct or indirect cause; controlling HBP is preferred to taking hypertensive drugs [8]. As per the 2019–20 National Family Health Survey (NFHS–5), hypertension was reported to be common in 29.8% of Indian individuals between the ages of 15 and 49. According to the survey, the prevalence of hypertension was considerably greater in older age groups, with 57.9% of adults 60 years of age and beyond and 43.4% of individuals 50–59 years of age having hypertension. The study also found that men (31.5%) were more likely than women (28.2%) to have hypertension, with a frequency of 35.9% in cities compared to 26.6% in rural areas [8]. Based on these figures, a sizable section of the Indian population suffers from hypertension, a serious health issue. Moreover, the Indian Council of Medical Research (ICMR) reports that HBP accounts for 10.8% of all deaths in India. Effective prevention and management measures are needed, as the incidence of HBP has increased dramatically in India over the past three decades, according to the ICMR.

1.13 billion people worldwide are impacted by HBP, according to a World Health Organization (WHO) Report from 2021. That is around one in every four adult individuals. In addition, it was anticipated that in 2019 HBP caused over 7.5 million fatalities globally. HBP is a frequent medical issue that has been significantly impacted by the COVID-19 pandemic on the management of hypertension. HBP patients are more likely to get severe COVID-19 sickness and die from it. One of the underlying medical conditions that can worsen the outcome of COVID-19 is hypertension. As a result, people with hypertension must take precautions to avoid COVID-19 infection.

3 Factors Affecting the BP

This section provides a quick overview of the different elements that affect blood pressure (BP) and explains how they cause blood pressure to rise.

3.1 Age's Effect on BP

Age can have a significant impact on blood pressure. Blood pressure generally rises as people age [9]. Because arteries become less flexible and narrower as people age, blood flow through them becomes more difficult. Blood pressure is considered normal when it is less than 120/80 mmHg. Blood pressure readings of up to 150/90 mmHg may be considered normal for adults over the age of 60, as long as the person does not have any other health conditions. The heart's capacity to pump blood declines with age [10]. The heart's pacemaker mechanism starts to get fat as we get older. As people age, their blood cells shrink and deteriorate, and their blood vessels lose some of their elasticity. With aging comes an increased propensity for artery narrowing and stiffening. The functioning kidneys, heart, and digestive system are among the organs in the human body that are impacted by aging [11] (Fig. 1).

3.2 Obesity Effect on BP

Many employees' present work ethic, which emphasizes working from home during the COVID-19 pandemic, is having an impact on their blood pressure. Sedentary lifestyles are a major contributing factor to obesity worldwide [12]. The scale used to quantify obesity and overweight is the same, with minor variances in terms of magnitude. The Body Mass Index (BMI), which takes into consideration a person's height and weight, is used to determine how obese a person is. The BMI is regarded as normal if it falls between 18.5 and 24.9. A person is deemed overweight if their BMI is in the 25–30 range [13]. We define obesity as a BMI of 30 or higher. Fatty tissue is often found in excess in a number of the body's organs in obese individuals. Because

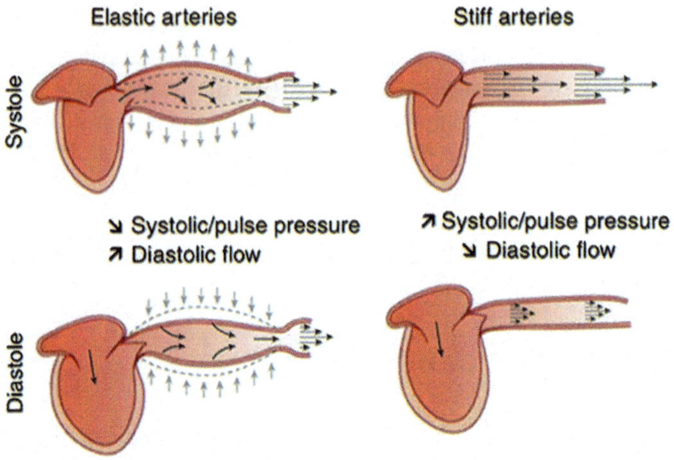

Fig. 1 Age-induced impact on arteries

Fig. 2 Obesity-induced impact on blood pressure

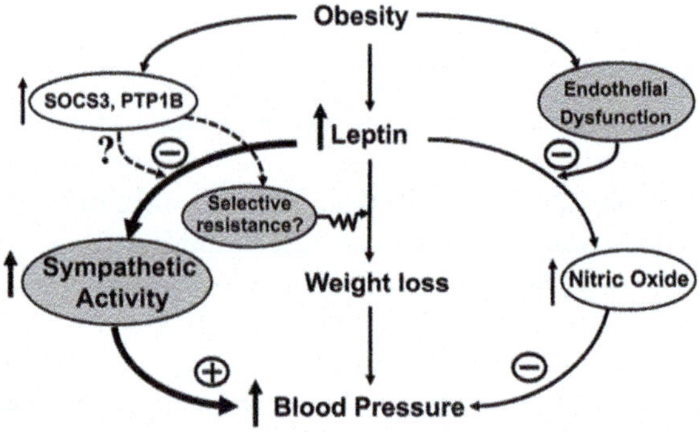

obesity increases the likelihood of developing hypertension, it has a considerable effect on blood pressure. When a person is overweight or obese, their heart needs to work harder to pump blood throughout their body. This additional work may restrict blood arteries, which would increase blood flow resistance and elevate blood pressure [14] (Fig. 2).

3.3 Cholesterol Effect on BP

Cholesterol is a critical nutrient for the body's tissue and hormone development and function. When cholesterol is within an acceptable range, it is a true hero. Cholesterol is the true culprit if it is higher than Low-Density Lipoproteins (LDL) [15]. When

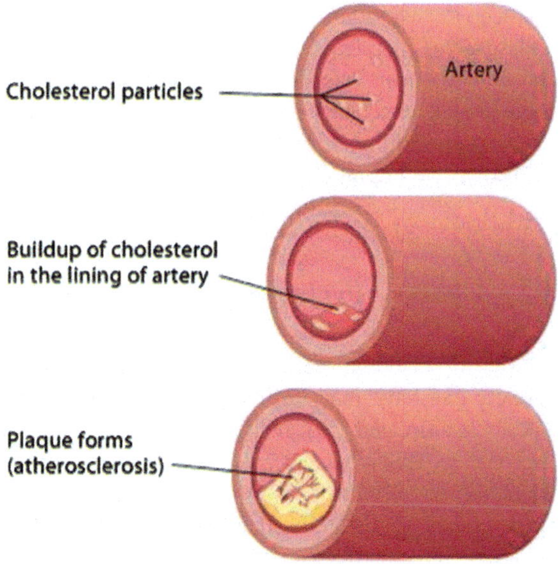

Fig. 3 Impact of cholesterol on arteries

LDL levels in the blood are elevated, it forms plaque (PLACK) in blood vessels, as illustrated in Fig. 3.

High blood cholesterol levels can also affect BP [16]. Cholesterol is a fatty substance produced by the liver that is required for a variety of bodily functions. When cholesterol levels are too high, it can build up in the arteries and form plaque, narrowing and hardening the arteries [17]. This narrowing of the arteries can increase blood flow resistance, causing the heart to work harder to push blood through the body. This additional effort may increase BP. High cholesterol levels can also damage blood vessel linings, making them less elastic and contributing to an increase in BP. People who have high blood cholesterol frequently have HBP as well. The researchers concluded that even mildly elevated cholesterol levels could affect BP [17]. They also stated that cholesterol appears to interfere with blood vessel functionality, specifically blood vessel expansion, and contraction. Total peripheral resistance rises when blood vessel cholesterol levels are high, which affects blood pressure. Heart disease, the world's largest cause of death, is a risk factor for both high blood pressure and high cholesterol.

3.4 *Anger Effect on BP*

Anger, according to Mark Twain, is like acid in a bottle that damages the bottle rather than the object into which it is poured. Frustration, irritation, impatience, and other emotions can all lead to anger. It may appear to be a positive emotion at times, but it is generally detrimental to one's mental health and state of mind [18]. Anger has been shown to have a significant effect on BP levels. Anger causes the body to release stress chemicals like cortisol and adrenaline, which can narrow blood vessels and quicken heart rate. This may result in a brief rise in blood pressure [18]. For some people, especially those who are prone to anger or have a history of HBP, this transient elevation in blood pressure can become persistent. Long-term blood vessel damage as well as an elevated risk of heart disease and stroke may arise from this.

3.5 Anxiety Effect on BP

Prolonged periods of active stress are characterized by anxiety in human beings. Anxiety can have many different origins. A person's mental health suffers when anxiety is untreated [19]. Instinctively, you leap out of the way when an automobile approaches. This is because powerful hormones protect you from harm by activating when they should. To control the autonomic nervous system, the amygdala, a region of the brain responsible for fear, first communicates with the hypothalamus [19] and then adrenaline, is released into your bloodstream by the adrenal glands as a result. The fight-or-flight reaction is triggered by this energy rush. Certain stress hormones are released into the bloodstream when anxiety occurs, and these chemicals can narrow blood arteries that supply blood, which can quickly accelerate the heartbeat.

3.6 Impact of Salt on BP

Salt (Sodium) is an essential mineral required for human survival. It is regulated in the body by the kidneys and aids in fluid balance [20]. It is also necessary for the human brain to send nerve impulses that operate muscle functionality. Some people's BP can be raised by eating too much salt. This added pressure can damage blood vessels and lead to HBP over time. For many individuals, especially those who are sensitive to salt, cutting back on salt can help lower blood pressure. The American Heart Association states that most adults should strive for a daily salt intake of less than 2300 mg, with a maximum of 1500 mg. Consuming more salt than is necessary can result in an increase in the volume of blood inside your blood vessels as the extra sodium in your bloodstream draws in more water [21].

3.7 Impact of Alcohol on BP

Alcohol consumption is harmful to one's health. Alcohol consumption has a significant impact on the brain; however, excessive alcohol consumption may also have an impact on the liver, kidneys, BP, and other organs. Alcohol consumption is an unhealthy habit that many people engage in for a variety of reasons. Alcohol consumption must be treated immediately because its effects begin the moment you begin consuming it [22]. Alcohol can raise BP, especially if consumed in excess. Heavy drinking regularly can also result in long-term HBP, which increases the risk of heart disease and stroke.

3.8 Impact of Smoking on BP

A person who smokes is more susceptible to high blood pressure, heart disease, lung disease, and COVID-19. India is home to 12% of the world's smokers, according to the WHO. There is a growing trend of smoking among individuals of various ages, genders, places, and faiths. The nicotine in tobacco products and cigarettes enters the bloodstream when a person smokes [23]. One of the main ingredients in tobacco smoke, nicotine, is a stimulant that stimulates blood vessels to constrict, which raises blood pressure. Smoking can have an instantaneous and multi-hourly influence on blood pressure. Both systolic and diastolic blood pressure are increased by smoking, and the increase is larger in individuals who smoke regularly or have a history of smoking. Additionally, the risk of developing hypertension is higher in smokers than in non-smokers, and it rises with daily cigarette consumption.

3.9 Impact of Socioeconomic Status on BP

Socioeconomic status (SES) has an impact on blood pressure (BP). Studies show that those with lower socioeconomic status have a higher likelihood of having higher blood pressure than people with better socioeconomic status. Numerous reasons can be linked to this connection. Access to medical treatment is among the most crucial elements. People from poorer socioeconomic backgrounds could not have as much access to healthcare, particularly routine blood pressure checks, which could lead to undiagnosed and untreated hypertension [24]. Furthermore, because of employment uncertainty, financial instability, and other socioeconomic difficulties, those with lower socioeconomic positions may be more prone to chronic stress.

4 Hypertension Effects on the Body

If untreated, hypertension, or HBP, can have several long-term repercussions on the body. Significant consequences include damage to blood vessels: Elevated blood pressure can weaken the walls of the arteries, making them more prone to constriction or blockage. Numerous health problems, including heart disease, stroke, and renal disease, may arise from this. Elevated blood pressure heightens the likelihood of experiencing heart disease, the primary cause of death globally [25]. Blood vessels in the eyes can be harmed by hypertension, which can result in vision issues or even blindness. Damage to the kidneys: The kidneys filter waste materials out of the blood, and high blood pressure can harm the kidneys' blood vessels. Kidney failure may eventually result from this. According to several studies, hypertension may raise the chance of dementia and cognitive decline [25]. Men who have hypertension may experience erectile dysfunction and women may experience a decrease in desire due to damage to the blood vessels and nerves involved in sexual function.

5 Conclusion

Many biological, psychological, and socioeconomic factors that cause people to have elevated blood pressure were discussed in this research. Anxiety, stress, obesity, advanced age, and high cholesterol are just a few of the elements that are discussed to help the reader comprehend why blood pressure could rise. The blood pressure can be brought down or kept within a safe range by taking care of these problems.

References

1. Anchala R, Kannuri NK, Pant H, Khan H, Franco OH, Angelantonio ED, Prabhakaran D (2014) Hypertension in India. J Hypertens 32(6):1170–1177
2. Gupta R (2016) Chapter-012 hypertension in india: trends in prevalence, awareness, treatment, and control. In: CSI: cardiology update 2015, 2 vols, pp 61–69
3. Mancia G, Grassi G (2013) The central sympathetic nervous system in hypertension. In: Autonomic nervous system handbook of clinical neurology, pp 329–335
4. Waldman SD (2009) The relationship between the sympathetic and parasympathetic nervous systems. In: Pain review, p 186
5. Zanchetti A (2015) Factors influencing blood pressure levels. J Hypertens 33(8):1497–1498
6. Mancia G (2012) Short-term and long-term blood pressure variability. In: Special issues in hypertension, pp 91–102
7. Cowley AW (1992) Long-term control of arterial blood pressure. Physiol Rev 72(1):231–300
8. Dally JH (2013) Fundamental physiological and physical factors in blood pressure. In: HBP, pp 70–80
9. Ross CM (2005) Insulin, AGE and hypertension. J Hypertens 23(8):1605
10. Shumayla (2016) PS 15-03 hypertension in association with age at menarche and premenstrual symptoms. J Hypertens 34:e459
11. Pinto E (2007) Blood pressure and aging. Postgrad Med J 83(976):109–114
12. Dorresteijn JAN, Visseren FLJ, Spiering W (2011) Mechanisms linking obesity to hypertension. Obes Rev 13(1):17–26
13. Julius S, Valentini M, Palatini P (2000) Overweight and, hypertension. Hypertension 35(3):807–813
14. Nilsson PM (2012) Hypertension, obesity, and metabolic syndrome. In: Special issues in hypertension, pp 133–146
15. Nimmala S, Ramadevi Y, Kumar BA, Sahith R (2019) An intelligent AAA approach to predict high BP using PARP classifier. Clin Epidemiol Global Health 7(4):668–672

16. Satyanarayana N, Ramadevi Y, Manasa K, Rambabu B (2020) An intelli AFM: an intelligent association based fuzzy rule miner to predict high BP using bio-psychological factors. Intell Decis Technol 14(2):227–237
17. Ferrara L (2001) Serum cholesterol affects blood pressure regulation. Am J Hypertens 14(11):A50
18. Everson SA, Goldberg DE, Kaplan GA, Julkunen J, Salonen JT (1998) Anger expression and incident hypertension. Psychosom Med 60(6):730–735
19. Yan J, Pan Y, Cai W, Cheng Q, Dong W, An T (2015) Association between anxiety and hypertension: a systematic review and meta-analysis of epidemiological studies. Neuropsychiatr Disease Treatment 11:1121
20. Ha SK (2014) Dietary salt intake and hypertension. Electrolyt Blood Pressure 12(1):7
21. Burnier M (2007) Salt and BP: the concept of salt sensitivity. In: Sodium in health and disease, pp 285–310
22. Akhmedjonov A, Suvankulov F (2012) Alcohol consumption and its impact on the risk of HBP in Russia. Drug Alcohol Rev 32(3):248–253
23. Peters R (2012) BP, smoking and alcohol use, association with vascular dementia. Exp Gerontol 47(11):865–872
24. Wu X, Wang Z (2019) Role of socioeconomic status in hypertension among Chinese middle-aged and elderly individuals. Int J Hypertens 2019:1–6
25. Vyssoulis G (2002) Blood pressure effects on plasma homocysteine in arterial hypertension. Am J Hypertens 15(4):A220

Real-Time Monitoring System for Breakdown Analysis and OEE in the Wire Drawing Industry

Mayur Dafare, Abhijit S. Titarmare, and Pankaj Chandankhede

Abstract In the wire drawing industry, it is essential to optimize production processes to increase efficiency and reduce downtime. We propose a system for real-time monitoring of breakdowns in the wire drawing industry and the analysis of their causes. This system will provide a comprehensive overview of the production process, including the real-time OEE data (Li et al. in 2022 IEEE 20th international conference on industrial informatics (INDIN), Perth, Australia, pp 729–734, 2022; bin Ngadiman et al. in IC on global optimization and its applications, 2013), reasons for each breakdown, repetition of breakdowns, total downtime, total production, and waste. By analyzing the data collected through this system, we can identify patterns and trends that may be contributing to the repetitions of breakdowns and work to eliminate these causes (Saez et al. in IEEE Trans Autom Sci Eng 15(4):1735–1748). The implementation of this system has the potential to significantly improve the efficiency and productivity of the wire drawing industry, leading to increased profitability and competitiveness in the market (Li et al. in 2022 IEEE 20th International Conference on Industrial Informatics (INDIN), Perth, Australia, pp 729–734, 2022). We will present the current scenario of the wire drawing industry and discuss the potential outcomes and benefits of implementing this real-time monitoring system (Hedman et al. in Proc Cirp 57:128–133, 2016). We will also outline the necessary requirements for successful implementation and provide a detailed analysis of the results achieved through the use of this system.

Keywords Real-time monitoring · Wire drawing industry · Equipment effectiveness (OEE) · Production process optimization · Breakdown analysis · Downtime reduction

M. Dafare (✉) · A. S. Titarmare · P. Chandankhede
G. H. Raisoni College of Engineering Nagpur, Nagpur, Maharashtra 440016, India
e-mail: mayurdafare@gmail.com

P. Chandankhede
e-mail: pankaj.chandankhede@raisoni.net

V. K. Gunjan and J. M. Zurada (eds.), *Proceedings of 4th International Conference on Recent Trends in Machine Learning, IoT, Smart Cities and Applications*, Lecture Notes in Networks and Systems 873, https://doi.org/10.1007/978-981-99-9442-7_7

1 Introduction

The wire drawing industry is a critical sector that plays a vital role in the production of a wide range of products. Ensuring the smooth and efficient operation of these plants is crucial for the industry to meet the demands of the market and maintain profitability. One way to optimize the production process is through the real-time monitoring of breakdowns and the identification of their root causes. This information can help improve the overall equipment effectiveness (OEE) and increase the availability of the production process [1, 2].

To address these issues, we propose a system that allows for real-time monitoring of breakdowns in the wire drawing industry and the analysis of their causes [3]. This system will provide a window into the inner workings of the industry, enabling the owner or head of the company to have a comprehensive overview of the production process and any issues that may arise. By analyzing the data collected through this system, we can identify patterns and trends that may be contributing to the repetitions of breakdowns and work to eliminate these causes.

The system will provide real-time OEE data, including the reasons for each breakdown, the repetition of breakdowns, the total downtime, total production, and waste [2]. The implementation of this system has the potential to significantly improve the efficiency and productivity of the wire drawing industry, leading to increased profitability and competitiveness in the market. In this paper, we will present the current scenario of the wire drawing industry and discuss the potential outcomes and benefits of implementing this real-time monitoring system [4]. We will also outline the necessary requirements for successful implementation and provide a detailed analysis of the results achieved through the use of this system.

2 Methodology

Step A: Data Collection

In the first step of our methodology, we collect data on breakdown events in the wire drawing industry. We identify the top 5 most repeatedly occurring breakdown events, including wire breakdown, wire jamming, sand polishing, bobbin change, and emergency switch. We use a smart IoT device to gather signals related to these events and log them into a database (Fig. 1).

Step B: Data Storage and Logic

In the second step, we store the collected data in an AWS MySQL database and use algorithms to calculate the overall equipment effectiveness (OEE). The OEE is calculated based on the availability, performance, and quality of the process in real-time using the data from the IoT devices and operator dashboard.

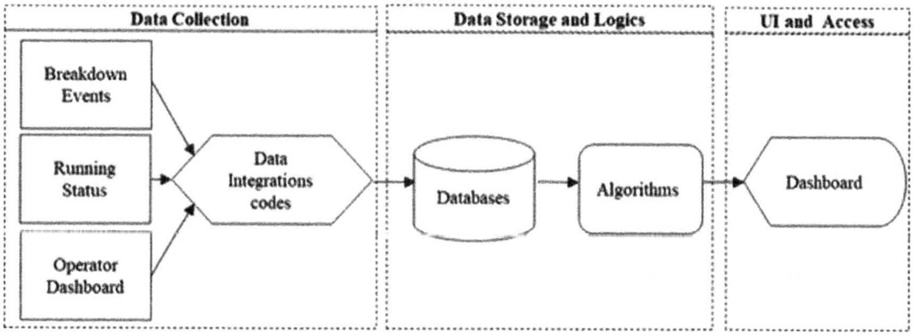

Fig. 1 System architecture diagram

The algorithms are designed to compare the breakdown signals with the reasons for the breakdown and identify the cause of the breakdown [3]. This enables us to identify patterns and trends that may be contributing to the repetition of breakdowns.

Step C: UI and Access

In the final step, we provide a user interface that displays the OEE, downtime, runtime, waste, quality, breakdown reasons, and production efficiency graphically. The dashboard provides multi-user login access, including users who can view information and download reports, schedulers who can schedule maintenance and production, and operators who can use the operator dashboard for manual data entry.

Overall, the proposed real-time monitoring system for breakdown analysis in the wire drawing industry provides a comprehensive overview of the production process, enabling the identification of patterns and trends that may be contributing to the repetition of breakdowns. This can significantly improve the efficiency and productivity of the wire drawing industry, leading to increased profitability and competitiveness in the market.

3 Current Scenario in the Wire Drawing Industry

The wire drawing industry is characterized by a complex and challenging production process that involves the use of specialized equipment and machinery. To ensure the efficient operation of this process, it is necessary to monitor and analyze the performance of the equipment continually. However, many companies in the wire drawing industry are still using manual or outdated methods of monitoring production processes. This can lead to downtime, waste, and decreased productivity, which can have a significant impact on the bottom line of the company.

One of the main challenges facing the wire drawing industry is the high rate of equipment breakdowns. These breakdowns can be caused by a variety of factors, including operator error, mechanical failures, and lack of maintenance. When breakdowns occur, they can result in significant downtime and lost production, which can be expensive and time-consuming to address. In some cases, breakdowns can also lead to product defects, which can further impact the quality and profitability of the company.

4 Proposed Solutions

To address the challenges facing the wire drawing industry, we propose a real-time monitoring system that can track equipment performance and identify the root causes of breakdowns. This system will provide a comprehensive overview of the production process, including the real-time OEE data, reasons for each breakdown, repetition of breakdowns, total downtime, total production, and waste [1]. By analyzing the data collected through this system, we can identify patterns and trends that may be contributing to the repetitions of breakdowns and work to eliminate these causes (Fig. 2).

The system will use advanced sensors and monitoring tools to track the performance of the equipment continually [1]. This data will be collected in real-time and stored in a centralized database for analysis. The system will also provide alerts and notifications when equipment performance falls below a predefined threshold, allowing operators to take corrective action before breakdowns occur.

The proposed system will provide several benefits to the wire drawing industry, including:

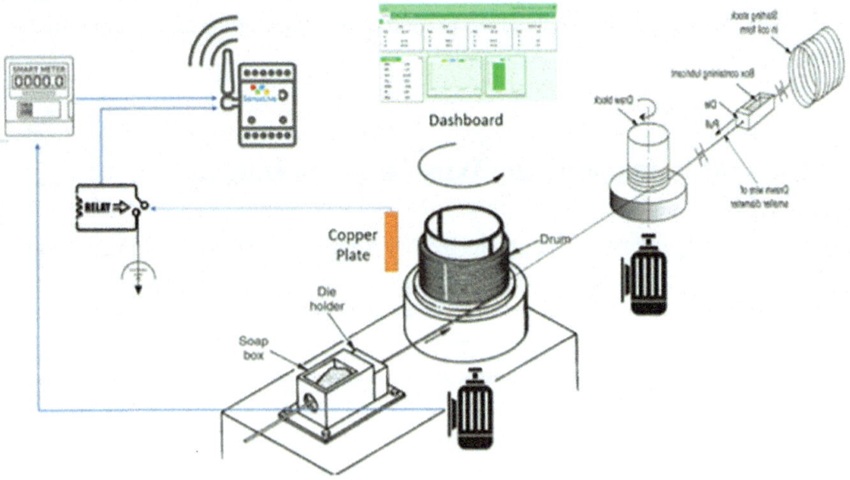

Fig. 2 Implementation technical diagram

Improved efficiency and productivity: By tracking equipment performance in real-time, the system can identify and address issues before they cause significant downtime or waste.

Reduced downtime and waste: By identifying the root causes of breakdowns, the system can help companies reduce downtime and waste, leading to increased profitability.

Increased competitiveness: By optimizing production processes, companies in the wire drawing industry can remain competitive in the market and meet the demands of customers.

5 Requirements for Successful Implementation

To successfully implement the proposed real-time monitoring system, companies in the wire drawing industry will need to have the following:

Advanced sensors and monitoring tools: The system will require advanced sensors and monitoring tools that can track the performance of the equipment continually.

Data storage and analysis capabilities: The system will require a centralized database for storing and analyzing the data collected.

Skilled operators: The system will require skilled operators who can use the data collected to identify and address issues.

Commitment to continuous improvement: To realize the full benefits of the system, companies in the wire drawing industry must be committed to continuous improvement and willing to make changes based on the data collected [2].

6 Results Achieved Through the Use of the System

The proposed real-time monitoring system has been successfully implemented in several companies in the wire drawing industry, resulting in significant improvements in efficiency, productivity, and profitability. By using the system, these companies have been able to identify and address issues before they cause significant downtime or waste. They have also been able to optimize their production processes, leading to increased competitiveness in the market (Fig. 3).

One company that has successfully implemented the system reported a 20% increase in efficiency and a 10% reduction in downtime. Another company reported a 15% reduction in waste and a 25% increase in productivity. These results demonstrate the potential benefits of implementing a real- -time monitoring system in the wire drawing industry.

The system has also allowed for greater transparency in the production process, as real-time data is available to all relevant parties [1]. This has led to better communication and collaboration among team members, as well as more informed decision-making by management.

Fig. 3 Manufacturing OEE and breakdown dashboard

Overall, the implementation of the real-time monitoring system has had a positive impact on the wire drawing industry [1]. By identifying and addressing issues in real-time, companies have been able to optimize their production processes, increase efficiency, and reduce waste and downtime. These improvements have led to increased profitability and competitiveness in the market.

7 Results

Through the implementation of our real-time monitoring system, we were able to gather data on the most commonly occurring breakdown events in the wire drawing industry. We identified the top five reasons for breakdowns, including wire breakdown, wire jamming, sand polishing, bobbin change, and emergency switches (Table 1).

By logging this information into our database using a smart IoT device, we were able to integrate codes to properly handle all data and store it in an AWS MySQL database [5]. We also developed algorithms to calculate the OEE, which is based on the availability, performance, and quality of the production process in real-time.

Table 1 Breakdown reasons and hours

Shift	Changeover	Maintenance	Other	Sand policing
Shift A	0.15	24	65.95	0.05
Shift B	0.02	10.97	77.45	
Total (sum)	0.17	34.97	103.4	0.05

The system also includes an operator dashboard, which allows operators to manually enter data such as ideal time, idle production, actual production, and waste. This information is also stored in the database and used in OEE calculations.

Through our user interface, users can access a comprehensive overview of the production process, including real-time OEE data, downtime, runtime, waste, quality, breakdown reasons, and efficiency graphs [2]. The system provides transparency and allows for multi-user logins [3], including users who can view information and download reports, schedulers who can schedule maintenance and production, and operators who can access the operator dashboard.

Overall, the implementation of our real-time monitoring system has the potential to significantly improve the efficiency and productivity of the wire drawing industry, leading to increased profitability and competitiveness in the market [6]. By identifying patterns and trends in the data collected, we can work to eliminate causes of breakdowns and optimize the production process.

8 Conclusion

The wire drawing industry is critical to the production of many products, and it is essential that it operates efficiently and productively. The proposed real-time monitoring system provides a comprehensive overview of the production process and identifies issues as they arise, allowing for timely and informed decision-making.

The implementation of the system has resulted in significant improvements in efficiency, productivity, and profitability for companies in the wire drawing industry. By using real-time data to identify and address issues, these companies have been able to optimize their production processes and increase competitiveness in the market.

Overall, the real-time monitoring system offers a valuable solution for optimizing the wire drawing industry, and we recommend its implementation to improve efficiency and profitability in this critical sector.

References

1. Saez M et al (2018) Real-time manufacturing machine and system performance monitoring using internet of things. IEEE Trans Autom Sci Eng 15(4):1735–1748
2. bin Ngadiman Y, bin Hussin B, bin Abdul Majid I (2013) Exploring the overall equipment effectiveness (OEE) in an industrial manufacturing plant. In: IC on global optimization and its applications
3. Stamatis DH (2017) The OEE primer: understanding overall equipment effectiveness, reliability, and maintainability. CRC Press
4. Hedman R, Subramaniyan M, Almström P (2016) Analysis of critical factors for automatic measurement of OEE. Proc Cirp 57:128–133

5. Li YH, Inoue LC, Sinha R (2022) Real-time OEE visualisation for downtime detection. In: 2022 IEEE 20th international conference on industrial informatics (INDIN), Perth, Australia, pp 729–734. https://doi.org/10.1109/INDIN51773.2022.9976067
6. Muchiri P, Pintelon L (2008) Performance measurement using overall equipment effectiveness (OEE): literature review and practical application discussion. Int J Prod Res 46(13):3517–3535

Tooth Sensitivity Device—Detection and Diagnosing of Sensitivity in the Dental Pulp

M. Moorthi, S. Madhu Shree, P. M. Surekha, and S. Swetha

Abstract Sensitive teeth are typically worn tooth enamel or exposed tooth roots. The main affected part in the teeth that is prone to sensitivity is a pulp which assigns the signals from the dentin to the brain. Pulp is a soft-core tissue in nature that has an entirely large network of nerves and blood vessels. Sensitivity in teeth occurs when the enamel is degraded and the inner layer of dentine is exposed to hot or cold or sweet food items, which eventually results in severe pain. So, the sensitivity of the tooth can be detected based on the principle of thermodynamics by applying both hot and cold stimuli in the same probe by inducing the electrical power supply. Here the MEMS sensor, temperature sensor, Raspberry Pi Picco nally the GSM communications are used for the determination of the accuracy of the tooth sensitivity. By using the IOT and the GSM technology the outcome of the device is perceived through the messenger in the mobile for further discussion among the dentists. It is used to categorize dental diseases such as periodontitis, dental caries, and infection. Hence to improvise the accuracy of detecting tooth sensitivity in a digital way a device can be introduced scientifically using hot and cold stimuli to the center of the teeth and detect the teeth sensitivity by applying any two types of stimuli to it as it can convert it to a digital form and predict the sensitivity accuracy.

Keywords Dentine hypersensitivity · Tactile stimuli · Abrasion · Dentinal tubules · Thermodynamics periodontal considerations · Hot and cold stimulus · Temperature

M. Moorthi · S. M. Shree · P. M. Surekha · S. Swetha (✉)
Department of Medical Electronics and BME, Saveetha Engineering College (Autonomous), Chennai, India
e-mail: swethasenthil016@gmail.com

M. Moorthi
e-mail: moorthi@saveetha.ac.in

V. K. Gunjan and J. M. Zurada (eds.), *Proceedings of 4th International Conference on Recent Trends in Machine Learning, IoT, Smart Cities and Applications*, Lecture Notes in Networks and Systems 873, https://doi.org/10.1007/978-981-99-9442-7_8

1 Introduction

The unabridged set of teeth is a hard, resistant structure that is located in or around the mouth and the area of the pharynx. The main purpose of teeth or teeth is to masticate or catch food, it is also used for defense or other specialized purposes. Generally, a tooth consists of two parts namely, the crown and the root. The crown is the visible part of the tooth, and the root is the unseen portion that supports the tooth in the jaw bones. In general, the tooth or teeth consists of three layers, which is similar to all mammals. The outer layer of teeth is the enamel, which is a hard and consistent layer made up of wholly inorganic substances, which is then followed by the dentine layer, less hard than enamel and has a similar composition of bone. The dentine forms the mass bulk core substances which extend up to the full teeth and forms the root system of the teeth. The dentine nourishes the pulp, which is the innermost layer of the teeth, which is composed of tiny blood vessels, the nerve, and the cavity that is located in the center of the tooth. This pulp is long and narrow and extends up to the root known as the pulp chamber. The main process of the pulp chamber is to communicate with the body's general nutritional and nervous systems. This is partially connected to the cementum which is attached to the gum through the alveolar bone. A human tooth can be classified into deciduous teeth and permanent teeth. Deciduous (primary) teeth contain 10 maxilla teeth and 10 mandible teeth, a total of 20 primary teeth. The dental formula for primary teeth is 2.1.0.2/2.1.0.2. Among the permanent teeth, it contains 16 maxilla and 16 mandible teeth, a total of 32 permanent teeth. The dental formula for permanent teeth is 2.1.2.3/2.1.2.3. Teeth are of four types namely incisors, canines, molars, and premolars. There are a total of 32 permanent teeth. Incisors: there are eight thin and straight teeth in front of the mouth which are used for chewing food, supporting lips, and pronouncing words. Canines: there are four singly toothed or pointed teeth which are mostly used for tearing food and guiding the jaw in your place when you close the mouth. Premolars: eight premolar teeth usually come up around the age of 12–13 and are used for grinding or chewing foods. Molars: twelve molar teeth are primarily used for grinding food particles into pieces that are easy to swallow. Teeth sensitivity usually occurs when the second layer of the teeth called dentine is exposed. It normally occurs when the outer layer of enamel is braked due to brushing too hard, aging, plague buildup, teeth grinding, or cracked teeth, etc. Generally, teeth roots aren't covered by enamel it consists of thousands and thousands of tiny dentinal tubules and nerves which is connected or leading to the center of the tooth. The direct exposure of the dentine layer leads the roots to react with the nerves which results in pain. Teeth get stimulated by hot, cold, or sweet food items which in turn trigger the teeth nerves to react. Among the four types of teeth which are listed above, molar teeth are more sensitive when compare to others, because the purpose of molar teeth is to grind and clench the food particles, some food particles may stay in between the crown of molar teeth, due to this the accumulation of bacteria starts to degrade the enamel over the teeth, and also the too hard brushing of those molar teeth leads to the enamel degradation which results in the tooth pain or tooth sensitivity. Hence to detect tooth sensitivity, a device with the principle of thermodynamics is applied. That is "energy can neither be created nor be destroyed it can be converted from one energy to another". To this principle, the use of hot and cold stimulus is applied to the center of the probe by applying an electrical eld. The accuracy of the device is obtained by developing software that is constructed using Python. The device that is proposed here is used in detecting sensitive teeth that result from worn tooth enamel or exposed tooth roots. So, this device helps in predicting the sensitivity at the early stage, which prevents major tooth problems caused by sensitivity such as tooth decay, and exposure of delicate nerves due to worn or broken, or cracked teeth. As a result, this sensitivity issue can be treated, avoiding excruciating procedures such as root canal treatment.

2 Literature Review

This paper involves the comparative performance of various methods of approximal caries detection at the primary molars. A survey was conducted among fty children, in which the examiners evaluated three important terms namely—visual inspection, radiographic method, and a pen-type laser fluorescence device. Further examination was done on the separated teeth. In both the threshold using the Mcnemar test, the parameters such as receiver operating characteristics curve (Az), sensitivity, accuracy, and specificity were compared and calculated. Parameters and methods used: Approximal surface, accuracy, sensitivity, specificity, visual inspection, radiographic method, caries detection, radiolucency, reproducibility, level of significance, LF pen, cavitation threshold [1].

This initial study deals with dentine hypersensitivity (DH) which is described by the pain that is caused by stimuli that often provoke zero symptoms. The concept includes the significance of the GaALAs diode laser along with the topical sodium fluoride gel (Naf). This study was conducted with 10 patients and dentine hypersensitivity was assessed by air and tactile stimuli, calculated with the help of a numeric rating scale (NRS) for several 115 teeth. The activation of the hydrodynamic process is assumed for dentinal sensitivity. The most familiar components included in dentine hypersensitivity are abstraction produced by teeth exion because of abnormal occlusal forces, occlusal disequilibrium anatomic predisposition caused by the structural deficiency of enamel-cement function, etc. As a result, the combination of GaALAs laser along with NaF gel produces better capability in improvising the DH—related pain, represents the action of combined laser and gel therapy. Parameter and methods used: Dentine hypersensitivity, numeric rating scale, GaALAs diode laser, sodium fluoride gel, dentinal tubules, analgetic effect, placebo group, desensitizing agent [2].

This paper deals with the collective study and compares effective methods to understand the characterization of dentin hypersensitivity, with various methods, and a device used to examine the tactile sensitivity in the clinical by small stimuli that is generally osmotic, thermal, tactile, evaporative, or chemical. The usual indication of dentine hypersensitivity is when the dentine gets exposed in response to stimuli, causing a short sharp pain that can't be able to assign any other dental pathology or defects. Visual analog sealed (VAS) and people probe on electronic pressure sensitive probes to assess the cold stimulus in response to the tactile stimulus are the methods used. The sensitivity sensor probe is used to examine tactile sensitivity in clinical settings. Parameters and methods used: visual analog scale (VAS), people probe, Schiff scale, tactile hypersensitivity, denitries [3].

In this paper, an evaluation of the DIAGNOdent device or DD is carried out which is a system used in the detection of caries. This detecting system's function is based on laser-stimulated fluorescence. Significantly, the evaluation performance was related to the observation of occlusal and the smooth surface carries present in the enamel and dentine involved several 25 studies which are classified into four categories, which expressed the result in assessing specificity and hypersensitivity value. Finally, the main idea was to produce a systemic analysis of the literature reporting the function of DD. Parameters and methods used: laser fluorescence device, DIAGNOdent device, carious lesions, occlusal dentinal caries, reliability, enamel lesions [4].

In this paper, the method and techniques used to reduce the onset of hypersensitivity, homecare, and professional management related to open dentinal tubule, and evaluation of the clinical support for different therapeutic agents are considered. Dentine hypersensitivity, the usual cause for this is gingival recession as a result of exposure to dentinal tubules. This condition predominantly occurs due to excessive brushing causing gingival abrasion/recession, dealing with the loss of enamel, and exposure to cervical dentine. By observation, studies have reported that dentine hypersensitivity widely occurs at the age of 30–39 [5].

This paper focuses on dentin hypersensitivity (DHS) which highlights pain mechanisms and etiology. The human dentinal tubule diameters are dependent on the age of the individual. This paper will also evaluate the etiology and risk factors related to dentine etiology including abrasion, erosion, and periodontal disease. Dentin detection, normally at the cervical margin, is caused by various procedures involving gingival recession or loss of enamel, predisposing factors such as periodontal disease and treatment, limited alveolar bone, thin biotype, erosion, and abrasion [6].

This technique proposes a device for deciding the pain sensitivity of the oral mucosa. Using the threshold of sensitivity or degree of compliance the working state of oral mucosa is determined. This device is composed of a rod along with a probe, a cylinder, a pressure sensor, a vinyl tube, wires, a battery, a microcontroller, and an LCD used for converting and processing the signal from the sensor and nally, it displays the information to them [7].

A survey was conducted on 10 patients with the absence of teeth. The device possessed benefits such as portability of the device and less consumption of material making it advantageous in the use of the dental chair and the feasibility of measuring at any position of the oral mucosa. This paper involves the examination of ceaseless and episodic twice–daily utilization regimens of a desensitizing dentifrice composed of 5% calcium sodium phospho-silicate (CSPS), and it involves subjects with dentinal hypersensitivity which are randomized to uninterrupted usage of 5% CSPS-composed dentifrice following two 8 weeks treatment periods along with the use of standard fluoride dentifrice. Using parameters such as tactile threshold, evaporative sensitivity was assessed. As a result, no treatment-related adverse events were reported. Finally, no performance variants were analyzed between the two usage regimens [8].

3 Proposed Method

To begin with the process, the device consists of a container where the exchange of stimulus takes place. While the stimulus is applied, it passes through the tube which is pushed to the end of the opening which is in contact with the tooth by the pressure of the water pump. As the tooth gets exposed to the stimulus at a certain degree, a MEMS sensor is a setup in the gums of the tooth which measures the slight movements by the individual due to the sensitivity, the individual produces a movement that is detected by the MEMS sensor (Figs. 1 and 2). This sensor transmits the signals to the microcontroller (see Fig. 3).

The microcontroller collects and stores the data of outcomes based on the application of temperature. The collection of measured data by the device is stored in the microcontroller and these are lined with an IOT platform called node MCU which contains memory of 128 Kbytes and 4 Megabytes which acts as an open source that is built around the system. The pipes connecting the mouthpiece and the stimulus cylinder are the water pump acting as a radiator, pumping the stimulus upwards. To set up the connections between the hardware and the software, the connecting wires send the signals from the device to the software for the accurate output display. Then

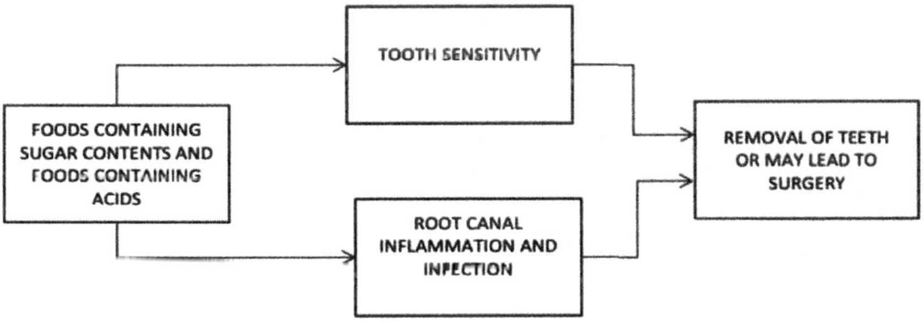

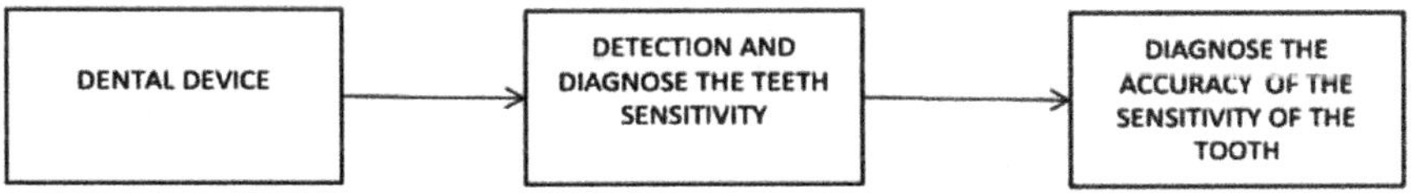

Fig. 1 Effects of enamel degradation

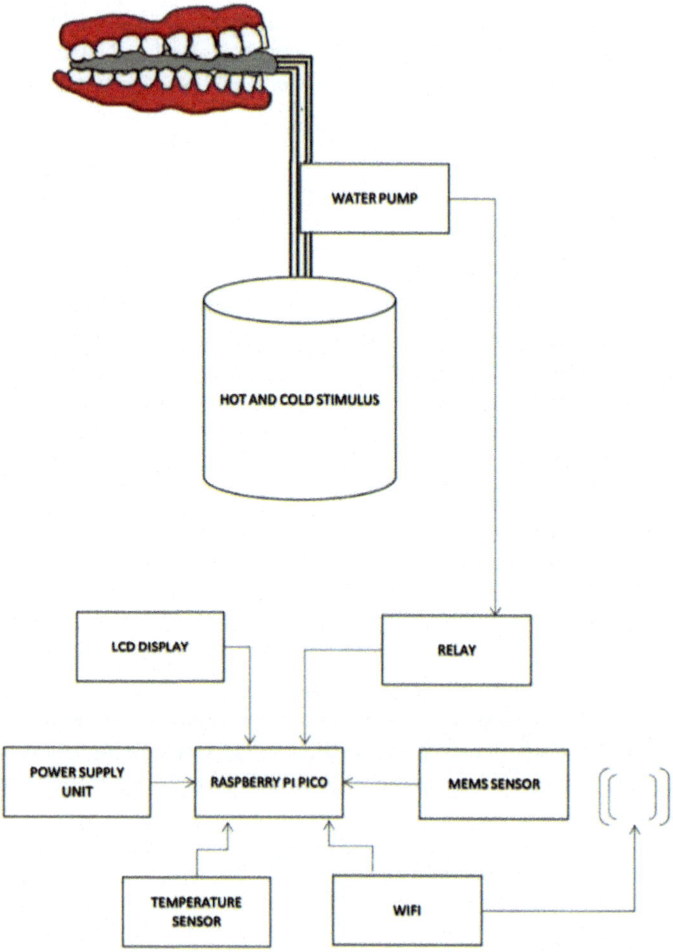

Fig. 2 Block diagram

Fig. 3 MEMS sensor (accelerometer)

Fig. 4 Raspberry Pi Pico

the power supply unit is used here to provide the desired power that activates the components present, and the connecting bers generate a speed of up to 940 megabytes per sec (Mbps) with relatively lesser lag time and stimulate the response faster by about 70% speed of light (see Fig. 4).

A temperature sensor helps in measuring the temperature, whether a hot or cold stimulus is given, and converts its data into an electronic form of data for the display. Followed by this CCD display uses the data from the temperature sensor along with the movements as a response to the stimulus measured by the MEMS sensor is displayed (see Fig. 5).

A breakout board is used to enable connection that provides easier form from larger circuit. Finally, the relay works to manage certain electrical parameters, acts as an ON/OFF switch with preset limits, which is worked without manual help in simpler terms generates a connection between sender and receiver. By training the datasets using micro python coding the software is developed to predict the accurate values of sensitivity are displayed and obtained. The Global System Mobile Communications (GSM) plays a foremost role in the kit by providing the telemedicine process, to make the action more convenient to the dentists (see Fig. 6).

Fig. 5 Temperature sensor

Fig. 6 GSM

Fig. 7 Prototype model

Figure 7 shows the system interface connected with the prototype model. In this phase, the GSM is connected through Wi-Fi to the cloud. Temperature readings are displayed on the LCD screen.

Figure 8 shows the sensing stimulus interfacing with the prototype model in order to detect the hot and cold stimulus that is displayed on the LCD screen (Fig. 9).

3.1 Hot Stimulus and Cold Stimulus

During the process, the stimulus produces heat with the following requirement needed. In this method, the hot stimulus is generated by using the burner which is in contact with the glass container of water and is pushed into the tube via the

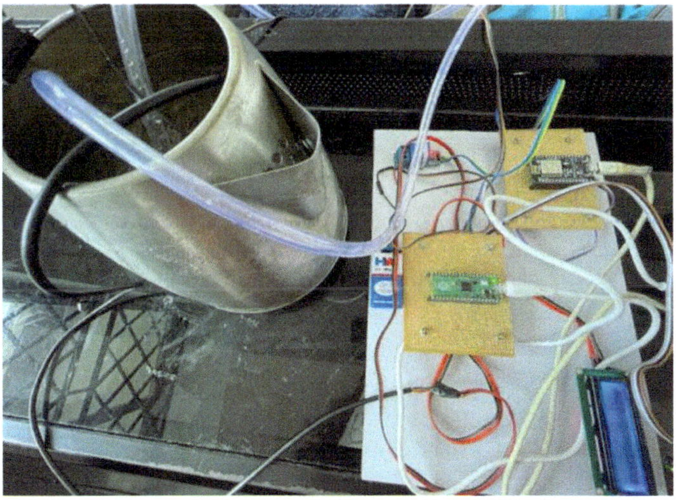

Fig. 8 Prototype model interfacing with sensing stimulus

Fig. 9 PC Interface with prototype model

water pump. And the temperature of the heat is sensed with the help of the temperature sensor. Similarly in the case of a cold stimulus cooling effect is produced by the use of ice cubes that is converted into vapor form and sent into the tube. And the temperature sensor tends to sense the temperature.

Fig. 10 LCD display output

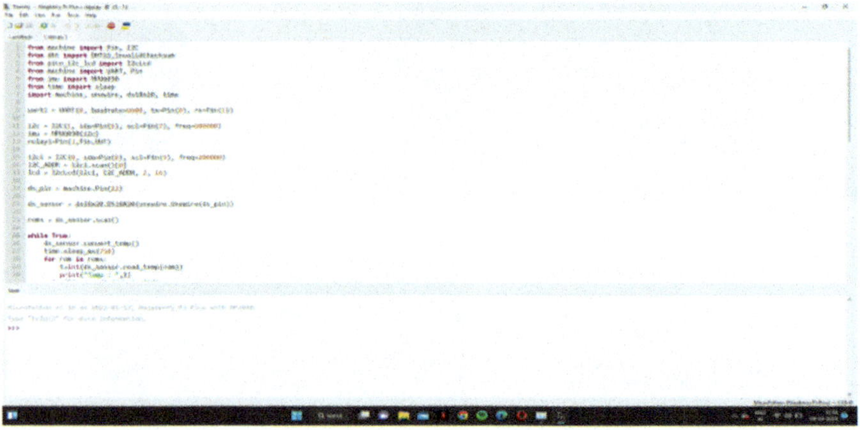

Fig. 11 Before compilation

4 Result and Discussion

This project is to provide a solution for the detection of accuracy in the enamel degradation of the tooth by using components such as MEMS sensor, and temperature sensor, Raspberry pi Pico plays a crucial role in sensitivity detection with preprocessed data collected. Thereby, contingent on early prediction of tooth decay and other dental complications.

According to the Fig. 10 the temperature level is 5, the pumping time is 1, and the sensor value is 70%. Similarly Figs. 11 and 12 shows the code before and after compilation.

Figures 13 and 15 shows the history of the temperature values during the hot and cold stimulus respectively and Fig. 14 shows the main page of the result, which contains the pump status, temperature value, date, and time information.

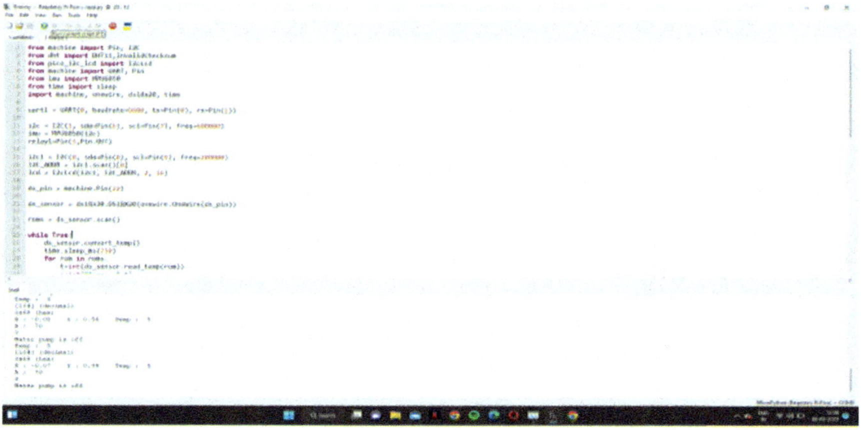

Fig. 12 After compilation

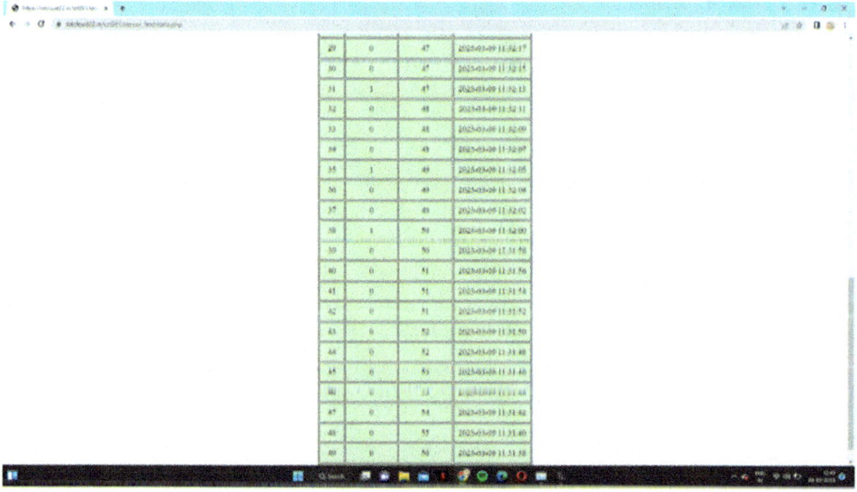

Fig. 13 Temperature during hot stimulus

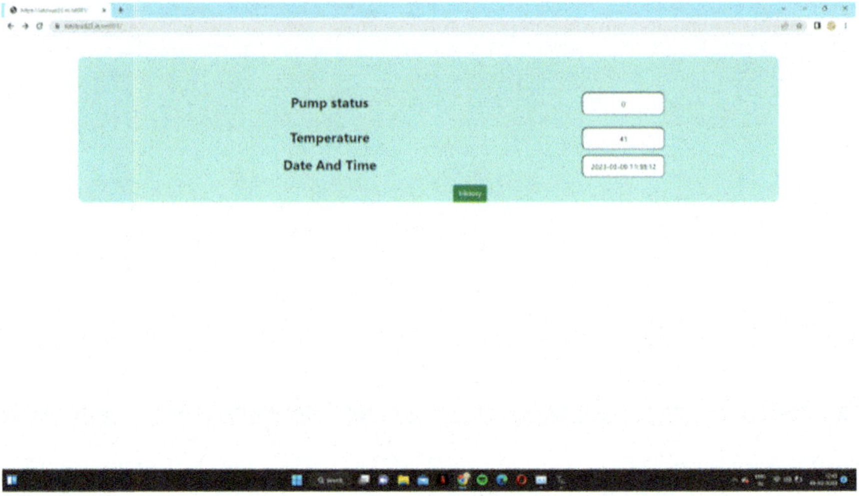

Fig. 14 Status of parameters

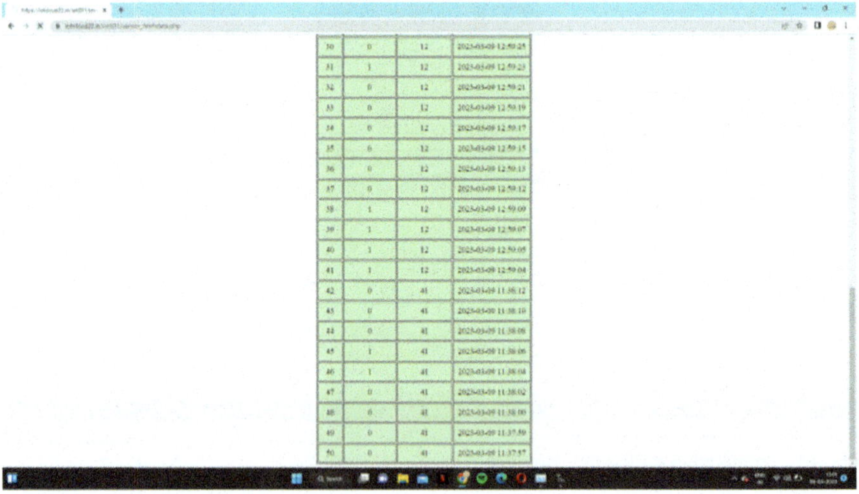

Fig. 15 Temperature during cold stimulus

5 Conclusion

As a result, this device predicts the sensitivity of the tooth with more accuracy. Due to this, we can reduce tooth decay and following root canal treatment because of the early detection of tooth sensitivity. Hence the sensitivity of the tooth can be diagnosed at an early period by this device.

References

1. Curro FA (1990) Tooth hypersensitivity in the spectrum of pain. Dental Clin North Am 34:429–437
2. Taani SD, Awartani F (2002) Clinical evaluation of cervical dentin sensitivity (CDS) in patients attending general dental clinics (GDC) and periodontal specialty clinics (PSC). J Clinic Periodontol 29:118–122
3. Bissada NF (1994) Symptomatology and clinical features of hypersensitive teeth. Arch Oral Biol 39(Suppl):31S–32S.
4. Rees JS (2000) The prevalence of dentine hypersensitivity in general dental practice in the UK. J Clin Periodonol 27:860–865
5. Fischer C, Fischer RG, Wennberg A (1992) Prevalence and distribution of cervical dentine hypersensitivity in a population in Rio de Janeiro. Brazil J Dent 20:272–276
6. Irwin CR, McCusker P (1997) Prevalence of dentine hypersensitivity in a general dental population. J Ir Dent Assoc 43:7–9
7. Kumar NG, Mehta DS (2005) Short-term assessment of the Nd: YAG laser with and without sodium fluoride varnish in the treatment of dentin hypersensitivity—a clinical and scanning electron microscopy study. J Periodontol 76(7):1140–1147
8. Moritz A, Gutknecht N, Schoop U et al (1996) The advantage of CO_2-treated dental necks, in comparison with a standard method: results of an in vivo study. J Clin Laser Med Surg 14(1):27–32

Recognition of Skin Cancer

G. Gurumoorthy, S. Ezra Deva Kirubakaran, S. Nitin Moses, S. Safiya Farheen, and V. Kabilan

Abstract Skin cancer is a condition that affects the skin, and it can manifest in either a malignant or benign form. Malignant melanoma is characterized by the presence of spots that may bleed, while malignant carcinoma is the deadliest form of skin cancer and is named after the cell from which it is believed to arise, the melanocyte. However, skin cancer can be cured if it is diagnosed at the right time. The diagnosis of melanoma is delicate and requires both a physical examination and laboratory tests, as it can spread throughout the body via the lymphatic system or blood. Unfortunately, laboratory testing often leads to inflammation or the spread of the lesion, which is why less dangerous and time-consuming methods are needed. Computer-based diagnosis can improve the speed and accuracy of skin cancer diagnosis by analyzing the symptoms of the condition. The common things between skin lesions make the diagnosis of malignant cells a challenging task. While there are several traditional methods available to detect skin cancer, early detection is always preferable. The medical community generally uses the biopsy system to diagnose skin cancer, which involves removing skin samples and subjecting them to various laboratory tests. However, this method is both painful and time-consuming. Therefore, the aim is to develop a computer-assisted diagnosis system for skin cancer, which would be safer and less time-consuming

Keywords Internet-of-Things · Skin cancer · Melanoma · MATLAB · CNNs · Arduino UNO · Biopsy · Malignant carcinoma

---

G. Gurumoorthy · S. E. D. Kirubakaran · S. N. Moses (✉) · S. S. Farheen · V. Kabilan
Department of Medical Electronics, Saveetha Engineering College, Chennai, India
e-mail: nitins0824@gmail.com

V. K. Gunjan and J. M. Zurada (eds.), *Proceedings of 4th International Conference on Recent Trends in Machine Learning, IoT, Smart Cities and Applications*, Lecture Notes in Networks and Systems 873, https://doi.org/10.1007/978-981-99-9442-7_9

1 Introduction

There are numerous types of fatal cancers, with skin cancer being the most common type. It affects the fair-skinned population in Europe, North America, and Australia, with two major types: malignant carcinoma and melanoma (such as basal cell, squamous cell, and Merkel cell lymphomas), and carcinoma is more dangerous and can be fatal if left untreated [1]. Early detection of carcinoma increases the chances of a cure, but advanced carcinoma is deadly. The increase in periodic rates of all skin cancer forms is concerning, with malignant carcinoma being the most dangerous and increasing rapidly [1, 2]. In Canada, more than 5500 people were affected by melanoma in 2010, with around 17.35% of them dying. In the United States, there were about 22,000 people affected with melanoma in 2012, and approximately 49.4% of them died. Based on the Cancer Trends Progress Report by the National Institute of Health of the United States (NIH), it is estimated that nearly half of all Americans who live to age 65 will develop skin cancer at least once [3]. A doctor cannot diagnose a pigmented lesion through its characteristics, even if they are a trained professional and using the naked eye. New criteria are necessary for a clinical diagnosis, and while it can be performed without any support using the naked eye, the result is not always reliable. Therefore, a demo copy was created, which involves using a device to take a picture of the skin lesion and analyze its features to determine whether it is benign or not [4]. As some people do not have access to a dermatologist, and even with a trained eye, the result can be false, it is necessary to develop automatic methods to increase the accuracy of diagnosis [5].

Early detection is the most effective tool for controlling this kind of cancer. The main criterion to distinguish between benign and malignant skin lesions is the ABCD rule, which evaluates the asymmetry, border, color, and diameter of the skin lesion to make a diagnosis [4, 6].

Our team has developed a self-learning model that we believe would perform better in addressing the probabilistic nature of the skin disease we are studying. This model is designed to learn the underlying patterns present in skin disease based on image and histopathology inputs.

In our project, we propose a new method for automated melanoma recognition in demo copy images, using a two-step process involving segmentation and classification. By using very deep convolution neural networks (CNNs), we can generate highly discriminating features, which improve the performance of both segmentation and classification tasks. Importantly, this approach requires no manual interaction.

There are, however, some potential disadvantages to this approach. One issue is the inference time speed; for non-trivial problems, it may be necessary to use a very large network, which can be extremely time-consuming to evaluate at inference time [7]. Additionally, because our model requires a large dataset, the training time can also be significant.

2 Proposed System

The methods used in this study involved several algorithms to analyze images of skin afflictions. These included converting the images to grayscale and applying various filters such as sludge, median sludge, smooth sludge, double mask, RGB birth, and histogram. The RGB values of the images were analyzed before being converted to grayscale to ensure accuracy. The sludge filter was used to enhance the details of the infected area, and the Euler value was used to determine the number of factors of the affliction present in the image. Contrast improvement was also applied to maintain intensity. Graphical representation was used to further analyze the numerical data obtained from the segmentation process. A threshold limit was heuristically assessed on the Euler value to identify the presence of a large number of afflictions, an important characteristic for diagnosis. The study also utilized deep learning models to improve the accuracy and sensitivity of the analysis, as evidenced by the confusion matrix obtained from the given image.

Proposed Methodology

See Fig. 1.

3 Literature Review

Melanoma is a type of skin cancer with increasing prevalence rates over the last three decades. The main risk factor for this type of cancer is prolonged exposure to UV radiation. However, early detection and treatment significantly increase the chances of a cure [1]. Therefore, there is a need for a real-time image analysis system to aid in the prevention and early detection of melanoma [2]. In this paper, we propose a system that utilizes image recognition technology to capture and analyze skin images of different types. The system can alert the user in real-time and prompt them to seek medical attention urgently. Our approach automates the process of melanoma prevention and detection, which is both accessible and effective [5]. To achieve this, we propose and evaluate six different styles for the segmentation of skin lesions in thermoscopic images [2]. This set includes some state-of-the-art methods [grade vector inflow (GVF) and the position set system of Chan et al. (C-LS)] and a set of styles developed by the authors [adaptive thresholding (AT), adaptive snake (AS), EM position set (EM-LS), and fuzzy-grounded split-and-combine algorithm (FBSM)] [3, 4]. We applied these segmentation styles to 100 thermoscopic images and evaluated them using four different criteria, with the segmentation results obtained by a dermatologist as the ground truth. The results show that the semi-supervised AS and EM-LS styles achieved the best segmentation results. The fully automatic system, FBSM, achieved results that were only slightly worse than those of AS and EM-LS [3]. In conclusion, our real-time image analysis system and segmentation styles have the potential to aid in the prevention and early detection of melanoma. The prevalence

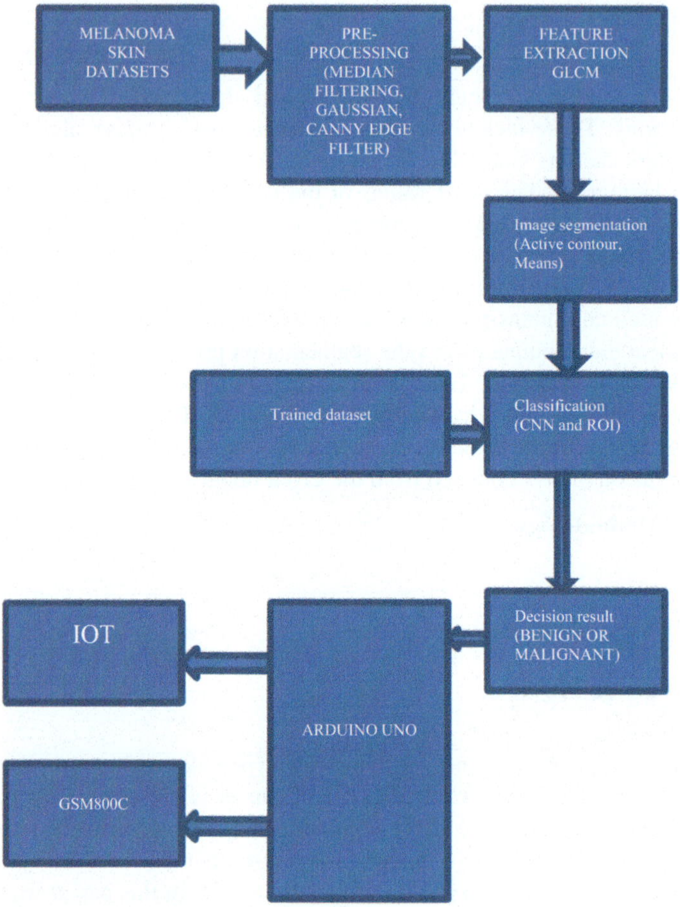

Fig. 1 Proposed methodology

of this type of skin cancer continues to increase worldwide, making our proposed approach an important step toward addressing this health issue.

This is a type of cancer that can affect individuals of any age and is one of the leading causes of death among young people. Fortunately, it is cancer that is visible on the skin, making it potentially detectable at an early stage when it is still curable. Recent advancements have made fully automated early detection of melanoma a real possibility. One such advancement is the introduction of dermo copy, which has greatly enhanced the clinical ability to detect skin cancer at the earliest stages [6].

The incidence of cancer can occur at any age, and it is considered one of the leading causes of death in young people. However, due to its visible characteristics on the skin, it can be detected at an early stage when it is still curable. With the introduction of dermo copy, there has been a significant improvement in the clinical diagnosis of skin cancer, allowing for early detection [2, 3]. The widespread adoption of this technology has led to the accumulation of large datasets of dermo copy images of both malignant and benign lesions, which have been validated through histopathology. Recent advancements in the fields of image processing and machine learning have made it possible to distinguish malignant lesions from benign ones with greater accuracy [3]. These new technologies have the potential to not only improve the early detection of skin cancer but also reduce the number of unnecessary and costly biopsies [4, 7].

Skin cancer is a leading cause of death, and early detection is crucial for successful treatment. Recent advancements in technology have made automatic early detection of carcinoma a possibility [5]. One of these advancements is dermo copy, which allows for the detection of skin cancer at an early stage. The development of machine learning algorithms, such as deep learning and support vector machines, has enabled the differentiation of malignant tumors from benign lesions [4]. This approach eliminates the need for unnecessary and expensive biopsies. While some of these technologies have shown promise in initial trials, further advancements are necessary to increase their sensitivity and reproducibility. This study proposes an approach for detecting skin cancer using a combination of deep learning, coding, and SVM algorithms [4, 6]. The proposed method eliminates the need for manual point transfer and allows the system to draw comparisons between the paragraph discusses the need for early detection and treatment of melanoma, a type of skin cancer that has seen increasing prevalence rates over the last three decades [1, 5]. The primary risk factor for melanoma is prolonged exposure to UV radiation. The paragraph proposes a real-time image analysis system that utilizes image recognition technology to detect and analyze skin images of various types [1]. The system can alert users in real-time and prompt them to seek medical attention urgently, which could aid in the prevention and early detection of melanoma [2, 5]. The paragraph also discusses the six different styles proposed for the segmentation of skin lesions in thermoscopic images, including state-of-the-art methods and new styles developed by the authors. The study evaluated the different styles and found that the semi-supervised adaptive snake and EM position set styles achieved the best segmentation results [4, 6]. The fully automatic system, fuzzy-grounded split-and-combine algorithm, achieved results that were slightly worse but still promising. The paragraph concludes that this proposed system has the potential to aid in the prevention and early detection of melanoma, an increasingly prevalent type of skin cancer worldwide [6, 7]. The system uses a neural network classifier to classify lesions as either carcinoma or nevus. The system is highly accurate and has been tested on a dataset of skin lesion images. The paragraph emphasizes the importance of safe sun practices and early detection of skin cancer [5, 7]. The ABCD and seven-point rosters are helpful tools for diagnosis, and the presented approaches offer promising solutions for automatic skin cancer detection, potentially reducing unnecessary biopsies [7].

4 Module Description Image Acquisition

The image accession step is the first stage of any vision system. After the image has been attained, colorful styles of processing can be applied to the image to perform the numerous different vision tasks needed moment. Still, the desired tasks may not be possible if the image has not been acquired adequately, indeed, with the aid of some form of image improvement. Digital imaging or digital image accession is the creation of a digitally decoded representation of the visual characteristics of an object, similar to a physical scene or the interior structure of an object. The term is frequently assumed to indicate or include the processing, contraction, storehouse, printing, and display of similar images. A crucial advantage of a digital image, versus an antilog digital imaging, also known as digital image acquisition, refers to the process of creating a digital representation of the visual properties of an object or scene, including the internal structure of an object. Processing, compressing, storing, printing, and displaying samples are just a few illustrations of the operations that may be included in this process. Processing, compressing, storing, printing, and displaying samples are just a few illustrations of the operations that may be included in this process. One of the key benefits of digital images is the ability to create exact copies of the image without any degradation in quality, which is not possible with analog images like film photographs (Fig. 2).

Digital imaging refers to the process of creating a digitally encoded representation of the visual characteristics of an object or scene, which can include processing, storage, compression, printing, and display of images. This procedure entails converting electromagnetic radiation or other waves—whose attenuation varies as they travel through or reflect off objects—that carry information into digital signals. After that, a computer processes these impulses and produces visible-light images. Based on the sort of electromagnetic radiation or waves that were employed to capture the image, many types of digital imaging can be categorized. For example, digital photography and videography use visible light, while digital X-ray imaging, fluoroscopy, and CT use X-rays, and digital scintigraphy, SPECT, and PET use gamma rays. Similarly, ultrasonography and sonar use sound, and radar uses radio waves. One key advantage of digital imaging is that it allows for indefinite duplication without any loss of image quality.

(a) *2D Image input*

Fig. 2 Input image

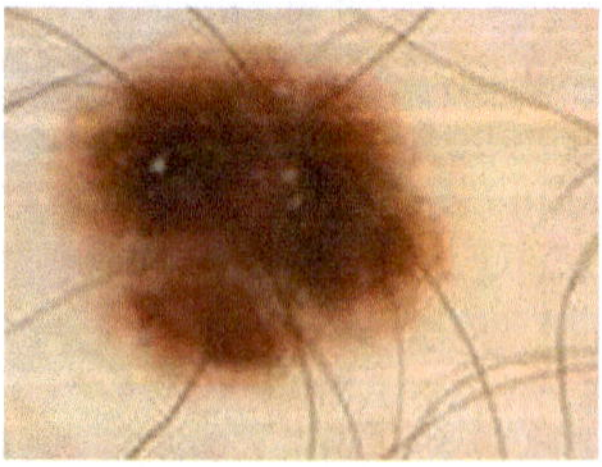

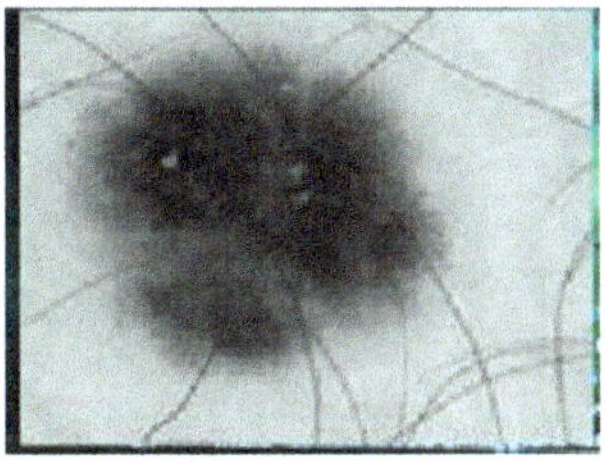

Fig. 3 Gray IMAGE

A digitized image is a digital representation of a two-dimensional grayscale image. It is created by converting an analog image into a digital format. The value of the function at any position (*x*, *y*) is proportional to the brightness or grayscale value of the image at that location. The image can be conceptualized as a function of two variables, *x* and *y*, which represent the spatial coordinates. In a digitized image, spatial and grayscale values are separated, and the intensity measured across a regularly spaced grid in the *x* and *y* directions is quantized to 8 bits (256 values). This allows the image to be stored as a matrix of numerical values, with each value representing the intensity of a particular pixel in the image. This matrix can then be processed using a variety of image processing techniques, such as filtering, segmentation, and feature extraction, to extract useful information from the image.

(b) *Gray IMAGE*

Yes, that is correct. One-bit-tonal black and white images are also called bi-level or binary images, and they only contain two colors: black and white. In contrast, grayscale images contain many shades of gray, ranging from black to white, and are typically represented using 8 bits per pixel, which allows for 256 different shades of gray to be represented. This makes grayscale images more suitable for applications that require more detailed intensity information, such as medical imaging, microscopy, and photography (Fig. 3).

Grayscale images consist of various tones between black and white. These images are created by measuring the light intensity at each pixel using a specific combination of frequencies or wavelengths. When only one frequency (usually a small band) is recorded, monochromatic grayscale images are the outcome. Any electromagnetic spectrum region, including infrared, visible light, and ultraviolet, can provide the frequency.

The stored numeric sample values of a photometric grayscale image are mapped to the achromatic channel of a common color space using a defined grayscale color space. Based on the quantified characteristics of human eyesight, this color space. There is no mapping from the grayscale image to the color image, though, if the original color image doesn't have a well-defined cooler space or if the grayscale image isn't meant to have the same perceived achromatic intensity as the color image.

(c) *Median Filter*

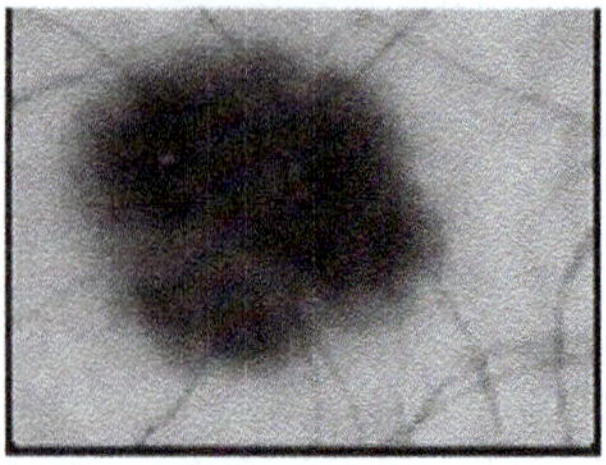

Fig. 4 Median filter

The Median Filter is a type of non-linear filter that is commonly used to reduce noise from signals or images. In contrast to linear filters, such as the mean filter, which replace each pixel's value with the average of its neighboring pixel values, the median filter replaces each pixel's value with the median value of its neighboring pixels. The median value is the middle value in a sorted list of values, and it is less sensitive to extreme values than the mean value.

Median filtering can effectively reduce noise in an image without blurring or distorting its edges, which is a common problem with linear filters. However, median filtering can also cause loss of detail and introduce artifacts, especially when the noise is correlated with the image features. Therefore, it is important to carefully choose the filter size and apply it selectively to the regions that are most affected by noise. Median filtering is widely used in digital image processing for tasks such as denoising, image restoration, and feature extraction (Fig. 4).

(d) *Histogram equalization*

Histogram equalization is a method commonly used to improve the contrast of images with limited intensity ranges. It works by redistributing the intensity values in an image, resulting in more even distribution and increased contrast. This is achieved by adjusting the histogram, which can improve the visibility of features that may have been difficult to discern previously. Unlike other contrast enhancement techniques, such as amplifying the most frequent intensity values, histogram equalization spreads out the histogram evenly. This method can be particularly useful when dealing with images that have closely spaced intensity values (Fig. 5).

(e) *Adaptive histogram equalization*

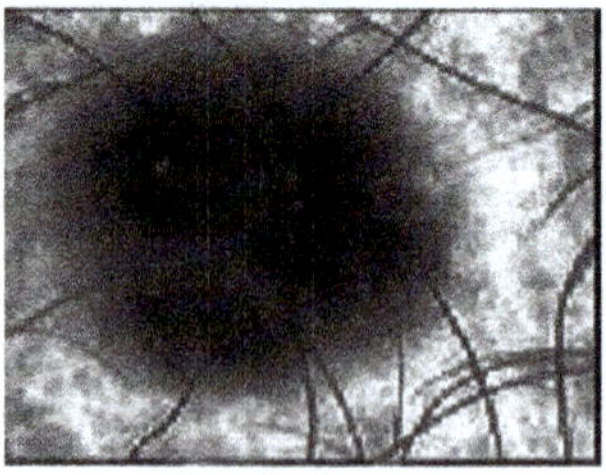

Fig. 5 Histogram equalization

Fig. 6 Adaptive histogram equalization

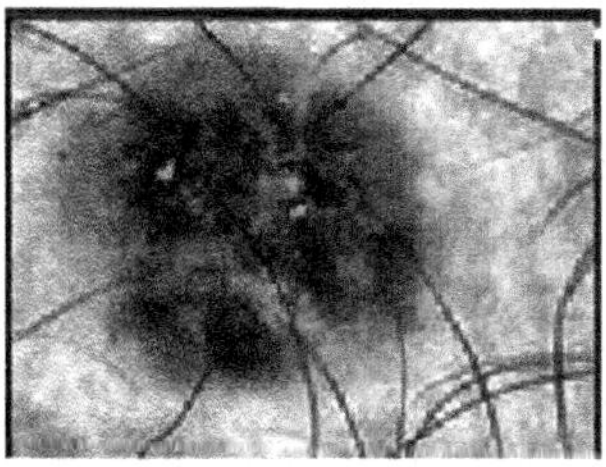

CLAHE limits the modification to prevent over-amplification of noise, while still improving the contrast and reducing inconsistencies in the image. This is done by dividing the image into small regions called tiles and applying histogram equalization separately to each tile. However, before equalization is performed, a clipping function is applied to limit the histogram range of each tile, which helps to prevent over-amplification of noise. The result is an image with improved contrast and reduced inconsistencies, while still preserving details and minimizing the amplification of noise (Fig. 6).

(f) *Black and White Image*

It converts an image to a double image, grounded on a threshold. im2bw produces double images from listed, intensity, or RGB images. The function "im2bw" applies a threshold to an input image and produces a double image from listed, intensity, or RGB images. Firstly, it converts the input image to grayscale format if it is not already an intensity image, and then it converts this grayscale image to double by thresholding. The resulting double image "BW" has pixel values of 0 (black) for all pixels in the input image with luminance lower than the specified threshold value, and 1 (white) for all other pixels. Note that the threshold value is specified in the range (0, 1), regardless of the input image class. The "im2bw" function can be used in different ways, such as converting an intensity image I to black and white using "BW = im2bw (I, threshold)", or converting a listed image X with colormap chart to black and white using "BW = im2bw (X, chart, threshold)", or converting an RGB image RGB to black and white using "BW = im2bw (RGB, threshold)". In computer vision, image segmentation is the process of dividing a digital image into multiple parts or sets of pixels, also known as super pixels. The purpose of segmentation is to simplify and/or change the representation of an image into a form that is more meaningful and easier to analyze (Fig. 7).

5 Architecture

(a) *Convolutional neural network*

Convolutional neural networks (CNN) is a type of neural network model that can extract advanced representations for image content. In contrast to classical image

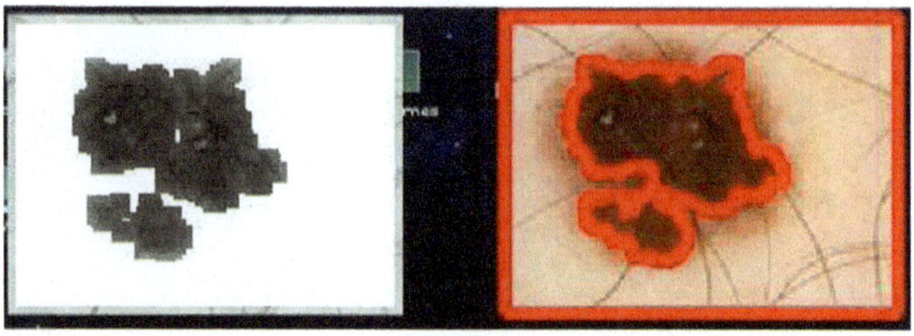

Fig. 7 Segmentation

recognition, where image features are defined manually, CNN takes the raw pixel data of the image, trains the model, and automatically extracts the features for better classification. CNNs are composed of neurons with learnable weights and biases, which receive multiple inputs and perform a weighted sum over them. The output passes through an activation function and results in an output. To teach an algorithm how to recognize objects in images, we use CNNs, which are a specific type of artificial neural network. CNNs get their name from one of the most crucial operations in the network, convolution. The output of a CNN is a 4D array, where the batch size is the same as the input batch size, and the other three dimensions of the image may change depending on the values of stride, kernel size, and padding used. A CNN generally consists of three layers: a convolutional layer, a pooling layer, and a fully connected layer.

6 System Perpetration

(a) *Data analysis and visualization*

MATLAB provides tools to acquire, dissect, and fantasize data, enabling you to gain sapience into your data in a time of bit, spreadsheets, or traditional programming languages are used. You can also validate and partake your results through plots and reports or as published MATLAB law.

(b) *Acquiring data MATLAB*

MATLAB provides a wide range of functionalities to access data from various sources, including files, databases, and external devices. Its I/O functions support popular file formats such as Microsoft Excel, text, double, image, sound, and video files, as well as scientific formats like netCDF and HDF. With MATLAB add-on products, users can even acquire data from hardware devices such as sound cards and live streams for real-time analysis and visualization. Additionally, MATLAB offers support for communication with instruments such as oscilloscopes, function generators, and signal analyzers.

(c) *Assaying data*

A variety of tools are available in MATLAB to help you manage, prepare, and modify your data. Exploratory data analysis can be used to find patterns, test theories, and build descriptive models. The programmer offers numerous filtering and smoothing, interpolation, regression, and quick Fourier transforms functions (FFTs). Additional modules extend its capabilities to include curve and surface fitting, multivariate statistics, spectral analysis, image processing, system identification, and other analytical tasks.

(d) *Imaging data*

MATLAB has built-in 2-D and 3-D deception functions, as well as methods for visualizing volumes. These features can be used to communicate results, comprehend data, and indulge in fantasies. Interactively and programmatically pots are customized. The MATLAB plot gallery provides exemplifications of numerous ways to display data graphically in MATLAB. To use with your MATLAB operation, you may view and download the source code for each illustration.

(e) *Programming and algorithm development*

High-level language and development tools from MATLAB enable quick construction and analysis of algorithms and operations.

7 The MATLAB Language

MATLAB is a programming language designed for engineering and scientific problem-solving. It offers native support for vector and matrix operations, which are commonly used in these fields. With MATLAB, you can write programs and develop algorithms more efficiently than with traditional languages because you don't need to perform low-level tasks such as declaring variables, specifying data types, and allocating memory. In many cases, the support for vector and matrix operations eliminates the need for loops, resulting in more concise code. Additionally, MATLAB provides features of traditional programming languages, including flow control, error handling, and object-oriented programming (OOP). You can use basic or advanced data structures, or even define custom data types. By interactively executing commands one at a time, you can produce immediate results, which allows for quick exploration of multiple options and iteration toward an optimal result. You can capture interactive workflows as scripts and functions to automate your work. MATLAB also offers add-on products with built-in algorithms for signal processing, image and video processing, control systems, and many other disciplines. By combining these algorithms with your own, you can create complex programs and operations.

8 Software Description MATLAB

MATLAB is a programming language developed by MathWorks. It started as a matrix programming language where direct algebra programming was simple. It can be run both under interactive sessions and as a batch job MATLAB (matrix laboratory) is a fourth-generation high-position programming language and interactive terrain for numerical calculation, visualization, and programming. MATLAB is developed by MathWorks. It allows matrix manipulations; plotting of functions and data; perpetration of algorithms; creation of stoner interfaces; uniting with programs written in other languages, including C, C, Java, and FORTRAN; dissecting data; developing algorithms; and producing models and applications. It has multitudinous built-in commands and calculation functions that help you in fine computations, generating plots, and performing numerical styles.

(a) *Features of MATLAB*

- MATLAB is a high-level programming language used for numerical computation, visualization, and application development.
- It provides tools for structure operations with custom visual interfaces.
- It provides functions for integrating.
- It features an interactive environment for iterative exploration, design, and problem-solving.
  MATLAB has a vast library of built-in functions for direct algebra, statistics, Fourier analysis, filtering, optimization, numerical integration, and solving ordinary differential equations.
- The platform includes built-in tools for imaging data and custom plot creation. MATLAB's programming interface provides development tools for enhancing code quality, maintainability, and maximizing performance.
- It offers tools for building user interfaces with custom visual interfaces. MATLAB provides functions for integrating MATLAB-based algorithms with external applications and programming languages, such as C, Java,.NET, and Microsoft Excel.

(b) *Digital image processing*

Digital image processing is a field that involves the manipulation of digital images using a computer. It is a subfield of signals and systems that is specifically focused on images. The goal of digital image processing is to develop computer systems that can effectively process digital images using algorithms and produce an output image. Popular examples of digital image processing software include Adobe Photoshop, which is widely used for editing digital images (Fig. 8).

The image in the figure was captured by a camera and then transferred to a digital system to remove extraneous details. By zooming in on the water, the image quality is maintained while focusing on this specific area.

Fig. 8 Image segmentation

9 Result and Discussion

See Figs. 9, 10, 11 and 12.

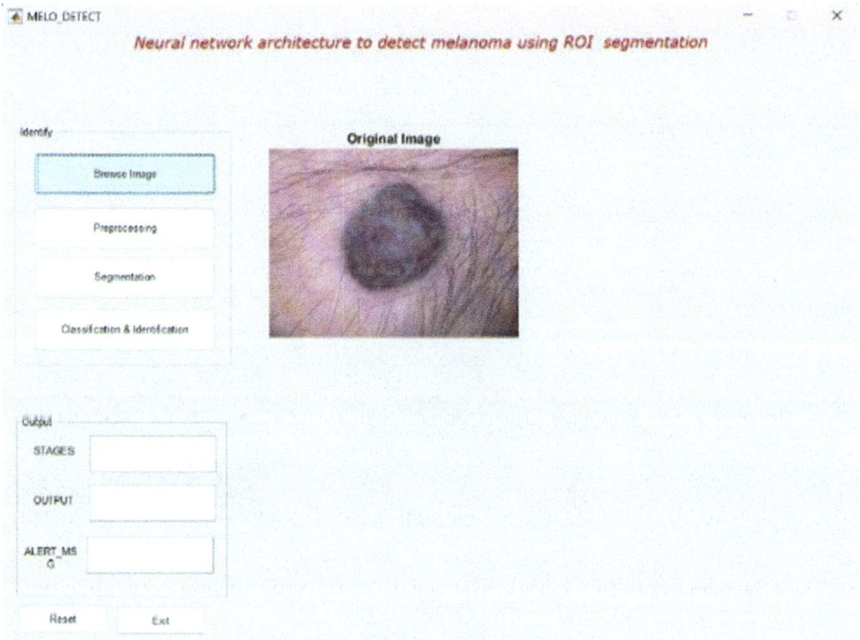

Fig. 9 Image selection

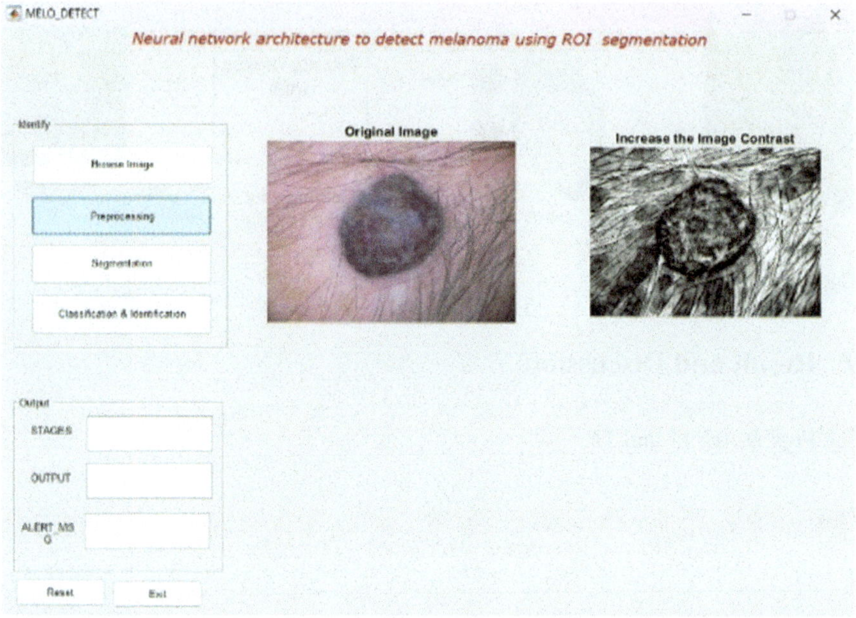

Fig. 10 I preprocessing

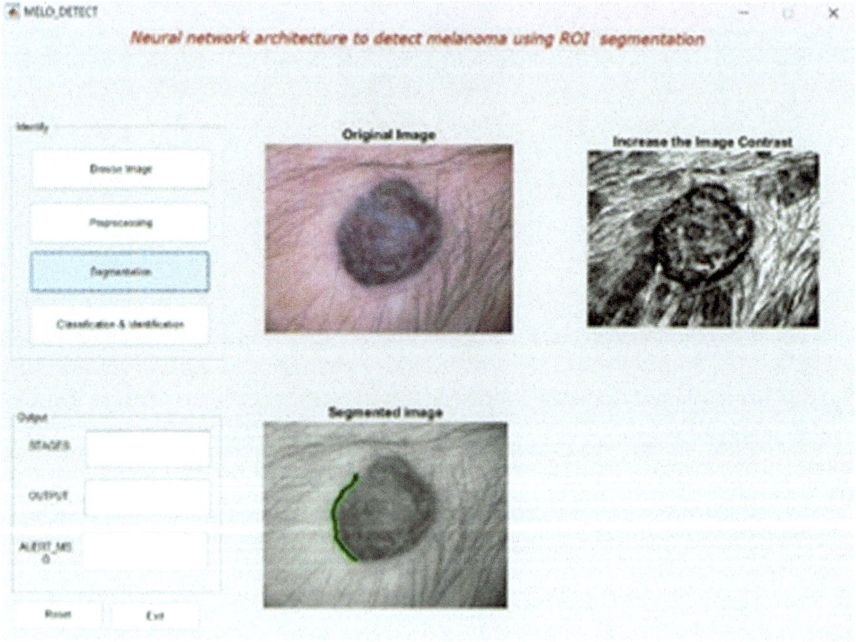

Fig. 11 ROI segmentation

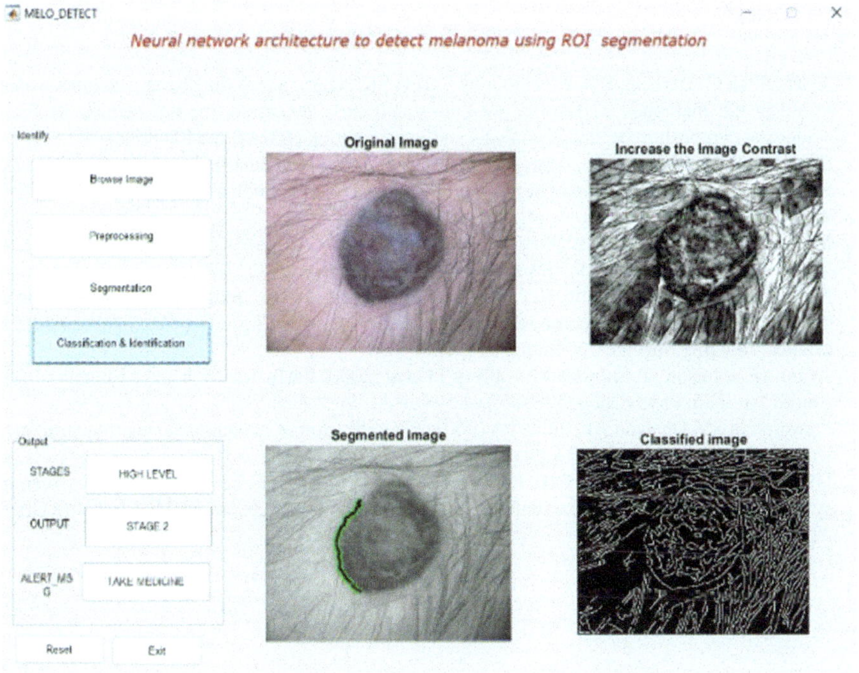

Fig. 12 Classification

10 Conclusion and Future Scope

We have developed a MATLAB algorithm that utilizes convolutional neural network (CNN) to automatically detect skin cancer from 2D images. Traditionally, skin cancer is diagnosed through a biopsy, which involves taking a small sample of tissue for microscopic examination. Our method examines an image of a skin lesion to detect whether skin cancer is present by using novel image processing techniques. Our methods for lesion image analysis look at a number of factors, including asymmetry, border, color, dimension, and structure, to quickly and accurately diagnose patients using photos from their medical records. Our goal is to simplify the theoretical methodology to a real-time system for practical operations. In the future, we plan to make the system more interactive and user-friendly for checking lesions and determining if they are cancerous.

References

1. Abuzaghleh O, Barkana BD, Faezipour M (2014) SKINcure: a real-time image analysis system to aid in the malignant melanoma prevention and early detection. In: Proceedings of IEEE southwest symposium on image analysis and interpretation (SSIAI), pp 85–88
2. Silveira M, Nascimento JC, Marques JS, Marçal AR, Mendonça T, Yamauchi S, Maeda J, Rozeira J (2009) Comparison of segmentation methods for melanoma diagnosis in dermoscopy images. IEEE J Selected Topics Sig Process 3(1):35–45
3. Mishra NK, Celebi ME (2016) An overview of melanoma detection in dermoscopy images using image processing and machine learning. arXiv preprint arXiv:1601.07843
4. N. Codella, J. Cai, M. Abedini, R. Garnavi, A. Halpern, and J. R. Smith, "Deep learning, sparse coding, and SVM for melanoma recognition in dermoscopy images," in Machine Learning in Medical Imaging.Springer, 2015, pp. 118–126.
5. Jerant AF, Johnson JT, Sheridan C, Caffrey TJ et al (2000) Early detection and treatment of skin cancer. Am Fam Phys 62(2):357–386
6. Patwardhan SV, Dhawan AP, Relue PA (2003) Classification of melanoma using tree-structured wavelet transforms. Comput Methods Programs Biomed 72(3):223–239
7. Iyatomi H, Oka H, Celebi ME (2008) An improved internet-based melanoma screening system with dermatologist-like tumor area extraction algorithm. Comput Med Imaging Graph 32(7):566–579

IoT-Based Smart Street Lighting Surveillance System

Inampudi Srinivas, N. Neelima, Jangam Pragathi, and Kisari Shyam Kumar Reddy

Abstract The control of street lights technology is one of the intelligent machines that greatly raise the cost of building a city's infrastructure. Without a doubt, intelligent street light regulation may lower the cost of public illumination. The increasing importance given to energy conservation and proper maintenance has resulted in the creation of reducing technology that allows for large power savings, the highest consideration for the ecology, and a decrease in traffic accidents. This concept presents a fresh approach to the issue of major disasters and street lighting. Wireless technology is used, in which a base server may control by merely sending data. For this project, we used the Arduino IDE, LDR sensors, and IR sensors. The reduction of energy waste and the protection of life from accidents are the key objectives of this research.

Keywords Intelligent machines · Reducing technology · Arduino IDE · LDR sensor · IR sensor · Power systems

1 Introduction

Street lighting is one of a city's top energy consumers. An adaptive lighting system could decrease municipal street lighting costs by up to 50–70%. The existing technique, once it is bright enough outside in the morning, works like switching on the lights just before dusk and turning them off the following day. Yet, the proper time to switch on these lights is when it is completely dark. There will be a huge electrical waste as a result. One of the primary issues with the actual street light systems is the evident difference between their on and off times on bright and wet days which replaces the traditional lighting system entirely. Due to population increase and economic expansion, the world's electricity consumption is rising at the quickest rates imaginable, but there are still several sources of energy that are severely limited.

I. Srinivas · N. Neelima · J. Pragathi (✉) · K. S. K. Reddy
Department of IT, Velagapudi Ramakrishna Siddhartha Engineering College, Vijayawada, India
e-mail: pragathirachel91@gmail.com

V. K. Gunjan and J. M. Zurada (eds.), *Proceedings of 4th International Conference on Recent Trends in Machine Learning, IoT, Smart Cities and Applications*, Lecture Notes in Networks and Systems 873, https://doi.org/10.1007/978-981-99-9442-7_10

There are still significant energy shortages since the development and expansion of resources in the power generation have lagged in expanding demand. Any expanding neighbourhood must have streetlights. They are present on all main highways and in the suburbs. Streetlights remain on from dark to dawn at full power, even when no one is around. These street illuminations cost hundreds of thousands of dollars every day to provide the required electrical energy on a worldwide basis. Traditional incandescent bulbs are extremely expensive to maintain and replace. They produce a lot of heat emissions and require a lot of electricity to run. All of this increases the need for electricity generation, which leads to increased carbon dioxide emissions from power plants. As a result, this behaviour affects the environment and contributes to unneeded light pollution. The project's primary goal is to develop an "IoT-based automatic street light system" which uses solar energy at night. We are using the word "smart" since the system helps identify pedestrian direction of movement and offers aid by lighting the pedestrian's line of movement until the next street light is nearby in addition to providing power for the street lights. In remote rural areas with serious issues with electric power supply, it is possible to methodically help people get to their destination through combining intelligent lighting systems with streetlights. Metropolitan areas can also adopt a similar strategy. Dimming the lights when it is not crowded would be a straightforward and practical solution to this. When presence is detected, the lighting in its direct proximity will turn on in regular (bright) mode. Massive energy saves and lower operational expenses for streetlights would follow from this. We may check the state of the street lights online. To handle any issues that can develop during processing, Internet of things (IoT) can be used from any place in real time.

2 Literature Survey

- Dheena et al. [1], Energy can be saved significantly by replacing sodium vapour lights with LEDs and adding an additional security feature. It ends senseless waste. Electricity is wasted manually switching on and off street lighting. It provides an efficient and intelligent automated street light control scheme with LDR. It is possible to reduce both the maintenance costs and the energy consumed. It can be applied in both urban and rural environments. The system is totally customizable to user needs and can be expanded. It creates a safe environment with the lightest intensity as necessary. Both system maintenance costs and system lifetime must be reduced.
- Abinaya et al. [2], The system's main objective is to minimise the two significant problems that our country is now working to resolve criminal detection and wasted energy. Using LED bulbs results in less heat production than using mercury lamps. This method reduces the cost of conventional methods about 50% to 60%, enhancing the economic growth and free up a sizable sum of money that may be utilised to support other advantageous projects. The gadget provides excellent

protection for women alone while yet protecting everyone's safety. Technology use can put a stop to threats, thievery, and harassment of women.

- Saifuzzaman et al. [3], The use and requirement of technological system are very crucial to construct a smart city in order to enhance our daily lives using IoT. We think that the smart city will flourish and become more established the more IoT is researched and developed. We now understand that our suggested system can satisfy this specific need. This project is simple, dependable, affordable, and important in day-to-day living. The components of the traffic lighting system in our proposal are excellent and can meet the need for energy conservation. Also, the traffic control aspect is excellent. We hope to create an Android app in future to monitor and assess traffic load. This software will be able to display the current state of any and all traffic jams on any streets or roads.
- Ouerhani et al. [4], The article displays a dynamic outdoor lighting management system powered by the Internet of things in operation. The system is currently reliant on the environment. For the situation in question, estimated energy savings when using our approach equal 56%. There have been no discernible modifications in the security impression of the intensity of light cycle according to qualitative tests done with people. This supports the idea that such technologies could aid in the growth of future low-carbon communities.
- Arjun et al. [5], We look at the way the Internet of Everything has ruined our daily life in this study. The smart streetlight system is one of the main elements that applies IoT concepts. A number of urgent issues, including energy waste, investigate crime, the disposal of incandescent lamps, maintenance expenses, etc., are directly addressed by the smart street illumination system. By preventing women from being harassed, having their homes broken into, and being further intimidated, this strategy ensures both public safety and traffic safety. If about 50 and 60 percentage of an electricity is saved and utilised for other crucial uses, the urban energy problem might be lessened. This system supports a secure environment and adequately meets user needs. This tactic makes use of straightforward hardware and software. Determinations governed the lighting of the streets.
- Prasad [6] Prasad ($n = 6$), In this study, a detailed comparison between the traditional illumination system with the ICT-enabled lighting has been made. It has been claimed that switching to new fluorescent bulbs and controllers from HPSV/ metal halide results in significant energy savings. This study demonstrates significant energy savings of roughly 55% over previous, conventional lighting systems without lowering residents' living standards. Safe environments are created by this smart lighting, which also requires extremely little operating maintenance. Similar models can be employed in India's rural areas and are now being utilised in other big cities around the world.
- Subramanyam et al. [7], Working on an improved capabilities street light monitoring and control system that incorporates modern technology and saves energy. Installation of solar panels on a lamppost A GUI programme that shows the state of lights in a road or highway light system allows us to monitor and manage the street lights as well as conserve more electricity and energy by utilising LDR

3 Proposed System

3.1 Description

A single stage circuit in the current system supplies power to the streetlight. To control power distribution, the proposed system adds five more elements. At the base of a street light, an infrared model usually detects a limited region surrounding it. Arduino, the circuit's brain, either receives or generates the data from the sensor. The Arduino then will issue instructions to change from low and bright modes in response to the circumstance, adjusting the brightness of the street light. The 5 V inputs towards the Arduino and detectors are supplied via the single stage line and batteries' eliminator, respectively.

Arduino IDE: The free freeware Arduino freeware makes it easy to write code and upload it to a board. (IDE). It works on different operating systems, including Windows, Mac, and Ubuntu. Since it was created in Java, any Arduino board may be used in the environment without the need to initially install Java on the PC (Fig. 1).

LDR: A photoresistor, also known as a light-sensitive load resistance (LDR), is an electrical component whose resistance is affected either by electromagnetic radiation that it is exposed to. Hence, they are light-sensitive technologies. They are also referred to as photoconductors, photoconductive cells, and photocells. They are constructed from incredibly robust semiconductor materials. The concept of photoconductivity helps to understand how devices like light-dependent resistors function (Fig. 2).

Fig. 1 Arduino IDE

Fig. 2 LDR sensor

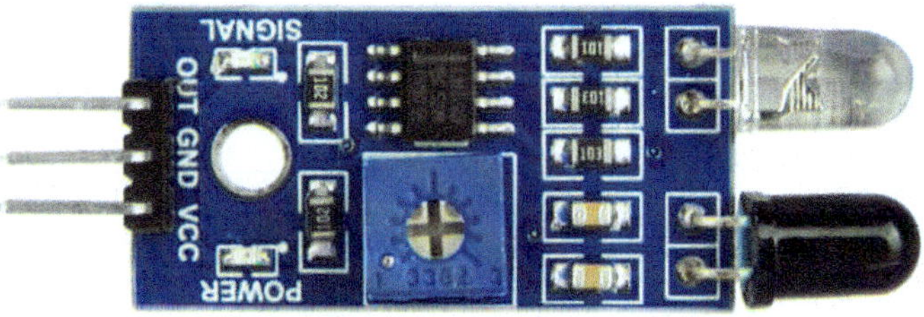

Fig. 3 Infrared sensor

IR sensor: Infrared radiation is used by an electronic appliance called like an infrared sensor to determine and/or emit specific aspects of its environment. Moreover, it measures an object's temperature and spot movement. Wavelengths of infrared light remain invisible to the naked eye (Fig. 3).

4 Design Methodology

The design in essence has three operating modes:

- **OFF mode**: As evident from the OFF mode, internal infrastructure will be switched off as the battery packs are billed whenever there is a direct sunlight in the area where it occurs frequently during the day.
- **Active mode**: As shown by the active mode, when the ambient light level drops below a predetermined threshold, the device automatically turns on and activates the motion sensors.
- **ON mode**: As evident from the ON mode, the sensors turn on whenever there are vehicles around, which leads to turning on the LED bulbs each one at a time. These illuminations soon dim.

Figure 4 shows the block diagram of the proposed system.

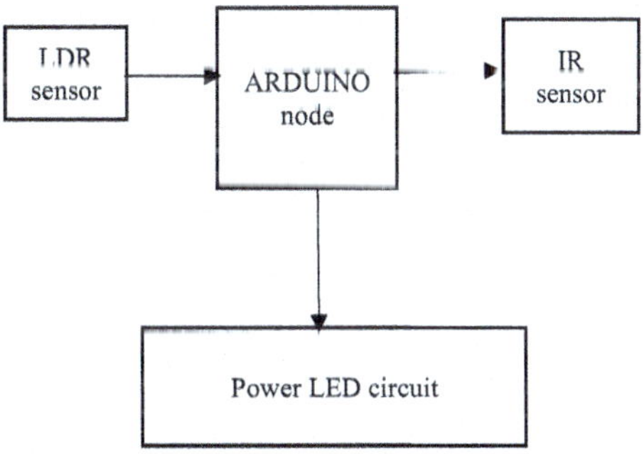

Fig. 4 Proposed system block diagram

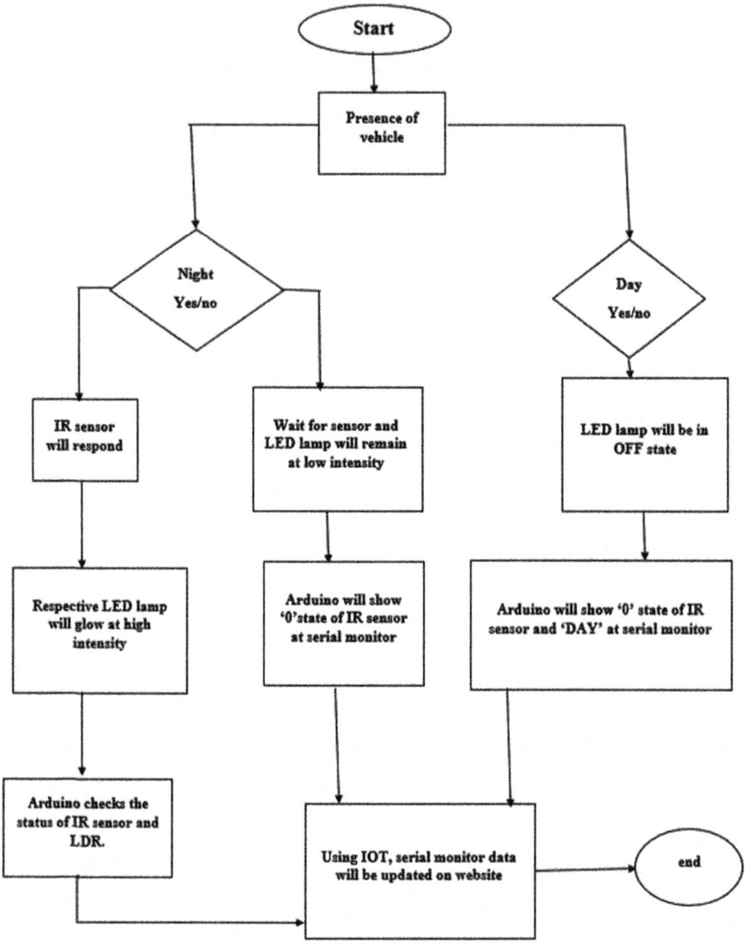

Fig. 5 Flow chart of working of smart street light

5 Flow Chart of Smart Lighting System

The flow chart of the smart street light system is shown in Fig. 5.

6 Advantages

- Lowering the cost of maintenance.
- The reduction of accidents and increased safety are two of the major benefits of street lighting.

- Conserve energy.
- Reduction in manpower.
- Additionally, the use of lightning systems significantly lowers crime such as theft, murder, and many other types.
- The purpose of street lighting is to make it possible for drivers to see the road clearly and easily at night, as well as their immediate surroundings.
- Artificial lighting at night improves vision and provides comfort while driving.
- Under good street lighting, drivers feel more at ease on the road.

7 Limitations

- Compared to traditional street lights, the automatic street light system demands a larger upfront cost.
- If the lighting is automated, the city, state, or county has probably done so to reduce the number of workers needed to maintain the street lighting. If a street light is automated, replacing it will take longer.
- Since automatic street light systems are expensive and non-wired, the risk of theft is substantially higher.
- These lights are primarily impacted by weather factors like air and thunder.
- The automated street light system's rechargeable batteries need to be changed several times, which takes time and is a challenging management process.
- Automatic street lights might not turn on and off at the appropriate times, which might lead to more accidents and more repairs.

8 Result

The smart streetlight system's functioning prototype seems to have been successful in cutting down on energy waste. When it becomes dark and a person is observed in the street, its lights will fully turn on. Otherwise, the streetlight will be turned on with much less brightness when it is night and no one can be seen. This significantly increases energy savings.

The overall connection setup for the smart street light system is shown in Fig. 6.

Figure 6 demonstrates how the setup's light bulb comes on and light will be subdued, i.e. stays dim.

Figure 7 demonstrates how the setup's light bulb brightens up right away after detecting an object.

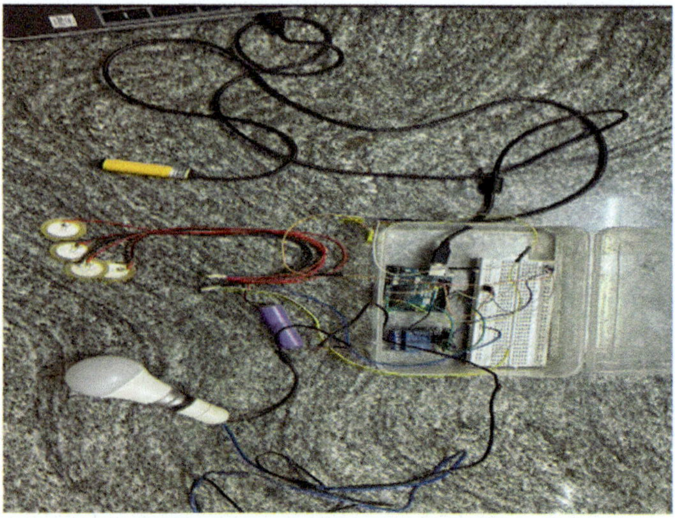

Fig. 6 Overall setup of the smart street light system

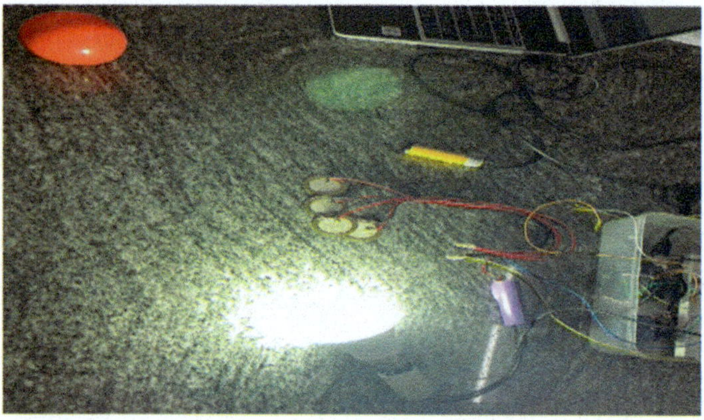

Fig. 7 Light gets brighter after detection of an object

9 Conclusion

We look at how the Internet of things has ruined our daily life in this study. The smart street light framework is one of the main elements that applies IoT concepts. Several urgent issues, including power loss, crime detection, the disposal of incandescent lamps, maintenance expenses, etc., are directly addressed either by smart street lighting system. By its ability to stop female harassment, cut, as well as other forms of intimidation, this system ensures both public safety and the safety of transportation. Cities may experience less energy issues since between fifty and sixty per cent of their electricity is conserved and used for other essential uses. This system fully accommodates user requirements and promotes a secure environment. Simple software and hardware are used in this strategy. The system makes decisions for managing the street lights. It will help make our city a smart city, which is advantageous.

References

1. Dheena PPF, Raj GS, Dutt G, Jinny SV (2017) IOT based smart street light management system. In: 2017 IEEE international conference on circuits and systems (ICCS), Thiruvananthapuram, India, pp 368–371. https://doi.org/10.1109/ICCS1.2017.8326023
2. Abinaya B, Gurupriya S, Pooja M (2017) Iot based smart and adaptive lighting in street lights. In: 2017 2nd international conference on computing and communications technologies (ICCCT), Chennai, India, pp 195–198. https://doi.org/10.1109/ICCCT2.2017.7972267
3. Saifuzzaman M, Moon NN, Nur FN (2017) IoT based street lighting and traffic management system. In: 2017 IEEE region 10 humanitarian technology conference (R10-HTC), Dhaka, Bangladesh, pp 121–124. https://doi.org/10.1109/R10-HTC.2017.8288921
4. Ouerhani N, Pazos N, Aeberli M, Muller M (2016) IoT-based dynamic street light control for smart cities use cases. In: 2016 international symposium on networks, computers and communications (ISNCC), Yasmine Hammamet, Tunisia, pp 1–5. https://doi.org/10.1109/ISNCC.2016.7746112
5. Arjun P, Stephenraj S, Kumar NN, Kumar KN (2019) A study on IoT based smart street light systems. In: 2019 IEEE international conference on system, computation, automation and networking (ICSCAN), Pondicherry, India, pp. 1–7. https://doi.org/10.1109/ICSCAN.2019.8878770
6. Prasad R (2020) Energy efficient smart street lighting system in Nagpur Smart City using IoT—a case study. In: 2020 fifth international conference on fog and mobile edge computing (FMEC), Paris, France, pp 100–103. https://doi.org/10.1109/FMEC49853.2020.9144848
7. Subramanyam BK, Reddy KB, Reddy PAK (2013) Design and development of intelligent wireless street light control and monitoring system along with GUI. Int J Eng Res Appl 3(4):2115–2119

Rock Segmentation of Real Martian Scenes Using Dual Attention Mechanism-Based U-Net

Shrani Sethy, Santi Kumari Behera, J. Ramadevi, Prabira Kumar Sethy, and Preesat Biswas

Abstract Segmentation of Martian rocks is essential for future space exploration and scientific research. In this study, we propose a dual attention mechanism-based U-Net for the segmentation of Real Martian scenes. The dual attention mechanism captures both spatial and channel-wise dependencies to improve segmentation performance. The proposed method is evaluated on a dataset of Real Martian scenes, and the results show that it outperforms state-of-the-art approaches. The dual attention mechanism-based U-Net segmentation technique successfully detected the Real Martian scenes of the rock segmentation using Mars images with an accuracy of 96.38%, sensitivity of 95.36%, specificity of 95.39%, and Jaccard index of 0.86. The proposed method is helpful for rovers to find a route without any obstacles.

Keywords Martian rock · Segmentation · Dual attention mechanism · U-Net · Deep learning

S. Sethy
Kendriya Vidyalaya No. 2, Sambalpur, Burla, Odisha, India

S. K. Behera
Department of CSE, VSSUT Burla, Burla, Odisha, India

J. Ramadevi
Department of CSE, PVP Siddhartha Institute of Technology, Vijayawada, India

P. K. Sethy (✉)
Department of Electronics and Communication Engineering, Guru Ghasidas Vishwavidyalaya, Bilaspur, Chhattisgarh, India
e-mail: prabirsethy.05@gmail.com

P. Biswas
Department of ET & T, GEC Jagdalpur, Jagdalpur, Chhatisgarh, India

V. K. Gunjan and J. M. Zurada (eds.), *Proceedings of 4th International Conference on Recent Trends in Machine Learning, IoT, Smart Cities and Applications*, Lecture Notes in Networks and Systems 873, https://doi.org/10.1007/978-981-99-9442-7_11

1 Introduction

It is conceivable that Mars, the fourth planet from the sun and the second smallest planet in the solar system, is responsible for the dry, toxic state of Earth today. Mercury is the only planet that is smaller than all planet. It is 2279 million kilometres from the sun, has a radius of 389.5 km, is one day and seven minutes long, and weighs 6.102 kg. Each orbit consisted of 687 days. Owing to Mars' scant atmosphere, the planet is dusty, chilly, and similar to a desert. Mars' seasons, polar ice sheets, canyons, and ancient volcanoes are indications of past or even greater activity. Our rovers have made this planet one of the most explored in the solar system.

On September 24, 2014, the Mars Orbiter Mission (MOM), also known as Mangalyaan, entered Mars orbit. It was launched in space by the Indian Space Research Organization on November 5, 2013. (ISRO). India is a first-time entrant in this space. ISRO became the fourth space agency to enter Martian orbit, following NASA, the Roscosmos (which collaborated with NASA on the mission), and the European Space Agency. Thus, India made history the first country in the world and the first country in Asia to reach Mars' orbit on its very first attempt. Craters, desiccated lake beds, and even dormant volcanoes dot desolate landscapes. It was obscured by red sediment, making it difficult to observe. Mars has the same atmosphere and winds as Earth. The Mars Exploration Program intends to explore Mars and provide ongoing scientific information and discovery using a carefully selected succession of robotic orbiters, landers, and mobile laboratories interconnected by a high-bandwidth Mar/Earth communication network.

It is difficult to segment rocks on Mars because of the vast variety of geological features prevalent there. Understanding the geological history of the Red Planet requires accurate classification of Martian materials. Using algorithms based on deep learning, significant progress has been made in the segmentation of Martian scenes in recent years. In this study, we developed a U-Net to classify Martian landscapes based on a dual attention algorithm. According to our study, the Martian landscape consists of numerous canyons, volcanoes, dried lake basins, and craters. A substantial amount of red pigment was visible on the exterior surface. Similar to Earth, it has an atmosphere and clouds. The Mars Exploration Program intends to explore Mars and provide ongoing scientific information and discovery using a carefully selected succession of robotic orbiters, landers, and mobile laboratories interconnected by a high-bandwidth Mar/Earth communication network. The scientific objectives of a Martian mission are to seek life, study the planet's history and surface, and prepare for future human exploration. Understanding whether life has existed anywhere else in the universe besides Earth is a foundational concern for humans. There may be minerals on Mars that were dissolved in water. According to Wang et al. [1], the capability of the planetary vehicle to drive autonomously is the most essential aspect of a trip to Mars. This training assists in the general naming of the Martian terrain and landforms, which is essential for the success of planetary vehicles. According to a study by Federico et al. [2], CNN is a variation of U-Net with fewer parameters that are optimized for rapid inference. Liu et al. [3] want to accomplish the ability to distinguish rock pixels from the background, which is essential for subsequent activities such as Mars rover exploration and geology research. With U-Net, neither the global setting nor long-distance spatial interconnectedness can be adequately modelled. Modern transformers provide a possible solution, but they still work in terms of observing and memorizing their immediate environment. When applied to stones of varying sizes, textures, and colours on Mars, the flaws in extant networks reduce their effectiveness. FRM makes rock formers more resistant to rocks with variable appearances on Mars by retaining more representative features and capturing global dependencies between multiscale properties. Furlán [4], and colleagues discovered that rocks do not function well with existing visual segmentation methods owing to their diverse morphologies and lack of a constant attribute that makes them stand out from the background soil features. Attributes may be local, such as texture; object-level, such as shading and 2-D shape; or scene-level, such as illumination direction. The photographic progression of the terrain is based on a digital model derived from the TianWen-1 dataset. Song et al. [5] developed and implemented a system for automatic boulder segmentation using texture-based image segmentation and edge-flow-driven active contours. For texture-based image segmentation, we utilized a hybrid of three techniques: a local transform based on wavelets, histograms at multiple resolutions, and combining results from various scales. Liu et al. [6] presented a hybrid attention semantic segmentation (HASS) network, with its local attention branch modelling the interactions between specific heterogeneous pixels under the direction of a carefully constructed loss function. The network was constructed with a representation-merging module, global intraclass attention, local interclass attention, and a representation-merging branch. Current methods for landscape segmentation on Mars rely on supervised learning, which requires a large quantity of data and is challenging to train using a small number of labelled samples. As suggested by Goh et al. [7], a deep segmentation network was initially trained unsupervised on unlabeled images before being converted to terrain segmentation and trained on a small number of labelled images. Li et al. [8] presented a one-of-a-kind autonomous framework for locating artefacts on Mars. He presented a method that takes a regional approach to solve the problem by dividing the image into groups with comparable intensity data and spatial arrangement. Xiao et al. [9] also addressed autonomous boulder detection by developing a novel method based on area contrast. Initially, the image was divided into comparable sections based on the intensity data and general image structure. Owing to memory constraints, only low-level features, average intensity, and variation within the super-pixel were identified; subsequently, a contrast map was created and pebbles were extracted using an adaptive threshold. The authors [10] used a mean-shift method to partition images into homogenous objects before classifying them as tiny rock candidates, rock shadows, and large objects in Navcam and Pancam images captured by Martian rovers. This study identified potentially large boulders by ground-plane fitting using 3D point-cloud data. Small and large candidates for rock extraction were combined and processed to obtain the final results.

2 Materials and Methodology

This section discusses the details of the dataset used and the adapted methodology.

2.1 About Dataset

We evaluated the performance of the dual attention mechanism-based U-Net segmentation method on a publicly available IEEE Data port. These images were collected using a Mastcam camera of the Curiosity rover on Mars between August 2012 and November 2018. There were 300 raw and 300 binary-labelled images with a resolution of 512 × 512 pixels. Samples of the neck ultrasound images and mask of the nerve are presented in Fig. 1.

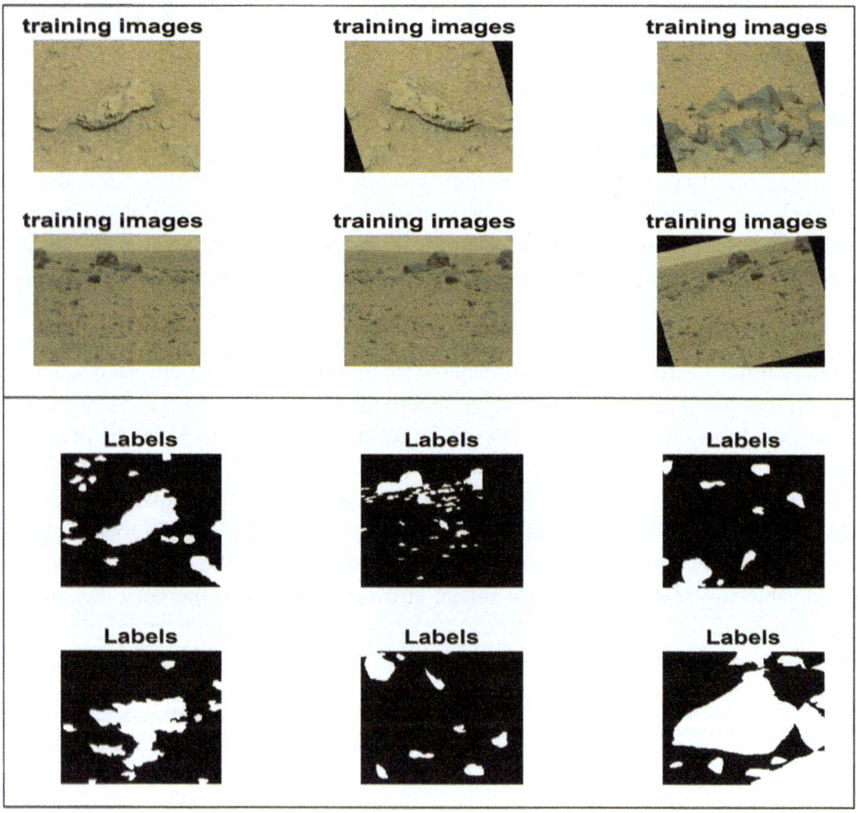

Fig. 1 Samples of train images and their ground truth

2.2 Methodology

The proposed method employs a U-Net architecture that was trained using a dual attention mechanism on a collection of authentic Martian images. The dual attention technique improves segmentation performance by simultaneously capturing spatial and channel-wise relationships. Because of its encoder-decoder structure with skip connections, the U-Net model can encompass both high-level and low-level characteristics.

The Dual Attention Mechanism U-Net (DAM-U-Net) is an enhanced variant of the widely used U-Net convolutional neural network (CNN) design. It is used for image segmentation tasks. By combining the Channel Attention Mechanism (CAM) and the Spatial Attention Mechanism (SAM), DAM-U-Net intends to increase the efficiency of the U-Net model. (SAM). The Channel Attention Mechanism (CAM) is designed to document the relationships between the channels of a feature. To achieve this, we must employ a self-attention mechanism to discover the weights to apply to the sum of the feature maps across channels. This mechanism's ability to suppress less informative channels while concentrating on more informative ones results in improved feature representation and increased segmentation accuracy. The objective of the Spatial Attention Mechanism (SAM) is to register the relationships between the positions of various feature maps. Self-attention is used to learn weights, which are then utilized to compute a weighted sum of feature maps across geographic locations. To further improve feature representation and segmentation precision, this strategy enables the network to focus on only the most informative spatial regions while disregarding the less informative regions. DAM-U-Net enhances the segmentation precision of the original U-Net architecture by using CAM and SAM techniques to more precisely capture spatial and channel-wise correlations between features (Fig. 2).

3 Result and Discussion

The results of applying the proposed method to Martian pebbles indicate that the dual attention mechanism could significantly improve segmentation outcomes. On an HP Victus with Windows 11, 16 GB RAM, and 8 GB GPU equipped with MATLAB 2021a, a public dataset of Real Martian scenes is used to verify the proposed method. The Real Martian dataset was used to train and evaluate the U-Net segmentation algorithm, which is based on a dual attention technique. The results show that it outperforms state-of-the-art approaches in terms of accuracy, sensitivity, specificity, Jaccard index, and the Dice coefficient. A total of 100 images were considered for testing. The proposed method achieved an average Dice coefficient of 0.85, which is higher than that of other state-of-the-art approaches. This method is robust to different lighting conditions and image noise. However, there are still limitations to the proposed method, such as the requirement for a large amount of training data

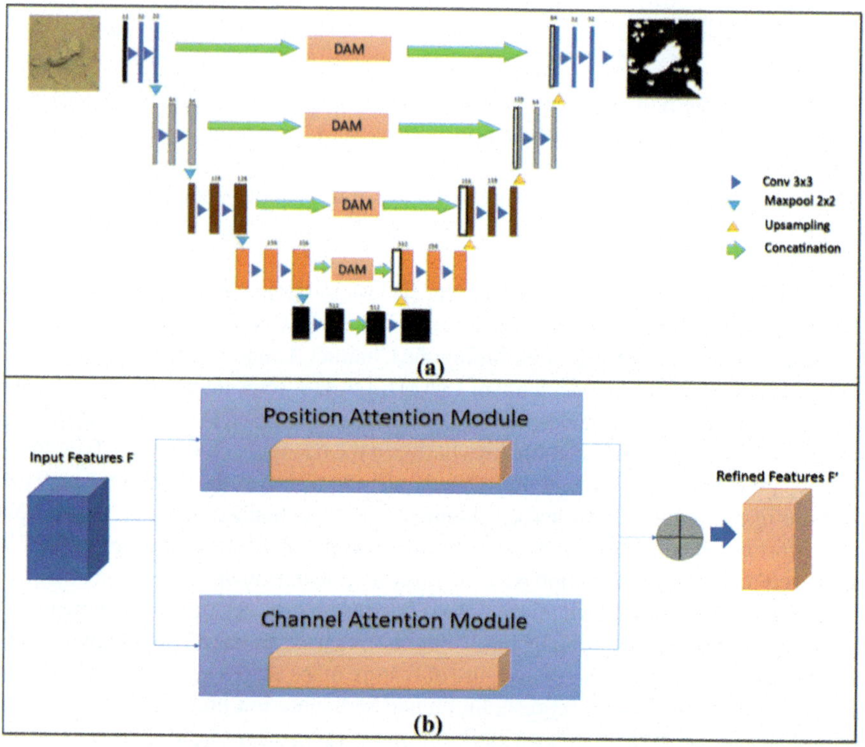

Fig. 2 Proposed model for Martian rock segmentation **a** Base architecture of dual attention mechanism-based U-Net and **b** dual attention model (DAM)

and further improvements in the computational efficiency. Qualitative analysis of the segmentation results also shows that the proposed method can accurately segment Martian rocks of different shapes and sizes. Furthermore, the proposed Dual U-Net Segmentation technique with an attention mechanism successfully detected the rock position, which was better than the baseline U-Net segmentation technique. To show the genuineness of our approach, we presented some outputs of Mars Rock Segmentation, as illustrated in Fig. 3.

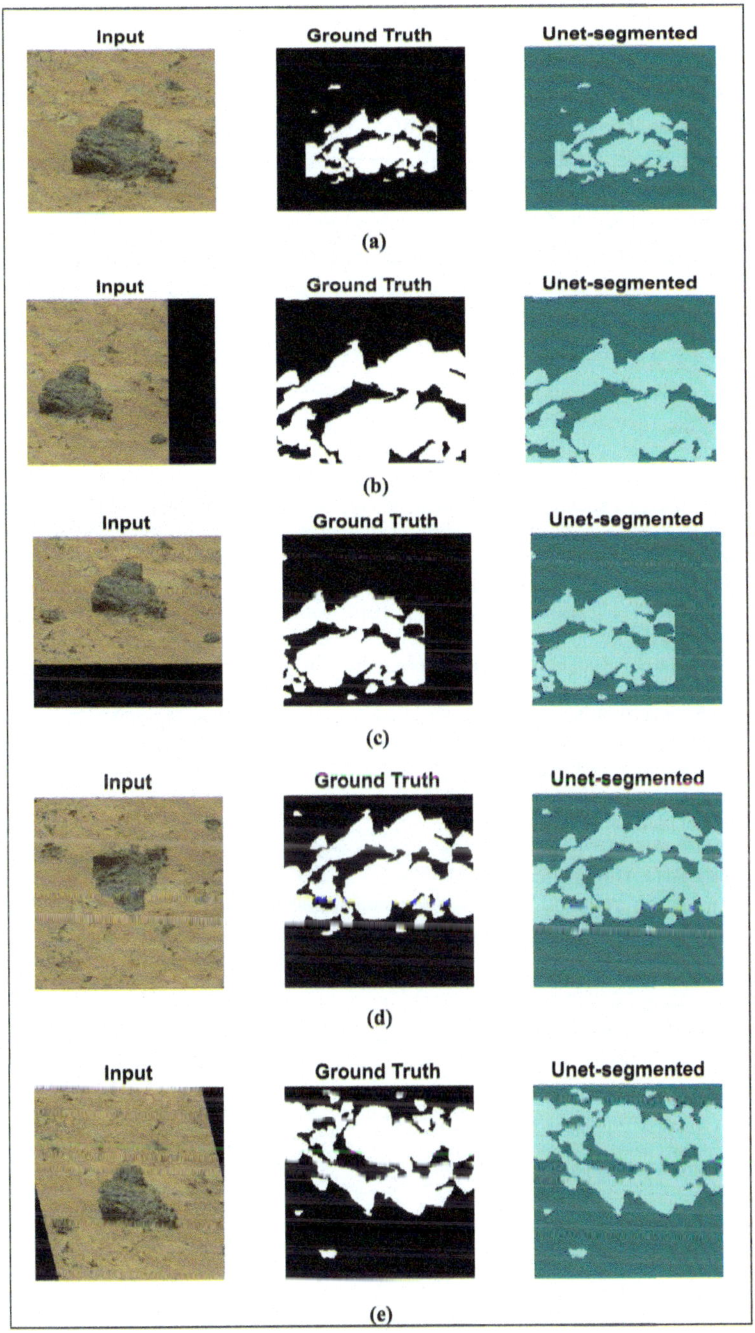

Fig. 3 Result after the DAM-based U-Net segmentation

4 Conclusion

In this study, we proposed a dual attention mechanism for the segmentation of Real Martian scenes. The experimental results demonstrate that the attention-based U-Net method outperforms other state-of-the-art methods for Rock Segmentation of Real Martian scenes. The Dual U-Net segmentation technique successfully detected the Rock martian scenes of Mars using images with an accuracy of 96.38%, sensitivity of 95.36%, specificity of 95.39%, and Jaccard index of 0.86. The proposed method is helpful for rovers to find route obstacles to any obstacle. Future work can focus on reducing computation and improving segmentation accuracy.

References

1. Wang Y, Yang L, Huang G (2022) Semantic segmentation of Martian Terrain based on dual-branch. In: ICETIS 2022; 7th international conference on electronic technology and information science, Harbin, China, pp 1–5
2. Dunlop H, Thompson DR, Wettergreen D (2007) Multi-scale features for detection and segmentation of rocks in mars images. In: 2007 IEEE conference on computer vision and pattern recognition. IEEE
3. Liu H et al (2023) RockFormer: a U-shaped transformer network for martian rock segmentation. IEEE Trans Geosci Remote Sensing 61:1–16
4. Furlán F et al (2019) Rock detection in a Mars-like environment using a CNN. In: Proceedings of pattern recognition: 11th Mexican conference, MCPR 2019, Querétaro, Mexico, vol 11. Springer International Publishing
5. Song YH, Shan J (2008) Automated rock segmentation for Mars exploration rover imagery. Lunar and planetary science conference XXXIX, Houston, USA
6. Liu H et al (2023) A hybrid attention semantic segmentation network for unstructured terrain on Mars. Acta Astronautica 204:492–499
7. Goh E, Chen J, Wilson B (2022) Mars terrain segmentation with less labels. In: 2022 IEEE aerospace conference (AERO). IEEE
8. Li G, Geng Y, Xiao X (2018) Multi-scale rock detection on Mars. Sci China Inf Sci 61:1–8
9. Xiao X et al (2017) Autonomous rock detection on mars through region contrast. Adv Space Res 60(3):626–635
10. Di K, Yue Z, Liu Z et al (2013) Automated rock detection and shape analysis from Mars rover imagery and 3D point cloud data. J Earth Sci 24:125–135. https://doi.org/10.1007/s12583-013-0316-3

IAAS: IoT-Based Automatic Attendance System with Photo Face Recognition in Smart Campus

D. Komala, Rishabh Kumar, Nikitha Ramya Sathish, and J. P. Sagar

Abstract A lecturer's most oppressive task is marking attendance. Manually marking the attendance of students also takes a lot of time. Although biometric systems have been used for automation in the past, it has not proven to be a very efficient method since it is much easier to fake a fingerprint today. When selecting a method for this task, accuracy and authenticity are also important considerations, something many proposed methods do not address. Nowadays, smart attendance monitoring systems such as biometric facial recognition systems are being proposed in order to eliminate manual paper-based attendance systems. There are thousands of applications for it, such as monitoring the classroom using CCTV, creating computer–human interaction, maintaining accurate attendance, and maintaining security. Students can be marked absent even when inside the classroom using this system. Auto-attendance systems rely primarily on facial recognition in contrast to other biometrics. In this study, three methods are applied: Convolution Neural Network (CNN), Viola–Jones (Haar-Cascade), and Histogram Oriented Gradients (HOG) are evaluated on the basis of face detection accuracy. Using an excel sheet, attendance information is automatically updated in the system. In the event of an absent student, a message will automatically be sent by using SMTP protocol to their parent's phone number.

Keywords Convolutional neural network (CNN) · Histogram oriented gradients (HOG) · SMTP · Fingerprint · Viola–Jones (Haar-Cascade)

1 Introduction

The use of face recognition to mark attendance could replace biometrics in an efficient manner. An innovative technique for identifying and recognizing a person is to use their facial features. Each person has a unique facial feature. Individuals in

D. Komala (✉) · R. Kumar · N. R. Sathish · J. P. Sagar
Don Bosco Institute of Technology, Bengaluru, Karnataka, India
e-mail: Komala.dbit@gmail.com

V. K. Gunjan and J. M. Zurada (eds.), *Proceedings of 4th International Conference on Recent Trends in Machine Learning, IoT, Smart Cities and Applications*, Lecture Notes in Networks and Systems 873, https://doi.org/10.1007/978-981-99-9442-7_12

the wild can be identified and detected more accurately and exquisitely thanks to the uniqueness and precision of facial features [1]. Security and surveillance systems can use facial recognition directly or indirectly, but it has not been used efficiently due to noticeable flaws and defects. There are many advantages and disadvantages to marking attendance in the most common way. In the traditional attendance marking system, the main snags are the manual effort and the time consumed. Any automation aims to eliminate those drawbacks. Biometric attendance marking is improved based on these limitations. The amount of time consumed remains the same, even though biometrics reduces manual effort. Additionally, biometrics isn't cheap. Students' attendance is monitored by capturing different facial expressions as an image when they enter the class or once all students are seated inside the classroom with the attendance system that incorporates Recognition of Human Face. Figure 1 illustrates two techniques for face recognition: Brightened and Featured. Brightened techniques use representations of faces with expressions, while Featured approaches use specific attributes on faces [2].

In increased brightness, the divisions of the captured image are evaluated and calculated. The technique is based on images. This results in the computation of a full image. There is an additional time commitment and a slight challenge associated with this method. The human face is a significant portion of which can be identified based on two types of evaluations. There are two types of verification: identification and verification. As part of the verification process, the system compares a person's stated identity with who they claim to be, and gives a yes or no answer. It is possible to improve face recognition accuracy by using good datasets for training and testing, as well as optimizing the algorithms used. The field of face recognition has been flooded with a number of techniques, many of which make use of improved data mining and statistics techniques, while others use standard algorithms. In some methods, microcontrollers and cameras are incorporated into mobile phones, while in others, these components are incorporated into hardware systems.

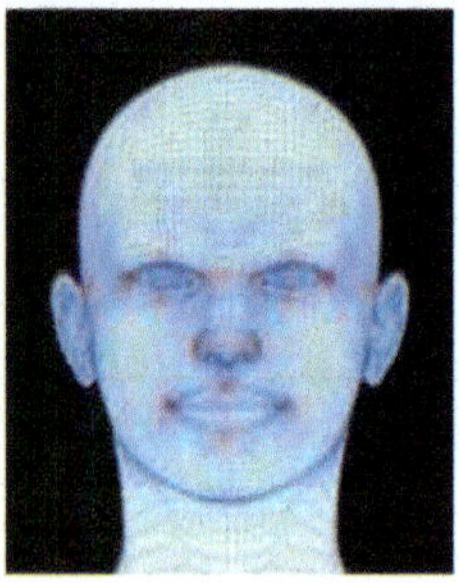

Fig. 1 Face recognition techniques

2 Literature Review

Kodali and Hemadri [1] conducted that, to provide a better solution to the problem, researchers analyze the solutions proposed by others and identify their shortcomings. As a solution to the attendance tracking problems, an approach based on face recognition was used. This consists of matching up the recent images with those in the attendance database. A solution to this problem has been presented by Naveed et al. in, where two databases are available [3]. In one database, there are pictures of the students, and in the other, there is information about attendance. It removes the background and noise associated with the image when taking a picture of the class. Consequently, the identification of the face is matched with the profile database and the presence is recorded in the database for attendance based on the classification of the skin. Fixed seating ensured that students' attendance was marked. Their proposal was to detect faces in real-time using a learning management system already in place. Students' faces were detected and attendance was recorded by fixing the order of the students. Through the front-facing camera, students were marked out of class for all classes. They were then photographed with a roof-mounted camera and their faces were compared with those in the record.

Several other solutions were proposed by Muthu Kalyani et al. in addition to the prior one mentioned in [4]. An enhanced 3D mobile device was mounted on a CCTV camera for facial recognition using the put forth system. To match the taken picture to the stored image and the image of the student, Canonical Techniques and 3D Modeling were applied after the image was captured. When the images do not match, the administration can use the image as a security measure against strangers.

Nasser et al., also proposed a system in [5]. It is difficult to mark attendance in large businesses due to the number of employees. Their solution proposed storing faces and attendance information in an SQL database after face detection and attendance marking. A picture is taken at the beginning of the meeting. An attendance mark is made by taking a picture with the camera, which is then processed to remove background and picture noise. In an article by Roshan et al., they provided the solution to the orthodox problem in [6] a situation in which the existing attendance management systems have been amended or where a situation has been identified that has been left unattended. During the Muslim/Islamic tradition, girls and women wear veils covering their faces, while men wear beards, which may alter their appearance from day to day. Attendance marking should be robust enough to handle different beard styles and lengths, as well as the presence of a veil.

Tan et al. [7], using the spatial transformer network, a new frame was proposed to improve the recognition of face attributes. On both CelebFaces Attributes Dataset (CelebA) and Labeled Faces in the Wild-a (LFWA), our method achieves state-of-the-art performance. Celebi and LFWA performance has been upgraded by 4% and 2%, respectively, on average for face attribute recognition.

Cherifi et al., proposed attendance system that will incorporate global and local methods of face recognition (PCA, FLD, DCT, DWT) [3]. Different conditions and working environments are tested with this method, such as pose variations, illumination, and facial expression changes, along with blur, motion, and noise effects. Due to this, DCT + LBP is more effective than other methods.

For facial recognition and association in facial identity, Nazare Kanchan Jayant et al use the equation Viola and Jones. Face recognition algorithm based on alignment-free partial face recognition. A real-time database capture is used to analyze the data using an iBall webcam CHD20.0. On average, 85% of faces are detected and 53.1% are recognized.

3 Algorithms

The purpose of our work is to utilize face recognition technology to create an automatic attendance system. By comparing and combining existing face recognition techniques, we are able to find the best results for our system based on an existing technique.

A. *SRGAN*

In order to enhance the resolution of a single image, a generative adversarial network (SRGAN) is used. According to this algorithm, its perceptual loss function incorporates adversarial loss as well as content loss. By using a discriminator network, a solution is pushed toward the natural image manifold when the adversarial loss is introduced. The authors also use a content loss based on perceptual similarity rather than pixel similarity. Remaining blocks for feature extraction are depicted in the figure to the right.

B. *MTCNN*

As a solution for both face detection and face alignment, multi-task cascaded convolutional networks (MTCNN) employ multi-task cascaded convolutional networks. Faces and landmark locations like eyes, noses, and mouths can be recognized through three stages of convolutional networks. A multi-task learning approach called MTCNN is presented in the paper as a way to integrate both tasks (recognition and alignment). A shallow CNN is used in the first stage to produce candidate windows quickly. A more complex CNN is used to refine the proposed candidate windows in the second stage. Third, it further refines the result and outputs facial landmark positions using a third CNN that is more complex than the other two.

C. *GoogleNet*

GoogleNet is a convolutional neural network with 22 layers. A wide range of pre-trained networks are available for loading, including those based on ImageNet or Places 365 data sets. Based on ImageNet, the network classifies images into 1000 categories, including keyboards, mice, pencils, and many animals. This network recognizes 365 different categories of places, such as fields, parks, runways, and lobby, similar to the one trained on ImageNet. Various feature representations have been learned for a wide range of images using these networks. Input images for both trained networks are 224 by 224 pixels.

D. *Haar-Cascade*

Object detection algorithms are used to recognize faces in images and videos. According to their research paper "Rapid Object Detection using a Boosted Cascade of Simple Features", a line or edge detection algorithm was published in 2001. Algorithms are trained by presenting them with many images with faces as well as many without faces.

E. *Viola–Jones*

Due to Viola–Jones' design, it can detect frontal faces better than ones looking sideways, upwards, or downwards. For faster processing of the image and less data to process, the image is converted to grayscale before detecting faces. In the Viola–Jones algorithm, the grayscale image detects the face first, then the colored image finds its location.

4 Problem Statement

In order to prevent illegal attendance, the goal of this project is to use face recognition technology to develop an automatic attendance system in university lectures.

To make university lectures more convenient for those using the system and to prevent proxy attendance, a form of illegal attendance, we intend to include a face recognition-based automatic attendance system. To increase convenience for those who use the system and to prevent proxy attendance, a type of illegal attendance, we would like to provide an automatic attendance system based on face recognition in university lectures.

Input: An Image.
Output: IAAS system notifies each user of the attendance completion by an email message.

5 Objective

People would have been baffled and perhaps even criticize overengineered products if they were told that your Geyser and bedroom lights were connected to the Internet a few years ago. Nowadays, with the advent of IoT, smart cities, and the Internet of Things, the idea is no longer strange. We have devices around us that communicate with the Internet and have become smarter as a result.

This project aims to enhance and upgrade a current attendance monitoring system with a face recognition system for an educational institution so that it becomes more efficient and effective. The old system provides a number of problems when the authority cannot enforce its regulations. A lack of enforcement of the old system poses many problems for the authority. In this way, by automating most of the tasks involved in attendance taking, the project will address the flaws in the current system. Face recognition technology will be used behind the scenes. Because there is little possibility of a face deviating or being duplicated, it is used to trace identity. Face databases will be created to feed data into the recognizer algorithm. In order to identify a person, faces will be compared against the database during the attendance-taking session [8].

Ten (10) students are in the training database of the system implemented. Students' faces are photographed under different levels of illumination and with different levels of composure and look. As low as 70% and as high as 90% accuracy was found in tests conducted on nine (9) students in the database. As Fisherfaces' database grows, its accuracy increases. As a real-time attendance monitoring tool, the database was integrated with the simple mail transfer protocol (SMTP) that sends identification messages (names and dates of student identification) to a lecturer's (in this case, the administrator's) email address. The result revealed that attendance fraud had been eliminated [9].

6 System Design and Architecture

A. *Multi-task Cascaded Convolutional Network*

An integrated framework for both face detection and alignment has been developed. Face recognition is accomplished by using convolutional networks in three stages that include eyes, noses, and mouths as landmarks. Multi-task learning is proposed in the paper as a method of integrating both tasks (recognition and alignment). To quickly produce candidate windows, a shallow CNN is used in the first stage. A more complex CNN refines the proposed candidate windows in the second stage. Lastly, a third CNN is used to refine the initial results and output facial landmark positions at the third stage.

The three stages of MTCNN:

Stage 1: The Proposal Network (P-Net)

An image pyramid is created by resizing the image to different scales as the first stage of the cascaded three-stage network. This process begins with an FCN. As a result of their architecture, FCNs do not utilize dense layers like CNNs. This proposal network is used to determine candidates' bounding boxes and their regression vectors. It is common to use bounding box regression to predict the location of boxes when detecting a certain class of objects, such as faces. After that, overlapping regions must be combined by refining the bounding box vectors. When all candidate windows have been refined to reduce the volume of candidates, they are created [10] (Fig. 2).

Stage 2: The Refine Network (R-Net)

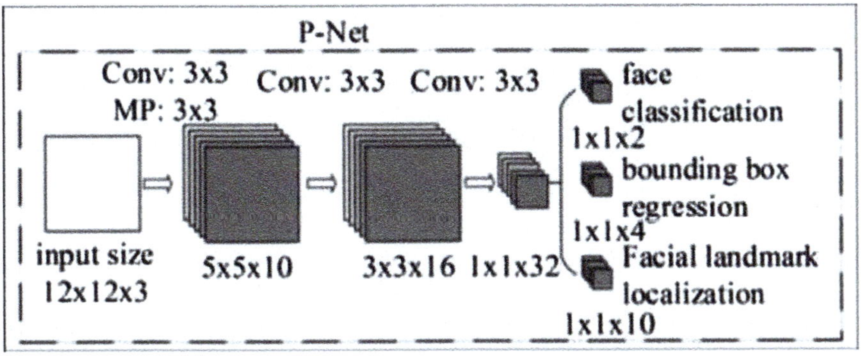

Fig. 2 P-Net

Candidates are refined from the P-Net using a Refine Network. This network is a CNN instead of a FCN because it has a dense layer at the end. After reducing the number of candidates, calibration with bounding box regression, and merge overlapping candidates using NMS, we employ R-Net to reduce the number of candidates. The algorithm generates a four-element bounding box vector, a ten-element vector indicating facial landmark locations, and a four-element vector indicating whether the input is a face (Fig. 3).

Stage 3: The Output Network (O-Net)

A similar approach is taken in this stage, however, this Output Network aims to provide more detail about the face and output the five facial landmarks, namely the eyes, nose, and mouth, and their positions (Fig. 4).

B. *GoogleNet and VGG16*

A major breakthrough in neural networks, especially CNNs, was the Inception Network. Inception Networks currently have three versions, called Inception

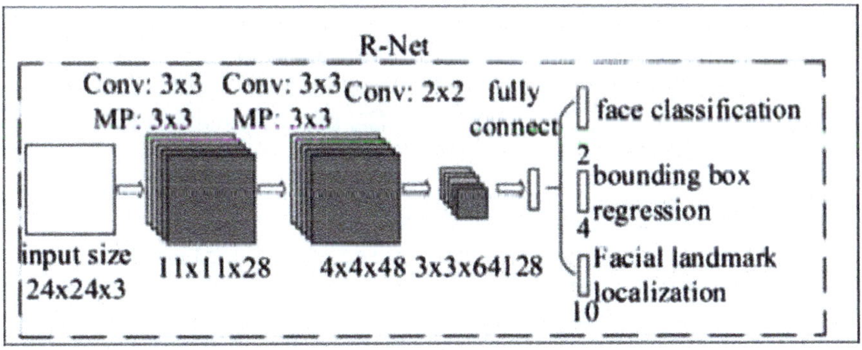

Fig. 3 R-Net

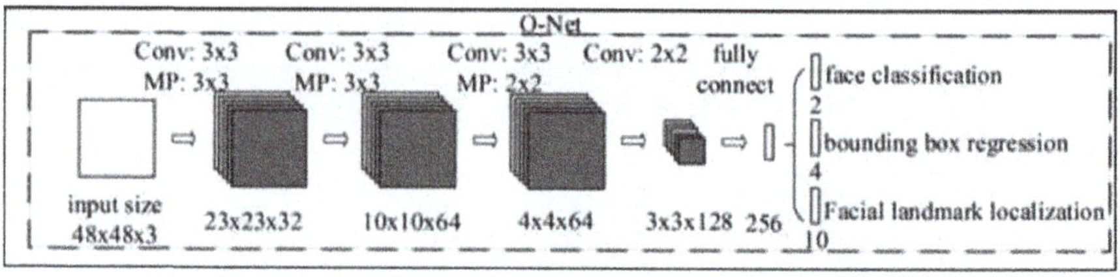

Fig. 4 O-Net

(a) Siberian husky (b) Eskimo dog

Fig. 5 Two different classes from the 1000 classes of the ILSVRC 2014 classification challenge

Versions 1, 2, and 3. This application was launched in 2014 and has been developed by a Google team as its name implies. In the ILSVRC, this network set a new standard for classification and detection. There is the first version of Inception known as GoogleNet, which is a precursor to the existing version [11] (Fig. 5).

Simonyan and Zisserman of the University of Oxford developed the VGG neural network architecture. This method is also based on CNNs and was considered for the ImageNet Challenge in 2014. An overview of their work is provided in the paper Very Deep Convolutional Networks for large-scale Image Recognition, in which the authors discuss their findings. Top-five test accuracy of 92.7% was achieved by the network on ImageNet. With VGG, the major improvements are the addition of multiple (3 * 3) kernel-sized filters, one after the other, in the first and second convolutional layers (sizes 11 and 5, respectively) (Fig. 6).

C. Simple Mail Transfer Protocol (SMTP)

In order to fully understand how SMTP works, we need to look at the sending process, the rules and commands that govern it, and the errors you might encounter along the way. Here's a warning: Things get technical. Nonetheless, we'll simplify this protocol into chunks that are easy to understand. SMTP servers can be connected to and used by email clients once they have been set up. A user's email client opens an SMTP

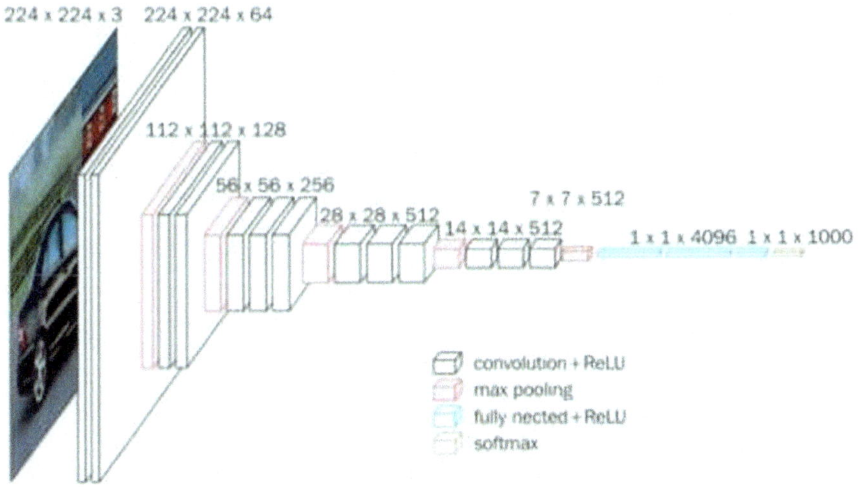

Fig. 6 VGG16 architecture

connection to the server when they click "send" on the email message. TCP stands for Transmission Control Protocol, which is the basis for SMTP connections. Senders and recipients' email addresses, as well as the message's contents, are provided to clients by servers in response to commands. Mail Transfer Agents or Message Transfer Agents (MTAs) verify that both email addresses come from the same domain [12] (Fig. 7).

7 System Requirements

7.1 Hardware Requirements

- System: Windows 7 or more
- Hard Disk: 40 GB
- Monitor: 15VGAColour
- Mouse: Logitech
- RAM: 4 GB.

7.2 Software Requirements

- OpenCV-Python Software
- Visual studio code Software
- Qt Creator Software

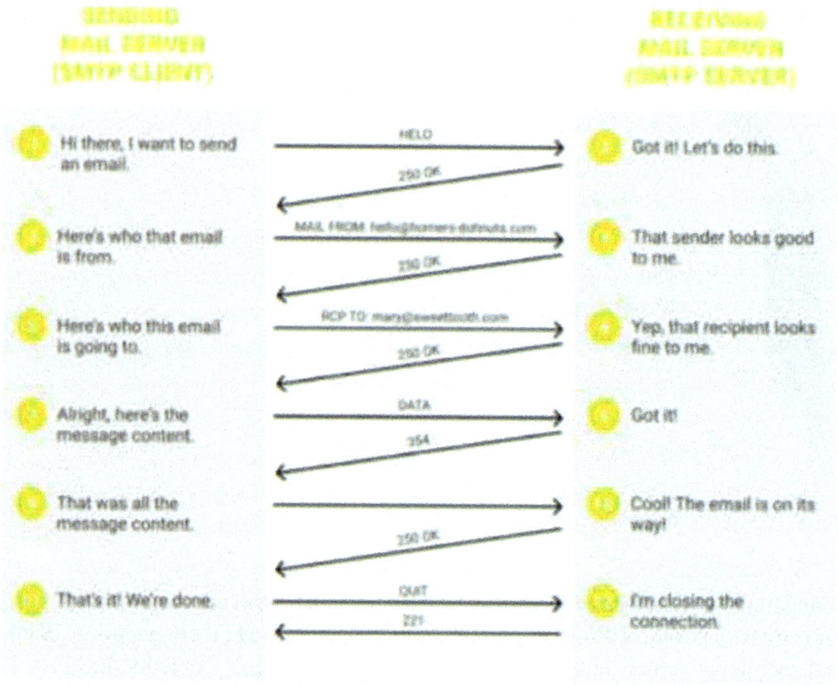

Fig. 7 SMTP client and SMTP server

- Anaconda.

8 System Implementation and Methodology

We will develop a possible solution to our problem based on the literature survey since we have studied several topics which are directly related to our project. As part of this part, we will present a method that will provide an overview of the approach to our project and how it should be accomplished. In this project, we developed it in the most feasible and efficient manner possible since the previous work was not enough.

The system architecture is shown in Fig. 8. Face recognition algorithms are used in the proposed automated attendance management system. The camera at the entrance captures a person's image as he enters the classroom. In order to process the face region further, the face region is extracted and pre-processed. Due to the limited number of people entering a classroom at once, face detection algorithms have less work to do. There are several advantages to using a face recognition system over other systems. Face recognition systems perform better with an efficient and accurate face detection algorithm. Pre-processing is performed on the detected face after it has

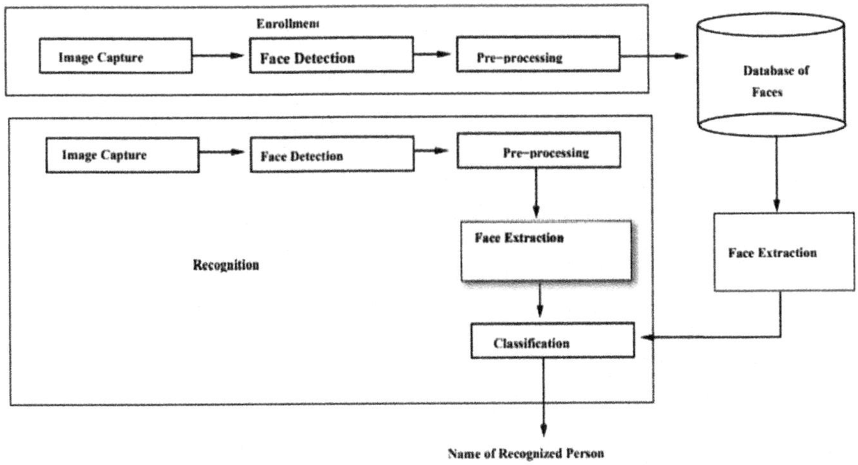

Fig. 8 System architecture

been extracted. After the extracted face image has been histogram equalized, it is resized to 100 × 100 pixels and pre-processed. After the extracted face image has been histogram equalized, it is resized to 100 × 100 pixels and pre processed. The image is made clearer by stretching the range of the intensities in it, which improves its contrast. In this stage, every individual's image is captured and the biometric feature is extracted. In our case, it is the face, which is then enhanced using pre-processing techniques and stored in the database. In order to get accurate results, the performance of a face recognition system is also affected by the feature extraction and classification process. Techniques such as feature-based extraction or holistic extraction are used to extract features. Dimensionality reduction can be used before classification in some holistic techniques. A real-time scenario was used to compare the results of different holistic approaches for feature extraction and classification.

9 Results

Output 1: In output 1 a bounding box vector is created around the face and it is detecting a face. As we can see in the figure it is not detecting face because datasets is not trained.

After capturing images we trained an image using Viola–Jones and Haar-cascade. There is a need of proper training algorithm for training model. Super Resolution GAN (SRGAN) is used to enhance the picture quality (Fig. 9).

Output 2: Test data is data which has been specifically identified for use in tests, typically of a computer program. Some data may be used in a confirmatory way,

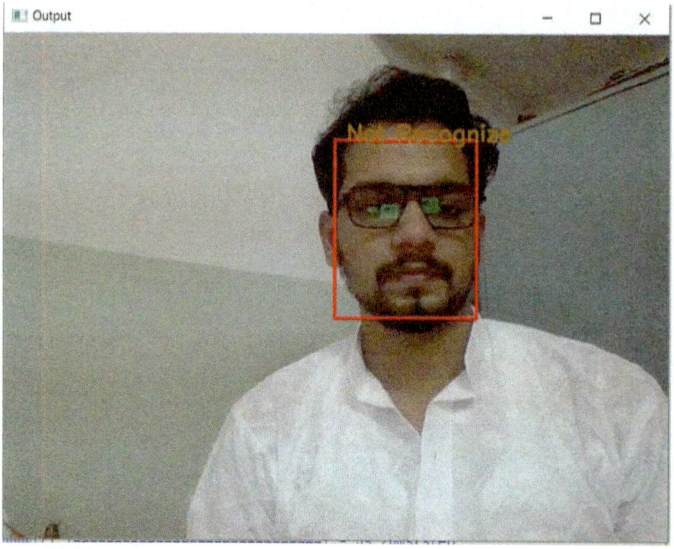

Fig. 9 Train the data

typically to verify that a given set of input to a given function produces some expected result (Fig. 10).

Fig. 10 Test the data

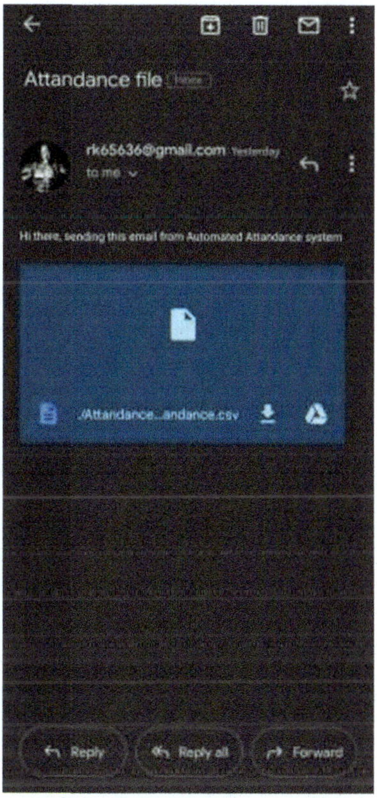

Fig. 11 Attendance mail

Output 3: We are proposing a model where the memory cost can be reduced and this model can run on legacy systems thereby reducing the cost of implementation. All attendance has been stored in excel sheet form and it mailed to faculty (Figs. 11 and 12).

ID	Name	Time	Data	Attandance
1	Aashil	11:26:45	2022-06-30	absent
2	Akash	11:26:45	2022-06-30	absent
3	Nikitha	11:26:45	2022-06-30	absent
4	Raj	11:26:45	2022-06-30	absent
5	Ranjeet	11:26:45	2022-06-30	absent
6	Rishabh	11:26:45	2022-06-30	present
7	Sanrab	11:26:45	2022-06-30	absent
8	Satish	11:26:45	2022-06-30	absent
9	Shubham	11:26:45	2022-06-30	absent
10	vivek	11:26:45	2022-06-30	absent

Fig. 12 Attendance in excel sheet

10 Conclusion

Automated attendance system based on Internet of Things (IAAS). With the help of face recognition technologies, we aim to provide a reliable and improved automatic attendance tracking system. Using Haar-cascade techniques and improved system structure, it is possible to extract a user's face data using training data and suitable face recognition technologies. Both students and professors are expected to find our IAAS system convenient for checking attendance. In this research, we will analyze how postures and appearances of users affect performance and develop a solution for dealing with such factors.

References

1. Kodali RK, Hemadri RV (2021) Attendance management system. 978-1-7281-5875-4/21/ $31.00 ©2021 IEEE. https://doi.org/10.1109/ICCCI50826.2021.9402659
2. Kumar A, Sharma M, Gautam SP, Kumar R, Raj S (2020) Attendance management system using facial recognition. https://doi.org/10.1109/DASA51403.2020.9317104
3. Prabowo H, Gunawan MA, Gaol FL (2020). 978-1-6654-0422-8/20/$31.00 ©2020 IEEE
4. Muthunagai R, Muruganandhan D, Rajasekaran P (2020) Classroom attendance monitoring using CCTV
5. Patil P, Shinde S (2020) Comparative analysis of facial recognition models using video for real time attendance monitoring system. 978-1-7281-6387-1/20/$31.00 ©2020 IEEE

6. Raj AA, Shoheb M, Arvind K, Chethan KS (2020) Face Recognition based smart attendance system. In: 2020 international conference on intelligent engineering and management
7. Hachad T, Sadiq A, Ghanimi F (2020) Student's attendance management using deep facial recognition. https://doi.org/10.1109/ICECOCS50124.2020.9314403
8. Yashwant CC, Sanjay GK, Bangal SP (2021) Development of IoT based smart attendance system using raspberry PI. 12(7). ISSN NO: 0377-9254
9. Okokpujie K, Noma-Osaghae E, John S, Oputa R (2017) Development of a facial recognition system with email identification message relay mechanism. 978-1-5090-4642–3/17/$31.00 ©2017 IEEE
10. https://medium.com/@iselagradilla94/multi-task-cascaded-convolutional-networks-mtcnn-for-face-detection-and-facial-landmark-alignment-7c21e8007923. By Rosa Gradilla 28 Jul 2020
11. https://blog.paperspace.com/popular-deep-learning-architectures-alexnet-vgg-googlenet/. By Vihar Kurama
12. https://postmarkapp.com/guides/everything-you-need-to-know-about-smtp. By Bettina Specht. 21 June 2022

Hardware Implementation of Moving Object Detection Using Background Subtraction Algorithm

Rajeshwari Mattimani and Saroja V. Siddamal

Abstract Moving object detection is used in many applications such as surveillance and vehicle navigation. Detection of moving object is an important process. The authors proposed adaptive background subtraction for detection of moving object. The proposed architecture is designed and implemented on FPGA board Spartan 6[xc6scx45-2csg324) FPGA. The proposed technique uses less area 2001 LUTs at an operating frequency which is 154 MHz.

Keywords Spartan 6 · FPGA board · RTL

1 Introduction

Object detection is very concerned nowadays for any moving object. Security is one of the concerns in today's modern society. In many places video camera is installed to address the security issue; one among the techniques is for surveillance. Moving object detection is a technique that is widely used in image processing. Recognizing an object's physical movement in a certain location or area is known as moving object detection. It has applications in different fields like visual and clinical surveillance, robotics, autonomous navigation, and collision avoidance. Otherwise a person should continuously monitor the video frame at the base station to detect objects in the video which may cause error intentionally or unintentionally. This necessitates the need for developing an algorithm for automatically detecting moving object present in any video frames [1]. The technique is widely used in image processing. Recognizing an object's physical movement in a certain location or area is known as moving object detection [2]. It has applications in different fields like visual and clinical surveillance, robotics, autonomous navigation, and collision avoidance. In this technique,

R. Mattimani (✉) · S. V. Siddamal
School of Electronics and Communication, KLE Technological University, Hubli, India
e-mail: rajeshwari_m@kletech.ac.in

S. V. Siddamal
e-mail: sarojavs@kletech.ac.in

V. K. Gunjan and J. M. Zurada (eds.), *Proceedings of 4th International Conference on Recent Trends in Machine Learning, IoT, Smart Cities and Applications*, Lecture Notes in Networks and Systems 873, https://doi.org/10.1007/978-981-99-9442-7_13

consecutive frames of video are extracted and compared using different methods to check if any moving object is detected [3]. The following are some of the well-known approaches for moving object detection:

1. Background subtraction
2. Temporal differencing
3. Frame differencing
4. Optical flow.

2 Literature Survey

One of the most important methods for processing images is moving object detection. For object detection and object recognition and analysis, numerous methods have been developed. The model uses background subtraction to detect the moving object followed by median filtering to remove the noise pixels over the entire image. The model is simulated using MATLAB tool [4].

A paper was published which describes a method for segmenting moving objects using Adaptive Block Partition Decision (ABPD). This method uses a fixed threshold value. The model was implemented on virtex6 board with 5 pipeline stages, and it achieves high throughput with less resource utilization. The effectiveness of this approach is verified using MATLAB tool [5].

Another such work in the field of object detection was carried out that basically has four main stages. The convolution filter, the background subtraction and segmentation process, the erosion operation, and the gravity center calculation. This model provides one processed pixel for each FPGA clock cycle. The hardware implementation is 32 times better than software implementation in terms of speed. The system has very less area consumption as it uses very less logic elements of FPGA. The EDA Quartus II Altera's tool was used for the system design [6].

For high-speed applications, real-time moving object detection on FPGA is suitable. A system with respect to this is modeled that uses the combination of background subtraction algorithm and OTSU algorithm (adaptive image segmentation algorithm) with floating point rounding method. OTSU algorithm is used to determine the threshold. This system makes use of the pipeline structure, which increases the speed of data processing and hence helps in real-time moving object detection. The simulation and testing is done using MATLAB and Lattice Diamond. This system is observed to be fast and stable in indoor scenes [7].

Utilizing MATLAB tool and Xilinx ISE, video object identification with background subtraction has been implemented on the Spartan 3 FPGA. In this work, first the video is converted to frames and then two frames are selected among which one is reference background frame and other is foreground frame with moving object to be detected. Sobel's edge detection approach is employed to obtain outline shape of the recognized moving object [8].

In many applications moving object detection is important task. In this paper they have used background subtraction method and implemented on FPGA board XC 3S 250e-4vq 100 [9].

In this paper, an efficient FPGA-based reconfigurable architecture of moving object detection using adaptive threshold method is proposed. The architecture is optimized and implemented on FPGA board using Spartan 6 [10].

3 Theoretical Aspects of Motion Detection

The proposed work involves background subtraction due to its advantages such as high accuracy, less processing time, and its expressive potential for other applications like edge detection.

The mathematical equations related to background subtraction are as follows:

$$D_t(x, y) = |I_t(x, y) - \mathrm{BG}_t(x, y)| \tag{1}$$

where x, y are the coordinates of the pixels, $I_t(x, y)$ is the current frame, $\mathrm{BG}_t(x, y)$ is the background frame at time 't', and $D_t(x, y)$ is the background subtracted frame at time 't'.

$$\begin{aligned} \mathrm{MD}_t(x, y) &= 1 \text{ if } D_t(x, y) \text{is greater than or equal to } T \\ \mathrm{MD}_t(x, y) &= 0 \text{ if } D_t(x, y) \text{is lesst han } T \end{aligned} \tag{2}$$

where T is a pre-defined threshold value and $\mathrm{MD}_t(x, y)$ is a binary image.

4 Proposed Hardware Architecture for Motion Detection Using Background Subtraction

The proposed motion detection architecture involves stages such as preprocessing, filtering, thresholding, and background subtraction (Fig. 1).

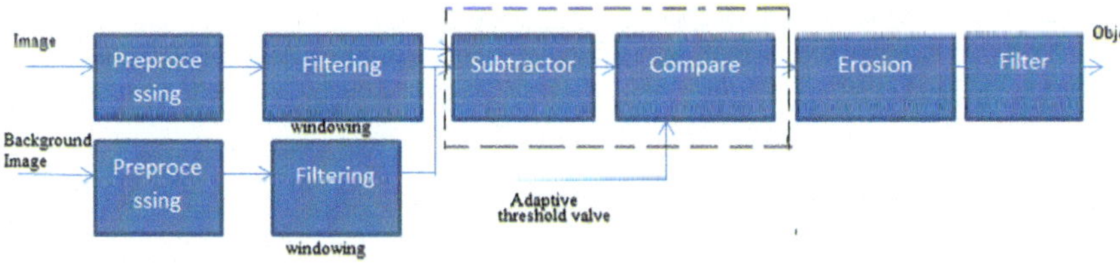

Fig. 1 Proposed background subtraction architecture

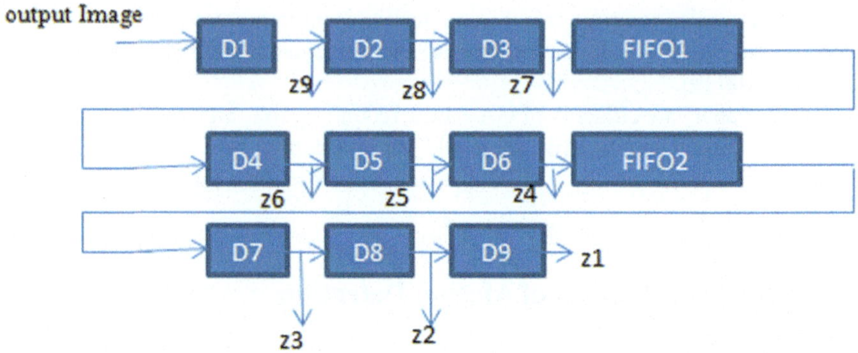

Fig. 2 Moving window architecture

Preprocessing: In this stage both the images input and background are converted from RGB to gray and resized. The conversion of image from 2 to 1D takes place in the stage. The gray image are converted to binary valves and stored in memory.

Filtering: In this stage filters are used to remove the noise, and image high-frequency edges moving window is used to obtain 3 * 3 matrixes of input valves. The Gaussian mask considered is 3 * 3 matrixes; hence the input image matrix also needs to be 3 * 3. Figure 2 shows the moving window architecture used to obtain 3 * 3 matrixes. The input image (256 * 256) is considered in the work. To obtain 3*3matrix two shift registers of size 256 which work as FIFO and 9 D flip flops are used. This acts as dynamic arrays.

After 515 clock cycles the kernel is obtained and represented as $\begin{bmatrix} z_1 & z_2 & z_3 \\ z_4 & z_5 & z_6 \\ z_7 & z_8 & z_9 \end{bmatrix}$.

The Gaussian mask 1/16 $\begin{bmatrix} 1 & 2 & 1 \\ 2 & 4 & 2 \\ 1 & 2 & 1 \end{bmatrix}$ is multiplied with obtained kernel which removes the noise occurs due to light intensity.

Adaptive Threshold Calculation: The adaptive threshold valve is calculated, and it is used to detect the object. To calculate the adaptive threshold valve following equations are used:

Weighted Mean Square Difference

$$\text{WMSD} = (I_{I0} - I_{R0})^2 + (I_{I1} - I_{R1})^2 + \cdots + (I_{\text{IN}} - I_{\text{RN}})^2/(K + M + N) \quad (3)$$

I_0, I_{I1}, I_{Rn}	Pixel valves of output image
$I_{R0}, I_{R1}\ I_{Rn}$	Pixel valves of background image
N and M	Dimension of image (256 * 256)
K	sensitivity factor $= \delta$.

$$\text{AdTh} = \text{WMSC} + [\text{Background Image pixel valve}] \quad (4)$$

Background Subtraction: The background subtraction calculates the difference of filtered output image with filtered background image pixel by pixel. The authors use modified background subtraction whose equation is given as

$$\text{MBS} = \left\{ \begin{array}{l} \text{SI} : \text{if SI} \geq \text{AdTH} \\ 0 : \text{if SI} \leq \text{AdTH} \end{array} \right\} \quad (5)$$

where SI: is subtracted Iiage = |filtered output image-filtered modified image| AdTh: is threshold valve.

5 Results

5.1 Hardware Utilization

The proposed object detection method is synthesized using Xilinx 14.5 ISE and ported to Spartan 6 FPGA board. Table 1 gives the hardware utilization of the proposed design (Fig. 3).

Table 1 Hardware utilization

Parameters	Module
Number of slice register	1507
Number of slice LUTs	2001
Number of LUT-FF pairs	2376
Max operating frequency (MHz)	154

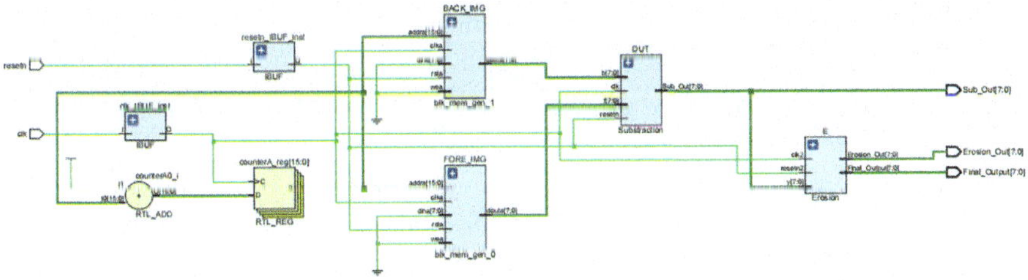

Fig. 3 Shows the RTL schematic representation of the proposed architecture

5.2 *Object Detection Output*

Figure 4a, b shows the foreground image (input image) and background image. Figure 4c shows the background subtracted image.

Figure 4c shows the edge detected image obtained from subtracting background subtracted Fig. 4a with eroded image Fig. 4b.

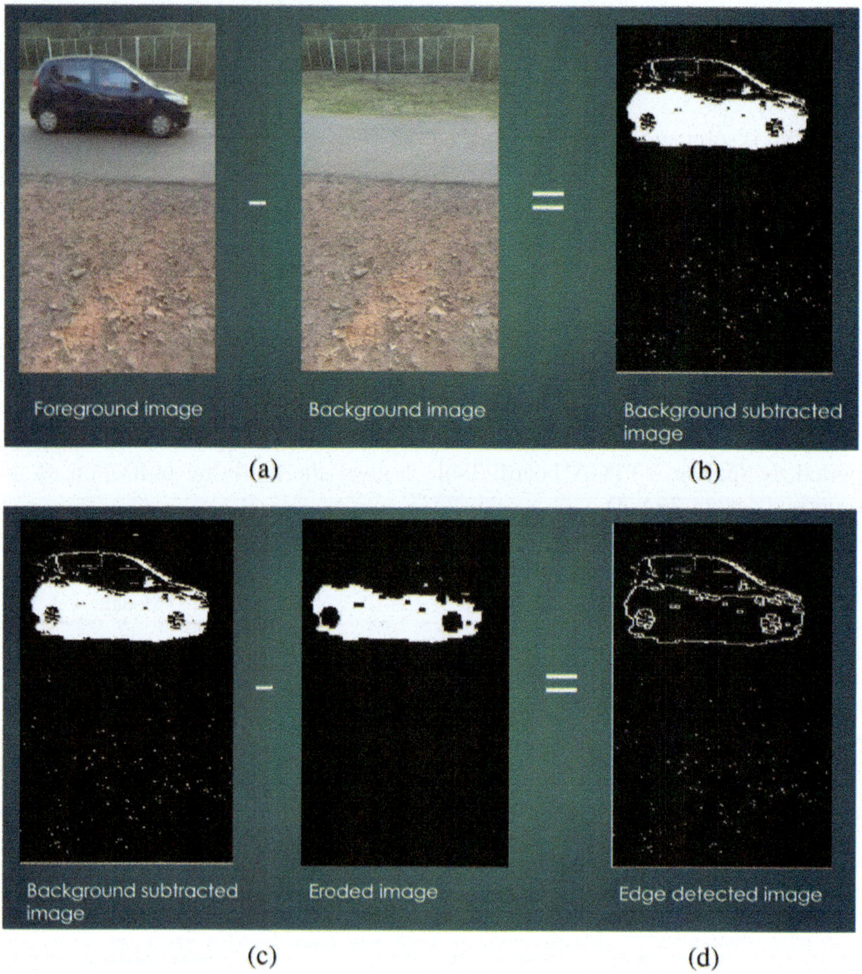

Fig. 4 **a** and **b** Simulation results of background subtraction image. **c** and **d** Simulation results of edge detected image

Table 2 Comparison

Existing work	Board	Area (no. of LUTs)	Operating frequency (MHz)
1	Virtex5	–	118
2	Spartan 3	3203	23
3	Spartan 6	2971	–
Proposed	Spartan 6	2001	154

5.3 Comparison with Existing Work

The hardware compared with existing works is as shown in Table 2. The method proposed in reference 2 uses 3203 LUTs, and the frequency is 23 MHz. The frequency is reduced since it uses block sets of Xilinx.

6 Conclusion

The modified background subtraction and adaptive threshold are used to design hardware architecture for moving object detection in this paper. The proposed model is implemented in Spartan 6 FPGA board. The proposed architecture consumes less area and it is multiplier architecture.

References

1. Chowdary MK, Babu SS, Babu SS, Khan H (2013) FPGA implementation of moving object detection in frames by using background subtraction algorithm. In: 2013 international conference on communication and signal processing, 2013
2. Gujrathi P, Priya RA, Malathi P (2014) Detecting moving object using background subtraction algorithm in FPGA. In: 2014 Fourth international conference on advances in computing and communications, 2014
3. Kupade RN, Meenpal T, Nath PK (2019) FPGA implementation of moving object segmentation using ABPD and background model. In: 2019 IEEE international conference on electrical, computer and communication technologies (ICECCT), 2019
4. Nivetha T, Kapali BS, Shally SP (2020) Video-object detection using background subtraction in Spartan 3 FPGA Kit. In: 2020 7th International conference on smart structures and systems (ICSSS), 2020
5. Pagire VR, Phadke AC (2019) Detection of moving object based on FPGA. In: 2019 IEEE Pune section international conference (PuneCon), 2019
6. Rahiminezhad A, Tavakoli MR, Sayedi SM (2022) Hardware implementation of moving object detection using adaptive coefficient in performing background subtraction algorithm. In: 2022 International conference on machine vision and image processing (MVIP), 2022
7. Ren X, Wang Y (2016) Design of a FPGA hardware architecture to detect real-time moving objects using the background subtraction algorithm. In: 2016 5th International conference on computer science and network technology (ICCSNT), 2016

8. Sridevi N, Meenakshi M (2018) An efficient FPGA based background subtraction algorithm for object detection. In: 2018 3rd IEEE international conference on recent trends in electronics, information communication technology (RTEICT), 2018
9. S´anchez-Ferreira C, Mori JY, Llanos CH (2021) Background subtraction algorithm for moving object detection in FPGA. In: 2012 VIII southern conference on programmable logic, 2012
10. Venkatesh A, Karanth P, Talawar K (2019) An efficient FPGA based reconfigurable architecture for object detection using adaptive threshold. In: 2019 4th International conference on electrical, electronics, communication, computer technologies and optimization techniques (ICEECCOT), 2019

Crime Pattern Identification and Prediction Using Machine Learning

Kavita Jhajharia, Gaurav Aggarwal, Mayank Goyal, and Ishank Rastogi

Abstract In today's world, crime rate is reaching the heights of sky, which calls for the help of technology to predict and hence control the crime before it even occurs, thus finally reducing the crime rate. This research's main aim is to predict whether the crime is going to occur or not based on certain inputs such as location, date, time, and previous crime records related to those particular attributes. This work uses several machine learning algorithms to detect and classify the crime and hence predicting the crime. In the end after comparing several machine learning algorithms, studying several plots and graphs, the algorithm with the highest accuracy is chosen. The result predicts whether the crime will occur or not.

Keywords Crime prediction · Statistics · Machine learning

1 Introduction

In today's modern world, technology has sunken its roots deeply in almost every aspect of our day-to-day lives. As the advice "Prevention is better than cure" goes, this research, viz., crime prediction using supervised machine learning, is created with the motivation to make the use of technology to its best for helping prevent and thus reducing the crime in our societies [1, 2]. Machine learning can be used for several purposes, and its algorithms are most commonly used for either predicting or classifying along with the prediction of the data [3]. Therefore, with the help of certain algorithms of supervised machine learning, this model works with the motive

K. Jhajharia (✉) · I. Rastogi
Manipal University Jaipur, Dehmi Kalan, Jaipur 303007, India
e-mail: Kavita.chaudhary@outlook.com; Kavita.jhajharia@jaipur.manipal.edu

G. Aggarwal
Amity University in Tashkent, Street Labzak, Building-70, Tashkent, Uzbekistan 100028

M. Goyal
University of Southern California, 3551 Trousdale Pkwy, Los Angeles, CA 90089, USA

V. K. Gunjan and J. M. Zurada (eds.), *Proceedings of 4th International Conference on Recent Trends in Machine Learning, IoT, Smart Cities and Applications*, Lecture Notes in Networks and Systems 873, https://doi.org/10.1007/978-981-99-9442-7_14

of predicting and classifying the crimes so as to prevent them even before they occur [4–6]. Some of the machine learning algorithms that are used in this model are random forest classifier, decision tree, logistic regression, K-nearest-neighbor, etc., each of which predicts the desired output with different accuracies, and the algorithm with the best accuracy, in the end, is selected to predict the crime [7, 8].

The main objective of this research is to predict whether the crime is going to occur or not and if yes then what type of crime is most probable to occur. This model is currently working on the criminal dataset of Chicago taken from the website of Chicago Police Department to predict the occurrence of crime in Chicago city and which crime is most probable to occur. The material presented in this work is organized into four sections. After the introductory section, Sect. 2 describes the background overview explaining the conceptual overview and technologies using which this model produces the required results. Section 3 summarizes the methodology of the project/model. It contains detailed description of the methodologies that are being used by the model. For better visualization and understanding of this model, it contains several circuit layouts and block diagrams to help the reader guide through. Section 4 provides an account of implementation of different methods adopted in the model and their respective results containing the modules and its prototypes.

This model is based upon the concept of machine learning using its different algorithms to predict and classify the provided data so as to produce the desired result with the best accuracy. Random forest classifier, decision tree, AdaBoost, linear regression, logistic regression, *K*-nearest neighbor, and support vector machine are used in the study.

2 Methodology

2.1 Logistic Regression

Logistic regression is the appropriate regression methodology to use when the dependent variable is dichotomous (binary), as in "two possible answers." A predictive analysis is provided by logistic regression, just as it is by any other type of regression analysis. If a dependent binary variable and one or more independent variables are present at the nominal, ordinal, interval, or ratio level, then data can be described and their relationship between them explained using logistic regression. Logistic regressions might be tough to understand at times, but using Intellects Statistics, you can do the analysis fast and then interpret the findings in simple English. A binary dependent variable should be used (e.g., present vs. absent). Outliers should be removed from the data by converting continuous predictors to standardized values. Large correlations between the predictors should be avoided (multicollinearity). This can be assessed by a correlation matrix which is generated between the predictors. For binary logistic regression, several bogus R2 values have been fixed. These should be used with caution because they include various mathematical flaws that lead them to be abnormally high or low [7].

2.2 Random Forest

To select the final class of test items, random forest aggregates the votes from numerous decision trees. Random forest may generate three decision trees utilizing subset inputs like $[X1, X2, X3]$ if the training set is $[X1, X2, X3, X4]$ and the labels are $[L1, L2, L3, L4]$. $[X1, X2, X4]$ $[X2, X3, X4]$. Finally, the prediction is generated using the decision trees' majority votes. Because a single decision tree is sensitive to noise, merging many decision trees lowers noise's influence and gives more accurate results. Alternatively, the random forest may account for the influence of each decision tree's output using a word called weight [9]. A low weight value is assigned to trees with a high mistake rate, and vice versa. As a result, low mistake rate trees' decision effect would be enhanced. The random forest classifier is a technique that uses an ensemble of methods to get more accurate results. Because it works in principle: turn a lot of weak estimators into a lot of strong estimators. Even if one or more decision trees are incorrect, the overall outcome is almost always right. We have already obtained a high accuracy of 97% with a minimal number of estimators = 30.

2.3 Linear Regression

Linear regression is a method used in statistics for modeling the association between a continuous response and one or more independent variables along a straight line. Linear predictor functions are used to describe the relationships, and the unknown model parameters are obtained from the data using linear regression. These are known as linear models. Given the values of the explanatory factors, the conditional mean of the answer is generally seen as an affine function of those values (or predictors). The conditional median or another quantile is used less frequently [4, 10].

2.4 AdaBoost Classifier

The AdaBoost classifier is a powerful classifier that combines two weak classifier algorithms. A single algorithm cannot accurately categorize the items. We can acquire a decent accuracy score for the total classifier if we combine many classifiers, pick the training set at each iteration, and apply the proper weights at the final vote. In other words, the algorithm is iteratively retrained by selecting the training set depending on the prior training's accuracy. At each iteration, the weighted age of each trained classifier is determined by the accuracy attained and retrains the algorithm iteratively by selecting the training set depending on the accuracy of prior training. At each iteration, the weighted age of each trained classifier is determined by the accuracy attained. A classifier with a 50% accuracy is given a weight of zero, whereas one with a less than 50% accuracy is given a negative weight. Because they rely on numerous weak classifiers to reach the final choice, AdaBoost-like random forest classifiers produce more accurate results. Facial recognition systems are one of the applications of AdaBoost.

2.5 K-Nearest-Neighbor

K-NN is a supervised learning algorithm used for classification and prediction [11]. Based on the training data, the unknown sample may be categorized into present classes using K-NN. It needs more calculation than other methods. However, it is preferable for dynamic data that vary or are updated often. The K-NN process is the dissociation among the full sample in the training data for new sample categorization. The Euclidian distance is used to measure distance. The samples that are closest to the new sample are referred to as K-NN. The primary principle behind the K-NN is to predict those observations that are closest to the intended output on a limited number of observations. It may be utilized in both discrete and continuous decision-making processes, such as classification and regression. If $k = 1$, the element is given the class of its single immediate neighbor. Alternative ways to building a growing tree from a point of interest have been devised to reduce runtime. By efficiently encoding aggregated distance information for the sample, these approaches try to decrease the amount of distance calculations necessary. The essential principle is that if point A is extremely far away from point B and point B is very near to point C, we may infer that points A and C are very far apart without having to explicitly compute their distance. (This is not always the case.) A closest neighbor search's computational cost can be decreased to O $[D\ N * \log(N)]$ or better in this fashion. For big N.K-D trees, this is a major improvement over brute force, which works well enough for D 20. With a bigger D, it takes much longer. This is known as the dimensionality curse.

2.6 Support Vector Machine

SVM is a well-known supervised machine learning that is also known as support vector networks. It is also one of the techniques that is considered as an unsupervised machine learning. It can be utilized in the study of data for both classification and regression purposes. A set of pre-labeled training data is produced by the system through the application of input–output mapping functions. The SVM approach can be utilized for the classification of datasets that are either linear or nonlinear. Through the use of nonlinear mapping, the SVM was able to elevate the original training data to a higher dimension. The new dimension was then looked for by the support vector machine (SVM). Finally, the SVM classifier's final but most significant characteristic. SVM to core aims to maintain a reasonable distance. A gap is the distance between the line and the nearest class points. A good distance is one in which both groups are separated by a significant distance. The photographs below show an example of a good and terrible margin [11].

2.7 Decision Tree

By detecting lines, the decision tree classifier splits the work area (plot) into subparts repeatedly. (This is repeated since two distant sections of the same class might be separated by others.) When we have a trail from one class division to the next, we call it impurity. This might arise because: there are not enough methods to partition the class. To achieve quicker performance, we tolerate a fixed number of contaminants (we halt further subdivisions). (There is always a trade-off between precision and speed.) In the second scenario, for example, we can stop dividing when there is x or less components remaining. Gini contamination is another name for this. Entropy: Entropy is a measure of impurity that describes the degree of unpredictability of components. The decision tree classifier's core feature is efficient sharing based on maximal information gain. Pure class splitting is not practicable in the real world with millions of data (it may take longer to train), and thus we stop at places in tree nodes when specific criteria (e.g., impurity %) are fulfilled [12, 13].

2.8 Data Pre-processing

Removal of all the null values and creating another cleaned dataset with zero null values. Logistic regression—this method uses logistic regression to classify whether an arrest is made or not and same to classify whether the crime is domestic or not. Random forest classifier—using a collection decision trees, RFC is, here, used to classify the same Boolean values of arrest and domestic as done by logistic regression, only with a better accuracy. Support vector machine—this algorithm divides our two features into two classes, i.e., true and false which again provides us with the information whether or not the arrest has been made. Similarly, linear regression, decision trees, AdaBoost, K-NN, and Gaussian naïve Bayes have been used to produce results with different accuracies. In the last, using folium library, this program also plots a map of Chicago State to provide a better visualization of the crimes with their various details.

Table 1 Comparison among the machine learning algorithms

Algorithms	Accuracy		Precision
	Training data	Test data	
Logistic regression	91.49	91.28	91.30
Random forest classifier	97.24	87.20	88.46
Linear regression	36.25	36.25	33.39
AdaBoost classifier	88.90	88.44	86.94
K-neighbor classifier	84.70	84.70	88.19
Support vector machine	89.06	88.56	87.12
Decision tree classifier	96.28	85.36	88.00

3 Results

We implemented seven different machine learning algorithms to compute the accuracy of test data and train data. Table 1 represents the results of implemented algorithms. The result observation indicates that logistic regression performs better than other algorithms with 91.28% with respect to test data accuracy but for training data accuracy random forest provides best results with 97.24%. Linear regression provides the least accuracy with 36.25%.

We have further conducted more analysis and identified the committed crimes with their frequency in months. As per Fig. 1, theft is the most frequent crime and battery is second most frequently committed crime with respect to other crimes. To get the insight arrests made each year from 2001 to 2019 a line graph has been generated (Fig. 2). Figure 2 indicates rise in arrests in percentage of successful arrest year-wise. The arrests rose from 2001 to 2007 and hit the peak in 2005, but it dropped in 2008 and again rose in 2009 to 2015. In 2016 the arrest had a big fall from 26% successful arrests to 19%.

Figure 3 explains the arrests and not arrests described in percentage with respect to all the filed cases. About 72.49% are not arrested from total number of cases (443,922), and 27.51% arrests were made which is 1684503 cases. On the basis of pattern and frequency of the crime we have predicted the hot spots of the crime. Figure 4 highlights the areas which are more prevalent to crime in the future.

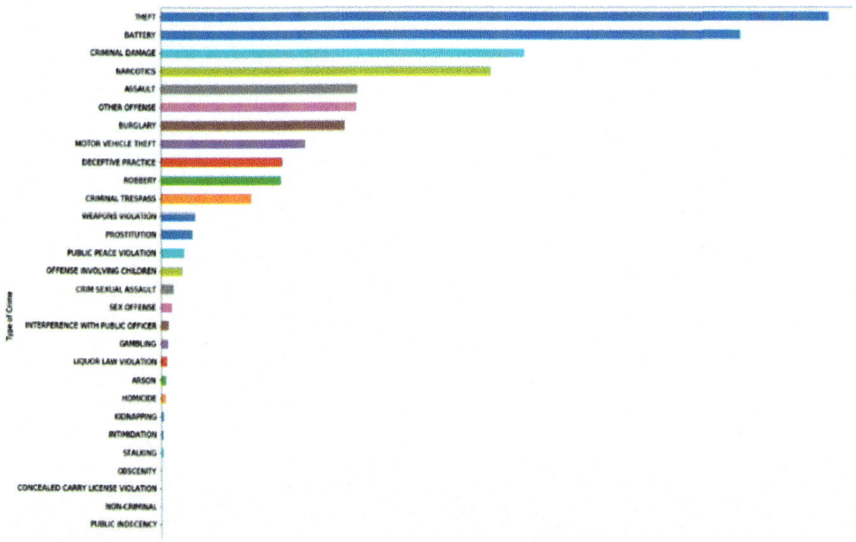

Fig. 1 Type of crime and frequency of crime correlation

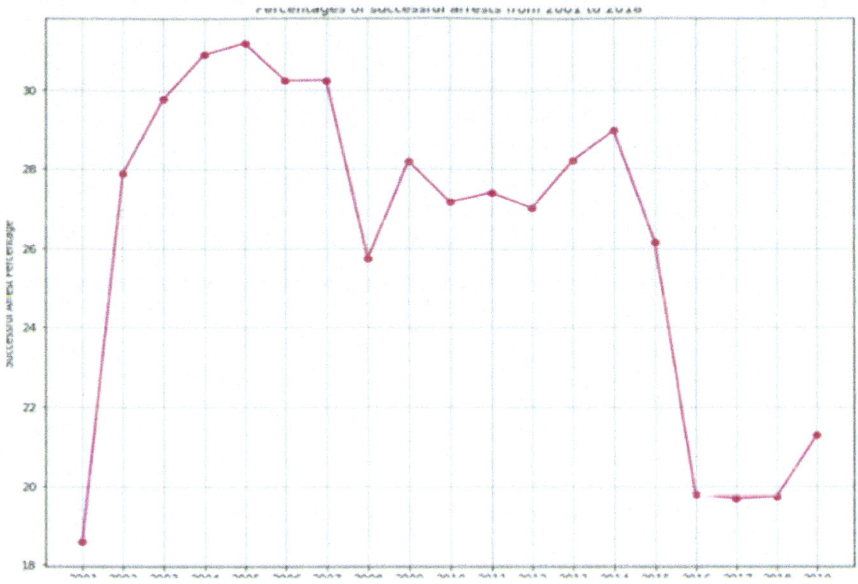

Fig. 2 Year-wise successful arrests

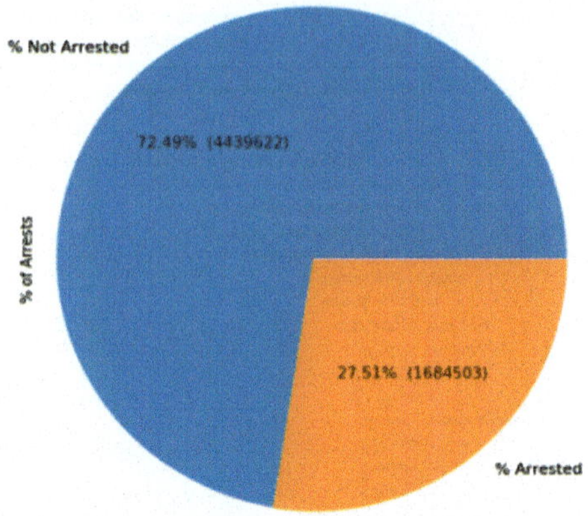

Fig. 3 Total percentage of arrests with respect to all the reported crimes

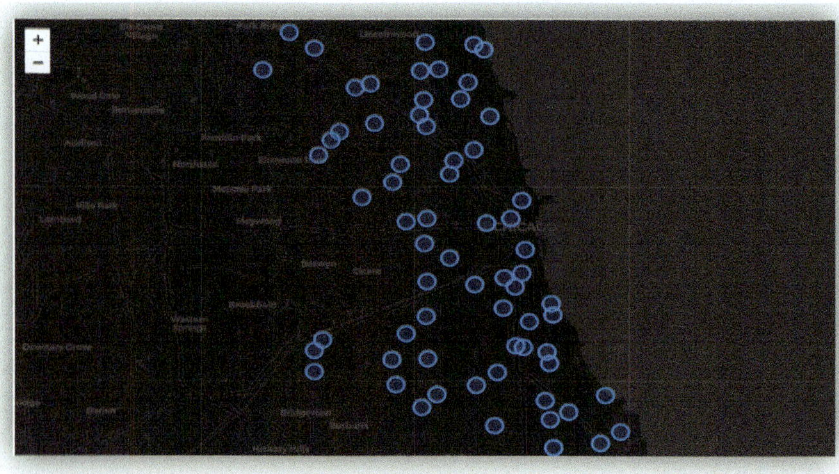

Fig. 4 Crime prevalent areas on Chicago State map

4 Conclusion

The purpose of this research is to analyze the crime statistics and find a pattern in Chicago with the help of machine learning algorithms. The research predicts crime using several machine learning algorithms with the accuracy of 91.28% in test data. The algorithms have obtained promising results for crime prediction with high accuracies for testing and training data. Logistic regression and random forest have been more viable than other algorithms. Machine learning is implementation important as it deals with data very efficiently. The research has also predicted the hot spots of the crime which might help the law enforcement agencies to design strategies. The study can be extended to a software or mobile-based application for better interactivity.

References

1. Liang W, Wang Y, Tao H, Cao J (2022) Towards hour-level crime prediction: a neural attentive framework with spatial–temporal-categorical fusion. Neurocomputing 486:286–297. https://doi.org/10.1016/j.neucom.2021.11.052
2. Wang H, Ma S (2022) Preventing crimes against public health with artificial intelligence and machine learning capabilities. Socioecon Plann Sci 80:101043. https://doi.org/10.1016/j.seps.2021.101043
3. Hussain FS, Aljuboori AF (2022) A crime data analysis of prediction based on classification approaches. Baghdad Sci J 19(5):1073
4. Shamsuddin NHM, Ali NA, Alwee R (2017) An overview on crime prediction methods. In: 2017 6th ICT international student project conference (ICT-ISPC), pp 1–5. https://doi.org/10.1109/ICT-ISPC.2017.8075335
5. Llaha O (2020) Crime analysis and prediction using machine learning. In: 2020 43rd international convention on information, communication and electronic technology (MIPRO), pp 496–501. https://doi.org/10.23919/MIPRO48935.2020.9245120
6. Bharati A (2018) Crime prediction and analysis using machine learning 05:6
7. Oh G, Song J, Park H, Na C (2021) Evaluation of random forest in crime prediction: comparing three-layered random forest and logistic regression. Deviant Behav 0:1–14. https://doi.org/10.1080/01639625.2021.1953360
8. Zhang X, Liu L, Lan M, Song G, Xiao L, Chen J (2022) Interpretable machine learning models for crime prediction. Comput Environ Urban Syst 94:101789. https://doi.org/10.1016/j.compenvurbsys.2022.101789
9. Aziz RM, Hussain A, Sharma P, Kumar P (2022) Machine learning-based soft computing regression analysis approach for crime data prediction. Karbala Int J Mod Sci 8:1–19. https://doi.org/10.33640/2405-609X.3197
10. Chen X, Cho Y, Jang SY (2015) Crime prediction using Twitter sentiment and weather. In: 2015 systems and information engineering design symposium, pp 63–68. https://doi.org/10.1109/SIEDS.2015.7117012
11. Bhavsar H, Panchal MH (2012) A review on support vector machine for data classification 1:5
12. Xiaohu W, Lele W, Nianfeng L (2012) An application of decision tree based on ID3. Phys Procedia 25:1017–1021. https://doi.org/10.1016/j.phpro.2012.03.193
13. Adhikary A, Murad SA, Munir MdS, Hong CS (2022) Edge assisted crime prediction and evaluation framework for machine learning algorithms. In: 2022 International conference on information networking (ICOIN), pp 417–422. https://doi.org/10.1109/ICOIN53446.2022.9687156

IMICE: An Improved Missing Data Imputation Using Machine Learning

Kokkerlapati Manoj Varma, Subhashish Nayak, and Pabitra Moha Khilar

Abstract When researchers use clinical data for studies, they often face the problem of missing data. Missing data can have a detrimental impact on analytical results and introduce bias into the study. Thus, the handling of missing data is required to obtain results from the forecast. In this paper, we have put forward a method for imputing missing values known as IMICE: Improved Multiple Imputation by Chained Equations which is improved from MICE along with Principle Component Analysis data. We also implemented the existing algorithms called Single Center Imputation from Multiple Chained Equations (SICE) and Multiple Imputation with Chained Equations (MICE) to impute missing values and compared our results. We have compared the performance of our method in Diabetes dataset with MICE and SICE algorithms using this dataset. We simulated missing data rates between 1 and 30%, with the missing data being fully random and the Error metrics are Mean Absolute Error and R squared value to examine the performance of the proposed method. The results of the experiments shows that our method is able to obtain less MAE value and also high R^2 value after missing data imputation than the other methods for all the missing data percentages. The findings of this study have important implications for researchers who are dealing with missing data in clinical datasets and may benefit from using our proposed method.

Keywords Imputation · MICE · SICE · IMICE · PCA

K. M. Varma (✉) · S. Nayak · P. M. Khilar
National Institute of Technology Rourkela, Rourkela, India
e-mail: mv888614@gmail.com

S. Nayak
e-mail: 520CS2008@nitrkl.ac.in

P. M. Khilar
e-mail: pmkhilar@nitrkl.ac.in

V. K. Gunjan and J. M. Zurada (eds.), *Proceedings of 4th International Conference on Recent Trends in Machine Learning, IoT, Smart Cities and Applications*, Lecture Notes in Networks and Systems 873, https://doi.org/10.1007/978-981-99-9442-7_15

1 Introduction

In the recent past, there has been a significant expansion in the production of digital data, which has been accompanied by an equally quick increase in the amount of processing capacity [1]. They make it possible to glean previously unknown insights from huge databases. When clinical data that was obtained digitally is utilized in research, the challenge of missing data commonly emerges. This might possibly introduce bias or have a negative influence on the analytical results. The absence of some data can be attributed to a wide range of factors, some of which are as follows [2]:

- When people in charge of an electronic medical record fail to document a value under a structured data element on a consistent basis.
- Problems of a technical nature that prohibit a device from fulfilling its intended function of tracking certain data.
- The vast majority of the data are stored in unstructured formats, so it is challenging to mine them electronically without resorting to human abstraction or preparation. This makes it difficult to access these data for analysis purposes.

Inconsistent data recording leads to missing data, which limits the capacity to analyze and understand in a systematic manner the relative effect of different data items [3]. Finding solutions to missing data needs first having an understanding of the types of data that are absent [4].

Unless there was a problem with the data used in order to train the algorithms, the predictions that these algorithms make are often spot on [5]. In the process of data analysis and mining, most essential step to do is refinement of the data that will be used to train the system. Data preparation, which is one of the steps involved in the data mining process, is regarded by data analysts as being the most challenging stage. Forecasts are frequently off because data are either absent or inputted improperly by humans, both of which are common occurrences [6]. One of the most problematic aspects of the quality data is the missing values in data. The null values in a dataset can increase the cost of computation, cause the findings to be inaccurate, and frustrate researchers [7].

Missing values in dataset can be mainly of three types:

1. Missing Completely at Random (MCAR): There is a lack of data regardless of whether or not there is visible or un-observable data [1].
2. Missing at Random (MAR): Data is missing regardless of unobserved data given the observed data [1].
3. Missing Not at Random (MNAR): The values of the unobserved data themselves are related to missing observations [1].

2 Related Work

Single Center Imputation from Multiple Chained Equation (SICE) has been proposed in [1] that outperforms Multivariate Imputation by Chained Equation in terms of accuracy and RMSE. But the algorithm may not perform well on datasets with different characteristics or distributions. The paper [2] introduces multiple imputation method that provides unbiased result based on the information from available data. And the discussions in the paper are about several methods used to impute missing values. These methods include multiple value imputation, complete case analysis, missing indicator method, single value imputation, and sensitivity analysis. Decision Trees with Support Vector Machine was proposed in [3] for classification. The proposed method outperforms other methods in terms of accuracy, F-measure, recall, precision. Random Forests with Multivariate Imputation by Chained Equation outperforms traditional algorithms like kNN, RF, etc. RF-MICE proposed in [4] that outperforms all other methods in terms of RMSE. But, all other methods are tested under similar conditions like same testing and training data. Improved kNN proposed in [5] outperforming traditional algorithms like PMM, kNN, LR in terms of Normalized RMSE. But, datasets considered are numeric data. A new model that comprises PCA, logistic regression, and *k*-means clustering algorithm that outperforming algorithms like *k*-means and logistic regression was proposed in [6]. But, dataset used is only Pima Indian Diabetes Dataset. So, it may not be generalized to other datasets. Two novel techniques was proposed in [7] using Decision Trees and Forests and compared with existing ones on nine publicly available datasets, and the results show that the techniques outperform other commonly used approaches based on evaluation criteria used. The study in [8] discusses the concept, application and qualitative advantages/ disadvantages of various methods. It also presents an approach to selecting an applicable technique based on available data. A non-parametric missing value imputation method called Clustering-based Missing value Imputation (CMI) based on clustering was proposed in [9] that outperforms other state-of-the-art methods such as KNNImpute, Soft-Impute, MICE. But, method assumes missing values are of type Missing Completely At Random.

3 Methodology

In this paper, we have developed a model called IMICE : An improved method for missing data Imputation which is improved from Multiple Imputation by Chained Equations (MICE). The aforementioned model is built and put into operation by combining the advantages gained from using PCA to IMICE algorithm. After that, a new strategy is suggested, which involves first employing PCA to convert the initial collection of features into a new set of features, and then use MICE to impute the missing values.

3.1 Principal Component Analysis

It can be exceedingly challenging to identify all the correlations between attributes during data analysis. Principal Component Analysis (PCA) allows for the conversion of a vast amount of information from originally correlated data into a new set of data containing orthogonal components. This transformation enables the identification of concealed relationships and enhances data visualization, outlier detection, and classification within the newly defined dimensions [6]. The dataset used in this study has n dimensions, and the dimensions chosen are $\log(n)$ because they best preserve data variation without causing any loss. Principal component analysis (PCA) delivers the variance from the feature subset along axis that is why it is required for the best performance of almost all machine learning algorithms. To begin, we standardize the data into a unit scale (variance = 1, mean = 0) to optimize the PCA result [6]. Our goal is to take our existing dataset X, which has P dimensions and convert it into a new set Y, which has some less dimensions than P. In this case, Y will be the principal component of X, i.e.,

$$Y = PC(X) \tag{1}$$

- Organize the dataset

  X has a collection of n vectors $(x_1, x_z, \ldots, x_n)$ where each individual (x_i) element represents a single record from our data.

- Calculate Mean

$$\overline{X} = \frac{\sum_{i=1}^{n} X_i}{n} \tag{2}$$

- Calculate Variance

$$\overline{X} = \frac{\sum_{i=1}^{n} \left(X_i - \overline{X}\right)^2}{(n-1)} \tag{3}$$

- Calculate Covariance

$$X^{n*n} = \left(x_{i,j}, x_{i,j} = \text{cov}(\text{dim}_i, \text{dim}_j)\right) \tag{4}$$

  where X^{n*n} is the data matrix and it contains n rows, columns and dim_i is the ith dimension

- Calculate EigenValues and EigenVectors

  The covariance matrix's eigenvector and eigenvalues form the basis of a PCA. The directions and magnitude of the new feature space will be determined by the eigenvectors and eigenvalues, respectively.

$$Ax = \lambda x \tag{5}$$

$$(\lambda I - A)x = 0 \tag{6}$$

- The eigenvectors are then arranged from highest to lowest eigenvalue after the Eigen space has been determined from the covariance matrix. Due to the removal of less important components, the major components which give major representation of original data. Obtained principal components is used as input for MICE algorithm.

3.2 IMICE

IMICE operates under the assumption that the observed data may be used to predict the probability of a variable being absent. IMICE generates a series of regression models, or other models that are appropriate for the situation, in order to supply many values in place of a single missing variable [1]. In IMICE, any variable that is missing is treated as dependant variable and the other variables are treated as independent variables. IMICE initially predicts missing values based on the data already present from other features. Then it created new dataset after imputation using projected values. In the same manner, it imputes missing values for all other features. Every dataset is then analyzed using statistical methods and results are used [4]. Multiple imputation methods may give better results because single imputation techniques like mean, median, mode are most likely to result in a biased imputation results. The regression methods used for imputation are Classification and Regression Trees (CART) for Categorical features and Multiple Linear Regression (MLR) for other features. Because for categorical data, CART performs better [7] and for numerical data, Linear Regression performs better [1].

3.3 Flowchart

Workflow of the proposed method IMICE is shown in Fig. 1

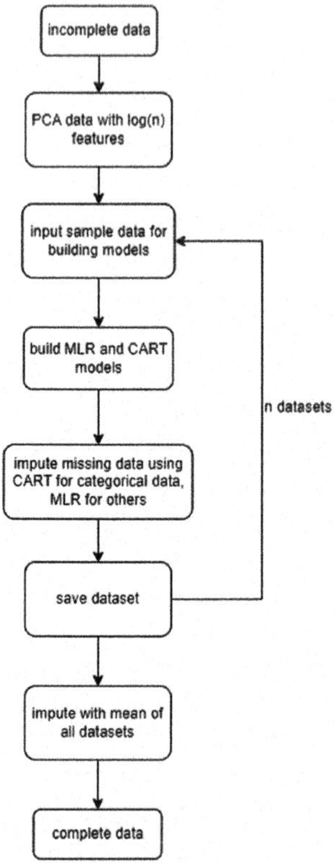

Fig. 1 IMICE flowchart

4 Experimental Setup

4.1 Diabetes Dataset

Diabetes dataset is taken in this paper is from National Institute of Diabetes and Digestive and Kidney Diseases. Description of dataset is shown in Fig. 2 and statistics of dataset is shown in Table 1.

Fig. 2 Diabetes dataset

	Data Type	Column Name
0	int64	Pregnancies
1	int64	Glucose
2	int64	BloodPressure
3	int64	SkinThickness
4	int64	Insulin
5	float64	BMI
6	float64	DiabetesPedigreeFunction
7	int64	Age
8	int64	Outcome

Table 1 Statistics table

	Mean	Variance	Standard deviation
Before imputation	3.845	3225.613	3.369
After imputation	3.834	3204.497	3.389

Diabetes dataset follows log normal distribution before and after imputation. Singular Value Decomposition method is used for data distribution. Distribution of dataset before imputation is shown in Fig. 3 and distribution of dataset after imputation is shown in Fig. 4.

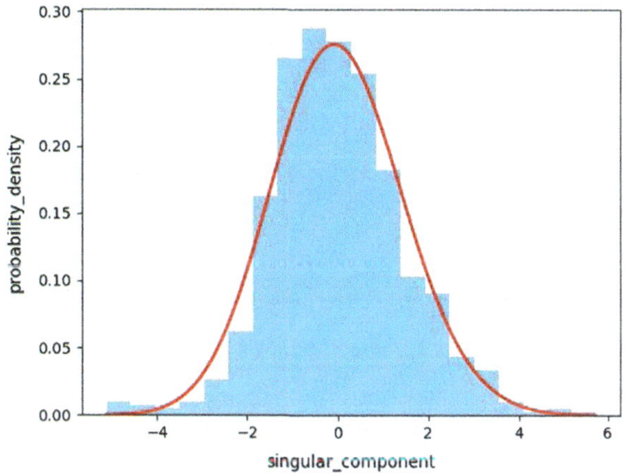

Fig. 3 Diabetes distribution before imputation

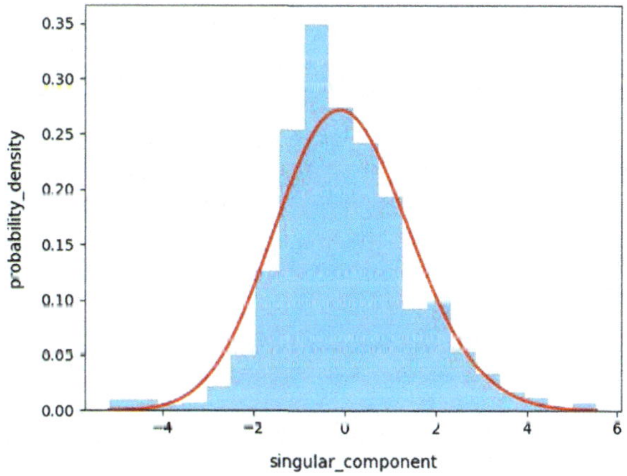

Fig. 4 Diabetes distribution after imputation

4.2 Results

Performance of IMICE is calculated based on Mean Absolute Error (MAE) and R squared (R^2) value. MAE value of IMICE is compared with MICE and SICE and shown in Fig. 5. (R^2) value of IMICE is compared with MICE and SICE and shown in Fig. 6.

Fig. 5 Mean absolute error

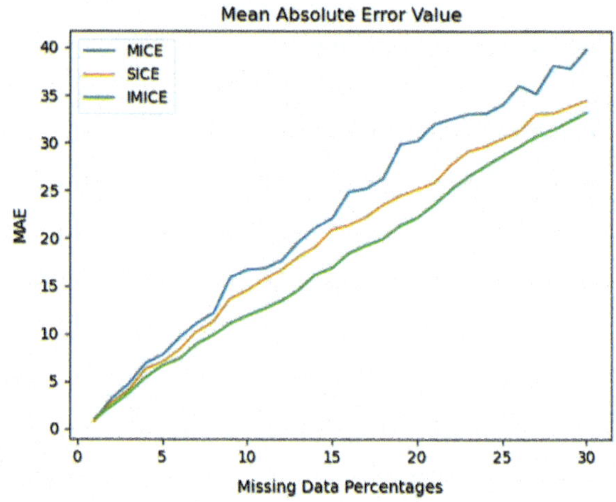

Fig. 6 *R* squared value

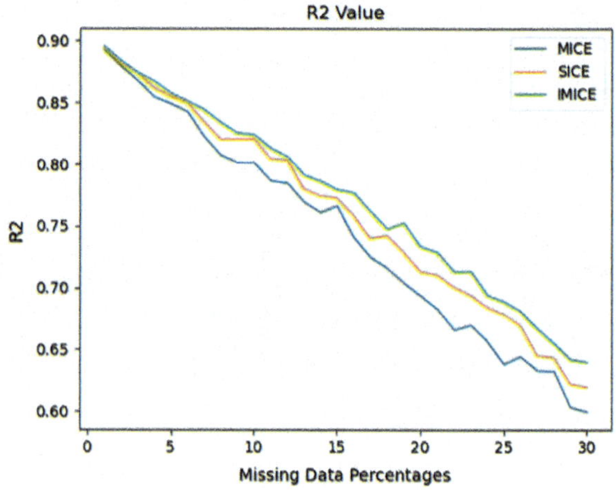

5 Conclusion and Future Work

In the field of data analytics, missing data imputation is given a very high level of importance. Finding the best method for imputing missing values that works for every type of dataset can be a great challenge. However, statisticians caution that handling missing values using single imputation techniques may result in biased imputations. On the other hand, approaches that are based on multiple imputation take into account, the uncertainties of the data and create new set of appropriate values for every missing value. In this paper, we have put forward a method called Improved MICE with PCA for missing data imputation. Diabetes and heart disease datasets were used to generate the data for this article. We simulated missing data rates between 1 and 30%, with the missing data being fully random and the Error metrics are (R^2), MAE to examine the effectiveness of the proposed method. It is observed from the results that our method achieves less MAE value and high R^2 value after missing data imputation than the SICE algorithm for all the missing data percentages in Diabetes dataset. In this work, experiments are conducted on Missing Completely At Random category only. In future, we will try to develop algorithm for Missing At Random category.

References

1. Khan SI, Md Latiful Hoque AS (2020) SICE: an improved missing data imputation technique. J Big Data 7(1):1–21
2. Pedersen AB, Mikkelsen EM, Cronin-Fenton D, Kristensen NR, Pham TY, Pedersen L, Petersen I (2017) Missing data and multiple imputation in clinical epidemiological research. Clin Epidemiol 157–166
3. Jena M, Dehuri S (2022) An integrated novel framework for coping missing values imputation and classification. IEEE Access 10:69373–69387
4. Phan Q-T, Wu Y-K, Phan Q-D, Lo H-Y (2022) A study on missing data imputation methods for improving hourly solar dataset. In: 2022 8th International conference on applied system innovation (ICASI). IEEE, pp 21–24
5. Jadhav A, Pramod D, Ramanathan K (2019) Comparison of performance of data imputation methods for numeric dataset. Appl Artif Intell 33(10):913–933
6. Zhu C, Idemudia CU, Feng W (2019) Improved logistic regression model for diabetes prediction by integrating PCA and k-means techniques. Informatics Med Unlocked 17:100179
7. Rahman MdG, Islam MdZ (2013) Missing value imputation using decision trees and decision forests by splitting and merging records: two novel techniques. Knowl Based Syst 53:51–65
8. Osman MS, Abu-Mahfouz AM, Page PR (2018) A survey on data imputation techniques: water distribution system as a use case. IEEE Access 6:63279–63291
9. Zhang S, Zhang J, Zhu X, Qin Y, Zhang C (2008) Missing value imputation based on data clustering. In: Transactions on computational science I. Springer, Berlin, pp 128–138

Analyzing Students' Opinion on E-Learning—Indian Students' Perspective

Shubhangi V. Urkude and C. Lakshmi Devasena

Abstract Online education has become an indispensable part of the teaching-learning process during the pandemic. Due to the vast benefits of online learning, it can be continued further. The appropriateness of digital education depends on the students' compassion and the available infrastructure. This research work aims to analyze the student's perception of online learning. The data is collected from various discipline students across India and analyzed using descriptive statistics. It is revealed that most female students prefer online learning, even though they are less privileged to have smart devices and internet facilities. Postgraduate students prefer online learning over other educational groups. After having practical experience of almost two years in online learning, only 58% of students prefer it in the future. The government needs to take necessary actions to supply infrastructure to rural students, especially the unprivileged female students from rural and urban areas, to support online learning.

Keywords Online learning · E-learning · Students' perception on online learning

1 Introduction

The coronavirus 2019 has threatened human lives. There was a drastic change in everything across the whole world. People started thinking about how to use digital technology in daily life to improve it. The traditional approach to doing business was changed completely to avoid the spread of viruses. Digitization of businesses creates various job opportunities for technical students. The scope of business in the software industries was increased. Different web-based and mobile applications

S. V. Urkude (✉) · C. L. Devasena
Department of Operations and IT, ICFAI Business School (IBS) Hyderabad, The ICFAI Foundation for Higher Education (IFHE), Hyderabad, India
e-mail: shubhangini@ibsindia.org

C. L. Devasena
e-mail: ldevasena@ibsindia.org

V. K. Gunjan and J. M. Zurada (eds.), *Proceedings of 4th International Conference on Recent Trends in Machine Learning, IoT, Smart Cities and Applications*, Lecture Notes in Networks and Systems 873, https://doi.org/10.1007/978-981-99-9442-7_16

were available in the market to satisfy the need of businesses like digital payment apps Paytm, UPI, PhonePe, and so on. E-commerce websites capture the market to fulfill the customer's needs, such as vegetables, daily necessities, and medicines.

The government announced a complete lockdown as the number of cases increased worldwide and in India. As an effect of this, all the businesses and educational systems were closed. But education should go on, and all the education systems change from face-to-face to online to provide students with classes. With this, other problems arose, such as infrastructure facilities and no awareness of using various digital platforms for education. Before also, these digital platforms were available, but their use was limited, and very few people were aware of them [2]. In online classes, instructors should understand the student's mentality and learning capacity and motivate them to attend the classes. The instructor has to be trained to use the digital teaching aids and give support and guidance to improve them to handle online classes [3].

COVID-19 is considered the longest pandemic till now on earth. The most dangerous thing about this was the nature of the virus, which mutated itself and produced different types of variants. Some of the variants were more dangerous, and some were less effective. Even then, it affects human life to too much extent. The common symptoms of the coronavirus were headache, running nose, fever, cough, cold, throat pain, body pain, and so on. There could be various preventive measures to avoid the infection, like regularly washing the hand, not anywhere face frequently, wearing a face mask, and staying away from crowded places. This pandemic also taught everyone how to use various technologies to handle different situations; robots and drones were used to avoid physical contact with patients, supply medicines, etc. [1]. E-learning gives a learning experience to the teacher to operate various digital devices such as smart phones and laptops with internet connections. These e-learning processes make students more creative and flexible and develop self-learning skills [4].

In the online learning system, the teacher plays multiple roles, such as trainer, facilitator, and guide. Teaching the students is the most crucial role of the teacher; along with that, he/she can design the course and assist the students in using the multiple facilities available on the platform. The teacher will guide the students and perform other activities, such as conducting exams and monitoring performance [24]. Many e-learning platforms are accessible freely, and students can use them anywhere globally with the internet facility. Using these platforms solely depends on the cognitive power of the student and willingness to use available facilities [22]. One of the significant drawbacks of the e-learning process was that the necessary infrastructure required to attend classes was not available in rural areas. Most of the time, students felt bored in the class without physical interaction with their peer groups.

This research work aims to understand the students' perception of adopting online education across disciplines in India. It will answer the following questions:

- Is online learning suitable for students of different economic statuses?

- How do the affordability and availability of internet facilities affect their preference?
- Whether students want to continue online education in the post-pandemic period?

2 Literature Survey

Karl [8] discussed the e-learning process and concluded that instructors should design the course carefully depending on the student's characteristics, syllabus, and learning environment. They have used data provided by the Information Technology Training Initiative (ITTI) of the Michigan Virtual University students. They also said that overall satisfaction with online classes is 4.76, and with the traditional method, it is 3.70 on a scale of 7. The primary issue with e-learning is that online content will be available temporarily, and performance evaluation is complex. Paechter [25] discussed the student's experience and preferences for online and face-to-face teaching for Austrian university students. They concluded that they prefer modes of instruction based on the type of knowledge students want to get. The classroom teaching was preferred for application-oriented subjects to get more practical experience, and conceptual subjects can be learned online. Bali [10] discussed students' perception of e-learning concerning social presence, interaction, and satisfaction in Indonesia. They concluded that there is not much difference between face-to-face teaching mode and online mode, but those who want to increase their technical skills are enjoying it.

Syauqi [9] analyzed students' perception of Yogyakarta State University students for vocational courses in the technical background. They analyzed the data on five different aspects of mentoring the students, knowledge transfer, materials used, resources available, and willingness to continue in the future. They finally concluded that most students disagree with the effectiveness of e-learning and need to improve from the teacher and institution sides. Dommett [26] analyzed 95 staff and 522 students' data to check the perception of capturing the lecture contents. They concluded that students would give more importance to capturing the contents, depending on many factors like demand of live lectures, student presence, and performance. However, attendance is indirectly related to the lecture capture. Thomas et al. [5] discussed the effectiveness of online teaching and students' perception. They analyzed the data collected from the symbiosis college in Pune. They concluded that 73% of the students don't have any opinion on the course design and content delivery in the online class. Half of the students are interested in searching for experts in an online class.

Kulal and Nayak [11] analyzed the student perception of online learning in a specific region such as Dakshina Kannada and Udupi district. They concluded that this study will help the colleges to build the necessary infrastructure to start online courses in the future and give better education. Rojabi [12] explored the student perception of the use of Microsoft Team for Online Education among university students of Indonesia. He found that students have a positive approach to online education. Biswas et al. [15] discussed mobile learning in Bangladesh. As per the analysis, they concluded that most of the students are happy with mobile learning and they want to continue in the future to minimize the learning gap due to COVID-19. They analyzed the data in different parts, basic details of the student, and others were the use of social media for learning purposes, last was the student's opinion about the e-learning process. They found that the necessary infrastructure will help the students attend online classes and improve themselves. Agarwal and Kaushik [20] analyzed the survey data to study students' perception of online education. As per the analysis, many students want online classes to be part of regular classes to save time. They also suggested better software can be used for online classes to get more facilities.

Nandal et al. [6] analyzed the perception of college students from Delhi. The chi-square test is used to analyze the relationship between gender and subject blocking period, concluding that the association is not considerable. Many students have faced internet problems in finishing the given courses. Avila [7] discussed the student's perception of e-learning and different learning techniques during the pandemic for the Polytechnic University of the Philippines' Ragay. They concluded that online classes are not very effective and that students and teachers should be given some practice to use them effectively before starting the semester. Khan [13] analyzed the students' perceptions and issues encountered by school students from 8 to 12th grade in Delhi. The data were analyzed using the logistic regression method to understand the influence of various factors such as financial status, parents' education level, internet facility, and support from the educational institutes. They found that the student's financial condition and internet facility were very important for attending online classes. The majority of the students prefer regular classroom teaching to avoid these problems. Kang [14] discussed the adoption of online education and the difficulties faced by students in Vietnam. They analyzed that online class preparation was needed for the students and teachers. The courses should be designed so that interaction between the students and the teachers is greater and use different teaching aids to engage the students.

Muthukrishnan [23] discussed the perception of online education and challenges faced by the students of agricultural universities. They analyzed the collected data and concluded that e-learning is beneficial for most students but the students are not happy without social interaction between them. They also suggested that students should be motivated to attend the classes and enhance their performance. Galgewela [16] discussed the perception of higher education students in Sri Lanka. Based on the data analysis, she concluded that due to the flexible nature of online classes, students are happy and enjoy the learning process to enhance their skills. Bączek [17] analyzed the data collected from eight hundred four Polish medical students and concluded that most students were interested in online classes. As per their studies, online classes were not efficient in transferring practical knowledge compared to face-to-face teaching. Students were not able to learn from the feedback received. Almahasees et al. [18] discussed the faculty and student perception of online learning during pandemics in Jordan. They analyzed the data collected from the faculty and students separately and concluded that the faculty's opinion, class preparation, knowledge of computers, effectiveness, and faculty approach were critical. According to the student's point of view, classroom learning, challenges faced, and advantages were considered. They have concluded that students faced many challenges in adopting digital classes and short of physical interaction in peer groups. One benefit is that it makes a student more responsible so face-to-face teaching is more efficient.

Bast [19] discussed the student perception of a pandemic at the central university of Punjab. They found that many students from higher class and economically strong have a positive perception of online education. Many students prefer recorded courses so they can refer to them whenever they want. Muthuprasad [21] discussed the agriculture student's perception. They analyzed the collected data on different aspects like the online class structure, duration, students responding to the queries, and available facilities. They considered the benefits of online classes and the various loopholes in the process to conclude that flexibility is the main thing for the successful conduction of classes. But the hybrid model was helpful for practical-based subjects in the future. Camilleri and Camilleri [27] discussed the remote learning via video conferencing technique by analyzing the collected data from 777 students in tertiary education. They used the partial least squares (PLS) approach to analyze the data and identified that students are willing to participate in the learning process through video conferencing. Facilitating condition positively impacts the student's perception of the technology they use. Performance expectancy and attitude toward mentioned technology have a significant weightage. They also considered demographic variables like age, experience, and gender for mapping the influence of the technology adaptation. Guo [28] discussed online learning in China during the pandemic. They analyzed the data collected from high school students in China to find the impacting factors for the digital break up. The first digital split was available equipment and network setup; from the analysis, they found that 95% of the students had digital equipment to take online learning, which was also used by other family members in schooling time and may damage also. The second digital split was the student adoption of online learning during COVID-19. The analysis shows that family background is the most affecting factor and the standard deviation between the families having one child versus two children is more. The third split was overall changes in the learning outcomes. The analysis shows that 46.95% of students said their outcome was reduced during online learning. They said information and communication technology can enhance performance, and the inequalities in the learning group should be considered during online learning.

2.1 Research Gap

Many researchers have analyzed the student's opinions about the e-learning process by collecting data from some limited regions, particularly domains outside India. However, very few of them concentrated on the practical subjects and challenges faced in handling and transforming the practical knowledge to the students.

However, this paper is based on students' perceptions of various educational streams from multiple institutions in India and rural and urban areas. Most students are doing junior college, undergraduate, and postgraduate studies. So let us assume some mental maturity among them, and they are grown up to understand and answer the questions.

3 Methodology

E-learning becomes a crucial part of the teaching and learning process. During the pandemic, online learning was introduced in India in the regular teaching and learning process, even before it was used only in MOOC courses. Online education has numerous benefits, so educational institutions may want to continue it in the future. Online learning might be continued in different situations like using it in pandemic situations, having full-time online courses for different degrees where students need not attend classes physically but can attend them from their place. For any of these situations, the institution and the students can get benefitted only based on the student's interests and infrastructural availability. This research work analyses students' perception in understanding whether this online education can be continued in feature or not, as well as their capability.

3.1 Data Collection

This research is based on the primary data collected from students of different disciplines across India and analyzed using descriptive statistical methods. A questionnaire was prepared and circulated through Google Forms to collect the data. Data was collected from students of various age groups and educational categories. The questionnaire contains around 15 questions for a deeper analysis of the student's perceptions. The Snowball sampling method is used to collect the data. Students who are currently studying are considered for this research work. A total of 228 data samples were collected. Table 1 summarizes the questions asked in the survey. The students presently studying various levels of education across India are considered for data collection.

4 Result and Discussion

After analyzing the data, numerous exciting facts are revealed. After having two long years of online learning experience, only 58% of them prefer online learning in the future. Around 42% of the students still prefer traditional face-to-face learning (Fig. 1) according to gender are shown in Fig. 2.

Table 1 Questionnaire included in the research work

Q.No.	Question
1.	Age Group: ◯ 15 and Below ◯ 16-20 ◯ 21 to 30 ◯ Above 30
2.	Gender: ◯ Male ◯ Female
3.	Education: ◯ School ◯ HSc or Junior College ◯ UG ◯PG ◯ PhD ◯ Diploma
4.	Family Income: ◯ Below 1 lakh◯ 1-3 Lakhs ◯ 3-5 lakhs ◯ 5 – 10 lakhs ◯Above 10 lakhs
5.	Area: ◯ Rural ◯ Urban
6.	Do you prefer online education? ◯ Yes ◯ No
7.	Do you have smart device to attend classes? ◯ Yes ◯ No
8.	Do you have sufficient Internet facility to attend classes? ◯ Yes ◯ No
9.	Which is your Native State? __________
10.	Is it easy to understand online classes? Strongly Disagree to Strongly Agree (5 scale)
11.	Is it easy to clarify doubts in online classes? Strongly Disagree to Strongly Agree (5 scale)
12.	Are there fewer disturbances than traditional classes, so that you can easily understand the subjects? Strongly Disagree to Strongly Agree (5 scale)
13.	Is it convenient to attend classes from home? Strongly Disagree to Strongly Agree (5 scale)
14.	Do you prefer on line education in future? Strongly Disagree to Strongly Agree (5 scale)
15.	Do you willing to attend online classes in future? Strongly Disagree to Strongly Agree (5scale)

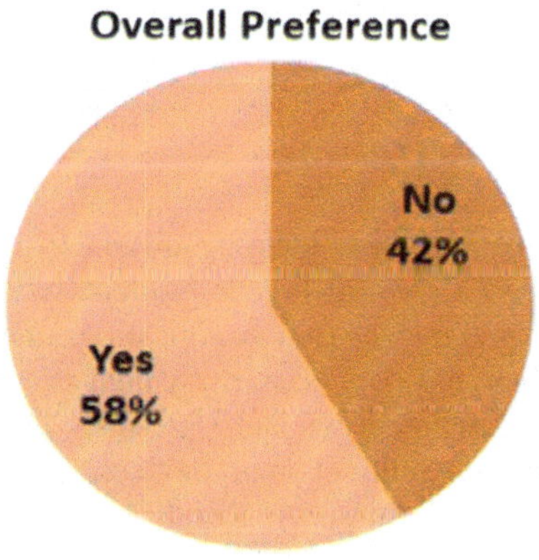

Fig. 1 Overall preference reference details

Amazingly, more female students (61%) wish to continue online classes than male students (only 55%). Preferences details according to different educational levels are depicted in Table 2. Most postgraduate students (73%) prefer online education over other groups. The reason may be that continuing their job and attending online classes might be their choice. Ph.D. students choose less (33%) than other students since they need constant communication with their faculty guides.

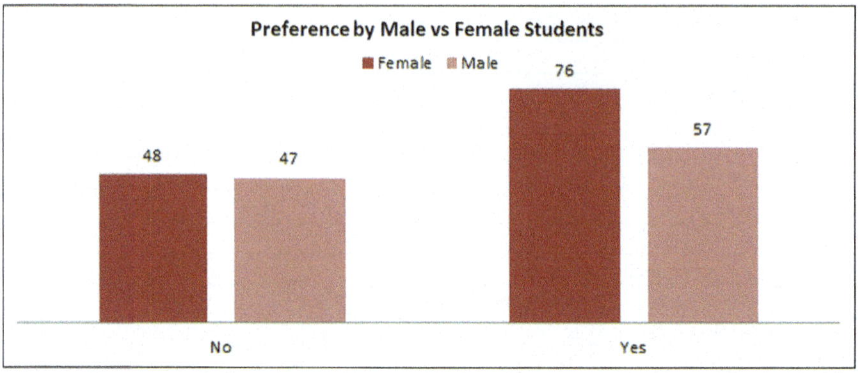

Fig. 2 Preference by male versus female students

Table 2 Preference details according to educational level

Educational level	Preference		Grand total
	No	Yes	
Diploma	5	8 (62%)	13
Higher secondary/junior college	17	23 (58%)	40
PG	3	8 (73%)	11
PhD	2	1 (33%)	3
School	3	8 (63%)	11
UG	65	85(56%)	150
Grand total	95	133	228

Preference to continue in online learning differs for different family income students. These details are shown in Fig. 3. Students who have a family income above 10 lakhs (Rs.) less prefer (55%) online education; since they might desire peer-group communications and physical class enjoyment over other students. Students who have a family income of below one lakh (Rs.) also show similar interest, but the reason may be the non-affordability of smart devices and internet facilities.

Preference details for different areas (Rural vs Urban) are shown in Table 3. Students from rural areas (57%) prefer online learning less than urban students (60%).

Preference for online learning in different age groups under area and gender level is shown in Table 4. Urban students prefer (60%) online education to rural students (57%). In that gender-wise preference is similar, not much difference observed in rural and urban students.

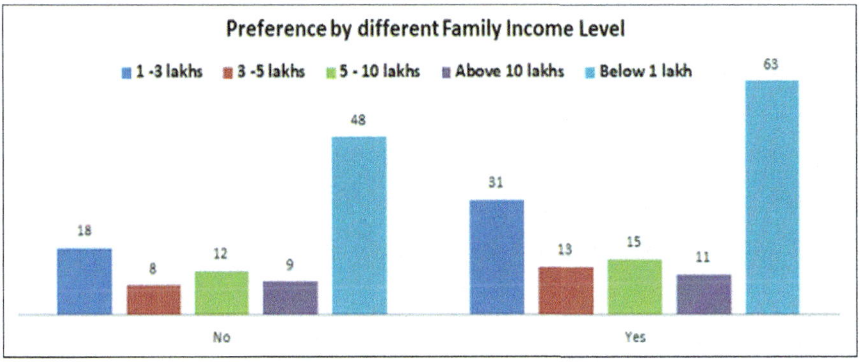

Fig. 3 Preference by different family income levels

Table 3 Preference details according to area

Area	Preference		Grand total
	No	Yes	
Rural	39	51 (57%)	90
Urban	56	82 (60%)	138
Grand total	95	133	228

Online learning is not possible without smart devices and sufficient internet facilities. Details about smart device holders and students having internet facilities are shown in Figs. 4 and 5, respectively. The survey reveals 84% of students could rent or own smart devices because of the affordable price of smart devices. Most students could buy them or could use their parent's devices during online classes. Only 16% of the overall students do not have smart devices. And most of the (7 + 10/90 = 0.18 = 18%) rural students do not have smart devices like urban students (15 + 5 = 20/138 = 0.14). Table 5 depicts that most female students (around 20% i.e., 25/124 = 20%) do not have smart devices than male students (12/104 = 12%). Even though 84% have smart devices, only 66% have sufficient internet facilities.

Smart device holder details for different age groups are shown in Table 6. 17% of students from the age group 16–20 do not have smart devices, and 13% of 21–20. School-going children might have used their parent's smart devices. Internet facility availability details of different age groups are shown in Table 7. Around 30–36% of the students do not have internet facilities. Internet facility might be a significant criterion for non-preferring online learning.

Table 4 Students' preference for online education in all age groups under area and gender level

Age	Online education preference														
	Female						Female total	Male						Male total	Grand total
	Rural		Rural total	Urban		Urban total		Rural		Rural total	Urban		Urban total		
	No	Yes		No	Yes			No	Yes		No	Yes			
15 and below	–	–	–	–	–	–	–	2	5	7	1	3	4	11	11
16–20	17	28	45	26	32	58	103	11	10	21	18	18	36	57	160
21–30	3	5	8	2	11	13	21	6	3	9	7	17	24	33	54
30 and above	–	–	–	–	–	–	–	–	–	–	2	1	3	3	3
Grand total	20	33	53	28	43	71	124	19	18	37	28	39	67	104	228

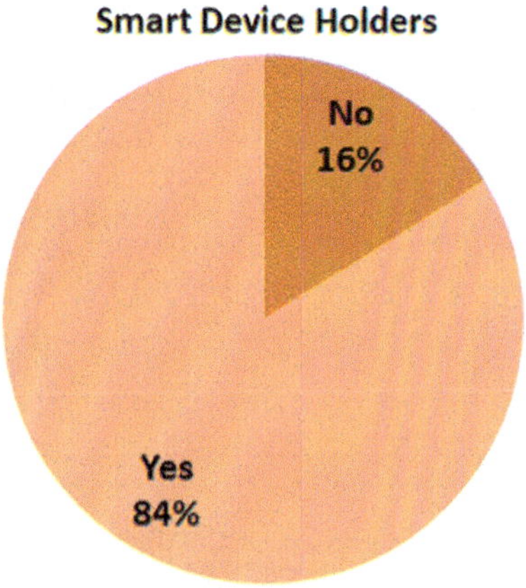

Fig. 4 Smart device holders

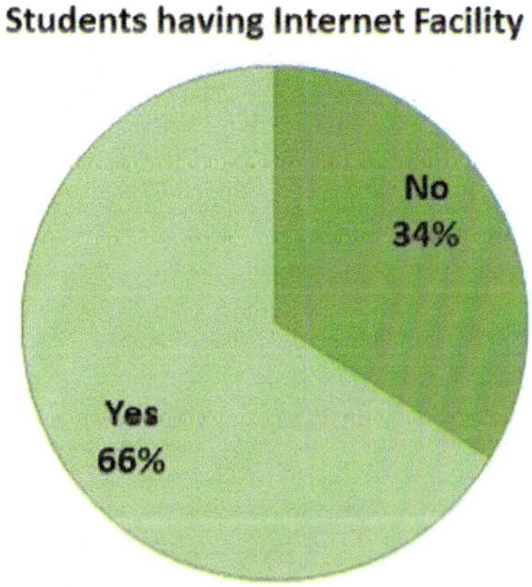

Fig. 5 Students having internet facility

Sufficient Internet availability in different age groups with area and gender level is shown in Table 8. More rural students (24 + 15 = 39/90 = 0.43 = 43%) do not have internet facilities over urban students (28 + 10 = 38/138 = 0.28 = 28%). Even in that, female students are more affected.

Some questions were asked to diagnose the satisfaction level of the students during online learning. Whether they can understand the concepts in the online classes? (Depicted in Fig. 6)—Around 21% of the students agreed that and 47% of the students do not agree. Whether they could clarify their doubts? (Depicted in Fig. 7)—Only 23% of students agreed. Whether are there fewer disturbances than in traditional classes? (Depicted in Fig. 8)—Only 30% of the students agreed. Is it convenient to attend classes from home? (Depicted in Fig. 9)—Around 44% of

Table 5 Students having smart devices in all age groups under area and gender level

Age	Having smart devices														
	Female							Male							
	Rural		Rural total	Urban		Urban total	Female total	Rural		Rural total	Urban		Urban total	Male total	Grand total
	No	Yes		No	Yes			No	Yes		No	Yes			
15 and below	–	–	–	–	–	–	–	2	5	7	–	4	4	11	11
16–20	9	36	45	13	45	58	103	3	18	21	2	34	36	57	160
21–30	1	7	8	2	11	13	21	2	7	9	2	22	24	33	54
30 and above	–	–	–	–	–	–	–	–	–	–	1	2	3	3	3
Grand total	10	43	53	15	56	71	124	7	30	37	5	62	67	104	228

Table 6 Age-wise and overall smart device holder details

Age	Having smart device		Grand total
	No	Yes	
15 and below	2	9	11
16–20	27	133	160
21–30	7	47	54
30 and above	1	2	3
Grand total	37 (16%)	191 (84%)	228

Table 7 Internet facility availability details for different age groups

Age	Having internet facility		Grand total
	No	Yes	
15 and below	3	8	11
16–20	57 (36%)	103 (64%)	160
21–30	16 (30%)	38 (70	54
30 and above	1	2	3
Grand total	77 (34%)	151 (66%)	228

the students agreed. Many of them feel it is convenient to participate in the online classes. Still, most students did not agree upon other parameters about understanding concepts, clearing doubts, and fewer disturbances.

Students who prefer online learning in the future and are willing to attend online classes in the future are shown in Figs. 10 and 11. Around 38% of students agree with that, and about 23% are confused. They can go with any decision, and about 41% of students do not agree with online education in the future.

Table 8 Internet availability of all age groups under area and gender level

Age	Internet facility available														
	Female						Female total	Male						Male total	Grand total
	Rural		Rural total	Urban		Urban total		Rural		Rural total	Urban		Urban total		
	No	Yes		No	Yes			No	Yes		No	Yes			
15 and below	–	–	–	–	–	–	–	3	4	7	–	4	4	11	11
16–20	21	24	45	24	34	58	103	7	14	21	5	31	36	57	160
21–30	3	5	8	4	9	13	21	5	4	9	4	20	24	33	54
30 and above	–	–	–	–	–	–	–	–	–	–	1	2	3	3	3
Grand total	24	29	53	28	43	71	124	15	22	37	10	57	67	104	228

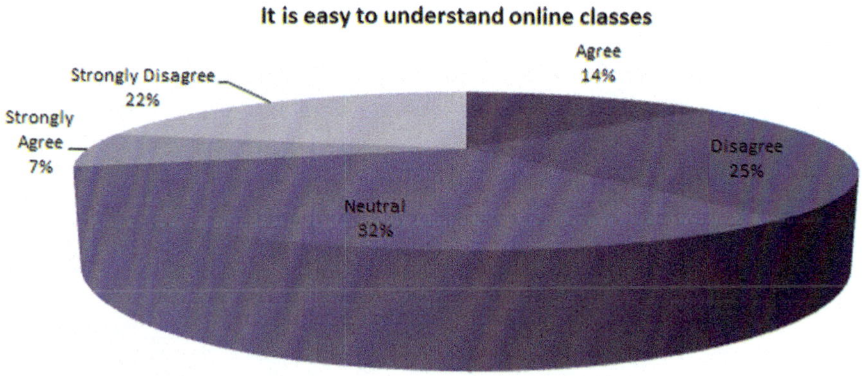

Fig. 6 Student's view on understanding online classes

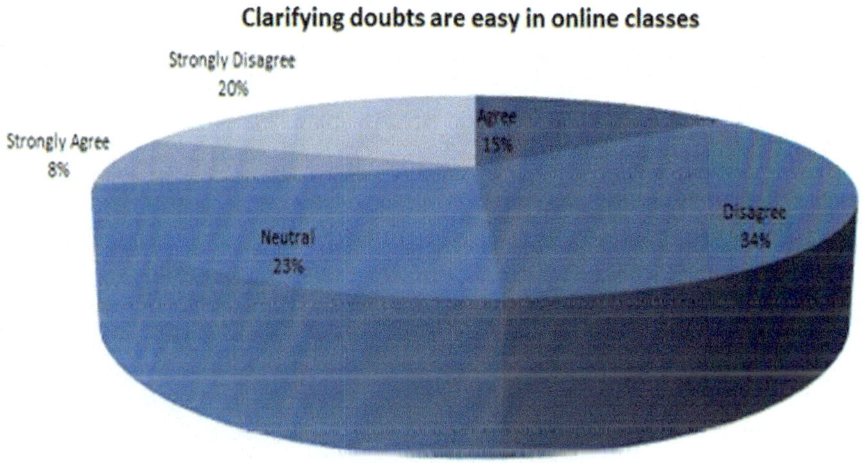

Fig. 7 Ease of clarifying doubts

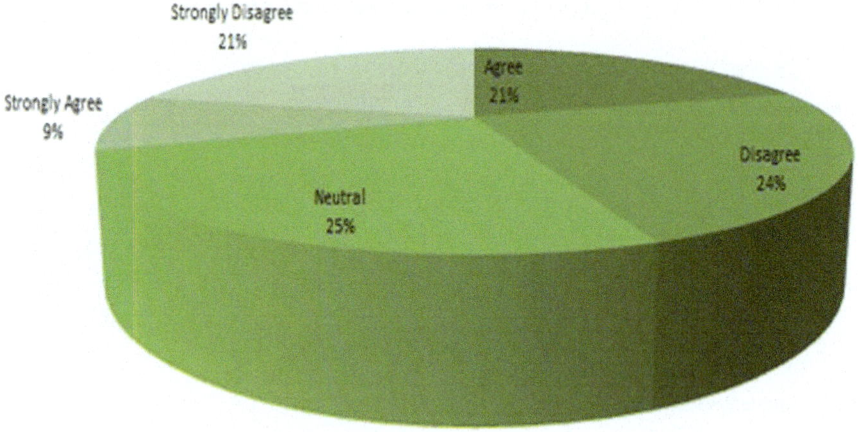

Fig. 8 Less disturbance than traditional classes

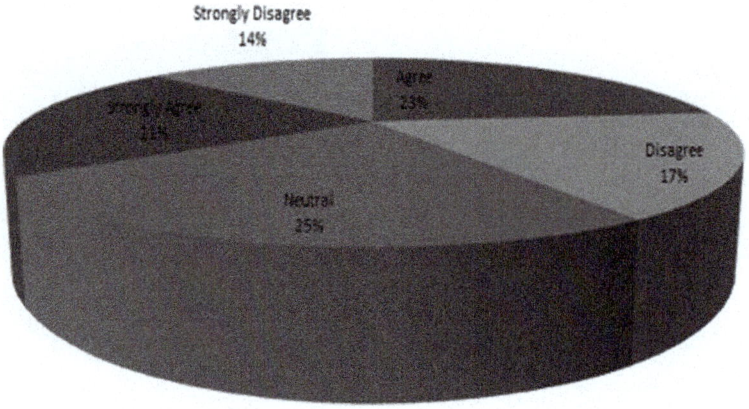

Fig. 9 Convenient to attend from home

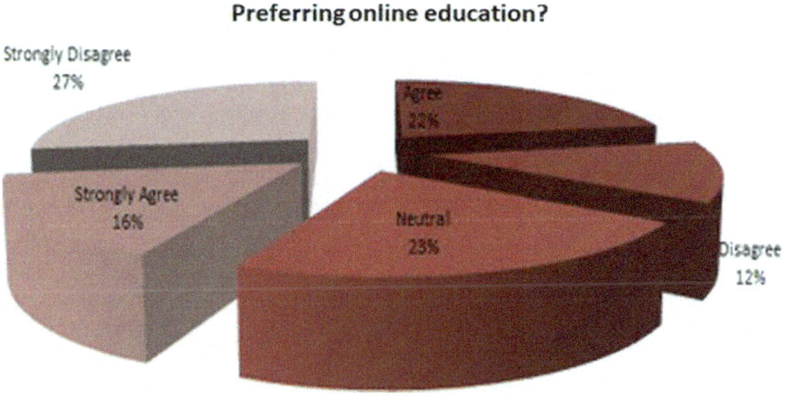

Fig. 10 Preferring online education

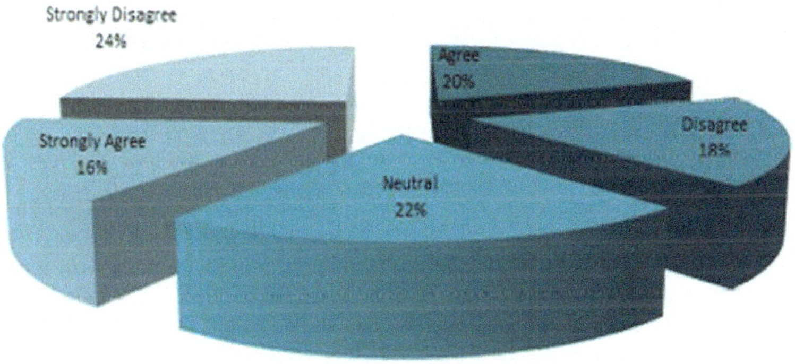

Fig. 11 Willingness to attend online classes

5 Conclusion

Online learning was imposed in almost all institutions in the past two years after the start of covid-19; even then, only 58% of the students expressed their willingness to prefer it in the future. The rest of the students might have their reasons: students from rural areas might not be able to afford smart devices and internet facilities, and urban students might not want to miss peer group interactions during academics. Female students from both rural and urban areas could not get the opportunity to afford smart devices and internet facilities; even then, their preference for online learning is more than male students. Final-year undergraduate and postgraduate students from various financial backgrounds could afford smart devices and internet facilities. Educational institutions intending to start online education programs in the future could target these groups. Government should initiate providing facilities (like smart devices and data packages) to students from rural and unprivileged female students to succeed in conducting online learning. Educational institutions can also support in this regard.

References

1. Urkude SV, Urkude VR, Vairachilai S, Panda SK (2022) An investigation into COVID-19 pandemic in India. New Adv Soc Artif Intell Indus IoT Paradigm. https://doi.org/10.1002/9781119884392.ch13
2. Baloran ET (2020) Knowledge, attitudes, anxiety, and coping strategies of students during COVID-19 pandemic. J Loss Trauma 1(8)
3. Moralista R, Oducado RM (2020) Faculty perception toward online education in a state college in the Philippines during the coronavirus disease 19 (COVID-19) pandemic. Univ J Educ Res 8(10):4736–4742
4. Singh V, Thurman A (2019) How many ways can we define online learning? A systematic literature review of definitions of online learning (1988–2018). Am J Distance Educ 33(4):289–306
5. Thomas LB, Suji M, Yadav P (2020) Students perception on the effectiveness of online teaching. Indian J Forensic Med Toxicol 14(4)
6. Nandal N, Nandal DN, Jora MN (2021) Perception of online learning among college students: a systematic. Int J Mod Agric 10(1):1142–1149
7. Avila EC, Abin GJ, Bien GA, Acasamoso DM, Arenque DD (2021) Students'perception on online and distance learning and their motivation and learning strategies in using educational technologies during COVID-19 Pandemic. J Phys 1933. https://doi.org/10.1088/17426596/1933/1/012130
8. Smart KL, Cappel JJ (2006) Students' perceptions of online learning: a comparative study. J Inf Technol Educ 5(1)
9. Syauqi K, Munadi S, Triyono MB (2020) Students' perceptions toward vocational education on online learning during the COVID-19 pandemic. Int J Eval Res Educ (IJERE) 9(4):881–886
10. Bali S, Liu MC (2018) Students' perceptions toward online learning and face-to-face learning courses. J Phys 1108:012094. https://doi.org/10.1088/1742-6596/1108/1/012094
11. Kulal A, Nayak A (2020) A study on perception of teachers and students toward online classes in Dakshina Kannada and Udupi District. Asian Assoc Open Univ J 15(3):285–296
12. Rojabi AR (2020) Exploring EFL students' perception of online learning via Microsoft teams: university level in Indonesia. Engl Lang Teach Educ J 3(2):163–173
13. Khan MA, Kamal T, Illiyan A, Asif M (2021) Schoolstudents' perception and challenges towards online classes during COVID-19 pandemic in India: an econometric analysis. Sustainability 13(1):4786
14. Kang M, Duong A (2021) Student perceptions of first-time online learningduring the COVID-19 pandemic in Vietnam. Inq Educ 13(1)
15. Biswas B, Roy SK, Roy F (2020) Students perception of mobile learning during COVID-19 in Bangladesh: university student perspective 4(2)
16. Galgewela N (2021) Students' perception of online education during the Covid- 19 pandemic: an analysis of students of higher education sector in Sri Lanka
17. Bączek M, Zagańczyk-Bączek M, Szpringer M, Jaroszyński A, Wożakowska-Kapłon B (2021) Students' perception of online learning during the COVID-19 pandemic—a survey study of Polish medical students. Medicine 100(7):e24821
18. Almahasees Z, Mohsen K, Amin MO (2021) Faculty's andstudents' perceptions of online learning during COVID-19. Front Educ 6(1):638470. https://doi.org/10.3389/feduc.2021.638470
19. Bast F (2021) Perception of online learning among students from India set against the pandemic. Front Educ 6:705013. https://doi.org/10.3389/feduc.2021.705013
20. Agarwal S, Kaushik JS (2020) Student's perception of online learning during COVID pandemic. Indian J Pediatr 87(7):554
21. Muthuprasad T, Aiswarya S, Aditya KS, Jha GK (2021) Students'perception and preference for online education in India during COVID-19 pandemic. Soc Sci Hum Open 3:100101

22. Koohang A, Durante A (2003) Learners' perceptions toward the web-based distance learning activi-ties/assignments portion of an undergraduate hybrid instructional model. J Inf Technol Educ 2:105–113
23. Muthukrishnan R, Janvi JS, Sourav S, Deepika G (2021) A study on students' perception on online learning during Covid-19 pandemic. Turk J Comput Math Educ 12(11):138–144
24. Rus M, Sandu ML, Salceanu C, Tanase T (2021) Studentperception of online education supported in the Covid-19 period. Technium Soc Sci J 15(1):71–79
25. Paechter M, Maier B (2010) Online or face-to-face? Students' experiences and preferences in e-learning. Internet Higher Educ 13:292–297
26. Dommetta EJ, Gardner B, van Tilburg W (2020) Staff and students perception of lecture capture. Internet Higher Educ 46:100732
27. Camilleri MA, Camilleri AC (2022) Remote learning via videoconferencing technologies: implications for research and practice. Technol Soc 68:101881
28. Guo C, Wan B (2022) The digital divide in online learning in China during the COVID-19 pandemic. Technol Soc 71:102122

Rash Driving Detection and Alerting System

Vishwas Dehankar and Surekha Tadse

Abstract The aim of this study is to depict the general system of rash driving detection whose goal is to detect the parameters of rash driving like over speeding, lane crossing, influence of drugs or alcohol, breaking traffic rules on driving-based mechanisms. Recently the traffic accidents as we can see, are frequently the result of driver carelessness. The drivers reckless and intoxicated behaviour revealed for such incidences. To overcome such accidents many research was going on which depends on abnormal driving behaviour of drivers, drunken driver detections, and many more. The rash driving detection and alerting system is one of the research work on the same domain which employs a Arduino UNO microcontroller to focus on the driver's unusual behaviour. The system can detect driving recklessness with the help of database and check whether the vehicles violating any rules.

Keywords Rash driving detection · Arduino UNO · Lane · Alarm · Alcohol detection

1 Introduction

Driving in Asian countries is regulated by various regulatory authorities and in some cases requires passing a driving test. The Ministry of Road Transport and Highways is a branch of the government of Asian countries and is the largest institution that develops and maintains the principles, rules, and laws related to transport, national roads, and transport inspections to expand and improve the efficiency of transport in Asia.

India is usually thought as one of the worlds riskiest nations in which to drive. According to the 2007 survey, 130,000 plus people died due to road accidents on the

V. Dehankar (✉) · S. Tadse
G. H. Raisoni College of Engineering, Nagpur, India
e-mail: falsevishwasdehankar@gmail.com

S. Tadse
e-mail: Surekha.tadse@raisoni.net

V. K. Gunjan and J. M. Zurada (eds.), *Proceedings of 4th International Conference on Recent Trends in Machine Learning, IoT, Smart Cities and Applications*, Lecture Notes in Networks and Systems 873, https://doi.org/10.1007/978-981-99-9442-7_17

roads in India. Although, this figure should be measured was not much in relation to the nation's population. The spokesman from Delhi's Institute of Road Traffic Education pointed out that major reason for it was the inadequate management of traffic, "we don't modify traffic management systems to split different streams," and poor driver training [1]. An article from Reuters from 2014 described that the driver's licence exam in Delhi were spent lesser than two minutes, as an examiner was testing ten people consecutively. The big variety of ways of transportation and what's typically depicted and neglecting the principles of the road contribute to the accidental deaths on road.

Our project is based on rash driving detection, which takes the database from the internet, checks whether the vehicle is violating any rules, and shows that the vehicle is rash driving.

This methodology aims to provide a standard model for diagnosing rash driving behaviour. The model's responsiveness reveals recklessness by collecting data from the database. The requirements for hardware are as follows: Arduino UNO hardware, LCD Display Module (16 * 2), buzzer, database, and GPS.

2 Literature Review

The accidental situations are steadily increasing as the number of vehicles and their movement on the road increased. Preventing or reducing such serious accidents has become a nightmare for the authorities and all efforts are in vain. According to the Road Traffic Accident Study of India, more than 135,000 traffic accidents occur each year. Most of these accidents are due to reckless driving. According to the Indian Constitution, IPC Section 279, rash driving is an crime [2]. So, the idea is to create a module that can detect the vehicle and send the information about it to the relevant authorities when the vehicle found to be driven rashly or exceeds the speed limit.

The study conducted by Beneret et al. (2010) in Qatar found that young drivers are more likely to drive recklessly than older drivers and with factors such as alcohol consumption, lack of experience, and peer pressure were significant predictors of this behaviour. Another study by Karthikeyan and Sastry (2012) in India found that drivers who were aggressive or had a sensation-seeking personality were more likely to engage in reckless driving. Several strategies have been proposed to prevent rash driving. One approach is to increase the penalties for this behaviour, such as through increased fines or licence suspensions (Rafique et al. 2016). Another approach is to implement educational programmes that target high-risk groups, such as young male drivers (Bener et al. 2010). In addition, technological solutions such as speed cameras and automated enforcement systems have been shown to be effective in reducing rash driving behaviour (Khattak et al. 2016).

Overall, impaired driving is a serious problem that requires a diverse approach to address. Through targeting the underlying causes of this behaviour and enacting effective prevention plan, it is possible to reduce the incidence of reckless driving

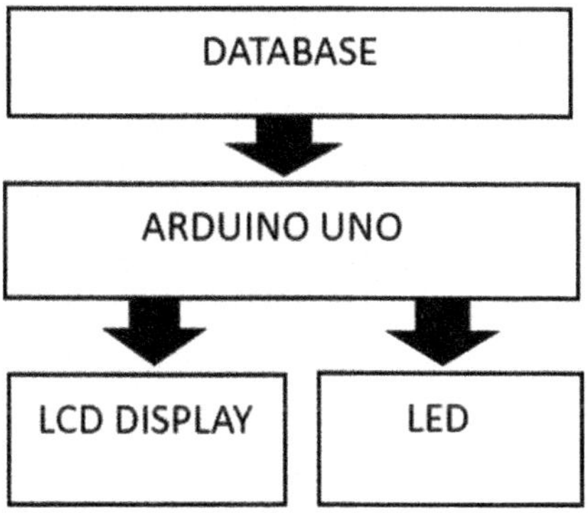

Fig. 1 Architecture of prototype interface

and improve public safety on the roads. Our prototype is one that works for reducing these incidents of recklessness [3] (Fig. 1).

3 Methodology

To detect rash driving, we have created a summarized database with different parameters such as vehicle number, lane, overtaking, speed limit, influence of drugs or alcohol, and traffic rules. With the help of cameras and sensors, we collect this information, and with a specific algorithm using Arduino UNO, if the driver is found guilty, we can state that he was a rash driver who was not appropriate to drive on the road. Our concept is the prototype of how the systems can be smart through data management to detect and reduce the incidences of rash driving situations and how we can take action over them in a smart way.

In India, most of the cities and roads are now under surveillance cameras. On every signal, we see cameras installed by the authorities for road safety concerns, and they are working those to check for signal breaks, helmets worn or not, and similarly many other things. If these same cameras were implemented in more diverse ways for collecting some data that they could track, then our prototype would become more useful in the future. And also, the burden of manual inspection has become less and more automated [4–6].

In the same sequence, we have created the alcohol detection drunk and driving system model, which works to detect the alcohol-consuming driver. In this model, the alcohol detection sensor is placed in the vehicle, and if the driver is drunk, the vehicle engine gets off automatically if the threshold value is crossed.

With a similar streak, we have studied one more model related to road safety that works for checking the drowsiness of the driver. In this drowsiness model, the driver's face and eyes were monitored [7, 8]. And factors like yawning, blinking of the eyes, and facial motion were detected. And with the help of image processing, these factors were compared. If the driver was found to be drowsy, a particular signal was given to him, like a blinking led or an alarm [9, 10].

After the research in different areas related to road safety concerns, we created this prototype after successful testing of those models.

4 Design and Implementation

This method is built in three stages.

Stage 1: Data collection.

Here we have summarized data on different parameters taken from foreign countries road data records.

The database has parameters that we wish to consider with respect to rash driving. The parameters were given in the sense of true (1's) and false (0's) (Table 1).

The hardware implementation of the model is given in Fig. 2. These are the combinations of the complete input and output setup with the CPU.

The workings of the prototype are as follows:

Stage 2: Tracing Stage

The system moves on to the standard tracing stage, in which, firstly, the data sheet of the driving record is there, which contains different parameters to detect to

Table 1 Database sample

ID	Vehicle_No.	Inlane	Overtaking	In speed limit	Drugs/alcohol influence	Following_ traffic_rules
1	TS 09 MD 7987	1	0	1	0	1
2	KA 01 AF 8192	0	0	1	1	0
3	DL 11 CA 1111	1	0	0	1	0
4	MH 29 AF 0217	1	0	1	0	1
5	MH 49 AF 5519	0	1	1	1	0

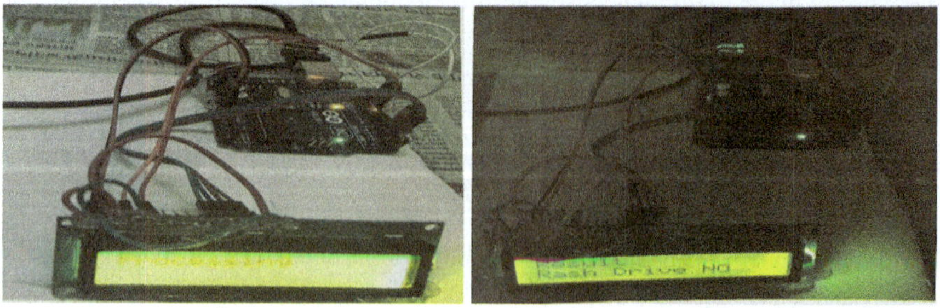

(i) (ii)

Fig. 2 Hardware implementation

categorize the vehicle, like whether the vehicle is in lane, overtaking, speed limit, or if the driver has taken alcohol, which acts as input data (a database) to the system (an Arduino).

The system goes through each row by first detecting the vehicle number along with the parameters it has to check. While moving through all the parameters, if the driver who is driving the vehicle is found guilty in any of the parameters, he is detected as the rash driver by the system, and the red led will glow to inform us on the led display, and action can be taken on him. Otherwise, a green light will glow, and the driver's safety will be considered [11, 12].

Stage 3: Warning and Action Stage

It is necessary to gauge the police's or traffic controllers attention regarding this rash driver so the warning notification will be forwarded to the concerned authority to take the required action on such drivers. To take action against this driver, all information regarding where the driver was found guilty will be shared with them. So, with the help of the respective driver's vehicle number, the authority can take the necessary actions so that the GPS tracking record of the vehicle can also be shared with them.

Please note that the first paragraph of a section or subsection is not indented. The first paragraphs that follows a table, figure, equation etc. does not have an indent, either.

Subsequent paragraphs, however, are indented.

5 Result

With this implemented system, we have designed the rash driving detection system, which will be useful in the world of automobiles to reduce reckless drivers and take action against them. It also helps to implement government plans to reduce road accident rates, as road accidents due to rash driving are a major issue in countries like India.

6 Conclusion

Through this research, we have discussed how to integrate different driving-related parameters and algorithms to create and utilize it for the detection of reckless driving behaviour. Currently, tremendous work is going on from the union ministry for road safety-related concerns. So, our prototype was a way to enhance drivers safety by monitoring driving behaviour, recording hostile driving events, and giving estimation over the recorded instances. Observing and regulating driving behaviour using database management has been less expensive, considering that many systems already posses it. This study explores various methods of detecting driving behaviour, which deliberatively helps the concerned authorities take action against drivers and punish them as per protocol.

References

1. Impaired Driving (2017) Get the facts page. Last updated: 16 June 2017. Available at https://www.cdc.gov/motorvehiclesafety/impaireddriving/impaired-drv_factsheet.html. Accessed 15 July 2017
2. Sivakumar, Krishnaraj R (2012) Road traffic accidents (RTAS) in India challenges in prevention. Int J Res Manage Technol (URMT) 2(4)
3. Prashanth KP, Padiyar K, Naveen Kumar PH, Santhosh Kumar K (2014) Road accident avoiding system using overspeeding sensing technique. Int J Eng Res Technol (IJERT) 3(10). Available at Road accident avoiding system using overspeeding sensing technique. http://www.ijert.org/download/11491/road-accident-avoiding-system-using-overspeeding-sensing-techniquef. Accessed 15 July 2017
4. Baru D, Jain P, Gupta J, Gadre DV (2013) Road accident prevention unit (RAPU). In: Texas instruments India Educators' conference
5. Bhuta D, Keni (2015) Vehicle controlling. Int J Eng Trends Appl (LETA) 2(2)
6. Phalak, Kowekar, Joshi (2015) Smartphone and sensor based drunk driving prevention system. Int J Res Emerg Sci Technol 2(9)
7. Ramanath, Sudharsan, Udhayara (2010) Rash driving prevention system. In: International conference on mechanical and electrical technology-ICMET
8. Comas J, Aspandi D, Binefa X (2019) End-to-end facial and physiological model for affective computing and applications, 10 Dec 2019
9. Mardi Z, Naghmeh S, Ashtiani M, Mohammad. EEG-based drowsiness detection for safe driving using chaotic features and statistical tests
10. Ahmed R, Emon KEK, Hussain MF (2014) Robust driver fatigue recognition using image processing. Khulna University of Engineering & Technology, Bangladesh
11. Pérez J, Milanés V, Onieva E (2011) Cascade architecture for lateral control in autonomous vehicles. IEEE Trans Intell Transp Syst 12(1):73–82
12. Mathur R, Sakarkar G, Kalbande K, Mathur R, Kolhe H, Rathi H (2021) Orthopantomogram (OPG) image analysis using bounding box algorithm. In: 2021 international conference on computational methods and data engineering (ICCMDE 2021)

Logistic-Based OVA-CNN Model for Alzheimer's Disease Detection and Prediction Using MR Images

Chinchu M. John and Prafulla Phalgunan

Abstract Alzheimer's disease (AD) is a form of dementia that diminishes a patient's cognition and memory. By predicting AD progression early, physicians can begin medication that may reduce the disease's progression. AD detection and prediction have been extensively performed using convolutional neural networks (CNNs). Despite this, CNN's use for AD detection is limited by the number of parameters and the extensive training data required. Consequently, this study proposes a logistically based OVA-CNN model for AD detection and prediction using Magnetic Resonance Images (MRI). OVA-CNN consists of the four binary CNN which classified four classes of AD and the logistic regressor predicts the time conversion of pMCI patient who converts AD. The experimental results on the Alzheimer's Disease Neuro Imaging Initiative (ADNI) demonstrate that the proposed model achieves a high level of accuracy of 93%.

Keywords Alzheimer's disease · Logistic regression · MRI images

1 Introduction

Alzheimer's disease (AD) is a type of dementia caused by a neurological process that gradually reduces mental capacity. By 2050, the Alzheimer's association predicts that there will be over 100 million Alzheimer's patients worldwide [1]. Because of this, Alzheimer's patients are a psychological and emotional burden to their families, society, and the health-care system. Memory loss, misplacing things, and difficulty focusing on a single task are signs of this irreversible brain disorder. The disease's

C. M. John (✉) · P. Phalgunan
Department of Computer Science and Engineering, College of Engineering, Trivandrum, Kerala, India
e-mail: Chinchumjohn2022@gmail.com

P. Phalgunan
e-mail: prafullap@cet.ac.in

V. K. Gunjan and J. M. Zurada (eds.), *Proceedings of 4th International Conference on Recent Trends in Machine Learning, IoT, Smart Cities and Applications*, Lecture Notes in Networks and Systems 873, https://doi.org/10.1007/978-981-99-9442-7_18

root cause is the improper accumulation and buildup of amyloid and tau protein surrounding brain cells in the form of plaques and tangles [2].

Alzheimer's disease is divided into three categories: Alzheimer's Disease (AD), Cognitive Normal (CN), and Mild Cognitive Impairment (MCI) depending on how quickly the disease is progressing. CN is at a normal stage of cognitive development, and the patient is free of depression, dementia, or memory loss. AD is the advanced stage of dementia, and MCI is the stage in-between AD and CN. Stable MCI (sMCI) and progressive MCI (pMCI) are the two subclasses of MCI also, the pMCI stage will eventually get AD, but patients with sMCI do not.

The biomarkers obtained from imaging modalities such as Positron Emission Tomography (PET), Cerebro Spinal Fluid (CSF), and Magnetic Resonance Imaging (MRI) are commonly used to diagnose the progression of AD [2]. MRI captures the images of the brain and detects the progression of AD by cortical thickness, the volume of the hippocampus, and Region of Interest (ROI) volume. The clinical assessment examination can be used to determine the subject's level of cognitive impairment. To conduct this examination, individual is given a collection of questionnaires, and the physician determines the AD based on the responses.

The traditional AD detection approaches reliably identify Alzheimer's disease utilizing a number of machine learning methods, including Recurrent Neural Networks (RNN), Support Vector Machine (SVM) [3], and Convolutional Neural Networks (CNN). CNN gives the hasty outcome of AD detection. However, it leads to a significant issue. The large amount of weight parameters in deep CNN models like Alex-Net, VGG-16, and ResNet-50 raises the memory requirement for their implementation. Moreover, a single CNN for multi-class classification results in more weight parameters which also causes more memory and high model complexity. As a result, when employing CNN in a memory-constrained environment, it is essential to reduce the amount of weight components.

The One-vs-All (OVA) CNN model is suggested in this paper to identify the several classes of Alzheimer's disease, including CN, AD, pMCI, and sMCI, using MRI images. To identify the various phases of AD, the four binary CNN classifier is utilized to analyze the MRI images from the ADNI dataset. To determine the precise conversion period of a patient from pMCI to AD, logistic regression is performed.

The following is the structure of this paper. The research that is pertinent to this work is described in Sect. 2. Section 3 discusses the methodology. The experiments and results are covered in Sect. 4. Conclusions are discussed in Sect. 5.

2 Related Works

The conventional AD diagnosis techniques used MRI modality images at a single time point and co-related features are extracted from these MR images. For example, A feature extraction technique based on anatomical landmarks from longitudinal MR images was proposed by Jun Zhang et al. [4]. High-level spatial features for MR images are extracted using the bag-of-words method, and contextual longitudinal features are created for MR images using normalized longitudinal deformations. These features are given into the linear SVM classifier after Z-score normalization, which assesses whether AD is present. Noisy features have an impact on the proposed technique, which reduces accuracy.

The MRI modality makes it easier to extract features than other modalities like PET and CSF. According to a model put forth by Mathew et al. [5], certain features are taken from the brain's MRI image. The Gray Level Co-occurrence Matrix (GLCM) is used to derive the statistical features in this instance. Based on the data gathered from the features collected, the images are categorized. Every image is given a set of features, which are then input into a probabilistic neural network (PNN) classifier, which divides the images into two categories: AD and CN. This method failed to predict mild cognitive impairment (MCI) class.

To solve the issues, Faisal and Kwon [6] proposed an end-to-end deep CNN architecture employing the entire image volume as input for multi-label AD biomarker identification. Here, CNN architectures like VGG-Net can be used for classification. The VGG model supports the use of down sampling, and enhanced feature extraction. The AD, CN, and MCI stages of AD are more effectively and precisely predicted using the Soft-Max activation function. Additionally, using data from multiple models, Srivardhini et al. [7] proposed a principal component analysis is used to extract the feature, which is then fed into a convolutional neural network (CNN) that can distinguish between mild, moderate, and normal levels of AD. The class imbalance issue with the MRI data is another issue that shows the limitations of this approach.

A model for early MRI image-based diagnosis of AD and dementia has been developed by Murugan et al. [8]. The SMOTE method addresses the issue of class imbalance in the dataset while reducing overfitting. MRI images that have been normalized and passed to the convolution layer, which creates the feature map or response by combining the input with bias and weight filters. The successive layers receive the feature response after that. By avoiding procedures like skull stripping during the pre-processing stage, the approach produces better results.

MCI has two subclasses: pMCI and sMCI. The pMCI patients will eventually transform into AD patients. A two- stage deep learning method is used by El-Sappagh et al. [9] for predicting Alzheimer's disease. This model combines the classification and regression tasks to determine whether a patient has MCI and then predict the progression time. A three-class classification model is proposed in the initial stage to identify the patient's class (CN, MCI, or AD), and a regression model is proposed in the second stage for predicting the conversion time for pMCI patients.

Mehmood et al. [10] proposed a Siamese convolutional neural network-based model for the detection of AD disease and ResNet-34 is used to classify the CN and AD. The proposed model's flaw is ResNet-34 deployment requires more memory. Shanmugamet al. [11] suggested using MR images to predict several kinds of AD using Google-Net. A VGG-Net was proposed by Kwon and Faisal [6] for the classification of AD. For the models to be established, deep CNN networks such as Res-Net, Alex-Net, Google-Net, and VGG-Net need a larger weight parameter. Additionally, having more weight parameters uses up more memory and makes the model

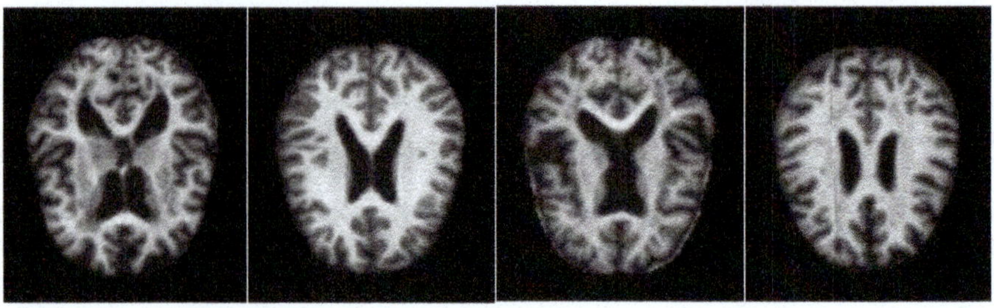

Fig. 1 Image of pMCI, CN, AD, sMCI after smoothing

more complex. Furthermore, a large amount of memory and weight parameters are produced when using a single CNN for multi-class classification.

3 Methodology

This research proposes a revolutionary OVA-CNN for the categorization of AD in various stages to alleviate the limitations of conventional CNN. Initially, this section introduces the pre-processing of MRI images. Next, describe the theoretical concepts of OVA-CNN. Toward the end of the section, gives the explanation of the OVA-CNN architecture.

3.1 Pre-processing

Digital Imaging and Communications in Medicine (DICOM) formatted brain MRI images are obtained from the ADNI database. First, resize the images to 128 × 128 and convert them from DICOM format to JPEG. After that, perform out digital image processing methods such as smoothing to eliminate the images' noise [12]. Over the images, a linear median box filter is used, with each output pixel being the weighted mean of the kernel neighbor. Figure 1 shows images of pMCI, CN, AD, and sMCI that were acquired after smoothing. To prevent over-fitting, the images are then split into three sets: train set, test set, and validation set.

3.2 OVA-CNN: Theoretical Concept

The training method becomes stalled when attempting to occur a single error function in a multi-class classification issue in AD. As a result, it might be challenging for

AD detection with limited medical data to identify the best hypothesis using a single classifier.

A CNN classifier named X was trained on samples $Z = (x_1, y_1), (x_2, y_2), \ldots, (x_m, y_m)$ for the multi-class classification problem. Where X_i signifies the r-dimensional vector, y_i denotes the label from $y = 1, \ldots, l$, and m denotes the number of training samples. In this proposed framework, the multi-class classification issue is divided into multiple binary classification sub-problems [13]. Every sub-problem p corresponding to an unidentified sample X to $Y = 1, 1$ can be resolved by the binary classifier, BCNN. The Vapnik–Chervonenkis (VC) dimension of a classifier is typically related to its generalizability. The capacity of each class in the classifier is measured by the VC dimension. A classifier that has a greater VC dimension produces data that is more complex and demands more training images [13]. Equation (1) describes the association between the error rate and VC dimension.

$$\frac{2e\gamma|Z|}{VC_{\text{dim}}} = 1 \tag{1}$$

where |Z| is the amount of training samples and e is the natural logarithm's base. Equation (1) demonstrates a direct relationship between the model's VC dimension and the γ. Therefore, decreasing the classifier's VC dimension can lower the error rate of the classifier for a preset amount of training images. Moreover, a classifier's VC dimension is inversely related to the model complexity S. The association between the model's complexity S and the l number of classes is stated in Eq. (2).

$$S = \log_2 *r \tag{2}$$

Equation (2) states that for a constant amount of training images, the number of classes to be discovered directly relates to the complexity of the classifier. Reduced model complexity with fewer classes and a reduced error rate result from lowering the VC dimension. As a result, the depth or width of the CNN layers can be decreased to reduce the VC dimension. By decreasing the weight values in the convolution layers, it is possible to reduce the CNN's layer depth or width. To calculate the amount of weight parameters in convolution layer and fully connected layer using Eqs. (3) and (4) respectively.

$$\text{wt}_{\text{conv}} = \sum_{r-1}^{R} f_{r-1}^2 n_r L^2 + n_r \tag{3}$$

$$\text{wt}_{\text{fully}} = \sum_{r-1}^{R} n_{r-1} n_r \tag{4}$$

where the total amount of parameters for the convolutional layer and the fully connected layer are denoted by wt_{conv} and wt_{fully}. The current layer's feature map size is f_r, whereas the previous layer's feature map size was f_{r-1}. The words n_r

and n_{r-1} denote the number of neurons in the current layer and the preceding layer, respectively. The entire number of weight parameters, W_{tot}, in CNN are shown in Eq. (5). Where n_{batch} denotes the total amount of neurons in the batch layer.

$$W_{tot} = wt_{conv} + wt_{fully} + n_{batch} \tag{5}$$

3.3 OVA-CNN: Architectural Design

Figure 2 illustrates the architecture design for the suggested OVA-CNN model. Classification and regression are used in this ensemble technique. In this framework, the OVA-CNN module classifies the four classes of AD, and the regression module predicts the conversion time. In OVA-CNN, there are two components: BCNN and Max-value Selector. Box filtering the MR images obtained from the ADNI database and the images are passed through the BCNN classifier. The framework contains four binary classifiers BCNN1, BCNN2, BCNN3, and BCNN4 for the four classes CN, pMCI, sMCI, and AD, respectively. Moreover, each BCNN is exactly predicting one class correctly.

Figure 3 demonstrated the structure of BCNN. The BCNN has three layers: a dense layer, a pooling layer, and a convolutional layer. In the first convolution layer, 16 neurons with a kernel size of 5 are present. The second convolution layer consist of 32 neurons with the activation function ReLU. Maxpooling 2×2 is used in each convolution layer. The feature vectors with a 1×288 size are flattened from the feature maps acquired from the convolution layers. Following that, a ReLu activation function feeds the feature vector into three completely connected layers. The total number of 64, 32, and 16 neurons are present in each fully connected layer. Finally, the output was generated by a dense layer composed of two neurons with soft-max activation functions. In order to identify the identified class, the max-value selector uses the BCNNs to decide which has the greatest probability of occurring.

As a next step, pMCI data extracted from the MR images are passed through the logistic regression module. As a result, the regressor can predict the precise time at which pMCI patients will change to AD. The progressive MCI will convert into AD after some point in time. Logistic regression is used to predict the exact conversion time. Logistic regression performs binary classification with the sigmoid activation function. However, if the function continues the result will be calculated from the anti-log of the regression coefficient. The extracted data from different time point are fed into the logistic regressor, which gives time of prediction.

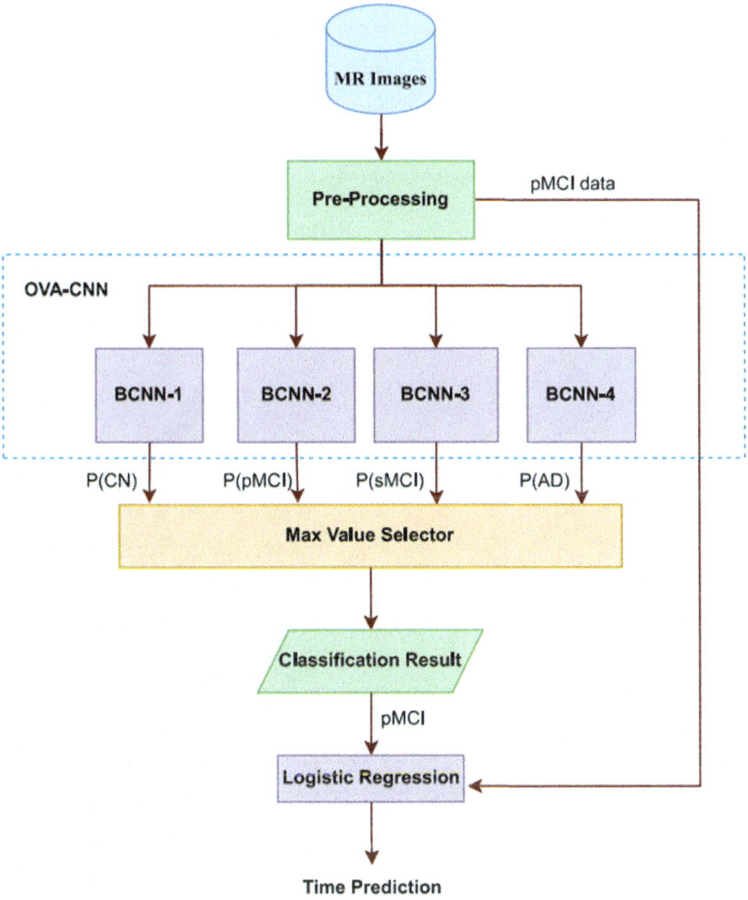

Fig. 2 Structure of OVA-CNN model

4 Results and Discussion

This section deals with the dataset and experimental setup followed by a discussion based on the result obtained.

4.1 *Dataset and Experimental Setup*

MR images are acquired from the ADNI database, and the dataset consists of 2279 MRI images. These 2279 images are divided into four different classes: AD CN, pMCI, sMCI, and AD and 70% of these images are training sets while 30% are

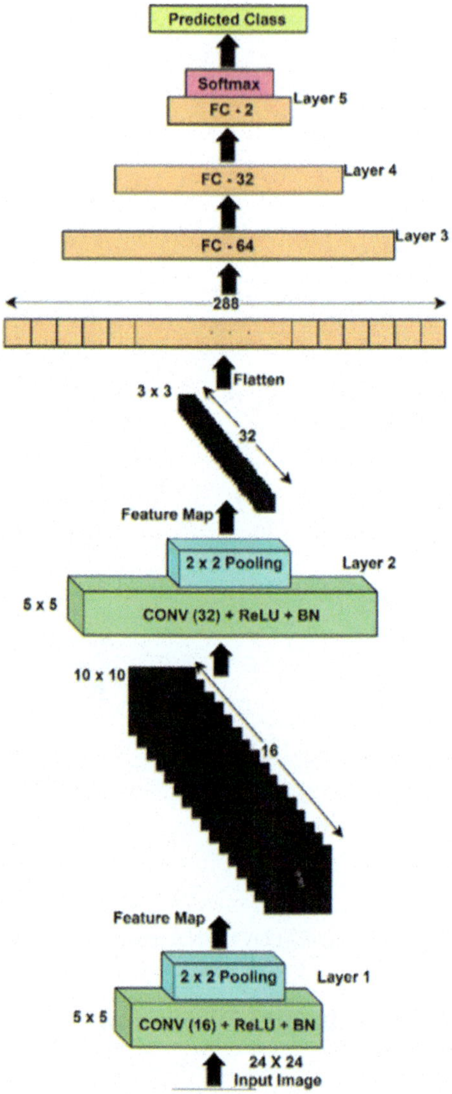

Fig. 3 Architectural diagram of BCNN

testing sets. The number of images in each class based on training, validation, and testing is tabulated in Table 1.

We conducted our experiments on a laptop running Windows 11 Pro, equipped with an Intel i5 processor and 4 GB of RAM. For the development of BCNN models, Python codes are used in conjunction with the Keras API. The Adam optimization algorithm is used to train each BCNN for 30 iterations, with a fixed learning rate of 0.02 and loss can be estimated using categorical cross-entropy.

Table 1 No. of images in each class based on training, testing, and validation

Class	Train	Validation	Test
AD	508	119	64
pMCI	437	104	58
sMCI	432	67	34
CN	256	127	76

5 Discussion

Performance evaluation of the proposed model based on the confusion matrix obtained. The graphical representation of loss and accuracy for each model in 30 epochs is shown in Fig. 5. The testing accuracy of pMCI versus Rest, AD versus Rest, CN versus Rest, and sMCI versus Rest are 86.21%, 96.65%, 96.12%, and 91.38% respectively. Accuracy of the model AD versus Rest gives high performance than the other individual models. The validation loss for pMCI versus Rest is 1.0293, for CN it is 0.1747, for AD it is 0.1165, and for sMCI it is 0.3622. From these values, the loss of AD is lesser than in the other individual models.

The performance of logistic regression for predicting the time conversion of pMCI to AD is compared with that of other conventional machine learning classifiers in Fig. 4. The results of the logistic regressor's accuracy, precision, recall, and $F1$-score are 94%. MRI data are used to train existing classifiers like Random Forest, SVM, KNN, and AdaBoost. The obtained values for Random Forest's accuracy, precision, recall, and $F1$-score are 81.5% and 81% for AdaBoost.

The accuracy and other performance metrics for the SVM and KNN were both 87.5%. This data demonstrates that, for this time prediction, the logistic regressor

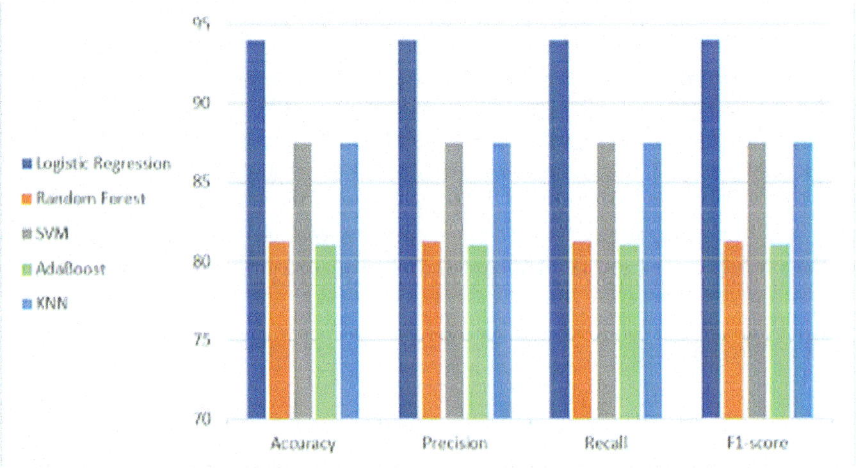

Fig. 4 Performance of logistic regressor versus other regressor

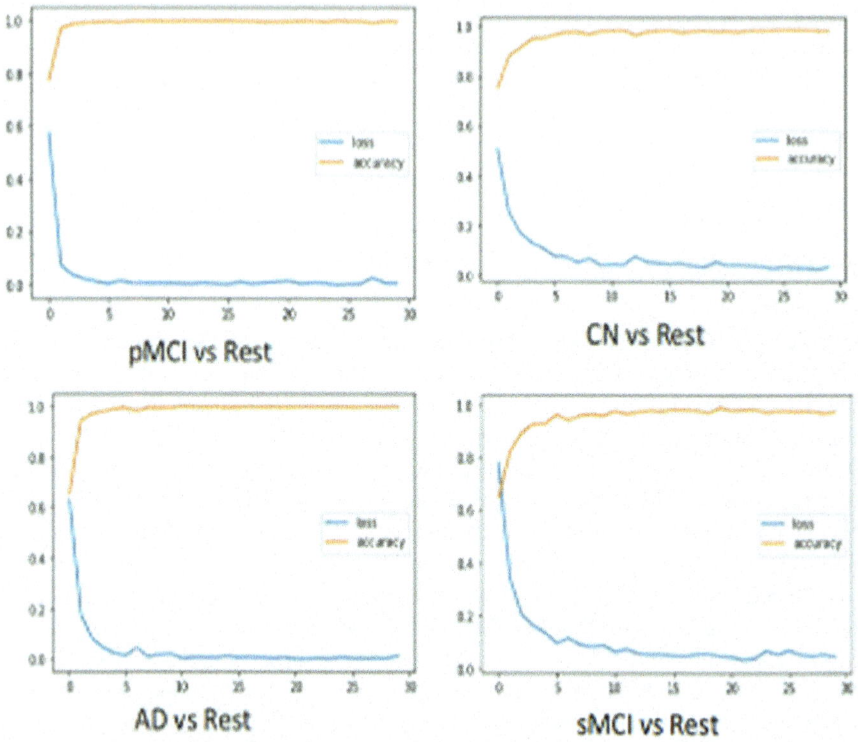

Fig. 5 Loss and accuracy of individual model

outperforms traditional machine learning classifiers. The performance matrices of OVA-CNN model are tabulated in Table 2. The model's 93% accuracy, precision, recall, and $F1$-score indicate outstanding performance for our prediction model.

Table 3 shows a comparison of the suggested model with two previous works which currently exist. Faith et al. [14] proposed an 83.27% accurate 3D CNN-based Alzheimer's disease diagnosis model. The second model put forth by Ebirami et al. [15] employs the TCN technique with Resnet-18 to achieve an accuracy of 91.78%. Compared to the other two models, the suggested OVA-CNN model provides 93% accuracy. According to Eqs. (4) and (5), there are a total of 1.89 M weight parameters in the proposed framework. This result indicates that the model contributes less

Table 2 Performance of OVA-CNN

Performance metrics	OVA-CNN
Accuracy	93
Precision	93
Recall	93
$F1$-score	93

Table 3 Comparison of OVA-CNN model versus existing model

Model	Method	Accuracy (%)
Faith et al. [14]	3D-CNN	83.27
Ebirahmi et al. [15]	Res-Net and TCN	91.27
Proposed model	OVA-CNN	**93**

weight parameters. The model's smaller weight parameter reduces to the width of the CNN layers, in accordance with the theoretical concept. The VC dimension and model memory are both reduced because of the CNN layers' decreased width. OVA-CNN model has less memory as a result than existing multi-class AD detection approaches.

6 Conclusion

The task of early AD detection has always been difficult, and relevant computer experts are continually investigating. Based on the ADNI dataset from MRI images, this study employs the One-vs-All model to detect Alzheimer's disease, and the conversion of pMCI patients is predicted by logistic regression. The four classifications of Alzheimer's disease in this study were divided into AD, CN, and MCI sub types such sMCI and pMCI by four BCNN models. By using logistic regression, the time can be predicted for patients who converted from pMCI to AD. The proposed model's overall accuracy is 93% and employing fewer weight parameters contributed to a reduction in the model's memory usage.

References

1. Brookmeyer R, Johnson E, Ziegler-Graham K, Michael Arrighi H (2007) Forecasting the global burden of Alzheimer's disease. Alzheimer's Dement 3(3):186–191. ISSN: 1552-5260
2. Coedert M, Sisodia SS, Price DL (1991) Neurofibrillary tangles, and P-amyloid deposits in Alzheimer's disease. Elsevier
3. Kavitha C, Mani V, Srividhya SR, Khalaf OI, Tavera Romero CA (2022) Early-stage Alzheimer's disease prediction using machine learning models. Front Public Health. https://doi.org/10.3389/fpubh.2022.853294. PMID: 35309200; PMCID: PMC8927715
4. Zhang J, Liu M, An L, Gao Y, Shen D (2017) Alzheimer's disease diagnosis using landmark-based features from longitudinal structural MR images. IEEE J Biomed Health Inform 21(6):1607–1616. https://doi.org/10.1109/JBHI.2017.2704614
5. Mathew NA, Vivek RS, Anurenjan PR (2018) Early diagnosis of Alzheimer's disease from MRI images using PNN. In: 2018 international CET conference on control, communication, and computing (IC4), pp 161–164. https://doi.org/10.1109/CETIC4.2018.8530910
6. Faisal FUR, Kwon G-R (2022) Automated detection of Alzheimer's disease and mild cognitive impairment using whole brain MRI. IEEE Access 10:65055–65066. https://doi.org/10.1109/AC-CESS.2022.3180073

7. Srivardhini V, Sankavi M, Sanjana S, Kalaivani N, Dinesh Kumar JR (2021) A deep learning based multi-model for early prognosticate of Alzheimer's dementia using MRI dataset. In: 2nd international conference on smart electronics and communication (ICOSEC), pp 1048–1055. https://doi.org/10.1109/ICOSEC51865.2021.9591936
8. Murugan S et al (2021) DEMNET: a deep learning model for early diagnosis of Alzheimer diseases and dementia from MR images. IEEE Access 9:90319–90329. https://doi.org/10.1109/AC-CESS.2021.3090474
9. El-Sappagh S, Saleh H, Ali F et al (2022) Two-stage deep learning model for Alzheimer's disease detection and prediction of the mild cognitive impairment time. Neural Comput Appl 34:14487–14509. https://doi.org/10.1007/s00521-022-07263-9
10. Mehmood A, Maqsood M, Bashir M, Shuyuan Y (2020) A deep Siamese convolution neural network for multi-class classification of Alzheimer disease. Brain Sci 10(2):84. https://doi.org/10.3390/brainsci10020084. PMID: 32033462; PMCID: PMC7071616
11. Shanmugam JV et al (2022) Alzheimer's disease classification using pre-trained deep networks. Biomed Signal Process Control 71(Part B):103217. ISSN: 1746-8094. https://doi.org/10.1016/j.bspc.2021.103217
12. Ilea et al (2014) Identification of body fat tissues in MRI data. In: OPTIM 2004—9th international conference on optimization of electrical and electronic equipment, 20–21, Brasov, Romania
13. Babu BP, Narayanan SJ (2022) One-vs-all convolutional neural networks for synthetic aperture radar target recognition. Cybern Inf Technol 22(3):179–197. https://doi.org/10.2478/cait-2022-0035
14. Altay F, Sánchez GR, James Y, Faraone SV, Velipasalar S, Salekin A (2020) Preclinical stage Alzheimer's disease detection using magnetic resonance image scans
15. Ebrahimi-Ghahnavieh A, Luo S, Chiong R (2019) Transfer learning for Alzheimer's disease detection on MRI images. In: Proceedings of the IEEE international conference on Industry 4.0, artificial intelligence, and communications technology, pp 133–138

Comparative Study of CNNs for Camouflaged Object Detection

Chinmay Avinash Javalagi, K. M. Medha, Neha T. Patil, Srujan Itagalli, Uday Kulkarni, and Satish Chikkamath

Abstract Convolutional neural networks (CNNs) have revolutionized the computer vision field, achieving remarkable performance in various visual recognition tasks, including object detection. One challenging problem in object detection is the detection of camouflaged objects in natural scenes. Camouflaged object detection (COD) is an essential task in various real-world applications, such as military surveillance, autonomous driving, and robotics. In recent years, numerous methods have been proposed for this task. The existing research on this issue has relied either on biological traits of the camouflaged objects or computer vision methods. In recent years, significant progress has been made in this field, with various techniques proposed for the detection of camouflaged objects. The detection and tracking of camouflaged objects using computer vision algorithms is discussed in this article from a theoretical standpoint. The proposed work presents a comprehensive review of the different methodologies used for camouflaged object detection, which includes an overview of the most recent cutting-edge techniques for COD and evaluate their performance on three popular Benchmark Datasets: COD10K, CAMO, and Chameleon. We analyse five different models, including SINet, SINet-V2, DGNet, ERRNet, and HITNet, based on two evaluation metrics: S-measure and E-measure. After evaluating the approaches in a common setting of dataset and computing system, the result is provided. Our results demonstrate that HITNet outperforms the other models across all datasets with an accuracy of 97%. Therefore, HITNet model underwent post-dynamic range quantization to reduce its size from 101 MB to 25.5 MB to enable deployment on edge device.

Keywords Convolutional neural networks (CNNs) · Camouflaged object detection (COD) · Benchmark datasets · Biological traits · Cutting-edge methods · Evaluation metrics · Edge device

N. T. Patil · S. Itagalli · U. Kulkarni
School of Computer Science, KLE Technological University, Hubballi, Karnataka 580021, India

C. A. Javalagi (✉) · K. M. Medha · S. Chikkamath
School of Electronics and Communication, KLE Technological University, Hubballi, Karnataka 580021, India
e-mail: chinmayjavalagi@gmail.com

V. K. Gunjan and J. M. Zurada (eds.), *Proceedings of 4th International Conference on Recent Trends in Machine Learning, IoT, Smart Cities and Applications*, Lecture Notes in Networks and Systems 873, https://doi.org/10.1007/978-981-99-9442-7_19

1 Introduction

In recent years, machine learning (ML) and deep learning (DL) have transformed the area of computer vision, enabling the development of sophisticated algorithms capable of detecting objects and patterns in images and videos [1]. Particularly since the advent of fully convolutional neural networks (FCNs), convolutional neural networks (CNNs) [2] have recently been able to surpass the limitations of traditional handmade features. Most COD methods based on CNN architecture that take picture patches as input use multi-scale or multi-context data to produce the final saliency map. As a result of the fully convolutional network [3] being suggested for pixel labelling issues, a number of end-to-end deep architectures for salient object detection emerge.

Early investigations are principally based on the assumption that contrast plays the most significant role in saliency detection [4] and are inspired by visual attention cognitive studies [5]. These techniques primarily profit from learnt fusion weights for either global or local contrast stimuli. Unfortunately, despite the fact that these manually created features can occasionally detect the most important items, the saliency maps they produce have irregular forms as a result of the unfavourable segmentation techniques and are unreliable when there is inadequate contrast between the foreground and the background [6].

Camouflaged object detection involves the identification and tracking of objects that are visually concealed within complex backgrounds [7]. It presents a unique challenge to machine learning algorithms, as these objects are intentionally or unintentionally concealed in the background, making them difficult to detect with traditional image processing techniques [8]. This is accomplished by augmenting the training data with various types of camouflage techniques, such as changing the texture, shape, or colour of objects [9]. The goal of camouflaged object detection (COD), which has lately received a lot of attention, is to identify the most visually identifiable things in an image [10]. It is extensively utilized in several fields related to vision and image processing, including content-aware image editing, object recognition, photosynth, non-photorealistic rendering, weakly supervised semantic segmentation, and image retrieval [11]. The purpose of this comparative study is to provide a complete evaluation of the most recent COD methodologies. It will examine the benefits and drawbacks of each technique, as well as provide a critical review of their performance on popular benchmark datasets. The study will also identify research obstacles and future goals in the field of COD. The paper is organized into the following sections. In Sect. 2, we give a summary of the literature that includes the most recent advances in the field of concealed object recognition. In Sect. 3, we present our proposed approach for deploying the best model on edge devices. In Sect. 4, we present a detailed description of the results of our suggested technique. In Sect. 5, we provide a comprehensive overview of the future scope of our work. Finally, we provide a conclusion in Sect. 6.

2 Related Work

In this section, we examine current works on camouflaged object segmentation, which share some similarities.

2.1 *Camouflaged Object Detection Using SINet*

In [12], the authors proposed the first complete benchmark on object identification from the perspective of camouflage. Paper, in particular, has supplied a new demanding and richly annotated COD10K dataset, carried out a large-scale assessment, developed a simple yet effective end-to-end SINet framework, and discovered many possible applications. SINet is competitive and produces better visually appealing outcomes when compared to existing cutting-edge baselines. The contributions listed above give the community the ability to create new models for the COD assignment. They intend to extend the COD10K dataset in future to give input in multiple formats, such as RGB-D camouflage object detection (similar to RGB-D salient object recognition, among others).

The loss function used to train SINet is cross-entropy loss as shown in equation (1).

$$L = L_{\text{CE}}^{s}(C_{\text{csm}}, G) + L_{\text{CE}}^{i}(C_{\text{cim}}, G) \tag{1}$$

where C_{csm} and C_{cim} are the two concealed object mappings.

2.2 *Camouflaged Object Detection Using ERRNet*

In [13], the authors proposed an edge-based reversible re-calibration network (ERRNet) for concealed object detection. The network uses a ResNet-50 backbone to extract features from the image. The traits are subsequently processed through an atrous spatial pyramid pooling (ASPP) module to acquire the global prior. It is composed of selective edge aggregation (SEA) modules and reversible re calibration units (RRUs) that work alongside neighbour prior, global prior, edge prior, and semantic prior (NEGS priors) at both the low and high layers. The RRU adjusts the coarse prediction, while the SEA aggregation approach mines the edge prior and prevents weak edges from fading. Furthermore, they performed extensive ablation tests illustrating that ERRNet might be a universal and robust solution to the COD challenge. However, under present conditions, the camouflage object identification job still encounters numerous difficult scenarios.

The loss function used to train ERRNet is shown in Equation (2)

$$L_{\text{total}} = L^{e} + L \sum_{i\varepsilon(3,4,5,g)} L_{i}^{c} \tag{2}$$

where $L^{c} = L_{\text{BCE}}^{W} + L_{\text{IoU}}^{W}$ and $L^{e} = L_{\text{BCE}}^{W}$.

2.3 *Camouflaged Object Detection Using DGNet*

In [14], DGNet, a unique deep gradient learning framework for effectively identifying concealed objects, was suggested by the authors in [15]. They recommended splitting the job into two branches to extract the camouflaged features: a context encoder and a texture encoder. To simultaneously learn features from these two branches, they developed a novel plug-and-play module termed gradient-induced transition (GIT), which functions as a soft grouping module.learn characteristics from these two branches.

In comparison with the 20 SOTA opponents, our simple and adaptable architecture demonstrated high generalization capabilities on three hard datasets. Furthermore, the efficient DGNet-S accomplished an exceptional performance-efficiency trade-off.

The overall optimisation goal is stated as the loss function in Equation (3):

$$L = L_{C}(P^{C}, Z^{C}) + L_{G}(P^{G}, Z^{G}) \tag{3}$$

where L_C is segmentation and L_G is object gradient loss function. L_C is expressed as $L_C = L_{\text{IoU}}^{w} + L_{\text{BCE}}^{w}$, where L_{IoU}^{w} and L_{BCE}^{w} represent the weighted intersection-over-union (IoU) loss and the weighted binary cross-entropy (BCE) loss, respectively [16].

2.4 *Camouflaged Object Detection Using SINet-V2*

In [17], the author proposes a framework consisting of two modules: the search module (SM) and the identification module (IM) just as in the previous version of SINet.

SINet employs a densely linked technique to save more information from multiple levels and a RF component to expand the receptive area. When compared to existing cutting-edge approaches in object recognition from a camouflage standpoint, the authors discovered that their methodology was competitive and produced visually favourable results. The suggested architecture holds potential in military, security, and wildlife conservation applications where spotting concealed objects is critical.

The loss function used for training SINet-V2 is shown in Eq. (4) as follows:

$$L = \lambda_{\text{edge}} L_{\text{edge}} + \lambda_{\text{CVAE}} L_{\text{CVAE}} + \lambda_{\text{ref}} L_{\text{ref}}, \tag{4}$$

where λ denotes the weight for each loss.

2.5 *Camouflaged Object Detection Using HITNet*

In [18], a high-resolution iterative feedback network (HITNet) for extracting informative and high-resolution illustrations for solving the degrading problem of segmentation characteristics on the COD challenge is being introduced. In an iterative feedback loop, HITNet may adaptively update low-resolution features with high-resolution information. More crucially, on four hard datasets, this technique produces substantial performance increases and considerably outperforms 29 cutting-edge models. Finally, they provide the cross-domain learning technique, which they utilize to build an application that turns a visible item to the concealed thing, possibly increasing the variety of the COD dataset.

HITNet consists of three different modules:

2.5.1 Transformer-Based Feature Extraction

The transformer-based feature extraction module uses the pyramid vision transformer (PVT) to retrieve multiple-scale characteristics and handle feature maps with substantially greater resolutions while using less memory. The progressive shrinking technique is used to divide the input feature into patches of varying sizes, and the spatial reduction attention mechanism is used to lower the computing cost.

The progressive shrinking strategy of PVT involves evenly splitting the input feature into patches of size Pi for the i-th stage. After linear projection, the height and width of the patches are reduced to 1/Pi of the input features. This strategy helps to reduce memory usage. By lowering the spatial dimension of the key K and query Q prior to the attentive action, the PVT's **spatial reduction attention mechanism** lowers computing costs even further. At the i-th stage, given an input sequence x, a reduction ratio of spatial dimension Ri is employed to rearrange the input x to the size $H_i W_i / R_i 2(R_i 2C_i)$. The dimension of the input sequence (i.e. $R_i 2C_i$) is then further reduced using a linear projection. The computational cost of spatial reduction attention is reduced to only $1/R_i 2$ of the normal multi-head attention layer using this method.

2.5.2 Multi-Resolution Iterative Refinement

The multi-resolution iterative refinement module uses a basic block BA(·) and an iterative feedback mechanism to improve the multi-scale features retrieved by the transformer backbone. To modify the output features, the iterative feedback mechanism carries feedback features created by the previous iteration into the feedback block FB(·).

2.5.3 Iteration Feature Feedback

The iteration feature feedback module increases segmentation accuracy even further by refining the output features using feedback features generated by the previous iteration. For improved cross-scale data flow, the feedback characteristics are specifically imported into the first and third upper scales.

The fundamental loss employed is $L_C = L_{\mathrm{IoU}}w + L_{\mathrm{BCE}}w$, where $L_{\mathrm{IoU}}w$ represents the weighted intersection-over-union (IoU) loss, and $L_{\mathrm{BCE}}w$ represents the weighted binary cross-entropy (BCE) loss, respectively. As indicated in Eq. (5), a technique is utilized to penalize the outcome of each iteration while forecasting an HR segmentation map.

$$L_{\mathrm{HIF}} = \sum_{in}^{N} (\mathrm{win}) L(Y^{\mathrm{in}}) + L(Y'), \quad (5)$$

where Y^{in} is the result of the *in*-th iteration, Y' is the output of graph-based resolution fusion, *in* is the current iteration number, N is the total iteration number, and w is the weight parameter.

3 Proposed Work

The proposed work will provide a comprehensive and systematic analysis of the cutting-edge models for camouflaged object detection, which can aid researchers and practitioners in selecting the most suitable model for their applications. As shown in Fig.1, we evaluate the performance of five different models: SINet and SINet-V2 are two models that successfully capture spatial information and lessen the impact of vanishing gradients during training by utilizing dilated convolutional layers and residual connections. A model called DGNet makes use of both global and local characteristics, efficiently capturing context information and enhancing object detection precision. The ERRNet model enhances the accuracy of object detection in cluttered and complicated surroundings by using an attention mechanism that selectively focuses on key areas of a picture. Finally, HITNet is a model that makes use of a hierarchical feature extraction strategy, enabling it to efficiently capture multi-scale information and enhance the precision of object detection. Thus, we choose HITNet model to deploy and test on the edge device.

The testing of all the models is performed on a system with configurations of Intel(R) Core(TM) i5-10210U CPU @ 1.60GHz with 8 GB Random Access Memory

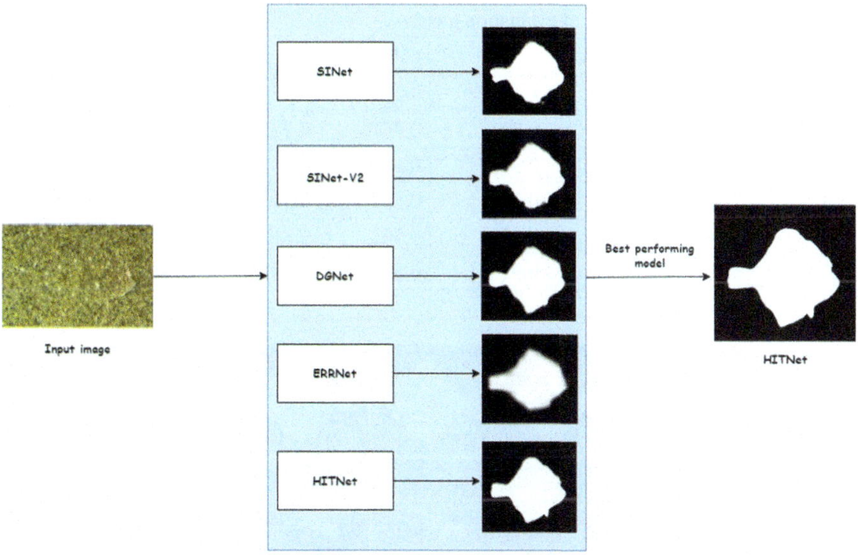

Fig. 1 Overview of testing the models

(RAM) and VS Code on Microsoft Windows 11. We used Android Studio to develop an Android application.

4 Results and Discussion

4.1 Comparison of Different Models

Several studies have been conducted to evaluate the performance of various deep learning models for camouflaged object detection on benchmark datasets such as Chameleon, CAMO, and COD10K, and the results obtained are shown in Table 1 and Fig. 2. Here is a summary of the results obtained using SINet, SINet-V2, DGNet, ERRNet, and HITNet on these datasets:

1. SINet: The original SINet architecture achieved cutting-edge outcomes on the Chameleon dataset. However, its performance was lower on the CAMO dataset and on the COD10k dataset. SINet uses a modified version of the MobileNet architecture as its backbone network, while SINet-v2 uses the EfficientNet architecture. The block design in SINet is based on depth-wise separable convolutions.

2. SINet-V2: The SINet-V2 architecture outperformed the original SINet architecture on all three datasets, i.e. on the Chameleon, CAMO, and COD10K datasets, respectively. SINet-V2 is a more powerful and efficient architecture than SINet,

Table 1 Comparative analysis of evaluation metrics

Baseline model	Chameleon		CAMO		COD10K	
	$S_\alpha \uparrow$	$E_\Phi \uparrow$	$S_\alpha \uparrow$	$E_\Phi \uparrow$	$S_\alpha \uparrow$	$E_\Phi \uparrow$
SINet	0.869	0.891	0.751	0.771	0.771	0.806
SINet-V2	0.888	0.942	0.820	0.882	0.815	0.887
DGNet	0.868	0.905	0.839	**0.915**	0.822	0.911
ERRNet	0.877	0.927	0.761	0.817	0.780	0.867
HITNet	**0.922**	**0.970**	**0.844**	0.902	**0.868**	**0.932**

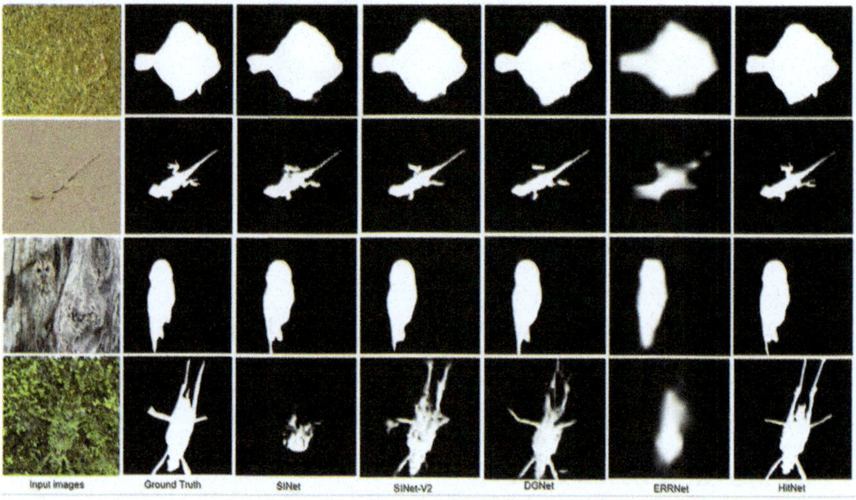

Fig. 2 Comparison of results obtained from different models

due to its use of the more advanced EfficientNet backbone network and the more efficient block design based on ECA (efficient channel attention) modules. Additionally, the inclusion of a spatial attention module further improves the accuracy and efficiency of SINet-v2 compared to SINet.

3. DGNet: The DGNet architecture uses a deep gradient mechanism to enhance the detection of concealed objects. Based on the evaluation metrics used, SINet-V2 performs better than DGNet in camouflaged object detection on Chameleon, CAMO, and COD10k datasets. SINet-V2 uses multi-scale feature learning to capture details at different levels, which helps in detecting camouflaged objects that may be hidden at different scales.
4. ERRNet: The ERRNet architecture uses an edge-enhanced residual block to capture edge information. While ERRNet performs better than some of the baseline models, it is outperformed by HITNet and SINet-V2 on all three datasets, indicating that HITNet and SINet-V2 are more effective in detecting camouflaged objects.

5. HITNet: The HITNet architecture uses a hierarchical feature fusion mechanism and hitmaps to detect concealed objects. The hierarchical fusion mechanism in HITNet refers to the process by which information from different scales of the feature maps is integrated to produce more accurate hitmaps. This is achieved through a series of fusion modules that combine feature maps from adjacent scales in a hierarchical manner. The resulting hitmaps are then used to locate concealed objects in the input image. Based on the results, HITNet consistently outperformed the other models across all three datasets in terms of S-measure and E-measure.

4.2 Analysis of Datasets Used

As shown in Table 2, the following are the datasets used for training and testing the models:

1. **Chameleon Dataset:** The Chameleon dataset was released in 2017 and consists of 1200 images categorized as camouflaged and non-camouflaged objects. The non-camouflaged class has 600 images with clearly visible objects against the background, while the camouflaged class includes 600 images with objects that are camouflaged. The dataset provides pixel-level ground truth annotations for camouflaged objects.

2. **Camo Dataset:** The Camo dataset, released in 2020, contains 1000 high-resolution photos divided into camouflaged and non-camouflaged groups. The non-camouflaged class comprises 500 images with clearly visible objects, while the camouflaged class has 500 images with camouflaged objects. Bounding box annotations are available for the camouflaged objects (Fig. 3).

3. **COD10K Dataset:** The COD10K dataset, released in 2021, is a benchmark dataset for camouflaged object detection. It consists of 10,000 images divided into camouflaged and non-camouflaged classes. The non-camouflaged class includes 5000 images with clearly visible objects, while the camouflaged class has 5000 images with camouflaged objects. The dataset provides pixel-level ground truth annotations and additional comments describing the type of camouflage used. It also incorporates various environmental variables such as lighting and weather.

Table 2 Analysis of datasets

Datasets	Number of images
Chameleon	1200
CAMO	1000
COD10K	10,000

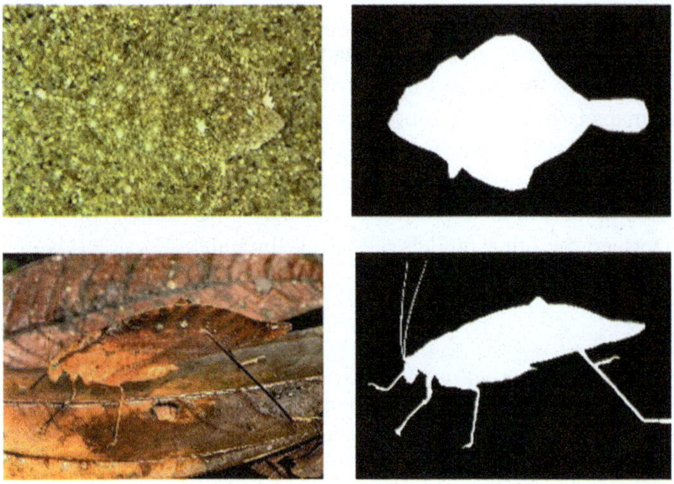

Fig. 3 Sample input images of camouflaged objects

4.3 Evaluation Metrics

In camouflaged object detection, evaluation metrics [19] are used to assess the performance of models. Two commonly used metrics for this task are the S-measure and the E-measure.

The **S-measure** (Structure measure) is a metric used to compare the structural similarity of two images. It considers factors such as brightness, contrast, and spatial frequencies to capture the spatial relationships and texture details between image regions. The S-measure has a score range of 0–1, with a higher score indicating a greater structural similarity between the images. The *S*-measure can be employed as a single metric to evaluate picture quality or as part of a composite metric [20].

The **E-measure** (Enhanced alignment measure) is a metric used to evaluate the accuracy of image segmentation results. It combines object-level similarity and region-level alignment measures. The object-level similarity measure assesses how well the segmented regions match the ground truth objects, while the region-level alignment measure analyses the alignment between the segmented and ground truth regions. The E-measure is a single value ranging from 0 to 1, with a higher score indicating higher segmentation accuracy.

These metrics provide quantitative assessments of model performance in camouflaged object detection tasks.

4.4 Quantization

In order to reduce memory usage and speed up computation without requiring a representative dataset for calibration, dynamic range quantization is a suggested first step. When converting from floating point to integer, this kind of quantization simply quantifies the weights statically, with an accuracy of 8 bits. To further reduce latency during inference, "dynamic range" [21] operators perform calculations using 8-bit weights and activations after dynamically quantizing activations based on their range to 8 bits. With this optimisation, inference latencies are almost entirely fixed-point. Dynamic range operations are faster than full fixed-point computations, but the outputs are still stored in floating point.

We used dynamic range quantization technique to reduce the size of the network while maintaining network accuracy. As a result, deployment is possible to hardware with lower memory footprints, freeing up more RAM for control logic and additional algorithms.

The model used is HITNet, having an accuracy of 97%. Quantization resulted in the reduction of our network size from 101 MB to 25.5 MB with the accuracy of 94%.

4.5 Deploying the Model on Edge Device

Deploying a quantized model on an edge device can significantly reduce the model size and memory usage without compromising on the performance [22]. This makes it feasible to deploy the model on edge devices with limited resources such as memory, computation, and power. Once the model is converted to the appropriate format, it can be deployed on the edge device. This typically involves writing code to load the model onto the device and running it to make predictions [15]. The results generated after deploying the quantized model on edge are shown in Fig. 4.

According to the results obtained, HITNet outperformed all other models, achieving a high accuracy and speed. So, we chose to deploy HITNet on an edge device as it can have several benefits.

Android Studio is used for deploying HITNet on edge devices. It has built-in support for integrating machine learning models into Android applications. The Android platform has a large user base, making it a popular choice for deploying machine learning models on edge devices.

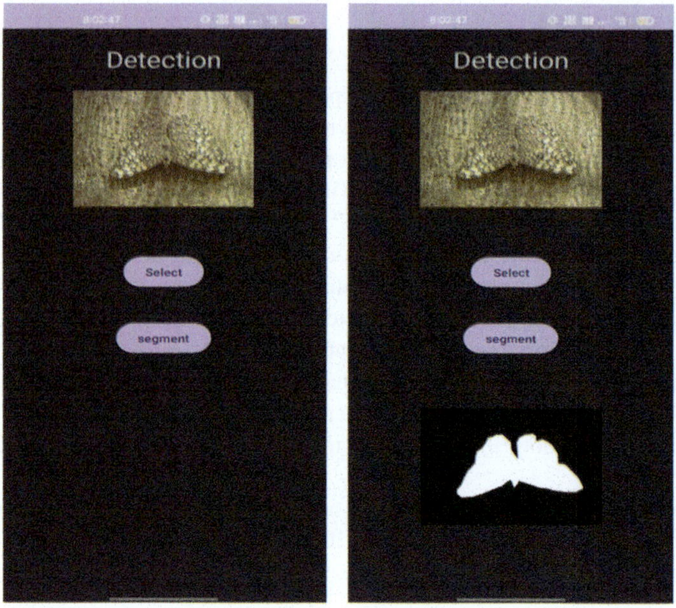

Fig. 4 Generated results after deploying dynamic range quantized model on edge device

5 Future Scope

Camouflaged object detection is an active topic of research with potential future possibilities. Improving the performance of existing models on complicated scenarios and datasets is an area of focus. This includes techniques for feature extraction, categorization, and object recognition. Additionally, there is a need for datasets that depict real-world circumstances with complicated backgrounds and lighting conditions. Detecting camouflaged objects in the spectral or thermal domains is another area of research. This involves exploring multi-spectral or hyperspectral imaging techniques and developing algorithms for detecting objects based on their thermal signatures. Machine learning techniques can be applied in various fields such as surveillance, wildlife preservation, and military applications. Camouflaged object recognition techniques can help identify and track wildlife in natural settings or improve the detection of suspicious objects in public spaces.

6 Conclusion

According to the results, HITNet consistently outperformed the other models in terms of *S*-measure and *E*-measure across all three datasets. This could be due to the use of hitmaps to highlight the hidden items in the image, resulting in better recognition and segmentation. While other models, such as SINet, SINet-V2, DGNet, and ERRNet, have also demonstrated promising results, they may have limits when dealing with complicated or deeply camouflaged objects. SINet-V2 and ERRNet, for example, may struggle with objects that have low contrast with their background, but DGNet may struggle with complicated or irregularly shaped objects. SINet, DGNet, and SINet-V2 performed well as well, while HITNet constantly outperformed them. Among the models tested, ERRNet performed the worst.

It's important to note that the performance of these models might differ depending on the application and dataset used. Furthermore, the performance of these models could be improved with additional optimisation and fine-tuning.

References

1. Kamran M, Rehman SU, Meraj T, Alnowibet KA, Rauf HT (2022) Camouflage object segmentation using an optimized deep-learning approach. Mathematics 10(22):4219
2. Nemade SB, Sonavane SP (2019) Image segmentation using convolutional neural network for image annotation. In: 2019 International Conference on Communication and Electronics Systems (ICCES), pp 838–843. https://doi.org/10.1109/ICCES45898.2019.9002121
3. Long J, Shelhamer E, Darrell T (2015) Fully convolutional networks for semantic segmentation
4. Borji A, Cheng M-M, Hou Q, Jiang H, Li J (2019) Salient object detection: a survey. Comput Vis Media 5(2):117–150. https://doi.org/10.1007/s41095-019-0149-9
5. Chikkamath SSN (2021) Melody generation using lstm and bi-lstm network. In: 2021 International Conference on Computational Intelligence and Computing Applications (ICCICA), pp 1–6. https://doi.org/10.1109/ICCICA52458.2021.9697286
6. Cheng M-M, Mitra NJ, Huang X, Torr PHS, Hu S-M (2015) Global contrast based salient region detection. IEEE Trans Pattern Anal Mach Intell 37(3):569–582. https://doi.org/10.1109/TPAMI.2014.2345401
7. Zhao Z-Q, Zheng P, Xu S-t, Wu X (2019) Object detection with deep learning: a review
8. Fan D-P, Ji G-P, Sun G, Cheng M-M, Shen J, Shao L (2020) Camouflaged object detection. In: IEEE Conference on Computer Vision and Pattern Recognition (CVPR)
9. Sun Y, Chen G, Zhou T, Zhang Y, Liu N (2021) Context-aware cross-level fusion network for camouflaged object detection. In: Zhou Z-H (ed.) Proceedings of the Thirtieth International Joint Conference on Artificial Intelligence, IJCAI-21, pp 1025–1031. International Joint Conferences on Artificial Intelligence Organization. https://doi.org/10.24963/ijcai.2021/142
10. Chu H-K, Hsu W-H, Mitra NJ, Cohen-Or D, Wong T-T, Lee T-Y (2010) Camouflage images. ACM Trans Graph 29(3):1–8
11. Cordts M, Omran M, Ramos S, Rehfeld T, Enzweiler M, Benenson R, Franke U, Roth S, Schiele B (2016) The cityscapes dataset for semantic urban scene understanding. In: 2016 IEEE Conference on Computer Vision and Pattern Recognition (CVPR), pp. 3213–3223. https://doi.org/10.1109/CVPR.2016.350
12. Fan D-P, Ji G-P, Cheng M-M, Shao L (2022) Concealed object detection. IEEE Trans Pattern Anal Mach Intell 44(10):6024–6042. https://doi.org/10.1109/TPAMI.2021.3085766
13. Ji G-P, Zhu L, Zhuge M, Fu K (2021) Fast camouflaged object detection via edge-based reversible re-calibration network. Pattern Recogn 123:108414
14. Ji G-P, Fan D-P, Chou Y-C, Dai D, Liniger A, Gool LV (2022) Deep gradient learning for efficient camouflaged object detection. Mach Intell Res 20(1):92–108
15. Kulkarni U, Meena SM, Gurlahosur SV, Bhogar G (2021) Quantization friendly mobilenet (qf-mobilenet) architecture for vision based applications on embedded platforms. Neural Netw 136:28–39. https://doi.org/10.1016/j.neunet.2020.12.022

16. Le T-N, Nguyen TV, Nie Z, Tran M-T, Sugimoto A (2019) Anabranch network for camouflaged object segmentation. Comput Vis Image Underst 184:45–56. https://doi.org/10.1016/j.cviu.2019.04.006
17. Fan D-P, Ji G-P, Cheng M-M, Shao L (2021) Concealed object detection
18. Hu X, Wang S, Qin X, Dai H, Ren W, Tai Y, Wang C, Shao L (2023) High-resolution iterative feedback network for camouflaged object detection
19. Padilla R, Netto SL, Silva EAB (2020) A survey on performance metrics for object-detection algorithms. In: 2020 International Conference on Systems, Signals and Image Processing (IWSSIP), pp 237–242. https://doi.org/10.1109/IWSSIP48289.2020. 9145130
20. Fan D-P, Cheng M-M, Liu Y, Li T, Borji A (2017) Structure-measure: a new way to evaluate foreground maps
21. Kulkarni N, Singh N, Joshi Y, Hasabi N, Meena SM, Kulkarni U, Gurlaho- sur SV (2022) Hybrid optimization for DNN model compression and inference acceleration. In: 2022 2nd International Conference on Intelligent Technologies (CONIT), pp 1–8. https://doi.org/10.1109/CONIT55038.2022.9847977
22. Kulkarni U, Meena SM, Gurlahosur SV, Benagi P, Kashyap A, Ansari A, Kar-nam (2021) Modern approaches in machine learning and cognitive science: a walkthrough: latest trends in AI. In: Gunjan VK, Zurada JM (eds) AI model compression for edge devices using optimization techniques. Springer, Cham, pp 227–240. https://doi.org/10.1007/978-3-030-68291-0_17

3D Avatar Reconstruction Using Multi-level Pixel-Aligned Implicit Function

Shreedhar I. Muttagi, Vaishnavi Patil, Pooja P. Babar, Ritvik Chunamari, Uday Kulkarni, Satish Chikkamath, and S. M. Meena

Abstract In recent years, there has been significant interest in 3D avatar reconstruction from images, driven by its applications in gaming, entertainment, and augmented reality. This paper introduces a novel approach that utilizes the Fast Super-Resolution Convolutional Neural Network (FSRCNN) and multi-level pixel-aligned implicit function (ML-PIFu) framework to reconstruct high-fidelity 3D avatars from a single input image. Our method integrates the FSRCNN with a two-module ML-PIFu pipeline, combining global geometric information and local fine details. The FSRCNN enhances the input image quality before it is processed by the ML-PIFu framework. The coarse-level module of ML-PIFu improves the image quality, capturing the overall structure and shape of the human form, while the fine-level module incorporates intricate details for enhanced output. Through the use of neural networks and an end-to-end training strategy, our approach achieves accurate and precise reconstructions without requiring multiple images or complex camera setups. The potential applications of our approach are extensive, including the creation of personalized avatars for telepresence, Virtual Reality (VR), Augmented Reality (AR), anthropometry studies, and virtual try-on experiences. By leveraging our approach, these applications can benefit from enhanced accuracy, realism, and versatility, thereby opening up new possibilities in the realm of digital human representation and interaction.

Keywords 3D avatar reconstruction · Multi-level pixel-aligned implicit function (ML-PIFu) · Neural networks · End-to-end training · Single image reconstruction · Coarse-level module · Fine-level module · Augmented reality (AR) · Virtual reality (VR)

S. I. Muttagi · V. Patil · P. P. Babar (✉) · R. Chunamari · U. Kulkarni · S. Chikkamath · S. M. Meena
KLE Technological University, Hubli, Karnataka, India
e-mail: babarpooja2002@gmail.com

U. Kulkarni
e-mail: uday_kulkarni@kletech.ac.in

S. Chikkamath
e-mail: chikkamath@kletech.ac.in

V. K. Gunjan and J. M. Zurada (eds.), *Proceedings of 4th International Conference on Recent Trends in Machine Learning, IoT, Smart Cities and Applications*, Lecture Notes in Networks and Systems 873, https://doi.org/10.1007/978-981-99-9442-7_20

1 Introduction

The detailed digitization of the human body has significant implications in various fields, including medical imaging and virtual reality. However, its widespread use has been limited due to the high cost and specialized requirements of professional capture systems, such as multiple cameras and controlled lighting conditions. Fortunately, there has been an increasing interest in leveraging advanced deep-learning models to overcome these limitations and enable reconstructions from just a single image. While these approaches have demonstrated promising results, there is still a need to enhance their performance further. Our objective is to achieve an exceptional level of fidelity in the 3D reconstruction of clothed individuals using just a single image, surpassing the performance of existing techniques and specialized capture devices. We strive to capture intricate details such as finger positions, facial expressions, and clothing folds with utmost accuracy, ensuring a resolution that allows for precise restoration. However, current methods have not fully harnessed the potential of high-resolution imaging and struggle to handle low-quality photographs typically taken with standard sensors on mobile devices. As a result, we aim to develop innovative approaches that take advantage of modern imaging technology while remaining effective even when dealing with sub-optimal image quality. By doing so, we aim to revolutionize the field of 3D reconstruction from a single image.

Prior methods often use down-sampled images due to memory limitations, relying on holistic reasoning to connect a human's 2D appearance to its 3D shape. Local image patches provide valuable details for precise 3D reconstruction, but their usage in high-resolution inputs is limited due to graphics technology constraints. Our approach aims to overcome these limitations, incorporating local patches for comprehensive 3D representation, despite memory constraints and graphics limitations (Fig. 1).

Local image patches are valuable for capturing crucial details in 3D reconstruction, yet their utilization in high-resolution inputs remains limited due to current graphics technology limitations. However, advancements in this field can potentially overcome these constraints, paving the way for comprehensive integration of local image patches in future high-resolution implementations. By harnessing the power of advanced graphics capabilities, the potential for leveraging local image patches in their entirety can be unlocked, enriching the accuracy and fidelity of 3D reconstructions.

The constraints that hinder the use of local image patches can be resolved in two ways. In the first category, the issue is solved by enhancing low-fidelity surfaces with high-frequency details in a coarse-to-fine fashion. With this method, a low-resolution image is used to first get a crude form. Subsequently, through post-processing methods like Shape From Shading or neural network-based composition [15], precise features in the form of surface normal or displacements [8] are added. The second category generates convincing details using high-fidelity human models like SCAPE [3]. While both methods produce reconstructions that appear to be detailed, they frequently fall short of accurately recreating the true details found

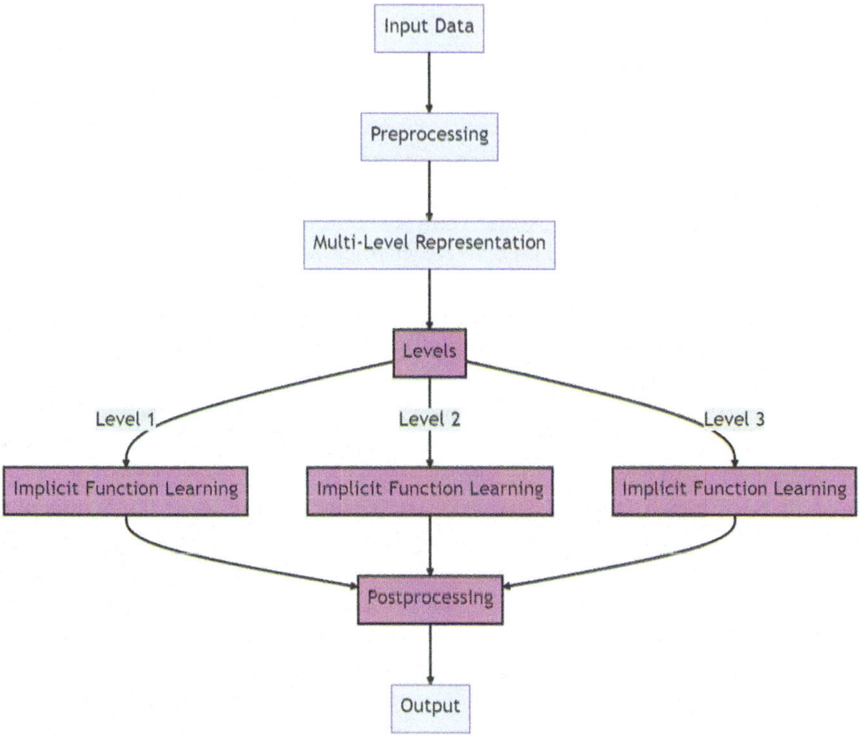

Fig. 1 Using a high-resolution image of an individual, we are able to reconstruct remarkably detailed 3D representations of clothed individuals

in the raw photos. By incorporating state-of-the-art algorithms and deep-learning models, we strive to overcome the limitations of existing methods and achieve accurate and detailed 3D reconstructions. The goal is to enable the creation of realistic and expressive 3D avatars from minimal input, revolutionizing applications such as virtual reality, gaming, and entertainment. In this paper, we present the following sections. Section 2 offers an overview of the background and context of our work on existing 3D human avatar digitization models. In Sect. 4, we examine the limitations of these models and propose an architecture that effectively overcomes these limitations. Section 4 presents the results of our proposed architecture, comparing it to other existing models. Finally, Sect. 5 summarizes the key findings and concludes our proposed architecture.

2 Related Works

1. **3D Human Avatar Digitization from a Single Image**. Research on 3D human reconstruction [16] encompasses various areas, including estimating body pose, single-image-based 3D human body reconstruction, video-based human body reconstruction, and inferring occluded or invisible texture color. The focus is on creating a 3D Human Avatar from a single image. A single image is used to reconstruct the human body. To reconstruct the final geometry, the proposed method utilizes 2D non-rigid registration. Infer Generative Adversarial Networks (GAN) [7] is used to extract texture based on foreground texture. In this method, a single image is used to reconstruct the entire geometry and intricate texture of a human body, and an animated model is generated based on the weights and joints of the parametric motion model. To demonstrate the capabilities of the reconstruction pipeline, a mobile application has been developed, showcasing the process and hosted on a server.
2. **Self-Recon: Self Reconstruction Your Digital Avatar from the Monocular Video** In this paper [4], the authors highlight the importance and challenges of reconstructing clothed bodies, particularly in industries such as film and gaming. Traditional methods for achieving high-quality human reconstructions often involve pre-captured templates, multi-camera setups, controlled environments, and manual labor by skilled artists. However, these requirements are impractical for general consumers who seek personalized avatars for applications like telepresence, AR/VR, anthropometry, and virtual try-on. This necessitates the development of direct high-fidelity digital avatar reconstruction methods from monocular videos to meet the growing demand. Additionally, a non-rigid ray casting algorithm is introduced, combining explicit representation for differentiable intersections with the deformed implicit surface. As a result, Self-Recon achieves high-fidelity reconstructions of detailed clothed body shapes without relying on pre-computed templates. The authors demonstrate the potential applications of Self-Recon through the generation of high-quality avatars based on tracking results.
3. **Multi-Level Pixel-Aligned Implicit Function for High-Resolution 3D Human Digitization** The purpose of this paper [8] is to construct high-resolution 3D reconstructions of clothed individuals that utilize the pixel-aligned implicit function (PIFu) [14] to facilitate the generation of 3D reconstructions of clothed individuals. This technique involves populating a dense 3D volume to determine the presence of human body points in 3D space. The performance of recent 3D human reconstruction methods was compared using the People Snapshot dataset, including multi-scale voxel representation (Deep Human), pixel-aligned implicit function (PIFu), and a model-based approach with texture mapping using displacements and surface normal (Tex2shape) [12]. The results suggest that Tex2shape and Deep Human methods show limited refinement effects, primarily

because their basic forms have limited representational power. Voxel representation introduces spatial resolution constraints, while model-based approaches face challenges in handling varying topologies and significant deformations.

3 Proposed Methodology

In order to improve the accuracy and resolution of 3D reconstructions from 2D images, we propose an enhancement to the pixel-aligned implicit function (PIFu) [8] framework. It utilizes 512 × 512 resolution images to generate low-resolution feature embeddings of size 128 × 128, which are then used to produce higher-resolution outputs. Two additional modules are introduced in this architecture to augment the PIFu framework. A higher-resolution image (1024 × 1024) is taken as input and encoded into 512 × 512 features. This module focuses on capturing fine details and enhancing the resolution of the reconstructed 3D object. The second module takes the high-resolution feature embedding from the first module, along with the 3D embedding from the PIFu framework, and utilizes them to estimate an occupancy probability field. The occupancy probability field represents the likelihood that a given point in 3D space is occupied by the object being reconstructed. By incorporating these additional modules, the proposed architecture extends the capabilities of the PIFu framework and improves the quality and resolution of 3D reconstructions from 2D images.

Overall the proposed architecture is designed to improve the accuracy and resolution of 3d reconstructions from 2d images by leveraging higher-resolution inputs and fine details. This approach has several potential applications in fields such as computer graphics virtual reality and robotics where accurate and detailed 3d reconstructions are essential for creating realistic and immersive environments. The proposed architecture may also have practical applications in fields such as medicine and architecture where detailed 3d models are essential for planning and analysis. In our proposed work we have improved the existing state of art PIFu framework. To enhance the quality of the reconstructed avatars, we integrate the PIFu framework with some advanced image processing techniques present in the open-source computer vision (OpenCV [1] (cv2)) library. These advanced image processing techniques preprocess the input images and improve the quality of the reconstructed avatars. Super-Resolution model, Fast Super-Resolution Convolutional Neural Network (FSRCNN) [11] has been used to improve the quality of the input image. By recovering a high-resolution (HR) image from a given low-resolution (LR) one, single image super-resolution (SR) attempts to improve image quality.

3.1 FSRCNN

FSRCNN: There are five fundamental steps in the development of the Fast Super-Resolution Convolutional Neural Networks (FSRCNN) [11]: removing the feature, shrinking the feature, mapping the feature, expanding the feature, and deconvolution the feature. As part of the network's architecture, convolutional and deconvolutional layers are used. A Fast Super-Resolution Convolutional Neural Network (FSRCNN) is a development of the Super-Resolution Convolutional Neural Network (SRCNN). It eliminates the need for interpolation by directly learning the transformation between the original low-resolution image and the high-resolution image during the mapping process of FSRCNN. In addition, the output is further refined by adding a deconvolution layer. Secondly, FSRCNN incorporates a shrink-expand approach, where the input feature dimension is initially reduced (shrunk) before the mapping step and then expanded again afterward. This technique captures and preserves essential information throughout the network. Thirdly, FSRCNN employs multiple mapping layers with smaller filter sizes. This design choice enables the network to learn more detailed and localized features. The output of FSRCNN is passed through the MobileNet model [10, 9], which is responsible for keypoint detection and annotation for human pose estimation. These modifications enhance the performance and efficiency of the FSRCNN model, allowing it to achieve fast and accurate super-resolution while maintaining significant image details.

As depicted in Fig. 2, the network utilizes preprocessed images as inputs for its image encoders. Two encoders are employed: a low-resolution and a high-resolution encoder. The coarse image encoder generates feature maps sized at $128 \times 128 \times 256$, which are then passed through a multi-layer perceptron (MLP). The coarse-level MLP consists of six layers with neuron counts of (257, 1024, 512, 256, 128, 1). To facilitate information flow, skip connections are implemented at the third, fourth, and fifth layers. These connections improve feature preservation and enhance the overall performance of the network.

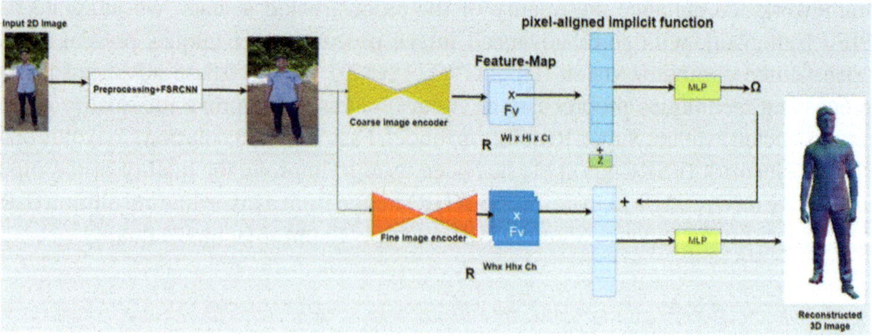

Fig. 2 Multi-level pixel implicit function architecture

The MLP layer produces 3D embedding as the final output. The pixel-based image encoder generates feature maps of 512 × 512 × 216. The coarse-level encoder uses the feature maps along with the 3D embedding to create the pixel-aligned implicit function, which is then used to create the fine-level encoder. The output generated by the pixel-aligned implicit function is then passed through the fine-level encoder MLP. This process ultimately produces a 3D avatar of the human depicted in the input image as the final output.

Multi-Level Pixel-Aligned Implicit Function The multi-level pixel-aligned implicit function [8] takes a 1024 × 1024 image as input and processes it through two levels of encoders.

3.2 The Coarse-Level Module

The multi-level approach in 3D human digitization is designed to follow the structure of the PIFu method described in previous research. It takes a down-sampled image of size 512 × 512 as input and generates backbone image features with a resolution of 128 × 128. The objective of this module is to incorporate global geometric information and establish a foundation for the subsequent fine-level module to add finer details to the 3D human digitization process. By processing the input image at a lower resolution, the module focuses on capturing largerscale geometric features that are crucial for accurately representing the human form in 3D space.

3.3 The Fine-Level Modules Fine-Level Module

The proposed approach, takes the original 1024 × 1024 resolution image as input and makes a backbone image feature with a resolution of 512 × 512. This resolution is four times higher than the resolution used in the previous PIFu research project [8]. The fine-level module incorporates more subtle details into the 3D human digitization process that were not captured by the coarse-level module. By using 3D embedded features rather than absolute depth values, the fine-level module avoids relying on absolute depth values. This allows the module to focus on refining the geometric features provided by the coarse level and incorporating finer details into the final output. By utilizing these 3D embedding features, our approach achieves enhanced fidelity and an increased level of detail in the reconstructed output. In our approach, we use a coarse-level module similar to that used in PIFu [16], but we modified it to enhance its performance. As a result of the modifications, the generated results should be more accurate and higher quality. The coarse-level module has been fine-tuned to better suit our approach by incorporating these adjustments, resulting in more precise and detailed results.

3.4 Multi-layer Perceptron (MLP)

It is a type of artificial neural network with many layers of interconnected neurons. Each neuron in an MLP [6] takes inputs, applies weights to those inputs, and then passes the weighted sum through an activation function to produce an output. In 3D avatar reconstruction, the MLP can estimate the avatar model parameters based on the extracted features. These parameters could include shape parameters (e.g., head size, body proportions), appearance parameters (e.g., skin color, hairstyle), or pose parameters (e.g., joint angles).

Predicting the exact shape and structure of the human back solely from images poses a challenge since direct observation is not possible. The reconstruction of the trunk using the MLP network adds another layer of difficulty to the task. Inherently ambiguous and multimodal problems result in back reconstructions that lack features. The occupancy loss function, designed to handle uncertainty, favors average reconstructions, further diminishing detail. The final layers of the MLP network face the challenge of learning a complex prediction function, adding to the overall complexity of the reconstruction process.

4 Experimental Results

4.1 Dataset Description

In this study, we obtained high-quality 3D geometry and images from the Render People dataset. In total, 500 photogrammetry scans were analyzed, divided into 450 training scans and 50 test scans. Each subject's renderings were generated from various viewpoints along the yaw axis while maintaining a fixed elevation of 0 degrees. To render the meshes, we employed pre-computed radiance transfer, utilizing 163s-order spherical harmonics derived from HDRI Haven1. This technique allowed us to accurately capture and simulate the lighting conditions, leading to improved realism and visual quality in the rendered images. The combination of the Render People dataset and the radiance transfer approach provided a strong foundation for our research, enabling us to effectively evaluate and enhance our proposed methods. This integration played a crucial role in facilitating our research objectives and ensuring the validity of our findings. As discussed earlier our proposed architecture is built upon the pixel-aligned implicit function (PIFu) framework [8, 14]. The loss values obtained after comparing our model with the baseline model are shown in 4.1. As compared to the baseline model, our model gives a less mean square error, a higher peak signal-to-noise ratio (PSNR), a lower mean square error, and a lower Structural Similarity Index Measure (SSIM).

In order to determine the similarity of two images, the Structural Similarity Index (SSIM) is used. It extracts three key features: luminance, contrast, and structure. The comparison between the two images is conducted based on these three features, enabling a comprehensive assessment of their similarity. The SSIM metric considers not only pixel values but also the structural information present in the images, providing a more robust evaluation of their resemblance.

SSIM The Structural Similarity Index (SSIM) is a metric utilized to measure the similarity between two provided images. It extracts three key features: luminance, contrast, and structure. The comparison between the two images is conducted based on these three features, enabling a comprehensive assessment of their similarity. The Structural Similarity Index (SSIM) [14] is calculated as follows:

The Structural Similarity Index (SSIM) is calculated as follows:

$$\mathrm{SSIM}(a, b) = \frac{(2\mu_a\mu_b + Z_1)(2\sigma_{ab} + Z_2)}{(\mu_a^2 + \mu_b^2 + Z_1)(\sigma_a^2 + \sigma_b^2 + Z_2)} \tag{1}$$

μ_a and μ_b represent the mean intensities of image patches a and b, respectively. σ^2 and σ^2 correspond to the variances of the intensities in image patches a and b, respectively. σ_{ab} indicates the covariance of intensities between the image patches a and b. Z_1 and Z_2 are constants introduced to stabilize the division and prevent division by zero. Typically, they are defined as $Z_1 = (k_1L)^2$ and $Z_2 = (k_2L)^2$, where L represents the dynamic range of the pixel values (e.g., $L = 255$ for 8-bit images). The constants k_1 and k_2 are small values, often set as 0.01 and 0.03, respectively.

MSE The Mean Square Error (MSE) is a metric used to quantify the cumulative squared error between a compressed image and its original counterpart. It provides a numerical measure of the overall discrepancy between the two images, with a lower MSE value indicating a higher level of fidelity in the compression process.

$$\mathrm{loss}_{\mathrm{MSE}} = \frac{1}{n}\sum_{i=1}^{n}(r_i - \hat{r}_i)^2 \tag{2}$$

PSNR An image's Peak Signal-to-Noise Ratio (PSNR) [2] measures how well it compares with its original image after compression or reconstruction. It is typically expressed in decibels (dB), and a higher PSNR value indicates better image quality. The PSNR formula is as follows:

$$\mathrm{PSNR} = 10\log_{10}\frac{R^2}{\mathrm{MSE}} \tag{3}$$

The comparison Table 1 reveals that our Improved PIfu model outperforms the original PIfu model in three crucial metrics: MSE, PSNR, and SSIM. The substantially lower MSE of 37.9 signifies its superior accuracy in reconstructing and preserving image details. Additionally, the higher PSNR of 11.244 dB and superior SSIM of 0.58 demonstrate the improved signal-to-noise ratio, visual quality, and maintenance of structural information compared to PIfu's performance (MSE:

Table 1 Comparison table

	PIfu	Improved PIfu (our model)
MSE/s	108.81	37.9
PSNR/ms	9.594	11.244
SSIM/ms	0.156	0.58

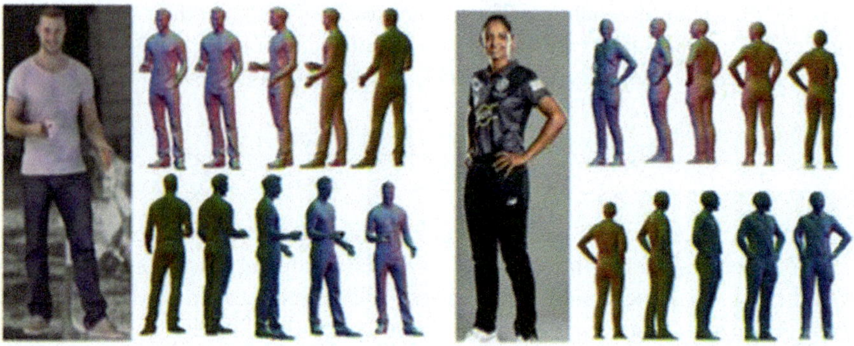

Fig. 3 3D representation of a man with realistic anatomical proportions, facial features, and customizable clothing

108.81, PSNR: 9.594 dB, SSIM: 0.156). These results confirm the effectiveness of our model for applications requiring high-quality images (Fig. 3).

5 Conclusion

This research paper introduces a groundbreaking approach for 3D avatar reconstruction called the multi-level pixel-aligned implicit function (ML-PIFU) framework. The method effectively addresses the challenges of reconstructing high-fidelity 3D avatars from a single input image by employing a two-module pipeline that integrates global geometric information and local fine details. Extensive experiments validate the effectiveness and innovation of the approach, showcasing its superiority over existing state-of-the-art techniques. The comprehensive nature of the experiments solidifies the credibility and reliability of the findings, establishing ML-PIFU as a cutting-edge solution in the field. In summary, the method presents an accurate and precise approach for 3D avatar reconstruction.

References

1. Abram ICD (2012) A brief introduction to opencv. https://doi.org/10.48550/arXiv.2107.03700
2. Alain Horé MOIVRE, Université de Sherbrooke, S.: Image quality metrics: Psnr vs. ssim (10 2010). https://doi.org/10.1109/ICPR.2010.579
3. Anguelov D, Srinivasan P, Koller D, Thrun S, Rodgers J, Davis J (2005) Scape: shape completion and animation of people, pp. 408–416
4. Jiang B, Hong Y, Bao H, Zhang J (2022) Selfrecon: self reconstruction your digital avatar from monocular video. https://doi.org/10.48550/arXiv.2201.12792
5. Dong C, Loy CC, Tang X (2016) Accelerating the super-resolution convolutional neural network. https://doi.org/10.48550/arXiv.1608.00367
6. Gurpreet Singh MS (2017) Mlp: multi-layer perceptron (mlp) neural network technique for offline handwritten gurmukhi character recognition. https://doi.org/10.1109/ICCIC.2014.7238334
7. Goodfellow I, Pouget-Abadie J, Mirza M, Xu B (2004) Gan: generative adversarial networks. https://doi.org/10.48550/arXiv.1406.2661
8. Saito S, Simon T, Saragih J, Joo H (2020) Pifuhd: Multi-level pixel-aligned implicit function for high-resolution 3d human digitization. https://doi.org/10.1109/CVPR42600.2020.00016
9. Kulkarni U, Meena SM, Gurlahosur SV, Bhogar G (2020) Quantization friendly mobilenet (qf-mobilenet) architecture for vision based applications on embedded platforms. https://doi.org/10.1016/j.neunet.2020.12.022
10. Michele A, Colin V, Santika DD (2019) Mobilenet convolutional neural networks and support vector machines for palmprint recognition. Procedia Comput Sci 157:110–117
11. Passarella LS, Mahajan S, Pal A, Norman MR (2022) Reconstructing high resolution esm data through a novel fast super resolution convolutional neural network (fsrcnn). Geophys Res Lett 49(4):e2021GL097571
12. Pons-Moll TAG (2019) Tex2shape: detailed full human body geometry from a single image. https://doi.org/10.1109/ICCV.2019.00238
13. Pumarola A, Sanchez J (2019) Choi: render people dataset
14. Saito S, Huang Z, Natsume R, Morishima S, Kanazawa A, Li H (2019) Pifu: pixel-aligned implicit function for high-resolution clothed human digitization, pp. 2304–2314
15. Thiemo Alldieck1 MM (2018) Video based reconstruction of 3d people models. https://doi.org/10.1109/{CVPR}.2018.00875
16. Li Z, Chen L (2019) 3d human avatar digitization from single image, pp 1–8. https://doi.org/10.1145/3359997.3365707

Helmet Detection Using YOLO-v5 and Paddle OCR for Embedded Systems

Uday Kulkarni, Satish Chikkamath, Apoorva K. Kasigavi, Smitha K. Venkatesh, Siri K. Venkatesh, Sindhu Bhat, and S. M. Meena

Abstract One of the best methods to lessen the chance of suffering brain injuries in a two-wheeler collision is to wear a helmet, but many riders choose not to do so for a variety of reasons, including comfort, convenience, or ignorance. By automatically detecting and identifying riders who are not wearing helmets, automatic helmet detection tools can assist in overcoming these difficulties. These tools may result in fines or other consequences that may serve as an incentive for riders to wear helmets. Thousands of images and video frames can be quickly scanned and analyzed using powerful image processing algorithms employed in automatic detection systems, yielding consistent, accurate results without the possibility of human mistake. The YOLO-v5 object detection technique utilizes deep learning to detect and identify objects in a frame of an image or video. With the help of this model riders without helmet can be detected, and their license plate can be extracted. Optical character recognition (OCR) technique called Paddle OCR can precisely extract and identify text from an image or video frame. With the aid of OCR, the license plates can be recognized and can be stored. The proposed tool accurately detects the license plate of riders without helmets by bringing together these two potent models. Both the YOLO-v5 model and Paddle OCR model were trained using custom datasets with 1000 images each and obtained accuracy by 91%.

Keywords YOLO-v5 · Paddle OCR · Deep neural network

U. Kulkarni (✉) · S. Chikkamath · A. K. Kasigavi · S. K. Venkatesh · S. K. Venkatesh · S. Bhat · S. M. Meena
KLE Technological University, Hubli, Karnataka, India
e-mail: uday_kulkarni@kletech.ac.in

V. K. Gunjan and J. M. Zurada (eds.), *Proceedings of 4th International Conference on Recent Trends in Machine Learning, IoT, Smart Cities and Applications*, Lecture Notes in Networks and Systems 873, https://doi.org/10.1007/978-981-99-9442-7_21

1 Introduction

Helmets [4] are the primary piece of safety gear for motorcycle riders. The motorcycle rider is protected from accidents by the helmet [3]. Although wearing a helmet is required in many nations, some motorcycle riders choose not to wear one or wear it incorrectly.

Deep learning [7] is now one of many artificial intelligence [10] methods utilized in a variety of fields [2]. The term "artificial intelligence" [5] is often used to describe algorithms designed for thinking and problem-solving. Applications using deep learning handle data more precisely and effectively and can quickly extract key features [2].

After training with a certain dataset, a helmet detection model can be created. The rider's license plate is cropped out and saved as an image based on the recognized classes as shown in Fig. 1. An optical character recognition (OCR) [12] model is given in this image and recognizes the text, and outputs are text encoded by a machine for the number plate number. A webcam can be used to implement it in real time.

In our paper, we discuss the literature study briefly in Sect. 2. The research's methodology is described in Sect. 3. The findings of the experiment are discussed in Sect. 4. Conclusions and future scope are discussed in Sect. 5.

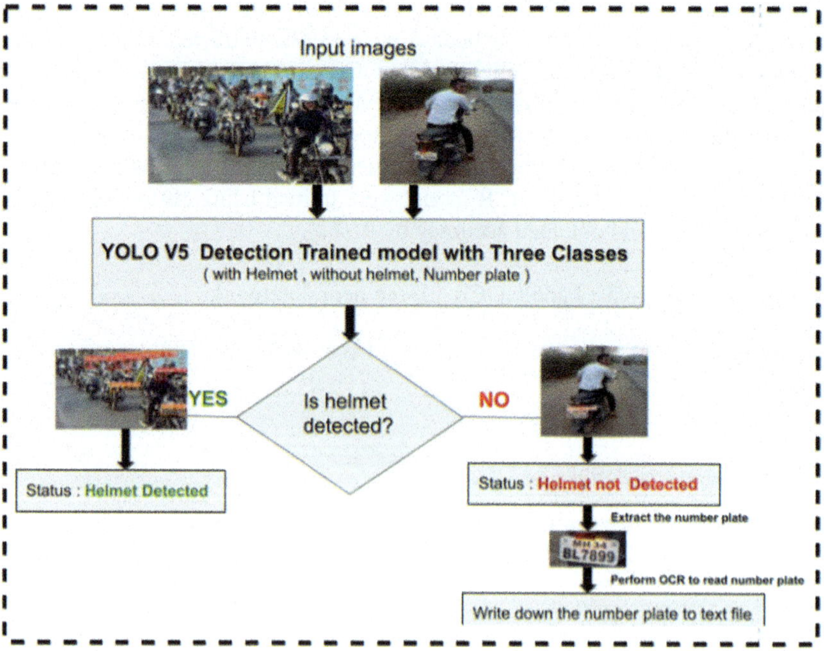

Fig. 1 Workflow of the proposed model

2 Related Works

In [1], authors introduced a brand-new Automatic Number Plate Recognition (ANPR) system, a cutting-edge machine vision technique used to identify automobiles by their number plates without overt human involvement. For Indian automobiles, this system will primarily do automatic number plate extraction, character segmentation, and identification. Sobel filter, morphological techniques, and linked component analysis are used for number plate extraction. Vertical projection analysis and connected component analysis are used for character segmentation. Support vector machine (SVM) are used for character recognition. The segmentation accuracy is 80%, and recognition rate is 79.84 %.

In [6], authors proposed unique approach based on texture and edge information. Candidate extraction and candidate verification make up both halves of the entire procedure. In the first section, numerous potential locations rather than one with the most texture information are extracted to verify that the real license plate is being extracted from a difficult context. The plate candidates are verified in the second section using a binary image and projection approach based on auto-correlation. The aforementioned traits paired with various binary techniques were used to validate each candidate's license plate.

In [11], authors address the issues of license plate detection in distorted images, varying background color, length of numbers, and character arrangement. A rectified license plate image is produced by the perspective rectification network (PRN), which automatically predicts the perspective transformation of an input license plate image. PRN includes perspective transformation prediction and grid sampling. Offset matrix T_{p} is predicted using rectification network, and the perspective transformation T is calculated. The sampling grid is then produced by T and features of perspective transformation. The 2D attention mechanism is used [8] in order to directly predict the sequence of license numbers from the perspective rectified license plates' 2D feature maps.

3 Proposed Methodology

For real-time helmet detection, accuracy and speed are crucial. The You Only Look Once (YOLO) DNN model was decided upon as a result. YOLO makes advantage of current, real-time object detection technology. The YOLO-v5 model is the most recent model in the YOLO series. When YOLO was first developed, it was the first object detection model to combine object categorization and bounding box prediction into a single end-to-end differentiable network. It was made in the DarkNet system, where it is currently kept up-to-date. YOLO-v5, the first of the YOLO models to be developed utilizing the PyTorch framework, is significantly more lightweight and user-friendly.

Even with antiquated technology, YOLO-v5's lightweight and speedy architecture allows for real-time object detection. Additionally, it makes use of a more intricate architecture with features like spatial pyramid pooling and self-attention, which can improve the accuracy of object detection. In contrast, faster R-CNN is slower and requires stronger hardware to achieve real-time performance. As a result, YOLO-v5 is better than Faster R-CNN.

Helmet detection and license plate detection involve the precise recognition of three objects: a rider with helmet, a rider without helmet, and a license plate. To find these items, a special object detection model must be created. The target objects from the target classes are represented by a dataset of images. Then, using this dataset, the custom model is trained. The model can be used to locate these particular objects after being trained.

3.1 Helmet Detection

We implement a deep learning classifier built on a convolutional neural network to extract the characteristics and store them, so we can identify them in other images. When an image is provided to this trained model, it must detect the untrained class. EfficientDet shown in Fig. 2 is a more complex design used by YOLO-v5 and is based on the EfficientNet network architecture. Because of the usage of a more intricate architecture, YOLO-v5 can achieve improved accuracy and better generalization to a wider range of item categories which is described in Fig. 3.

Applying this approach has significant benefits for YOLO-v5 as shown in Fig. 4, as it requires fewer parameters and less processing power (FLOPS), which boosts inference speed, a crucial component of real-time object identification models. YOLO-v5 uses the same head as YOLO-v3 and YOLO-v4. It consists of three convolutional layers which predict the locations of scores, object classifications, and bounding boxes (x, y, height, and breadth). In contrast to previous iterations, Eqs. (1)–(4) are now employed to calculate the target coordinates of the bounding boxes.

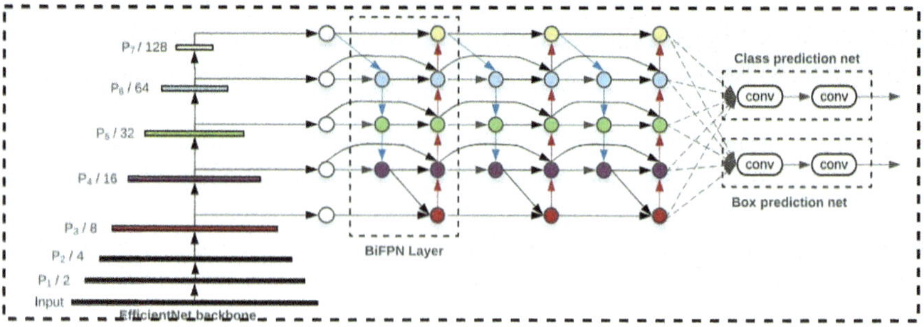

Fig. 2 EfficientDet architecture

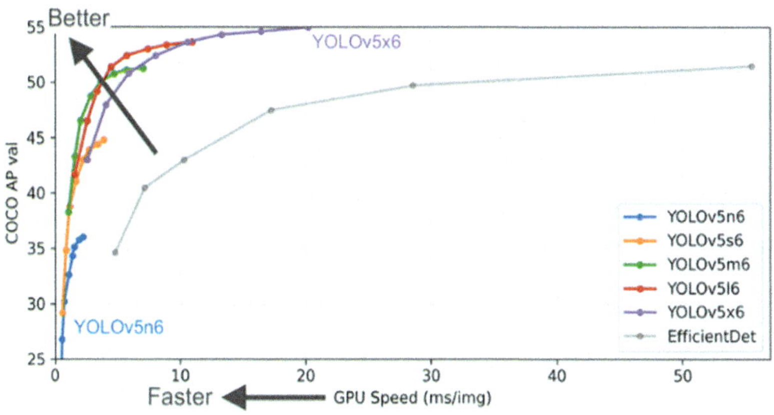

Fig. 3 Performance of YOLO-v5 different size models

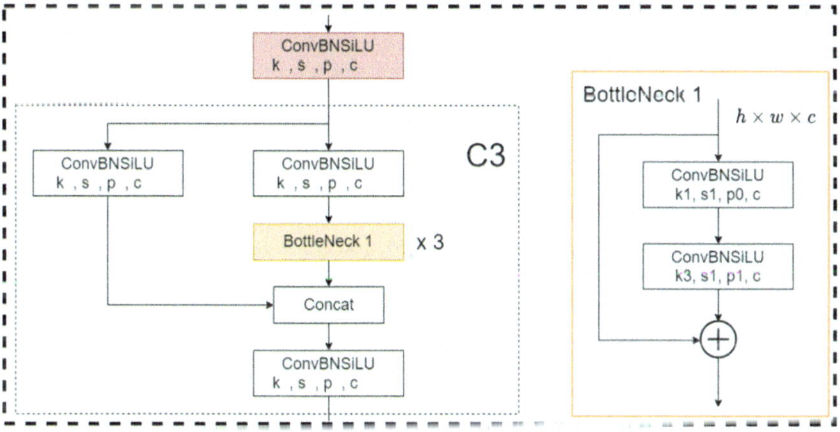

Fig. 4 BottleNeckCSP module architecture[21]

$$Bx = (2\sigma(Tx) - 0.5) + Cx \quad (1)$$

$$By = (2\sigma(Ty) - 0.5) + Cy \quad (2)$$

$$Bw = Pw \cdot (2\sigma(Tw))^2 \quad (3)$$

$$Bh = Ph \cdot (2\sigma(Th))^2 \quad (4)$$

All deep learning model has to choose an activation function, and the YOLO-v5 activation functions used are SiLU and Sigmoid as shown in Fig. 5. The classes of the learned objects, their bounding boxes, and the objectness scores are the three outputs that YOLO-v5 generates. Therefore, the classes loss and the objectness loss

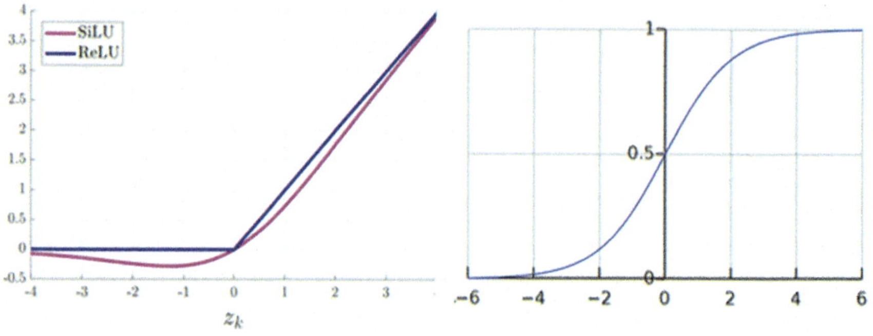

Fig. 5 SiLU function graph and Sigmoid function graph [21]

are calculated using Binary Cross-Entropy (BCE). To calculate the location loss, one uses the Complete Intersection over Union (CIoU) loss. Equation (5) provides the final loss equation.

$$\text{Loss} = \lambda_1 L_c + \lambda_2 L_o + \lambda_3 L_\ell \tag{5}$$

where λ_1, λ_2, and λ_3 are regularization hyperparameters, L_c is the classes loss, L_o is the objectness loss, and ℓ is the location loss.

Following the identification of the rider without a helmet, the associated person class is identified, the license plate is cropped, and a new image is saved. The number plate is recognized with the aid of Paddle OCR.

3.2 *License Plate Detection*

OCR uses computer vision and deep learning methods [9]. Using neural networks, OCR systems understand the text's basic characteristics and forecast the related output. A number of outstanding pretrained models are available from Paddle OCR. It provides text detection, text recognition, and text direction classification. Paddle OCR also provides a range of variants according to size.

Lightweight models are models that use less memory, run more quickly, but sacrifice accuracy. Server models (heavyweight) are slower models that use more memory but are more accurate.

One of the top OCR tools is the industry standard PP-OCR algorithm. Convolutional recurrent neural network (CRNN) architecture is the backbone of Paddle OCR as shown in Fig. 6. There are a total of three layers in this network, namely the transcription layer, RNNs, and CNNs. CTC, or Connectionist Temporal Classification Loss, is used by CRNN and is in charge of connecting anticipated sequences. The uppermost tier comprises the convolutional neural network (CNN), which is composed of convolutional layers and max-pooling layers. These are in charge of sorting

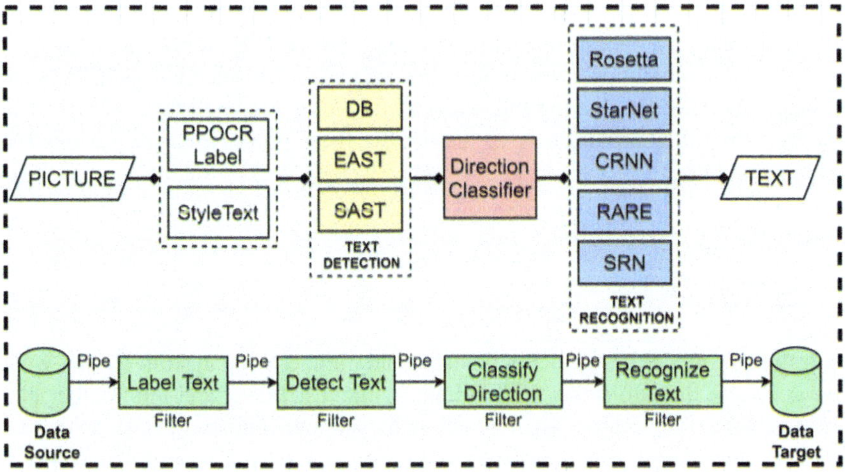

Fig. 6 Architecture of Paddle OCR

through the input images to extract features, and the outputs are feature maps. In order to output to the subsequent layer, feature maps are first converted into a number of feature vectors. Detailed architecture of Paddle OCR is shown in Fig. 6.

After the feature extraction process, each column within the feature maps represents a rectangular area within the input image, and this area is known as the receptive field. The convolutional neural network (CNN) serves as the foundation for the recurrent neural network (RNN) layer in this architecture. To address the vanishing gradient problem and create a more intricate network, two bidirectional long short-term memory (LSTM) units are utilized within the CRNN architecture. These recurrent layers are responsible for predicting labels for each feature vector or frame within the feature sequence derived from the CNN layers. For each frame x in the feature sequence $x = x1, \ldots, xt$, the layers theoretically forecast the label y. Transcription layer is in charge of generating the final sequence based on the per-frame predictions with the highest probability. The model learns to decode the output by calculating CTC, or Connectionist Temporal Classification Loss, using these predictions.

A tensor containing the probabilities of each label for each receptive field is the output that is received from the RNN layer. The decoding of the output tensor by the inference as well as training the network is both carried out via Connectionist Temporal Classification Loss (CTC loss). This problem is fixed by CTC by combining all the repetitive characters into one. Additionally, a blank character "–" is added when that word ends. For additional characters, this continues. Calculating loss and back-propagating it into the network is necessary for a model to learn. The probability of the output sequence is obtained in this case by summing each time step's scores for all possible alignments, from which the loss is determined.

The probability is negatively logarithmically calculated, and the loss is then calculated and used for back-propagation into the network. We require an output that is correct and unambiguous at the moment of inference. By selecting the CTC select the best sequence by selecting the characters with the highest probability for each time step from the output tensor or matrix. So, these processes have produced the finished product.

4 Results

We tested our model on the helmet dataset and number plate custom dataset which have 3 classes that are riders with helmet, without helmets, and the number plate of the vehicles. The Paddle OCR model achieved an accuracy of 0.91, which is a commendable improvement over the attempts on this issue. The complete workflow with results is described in Figs. 7, 8 and 9. The overall results are shown in Fig. 10.

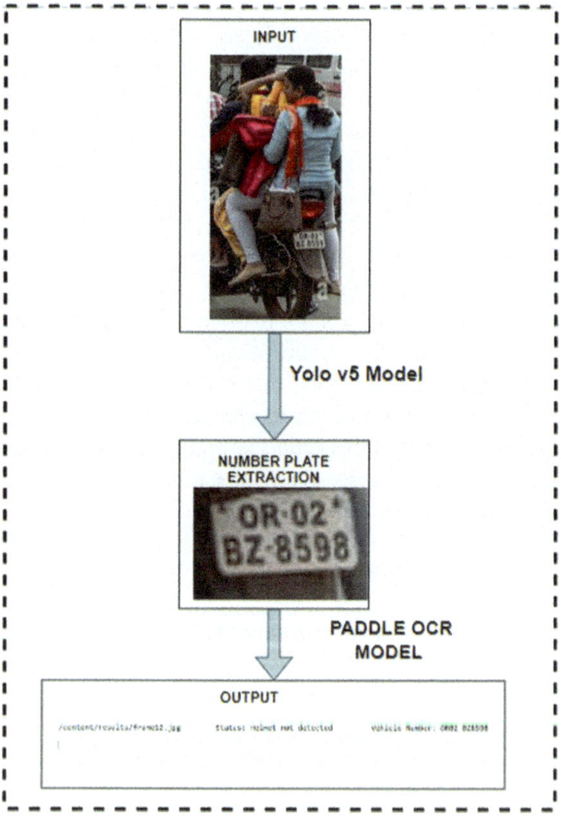

Fig. 7 Results of the proposed methodology

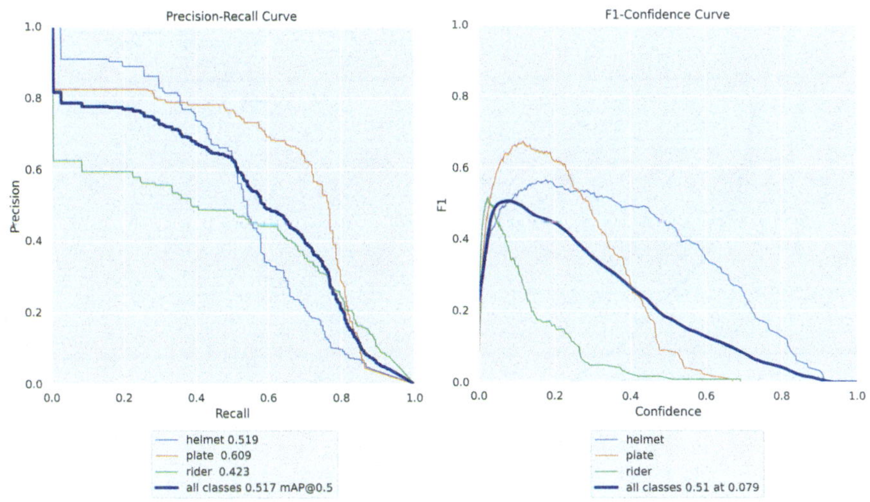

Fig. 8 Precision versus recall graph and $F1$ confidence curve of the proposed work

4.1 Dataset Description

The custom dataset used for training the YOLO-v5 model consisted of 1000 images specifically focused on helmet and license plate detection. The images were carefully curated to include a diverse range of scenarios and variations, such as different lighting conditions, angles, backgrounds, and helmet types. Each image was annotated with bounding boxes around the helmets and license plates to provide ground truth information for training the model. The dataset aimed to capture real-world scenarios commonly encountered in helmet detection applications. The inclusion of license plates in the dataset further enhanced the model's ability to accurately detect and localize helmets in various contexts.

4.2 Evaluations Metrics

X = True positive, Y = True Negative, P = False Positive, Q = False Negative.

Accuracy Accuracy is a measure of how many correct predictions were made out of the total number of predictions as shown in Eq. (6). Its range spans from 0 to 1, and it serves as a crucial metric for evaluating whether a model has either underperformed or overperformed on the given datasets.

$$\text{Accuracy} = \frac{X + Y}{X + Y + P + Q} \tag{6}$$

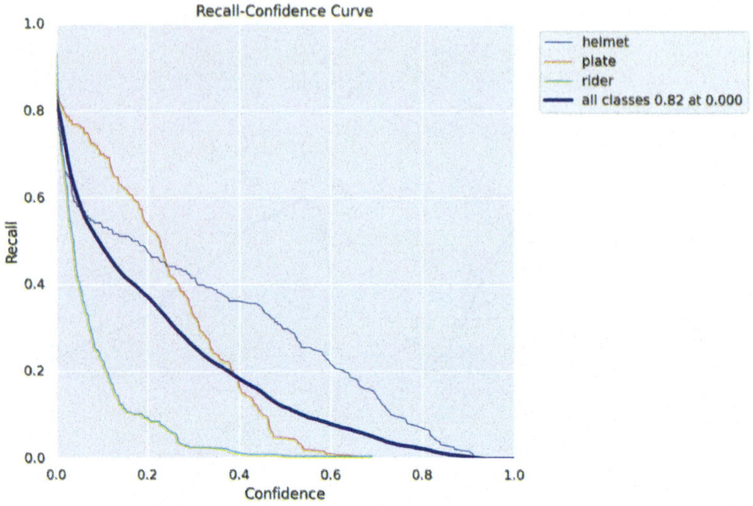

Fig. 9 Recall versus confidence graph of the proposed work

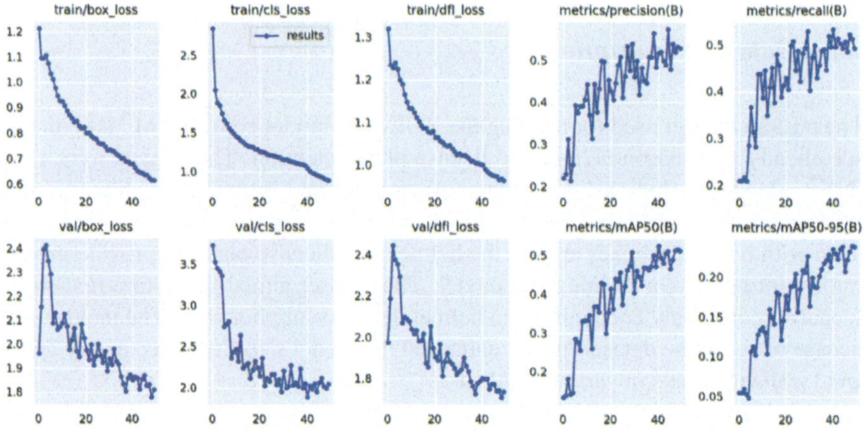

Fig. 10 Overall results of the proposed work

Precision Precision is the proportion of correctly identified positive instances out of the total predicted positive instances, as represented by the provided Eq. (7). A higher precision indicates a greater efficiency of the model. The graph in figure illustrates the trade-off between precision and recall Fig. 8.

$$\text{Precision} = \frac{X}{X + P} \tag{7}$$

Recall

$$\text{Recall} = \frac{X}{X + Q} \tag{8}$$

FL-Score The FL-score, as depicted in the equation, is the average of precision and recall Eqs. (8) and (9). $F1$ confidence curve is shown in Fig. 8. Recall versus confidence graph is shown in Fig. 9

$$F1\text{Score} = 2 \cdot \text{Precision} \cdot \text{Recall} \tag{9}$$

5 Conclusion

Lastly, our proposed work shows the efficiency of implementing YOLO-v5 and OCR for helmet detection. The model achieved an accuracy rate of 87%, which shows that object detection and character recognition can be a powerful approach to solving this problem. We believe that our research can help to improve the safety of individuals who travel by vehicle, especially those who work in construction, mining, and other high-risk industries. In the future, we plan to explore other applications of this model, such as detecting other types of safety gear in different settings.

During this study, the riders with and without a helmet were accurately classified, and through the aid of OCR, we could even read the number plates of the vehicles where the riders did not wear helmets.

References

1. Abdella AAES (2016) Libyan licenses plate recognition using template matching method. J Comput Commun 4
2. Allamki L, Panchakshari M, Sateesha A, Pratheek KS (2019) Helmet detection using machine learning and automatic license plate recognition
3. Chiverton J (2012) Helmet presence classification with motorcycle detection and tracking. Intell Transp Syst IET 6:259–269. https://doi.org/10.1049/iet-its.2011.0138
4. Hayat A, Morgado-Dias F (2022) Deep learning-based automatic safety helmet detection system for construction safety. Appl Sci 12:8268. https://doi.org/10.3390/app12168268
5. Kulkarni U, Meena SM, Gurlahosur SV, Bhogar G (2020) Quantization friendly mobilenet (QF-mobilenet) architecture for vision based applications on embedded platforms. https://doi.org/10.1016/j.neunet.2020.12.022
6. Kurpiel F, Minetto R, Nassu B (2017) Convolutional neural networks for license plate detection in images. https://doi.org/10.1109/ICIP.2017.8296912
7. Li Y, Wei H, Han Z, Huang J, Wang WD (2020) Deep learning-based safety helmet detection in engineering management based on convolutional neural networks. Adv Civil Eng 2020:1–10. https://doi.org/10.1155/2020/9703560
8. Lin Z, He Z, Tan B, Shen Y, Wang P, Liu T (2022) An attention-based network for serial number recognition on banknotes. Image Commun 106(C). https://doi.org/10.1016/j.image.2022.1167322

9. Salimah U, Maharani V, Nursyanti R (2021) Automatic license plate recognition using optical character recognition. IOP Conf Ser Mater Sci Eng 1115:012023. https://doi.org/10.1088/1757-899X/1115/1/012023
10. Yi X, Wu J (2020) Research on safety management of construction engineering personnel under big data + artificial intelligence. Open J Bus Manag 8
11. Xu H, Guo Z, Wang DH, Zhou X, Shi Y (2021) 2d license plate recognition based on automatic perspective rectification. In: 2020 25th international conference on pattern recognition (ICPR), pp 202–208
12. Xu Z, Yang W, Meng A, Lu N, Huang H, Ying C, Huang L (2018) Towards end-to-end license plate detection and recognition: a large dataset and baseline. In: 15th European conference, Munich, Germany, September 8–14, 2018, proceedings, part XIII, pp 261–277. https://doi.org/10.1007/978-3-030-01261-8_16

DefogNet: A Residual Network for Removal of Fog Using Weighted Combination Loss

Uday Kulkarni, Satish Chikkamath, Vinod Patil, Ankit Raichur, M. Siddesha, and Supreet Math

Abstract Fog is a common atmospheric phenomenon that significantly affects outdoor images and videos, causing reduced visibility and image quality. Various image processing techniques have been proposed for fog removal, but most of them require prior knowledge about the scene or complex models, which can be computationally expensive. In this paper, we propose a deep learning-based technique called DefogNet for removing fog from images. DefogNet is a residual network that uses a weighted combination loss function to generate visually pleasing and perceptually accurate results. The network is trained end-to-end using a dataset of foggy and clear images, and it can efficiently restore the visibility of images without requiring any prior knowledge of the scene. Experimental results show that the proposed method outperforms state-of-the-art methods in terms of both quantitative metrics and visual quality. The proposed approach can be useful in various applications, including surveillance, self-driving cars, and outdoor photography.

Keywords Fog removal · Residual network · Deep learning · Weighted loss combination

U. Kulkarni · A. Raichur
School of Computer Science Engineering, KLE Technological University Hubballi, Hubballi, India
e-mail: uday_kulkarni@kletech.ac.in

S. Chikkamath · V. Patil (✉) · M. Siddesha · S. Math
School of Electronics and Communication Engineering, KLE Technological University Hubballi, Hubballi, India
e-mail: vinodnspatil@gmail.com

S. Chikkamath
e-mail: chikkamath@kletech.ac.in

V. K. Gunjan and J. M. Zurada (eds.), *Proceedings of 4th International Conference on Recent Trends in Machine Learning, IoT, Smart Cities and Applications*, Lecture Notes in Networks and Systems 873, https://doi.org/10.1007/978-981-99-9442-7_22

1 Introduction

Fog is a common atmospheric phenomenon that can severely affect the visibility and quality of outdoor images. This degradation can pose significant challenges for computer vision applications, where high-quality images are essential for accurate detection, recognition, and analysis.

The presence of fog and haze may dramatically reduce the quality of outdoor pictures. It may be difficult for computer vision algorithms to execute their functions correctly due to the deteriorated pictures. Defogging solutions have therefore been suggested to clear the fog and haze and return the image's visibility. Defogging aims to improve the image's visual quality by minimizing light scattering and absorption brought on by fog and haze. Aerial photography, security systems, the automobile sector, and other fields all use defogging techniques. We present a thorough analysis of the most recent methods for defogging photographs in this work.

For fog removal, several methods have been proposed, including picture dehazing, color constancy, and dark channel prior methods. Modeling the physical features of fog and utilizing these models to determine the underlying scene radiance are standard strategies. However, fog removal continues to be a difficult challenge due to the intricate interaction of light, environment, and scene content.

Machine learning has gained considerable popularity in recent years due to its deployment on edge devices, and deploying it on edge devices is a complicated process as these devices have limited computational facilities and storage capabilities. Techniques like quantization [5] are used in this situation. With the help of quantization, we can make models to express parameters that use low-process data types such as 8-bit integers or floating point numbers, and this technique allows us to develop more and more complex models [4]. When deploying quantized machine learning models on edge devices, there are a number of important considerations that must be properly taken into account, including hardware constraints, network capacity, and latency requirements [3]. The precision and effectiveness of the deployed model may be impacted by the often occurring variations in the hardware capabilities and configurations of edge devices. To deploy a model successfully [6], the correct quantization strategy must be employed, and the model's parameters must be customized for a particular hardware configuration.

We give a thorough analysis of fog removal approaches in this study, including a description of the underlying physical models and computational methodologies. We also cover current issues and future objectives for fog removal research, as well as an examination of the performance of various approaches on benchmark datasets. Our objective is to present a complete review of the state-of-the-art in fog removal, as well as to identify prospective future research areas in this field. The contributions are summarized as follows. We propose a DefogNet, a network for removal of fog from indoor and outdoor images 3. We propose DeFog loss L_{DeFog} as a weighted combination of L_1 loss, feature loss L_{feature}, and SSIM loss L_{SSIM} toward removal of fog (Sect. 3.1). We demonstrate the results of proposed methodology on benchmark datasets in comparison with state-of-the-art methods using appropriate quantitative metrics 4.

2 Related Works

The author of this paper [2] proposes a learning-based defogging technique that focuses on thick and/or irregular fog. As a feature regularization technique, this method combines the structural representations from VIT and the CNN features to help networks recover background data. A grayscale network and an RGB network make up the pipeline. The debut of the grayscale feature multiplier, a tool for enhancing feature. They included uncertainty feedback learning in addition to the new structure loss a multiplier consistency loss, which improved the efficiency of the RGB generator network. According to experimental findings, this approach is effective for thick or irregular fog.

The author of this paper [9], despite the use of tiny training datasets, suggested that single picture dehazing approach is an end-to-end trainable architecture that can be applied in a variety of circumstances and non-homogeneous haze. Although the FFANet structure is straightforward, the network excels in restoring color fidelity and picture detail, and it is anticipated that it will also be able to handle other low-level vision tasks like deraining and super-resolution. The picture restoration algorithms heavily rely on FFA and other powerful modules in FFANet.

The author of this paper [10] presented single picture dehazing technique which is a fully trainable approach. The architecture can be used in various situations and non- homogeneous haze, despite the use of small training datasets. The method outperforms existing approaches in terms of SSIM and PSNR for all three datasets with available results. While the model preserves colors well in the reconstructed image for indoor images in the I-Haze dataset, the color reconstruction is not as for other datasets.

The author of this paper [8] presented a dehazing network called DHSGAN which was created particularly to clear photographs of smoke and fog. In contrast to conventional approaches, DHSGAN uses deep learning methods to do this. The network is trained from beginning to end, which enables it to immediately map hazy pictures to their equivalent haze-free versions without the need for further steps. By doing this, DHSGAN hopes to increase dehazing's efficiency and efficacy in comparison with more conventional methods. Fog and smoke may dramatically reduce image quality and visibility, and the network is particularly built to manage these problems. Overall, DHSGAN presents a potential method for improving photographs impacted by smoke and fog, resulting in clearer and more esthetically pleasing outcomes.

3 DefogNet: A Residual Network for Removal of Fog Using Weighted Combination Loss

Given a foggy image I_{fog} as input, the aim of the proposed DefogNet is to learn fog artifacts and generate a fog-free image. Fog removal module M_{FOG} detects and removes flare by masking the region as given as,

$$X_{\text{fog}} = M_{\text{FOG}}(I_{\text{fog}}) \tag{1}$$

where M_{FOG} learns removing of fog from I_{fog} to X_{fog} and X_{fog} is the fog-free image. M_{FRN} implicitly removes the fog from the image. The removal of fog from the scene creates additional artifacts resulting in unpleasant observations. Toward this we introduce a weighted loss function, to retain the image in accordance with ground truth I_{GT}.

Figure 2 shows the overall architecture of our proposed framework. We propose an end-to-end trainable architecture to simultaneously learn removing fog in the scene to generate fog-free observations. To learn the removal of fog efficiently, we incorporate L_1 loss, which is computed between the input flare image and the ground truth image.

The input to the network is a foggy image with 3 input channels, and the output is a clear image with 3 output channels. The network consists of 8 residual blocks, where each block contains two 3×3 convolutional layers with batch normalization and ReLU activation.

In a residual block, the input passes through a convolutional layer, followed by batch normalization and ReLU activation. The resulting output from the first convolutional layer is then fed into another convolutional layer, followed by batch normalization without an activation function. The output from the second convolutional layer is added to the input of the residual block, creating the final output of the block. This structure is commonly used in convolutional neural networks (CNNs) like ResNet to address the vanishing gradient problem and enable the training of deep networks. The last layer of the network is a convolutional layer with 3 output channels to produce the final clear image.

3.1 *Loss Functions Used*

L_1 Loss L_{L1}: Initially, we used L_1 loss as the loss function for our proposed architecture, and L_1 loss L_{L1} is computed between the predicted image X_{fog} and the ground truth I_{GT}. As shown in Fig. 1, the input fog image I_{fog} is fed to the network, for removing the fog. L_1 loss measures the pixel-wise difference between the predicted fog image and the ground truth image and is given by

$$\mathbb{L}_{L_1} = \sum \left\| I_{\text{GT}} - \hat{X}_{\text{fog}} \right\|_1 \tag{2}$$

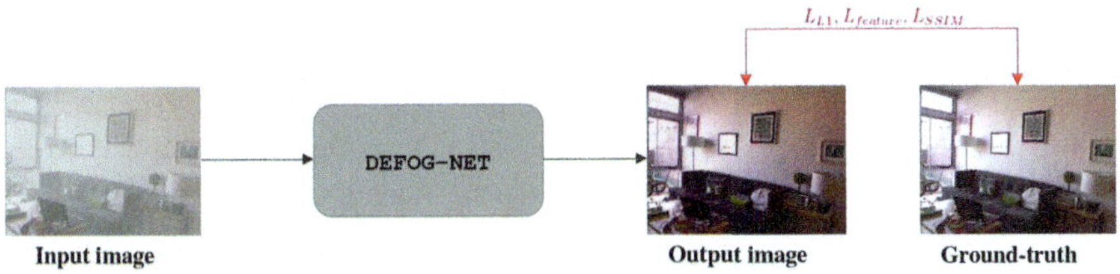

Fig. 1 Proposed framework of DefogNet

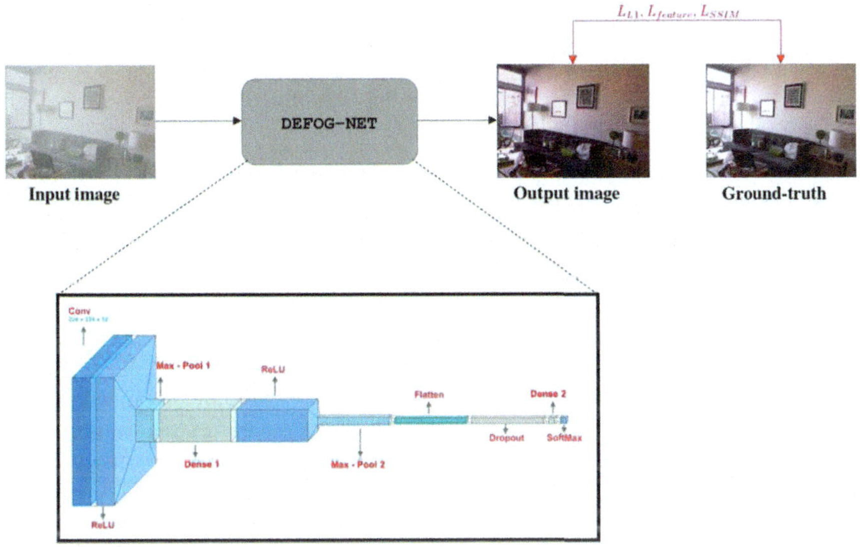

Fig. 2 Proposed framework of DefogNet

where I_{GT} is the ground truth corresponding to the input image and $\hat{X}_{\mathrm{fog}}$ is the generated image. The generated image is subtracted from the ground truth image I_{GT} for rendering the fog-free image I_{fog}; however the results we obtained using L_1 loss were not satisfactory because the model did not understand the structure of the image, resulting in poor image quality.

SSIM Loss L_{SSIM}: To improve the model's understanding of image structure, we then implemented DefogNet using SSIM loss. This allowed the model to understand the structure of the images more precisely, and we incorporate SSIM loss between the fog-free $\hat{X}_{\mathrm{fog}}$ image and the ground truth I_{GT}. L_{SSIM} measures the pixel-wise similarity between the X_{fog} and I_{GT} and is given by,

$$\mathrm{SSIM}(x, y) = \frac{(2\mu_x\mu_y + C_1) + (2\sigma_{xy} + C_2)}{\left(\mu_x^2 + \mu_y^2 + C_1\right)\left(\sigma_x^2 + \sigma_y^2 + C_2\right)} \tag{3}$$

where x and y are the predicted and ground truth images, respectively, μ_x and μ_y are the mean intensities of x and y, σ_x and σ_y are the standard deviations of x and y, and σ_{xy} is the covariance between x and y. The constants C_1 and C_2 are regularization terms that prevent division by zero. However, we found that the perceptual quality of the enhanced image was still not good enough as compared to ground truth as it still had artifacts in it.

Perceptual Loss: In order to further enhance the image quality, we turned to perceptual loss as the loss function. The VGG-16 perceptual loss is a measure of difference between the feature representations of a foggy input image and a ground truth image, as computed by the VGG-16 convolutional neural network.

$$L_{\mathrm{VGG}}(I, I') = \sum_{l=1}^{L} \lambda_l \frac{1}{w_1 h_1} \sum_{i=1}^{w_l} \sum_{j=1}^{h_1} \|\phi_l(I)_{i,j} - \phi_l(I')_{i,j}\|_2^2 \tag{4}$$

In this equation, I represents the ground truth image, I' represents the predicted image, L represents the number of layers used for feature extraction, w_l and h_l represent the width and height of the feature maps at layer l, respectively, and λ_l is a hyperparameter that controls the weighting of the feature loss at layer l. This helped to improve the perception quality of the enhanced images. However, we soon found out that to remove fog from the images, we needed all three loss functions (L_1, SSIM, and perceptual loss) to contribute to the process.

Proposed weighted combinational loss L_{DeFog}. To address this issue, we propose L_{DeFog} as a weighted combination of L_{L1}, L_{feature}, and L_{SSIM} by adding weights of 0.5 to each loss function. This allowed each loss function to contribute to the removal of foggy images, resulting in improved image quality, and it is given as

$$L_{\mathrm{DeFog}} = \alpha * L_{L_1} + \beta * L_{\mathrm{feature}} + \gamma * L_{\mathrm{SSIM}} \tag{5}$$

where α, β, and γ are weights, and we heuristically set $\alpha = \beta = \gamma = 0.5$. The proposed weighted combination loss function resulted in satisfactory fog removal results without any image artifacts. The combination of L_1, SSIM, and perceptual loss functions allowed the model to consider multiple aspects of the image simultaneously, resulting in a balanced approach to fog removal. By combining these loss functions with appropriate weights, we achieved optimal results in fog removal without sacrificing image quality or introducing any artifacts.

4 Results and Discussions

In this section, we discuss on the datasets used for training the proposed DefogNet. We demonstrate the results of DefogNet on benchmark datasets in comparison with SOTA methods using appropriate quantitative metrics.

4.1 Dataset Description

We train the proposed DefogNet on Indoor Training Set (ITS) [RESIDE-Standard] dataset [7]. RESIDE is a comprehensive benchmark containing a large-scale collection of synthetic and real-world hazy images and is divided into five subsets that cater to different training or evaluation purposes. Specifically, RESIDE-Standard's Indoor Training Set (ITS) comprises 13,990 clear images and their corresponding 13,990 hazy images and 13,990 trans images.

4.2 Experimental Setup

The proposed DefogNet is implemented using the PyTorch framework, coupled with Python (v3.8), and trained on NVIDIA RTX 3090 with 24 GB VRAM. In Sect. 2, we provide a detailed description of our architecture, which is designed to effectively learn from synthetically generated foggy images. We use Adam optimizer, with $\beta_1 = 0.9$ and $\beta_2 = 0.9999$, and ϵ set to $8e^{-2}$ we train the model for 100 epochs with learning rate set to lr = 0.0001. We heuristically set the loss weights $\alpha = \beta = \gamma = 0.5$.

4.3 Results

Results on Indoor Training Set (ITS) [RESIDE-Standard] dataset [7]. We compare the performance of the proposed DefogNet on benchmark dataset using appropriate quantitative metrics. In Fig. 3, we show the defogging performance on ITS dataset [7].

Quantitative Evaluation: We use the peak signal to noise ratio (PSNR) [1] and Structural Similarity Index Measure (SSIM) as evaluation metrics to assess the outcomes produced by the suggested architecture. SSIM [11] is a comprehensive metric and measures the overall image quality loss for retaining the overall structure of the image. SSIM is a complete reference metric and requires the ground truth and the corresponding processed image. The quality of its representation can be affected by corrupting noise, which can be measured using the PSNR metric. According to PSNR, the ratio of the signal's highest strength to the picture-degrading noise determines how well the image is rendered. PSNR and SSIM are defined as:

$$\mathrm{PSNR} = 10 \log_{10}\left(\frac{P_{\mathrm{signal}_{\max}}}{\mathrm{MSE}}\right) \tag{6}$$

where $p_{\mathrm{signalmax}}$ is the maximum power of the signal

Fig. 3 Qualitative results of the proposed DefogNet. The first row corresponds to input fog images, second row corresponds to results of proposed DefogNet, and the third row corresponds to ground truth images. We observe that the proposed DefogNet preserves contextual information well and removes fog

Table 1 Performance comparison of DefogNet with SOTA methods. The first highest value is highlighted in bold, and second highest value is highlighted in italics

Methods	PSNR	SSIM
Ayush et al. [10]	17.01	0.61
Jin et al. [2]	18.83	0.62
Xu et al. [9]	*22.47*	*0.78*
Ours	26.51	0.84

$$\text{SSIM} = \frac{(2\mu_x\mu_y + C_1) + (2\sigma_{xy} + C_2)}{\left(\mu_x^2 + \mu_y^2 + C_1\right)\left(\sigma_x^2 + \sigma_y^2 + C_2\right)} \tag{7}$$

where μ_x, μ_y, σ_x, σ_y, and σ_{xy} are means, standard deviation, and covariance for two images x and y (Table 1).

Acknowledgements We would like to express our sincere gratitude to Professor Uday Kulkarni and Professor Satish Chikkmath, for their invaluable guidance and support throughout this research project. Their expertise and mentorship have been instrumental in shaping our ideas and enhancing the quality of our work.

We would also like to thank the Heads of the ECE and CSE departments, whose encouragement and support were crucial to the successful completion of this project. We are grateful for the resources provided by the department that helped us carry out our research.

Finally, we would like to extend our heartfelt thanks to all those who directly or indirectly contributed to this project. We are grateful for the insightful feedback and constructive criticism provided by our peers and colleagues. Without their support, this research would not have been possible.

References

1. Huynh-Thu Q, Ghanbari M (2008) Scope of validity of PSNR in image/video quality assessment. Electron Lett 44(13):800–801
2. Jin Y, Yan W, Yang W, Tan RT (2023) Structure representation network and uncertainty feedback learning for dense non-uniform fog removal. In: Computer vision–ACCV 2022: 16th Asian conference on computer vision, Macao, China, December 4–8, 2022, proceedings, Part III. Springer, pp 155–172
3. Kulkarni N, Singh N, Joshi Y, Hasabi N, Meena SM, Kulkarni U, Gurlahosur SV (2022) Hybrid optimization for DNN model compression and inference acceleration. In: 2022 2nd International conference on intelligent technologies (CONIT). IEEE, pp 1–8
4. Kulkarni U, Meena SM, Gurlahosur SV, Benagi P, Kashyap A, Ansari A, Karnam V (2021) Ai model compression for edge devices using optimization techniques. In: Modern approaches in machine learning and cognitive science: a walkthrough: latest trends in AI, vol 2. Springer, pp 227–240
5. Kulkarni U, Meena S, Gurlahosur SV, Bhogar G (2021) Quantization friendly mobilenet (QF-MobileNet) architecture for vision based applications on embedded platforms. Neural Netw 136:28–39
6. Kulkarni U, Meena SM, Joshua P, Kruthika K, Platel G, Jadhav K, Singh A (2021) Performance improvements in quantization aware training and appreciation of low precision computation in deep learning. In: Advances in signal processing and intelligent recognition systems: 6th international symposium, SIRS 2020, Chennai, India, October 14–17, 2020, revised selected papers 6. Springer, pp 90–107
7. Li B, Ren W, Fu D, Tao D, Feng D, Zeng W, Wang Z (2019) Benchmarking single-image dehazing and beyond. IEEE Trans Image Process 28(1):492–505
8. Malav R, Kim A, Sahoo SR, Pandey G (2019) Dhsgan: an end to end dehazing network for fog and smoke. In: Jawahar C, Li H, Mori G, Schindler K (eds) Computer vision—ACCV 2018. Springer, Cham, pp 593–608
9. Qin X, Wang Z, Bai Y, Xie X, Jia H (2020) FFA-Net: feature fusion attention network for single image dehazing. In: Proceedings of the AAAI conference on artificial intelligence, vol 34, pp 11908–11915
10. Singh A, Bhave A, Prasad DK (2020) Single image dehazing for a variety of haze scenarios using back projected pyramid network. In: Computer vision–ECCV 2020 workshops: Glasgow, UK, August 23–28, 2020, proceedings, part IV 16. Springer, pp 166–181
11. Wang S, Rehman A, Wang Z, Ma S, Gao W (2011) Ssim-motivated rate-distortion optimization for video coding. IEEE Trans Circuits Syst Video Technol 22(4):516–529

Text-to-Image Generation Model with DNN Architecture and Computer Vision for Embedded Devices Using Quantization Technique

Uday Kulkarni, Satish Chikkamath, Rohit Devaranvadagi, Apoorva Angadi, Vishal Farande, and Nishanth R. Negali

Abstract Text-to-image generation is a type of deep learning task where the goal is to generate realistic images from textual descriptions. The model takes in a textual description as input and produces an image that closely matches the description. Quantization techniques have been used to reduce the size of the models while minimizing accuracy loss due to the high computational and memory requirements of these models. These models suffer from a number of problems, including a large model size and a high-accuracy model, which has led to a significant increase in computation and model storage requirements as well as increased power consumption. The study addresses these issues and concentrates on increasing model speed, lowering computational cost, compressing the size of the model, and improving the model's energy efficiency. We can achieve our objectives by using quantization techniques. This quantized model is later deployed on the embedded devices.

Keywords Quantization · Text-to-image · Embedded devices

U. Kulkarni · R. Devaranvadagi (✉) · A. Angadi · V. Farande · N. R. Negali
Department of Computer Science, KLE Technological University, Vidyanagar, Hubballi, Karnataka 580021, India
e-mail: falserohitbd999@gmail.com

U. Kulkarni
e-mail: kulkarni@kletech.ac.in

S. Chikkamath
Department of Electronics and Communication, KLE Technological University, Vidyanagar, Hubballi, Karnataka 580021, India
e-mail: chikkamath@kletech.ac.in

V. K. Gunjan and J. M. Zurada (eds.), *Proceedings of 4th International Conference on Recent Trends in Machine Learning, IoT, Smart Cities and Applications*, Lecture Notes in Networks and Systems 873, https://doi.org/10.1007/978-981-99-9442-7_23

1 Introduction

Text-to-image (T2I) generation is a form of deep learning challenge that demands the creation of realistic images from textual descriptions. Visualization is the process of transforming abstract scientific or business data into pictures that can help the viewer to comprehend the meaning of the data [1]. Here, on the basis of text prompts, novel visuals are created. As the author mentions in [2], the state-of-the-art in this work has recently advanced dramatically thanks to two crucial concepts. First off, it is possible to condition the synthesis [3] based on a user's level of general language comprehension by employing a sizable pre-trained language model as an encoder for the prompts [4]. Second, the models [5] can create virtually anything by employing large-scale training data made up of hundreds of millions of image-caption pairs [6]. Despite the advancements made in the field, there are still issues that need to be resolved, such as the difficulty in producing diverse, excellent images that reliably represent the textual descriptions. The attribute representations' great generalization capabilities and discriminative strength are appealing, but obtaining them can be difficult because they sometimes call for domain-specific expertise [7].

Here, we concentrate on T2I synthesis, which tries to create an image that accurately captures the meaning of a written description. T2I can be viewed as the reverse of image captioning, where an image is an input and a textual description of the picture is the output [2]. The overwhelming majority of T2I research focus on methodologies that generate aesthetically realistic, photographic, and natural images, despite the fact that the techniques discussed in this study can be utilized across a wide article range of image domains [8].

The challenge emerges as we try to deploy such DNN models [9] onto edge devices, given their limited memory and resources. Consequently, the model must be optimized, which can be accomplished through the techniques stated in Fig. 1.

Tensor and matrix decomposition approaches have been employed for low-rank factorization to identify repetitive deep neural network parameters. Whenever needed, a low-rank factorization approach helps to minimize the size of the model by simply splitting a large matrix into smaller matrices [10]. A huge, complicated model is trained using a large dataset in knowledge distillation. If this model generalizes

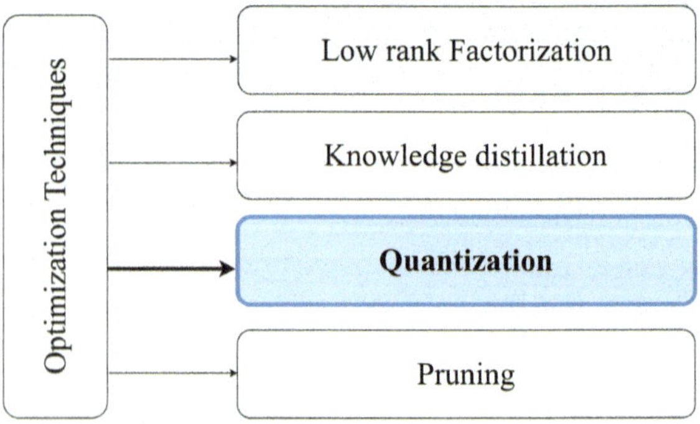

Fig. 1 Model optimization techniques

and performs satisfactorily with untested data, its abilities are passed on to a smaller network [11]. The larger model is referred to as the instructor model, and the less complex model is referred to as the student network. Pruning is an approach used to get rid of features that fail to help the model to make decisions adequately [12].

The two basic methods are weight pruning as well as filter pruning, which means that those weights which contribute the least to the final result are eliminated [13]. The 32-bit floating-point model's weights as well as activations are quantized within the confines of the optimization process that yields lesser n-bit precision while suffering little accuracy loss. The thought process of quantization is covered in the current piece [14]. Quantization [15] is one of the most effective techniques for reducing memory as well as computational costs, therefore allowing the use of DNN models on a fixed-point pipeline [16].

Deploying machine learning models on embedded end devices is becoming increasingly common as the demand for real-time data analysis grows. The advancement of laptops, smartphones, and other new IoT devices has further promoted the development of AI applications on embedded devices [17]. In this paper, we define a deep model that has been trained and applied to a specific service and is designed to run on mobile devices as Mobile Deep Learning Applications (MDLA) [18].

This paper is organized into five sections. Section 1 consists of the introduction to the problem domain. Section 2 mentions the different approaches that are used in generating images from the textual description. The baseline model is discussed in Sect. 2.2. The methodology we propose is presented in Sect. 3. Sections 4 and 5 make up the results and conclusion, respectively.

2 Literature Survey

2.1 Text-to-Image Generation by Using GAN

DALL-E is an artificial intelligence program developed by OpenAI that can create images from textual descriptions. DALL-E1 uses discrete (dVAE) along with Contrastive Language-Image Pre-training (CLIP) and DALL-E2 uses CLIP embedding directly. A substantial body of research on zero shot transfer, natural language super-vision, and multimodal learning is the foundation of CLIP [8]. The CLIP model has already been pre-trained on a sizable dataset of pictures and the text captions that go with them [14]. This pre-training step enables the model with the ability to carry out a variety of tasks that require it to understand how images and text relate to one another [1]. The VAE is a generative model that learns to convert a latent code into an image in DALL-E [19].

Text-to-Image Generation Using VAE and CLIP

A novel technique known as stable diffusion is being used to generate images for a number of purposes, including text-to-image conversion [20]. The stable diffusion process creates an image by repeatedly applying a series of diffusion steps to a given noise vector. A scaled Gaussian noise vector is introduced to the noise vector in each diffusion step, and a series of reversible neural network transformations are then used to update the noise vector [21]. A decoder network is used to convert the noise vector that results after a set number of iterations into an image. With the goal to produce images from textual descriptions by means of stable diffusion, the text that is given as an input is often encoded into the latent space using a pre-trained language model [7].

The ability to fine-tune the generated images is one advantage of making use of stable diffusion for text-to-image conversion. It is possible to generate images with varied levels of depth and complexity and additionally regulate the style and substance of those generated images, by modifying the parameters of the diffusion process [1].

Text-to-Image Generation Using Stable Diffusion

A novel technique known as stable diffusion is being used to generate images for a number of purposes, including text-to-image conversion [20]. The stable diffusion process creates an image by repeatedly applying a series of diffusion steps to a given noise vector. A scaled Gaussian noise vector [22] is introduced to the noise vector in each diffusion step, and a series of reversible neural network transformations are then used to update the noise vector [21]. With the goal to produce images from textual descriptions by means of stable diffusion, the text that is given as an input is often encoded into the latent space using a pre-trained language model [7].

The ability to fine-tune the generated images is one advantage of making use of stable diffusion for text-to-image conversion. It is possible to generate images with varied levels of depth and complexity and additionally regulate the style and substance of those generated images, by modifying the parameters of the diffusion process [1].

2.2 *Quantization Techniques*

For a 32-bit floating-point model, quantization is an optimizing approach that makes sure that the weights and activations are quantized in order to reduce the amount of n-bit precision while maintaining accuracy. The DNN model can now be used with a fixed-point pipeline owing to one of the most efficient methods for reducing memory and computational costs [20]. Quantization [23] involves training the model with lower-bit representation, such as 4-bit instead of full precision, without sacrificing accuracy. At various training stages, the quantization can happen, but methods like Parameterized Clipping Activation (PACT) [24] consistently degrade accuracy less than other approaches [25]. Another method, known as training with simulated quantization, suffers little accuracy loss by only quantizing weights during forward passes. In general, QAT seeks to create models that can be accurately deployed with low-bit representations while still being efficient [26] (Fig. 2).

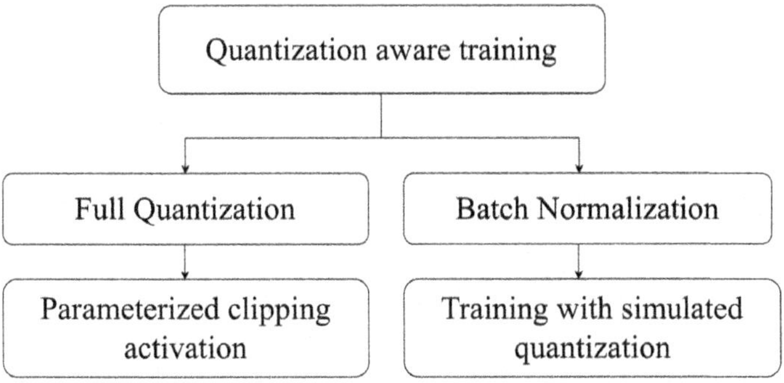

Fig. 2 Quantization techniques

PACT: Parameterized Clipping Activation Function

The ReLU activation function, which is commonly used in Convolutional Neural Networks (CNNs), can make activation quantization more difficult because it produces unbounded output [2]. While ReLU improves accuracy by allowing gradients to flow down into deep layers, it requires a higher dynamic range for quantization, which means that more bit precision is needed to avoid accuracy degradation due to quantization. To address this challenge, the Parameterized Clipping Activation (PACT) technique has been developed [3].

PACT generates a parameterized clipping level that is adjusted dynamically during training using gradient descent. This approach helps to quantize the activations more effectively while minimizing accuracy degradation due to quantization [7].

Training with Simulated Quantization

The weights of quantized networks are frequently quantized during or after training by first training them in floating-point format. This approach works well for larger models with large capacities for data representation, but it can significantly reduce the accuracy of smaller models. The weights are quantized before the input is subjected to convolution [11]. If batch normalization is used by the layer, the parameters are folded into the weights prior to quantization [1].

Post-quantization

A pre-trained model can be quantized using the post-training quantization technique to decrease CPU latency and model size without suffering a sizable accuracy loss. This method can quantize weights and activations as well as compress weights, leading to faster inference without the need for retraining [19]. Accuracy loss from post-training quantization is typically minimal. The value of weights can be restricted in Analytical Clipping for Int Quantization (ACIQ) without compromising accuracy. The Qdrop method and adaptive rounding both round significant activations and drop less significant activations to produce better results. By limiting the bit representation of parameters, per channel bit allocation and Adaptive Floating-Point Quantization offer nearly the same accuracy as the full precision model [14] (Fig. 3).

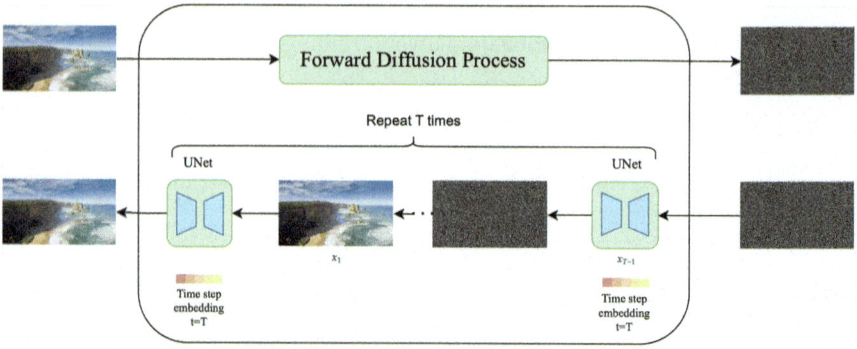

Fig. 3 Latent diffusion model architecture

2.3 *Stable Diffusion Model Architecture*

Stable diffusion models have emerged as a new state-of-the-art family of deep generative models. It is a latent text-to-image diffusion model and conducts the diffusion process in the latent space; thus, it is much faster than a pure diffusion model. The backbone diffusion model is modified to accept conditioning inputs such as text, images, semantic maps [7].

A diffusion probabilistic model (which we will call a "diffusion model" for brevity) is a parameterized Markov chain trained using variational inference to produce samples matching the data after a finite time. In order to reverse a diffusion process—a Markov chain that gradually adds noise to the data in the opposite direction of sampling until the signal is destroyed—transitions of this chain are memorized [25]. It is sufficient to set the sample chain transitions to conditional Gaussians as well when the diffusion only contains minor amounts of Gaussian noise, which enables a relatively straightforward parameterization of the neural network [17].

By adding noise, the original image's structure (distribution) is gradually destroyed. The image is then rebuilt, with the noise removed at each step, using a neural network model. The model eventually learns to estimate the underlying (original) data distribution by repeating this process often enough with high-quality data. The trained neural network can then be used to create a new image that is a representation of the initial training dataset, starting with just noise [21].

The training of the diffusion model can be divided into two parts:

(A) Forward diffusion process—add noise to the image.
(B) Reverse diffusion process—remove noise from the image [8].

(A) Forward Diffusion

The original image ($x0$) is slowly corrupted iteratively (a Markov chain) by adding (scaled Gaussian) noise. This process is done for some T time steps, i.e., xT. Image at timestep t is created by: $xt - 1 + t - 1$(noise) xt. No model is involved stage. At the end of the forward diffusion stage xT, due to the iterative addition of noise, we

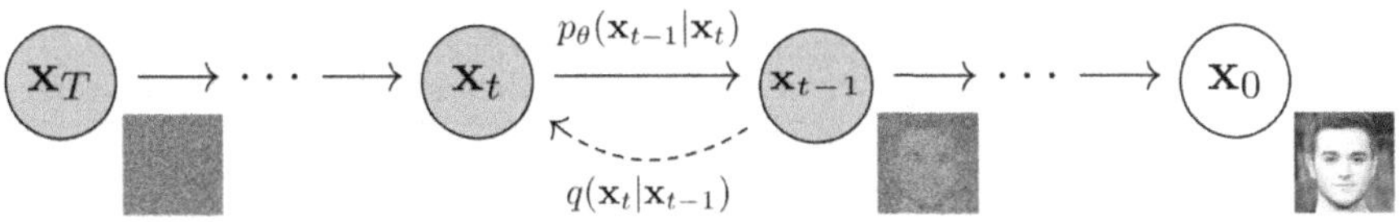

Fig. 4 Forward and backward diffusion

are left with a (pure) noisy image [27] representing an "Isotropic Gaussian." This is just a mathematical way of saying that we have a standard normal distribution and the variance of the distribution is the same along all dimensions [17]. We have converted the data distribution to a Gaussian distribution.

(B) Backward/Reverse Diffusion

In this stage, we undo the forward process [28]. The task is to remove the noise added in the forward process, again in an iterative fashion (a Markov chain). This is done using a neural network model [29]. The model is tasked with the following: Given a timestep t and the noisy image xt, predict the noise ($\acute{\varepsilon}$) added to the image at step $t - 1$.

$xt \rightarrow$ Model $\rightarrow \acute{\varepsilon}$ (predicted noise). The model predicts (approximates) the noise added to $xt - 1$ in the forward pass [10].

For the base 64 × 64 text-to-image diffusion model, we modify the U-Net architecture from Fig. 4. Similar to the class embedding conditioning method, a pooling embedding vector is added to the diffusion time step embedding and used to condition the network on text embeddings. By applying cross-attention over the text embeddings at various resolutions, we further condition the full sequence of text embeddings. We research several text conditioning techniques [7].

3 Proposed Methodology

3.1 Text-to-Image Generation

The technique for generating an image given a textual description is commonly referred to as T2I generation. It is a type of generative modeling that teaches how to transform a text input into an associated image output.

A parameterized Markov chain trained with variational inference is a diffusion model, which eventually generates samples that fit the data. Transitions of this Markov chain are learned to reverse a diffusion process that gradually introduces noise into the data in the opposite direction of sampling until the signal is eliminated. When the diffusion only contains weak Gaussian noise, it is sufficient to set the number of sample chain transitions to conditional Gaussians as well, enabling a relatively simple parameterization of the neural network.

The distribution that constitutes the original image's structure has gradually deteriorated through the addition of noise. The noise is then eliminated from the image. By repeating this procedure often enough with high-quality data, the model ultimately learns to estimate the fundamental (original) distribution of the data. Then, starting with just noise, the trained neural network is able to be used to produce a new image that is a representation of the initial training dataset.

3.2 *Quantization of the Baseline Model*

To deal with some of the above text-to-image generation mechanisms, DNN requires significant computational costs as well as memory needs. Because of the size of the model that was produced and the large number of float parameters, more memory is utilized. The model makes use of text-to-image conversion to get the desired output. It makes use of the PyTorch 8-bit post-training quantization technique to produce a quantized model. Multiple quantization techniques are supported by PyTorch for deep learning models. The model is often trained using FP32 before being transformed into an INT8 model.

The following mapping equation is used for the transformation from a floating point to a fixed point, wherein an is the input value, Z is the quantization parameter, and fit is utilized to fit the input value into the quantization range.

$$Z(a, \text{fit}, \text{zeropoint}) = \text{roundoff}(a/\text{fit} + \text{zeropoint}), \tag{1}$$

where Eq. (1) is used to calculate fit

$$\text{fit} = (2c - 1 - 1)/p - q. \tag{2}$$

Here, in Eq. (2), c denotes the bit width, and p and q denote the lower and upper bounds for the quantization range.

It is significant to observe that the visualization of the floating-point zero both prior to and following quantization is the same. Because the zero baselines before quantization correspond to a fixed-point value soon after quantization, it assures that quantization errors cannot occur in particular operations, like zero padding. The quantized tensor offers the capability to be used for numerous operations, which makes it simple for it to perform various operations. It can also be serialized and parallelized.

The original model is quantized in order to reduce its size by 4 $\times$ and its memory bandwidth desired by 4 $\times$ as compared with floating point 32. Any kind of computer vision-based application is effortlessly ported to edge devices thanks to hardware support for INT8 operation, which usually performs calculations 2–4 times quicker compared with floating-point operations.

Quantization does end up resulting in a tiny amount of loss of data termed quantization error, which corresponds to the difference between the input value and then the value that remains after quantization, regardless of the fact that it cuts down model size and the cost of computation. The impact of quantization loss on the model can be assessed by applying a measure of the parameter that is known as the signal-to-quantization noise ratio (x) (3).

$$x = 10\log_{10}\left(\frac{\sigma_a^2}{\sigma_b^2}\right). \tag{3}$$

Here, σa is the input value and σb is the quantized value. It is obvious that when it comes to preserving the model's accuracy, the x must be lowered. Decreasing the FLOPS, which is usually calculated by the amount of both addition and multiplication operations the model operates, would be another significant aspect of quantization provided that the x parameter is put into account. With quantization, MAC addresses (Multiplication Accumulate Operations) decrease substantially and the total number of operations (TO) in each layer is obtained using Eq. (4).

$$\text{TO} = \text{kernels}(\text{channels} * \text{width} * \text{height} + 1) * M * N. \tag{4}$$

In Eq. (4), kernels denote the number of kernels, channels denote the number of input tensor channels, width is the convolution kernel's width, and height is the convolution kernel's height. M and N are represented in Eqs. (5) and (6), respectively.

$$M = \left(\frac{\mu - \text{width} + 2p_\omega}{s_w} + 1\right), \tag{5}$$

where u is the width of input tensor, width is the convolution kernel's width, pw is the width of input tensor's padding.

$$N = \left(\frac{v - \text{width} + 2p_h}{s_h} + 1\right), \tag{6}$$

where v is the height of the input tensor, width is the convolution kernel's width, ph is the height of input tensor's padding.

3.3 *Porting onto Edge Devices*

Edge machine learning (Edge ML) has attracted an enormous amount of focus since the Internet of Things (IoT) for a good reason. The rapid growth of smart devices connecting across the cloud that occurred accompanied by the growing popularity of IoT was excessive for the network to handle at that point in time. In light of congested networks in the cloud, companies underestimated critical cloud computing concerns such as security. The suitable solution to the aforementioned issue is Edge ML.

Deploying ML models on edge devices has grown increasingly important in order to decrease latency, safeguard privacy, and enable innovative interactive applications. We used PyTorch Mobile, a tool that allows users to seamlessly switch between developing and delivering a model while remaining inside the PyTorch environment. In order to render the model simpler and less costly for computing, it has been quantized. The model is subsequently passed into the script function, which examines the source code, compiles it into the Torch Script code via the Torch Script compiler, and eventually produces a script function. Before being sent to the process that saves the optimized model in .ptl format, the model is further optimized for usage on mobile devices. Then, in order to get an inference from an Android application, the PLT model is deployed in the application that was developed using Android Studio.

4 Results and Analysis

4.1 Dataset Description

The Microsoft Common Objects in Context (MS COCO) dataset is a large-scale image recognition, segmentation, and captioning dataset [20]. Along with captions and 80 object categories, has more than 330,000 images and over 2.5 million instances of various objects. The photos in the collection are varied and intricate, showing both more concrete notions like events and emotions as well as situations from ordinary life like people, animals, and objects.

The dataset was developed as a standard for the analysis of image recognition, segmentation, and captioning problems. Numerous challenges and competitions are organized around it, and it is widely used in research on computer vision and machine learning. The dataset is a difficult testbed for creating new algorithms and methods in computer vision and natural language processing due to its size and complexity (Fig. 5).

4.2 Experimental Results

Quantization greatly reduces the memory footprint needed to store the model by encoding the parameters and activations of the model using lower-precision data formats. This is especially helpful when deploying models on memory-constrained devices like mobile phones or edge computing equipment. Quantized models are more effective since lower-precision computations during inference lead to quicker execution durations and better latency. They are useful for situations where there is a shortage of electricity since they use less power. Quantization makes use of

Fig. 5 Coco dataset [14]

the parallel processing capability of contemporary hardware designs, enhancing throughput and performance as a whole.

Our work mainly focuses on quantizing a given model to make the model more efficient. Quantization reduces the model size by around 40%, which is a big improvement in the model size. The main objective is to embed the machine learning model into an edge device, meaning that there is a constraint of space and computational power. By quantizing a model, we reduce the model size and make it more adaptive to an edge device.

Also, the downside to it is that the accuracy of the model reduces by around 30%. We tested the models by giving 13 prompts each. While the baseline model gives 12 out of 13 accurate results, the quantized model reduces it to 8 out of 13 as seen in Table 2. The accuracy of the prompts is subjective to the viewer's requirements. We summarize representative benchmarks in Table 1 (Fig. 6).

Table 1 Benchmarks for text-to-image generation

Benchmark	Measurement	Auto-eval.	Human-eval.
DrawBench [6]	Fidelity, alignment	N	Y
UniBench [5]	Fidelity, alignment	N	Y
PartiPrompts [9]	Fidelity, alignment	N	Y
PaintSkills [29]	Visual reasoning skills, social biases	Y	Y
EntityDrawBench [15]	Entity-centric faithfulness	N	Y
Multi-task benchmark [24]	Various capabilities	N	Y

Table 2 Indicates the model size and accuracy of the models before and after quantization

Models	Model size (MB)	Avg. model accuracy (13 prompts)
Baseline diffusion	923.75	92.3% (12/13 prompts)
After quantization	590.20	76.92% (10/13 prompts)

5 Conclusion and Future Scope

The key objective of the article is to quantize a given model to make it more efficient and reduce memory and computational requirements. The baseline model uses a diffusion model, which uses the U-Net architecture. The post-training quantization method is used to quantize the baseline model. As an objective with future scope in perspective, the model can be reduced more. The only drawback is that the accuracy takes a hit. Using different techniques and strategies can be challenging and novelty can be bought in this aspect. In future, this work can be extended for multiple language models and to make much more fine grained creative images by training the model with larger dataset.

Fig. 6 Column on the left consists of images generated from the baseline diffusion model and that on the right consists of images generated from the quantized model

References

1. Alcantara KD, Calandria JP, Calupas JS, Echas JPR, Sagum RA (2014) Storvi (story visualization): a text-to-image conversion. Int J Future Comput Commun 3(5):363
2. Sauer A, Karras T, Laine S, Geiger A, Aila T (2023) StyleGAN-T: unlocking the power of GANS for fast large-scale text-to-image synthesis. arXiv preprint arXiv:2301.09515
3. Gilboa G, Sochen N, Zeevi YY (2002) Forward-and-backward diffusion processes for adaptive image enhancement and denoising. IEEE Trans Image Process 11(7):689–703. https://doi.org/10.1109/TIP.2002.800883
4. Ramesh A, Dhariwal P, Nichol A, Chu C, Chen M (2022) Hierarchical text-conditional image generation with clip latents. arXiv preprint arXiv:2204.06125
5. Li W, Xu X, Xiao X, Liu J, Yang H, Li G, Wang Z, Feng Z, She Q, Lyu Y et al (2022) Upainting: unified text-to-image diffusion generation with cross-modal guidance. arXiv preprint arXiv:2210.16031
6. Saharia C, Chan W, Saxena S, Li L, Whang J, Denton E, Ghasemipour SKS, Ayan BK, Mahdavi SS, Lopes RG et al (2022) Photorealistic text-to-image diffusion models with deep language understanding. arXiv preprint arXiv:2205.11487
7. Reed S, Akata Z, Yan X, Logeswaran L, Schiele B, Lee H (2016) Generative adversarial text to image synthesis. In: International conference on machine learning. PMLR
8. Singh A, Anekar S, Shenoy R, Patil S (2022) Text to image using deep learning. Int J Eng Techn Res
9. Yu J, Xu Y, Koh JY, Luong T, Baid G, Wang Z, Vasudevan V, Ku A, Yang Y, Ayan BK et al (2022) Scaling autoregressive models for content-rich text-to-image generation. arXiv preprint arXiv:2206.10789
10. Nagel M, Fournarakis M, Amjad RA, Bondarenko Y, Baalen M, Blankevoort T (2021) A white paper on neural network quantization
11. Liu F, Zhao W, He Z, Wang Y, Wang Z, Dai C, Liang X, Jiang L (2021) Improving neural network efficiency via post-training quantization with adaptive floating-point. In: Proceedings of the IEEE/CVF international conference on computer vision (ICCV), pp 5281–5290
12. Zhao X, Wang Y, Cai X, Liu C, Zhang L (2020) Linear symmetric quantization of neural networks for low-precision integer hardware. In: International conference on learning representations. https://openreview.net/forum?id=H1lBj2VFPS
13. Rombach R, Blattmann A, Lorenz D, Esser P, Ommer B (2022) High-resolution image synthesis with latent diffusion models
14. Luo C (2022) Understanding diffusion models: a unified perspective
15. Chen W, Hu H, Saharia C, Cohen WW (2022) Re-imagen: retrieval-augmented text-to-image generator. arXiv preprint arXiv:2209.14491
16. Przewlocka-Rus D, Sarwar SS, Sumbul HE, Li Y, Salvo BD (2022) Power-of-two quantization low bitwidth & hardware compliant neural networks
17. Wang Q, Wang Z (2019) Research on deploying the deep learning models with embedded devices. In: 2019 IEEE 9th annual international conference on cyber technology in automation, control, and intelligent systems (CYBER), pp 1337–1341. https://doi.org/10.1109/CYBER46603.2019.9066643
18. Wang Y, Wang J, Zhang W, Zhan Y, Guo S, Zheng Q, Wang X (2022) A survey on deploying mobile deep learning applications: a systemic and technical perspective. Digital Commun Netw 8(1):1–17
19. Croitoru F-A, Hondru V, Ionescu RT, Shah M (2023) Diffusion models in vision: a survey. IEEE Trans Pattern Anal Mach Intell: 1–20. https://doi.org/10.1109/tpami.2023.3261988
20. Lin T-Y, Maire M, Belongie S, Hays J, Perona P, Ramanan D, Dollár P, Zitnick CL (2014) Microsoft coco: common objects in context. In: Computer vision—ECCV 2014: 13th European conference, Zurich, Switzerland, 6–12 Sept 2014, proceedings, part V 13. Springer, pp 740–755
21. Ho J, Jain A, Abbeel P (2020) Denoising diffusion probabilistic models
22. Wei X, Gong R, Li Y, Liu X, Yu F (2023) QDrop: randomly dropping quantization for extremely low-bit post-training quantization

23. Tao M, Tang H, Wu F, Jing X-Y, Bao B-K, Xu C (2022) DF-GAN: a simple and effective baseline for text-to-image synthesis. In: Proceedings of the IEEE/CVF conference on computer vision and pattern recognition (CVPR)
24. Petsiuk V, Siemenn AE, Sur Behera S, Chin Z, Tyser K, Hunter G, Raghavan A, Hicke Y, Plummer BA, Kerret O et al (2022) Human evaluation of text-to-image models on a multi-task benchmark. arXiv:2211.12112
25. Kulkarni U, Hosamani AS, Masur AS, Hegde S, Vernekar GR, Siri Chandana K (2022) A survey on quantization methods for optimization of deep neural networks. In: 2022 international conference on automation, computing and renewable systems (ICACRS), pp 827–834. https://doi.org/10.1109/ICACRS55517.2022.10028742
26. Ramesh A, Pavlov M, Goh G, Gray S, Voss C, Radford A, Chen M, Sutskever I (2021) Zero-shot text-to-image generation
27. Li B, Qi X, Lukasiewicz T, Torr P (2019) Controllable text-to-image generation. In: Wallach H, Larochelle H, Beygelzimer A, Alche-Buc F, Fox E, Garnett R (eds) Advances in neural information processing systems, vol 32. Curran Associates, Inc. https://proceedings.neurips.cc/paperfiles/paper/2019/file/1d72310edc006dadf2190caad5802983-Paper.pdf
28. Banner R, Nahshan Y, Hoffer E, Soudry D (2019) ACIQ: analytical clipping for integer quantization of neural networks. https://openreview.net/forum?id=B1x33sC9KQ
29. Cho J, Zala A, Bansal M (2022) Dall-eval: probing the reasoning skills and social biases of text-to-image generative transformers. arXiv:2202.04053

One-Shot Learning for Archaeological Site Data Using Deep Neural Network on Embedded Systems

Anoop Patil, Palak Jain, Shastransh Roy, Shri Nagahari Savanur, Uday Kulkarni, and Satish Chikkamath

Abstract Training Machine Learning models with effective features can be challenging and computationally expensive, especially when dealing with limited data. Conventional gradient-based approaches for target recognition require a large amount of data and extensive training. However, this study investigates a method for training Siamese Neural Networks using One-Shot learning, which reduces training time. By leveraging a unique structure, these networks can rank similarities between inputs and generalize to entirely new classes from unknown distributions. The Indian Archaeological Dataset is used to train and test the model, and after fine-tuning the network, its powerful discriminative features can be utilized to extend the predictive capabilities of the model beyond new data. In contrast, the Siamese Neural Network approach allows for effective feature extraction and recognition with limited data and reduces the risk of misconfiguring feature structures. Using this approach, we are able to achieve better accuracy of 86% and 66%, respectively, for One-Shot classification tasks for MNIST and Indian archaeological dataset by usage of triplet loss function. Using quantization, the model size is reduced to make it more compatible for working on embedded devices for classification tasks.

Keywords Siamese networks · One-shot learning · Triplet loss · Contrastive loss · Quantization

A. Patil (✉) · P. Jain · S. Roy · S. N. Savanur · U. Kulkarni · S. Chikkamath
K.L.E. Technological University, Vidyanagar, Hubballi, Karnataka 580031, India
e-mail: anooppatil500@gmail.com

U. Kulkarni
e-mail: uday_kulkarni@kletech.ac.in

S. Chikkamath
e-mail: chikkamath@kletech.ac.in

V. K. Gunjan and J. M. Zurada (eds.), *Proceedings of 4th International Conference on Recent Trends in Machine Learning, IoT, Smart Cities and Applications*, Lecture Notes in Networks and Systems 873, https://doi.org/10.1007/978-981-99-9442-7_24

1 Introduction

Deep Neural Networks (DNNs) [1, 2] have completely changed the way that things are detected and recognized. Since 2012, many other DNN architectures have been created and deployed for a range of tasks, including the classification of plants and animals, the recognition of faces and objects, image synthesis and reconstruction, and more. The Transfer Learning (TL) [3] strategy, which uses a well-structured DNN architecture built for an identification task as a feature extractor for new model, is a general technique for handling classification jobs. To obtain better results in a specific area, one must select a suitable base network, improve the model, set up an optimizer, and choose a loss function.

Even though Machine Learning (ML) [4] has attained cutting-edge performance in many applications, including speech recognition, image recognition, email spam detection, etc., these algorithms frequently struggle to perform predictions regarding data for which either small or no supervised information is available [5]. We require a more effective strategy to generalize to a new category without lengthy retraining.

Classification is a problem that is particularly intriguing when we are only allowed to examine a single case of each potential class before predicting a test occurrence. One-Shot learning is what it is referred to as and it is the main focus of the approach we offer in this study [5–8]. Contrast this with zero-shot learning [9, 10], in which the model is prohibited from looking at any examples from the target classes.

The classification job of One-Shot learning is highly domain-specific. Models that have been trained on a single dataset only provide the correct data for that specific dataset. For the target problem, One-Shot learning models often have strong discriminative features.

We build a architecture upon the Deep Learning(DL) [1] framework, which uses multiple layers of ResNet50 [11] pre-trained on the facial dataset followed by the adoption of the final few layers to adjust weights by training them on the archeological dataset. This is typically achieved by employing a model with a lot of parameters and then a lot of data to avoid overfitting. Though the learning process itself may be expensive, these characteristics are particularly powerful since we may learn them without imposing significant priors.

Our suggested method makes use of Siamese Neural Networks distinctive structure to rate similarities between inputs organically. The Indian Archaeological Dataset [12] is the dataset that is utilized for model training and testing. After the network is tuned, we may take advantage of its potent discriminative properties to expand the model's predictive ability beyond fresh data and generalize to whole new classes with unknown distributions. This approach is a potential direction for future research in the area of One-Shot learning because it is particularly beneficial in cases where data is scarce and when working with novel, previously unexplored classes. ResNet50's Deep Convolutional Neural Network [11] topology allows for the layer and weights to be combined to make the model more weighable. The model is made portable to edge devices, however, by utilizing certain quantization techniques [13–16] as described and removing additional weights for the involved classification tasks.

The paper is composed of several sections, and Sect. 2 explores the different approaches for One-Shot learning proposed by different authors. Section 3 discusses the detailed proposed solution for One-Shot learning using Siamese Network with triplet loss and contrastive loss. Section 4 discusses the outcomes of using the proposed model and results of the edge device deployment. The paper is concluded in Sect. 5, which also discusses achieved results.

2 Related Work

The Machine Learning community has not paid much attention to research into One-Shot learning algorithms, which is still in its early phases. Despite this, this field has been built on several important research areas.

The paper [17] conducted one of the earliest attempts at One-Shot learning in the early 2000s. For One-Shot picture classification, they created a variational Bayesian framework that used previously learned classes to predict new ones when there were insufficient examples of a given class.

The papers [18] developed a method called Hierarchical Bayesian Program Learning (HBPL) to address the issue of One-Shot learning for character recognition from the perspective of cognitive science. The method of generating character drawings to break the image up into smaller bits was modeled by the authors of various works. Finding a structural explanation for the observed pixels was the aim of HBPL. However, because of the size of the joint parameter space under HBPL, which created an insoluble integration problem, the inference was challenging.

The paper [5] proposed the idea of Siamese Neural Networks for One-Shot learning. A Deep Neural Network design that learns a similarity function between pairs of photos was suggested by the authors. The model is trained on pairs of images from several classes, with one serving as the anchor and the other as a good or bad example.

The formula for loss function is

$$\begin{aligned} L\, y_1^{(i)}, y_2^{(i)} &= x\, y_1^{(i)}, _2^{(i)} \log a\, y_1^{(i)}, y_2^{(i)} \\ | 1 \quad & x\, y_1^{(i)}, y_2^{(i)} \log 1 - u\, y_1^{(i)}, y_2^{(i)} + \lambda^F |p|^2. \end{aligned} \tag{1}$$

The model then gains the ability to predict whether or not the two photos are members of the same class. On numerous benchmark datasets, the authors attained cutting-edge findings. The paper [4] for One-Shot learning developed a new neural network design called matching networks. When producing a prediction about a new example, the model uses attention processes to dynamically weigh the significance of each training sample.

They calculate the probability using the formula:

$$P(\hat{x}|\hat{y}, T) = \sum_{i=1}^{k} w(\hat{y}, y_i)x_i. \tag{2}$$

On several benchmark datasets, the authors showed that the matching networks beat the previous state-of-the-art techniques.

The paper [19] proposed a triplet network that can be trained to map images into a metric space where the distance between images reflects their similarity. To train the network to produce smaller distances between similar images and bigger distances between dissimilar photos, triplets of images are used, with the first two image samples belonging to same class and the third sample image belonging to a separate class.

$$\begin{aligned} \text{TripletNet}y, y^-, y^+ = {} & \left\|Net(y) - \text{Net}(y^-)\right\|_2 \in A^2 \\ & + \|Net(y) - \text{Net}(y^+)\|_2. \end{aligned} \tag{3}$$

By calculating the distance between a new image and a set of reference images, the authors show that the learned embedding may be applied to One-Shot learning tasks.

The paper [20] introduces Relation Network (RN), a new deep learning architecture that is intended for few-shot learning. Nurses are trying to learn a function that can classify a new example by comparing it to a group of supporting instances. The relation score for each query is then determined using a formula. The equation is:

$$p_{i,j} = h_\phi\big(D\big(h_\varphi 1(w_d), h_\varphi 2(y_j)\big)\big), \quad i = 1, 2, \ldots, D. \tag{4}$$

In conclusion, Siamese Neural Networks have demonstrated success in One-Shot learning across a range of applications, including natural language processing and image recognition.

By learning a similarity metric between pairs of inputs, these networks can generalize to new categories effectively even with sparse data.

3 Proposed Work

In this research, by quantizing the models for deployment in an Android application, we explore One-Shot learning in greater depth using several methodologies. One-Shot image identification describes a Machine Learning model's capacity to identify an object or image from just one sample or instance. For a model to learn to recognize an object accurately in classical image recognition tasks, a sizable dataset of labeled images must be used for training.

The architecture used for our model is ResNet50. ResNet50 is a convolutional neural network unveiled by Microsoft Research Asia in 2015 [11]. ResNet50, in particular, has 50 layers made up of a mix of convolutional, pooling, batch normalization, and fully connected layers.

Using various loss functions, we further examine the Siamese Neural Network for One-Shot picture learning classification to conclude how to improve classification outcomes. One method for putting One-Shot learning into practice is the Siamese Neural Network. A particular kind of neural network design known as a "Siamese Network" is made up of two or more identical sub-networks that share the same weights. Input similarity measures are taught to these sub-networks during training.

Siamese Networks are commonly used in tasks that involve determining how similar two inputs are, such as face recognition, image matching, and text similarity. When there is little data, it is noisy or it is highly variable, and traditional classification methods would not be effective, they are especially useful. Triplet loss and contrastive loss functions [21] are the used loss functions. In single-shot learning, the model must distinguish images even when there is only one base image to compare. It does not have the advantage of learning from several photos, in contrast to typical Machine Learning techniques. However, loss functions play a crucial role in how well these models perform. The goal of the Siamese Network is to discriminate between the two inputs using the similarity score. The contrastive loss or triplet loss methods can be used to obtain the similarity score for the generic distance metric learning methodology.

When considering how triplet loss works, its main objective is to increase the distance between the anchor and the positive picture while decreasing it between the anchor and the negative one. Therefore, creating an appropriate dataset is the key challenge. A triplet pair of images is necessary when feeding input to a network that employs triplet loss, anchor, positive, and negative. Using triplet loss, a machine learning model can be trained to identify similarities and differences between objects. It makes use of triplets, which are groups of three related things made up of an anchor item, a positive item, and a negative item. The basic goal is to get the model to realize that the anchor is more strongly associated with the positive item than the negative item. The model can distinguish between similar and dissimilar things with greater accuracy as a result.

$$\text{Loss} = \sum_{i-1}^{N} \left[\| h_i^x - h_i^y \|_2^2 - \| h_i^x - h_i^z \|_2^2 + \alpha \right]_+, \tag{5}$$

where loss function is defined, superscript x indicates anchor image, y indicates positive image, z indicates negative image, i denotes ith number of input. Between the anchor and positive example images, the distance in Euclidean terms is $d(x, y)$, while between the anchor and negative example images, it is $d(x, z)$ and is the margin hyperparameter. The function takes the maximum of zero and the difference between the distances, penalizing the model only if the gap between the positive example and anchor is greater than the gap between the negative example and anchor, by at least the margin, where the margin parameter is configured as the minimum offset between distances of similar vs. dissimilar pairs.

Algorithm 1: Siamese network using triplet loss

Require: Generating Triplets from Dataset #[A,P,N]
Ensure: Images (I_A, I_P, I_N) where I_A and I_P belongs to same class and I_N is from different class.
1: Creating Siamese Network with a Base network as ResNet50 R_T
2: Get Embeddings of images
  (E_A, E_P, E_N)
Require: Adjust Weights using Triplet Loss where
Ensure: Distance (D_{AP}, D_{AN}) updates by time during training, such that D_{AP} will be less and D_{AN} is more.
  3:
4: **return** Similarity $_sScore$

To select the strategy with the best performance for further use, we also used Siamese Networks to approach One-Shot learning. We have discussed the proposed systems flowchart in Fig. 1. To learn a representation space where similar data points are mapped closely together and different data points are mapped widely apart is the goal of contrastive loss. Typically, a loss function based on margin is used to do this.

To educate feature representations that can distinguish between pairs of similar and different inputs, the contrastive loss function is described as being meant to optimize the model's parameters. Maximizing the distance between dissimilar pairings and limiting the distance between similar pairs are the aim of contrastive loss. By decreasing the contrastive loss, the Siamese Network learns to construct feature representations that are close to similar inputs and far apart for different inputs. As a result, the network can successfully acquire a similarity metric that can be used for tasks including face recognition, image matching, and text similarity.

$$\text{Loss} = X * C^2 + (1 - X) * \max(m - C, 0)^2, \tag{6}$$

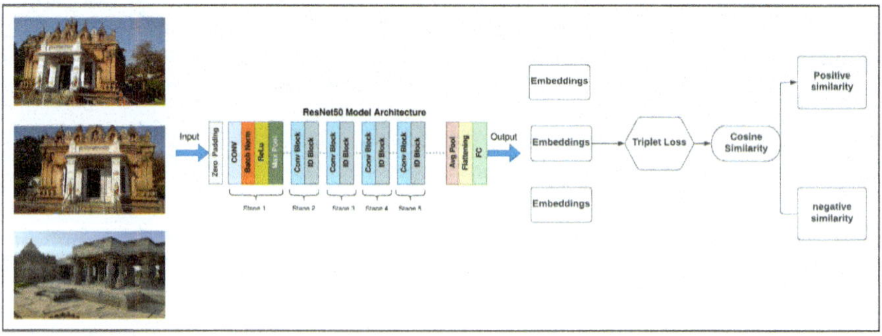

Fig. 1 System flowchart with triplet images as input, ResNet50 as the architecture, and similarity function as the output

where loss function defined, X, is the binary label (1 for similar pairs, 0 for dissimilar pairs), C is the distance between the two data points, and m is the margin hyperparameter.

Algorithm 2: Siamese network using contrastive loss

Require: Generating Similar/Dissimilar pairs from Dataset #[A,P] and [A,N],
Ensure: Images (I_A, I_P, I_N) where I_A and I_P same class and I_N is different class.
1: Creating Siamese Network with a Base network as ResNet50 R_T
2: Get Embeddings of images
(E_A, E_P, E_N)
Require: Adjust Weights using Contrastive Loss where
Ensure: Distance (D_{AP}, D_{AN}) updates by time during training, such that distance of similar pairs D_{AP} will be less and distance of dissimilar pairs D_{AN} is more.
3:
4: **return** Similarity $_sScore$

4 Results and Discussion

Experimental Settings: We follow different settings of datasets for different approaches used under One-Shot learning. Usage of triplet loss function in Siamese Networks requires input of triplets, i.e., anchor, positive, and negative images, whereas contrastive loss function usage in Siamese Network requires similar/dissimilar pair of images as input to model.

Dataset: For the experiments, used two publicly available datasets, i.e., MNIST and archaeological sites dataset.

MNIST dataset: The MNIST dataset consists of 70000 images with 28 × 28 pixel size. To evaluate the suggested loss functions, we have generated a dataset of 10,000 triplets from the data.

Archaeological dataset: The dataset used is a subset of the dataset generated by the authors of paper [12]. It has a total of 108,080 images of size 1920 × 1080 pixels. We created 10,000 similar–dissimilar pairs from 150 images belonging to 15 different classes for contrastive loss experiment. Similarly, 10,000 triplets containing anchor, positive, and negative were created for triplet loss experiment.

When testing a Siamese Network for One-Shot learning approach for both contrastive loss and triplet loss functions on MNIST dataset, the performance of the model on different datasets is different. To make a model understand similarity scores between two images, we need to train it on several pairs of similar/dissimilar images in the form of triplets and pairwise. Figure 4 is the vector representation as points in a two-dimensional space. The points are randomly chosen from a normal distribution. This is how two distinct loss functions organize vector space.

Table 1 Accuracy achieved from different approaches on different datasets

No.	Dataset	Triplet loss (%)	Contrastive loss (%)
1	MNIST	86.53	79.55
2	Archaeological dataset	66.22	59.52

Table 2 Model size before and after quantization

No.	Dataset	Before quantization (MB)	After quantization (MB)
1	MNIST (TL)	220	65
2	MNIST (CL)	220	65
3	Archaeological dataset (TL)	390	93
4	Archaeological dataset (CL)	322	87

Figure 3 shows the images captured from a similar angle of the same archaeological site. The test results indicate that the complexity of images has a significant impact on the performance of the Siamese Network model, and the performance varies depending on several parameters. The model performs better on the MNIST numerical dataset images, which have lower complexity, than on archaeological images. The model also performs better on images captured from the same angle of the given archaeological site than captured from different angles. Table 1 shows a comparison between the accuracies of the model on MNIST and archaeological dataset using triplet loss and contrastive loss.

Building a Siamese Network model, however, requires parallel networks with pre-trained weights derived from a huge dataset, which makes the model incompatible with embedded devices for One-Shot learning-based classification. To make the model suitable for deployment on end devices [22], essential to reduce its size by shrinking it [14, 16]. Therefore, quantization is performed on the models built using triplet loss and contrastive loss. As a result, the size of the models varies after the quantization process, which affects their performance to some extent. Table 2 shows a comparison between the model sizes before and after quantization of MNIST and archaeological dataset using triplet loss and contrastive loss. However, this trade-off between model size and performance can be compromised to achieve portability on edge devices. Figure 2 is the screenshots of the application deployed on the embedded system.

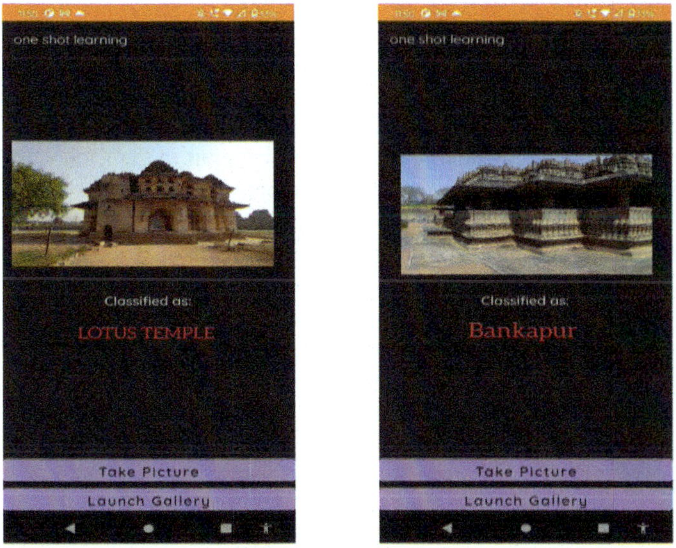

(a) (b)

Fig. 2 User interface of the Android application, that receives input image and classifies archeological sites

(a) Images captured from similar angle of same archaeological site.

(b) Images captured from different angle of same archaeological site.

Fig. 3 Example images of archaeological sites from dataset

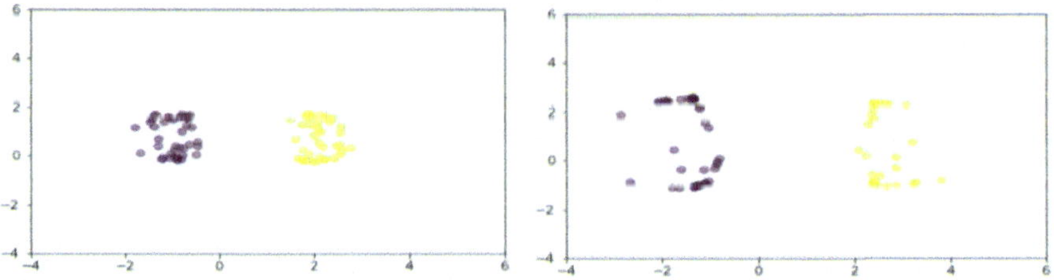

(a) Vector space organization by triplet loss function.

(b) Vector space organization by contrastive loss function.

Fig. 4 Vector space organizations

5 Conclusion

It is only necessary to update the weights in contrastive loss to either maximize or minimize the similarity of the same class. On the contrary, if triplet loss was used, the model would simultaneously pull the positive input to the anchor and push the negative image away from it. From the experiments, it is observed that the results on MNIST dataset are better compared to archaeological dataset, due to the difference in image complexities. On quantization, model size is lowered to one third of its original size with a little drop in performance. There is a lot of potential for identification problems with relatively little training data when the Siamese Neural Network and triplet loss are combined rather than contrastive loss.

References

1. LeCun Y, Bengio Y, Hinton G (2015) Deep learning. Nature 521(7553):436–444
2. Chikkamath S, Nirmala SR (2022) Music detection using deep learning with tensorflow. In: ICDSMLA 2020: proceedings of the 2nd international conference on data science, machine learning and applications. Springer Singapore, pp 283–291
3. Yang Q et al (2020) Transfer learning. Cambridge University Press
4. Mitchell TM (2007) Machine learning, vol 1. McGraw-Hill, New York
5. Koch G, Zemel R, Salakhutdinov (2015) Siamese neural networks for one-shot image recognition. In: ICML deep learning workshop, vol 2, no 1
6. Uzhinskiy AV et al (2021) One-shot learning with triplet loss for vegetation classification tasks. 45(4):608–614
7. Vinyals O et al (2016) Matching networks for one shot learning. Adv Neural Inf Process Syst 29
8. Thakurdesai N, Raut N, Tripathi A (2018) Face recognition using one-shot learning. Int J Comput Appl 182:35–39
9. Zhang L, Xiang T, Gong S (2017) Learning a deep embedding model for zero-shot learning. In: Proceedings of the IEEE conference on computer vision and pattern recognition
10. Sucholutsky I, Schonlau M (2021) Less than one-shot learning: learning N classes from M¡ N samples. In: Proceedings of the AAAI conference on artificial intelligence, vol 35, no 11
11. He K et al (2016) Deep residual learning for image recognition. In: Proceedings of the IEEE conference on computer vision and pattern recognition
12. Kulkarni U et al (2019) Classification of cultural heritage sites using transfer learning. In: 2019 IEEE fifth international conference on multimedia big data (BigMM). IEEE
13. Kulkarni U et al (2021) AI model compression for edge devices using optimization techniques. In: Gunjan VK, Zurada JM (eds) Modern approaches in machine learning and cognitive science: a walkthrough. Studies in computational intelligence, vol 956. Springer, Cham
14. Krishnamoorthi R (2018) Quantizing deep convolutional networks for efficient inference: a whitepaper. arXiv preprint arXiv:1806.08342
15. Kulkarni U, Meena SM, Gurlahosur SV, Bhogar G. Quantization friendly MobileNet (QF-MobileNet) architecture for vision based applications on embedded platforms. J Neural Netw
16. Kulkarni U, Meena SM et al (2022) Hybrid optimization for DNN model compression and inference acceleration. In: 2022 2nd international conference on intelligent technologies (CONIT), Hubli, India, pp 1–8
17. Fe-Fei L (2003) A Bayesian approach to unsupervised one-shot learning of object categories. In: Proceedings ninth IEEE international conference on computer vision. IEEE

18. Lake BM, Russ RS, Tenenbaum J (2013) One-shot learning by inverting a compositional causal process. Adv Neural Inf Process Syst 26
19. Hoffer E, Ailon N (2015) Deep metric learning using triplet network. In: Similarity-based pattern recognition: third international workshop, SIMBAD 2015, Copenhagen, Denmark, 12–14 Oct 2015. Proceedings 3. Springer International Publishing
20. Sung F et al (2018) Learning to compare: relation network for few-shot learning. In: Proceedings of the IEEE conference on computer vision and pattern recognition
21. Ghojogh B et al (2020) Fisher discriminant triplet and contrastive losses for training siamese networks. In: 2020 international joint conference on neural networks (IJCNN). IEEE
22. Nadeski M (2019) Bringing machine learning to embedded systems. Texas Instr: 1–5
23. Kulkarni U, Meena SM, Joshua P, Rodrigues K, Gurlahosur SV (2020) Integrated crowdsourcing framework using deep learning for digitalization of Indian heritage infrastructure. In: 2020 IEEE sixth international conference on multimedia big data (BigMM), New Delhi, India, pp 200–208. https://doi.org/10.1109/BigMM50055.2020.00036

EnhanceNet: A Deep Neural Network for Low-Light Image Enhancement with Image Restoration

Rahul R. Chheda, Khushi Priyadarshi, Shreyas M. Muragodmath, Farheen Dehalvi, Uday Kulkarni, and Satish Chikkamath

Abstract An important task in computer vision is improving low-light images, which aims to increase the quality of images collected in low light. Our study has introduced EnhanceNet, a deep learning-based technique, in this area. The DCE-Net low-light image enhancer architecture and the NAFNet baseline architecture make up this model. The model increases pixel intensity from a low-light image as input to enhance image quality, reduce grain, and remove blur. Altering the exposure, contrast, and gamma as well as adding filters like a histogram equalization or a simple linear transformation may all be used to achieve this. For image denoising and deblurring, we pass the enhanced image with a nonlinear activation-free network or NAFNet. The proposed combination model is trained on our own large dataset to uncover the association between low-light photographs and their corresponding enhanced versions. According to experimental results on numerous datasets like LOL, SICE, LIME, and NPE, the proposed approach provides state-of-the-art performance in terms of visual quality and quantitative measurements like PSNR value 60% more efficient than previous state-of-the-art model. The proposed EnhanceNet is computationally efficient since it has a small model size and a powerful design. This network can efficiently be implemented on edge devices for various applications like surveillance, automobile, and photographic camera. In this paper, the performance of the model on Android devices has been presented.

Keywords Deep curve estimation · NAFNet · EnhanceNet · Picture denoising · Image restoration

R. R. Chheda (✉) · K. Priyadarshi · S. M. Muragodmath · F. Dehalvi · S. Chikkamath
School of Electronics and Communication Engineering, KLE Technological University, Hubballi, India
e-mail: rrchbl3@gmail.com

S. Chikkamath
e-mail: chikkamath@kletech.ac.in

U. Kulkarni
School of Computer Science and Engineering, KLE Technological University, Hubballi, India
e-mail: uday_kulkarni@kletech.ac.in

V. K. Gunjan and J. M. Zurada (eds.), *Proceedings of 4th International Conference on Recent Trends in Machine Learning, IoT, Smart Cities and Applications*, Lecture Notes in Networks and Systems 873, https://doi.org/10.1007/978-981-99-9442-7_25

1 Introduction

Lack of light can be a challenge for low-light photography, resulting in images with poor brightness, high noise, and lost information [1–3]. However, machine learning algorithms have shown promising results in enhancing low-light photos [4]. By training models with large datasets of low-light photographs, these algorithms can effectively reduce noise, increase brightness, and enhance overall image quality.

This project aims to develop a machine learning-based low-light photo enhancement method that utilizes neural networks and is trained on several photographs with poor lighting. The resulting solution could have various practical applications, such as improving the performance of security cameras, enhancing the quality of smartphone photos taken in low light, and improving the accuracy of surveillance systems.

Three natural photographs obtained in dim illumination are shown in Fig. 1. Particularly in the first scenario, there is not much light. The darkness obscures significant noise and color distortion. Just increasing the image's brightness makes the top-right corner degeneration visible. The second photo was taken at night with low ambient light, which is why the majority of the objects are backlit. Imaging at midday while facing the light source (the sun) does not address the issue as effectively as the second instance, despite the scene being more obvious and the ambient light being stronger. Direct amplification will oversaturate the relatively light regions in the last two images. The bulk of low-level numerical vision tasks require ground truth training data, such as denoising [3] and super-resolution, and deep learning-based algorithms have outperformed conventional methods. Although the target issue, such as low-light picture improvement, lacks actual ground truth data, light intensity can be arranged in a certain order. Therefore, from the standpoint of the consumer, the ideal light levels for various people/requirements may vary slightly. In other words, it can be challenging to identify the ideal or actual lighting situation. Therefore, mapping an image to a specific degree of light is not ideal.

We list the problems with low-light picture improvement as follows, based on the previous investigation:

1. To determine the lighting component, estimate one photo, then adjust the light levels as needed. The data collection of photos that our team collected and processed serves as the model's training set.
2. After illuminating the dark areas, degradations such noise and color distortion that were previously hidden in the shadows should be eradicated.
3. To prevent overfitting, we introduce a low-light augmentation network that remains unaffected by either paired or unpaired training data. Consequently, our approach effectively adjusts to changing lighting conditions.

In this paper, we propose a deep neural network that can simultaneously consider the aforementioned problems.

Fig. 1 Left column: three natural photos acquired in various lighting situations. Our improved findings are shown in the right column

2 Related Works

2.1 Zero-Reference Deep Curve Estimation for Low-Light Image Enhancement Author: Chunle Guo et al.

To improve low-light images, Zero-DCE [5] can work in a variety of lighting situations, including uneven and low-light scenarios. In the case of Zero-DCE, instead of doing image mapping, the problem is framed as an image-specific curve estimation task. The suggested method specifically generates high-order curves from the input image in dim lighting. These curves are then used to build an improved image by pixel-by-pixel adjusting the dynamic range of the input. It differs from existing methods based on CNNs (like Mirnet [6]) and GANs (like EnlightenGAN [7]) in that it uses a null reference.

The Zero-DCE [5] model can effectively improve low-light images by reducing noise and increasing brightness while preserving important image details. It has

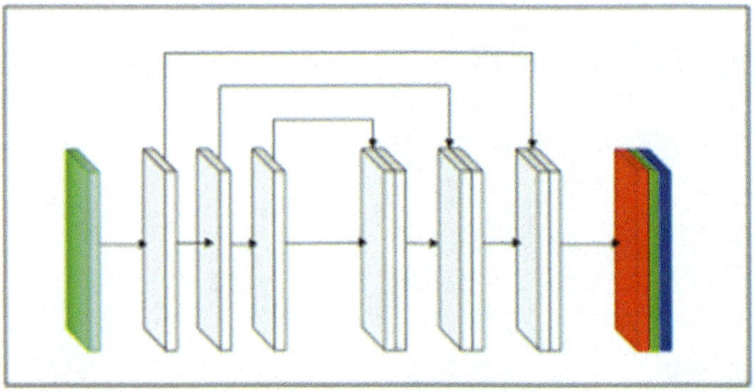

Fig. 2 Architecture of Zero-DCE [5]

shown promising results in a numerous tasks for improving low-light images, including object detection and recognition. Zero-DCE has potential for applications in a variety of fields, including security, surveillance, and mobile photography, where low-light image quality is a common concern (Fig. 2).

The Zero-DCE dataset is a collection of photos taken in low light and their corresponding well-lit reference images. It was created to train and evaluate image enhancement models, especially those based on the Zero Reference Curve (Zero-DCE) method. The dataset contains 2002 paired images of various scenes and objects and has been used in several studies on image enhancement and low-light image processing. On the other hand, the LOL (Low-Light Image Enhancement Benchmark) dataset is another dataset for evaluating low-light image enhancement algorithms. It contains 500 pairs of low-lit and well-lit reference images of natural scenes and has become a widely used benchmark for assessing the effectiveness of low-light image enhancing techniques. The LOL dataset has been instrumental in the creation of numerous cutting-edge low-light image enhancement algorithms and is frequently used in image enhancement and computer vision research.

2.2 Kindling the Darkness: A Practical Low-light Image Enhancer Author: Yonghua Zhang et al.

This research [8] proposes a deep neural network-based method for low-light picture enhancement.

The suggested technique begins by taking a series of dimly lit photographs at various exposures. The final step is to integrate these photos into a high-quality augmented image using a new neural network architecture. A new fusion block in the network averages input images using weights depending on their exposure levels. This method enhances contrast in the augmented image and aids the network in handling dynamic range compression.

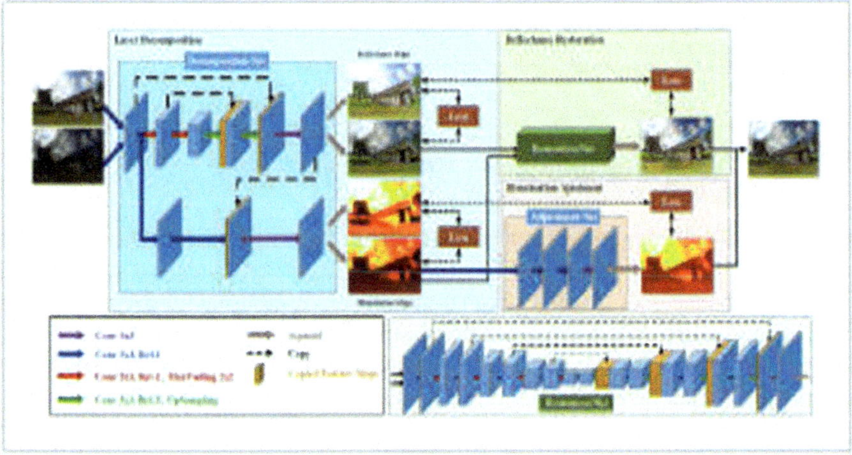

Fig. 3 Architecture of Kindling the Darkness [8]

The practicality of the suggested approach is one of its main benefits. It can function on a typical smartphone, making it appropriate for practical uses like tracking and mobile photography. The approach is also efficient because it may run in real-time on a mobile device and does not need any pre or post-processing steps.

According to experimental results, the proposed method outperforms a number of cutting-edge methodologies in terms of both objective and arbitrary evaluation criteria. In a variety of situations and applications, it might improve the quality of low-light photographs (Fig. 3).

The Kindling the Darkness (KtD) dataset is a practical low-light image enhancer dataset that was introduced in a research paper by Zhang et al. in 2021. The dataset consists of 200 low-light images that were taken in actual situations using a variety of tools and settings, together with 16 reference photos of the highest quality. The KtD dataset is made to assess and contrast several low-light image enhancement techniques in terms of their capacity to improve image quality, recover features, and reduce noise.

2.3 *Picture Denoising with Deep Neural Networks*

Deep neural networks have been used in a number of attempts to solve the denoising problem. When employed for image denoising, convolutional neural networks (CNNs) offer representational capabilities that are on par with or better than MRF models [9]. The multi-layer perceptron (MLP) technique developed by Bishop (1995) Neural [10] was successful in reducing noise from images. A trained nonlinear reaction-diffusion (TNRD [11]) opens a predetermined number of gradient descent inference steps, which can be used to define it as a feed-forward deep network. The results of this technique, which used stacked sparse denoising auto-encoders to get rid of Gaussian noise, are similar to those of K-SVD [12].

2.4 Batch Normalization and Residual Learning

Thanks to advancements in deep learning techniques, convolutional neural networks have shown to be highly successful in resolving a range of visual problems. Examples of major successes in CNN model training include adjusted linear unity (ReLU) [13], depth-width trade-off, parameter initialization, gradient-based optimisation approaches, batch regularization, and the rest of the training. Fast training on current, strong GPUs is another element influencing CNN performance. It is simple to create enough Gaussian's training data from a collection of high-quality photos. The primary goal of this investigation is to examine CNN's layout and imagery. In this part, we examine two DnCNN-related methods: batch regularization and recurrent learning [14].

2.5 Simple Baselines for Image Restoration-NAFNet Author: Liangyu Chen et al.

The authors of [15] have made significant contributions by presenting a novel approach for developing high-performing computer vision systems. By analyzing state-of-the-art (SOTA) methods and identifying their fundamental components, the authors have shown that a simpler baseline can achieve comparable or better results with lower computational costs, which could be beneficial for academic researchers looking to explore new ideas.

A nonlinear activation-free network (NAFNet) has also been developed by the authors by eliminating or substituting nonlinear activation functions such as Sigmoid [16], ReLU [13], and GELU [17]. The authors have also demonstrated the relationship between GELU and channel attention to gated linear unit. NAFNet has been shown to perform at least as well as the baseline, if not better. This study is the first to propose that SOTA performance in computer vision tasks may not always require nonlinear activation functions, which may open up new design possibilities for subsequent studies in this field.

2.6 *Deblurring using Analysis-Synthesis Networks Pair Author: Adam Kaufman et al.*

Some of the approaches used for image deblurring include subspace analysis, blind deconvolution techniques, deconvolution using statistics, local phase quantization, linear ternary patterns, and set theoretic characterization [18]. Using a training set of blurred images, subspace analysis includes building a model of each point spread function or blur kernel in the feature space. To sharpen both the query and target photos, blind deconvolution techniques employ learned models of face appearance fluctuation under various degrees of blur. There are two types of blind deconvolution procedures—blind deconvolution based on projections and maximum likelihood restoration. Blind deconvolution based on projections recovers both the real image and the point spread function simultaneously, while maximum likelihood restoration has low processing complexity and helps obtain the genuine image's blur, noise, and power spectra.

Estimating the blur kernel is necessary for the deblurring process using the blur estimate algorithm. The blurred input image is first re-blurred using Gaussian blur kernels with various blur radii, and the unknown blur radius is determined using the difference ratios between the various re-blurred images. Using both blurred and noisy photos, deblurring with noisy image pairings entails identifying an accurate blur kernel. An iterative technique has been developed to estimate a decent starting kernel and minimize deconvolution artifacts. However, there are drawbacks such as an unchanging spatial point spread function.

3 Methodology

We present the EnhanceNet framework for improving low-light images with reduced grain, which expands the NAFNet architecture for enhancing/restoring picture quality by denoising and reducing blurriness in the input image. For each input dark or low-light image, EnhanceNet estimates a collection of best-fitting light-enhancement curves using the (DCE-Net) a Zero-DCE framework. A low-light image might be automatically converted to its enhanced equivalent using a light-enhancement curve whose parameters are self-adaptive to the input image. When creating such a curve, three objectives should be taken into account: To avoid information loss due to overflow truncation, each pixel value in the enhanced image should fall inside the normalized range [0, 1]. To preserve the contrast between close pixels, it must be monotone (Fig. 4).

The shape of this curve must be straightforward, practical, and differentiable in order to support backpropagation.

Instead of merely the illumination channel, the light-enhancement curve is applied to all three RGB channels individually.

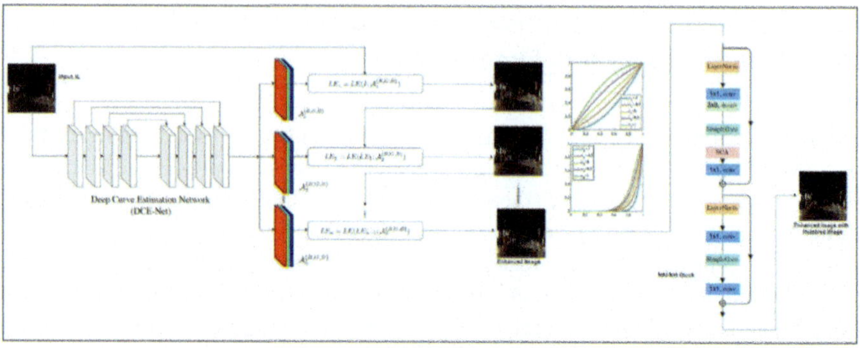

Fig. 4 **a** Zero-DCE [5] framework. A DCE-Net seeks to forecast accurate light-enhancement curves or LE-curves that progressively improve an image given as input. LE-curves in **b**, **c** with various iteration counts (n) and adjustment parameters. In **c** $\alpha 1$, $\alpha 2$, $\alpha 3$ all equal $-$ 1, while n equals 4. A V axis showing the output pixel values and a H axis showing the input pixel values are both present in each subfigure. **d** Free network block and nonlinear activation. With simplified channel attention (SCA) and SimpleGate, it replaces CA/GELU [17]. This figure depicts the specifics of various NAF-Net components [15]

A shallow deep neural network called the DCE-Net learns how to convert an input image into the best curve parameter mappings. A low-light image serves as the DCE-Net's input, and a series of pixel-by-pixel curve parameter translations for related higher-order curves are the result of this process. A simple CNN is made by symmetrically concatenating nine convolutional layers. After the ReLU activation function in each layer, 32 convolutional kernels with a size of 33 and a stride of 1 are added. After the final convolutional layer, the Tanh activation function generates 24 parameter maps over the course of eight iterations, requiring three curve parameter maps for each of the three channels.

3.1 Light-Enhancement Curve (LE-Curve)

In order to automatically improve low-light images, we intend to create a type of curve that is motivated by the curve alterations seen in photo editing software. The sole variable on which the self-adaptive curve parameters of this kind depend is the input image. When designing such a curve, it is crucial to consider three key factors. First, it is essential to guarantee that all pixel values in the improved image stay within the standardized range of [0, 1]. This precaution is vital to prevent information loss caused by overflow truncation. Second, the curve should exhibit monotonic behavior to maintain contrast between adjacent pixels. Lastly, the curve's shape should be as straightforward and differentiable as feasible to facilitate gradient backpropagation during the process.

The quadratic curve, depicted below, serves to accomplish the following three goals:

$$\mathrm{DE}(P(x); \alpha) = P(x) + \alpha P(x)(1 - P(x)). \tag{1}$$

DE($P(x)$; α), improved rendition of the designated input, where x depicts pixel coordinates. $P(x)$ is a trainable curve parameter that regulates both exposure volume and LE curve size. Each operation is carried out pixel by pixel, with each pixel's range set to [0, 1]. Instead of focusing primarily on the lighting channel, we apply the LE curve to each of the three RGB channels separately. The three-channel correction reduces the possibility of over saturation while preserving natural color. More information is provided in the supporting materials. The LE curve accomplishes the three previously mentioned goals. It effectively addresses the dynamic range of an input image and it can also be changed using the LE curve. With the help of this ability, you can improve areas with poor lighting while also eliminating overexposure artifacts.

The Higher Order Curve. Eq. (1)'s LE curve can be used repeatedly to allow for more variable modifications in low-light circumstances.

$$\mathrm{En}(x) = \mathrm{En} - 1(x) + \alpha n \mathrm{En} - 1(x)(1 - \mathrm{En} - 1(x)). \tag{2}$$

As an example, we show the maps of the curve parameters generated by the three channels. Comparable but different adjustment patterns may be seen in the parameter maps of different channels with the best-fit values that show how the three low-light picture channels differ and are relevant. The brightness of numerous spots, including the two glitters on the wall, is precisely depicted by the curve parameter map. The updated version picture may be produced right away by pixel-wise curve mapping using the fitted maps. The updated version preserves content in light parts while displaying stuff in dark areas.

3.2 DCE-Net

Using the input image's best-fitting curve parameter translations, the DCE-Net deep learning model aims to improve low-light photos. A straightforward CNN with seven convolutional layers and symmetric concatenation makes up the DCE-Net architecture [5]. Following the ReLU [13], activation function is a set of 32 convolutional kernels, each with a size of 33 and a stride of 1. The associations between adjacent pixels are preserved by removing the layers used for batch normalization and down-sampling. The Tanh activation function is used after the last convolutional layer to produce 24 parameter maps throughout the course of eight repetitions ($n - 8$). The model produces three curve parameter maps for each iteration for the three channels.

3.3 Unsupervised Loss Functions

To assess the upgraded images quality and in order to facilitate zero-reference learning within the DCE-Net, we utilized various differentiable non-reference loss functions during the training process of our DCE-Net [5], and we employed four types of losses as follows:

The contrast between adjacent regions in the input picture and their corresponding augmented regions is kept by the spatial consistency loss *Is*:

$$I_s = \frac{1}{M}\sum_{i=1}^{K}\sum_{j\in\Omega(i)}(|(Xi - Xj)| - |(Li - Lj)|) \tag{3}$$

(i) in Eq (3). The local regions in our images are the four regions (top, bottom, left, and right) that are centered on region i and are represented by 4×4 pixel squares. The average intensity value of a local region is denoted by X and L in both the input image and the enhanced version, and M denotes the overall number of local regions. The loss of spatial consistency quantifies the variation between nearby regions in the input image and their corresponding parts in the enhanced image. To prevent overexposure, we additionally use an exposure control loss. *I*ep to regulate exposure to prevent areas that are too or too little exposed. The difference between the average intensity value for a given area and the well-exposedness level J is the exposure control loss.

We use conventional methods to set J as the RGB color space's gray level. Although J was set at 0.6 for our trials, there was not much of a performance difference when J was varied between [0.4, 0.7] [5]. *I*ep loss can be expressed as follows:

$$I_{\text{ep}} = \frac{1}{K}\sum_{k=1}^{M}|(X_k - J)\,. \tag{4}$$

This loss in Eq. (4) is calculated with a 4×4 local area size utilizing the local region's average intensity values from both the upgraded version and the input image. The suggested technique also uses a color constancy loss based on the Gray-World theory of color constancy [19], which corrects any potential color discrepancies in the image by guaranteeing that the average color of each sensor channel is gray throughout the entire image.

Links are built between the three changed channels and enhanced the picture. It is possible to evaluate the loss of color *I c* constancy as follows:

$$I_c = \sum_{\forall(r,s)\in\epsilon}(T^r - T^s)^2, \epsilon = (R, G), (G, B), (B, R) \tag{5}$$

In Eq. (5), T_r represents the improved image's r channel's average intensity value, and (r, s) stands for a pair of channels. We implement a smoothness loss for illumination. B keeping the monotonicity linkages between adjacent pixels in each curve parameter map ItB stands for lighting smoothness loss.

$$I_{t_B} = \frac{1}{Z}\sum_{n=1}^{Z}\sum_{d\in\xi}\left(\left|\Delta_a A_n^d\right| + \left|\Delta_b Z_n^d\right|\right)^2, \xi = R, G, B. \tag{6}$$

The horizontal and vertical gradient methods are denoted by Δ_a and Δ_b in Eq. (6) above, where N represents the number of iterations, and below is a compilation of the overall loss functions:

$$IL_{\text{total}} = IL_s + IL_{\text{ep}} + WT_d IL_d + WT_{tB} I_{tB}. \tag{7}$$

In Eq. (7), W_{Tc} and WT_{tB} are the weights of the losses.

3.4 NAFNet

The inter-block complexity is decreased by using a conventional single-stage U-shaped layout with skip connections. We believe that the architecture would not be a bottleneck in terms of performance. Neural networks are constructed in modules. A fundamental building block with convolution, ReLU, and shortcut, which are most commonly used, organized. To make things simple, we will simply refer to it as PlainNet. We chose a transformer over a convolution network because of the following reasons. First, although another study suggests that transformers are not necessary to obtain SOTA results, they do function well for computer vision. Furthermore, depthwise convolution is less challenging to understand than self-attention. Third, rather than discussing the pros and cons of transformers and convolutional neural networks. This study aims to provide a fundamental framework. The attention process is described in Fig. 5 that follow.

In high-level computer vision applications, normalization is often employed, and it is becoming more popular in low-level vision. While instance normalization was

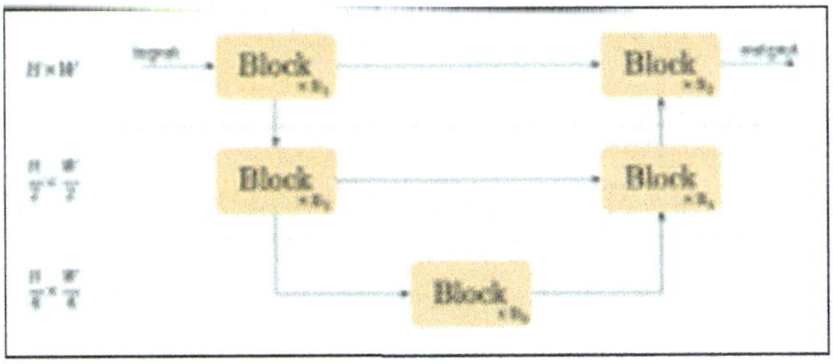

Fig. 5 UNet architecture of NAFNet [14]

revived and overcame the short batch size issue, batch normalization was abandoned due to the risk of unstable statistics resulting from tiny batch sizes. On the other hand, adding instance normalization necessitates human interaction and may not always increase performance. In contrast, as the usage of transformers grows, a rising number of solutions, including SOTA methodologies, are employing layer normalization. These findings led us to believe that layer normalization is critical for SOTA restorers, and we include it in the basic block outlined above. This tweak aids in easing training even with a 10% increase in learning pace.

Rectified Linear Unit (ReLU), the plain block's activation function, is commonly employed in computer vision. ReLU is typically swapped out with GELU in SOTA methods, though. This model does not account for this substitution. In line with the result, the performance on SIDD remains the same (from 39.73 to 39.71 dB), while it rises 0.21 dB (from 31.90 to 32.11 dB) on GoPro. To keep image denoising efficiency while providing a non-trivial increase in picture deblurring, we switched ReLU for GELU in basic block.

Transformer's attention mechanism is an inescapable issue in the construction of the block's internal structure given its recent popularity in computer vision. We will only touch on a few of the multiple attentional processes. Using the fundamental self-attention method, the target feature is formed by linearly combining all qualities, which are weighted based on how similar they are. As a result, while each feature contains global information, the computing cost climbs quadratically with size of the feature map. High-resolution data handling is used in some photo restoration tasks, rendering regular self-attention useless. To prevent the problem of excessive computation, only use self-attention within a fixed-sized local window, while having little global information. Since the depthwise convolution of the plain block may effectively capture local information, we do not utilize window-based attention.

4 Results

In this paper, we have presented the results of our experimental evaluation of the EnhanceNet algorithm on several benchmark datasets, including SICE, LIME, and NPE (Fig. 6).

In terms of objective measures such as PSNR [20] and SSIM [21] index, our studies reveal that EnhanceNet outperforms the original Zero-DCE method and other cutting-edge approaches. Our subjective judgment based on human perception study provides additional evidence that EnhanceNet produces images that are more visually pleasant and realistic than those produced by other approaches. These results demonstrate that the recommended modifications to the Zero-DCE algorithm can successfully address the drawbacks of the original approach and enhance its efficacy in enhancing low-light images. For a number of real-world uses requiring excellent picture quality, such as surveillance, autonomous driving, and medical imaging, EnhanceNet may be a viable choice. Overall, the recommended modifications to the

Fig. 6 Images in the left column depict the input images used, and the resulting images on the right showcase the improvements achieved by the EnhanceNet model

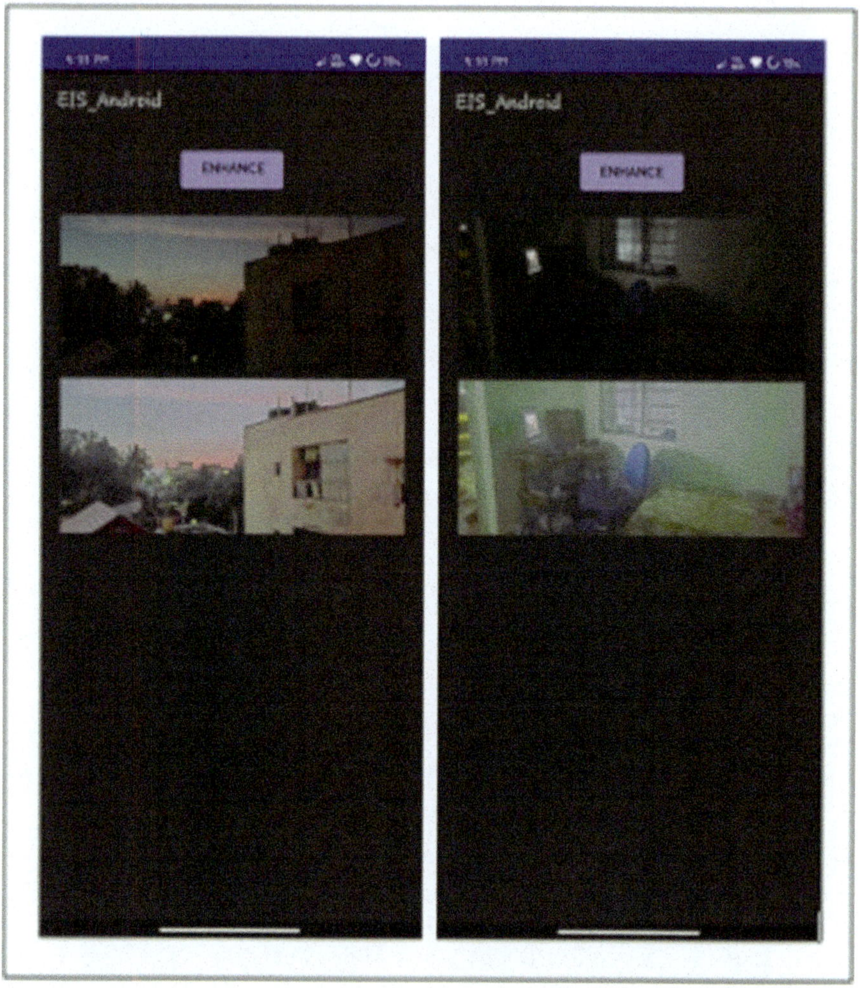

Fig. 7 Images in the top depict the input images used, and the resulting images on the bottom showcase the improvements achieved by the EnhanceNet model deployed on Android Platform

Zero-DCE approach have the potential to significantly advance the field of image processing and enhancement (Fig. 7).

4.1 Datasets

In order to advance low-light picture improvement, a collection of 600 unique photographs was created in a variety of lighting circumstances and brightness fluctuations. Low-light photography often yields photographs that are excessively dark

Fig. 8 Four natural photos acquired in various lighting situations, variations in brightness levels

and lack detail. This makes it difficult for image processing algorithms to extract important information and characteristics from the image.

The dataset consists of 4200 images with varying brightness levels to mimic various lighting conditions seen in real-world settings. The brightness levels were adjusted using common image processing techniques including gamma correction and histogram equalization (Fig. 8).

The dataset was created with the goal of providing a diverse collection of pictures for use in computer vision and machine learning applications, particularly those involving image processing and analysis. By providing multiple versions of each image, we want to boost the dataset's value for testing and training algorithms designed to handle a range of lighting circumstances. The collection may enhance low-light image quality, which could result in enhanced understanding and judgment in a variety of sectors.

Table 1 Comparative analysis using full-reference metrics for evaluating image quality considering PSNR, SSIM and MAE

Method	PSNR ↑	SSIM ↑	MAE ↓
SRIE	14.41	0.54	127.08
LIME	16.17	0.57	108.12
Li et al.	15.19	0.54	114.21
RetinexNet	15.99	0.53	104.81
Wang et al.	13.52	0.49	142.01
EnlightenGAN	16.21	0.59	102.78
Zero-DCE	16.57	0.59	98.78
EnhanceNet	27.59	0.58	149.05

Table 2 Comparative analysis using full-reference metrics for evaluating image quality

Method	RT	Platform
SRIE	12.1865	MATLAB (CPU)
LIME	0.4914	MATLAB (CPU)
Li et al.	90.7859	MATLAB (CPU)
RetinexNet	0.12	TensorFlow (GPU)
Wang et al.	0.021	TensorFlow (GPU)
EnlightenGAN	0.0078	PyTorch (GPU)
Zero-DCE	0.0025	PyTorch (GPU)
EnhanceNet	0.0134	PyTorch (GPU)

4.2 Quantitative Comparisons

We use the Peak Signal-to-Noise Ratio (PSNR, dB) [20], Structural Similarity (SSIM) [21], and Mean Absolute Error (MAE) [22] metrics to compare the effectiveness of various approaches for evaluating full-reference picture quality (Tables 1 and 2).

5 Conclusion

This study proposes an approach that can learn to decompose an observed image into reflectance [23] and illumination in a data-driven manner without using the decomposed reflectance and illumination as the ground truth. The introduction of subsequent light enhancement for illumination and denoising processes for reflectance. The disintegration network and the low-light augmentation network are trained from beginning to end. According to experimental findings, our technique offers enhanced results that are both aesthetically pleasing and accurately depict image decomposition.

NAFNet, a nonlinear activation-free network, is what we present as a consequence. Its performance is on par with or better than the baseline while being simplified. Our method performs better than other light-enhancement systems, according to experiments.

References

1. Fu X, Huang J, Zhang X, Ding X, Paisley J (2019) Exposure fusion using deep illumination estimation. In: IEEE conference on computer vision and pattern recognition (CVPR)
2. Li X, Guo Q, Li Y, Li H (2020) A low-light image enhancement method based on an improved deep convolutional neural network. J Electron Imaging 29(1):013011
3. Zhang K, Zuo W, Gu S, Zhang L. Learning deep cnn denoiser prior for image restoration. In: IEEE conference on computer vision and pattern recognition (CVPR)
4. Chen C, Li W, Sakaridis C, Dai D, Van Gool L (2018) Learning to see in the dark. In: IEEE conference on computer vision and pattern recognition (CVPR)
5. Guo C, Li C, Chen Q, Loy CC (2020) Zero-reference deep curve estimation for low-light image enhancement. In: IEEE conference on computer vision and pattern recognition (CVPR)
6. Zamir SW, Arsalan M, Khan FS, Munir H, Liu J, Zhang X, Shao L (2020) Learning enriched features for real image restoration and enhancement. In: European conference on computer vision (ECCV). Springer, pp 740–757
7. Jiang Y, Liu X, Cheng T, Li X, Yang C, Wang S (2019) Enlightengan: deep light enhancement without paired supervision. In: IEEE international conference on computer vision (ICCV). IEEE, pp 8100–8109
8. Zhang Y, Sun W, Huang Q, Jiang Y, Lin Z (1992) Kindling the darkness: a practical low-light image enhancer. In: International conference on computer vision (ICCV)
9. Geman S, Geman D (1992) Markov random field modeling in image analysis. IEEE Trans Image Process 1(3):312–321
10. Bishop CM (1995) Neural networks for pattern recognition. Oxford University Press, Oxford
11. Chen Y, Pock T, Ranftl R (2016) Trainable nonlinear reaction diffusion: a flexible framework for fast and effective image restoration. In: Proceedings of the IEEE conference on computer vision and pattern recognition, pp 3867–3875
12. Aharon M, Elad M, Bruckstein A (2006) K-SVD: an algorithm for designing overcomplete dictionaries for sparse representation. IEEE Trans Signal Process 54(11):4311–4322
13. Nair V, Hinton GE (2010) Rectified linear units improve restricted boltzmann machines. In: International conference on machine learning (ICML)
14. Zhang K, Zuo W, Chen Y, Meng D, Zhang L (2017) Beyond a gaussian denoiser: residual learning of deep CNN for image denoising. IEEE Trans Image Process
15. Chen L, Yuan S, Wang J, Hu W, Zeng B (2020) Simple baselines for image restoration-NAF-net In: European conference on computer vision (ECCV)
16. Hinton GE (2006) A fast learning algorithm for deep belief nets. Neural Comput 18(7):1527–1554
17. Hendrycks D, Gimpel K (2017) Gaussian error linear units (gelus). In: International conference on learning representations (ICLR)
18. Sun D, Yang X, Liu M-Y, Kautz J (2017) Deblurring using analysis-synthesis networks pair. In: IEEE conference on computer vision and pattern recognition (CVPR)
19. Buchsbaum G (2000) A "gray world" assumption for color constancy. J Opt Soc Am A 17:2015–2023
20. Hore A, Ziou D (2002) Peak signal-to-noise ratio as an image quality metric. Can J Remote Sens 28(1):45–62
21. Wang Z, Bovik AC, Sheikh HR, Simoncelli EP (2004) Image quality assessment: from error visibility to structural similarity. IEEE Trans Image Process 13(4):600–612

22. Willmott CJ, Matsuura K (2002) Mean absolute error. Encyclopedia Environmetr: 1–4
23. Nicodemus FE, Richmond JC, Hsia JJ (1964) Reflectance. Science 146(3642):1318–1320

Indian Music Instrument Classification Using Deep Learning on Embedded Platforms

Satish Chikkamath, Uday Kulkarni, Mahek Jain, Guruprasad Kamat, Rochan Bachari, and Sangamesh Mainale

Abstract Indian musical instruments are an integral part of the rich cultural heritage of India. These instruments have evolved over centuries and are unique in their sound, construction, and playing technique. The popularity of Indian music in Western countries has given rise to the need of identifying Indian musical instruments. In this study, we aim to classify Indian musical instruments by extracting Mel-Frequency Cepstral Coefficients (MFCCs) from the audio signals and training the Recurrent Neural Networks' (RNNs) architecture on these features. We compare the performance of the RNN with various Machine Learning (ML) techniques and find out that RNN outperforms all the other techniques. We quantize the trained RNN model to decrease latency by 13% and model size by 87.2% with negligible change in the model inference accuracy compared to the baseline model and deploy it on an Android device making a user-friendly application. This application allows users to record audio clips and classifies the audio into five different Indian musical instruments and other sounds as background noise. The app helps users to recognize Indian musical instruments in a convenient manner.

Keywords Long short-term memory · Mel-Frequency Cepstral Coefficients · Recurrent Neural Network · Android

---

S. Chikkamath (✉) · M. Jain · G. Kamat · R. Bachari
School of Electronics and Communication Engineering, KLE Technological University Hubballi, Hubballi, India
e-mail: chikkamath@kletech.ac.in

U. Kulkarni · S. Mainale
School of Computer Science and Engineering, KLE Technological University Hubballi, Hubballi, India
e-mail: uday_kulkarni@kletech.ac.in

V. K. Gunjan and J. M. Zurada (eds.), *Proceedings of 4th International Conference on Recent Trends in Machine Learning, IoT, Smart Cities and Applications*, Lecture Notes in Networks and Systems 873, https://doi.org/10.1007/978-981-99-9442-7_26

1 Introduction

Indian music is a vast and diverse art form that has a rich history dating back several millennia. Indian classical music is particularly famous for its depth, elegance, and complexity. It has a strong cultural and traditional foundation and is distinguished by its distinctive melody, rhythm, and improvisational skills. Indian music is made up of a variety of musical instruments, each having a distinctive sound and playing style. Essential components of this rich legacy are the different instruments employed in Indian music, each of which has a distinctive sound and personality [1].

Identifying and classifying these instruments can be a challenging task, especially for those who are not familiar with the Indian music tradition. Classification can be a significant tool in the process of teaching and learning about Indian music. By classifying the instruments in terms of their physical properties, playing techniques, and cultural significance, one can better convey the importance of the diversity and complexity of Indian music. Classification is also able to facilitate research on Indian music and instruments. By categorizing, one is able to study their historical and cultural context and also their acoustic properties and playing techniques. By using standardized terminology to describe different types of instruments, one can easily share information and knowledge about Indian music and the instruments [2].

We use ML and deep learning (DL) methods to tackle the classification problem for the instruments. The realms of music analysis and classification have seen great success using ML and DL methods. RNNs [3] and Deep Neural Networks (DNNs) [4] are examples of algorithms that have been successfully used to recognize and categorize musical instruments based on their acoustic characteristics.

In the proposed work, we opt for RNN [3] as RNNs are often used for audio-related deep learning tasks because they can effectively capture the temporal dynamics and dependencies of audio signals [5]. Audio signals are essentially time-series data, where the value of each sample is dependent on the values of the preceding samples. RNNs are designed to work with sequential data and have a memory of the previous inputs that they have processed. RNNs use their internal state, which is updated at each time step, to capture the context of the previous inputs, and this context is used to influence the processing of the current input [6]. For example, when processing musical audio, RNNs can capture the subtle changes in pitch, intonation, and rhythm that are characteristic of the instrument. This allows them to effectively model the complex structure of speech signals and to perform tasks like speech recognition or speaker identification.

In this study, we use audio signals as inputs. These input audio signals are first subjected to sampling. Once they are sampled, preprocessing of these sampled audio signals is performed by extracting a few features from them, which act as input to our RNN architecture. The architecture is then trained on the features extracted to classify the musical instruments, namely—Dhol-Tasha, Flute, Harmonium, Sitar, and Tabla, as shown in Fig. 1. To classify sounds apart from these instruments, we created a class called background noise.

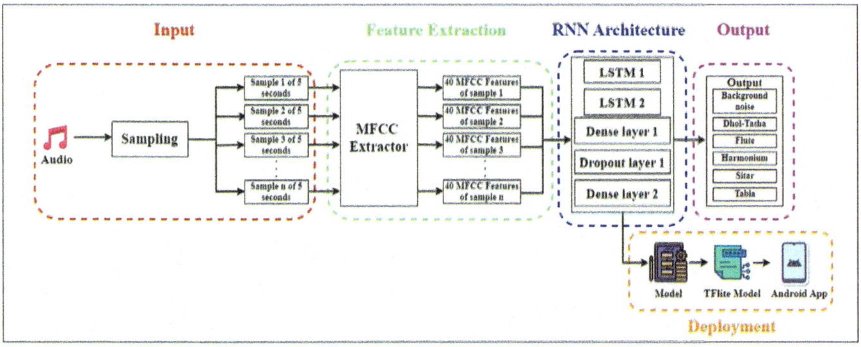

Fig. 1 Flow diagram of proposed model

The paper is organized as follows—Sect. 2 describes the background study. Section 3 talks about related works. Section 4 describes our proposed methodology. Section 5 describes the experimentations conducted and their results. Section 6 draws the conclusion of this study.

2 Background Study

This section covers the background research necessary for the RNN model's implementation.

2.1 MFCC

MFCC [7] is frequently applied to speech and audio processing. They are predicated on the findings that lower frequencies have a greater effect on the human auditory system's sensitivity to frequency fluctuations than higher frequencies.

The window function is then performed to each frame to calculate the MFCC after the audio stream is originally divided into brief frames, typically 20–40 ms long. Then, for every frame, a Fourier transform is used to determine the frequency spectrum. A Mel-scale filter bank [8] is then used to alter the frequency spectrum, which divides neighboring frequencies into bins that are more closely related to how people perceive pitch. The energy in each bin's logarithm is then calculated, and the vector of Mel-scale log energies that is produced is transformed using a Discrete Cosine Transform (DCT) [9] to produce the MFCC. The first 10–13 MFCCs are often kept as features because they tend to capture the most essential information about the spectral envelope of the audio source.

For a variety of audio processing applications, such as speech recognition, speaker identification, and music genre categorization, MFCC has been demonstrated to be incredibly effective. They are regarded as one of the most significant audio aspects in the area and are frequently employed in industry and research.

The equation for calculating the MFCC can be written as follows:

$$\mathrm{MFCC}_i = \sum_{j=1}^{N} \log\left(\sum_{n=1}^{N} |X_n| \cdot H_m(n) \cdot \cos\left[\frac{\pi}{N}\left(n + \frac{1}{2}\right)j\right]\right) \cdot \cos\left[\frac{\pi}{N}\left(i - \frac{1}{2}\right)j\right], \tag{1}$$

where N is the number of samples in each frame, X_n is the magnitude spectrum of the nth sample, $H_m(n)$ is the filter bank response for the mth Mel filter at frequency n, and i and j are the indices of the MFCC coefficients.

2.2 RNNs and LSTM

RNNs [3] are more effective than feedforward neural networks at modeling audio sequence data because they have cyclic connections. Due to their reliance on an acoustic frame's sliding window with a predetermined size, DNN can only give a limited amount of temporal modeling. They are inadequate to manage varying speech rates and longer-term dependencies since they can only simulate the data contained in the window.

Conceptually appealing for the task of acoustic modeling is the Long Short-Term Memory (LSTM) architecture as shown in Fig. 2, which addresses some modeling shortcomings of RNNs [3].

Memory blocks are distinct components found in the LSTM's recurrent hidden layer. The memory blocks are made up of unique multiplicative units called gates that regulate information flow and memory cells with self-connections that store the network's temporal state. Each memory block in the original architecture featured input and output gates. The input gate controls the flow of input activations into the memory cell. How cell activations leave the cell and enter the rest of the network is controlled by the output gate.

An LSTM network computes a mapping from an input sequence $x = (x_1, \ldots, x_T)$ to an output sequence $y = (y_1, \ldots, y_T)$ by calculating the network unit activations using the following equations iteratively from $t = 1$ to T:

$$i_t = \sigma(W_{ix}x_t + W_{im}m_{t-1} + W_{ic}c_{t-1} + b_i), \tag{2}$$

$$f_t = \sigma(W_{fx}x_t + W_{fm}m_{t-1} + W_{fc}c_{t-1} + b_f). \tag{3}$$

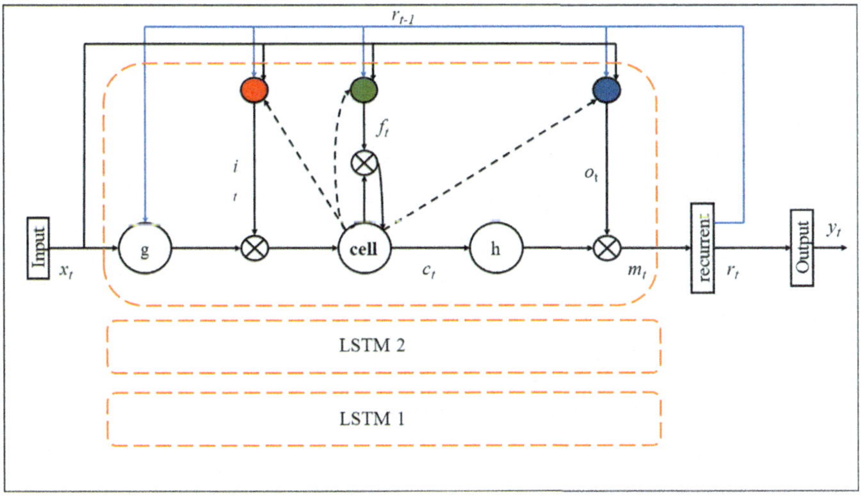

Fig. 2 RNN architecture with LSTM. For clarity, only one memory block is displayed [10]

- The W terms denote weight matrices (e.g., W_{ix} is the matrix of weights from the input gate to the input).
- W_{ic}, W_{fc}, and W_{oc} are diagonal weight matrices for peephole connections.
- The b terms denote bias vectors (e.g., b_i is the input gate bias vector).
- σ is the logistic sigmoid function.

$$c_t = f_t c_{t-1} + i_t g(W_{cx} x_t + W_{cm} m_{t-1} + b_c), \tag{4}$$

$$o_t = \sigma(W_{ox} x_t + W_{om} m_{t-1} + W_{oc} c_t + b_o), \tag{5}$$

$$m_t = o_t h(c_t), \tag{6}$$

$$y_t = \phi(W_{ym} m_t + b_y). \tag{7}$$

- i_t, f_t, o_t, and c_t are respectively the input gate, forget gate, output gate, and cell activation vectors, all of which are the same size as the cell output activation vector m_t.
- is the element-wise product of the vectors.
- g and h are the cell input and cell output activation functions, generally and in this paper tanh.
- ϕ is the network output activation function, usually softmax.

3 Related Works

Solanki et al. in [11] proposed a CNN [4] model using Alexnet [12] with some additional functionality of CNN [4] used to analyze spectrogram and detect the music. The dataset used is the Instrument Recognition in Musical Audio Signals (IRMASs) [13] contains 3 s of audio clips of various musical instruments. These audio clips are then converted to Mel-spectrograms, passed to the neural network, and trained in the model to detect Mel-spectrograms of different musical instruments. They also compared the results with different pooling and activation functions and came to the conclusion that the combination of ReLU [14] and Max Pooling [15] functions gave the best result with 92.80% accuracy to detect and classify the Western musical instruments.

Chikkamath et al. in [5] used the idea of transfer learning [16] in their work to identify the presence of music in a specific audio recording. According to the literature, it is possible to identify music in an audio file by manually extracting features like Zero Cross Rating (ZCR), entropy, and Adaptive Multi-rate (AMR), then training them using classifiers like Support Vector Machines (SVMs) [17] and Random Forest [18].

Racharla et al. in [19] found that when working with an audio dataset where, despite identical sound notes, the spectrogram varies depending on the instrument from which the note originated. IRMAS [13] data has been used in their study. The data includes 3-s audio recordings of eleven instruments. They looked at standard libraries like Librosa [20] and Essentia [21] in order to identify features and extract them from the audio files. They extracted the first 13 MFCC features using the libraries.

Chikkamath et al. in [6] have used RNNs [3] from the classes of LSTM [22] and Bidirectional Long Short-Term Memory (BI-LSTM) [22] to create melodies. To ensure that the generated music piece is distinct and not merely a replica of the training data, the authors of the study utilized the Dynamic Time Warping (DTW) [23] technique. The generated music sample and the training music sample are the two temporal sequences that are compared by DTW. The authors employed the DTW [23] technique to determine the dissimilarity between the data samples and the generated music pieces. The evaluation resulted in a 78.32% uniqueness score.

Kulkarni et al. in [24] proposed a method for compressing large Artificial Intelligence (AI) models and increasing inference speed to enable deployment on edge hardware. Trimming and quantization on the seven-layer CNN [4] architecture only result in a 0.44% accuracy loss for the suggested method. They recommended a hybrid optimization strategy. The two phases of hybrid optimization are pruning and quantization. Pruning is the process of removing unused weights and connections from the model to make it smaller. After removing the unnecessary parameters, the weights of the remaining parameters are quantized into 8-bit integer values. The difference between the specifications and the number of flops is 40%.

4 Proposed Methodology

In the context of classifying Indian musical instruments, there have been several works on musical instrument classification. However, there are certain drawbacks:

- Some musical instruments produce sounds that blend with those of other instruments, making it challenging to tell them apart.
- Different artists can use the same instrument in various ways, changing the sound that is generated. Because of this, it can be difficult to recognize instruments purely by their sound.
- Depending on where they are struck, some instruments, like drums, can produce different sounds. Because of this, it could be challenging to categorize the instrument using just one sound file.
- Obtaining substantial and varied training datasets for the categorization of musical instruments can be difficult, particularly for uncommon or unusual instruments.
- Since the nomenclature and classification of instruments can alter across different countries and eras, cultural and historical considerations can also have an impact on how musical instruments are categorized.

To counter a few drawbacks, the following are our contributions toward musical instruments classification:

- Dataset: As mentioned before, the scarcity of extensive annotated and well-structured datasets is a challenge for Music Information Retrieval tasks. Hence to tackle this, we performed data scraping from different open sources to create a dataset that is exclusively for Indian musical instruments.
- RNNs: We use RNN architecture to train our model as they are able to handle variable-length inputs by maintaining a hidden state that is updated at each time step, allowing them to process audio signals of different lengths.
- Implementation on Embedded Platforms: The application of the classification model on embedded platforms is another contribution of our research. Mobile phones and other embedded platforms have a finite amount of memory and processing capacity. Our research enables real-time or on-device classification of Indian musical instruments by creating a model that can be used on such platforms, making it more accessible and useful for end users.

4.1 Feature Extraction

To extract characteristics from the audio files, we looked at the common libraries. Both Librosa [20] and Essentia [21] were the available options. A Python module called LIBRIS is used to examine and modify audio signals. It provides a wide range of functions for loading, manipulating, and visualizing audio data, as well as for extracting a variety of features from audio signals including MFCC. Functions from Librosa are used to compute the MFCC features of the audio signal using a specific parameter to specify the number of coefficients to compute. Each audio sample is segmented into a constant length of 5 s. MFCC features are then extracted from those segments and saved in a JSON file. The frequency-to-Mel transform mathematical formula is as follows:

$$10m = 2595\log\left(1 + \frac{f}{700}\right). \tag{8}$$

The equation eq: Einstein specifies that the Mel value (m) corresponds to a given frequency value f in Hertz.

4.2 *Classifier Training*

As part of the preprocessing, the MFCC characteristics were extracted from the audio dataset of the instrument, normalized, and partitioned into training and test sets. The RNN model was optimized using the Adam optimizer after 150 training iterations with a learning rate of 0.0001 and a loss function of sparse categorical cross-entropy. Observing the model's performance while it was being trained, we adjusted the hyper-parameters to increase the model's precision. The model was trained and then quantized to make it a lightweight model that could be used on an edge device. The effectiveness of our approach was then contrasted with that of a number of ML algorithms, such as classifiers from Random Forests, SVM, Logistic Regression, and Decision Trees.

4.3 *Quantization of the Model*

The memory and processing power of edge devices is frequently constrained. Models can be optimized in a variety of ways to operate within these restrictions. Tools are available to reduce the complexity of optimizing inference, such as TensorFlow Lite [25] and the TensorFlow Model Optimization Toolkit. The size of a model can be decreased using some optimization techniques.

We quantize the trained RNN model to decrease latency by 13% and model size by 87.2% with negligible change in the model inference accuracy compared to the baseline model and deploy it on an Android device. Model accuracy may alter as a result of optimizations. By post-quantization, our model size of 20 MB is compressed to 2.56 MB. We want to emphasize that maintaining audio quality was a top priority during the post-quantization process. Our findings show that despite a slight loss of accuracy owing to compression, audio classification performance is still adequate even when the model size is shrunk.

5 Experimentation and Results

The training of classification of Indian music instruments is done on Google Colab which has NVIDIA K80/T4 @1.59 GHz GPU with 16 GB RAM. We have employed Android Studio to create a mobile application.

5.1 Dataset

The major problem of the Indian music instruments dataset is its availability. Thus, we have collected a total of 1177 audio samples from various sources on the internet. Figure 3 represents the audio spectrogram plots of six different classes. The dataset consists of six classes which include classes, namely background noise, Dhol-Tasha, Flute, Harmonium, Sitar, and Tabla. Background noise class consists of two samples with each of length 15 minutes. Similarly, class Dhol-Tasha consists of three samples each of 10 minutes in length, Flute consists of 227 samples each of 5 seconds in length, Harmonium consists of three samples each of 22 minutes in length, Sitar consists of 379 samples each of 5 seconds in length, and Tabla consists of 561 samples each of 10 seconds in length.

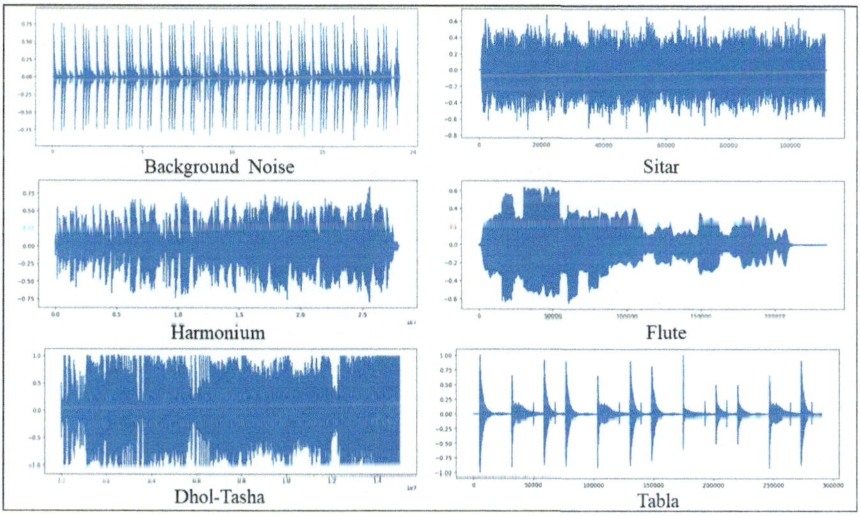

Fig. 3 Audio spectrogram plot of Indian music instruments and background noise

5.2 Comparing the Results

The experimental findings showed that in terms of accuracy and loss, RNN models outperformed traditional machine learning models. RNN models' superiority can be linked to their ability to simulate complex and nonlinear interactions between input features and output labels as well as their capability to accurately represent the temporal dependencies of sequential data.

Table 1 presents the outcomes of the RNN model along with other models that were trained using our data which concludes that RNN architecture outperforms the classification task because of its high accuracy and low loss.

Figure 4a shows the test and train accuracy variations with respect to the number of epochs, while Fig. 4b shows the variation of test and train errors respective to the number of epochs. Thus, we can infer that model accuracy and error saturate after a certain number of epochs.

Table 1 Accuracy results on different architectures

Architecture	Accuracy ↑	Loss ↓
KNN	0.9687	0.4834
Random Forest	0.9829	0.3124
RNN	**0.9872**	**0.2834**
XGBoost	0.9617	0.4734
Logistic Regression	0.9574	0.4989
Decision Tree classifier	0.9574	0.4899
SVM	0.9617	0.4317

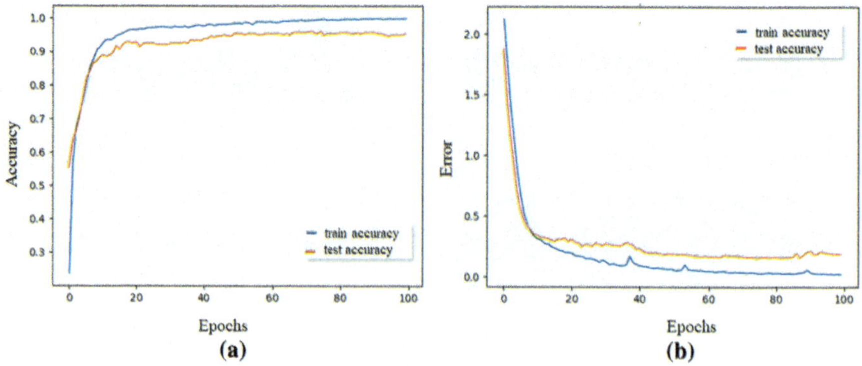

Fig. 4 Variation of accuracy and error with respect to number of epochs

5.3 Deploying the Model on an Edge Device

After being trained, the model is converted into a TensorFlow Lite (tflite) model. Tflite is a collection of tools that makes it possible for developers to run their models on mobile, embedded, and edge devices, enabling on-device machine learning. With the help of this tflite model [26], we developed an Android application using the Android Studio editor, a platform for developing Android applications.

The app continuously accepts audio input. An analysis of the supplied audio looks for a specific time stamp. The input label is retrieved by the RNN model. The user-friendly Android app lets the user select how many outputs will be shown. The threshold value can also be changed in the application. Thus, the user can quickly identify the Indian musical instrument being played in this way. Figure 5 shows the screenshots of the Indian music instruments classifier app.

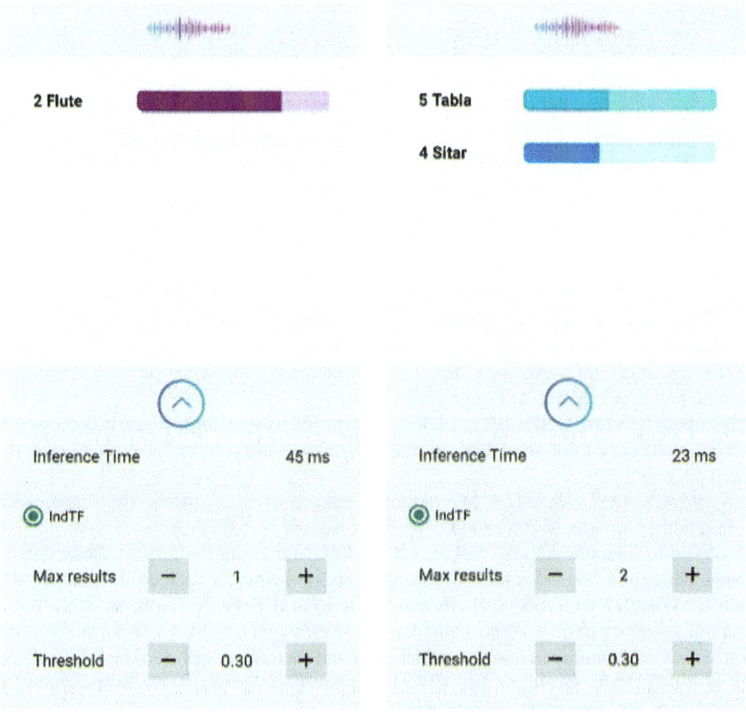

Fig. 5 Implementation on Android device which displays the probability of music instrument

6 Conclusion

This study looks at numerous classification schemes for Indian musical instruments. The method entails a number of phases, starting with audio sampling and ending with model distribution on an edge device. In conclusion, it has been demonstrated that utilizing RNN to categorize Indian musical instruments on mobile devices is a successful method. According to the study's findings, the model performed at a high level, with an accuracy of 98.7%. The model's deployment on an Android device serves as another example of how this method could work in the real world.

We have classified only five Indian musical instruments. The model can be trained on several other instruments with a much larger dataset. It would be advantageous for the end users if a model could close the current gap.

References

1. Sankaye SR, Mehrotra SC, Tandon US (2015) Indian musical instrument recognition using modified LPC features. Int J Comput Appl 122(13)
2. Patil SR, Machale SJ (2020) Indian musical instrument recognition using gaussian mixture model. In: Techno-societal 2018: proceedings of the 2nd international conference on advanced technologies for societal applications, vol 2. Springer, pp 51–57
3. Saon G, Tüske Z, Bolanos D, Kingsbury B (2021) Advancing RNN transducer technology for speech recognition. In: ICASSP 2021—2021 IEEE international conference on acoustics, speech and signal processing (ICASSP). IEEE, pp 5654–5658
4. Aldahdooh A, Hamidouche W, Fezza SA, Déforges O (2022) Adversarial example detection for DNN models: a review and experimental comparison. Artif Intell Rev 55(6):4403–4462
5. Chikkamath S, Nirmala SR (2022) Music detection using deep learning with tensorflow. In: ICDSMLA 2020: proceedings of the 2nd international conference on data science, machine learning and applications. Springer, pp 283–291
6. Chikkamath S, Nirmala SR (2021) Melody generation using lstm and bi-lstm network. In: 2021 international conference on computational intelligence and computing applications (ICCICA). IEEE, pp 1–6
7. Jin S, Wang X, Du L, He D (2021) Evaluation and modeling of automotive transmission whine noise quality based on MFCC and CNN. Appl Acoust 172:107562
8. Upadhya SS, Cheeran AN, Nirmal JH (2019) Discriminating Parkinson diseased and healthy people using modified mfcc filter bank approach. Int J Speech Technol 22(4):1021–1029
9. Shen X, Yang J, Wei C, Deng B, Huang J, Hua X-S, Cheng X, Liang K (2021) DCT-mask: discrete cosine transform mask representation for instance segmentation. In: Proceedings of the IEEE/CVF conference on computer vision and pattern recognition, pp 8720–8729
10. Sak H, Senior AW, Beaufays F (2014) Long short-term memory recurrent neural network architectures for large scale acoustic modeling
11. Solanki A, Pandey S (2019) Music instrument recognition using deep convolutional neural networks. Int J Inf Technol 14:01
12. Chen J, Wan Z, Zhang J, Li W, Chen Y, Li Y, Duan Y (2021) Medical image segmentation and reconstruction of prostate tumor based on 3d alexnet. Comput Methods Programs Biomed 200:105878
13. Solanki A, Pandey S (2022) Music instrument recognition using deep convolutional neural networks. Int J Inf Technol 14(3):1659–1668
14. Timor N, Vardi G, Shamir O (2023) Implicit regularization towards rank minimization in ReLU networks. In: International conference on algorithmic learning theory. PMLR, pp 1429–1459

15. Akhtar N, Ragavendran U (2020) Interpretation of intelligence in cnn-pooling processes: a methodological survey. Neural Comput Appl 32(3):879–898
16. Pinto G, Wang Z, Roy A, Hong T, Capozzoli A (2022) Transfer learning for smart buildings: a critical review of algorithms, applications, and future perspectives. Adv Appl Energy: 100084
17. Kurani A, Doshi P, Vakharia A, Shah M (2023) A comprehensive comparative study of artificial neural network (ANN) and support vector machines (SVM) on stock forecasting. Ann Data Sci 10(1):183–208
18. Genuer R, Poggi J-M, Genuer R, Poggi J-M (2020) Random forests. Springer
19. Racharla K, Kumar V, Jayant CB, Khairkar A, Harish P (2020) Predominant musical instrument classification based on spectral features. In: 2020 7th international conference on signal processing and integrated networks (SPIN). IEEE, pp 617–622
20. Shanmuga Sundari M, Priya KSS, Haripriya N, Sree VN (2023) Music genre classification using librosa implementation in convolutional neural network. In: Proceedings of fourth international conference on computer and communication technologies. Springer, pp 583–591
21. Correya A, Alonso-Jiménez P, Marcos-Fernández J, Serra X, Bogdanov D (2021) Essentia tensorflow models for audio and music processing on the web. In: Web audio conference (WAC 2021)
22. Alhussein M, Aurangzeb K, Haider SI (2020) Hybrid CNN-LSTM model for short-term individual household load forecasting. IEEE Access 8:180544–180557
23. Gella GW, Bijker W, Belgiu M (2021) Mapping crop types in complex farming areas using sar imagery with dynamic time warping. ISPRS J Photogr Remote Sensing 175:171–183
24. Kulkarni U, Meena SM, Gurlahosur SV, Benagi P, Kashyap A, Ansari A, Karnam V (2021) AI model compression for edge devices using optimization techniques. In: Modern approaches in machine learning and cognitive science: a walkthrough: latest trends in AI, vol 2. Springer, pp 227–240
25. David R, Duke J, Jain A, Reddi VJ, Jeffries N, Li J, Kreeger N, Nappier I, Natraj M, Wang T et al (2021) Tensorflow lite micro: Embedded machine learning for tinyml systems. Proc Mach Learn Syst 3:800–811
26. Varur S, Mainale S, Korishetty S, Shanbhag A, Kulkarni U et al (2023) Classification of maturity stages of coconuts using deep learning on embedded platforms. In: 2023 3rd international conference on smart data intelligence (ICSMDI). IEEE, pp 343–349

Facial Image Inpainting Using Vision-Based Quantized Conditional Generative Adversarial Network (QCGAN) on Edge Device

Viju Rangannavar, Swati Mudavi, Sushmita Korishetty, Mohit Gumber, Uday Kulkarni, and Satish Chikkamath

Abstract Image inpainting is considered a task in the field of computer vision aimed at filling in missing or corrupted parts of an image. By using deep learning techniques, they learn a mapping among the distorted image and its corresponding counterpart undistorted image. Applications for conditional image inpainting include enhancing, restoring, and creating new images. The goal of this task is to generate a plausible completion of the missing parts that is consistent with the visual aspect and structure of the surrounding image regions. The conventional method of image inpainting entails using information from the surrounding areas to fill in any damaged or missing portions of images. We proposed a QCGAN by training two neural networks parallelly: the discriminator network, which has a loss of 0.10833, and the generator network, which has a loss of 3.77887. The discriminator network assesses the generated pictures to determine if they are real or fraudulent, and the generator network attempts to generate a complete image of an incomplete image. Through this process, the generator network learns to generate realistic and accurate images that can rectify or restore incomplete or impaired portions of an image. Experiments using a customized dataset demonstrate the stability and effectiveness of our suggested model, as well as its higher naturalness, semantic similarity, and validity in terms of visual experience. The model underwent post-dynamic range quantization, and its size was reduced by 74.8% to enable deployment on edge devices. We achieved the PSNR value of 30.76 and the SSIM of 0.86.

V. Rangannavar · S. Mudavi · S. Korishetty · M. Gumber (✉) · U. Kulkarni
Department of Computer Science, KLE Technological University, Vidyanagar, Hubballi, Karnataka 580021, India
e-mail: mgpersonal.me@gmail.com

U. Kulkarni
e-mail: uday_kulkarni@kletech.ac.in

S. Chikkamath
Department of Electronics and Communication, KLE Technological University, Vidyanagar, Hubballi, Karnataka 580021, India
e-mail: chikkamath@kletech.ac.in

V. K. Gunjan and J. M. Zurada (eds.), *Proceedings of 4th International Conference on Recent Trends in Machine Learning, IoT, Smart Cities and Applications*, Lecture Notes in Networks and Systems 873, https://doi.org/10.1007/978-981-99-9442-7_27

Keywords QCGANS · Deep learning · Image inpainting

1 Introduction

Image inpainting [1] represents a field within the realm of image processing. The restoration of old books, the processing of medical images, and PhotoShop processing are a few of the instances where picture inpainting technology has been extensively applied. Therefore, it is worthwhile to research image inpainting. Image inpainting is a challenging task which encompasses the task of repairing absent or deteriorated sections in an image. It can be difficult to seamlessly blend the restored regions with the surrounding image content. As a result, there can be visible artifacts or blurring at the boundary between the restored and original regions of the image, which is commonly referred to as the fuzzy phenomenon. Existing techniques for solving these issues with picture inpainting has been divided into two groups: texture synthesis techniques based on patches, where the goal is to locate the border of distorted area in order to fill in the image's distorted area. The alternative method uses deep neural networks (DNNs) [2], and its fundamental idea is to interpret an image by first extracting its characteristics using a deep neural network [3] before filling in the unoccupied area.

Our demonstrated method, as shown in Fig. 1, uses the generator–discriminator [4] and it is able to produce convincing and logically consistent images during the process of inpainting. Our model takes in a damaged image as input and uses a generator to produce a fake image. The discriminator then examines the authenticity of the generated image, identifying real from fake. Our model utilizes a feedback mechanism to generate accurate results. Specifically, the generator loss produced by the discriminator is given back into the generator, which learns to produce images that are more similar to real images. This iterative process helps the generator to learn and improve over time, leading to increasingly accurate and realistic results. The generator also employs the skip connections [5] to enhance its ability to avoid the issue of gradient vanishing [6].

The following is the paper's workflow. Section 2 says about the related research. The outlined approach is elucidated in Sect. 3. The experimentation is detailed in Sect. 4. The findings of the experiment are detailed in Sect. 5. Lastly, the ultimate conclusions are formulated in Sect. 6.

2 Related Works

Isola et al. [7] proposed Pix2Pix transforming one image into another model which is leveraged on deep learning. Model converts an input image into high-quality output images by using a Conditional Generative Adversarial networks (conditional GANs). The generator network creates a restored image as the output from an input image

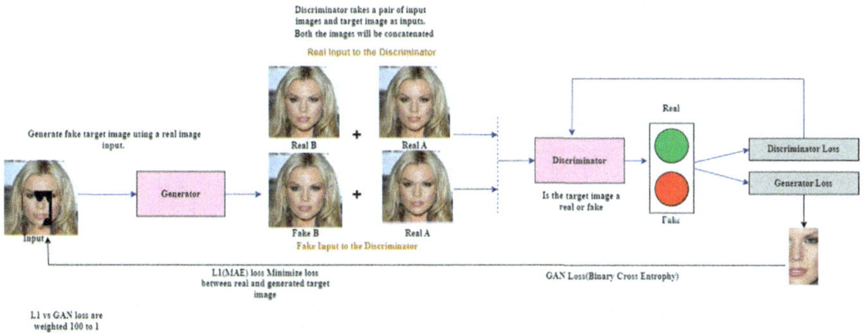

Fig. 1 Architecture of the network consisting of the generator, the global discriminator, the generator learn the loss to give the accurate results, and the global discriminator which tells whether the generated image is fake or real

that contains missing or corrupted parts. To differentiate between the produced and real-world images, the discriminator network is trained.

Barnes et al. [8] introduced Patch-Match for image completion by finding matching patches in the remaining portion of the image. While it improves background inpainting, it struggles with complex images like real faces, resulting in highly blurry outcomes. In our model, the generator utilizes skip connections to overcome gradient vanishing and enhance predictive ability in deep networks.

Pathak et al. [9] leveraged a deep neural network (DNN) approach for picture inpainting, which can learn to complete the missing parts of an image by converting the input image into an encoded representation a lower-dimensional feature representation and then decoding it to generate the missing parts.

Kaiming et al. [10] introduced skip connections in ResNet to mitigate the vanishing gradient problem in neural networks. While increasing the depth of the model can enhance recognition accuracy, it may eventually lead to decreased accuracy. This is due to the gradient vanishing issue, where gradients diminish as they propagate through numerous layers, preventing effective adjustments in the previous layers and resulting in blurred images.

Auto-encoders are a method of unsupervised learning introduced by Zhao et al [11]. Modeling is made up of an encoder and a decoder. The model's goal is to reduce the variation between the input image provided to the encoder and the resulting output image generated by the decoder. We include connectivity links that bypass the encoder and connect to the decoder in this paper as shown in Fig. 2. As in [12], authors do utilization of an auto-encoder architecture as the basic architecture for the generator in proposed model, just as one was utilized as the generative network for picture inpainting.

Yu et al. [14] introduced a model rooted in Generative Adversarial Networks (GANs), incorporating a contextual attention mechanism. This mechanism enables the model to selectively emphasize important areas of the image while generating the parts that are missing.

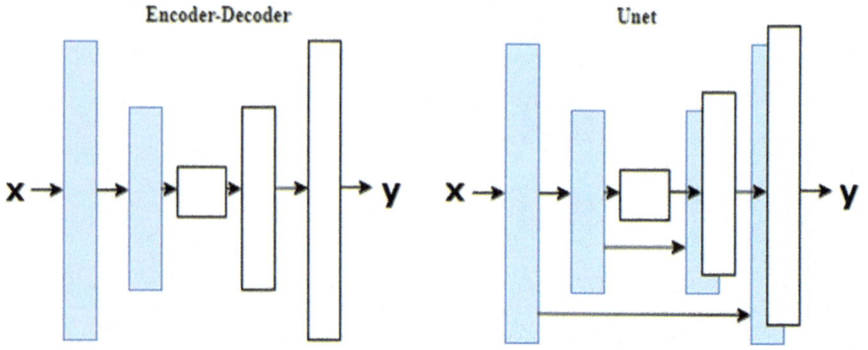

Fig. 2 Encoder–decoder architecture and encoder–decoder with skip connections [5] called U-net [13] architecture

3 Proposed Methodology

The proposed work will provide a comprehensive and systematic analysis of the QCGAN.

3.1 Method

The process of image inpainting can be described as follows: Given an input consisting of a random noise vector (z) and an image (x), the generator works with conditional GANs to create an output image (y). This is achieved by learning a mapping from the observed image (x) and the random noise vector (z). Meanwhile, a discriminator (D) is trained adversarially to excel in distinguishing between authentic images and the generated outputs produced by the generator (G), which are sometimes referred to as "fakes."

The goal of a conditional GAN is illustrated in the following Eq. 1:

$$L_{\text{cGAN}}(G, D) = E_{x,y}\big[\log D(x, y)\big] + E_{x,z}\big[\log(1 - D(x, G(x, z))\big], \quad (1)$$

where G represents the generator and the D represents the discriminator, where G strives to reduce this goal and a hostile D seeks to increase it as shown in Eq. 2:

$$G^* = \arg\min_G \arg\max_D L_{\text{cGAN}}(G, D). \quad (2)$$

The generator's job is to be close to the truth while simultaneously deceiving the discriminator. Since $L1$ encourages less blurring than $L2$, we are utilizing $L1$ distance instead of $L2$ as shown in Eq. 3.

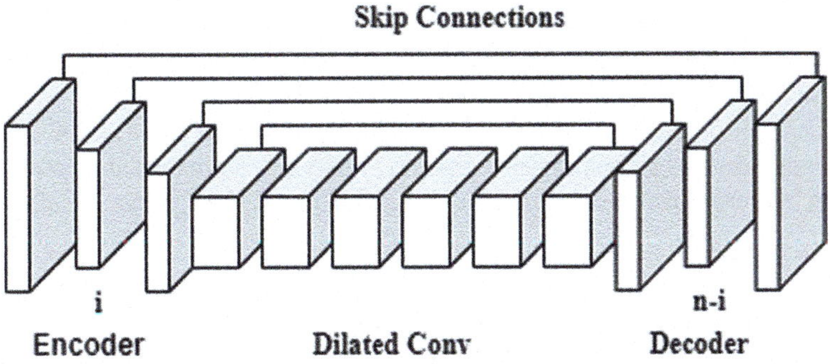

Fig. 3 Encoder–decoder architecture and encoder–decoder with skip connections are called U-Net architecture which is a generator of the conditional GAN

$$L_{L1}(G) = E_{x,y,z}\left[\|y - G(x, z)\|_1\right]. \tag{3}$$

3.2 *Generator with Skip Connections*

One of several issues with image generation is restoration. To tackle this problem, we use U-Net architecture for the generator, which contains two networks. An encoder network and a decoder network are used to repair images as shown in Fig. 2. The encoder is used to produce an encoded version of the given image, which is then decoded by the decoder to reconstruct and generate the repaired image. We add skip connections as illustrated in Fig. 3 in order to prevent the network layers from deteriorating as a result of the increased depth, and there should be a space in middle of each layer i and layer $n - i$, where n represents the total number of layers. Each skip connection merely combines all layer i channels with layer $n - i$ channels. It improves the generator's ability to predict structure.

The encoder employs a convolutional layer architecture [15] with multiple layers, and transposed convolution layers are used to provide an architecture for the decoder that is symmetric to the encoders.

3.3 *Discriminator*

The generator is tasked with filling in the missing or obscured parts of a picture, but it cannot guarantee the accuracy or consistency of the generated area with respect to the real image. As a binary classifier, the discriminator can tell real images from those created by the generator. The discriminator also aids the generator become more adept at producing convincing visuals that can fool the discriminator.

In our model, discriminator is composed of a number of convolutional layers, followed by a layer that is completely coupled and generates a single output. The given input to the discriminator is a pair of pictures the original picture with the missing parts, and the generated image with the missing parts filled in. The discriminator compares the two images and outputs a probability score indicating how similar they are. Discriminator uses the sigmoid function to tell if the picture is fake or real.

4 Experimentation

The process of training the conditional GAN for inpainting took place on a laptop equipped with an Intel(R) Core(TM) i5-10210U CPU running at 1.60 GHz, accompanied by 16 GB of random access memory (RAM), all within the framework of Microsoft Windows 11. For the development of a mobile application, we employed Android Studio.

4.1 Dataset and Preprocessing

We have picked the CelebFaces Attributes Dataset (CelebA), which has more than 200 K images of famous people annotated with 40 different attributes. The large-scale facial attributes' dataset is made up of each. This photo collection features a variety of positions and crowded settings, including 202,599 face images and 10,177 IDs. We have used 2700 images of CelebFaces dataset for training our model.

Preprocessing serves the purpose of enhancing image data by minimizing unintended distortions or amplifying specific elements essential for subsequent processing, despite potential geometric alterations in the images. In our research, the input images undergo resizing, transitioning from various dimensions to a uniform 512 $\times$ 512 pixel size, with the addition of a randomly positioned black mask measuring 256 $\times$ 256 pixels. Subsequently, the entire image is resized to 256 $\times$ 256 pixels. This dataset is then divided into separate training and testing datasets.

4.2 Training of the Model

To perform image inpainting, we employed a conditional GAN as our foundational model. Through the learning process of the generator, this task was accomplished. Conditional GAN was trained on CelebA dataset which consists of 200 k images in which we have used 2700 images with different attributes. We have trained our model using CelebA dataset for 200 epochs. Typically, the model accepts 256 $\times$ 256 pixel images as the input. We got the good results for various damaged images, further quantized [16] our model from 207 to 52 MB for deploying in android.

4.3 Post-quantization

Post-quantization is a technique used in the quantization of neural network models where the weights and activations of a pre-trained model are quantized after the model has been trained. In post-quantization, the quantization process does not affect the training process itself but is applied to the model after the training is complete.

The model underwent post-dynamic range quantization to reduce its size from 207 to 52 MB to enable deployment on edge devices without comprising the results.

5 Results and Discussion

We employed the generator learning method in this study with conditional GAN to inpaint the images with better accuracy on the test dataset. Conditional GAN is the different architectures which can be used for training the model to inpaint the images using the loss functions. It also gives good results for different images from different datasets.

5.1 Visual Analysis

Figure 4 illustrates a comparison between our proposed method for random regular and irregular masks and the input images from the CelebA dataset. The results show that our method performs good for smaller mask regions as compared to larger ones. The images generated for smaller masks are more realistic and accurate, demonstrate a higher level of detail and coherence, and closely resemble the features and textures of the input images, while the ones generated for larger masks larger masks may appear less realistic, with some parts of the image appearing distorted or blurred. To address this issue, our model requires more training epochs to refine its ability to reconstruct larger mask regions with higher accuracy.

To assess the impact of image recovery, the model is compared with GLCIC [17], Fast Marching Method (FMM) [18], and DIP [19].

The model is compared to GLCIC [17], Fast Marching Method (FMM) [18], and DIP [19] in order to examine the impact of image restoration.

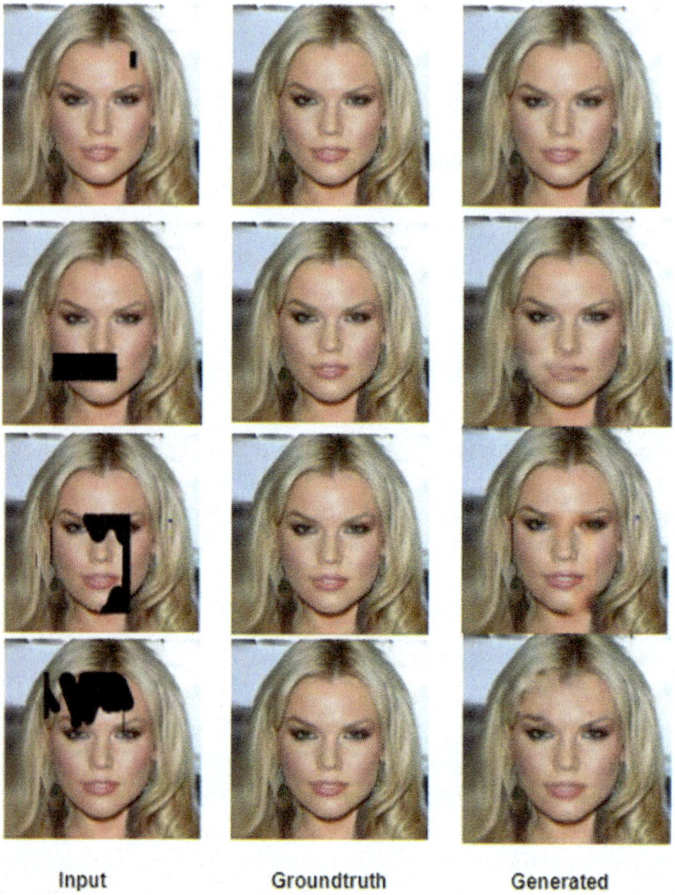

Fig. 4 Comparing the output for varying irregular masks

5.2 *Quantitative Comparison*

5.2.1 Comparison with PSNR

The Peak Signal-to-Noise Ratio (PSNR) is a commonly employed objective assessment method index for image quality assessment. It measures the error between corresponding pixels in an image. More PSNR value indicates that the picture has less distortion compared to the original or ground truth image.

PSNR is calculated by:

$$\text{PSNR} = 10 \log_{10} \frac{255^2}{\text{MSE}}, \tag{4}$$

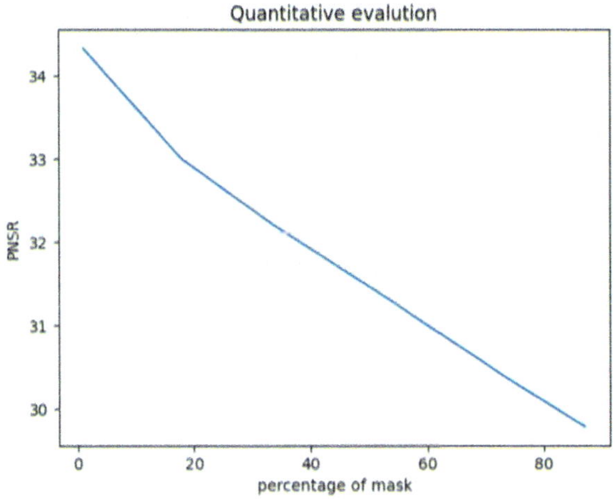

Fig. 5 Quantitative analyses of PSNR at various mask opacity levels

where MSE indicates the mean square deviation among the corrected image and the original image. The PSNR value at various mask opacity levels is being analyzed in Fig. 5.

5.3 *Comparison with SSIM*

Structural Similarity Index (SSIM) is a widely used an image quality evaluation metric that gauges the overall likeness, between two images by considering their brightness, contrast, and structure. The SSIM value lies in between 0 and 1, where a value of 1 indicates that the two images are identical. A higher SSIM indicates a higher degree of similarity among the two images. Unlike PSNR, SSIM considers the perceptual quality of the image, and it is more closely belonged to the perceived picture quality by human viewers.

We calculate SSIM by:

$$\mathrm{SSIM}(x, y) = \frac{(2\mu_x\mu_y + C_1)(2\sigma_{xy} + C_2)}{(\mu_x^2 + \mu_y^2 + C_1)(\sigma_x^2 + \sigma_y^2 + C_2)}, \tag{5}$$

where x and y are the input images, μ_x and μ_y are the mean pixel rates of x and y respectively, σ_x^2 and σ_y^2 are the difference of x and y, and σ_{xy} is the covariance of x and y. C_1 and C_2 are constants added for numerical stability. The SSIM value at various mask opacity levels is being analyzed in Fig. 6.

Table 1 provides the assessment outcomes for our proposed approach and alternative methods, utilizing the CelebA dataset, with a focus on PSNR and SSIM. The data in Table 1 illustrates that our proposed method exhibits slightly superior performance

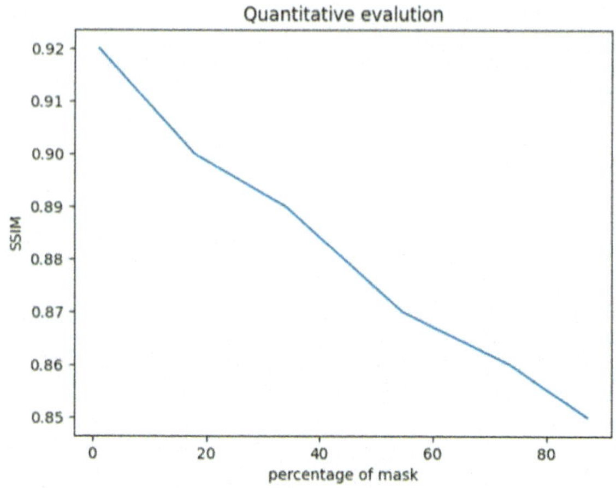

Fig. 6 Quantitative analyses of SSIM at various mask opacity levels

Table 1 Result comparison of different approaches applied quantitatively on the CelebA dataset

Model	PSNR ↑	SSIM ↑
FMM	23.52	0.84
GLCIC	27.81	0.87
DIP	25.46	0.85
Ours	30.76	0.86

when compared to all the other techniques, all of which fall under the category of image restoration methods.

5.4 *Generator and Discriminator Losses*

5.4.1 Generator Loss

The generator is trained using the loss function and binary cross entropy that measures how well the generated samples match the original data. The generator loss is the value of this loss function, which is minimized as illustrated in Fig. 7 during training to improve the quality of the generated samples.

5.4.2 Discriminator Loss

For the discriminator loss, we make use of the binary cross entropy (Fig. 8) which is calculated by comparing the output of the discriminator network for the real data and the generated samples. The goal of the discriminator is to correctly identify the real data as real and the generated samples as fake.

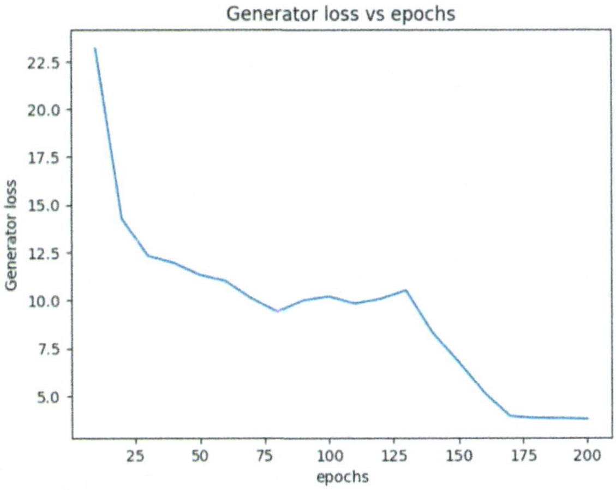

Fig. 7 Shows the generator loss at various epochs

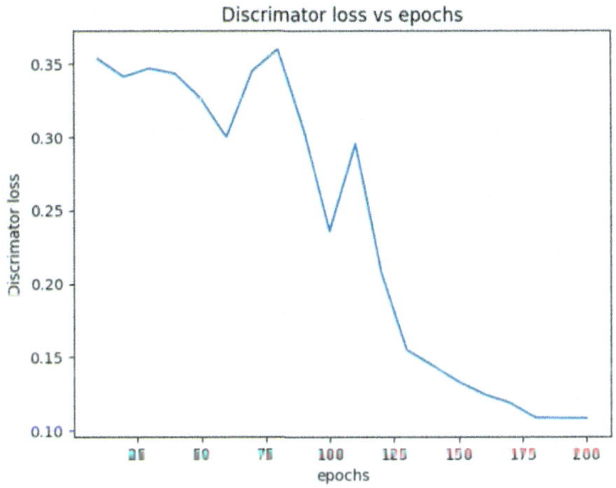

Fig. 8 Shows the discriminator loss at various epochs

5.5 *Implementing the Model on an Edge Device*

After the model is trained, it is converted into a tflite model. TensorFlow Lite comprises a set of utilities that empower developers to run their models on mobile, embedded, and edge devices, facilitating on-device machine learning, as elaborated in [20]. Leveraging this tflite model, we proceeded to create an edge device application using the Android Studio development environment, specifically designed for Android app development.

Our application offers a streamlined and intuitive user experience, making it easy for individuals to restore their appearance and regain their confidence. With our application, users can easily identify and select any damaged areas of their face and apply inpainting techniques to repair irregular and regular damage.

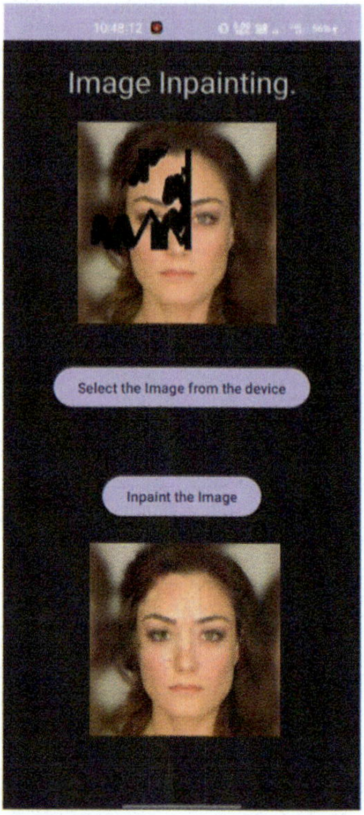

Fig. 9 Porting on edge device

Figure 9 demonstrates outcomes achieved by our method an edge device, which was constructed using the tflite model within the Android Studio environment

6 Conclusion and Future Scope

Conditional Generative Adversarial networks (cGANs) have illustrated promising results in the task of image inpainting. By using a combination of a generator and a discriminator, cGANs can effectively learn to fill in distorted area of a picture based on contextual details provided in the form of a conditioning image. The generator network in a cGAN takes the incomplete picture and the conditioning image and produces a completed image that is expected to be visually similar to the original image. The discriminator network is distinguished between the completed image generated by the generator and the real, complete image. Training a cGAN for image inpainting requires a carefully designed loss function that balances the content loss, which guarantees that the finished images are accurate to the original images, and the adversarial loss, which motivates the generator to generate realistic-looking images.

Despite the promising results, conditional GAN-based image inpainting still faces some challenges, such as handling large missing regions, preserving fine details and textures, and avoiding mode collapse, where the generator produces limited variations of the completed images. Future work is needed to address these challenges and to improve the performance [21] and efficiency of conditional GANs for image inpainting.

References

1. Bertalmio M, Sapiro G, Caselles V, Ballester C (2000) Image inpainting. In: Proceedings of the 27th annual conference on computer graphics and interactive techniques. SIGGRAPH'00. ACM Press/Addison-Wesley Publishing Co., USA, pp 417–424. https://doi.org/10.1145/344779.344972
2. Yang C, Lu X, Lin Z, Shechtman E, Wang O, Li H (2017) High-resolution image inpainting using multi-scale neural patch synthesis
3. Kulkarni N, Singh N, Joshi Y, Hasabi N, Meena S, Kulkarni U, Gurlahosur SV (2022) Hybrid optimization for DNN model compression and inference acceleration. In: 2022 2nd international conference on intelligent technologies (CONIT). IEEE, pp 1–8
4. Radford A, Metz L, Chintala S (2016) Unsupervised representation learning with deep convolutional generative adversarial networks
5. Tong T, Li G, Liu X, Gao Q (2017) Image super-resolution using dense skip connections. In: Proceedings of the IEEE international conference on computer vision, pp 4799–4807
6. Pascanu R, Mikolov T, Bengio Y (2013) On the difficulty of training recurrent neural networks. In: International conference on machine learning. PMLR, pp 1310–1318
7. Isola P, Zhu J-Y, Zhou T, Efros AA (2018) Image-to-image translation with conditional adversarial networks
8. Barnes C, Shechtman E, Finkelstein A, Goldman DB (2009) PatchMatch: a randomized correspondence algorithm for structural image editing. ACM Trans Graph (Proc SIGGRAPH) 28(3)
9. Pathak D, Krahenbuhl P, Donahue J, Darrell T, Efros AA (2016) Context encoders: feature learning by inpainting. In: Proceedings of the IEEE conference on computer vision and pattern recognition, pp 2536–2544
10. He K, Zhang X, Ren S, Sun J (2016) Deep residual learning for image recognition. In: 2016 IEEE conference on computer vision and pattern recognition (CVPR), pp 770–778. https://doi.org/10.1109/CVPR.2016.90
11. Zhao J, Mathieu M, Goroshin R, Lecun Y (2015) Stacked what-where auto-encoders. arXiv preprint arXiv:1506.02351
12. Li Y, Liu S, Yang J, Yang M-H (2017) Generative face completion
13. Ronneberger O, Fischer P, Brox T (2015) U-Net: convolutional networks for biomedical image segmentation
14. Yu J, Lin Z, Yang J, Shen X, Lu X, Huang TS (2018) Generative image inpainting with contextual attention. In: Proceedings of the IEEE conference on computer vision and pattern recognition, pp 5505–5514
15. Bezdan T, Bacanin N (2019) Convolutional neural network layers and architectures, pp 445–451. https://doi.org/10.15308/Sinteza-2019-445-451
16. Kulkarni U, Meena S, Gurlahosur SV, Bhogar G (2021) Quantization friendly MobileNet (QF-MobileNet) architecture for vision based applications on embedded platforms. Neural Netw 136:28–39
17. Iizuka S, Simo-Serra E, Ishikawa H (2017) Globally and locally consistent image completion. ACM Trans Graph 36:1–14. https://doi.org/10.1145/3072959.3073659

18. Barnes C, Shechtman E, Finkelstein A, Goldman D (2009) Patchmatch: a randomized correspondence algorithm for structural image editing. ACM Trans Comput Syst 28(3). https://doi.org/10.1145/1531326.1531330. ACM SIGGRAPH 2009, SIGGRAPH'09; conference date: 03-08-2009 Through 07-08-2009
19. Ulyanov D, Vedaldi A, Lempitsky V (2020) Deep image prior. Int J Comput Vis 128(7):1867–1888. https://doi.org/10.1007/s11263-020-01303-4
20. Kulkarni U, Meena S, Gurlahosur SV, Benagi P, Kashyap A, Ansari A, Karnam V (2021) Ai model compression for edge devices using optimization techniques. In: Modern approaches in machine learning and cognitive science: a walkthrough: latest trends in AI, vol 2. Springer, pp 227–240
21. Kulkarni U, Meena S, Joshua P, Kruthika K, Platel G, Jadhav K, Singh A (2021) Performance improvements in quantization aware training and appreciation of low precision computation in deep learning. In: Advances in signal processing and intelligent recognition systems: 6th international symposium, SIRS 2020, Chennai, India, 14–17 Oct 2020, revised selected papers 6. Springer, pp 90–107

Intelligent Prediction of Cardiac Abnormality

V. Kiruthika, Muthu Natesan, E. Sathish, and G. Vimalarani

Abstract Cardiac abnormalities are very common in today's world. Efficient and early prediction will help in saving the life of several individuals. ECG is a standard technique for prediction of any kind of abnormality prevailing in the heart. An intelligent cardiac abnormality prediction system is developed in this study. The study utilizes the real-time ECG data collected from a hospital. The contributing attributes are selected. The five chosen attributes include QRS value, heart rate, *P*-axis, *R*–axis, and *T*-axis. Machine learning techniques are explored for developing the system. Performance metrics are computed for each classifier and best performing classifier is selected for deploying the model. Naive Bayes has performed well by giving 93.50% accuracy.

Keywords Machine learning · ECG · Naive Bayes · Linear discriminant analysis

1 Introduction

Cardiovascular disease (CVD) is one of the major problems found in most of the adults after a certain age. The four main types of CVDs are aortic disease, coronary heart disease, peripheral arterial disease, and stroke. There are multiple reasons for the cause of a CVD which includes stress, unhealthy food habits, cholesterol, obesity, and blood pressure. Cardiovascular diseases are prevalent worldwide and it is predicted that 17.9 million people lose their life because of this disease. Heart attacks and strokes account for 85% of deaths from CVD, with more than 75% of those deaths occurring in low and middle-income nations [1]. Over the past several decades, estimates of the prevalence of coronary heart disease in India have ranged from

V. Kiruthika (✉) · E. Sathish
School of Electronics Engineering, Vellore Institute of Technology, Chennai 600127, India
e-mail: krithi.23.1985@gmail.com

M. Natesan · G. Vimalarani
Department of Electronics and Communication Engineering, Hindustan Institute of Technology and Science, Chennai 603103, India

V. K. Gunjan and J. M. Zurada (eds.), *Proceedings of 4th International Conference on Recent Trends in Machine Learning, IoT, Smart Cities and Applications*, Lecture Notes in Networks and Systems 873, https://doi.org/10.1007/978-981-99-9442-7_28

1.6 to 7.4% in rural populations and from 1 to 13.2% in urban populations [2]. Diabetics have a CAD prevalence of 21.4%, while non-diabetics have a prevalence of 11%. Rural areas of the country have a CAD prevalence that is almost half that of urban areas [3]. In western countries, only 23% of populations are threatened by CVD, but in India, 52% of population are threatened and empowered to death. Early detection of CVD will serve as a boon to save many individuals. There are many ways to detect CVD, and one such way is by observing and analyzing the electrocardiography (ECG).

ECG is a tool, which is based on electric signals, that detect the pace of the heart and return results as normal or abnormal. Twelve different leads are positioned on the surface of the human body and waves are generated based on the beats of the heart. The ECG procedure is usually performed in any designated laboratory or even in a primary health center. But interpretation of the obtained ECG is facilitated completely by medical experts only in a hospital. This creates a time delay in further diagnosis or treatment and even certain types of CVD require immediate attention too. So, a robust automatic system is required for immediate interpretation and analysis of ECG to diagnose the CVDs.

The ECG signal has lot of attributes which include heart rate, QRS complex, QT-interval, QTC-interval, *P*-interval, PR-interval, RR-interval, *R*-axis, and *T*-axis. Some attributes are very important as they contribute in predicting the functionality of the heart. Machine learning algorithms have been extensively used in medical field for diagnosis and treatment. Selection of suitable parameters is vital in developing a robust model for better prediction and classification. Best models with higher accuracies can aid medical diagnosis and serve as a decision support system in determining the prognosis of the treatment. In this study, it is proposed to develop an intelligent system that helps in classifying the condition of the heart as normal or abnormal with the ECG signal. Three intelligent algorithms such as Naive Bayes, Linear Discriminant Classifier, and Generalized Additive model were used to develop the system and the algorithm with maximum accuracy was selected as the best performing model for the developed system. The developed intelligent system can be used as a better testing approach, especially, in rural areas so as to facilitate immediate medical attention of patients suffering from cardiac abnormalities.

2 Related Study

Computer-aided diagnostic systems (CAD) which employ intelligent techniques help the medical experts in effective diagnosis and prompt treatment of the disease. An extensive survey was carried out on various algorithms used for heart disease prediction.

Various human health parameters were used as inputs to the machine learning classifiers for prediction of heart diseases. Appropriate feature selection is not carried out and the algorithms used may end up in computational complexity [4]. Heart disease was predicted using machine learning algorithms such as k-Nearest Neighbor,

Decision Tree (DT), Support Vector Machine (SVM) and Linear Regression. Support Vector Machine performed better, but only few input parameters were considered [5]. The rate of heart failure was predicted using univariate and multivariate analyses. Hyperparameters were tuned based on search techniques. But the limitation of the study was that only a single machine learning algorithm was used for prediction [6].

Hybrid dataset was used from two repositories such as UCI MI repository and Kaggle for predicting the heart disease. One dataset was utilized for training and the other was utilized for testing. Classifiers such as logistic regression, SVM, and Naive Bayes were used for the study. Since datasets from two different repositories were used, the accuracy obtained was very low. The study also proves that implementing a model in real-time dataset would be more proven and realistic for the development of heart disease prediction models [7].

Singh et al. [8] was able to extract information about different arrhythmias and concluded that the most important arrhythmia was the cardiovascular arrhythmia. Five different algorithms such as Artificial Neural Network (ANN), SVM, DT, Naive Bayes, and Random Forest were trained and ANN has turned to be the best algorithm to give 99.59% accuracy. The main disadvantage was that only three features were used as input to the classifier. But in general, prediction of arrhythmias depends upon more factors [8].

Hybrid Random Forest with a linear model was used to create an upgraded performance level model with an accuracy of 88.7%. However, predominant features were not selected in this study, and the accuracy obtained was not satisfactory [9]. An application was developed using machine learning techniques to predict the heart disease using basic symptoms like gender, age, and pulse rate. But the model did not incorporate the parameters in the ECG signal which is the most vital factor for heart disease prediction [10].

A method that combines classic machine learning (ML) and end-to-end deep learning (DL) was used. With a score of 89.3, the suggested approach outperformed the challenge's baseline approach by 9.1. Although the experimental findings are not directly comparable, the method's overall accuracy is 92.9% (error of 7.1%); this error rate is reasonably similar to the percentage of recordings classified as "unknown" (9.7%). However, the project uses existing dataset from internet [11]. Machine learning techniques have been used for prediction of heart diseases in various studies [12–15].

Existing studies focus on developing a system for heart disease prediction which either takes in some basic symptoms or demographic parameters as input to the system. It does not use the ECG parameters for prediction. Such a system may not yield reliable results. Moreover, most of the studies use the datasets available in the repositories for developing the prediction system where there is a chance for misinterpretation and wrong diagnosis. In certain studies, either multiple features or very few features are considered for developing the models. So, developing an intelligent system incorporating selective contributing features extracted from a real-time ECG signal would be more consistent and robust. Hence, in this study, selection of attributes that contribute more toward the prediction of the functionality of the heart is taken for developing the model.

3 Methods

A new automated system called as Intelligent Prediction for Cardiac Abnormality (IPCA) that incorporates efficient feature selection and intelligent classification of the condition of the heart is developed in this study. The flow diagram of the IPCA methodology is shown in Fig. 1.

3.1 Dataset Preprocessing

Dataset was collected from the Department of Cardiology, Oxymed Hospitals, Chennai. A total of 1000 ECG datasets comprising 341 normal and 659 abnormal were used in this study. The methodology was developed in MATLAB R2021a. The collected dataset had thirteen attributes as given in Table 1.

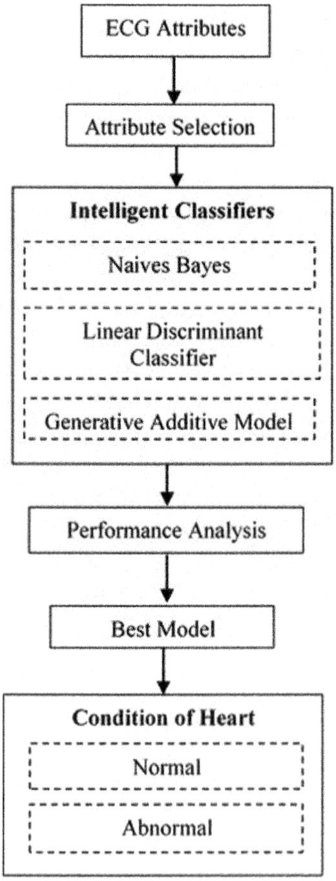

Fig. 1 Flow diagram of the IPCA methodology

Table 1 Attributes in the dataset

S. No.	Attribute	Description
1	Name	Name of patients
2	Age	Age of respective patient
3	Gender	Male or female
4	QR-interval	Time period taken to cover QR peak (ms)
5	QTC-interval	Time period taken to cover QT peak (ms)
6	QT-interval	RMS of QT-interval (ms)
7	PR-interval	Time taken to cover PR-interval (ms)
8	P-interval	Time taken to cover P-interval (ms)
9	RR-interval	Time taken to cover RR-interval (ms)
10	*P*-axis	P position of heart measured in degrees
11	*R*-axis	R position of heart measured in degrees
12	*T*-axis	T position of heart measured in degrees
13	QRS	QRS peak (R-peak) measured in ms

3.2 Dataset Collection

Careful investigation was done in selecting the attributes. Five different attributes were chosen from the thirteen ECG attributes. The five chosen attributes include QRS value, heart rate, *P*-axis, *R*-axis and *T*-axis. These features play a significant role in influencing the ECG outcomes of an individual. When it comes to heart disease prediction, QRS value becomes a significant factor as it is directly related to the heart beat. The pumping of blood through the arteries which causes to and fro oscillatory motion is determined by the QRS segment commonly called as the *R*-peak. The time period and the length of the *R*-peaks are studied to diagnose the condition of an individual. *P*-axis, *R*-axis, and *T*-axis are required to study about the positions of heart in degrees. *P*-axis and *T*-axis give the information pertaining to left atrial fibrosis and ventricular repolarization, respectively. *R*-axis deviation toward right indicates abnormalities such as right-sided heart strain, right bundle branch block, and right ventricular hypertrophy whereas deviation of *R*-axis toward left indicates ventricular contraction. Heart rate helps to predict the efficient pumping of blood by the heart throughout the entire body. Nine features were ignored as they had outliers or some ambiguities like overlapping between the normal and abnormal values. Moreover, they were not significant in contributing toward the desired outcome. With respect to age, cardiac abnormalities are common in the age group of 45–65 and so this feature was not considered as an input to the classifiers.

3.3 Machine Learning Algorithms

Machine learning (ML) is a concept that involves the opportunity to teach or to guide machine and enable machines to think like humans do. There are multiple techniques that can be used to teach machines. Choosing a ML model merely depends on the type of the dataset. In this study, three algorithms such as Naive Bayes, Linear Discriminant Classifier, and Generalized Additive Model were compared and studied. Few ML algorithms were not considered as they involved kernel tricks, rules, or clusters for its functionality.

Naive Bayes

It is a machine learning algorithm that relies on Bayes Theorem and is one among the simplest of the Bayesian Network Models. Bayes Theorem was introduced to identify the probability of an event A, given B had already occurred. When coupled with kernel density distribution, Bayes Theorem is said to give higher percentage of accuracies. When multiple classes are present, Naive Bayes plays a vital role in solving such problems. Naive Bayes interprets an outcome independently irrespective of correlations between the features. In this study, the event A which has already occurred is the input argument, that comprises the heart rate, QRS value, *P*-, *R*-, *T*-axes, and the event B that is to occur is the "normal/abnormal" variable.

Linear Discriminant Analysis (LDA)

Linear discriminant classifier is an algorithm that uses the help of linear curve and works on graphs and distribution of values. Linear discriminant analyses are used to reduce the dimensionality of the classes. When a model with more than two classes is to be trained, complexities are a natural part of such classes. In order to reduce complexities, the reduction in dimension is an apt and a proven technique in LDA. When reducing the dimension, the accuracies can also improve thereby enabling a better view of the clustered samples in the graph. For example, when a 2D graph is viewed without proper distribution of values, confusions can occur for other algorithms, whereas in case of LDA, the confusion can be solved by reducing 2D to 1D and visualizing the graph in an efficient manner.

Generalized Additive Model (GAM)

It is a type of linear model that deals with unknown smooth functions on the predictor variables. The point of focus is on the inference of these functions. It is generally used to fit nonlinear data in a linear surface. It is a type of Linear Regression which can be useful for performing analysis and predicting accurate results on nonlinear data. In the dataset used, distribution of values in graph is not clustered and uneven. As the distributions are uneven and require multiple inconsistent connections, GAM is used to solve linearity problem.

4 Performance Analysis

The condition of the heart was identified as normal or abnormal for the collected dataset by the physician. The proposed methodology has to be tested for its efficiency. Therefore, performance measures like sensitivity, specificity, accuracy, precision, *F*-scores, False Acceptance Rate (FAR), False Rejection Rate (FRR), misclassification rate, and area under the curve are calculated. These indices are computed with the true positive (TP), true negative (TN), false positive (FP), and false negative (FN) values available in the confusion matrix of the classifiers.

Sensitivity: It defines the percentage of true positives

$$\text{Sensitivity} = \frac{\text{TP}}{\text{TP} + \text{FN}}. \tag{1}$$

Specificity: It defines the percentage of true negatives.

$$\text{Specificity} = \frac{\text{TN}}{\text{TN} + \text{FP}}. \tag{2}$$

Accuracy: It defines the degree of closeness of the measured value to the true value.

$$\text{Accuracy} = \frac{\text{TN} + \text{TP}}{\text{TN} + \text{FP} + \text{TP} + \text{FN}}. \tag{3}$$

Precision: It defines the degree of closeness of two or more measeurements to each other.

$$\text{Precision} = \frac{\text{TP}}{\text{TP} + \text{FP}}. \tag{4}$$

F-scores: *F*-scores or *F*-test is used while comparing statistical models with overfitting data

$$F\text{-Score} = 2 * \left(\frac{\text{Precision} * \text{Recall}}{\text{Precision} + \text{Recall}}\right). \tag{5}$$

Misclassification rate (MCR): It is referred to as the error obtained from the predicted model.

$$\text{MCR} = \frac{\text{FP} + \text{FN}}{\text{TP} + \text{TN} + \text{FP} + \text{FN}}. \tag{6}$$

False Rejection Rate (FRR):

$$\text{FRR} = \frac{\text{FP}}{\text{FP} + \text{FN}}. \tag{7}$$

False Acceptance Rate (FAR):

$$\text{FAR} = \frac{\text{FN}}{\text{TP} + \text{FN}}. \quad (8)$$

When performance indicators like sensitivity, specificity, and accuracy are close to 100%, the classification algorithm's efficiency is considered to be good.

5 Results and Discussions

5.1 Results

The study data represented samples from patients of both genders in equal proportion between the age groups of 45 and 65. A total of 1000 datasets were divided into training and testing data in 80:20 ratio. The dataset distribution is given in Table 2.

In IPCA methodology, three algorithms such as Naive Bayes, Linear Discriminant Classifier, and Generalized Additive Model were tried and tested.

Naive Bayes: Naive Bayes algorithm was used for classifying the functionality of the heart. Naive Bayes algorithm works on the basis of likelihood between data samples. In this study, there was a likelihood between the data samples obtained for various abnormal conditions of the heart. This helped in segregating the abnormal data samples from the normal samples which in turn helped in classification of cardiac functionality. Confusion matrix and Region of Convergence (ROC) Curve was computed to determine the effectiveness of the classifier. The confusion matrix reveals the extent to which each classifier gives the accurate classification and helps to find the best performing classifier. Two hundred samples were used for testing the classifier accuracy. The obtained confusion matrix for Naive Bayes classifier reveals that 94 normal samples were classified as normal and 93 abnormal samples were classified as abnormal. Meanwhile, 13 normal samples were classified as abnormal. The classifier provided an accuracy of 93.5%. The ROC curve is plotted to understand the nature of the classifier. When the curve inclines itself toward the top left corner, it indicates a better efficiency of the classifier. In IPCA using Naive Bayes classifier, the ROC curve hugs top left indicating better performance of the classifier. Figure 2 shows the confusion matrix and ROC of Naive Bayes classifier.

Linear Discriminant Analysis: LDA works on the basis of a linear combination of input samples for prediction and classification. In the IPCA methodology, LDA was useful to predict the linear combination of the QRS values, axes information,

Table 2 Dataset distribution

Type of dataset	Training datasets	Testing datasets
Normal	234	107
Abnormal	566	93

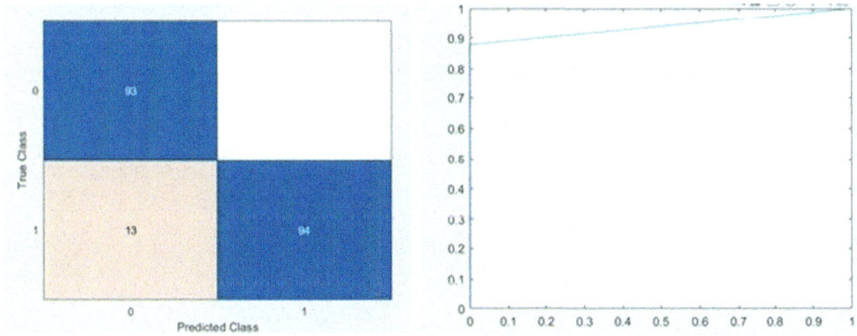

Fig. 2 Confusion matrix and ROC of Naive Bayes classifier

and heart rate for classification of cardiac functionality as normal and abnormal. The obtained confusion matrix for Linear Discriminant Classifier reveals that 94 normal samples were classified as normal and 67 abnormal samples were classified as abnormal. Meanwhile, 13 normal samples were classified as abnormal and 26 abnormal samples were classified as normal. The classifier provided an accuracy of 80.5%. In IPCA using Linear Discriminant Classifier, the ROC curve is deviated from the top left corner thus depicting a moderate performance efficiency of the classifier. Figure 3 shows the confusion matrix and ROC of Linear Discriminant Classifier.

Generalized Additive Model: The obtained confusion matrix for GAM classifier reveals that 81 normal samples were classified as normal and 67 abnormal samples were classified as abnormal. Meanwhile 26 normal samples were classified as abnormal and 26 abnormal samples were classified as normal. The classifier provided an accuracy of 74%. Figure 4 shows the confusion matrix and ROC of Generalized Additive Model.

In IPCA using GAM, the ROC curve is highly deviated from the top left corner thus depicting a poor performance efficiency of the classifier. Out of the three algorithms,

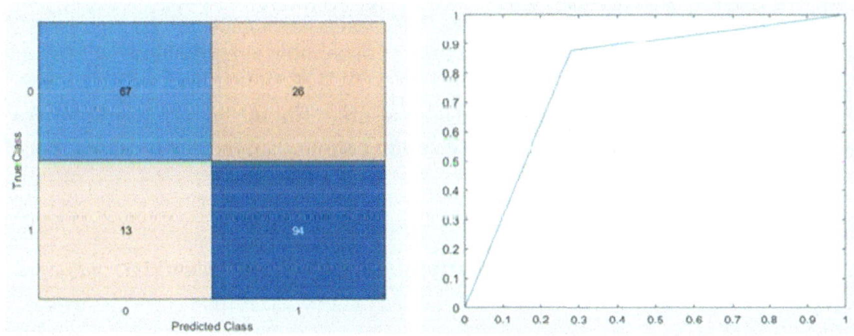

Fig. 3 Confusion matrix and ROC of Linear Discriminant Classifier

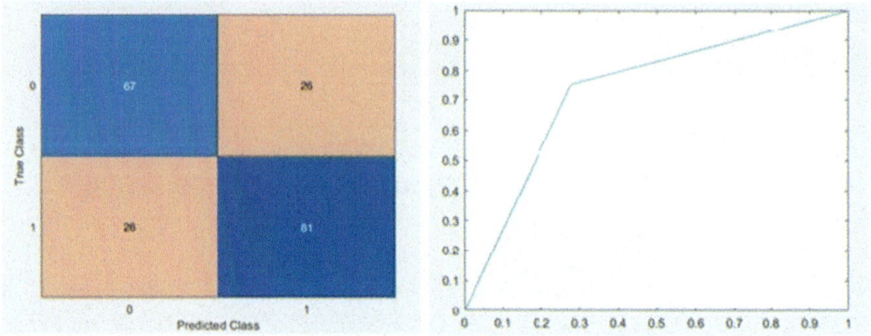

Fig. 4 Confusion matrix and ROC of Generalized Additive Model

Naive Bayes performed better yielding a better accuracy of 93.5% compared to LDA and GAM. Hence, Naive Bayes algorithm was predicted to be a suitable classifier for the IPCA study.

5.2 *Discussion*

Comparison was made between the various classifiers using performance indices like specificity, sensitivity, accuracy, precision, and *F*-score in order to find the overall performance of the methodology. This helped to identify a suitable classifier. Table 3 shows the results of various performance indices in percentages.

From Table 3, it is revealed that Naive Bayes Classifier has the highest values with respect to all the performance metrics as compared to LDA and GAM. This indicates that Naive Bayes algorithm would be the best classifier for the IPCA system. In addition to this, misclassification rate, FAR, FRR, and AUC were computed for all the three classification methodologies such as GAM, LDA, and NB which are given in Table 4.

From Table 4, it is observed that the misclassification rates obtained for Naive Bayes algorithm are acceptable and there is a considerable agreement in the FAR and FRR values obtained for Naive Bayes algorithm. FAR is slightly high which indicates that some normal samples are classified as abnormal. But this in turn would provoke further investigation to test the condition of the patient. But in case of LDA, though

Table 3 Performance metrics

Classifier	Sensitivity (%)	Specificity (%)	Accuracy (%)	Precision (%)	*F*-score
Naive Bayes	87.85	100	93.5	100	93.53
LDA	87.85	72.04	80.5	78.33	82.81
GAM	75.70	72.04	74	75.70	75.7

Table 4 Computation of classifier performance

Classifier	FAR (%)	FRR (%)	Misclassification rate (%)	AUC (Sq. units)
Naive Bayes	12.14	0	6.5	93.93
LDA	12.14	27.95	19.5	79.95
GAM	24.29	27.95	26	73.87

the FAR values are similar to Naive Bayes, the FFR values are very high indicating that some of the abnormal samples are classified as normal. This would create misinterpretation and wrong diagnosis. The misclassification rate is also moderately high. In case of GAM, both FAR and FRR are high depicting that the efficiency of this classifier is poor as compared to other classifiers. The misclassification rate is also very high. The AUC is also high for Naive Bayes classifier. On comparison with all the other methods, Naive Bayes algorithm can be concluded as a best classifier for the prediction of cardiac abnormalities.

6 Conclusion

A predictive algorithm was developed in this study to detect the functionality of the heart. Features were obtained from the ECG signal and contributing features were selected and were fed as input to the classifiers. Three classifiers such as Naive Bayes, Generalized Additive Model, and Linear Discriminant Classifier were tried and Naive Bayes classifier outperformed the other classifiers. The developed intelligent predictive algorithm will be useful in detecting the presence and absence of cardiac abnormalities. The intelligent system will serve as a decision support system and help the physician in efficient diagnosis. The future scope of the study includes classification of different cardiac abnormalities to aid better diagnosis and treatment. A graphical user interface can be created to facilitate easy and prompt diagnosis.

Acknowledgements The authors are thankful to the Department of Cardiology, Oxymed Hospitals, Chennai, India, for giving the dataset, for providing suggestions, and for validating the study.

References

1. Sheet F (2020) World Health Organization: cardiovascular diseases. WHO, Geneva, Switzerland
2. Gupta R, Joshi P, Mohan V, Reddy SK, Yusuf S (2008) Epidemiology and causation of coronary heart disease and stroke in India. Heart (British Cardiac Society) 94(1):16–26. https://doi.org/10.1136/hrt.2007.132951
3. Kumar AS, Sinha N (2020) Cardiovascular disease in India: a 360° overview. Med J Armed Forces India 76(1):1–3. https://doi.org/10.1016/j.mjafi.2019.12.005

4. Mohan N, Jain V, Agrawal G (2021) Heart disease prediction using supervised machine learning algorithms. In: 2021 5th international conference on information systems and computer networks (ISCON). IEEE, India, pp 1–3. https://doi.org/10.1109/ISCON52037.2021.9702314
5. Singh A, Kumar R (2020) Heart disease prediction using machine learning algorithms. In: 2020 international conference on electrical and electronics engineering (ICE3). IEEE, India, pp 452–457. https://doi.org/10.1109/ICE348803.2020.9122958
6. Ambesange S, Vijayalaxmi A, Sridevi S, Yashoda BS (2020) Multiple heart diseases prediction using logistic regression with ensemble and hyper parameter tuning techniques. In: 2020 fourth world conference on smart trends in systems, security and sustainability (WorldS4). IEEE, London, UK, pp 827–832. https://doi.org/10.1109/WorldS450073.2020.9210404
7. Khan MIH, Mondal MRH (2020) Effectiveness of data driven diagnosis of heart disease. In: 2020 11th international conference on electrical and computer engineering (ICECE). IEEE, Dhaka, Bangladesh, pp 419–422. https://doi.org/10.1109/ICECE51571.2020.9393055
8. Singh V, Tewary S, Sardana V, Sardana HK (2019) Arrhythmia detection—a machine learning based comparative analysis with mit-bih ecg data. In: 2019 5th international conference for convergence in technology (I2CT). IEEE, Bombay, India, pp 1–5. https://doi.org/10.1109/I2CT45611.2019.9033665
9. Mohan S, Thirumalai C, Srivastava G (2019) Effective heart disease prediction using hybrid machine learning techniques. IEEE Access 7:81542–81554. https://doi.org/10.1109/ACCESS.2019.2923707
10. Gavhane A, Kokkula G, Pandya I, Devadkar K (2018) Prediction of heart disease using machine learning. In: 2018 second international conference on electronics, communication and aerospace technology (ICECA). IEEE, Coimbatore, India, pp 1275–1278. https://doi.org/10.1109/ICECA.2018.8474922
11. Gjoreski M, Gradisek A, Budna B, Gams M, Poglajen G (2020) Machine learning and end-to-end deep learning for the detection of chronic heart failure from heart sounds. IEEE Access 8:20313–20324. https://doi.org/10.1109/ACCESS.2020.2968900
12. Parthiban G, Srivatsa SK (2012) Applying machine learning methods in diagnosing heart disease for diabetic patients. Int J Appl Inf Syst 3:25–30
13. Bharti R, Khamparia A, Shabaz M, Dhiman G, Pande S, Singh P (2021) Prediction of heart disease using a combination of machine learning and deep learning. Comput Intell Neurosci 2021:8387680. https://doi.org/10.1155/2021/8387680
14. Shah D, Patel S, Bharti SK (2020) Heart disease prediction using machine learning techniques. SN Comput Sci 1(6):345. https://doi.org/10.1007/s42979-020-00365-y
15. Nayeem MJN, Rana S, Islam MR (2022) Prediction of heart disease using machine learning algorithms. Euro J Artif Intell Mach Learn 1(3):22–26. https://doi.org/10.24018/ejai.2022.1.3.13

Innovative Design of a Solar-Powered Wireless Soil Moisture Sensor for Maximizing the Efficiency of IoT-Based Systems

Anjali Gangone, G. J. Bharat Kumar, and Swapna Gangone

Abstract This paper presents a comprehensive design and implementation of a Sunward Solar-Powered Wireless Soil Moisture Sensor, which combines cutting-edge technology to enable energy-efficient soil moisture measurement and wireless data transmission. The proposed sensor system employs capacitance-based soil moisture measurement principles and incorporates innovative design elements to reduce costs and simplify installation. It also integrates solar power technology to ensure autonomous and sustainable operation. The core components of the system include a microcontroller (MSP430F149), a Wireless Transceiver Unit (nRF905), a Soil Moisture Sensor Unit with annular and conical electrodes, and a servo motor-based solar panel module for optimal solar energy capture. The sensor's power consumption is meticulously analyzed, and its energy requirements are met by a combination of solar panels and rechargeable batteries. The design of the Sunward Solar Panel Module, featuring a spherical servo motor, is a highlight of this system. The servo motor ensures precise tracking of the sun's position, optimizing solar panel orientation for maximum energy generation. Four Light-Dependent Resistors (LDRs) strategically placed on the solar panel aid in this tracking process. Through this innovative system, real-time soil moisture data can be collected and wirelessly transmitted over long distances, making it a valuable tool for applications in agriculture and environmental monitoring. The paper also discusses the dimensions and efficiency of the solar panel, ensuring that it meets the energy needs of the sensor. It achieves the main objective of this research, that is, not only in day time when sun is with more intensity but also in the low-intensity times like morning and evenings, the proposed design exhibits better results.

Keywords Soil moisture · Soil moisture sensor · Wireless soil moisture sensor · Solar energy · Sunward

---

A. Gangone
Department of Computer Science and Engineering, CMR Institute of Technology, Hyderabad, Telangana, India

G. J. Bharat Kumar (✉) · S. Gangone
Department of Computer Science, Mattu University, Mattu, Ethiopia
e-mail: drkumar2018@gmail.com

V. K. Gunjan and J. M. Zurada (eds.), *Proceedings of 4th International Conference on Recent Trends in Machine Learning, IoT, Smart Cities and Applications*, Lecture Notes in Networks and Systems 873, https://doi.org/10.1007/978-981-99-9442-7_29

1 Introduction

Without agriculture and farmer, there is no life. The entire human kind is dependent on agriculture and most respectable farmer. For centuries together, the production of the agriculture is highly dependent on nature, climatic conditions, and natural resources. Maintaining a very good soil quality is very sentential for irrigation scheduling and nutrient management and the economic and environmental sustainability of the yield of the cropping; if it declines, it impacts production, growth of the plants, quality of the grains, cost of production and even it may increase the risk of erosion in soil [1]. To keep track of such a vital parameter like soil moisture, the sensors that are available need power supply to operate. But as per our traditional agricultural system, which depends mostly on natural resources, and also due to economic viability of the formers, most of the agricultural fields are not equipped with adequate power supplies and long-distant cables. As a necessity to monitor the soil moisture, this paper presents a design proposal for a soil moisture sensor based on the impedance transform of the frequency-domain technique with wireless communication capabilities to avoid the long-distance cables; solar energy as an operational power to the sensor avoiding the necessity of installing power supplies in the agriculture fields; always Sunward that is rotated according to sun direction in the day time to maximize the extraction of energy from the sun; the power consumption is also optimized so that the sensor can operate for longer time even when a cloudy day and also night. The physical design of the sensor is in the way that the sensor electrodes are embedded at the bottom of the sensor so as to measure the moisture content value at different depths of the soil.

2 Literature Survey

The selection of an appropriate soil moisture sensor tailored to specific soil types and crop varieties is paramount to ensure accurate data collection and informed decision-making on farms [2]. The selection and implementation of appropriate soil moisture sensor technologies are vital considerations for optimizing agricultural practices, enhancing crop yield, and ensuring sustainable water resource management [3]. The state-of-the-art and smart agricultural techniques have been proposed to get the appropriate information related control of irrigation according to crop periods [4]. The electrical based soil moisture measurement techniques have been proposed for the field of agronomy [5, 6]. Some resistive-based soil moisture sensing [7]; Capacitive soil moisture sensing using Frequency-Domain Reflectometry [8] and Time Domain Reflectometry [10]. Some widely in use sensors with wireless communication capabilities to avoid long-distance cables in the field with data acquisition and processing capabilities [9]. Long-distance power cables are not a viable solution for huge farmlands and the life of the batteries as a substitute is very limited. The solar power is the best solution and long-lasting energy to power the soil moisture sensor [11].

3 Design Methodology

3.1 Working Principle of the Soil Moisture Sensor

Figure 1 shows capacitance-based soil moisture sensors which work on the principle that the dielectric constant (permittivity) of a material, such as soil, changes with its moisture content. These sensors typically consist of two electrodes separated by a dielectric medium, which is the soil being measured.

The sensor measures the capacitance (C) between the two electrodes. Capacitance is a measure of the sensor's ability to store electrical charge and is influenced by the properties of the dielectric material between the electrodes, which, in this case, is the soil. Dielectric constant (permittivity) which is (ε_r) of a material is a measure of its ability to store electrical energy in an electric field. It changes with the moisture content level of the soil. Dry soil has a lower dielectric constant, while wet soil has a higher dielectric constant. The capacitance of the sensor can be expressed using the formula

$$C = \varepsilon_r \times \varepsilon_0 \times \frac{A}{d}, \tag{1}$$

where C is the capacitance. ε_r is the relative permittivity (dielectric constant) of the soil. ε_0 is the permittivity of free space (approximately 8.854×10^{-12} F/m). A is the surface area of the sensor's electrodes. d is distance between two electrodes.

Calibration

To use the sensor for soil moisture measurement, it needs to be calibrated. This involves taking measurements of capacitance (C) at different known moisture levels

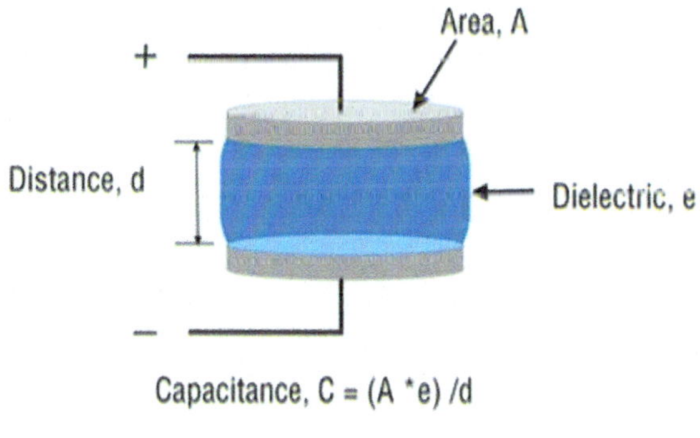

Fig. 1 Working principle of the soil moisture sensor

to establish a relationship between C and soil moisture content. Moisture Content Calculation method involves that, once calibrated, the sensor can estimate soil moisture content (θ) based on the capacitance reading (C) and the known calibration parameters. The equation for calculating soil moisture content is often linearized as

$$\theta = a \times C + b, \tag{2}$$

where θ is the soil moisture content (typically expressed as a percentage). a and b are calibration constants determined during calibration.

3.2 *Proposed Electrode Design*

The single probe soil moisture sensor proposed, with an annular electrode and a conical electrode, is an alternative design method that aims to achieve the advantages of a dual-probe sensor while potentially reducing costs. This potential advantages of this design. The annular electrode is ring-shaped and typically positioned closer to the surface of the soil. Its purpose is to measure moisture content at a shallower depth within the soil profile, such as the topsoil layer. The conical electrode has a cone-like shape and extends deeper into the soil. It is designed to measure moisture content at a greater depth within the soil profile, penetrating deeper into the root zone or beyond. It is important to note that while this alternative design can offer cost savings and simplify installation, it may have some limitations compared to traditional dual-probe sensors, such as potential differences in the accuracy of shallow and deep measurements. However, these limitations can often be addressed through careful sensor calibration and validation in specific soil conditions. This design is shown in Fig. 2.

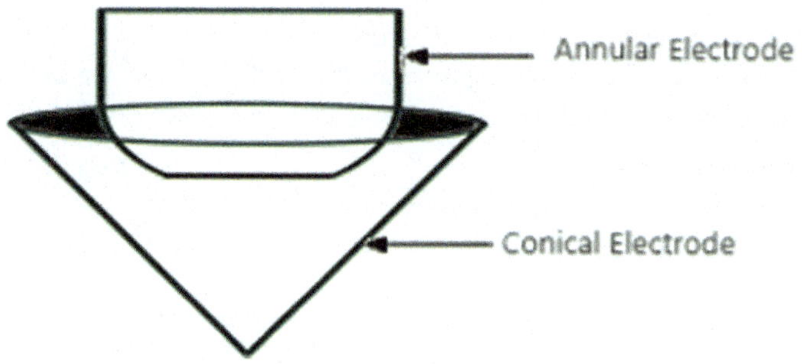

Fig. 2 Proposed electrode design

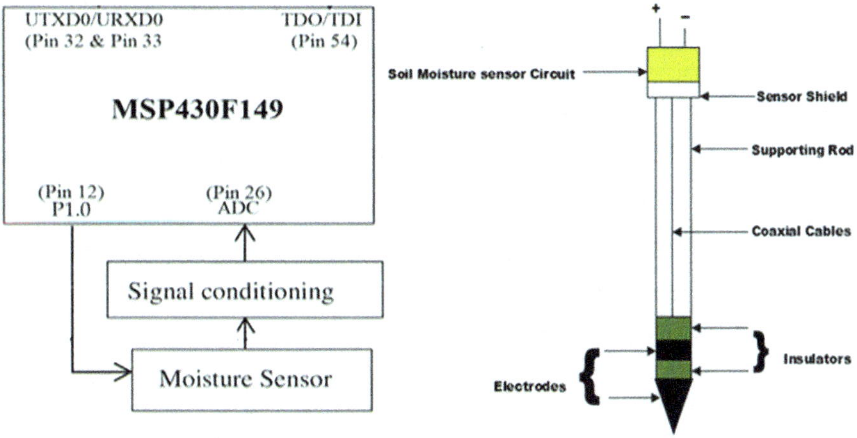

Fig. 3 Proposed soil moisture sensor design

3.3 *Proposed Soil Moisture Sensor Design*

As shown in Fig. 3, to find the most suitable microcontroller for our research, we explored various alternatives, including popular choices like Arduino. However, our primary objective is centered around achieving exceptional energy efficiency. After thorough evaluation, the MSP430F149 microcontroller is emerged as the ideal choice. Its standout feature lies in its remarkable ability to operate on extraordinarily low-power consumption levels. This capability aligns seamlessly with the core focus of our research, enabling us to design and implement energy-efficient solutions that address critical challenges in our field. The soil moisture sensing unit, equipped with the MSP430F149 microcontroller, serves the purpose of gauging soil moisture levels through the utilization of a specialized electrode and a coaxial cable. A stainless-steel supporting rod, boasting a diameter of 10 mm and extending to a length of 590 mm, has been ingeniously designed with a hollow configuration, housing the coaxial cable that links two distinct electrodes. Situated at the extremity of this supporting rod, the sensor components are strategically embedded to facilitate soil moisture measurements. These two electrodes exhibit contrasting shapes to enable easy insertion into the soil; one takes on an annular form, while the other adopts a conical structure. Additionally, a plastic shield is expertly interposed between the two electrodes to ensure their insulation from each other.

3.4 *Proposed Wireless Soil Moisture Sensor Design*

In Fig. 4, we present the wireless soil moisture sensor design, featuring a sensor shield crafted from multi-crystalline silicon. This shield serves the critical role of

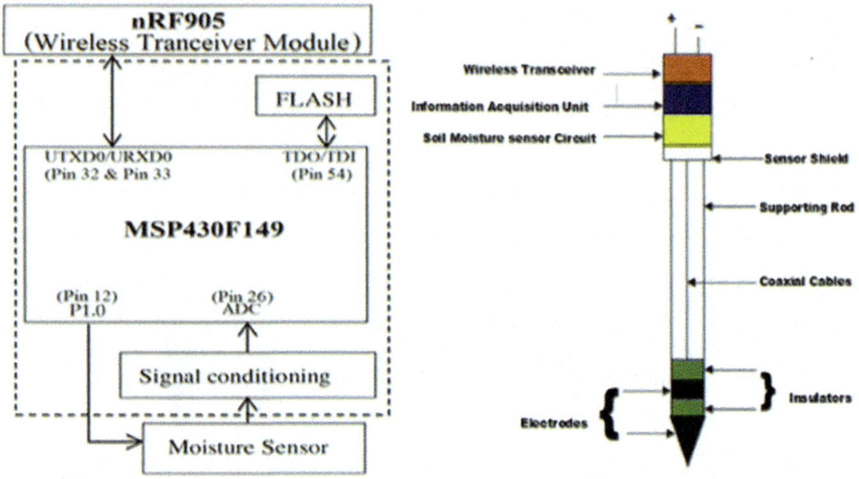

Fig. 4 Proposed wireless soil moisture sensor design

safeguarding the internal modules against water damage and rust when deployed in agricultural fields. The sensor shields house three essential components: the Wireless Transceiver Unit nRF905, the Information Acquisition Unit, and the Soil Moisture Sensor Unit. After experimenting with various wireless transceiver modules, such as ESP32 and others, the nRF905 has emerged as the optimal choice due to its exceptional energy efficiency and an impressive range of 100 m operating at 433 MHz. The Wireless Transceiver Unit functions as the communication hub, facilitating the transmission and reception of soil moisture data to and from other devices. Meanwhile, the Information Acquisition Unit is equipped with a 1 MB memory chip and FLASH memory, boasting a range of 100 m. Its primary role is to gather and store the soil moisture values detected by the Soil Moisture Sensor Unit.

3.5 *Proposed Design of Sunward Solar-Powered Wireless Soil Moisture Sensor*

Figure 5 shows the wireless soil moisture sensor designed above which is equipped with a solar panel with a battery of 2 Wh (Watt-hour) to store the power and to power up with solar energy, and also, we have added a backup battery of capacity 6 Wh to overcome the situation to handle solar circuit failures.

Power Consumption of Each and Every Module

Table 1 gives the information related to power consumption of each and every module.

Based on the above power consumption parameters, the required energy for the Sunward Solar-Powered Wireless Soil Moisture Sensor is calculated as 2 W-h (Wh).

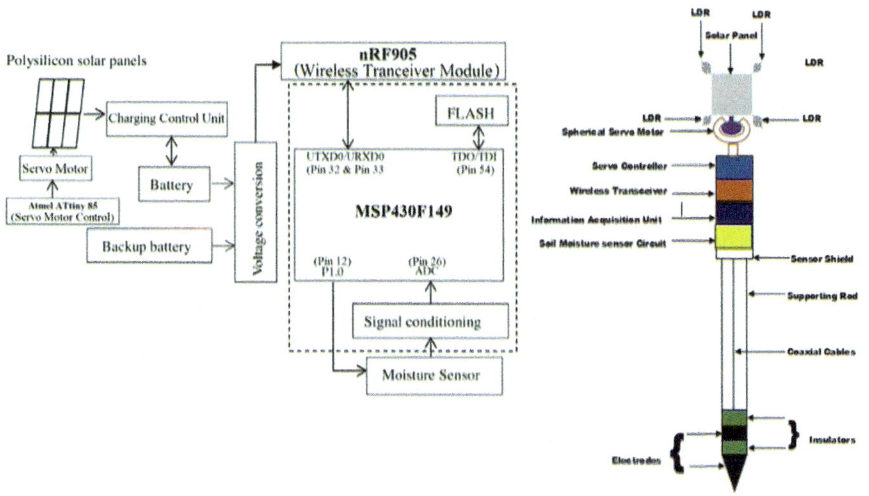

Fig. 5 Design of solar-powered wireless soil moisture sensor

Table 1 Power consumption of components

	Voltage (V)	Current (mA)
MSP430F149	3	0.02
nRF905	3	12.5 (transmit) 9 (receive)
Memory flash	3	8 (write) 3 (read)
Soil moisture sensor	5	18
Servo motor	5	220
LDR	3	3
Recharge control unit	5	2
Regulator DC	5	1

Solar Panel Size and Capacity and Dimension Calculation

To calculate the dimensions of a solar panel that can generate 2 Wh of energy per day under certain conditions, you can follow these steps. Please note that this is a simplified calculation, and real-world factors such as temperature, shading, and angle of the panel can affect performance. Additionally, solar panels are typically rated in peak watts (W_p) at standard test conditions (STCs), so efficiency is a crucial consideration. The required daily energy output is 2 Wh and we assumed that average daily sunlight hours are 6 AM–6 PM and a duration of 12 h and the Solar panel efficiency we assumed as 15%, that is 0.15 in decimal form, which is a typical efficiency for solar panels in the we used. Required Solar Panel Capacity (in watts) is the daily energy requirement (2 Wh) divided by the average daily sunlight hours (12 h) gives you the wattage needed is

$$\text{Required Solar Panel Capacity(in watts)} = \frac{\text{Daily Energy Requirement (Wh)}}{\text{Average Daily Sunlight Hours (h)}},$$

$$\text{Required Solar Panel Capacity(in watts)} = \frac{2\,\text{Wh}}{12\,\text{h}} = 0.167\,\text{W}.$$

Adjust for Panel Efficiency: Since solar panels are typically rated in peak watts at a certain efficiency, you will need to adjust for the panel's efficiency. To do this, divide the required capacity by the panel efficiency which is

$$\text{Adjusted Solar Panel Capacity(in watts)} = \frac{\text{Required Solar Panel Capacity (W)}}{\text{Panel Efficiency}},$$

$$\text{Adjusted Solar Panel Capacity(in watts)} = \frac{0.167\,\text{W}}{0.15} = 1.113\,\text{W}.$$

To calculate the area of the solar panel (in square centimeters), we need to calculate the area of the solar panel required using the adjusted capacity and assuming a specific power density for the panel which is

$$\text{Area (in square centimeters)} = \frac{\text{Adjusted Solar Panel Capacity (W)}}{\text{Panel Power Density}\left(\text{W/cm}^2\right)}.$$

The solar panel we used has power densities in the range of 1.5 W per square meter (W/cm^2).

$$\text{Area (in square centimeters)} = \frac{1.113\,\text{W}}{1.5\,\text{W/cm}^2} = 0.742\,\text{cm}^2 = 0.00798\,\text{feet}^2.$$

After calculating the surface area of the solar panel, we have taken shape of the solar panel as square with a dimension of each side is 0.861 cm that is 8.61 mm.

Design of Sunward Solar Energy Panel Module

Figure 6 schematically shows the design of a Sunward Solar Panel Module, which contains a solar panel attached to DC spherical stepper servo motor with its controller circuitry of operating voltage of + 5 V for rotating according to sun and it has also been attached with two Light-Dependent Resistors (LDR), LDR1 and LDR2. LDRs are photoresistors with operating voltage of + 5 V which has a capability changing its resistance with the light intensity that falls up on it. It has very high resistance value with it which indicates no light or dark conditions, when light falls on it has capability to reduce it resistance value according to intensity. By calculating the resistance value difference between LDR1 and LDR2, the servo controller circuitry take decision of in which direction the Solar Panel has to be rotated.

If the difference of resistance of LDR1 and LDR2 is positive, which means that LDR1 is at relatively darker side than LDR2, then the servo controller rotates the solar panel toward LDR2 side as it is at brighter side which means facing toward the

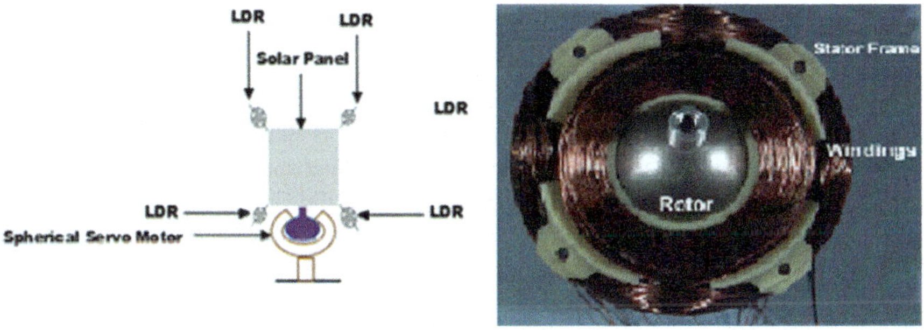

Fig. 6 Sunward Solar Panel and spherical servo motor

sun at that point of time, otherwise the servo controller will rotate the Solar panel to LDR1 side.

Figure 6 shows the picture of a Spherical Servo Motor which can rotate in 3D space in 360°. This tiny motor is capable of handling maximum of 100 g weight with the torque 0.15 Nm continuously and 0.40 Nm peak value. The servo controller Atmel AT Tiny 85. In addition to these batteries, this module also contains a dedicated backup battery to handle the power needs when solar power panel is under maintenance and repair.

Figure 7 shows the different placements of the LDRs on the solar panel. The installation of the direction of the sensor solar panel must be in such a way that the resistance value of four LDRs must be same; to achieve this, we tried with so many different placements mainly considered the angle with diagonal of the solar panel to LDR. When four LDRs are fixed at an angle of 52.5° from the central axis of the Solar panel to diagonal the produced best results.

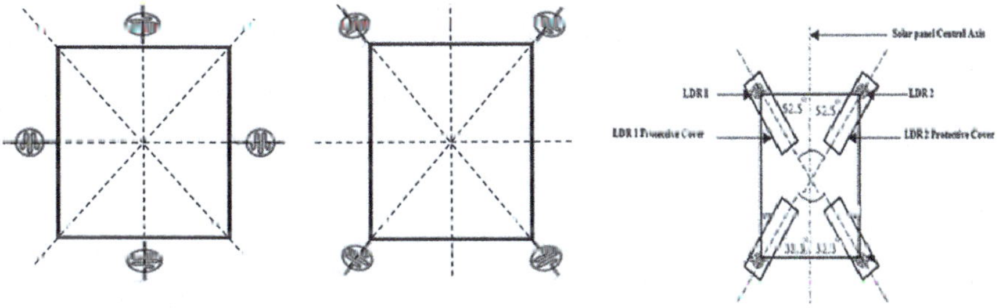

Fig. 7 LDR placement on the solar panel

3.6 Results and Discussion

The data in Table 2 is a 30-day average Wh of 12 h of the day from 6 AM to 6 PM. The results show that in that whenever sun intensity is like middle of the day from 10 AM to 3 PM, both the sensors are producing almost same amount of power, but the proposed sunward variant produces not only in the mid of the day also in the mornings and evenings, even in low light, the efficiency is high.

Table 2 Comparative results of sunward and regular solar-powered wireless soil moisture sensor

Time (h)	Regular solar-powered soil moisture sensor (Wh)	Sunward Solar-Powered Soil Moisture Sensor (Wh)
6:00 AM	0.1	0.3
7:00 AM	0.1	0.4
8:00 AM	0.2	0.6
9:00 AM	0.7	1.1
10:00 AM	1.3	1.4
11:00 AM	1.5	1.6
12:00 PM	1.9	1.9
1:00 PM	1.9	1.9
2:00 PM	1.7	1.8
3:00 PM	1.5	1.6
4:00 PM	0.7	1.2
5:00 PM	0.3	0.6
6:00 PM	0.1	0.3

4 Conclusion and Future Work

In this paper, we have presented a novel Sunward Solar-Powered Wireless Soil Moisture Sensor system that integrates advanced technologies to address critical challenges in soil moisture monitoring. The system's design encompasses capacitance-based soil moisture measurement, wireless data transmission, and solar power generation to achieve energy efficiency and sustainability. Key design elements include the use of annular and conical electrodes for versatile depth measurements, a highly energy-efficient MSP430F149 microcontroller, and the nRF905 wireless transceiver for long-range communication. The Sunward Solar Panel Module, equipped with a spherical servo motor and strategically placed LDRs, ensures optimal solar panel orientation and energy capture. By carefully considering power consumption and solar panel efficiency, the system achieves autonomous operation with the capacity to handle maintenance and repair situations through backup batteries. The proposed system is a valuable tool for various applications, including agriculture and environmental monitoring, where real-time soil moisture data can inform decisions and optimize resource utilization. Its cost-effectiveness, reduced installation complexity, and energy efficiency make it a promising solution for remote and off-grid locations. In summary, our Sunward Solar-Powered Wireless Soil Moisture Sensor represents a significant step forward in the field of soil moisture monitoring, offering a sustainable and data-rich solution for addressing critical challenges in agriculture and environmental science. Further research and field testing will help to refine and validate its performance in diverse soil and environmental conditions. In the future research, we will try to implement the same with GPS-based solar tacking mechanism.

References

1. Ojha T, Raghuwanshi NS, Misra S (2015) Wireless sensor networks for agriculture the stateoftheart in practice and future challenge. In: Computer and electronics in agriculture
2. Bharat Kumar GJ (2018) Selection of soil moisture sensor according to soil and crop type: a selection guide for farms. Int J Multi-Disc Educ Res (IJMER) 7(12(2)). ISSN 2277-7881
3. Bharat Kumar GJ (2018) Implications of soil moisture sensor technologies: a categorical review. Int J Multi-Disc Educ Res (IJMER) 7(12(2)). ISSN 2277-7881
4. Tarange PH et al (2015) Web based automatic irrigation sys using wireless sensor netw and embedded Linux boards In: Circuit, power and computing tech. IEEE
5. Gambazzi F, Suski B, Baron L et al (2014) Evolution of soil wetting patterns hydrologically induced landslide inferred from electrical resistivity survey of volumetric water content and pore water pressure. Water Resour Res
6. Dubois-Fernandez P et al (2014) Sensitivity of main parameters of multifrequency SAR data to soil moisture and surface roughness over bare agricultural soils. IEEE Geosci Remote Sensing Lett 10(4):731–735
7. Mouazen AM et al (2014) Combining frequency domain reflectometry for assessment of soil bulk density. Soil Tillage Res
8. Rhoades JD (1979) A compact low-cost circuit forreading fourelectrode sensors. Soil Sci Soc Am J

9. Rehmani MH (2016) Emerging common techs based on wireless sensor nets. Cur Res Future App J Anatomical Rec 112(2):195–215
10. Cope RE et al (2007) A matrix sensor to measure the soil water in the field. Euro J Soil Sci 58:18–23
11. Jiang M, Lv M, Deng Z, Zhai G (2017) A wireless soil moisture sensor powered by solar energy. PLOS One

Image Inpainting on Archeological Dataset Using UNet Architecture on Embedded Platform

Uday Kulkarni, Satish Chikkamath, James Samuel Mirajkar, Yashas Hittalmakki, Venkatpavankumar Thota, and Fardin Khan

Abstract Image inpainting is a widely used technique for reconstructing damaged or missing portions of images, with applications in image editing, object removal, and image restoration. This paper proposes a novel approach to enhance the quality of image inpainting specifically in archaeological datasets. Our method focuses on enhancing the UNet architecture by incorporating the MS-SSIM loss function, which effectively denoises the output. We conducted experiments using datasets from archaeological sites, which often contain photographs with missing or damaged areas due to the passage of time or excavation procedures. Our results demonstrate that the proposed approach significantly improves the SSIM score, increasing it from 0.9026 to 0.9590. Additionally, we employed post-training quantization, resulting in a remarkable 77.1% reduction in the model size, making it more suitable for deployment in mobile applications. The suggested approach enhances image restoration and analysis for archaeological data, providing academics and archaeologists with a valuable tool for their work. Furthermore, by integrating the model into mobile apps, we aim to expand its usability and accessibility, empowering users to examine and restore their own archaeological images effectively.

Keywords Image inpainting · MS-SSIM · UNet architecture · Quantization

Supported by organization KLE technological university.

U. Kulkarni (✉)
School of Computer Science, KLE Technological University, Hubballi, India
e-mail: uday_kulkarni@kletech.ac.in

S. Chikkamath · J. S. Mirajkar · Y. Hittalmakki · V. Thota · F. Khan
School of Electronics and Communication, KLE Technological University, Hubballi, India
e-mail: chikkamath@kletech.ac.in

V. K. Gunjan and J. M. Zurada (eds.), *Proceedings of 4th International Conference on Recent Trends in Machine Learning, IoT, Smart Cities and Applications*, Lecture Notes in Networks and Systems 873, https://doi.org/10.1007/978-981-99-9442-7_30

1 Introduction

Image inpainting is the process of reconstructing damaged or missing portions of an image with plausible content [14, 17]. As shown in Fig. 1, this method is used in image processing and computer vision to repair damaged images, get rid of extraneous items like text or objects, and complete images that are missing parts. The procedure entails examining the pixels that surround the missing or damaged area and then creating new pixel values based on the analysis to fill in the void [17]. To accomplish this, a variety of strategies can be used, including deep learning-based methods, patch-based approaches, and texture synthesis [4, 6]. There are several uses for image inpainting, including the digital restoration of damaged artworks, the elimination of undesirable objects or text from photographs, and the enhancement of image quality in surveillance and medical imaging. The created material must match the original image and appear realistic to the human eye, making it a difficult assignment. Image inpainting is becoming more efficient and accurate thanks to recent developments in deep learning algorithms [3], and it is being employed more frequently in a variety of real-world applications.

The use of the deep convolutional GAN (DCGAN) [3], a Generative Adversarial networks' (GANs) variation, was one of the first deep learning-based methods for image inpainting that was proposed in 2015. When creating new images, DCGANs use a generator network; when telling the difference between real and created images, they use a discriminator network. In image inpainting, the discriminator network is taught to recognize the difference between the filled image and the original image, while the generator network is trained to fill in any missing areas of an image. DCGANs do frequently generate hazy and inaccurate results, though, which is a significant drawback. The Partial Convolutional Neural Networks' (PCNNs) [16] technique was introduced by researchers in 2018 as a means of overcoming this

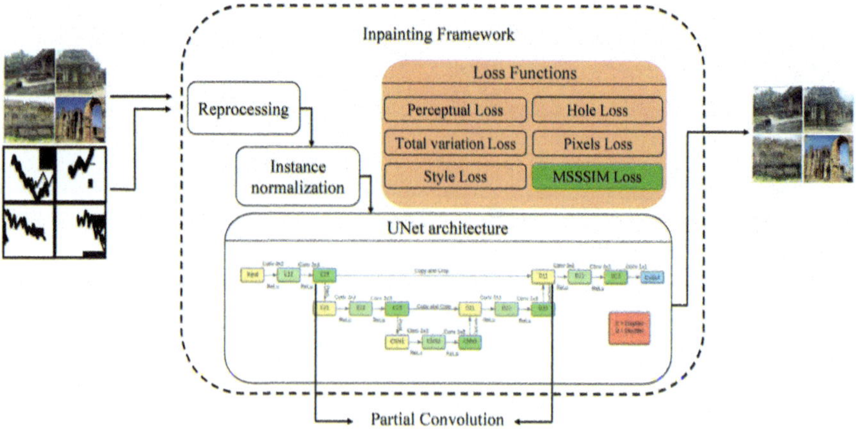

Fig. 1 Framework for image inpainting

restriction. While DCGANs employ regular convolutions in the generator network, PCNNs substitute partial convolutions. By modifying their normalization factors in accordance with the number of valid pixels in their receptive field, partial convolutions may manage missing areas in a picture. By just updating the missing portions of a picture, the generator network is able to update an image selectively, yielding results that are sharper and more lifelike [5].

For consumers, the study of image inpainting has always been a difficult and constant endeavor. Much of work has gone into developing improved tools for users to produce aesthetically pleasing images in response to the growing demand for shooting great pictures. Yet, occasionally users could capture unfavorable images when their subject of interest is obscured by an unrelated object.

The paper is organized as follows: The works on picture inpainting using different methods are in Sect. 2. The model's implementation is covered in Sect. 3, which also involves choosing the ideal approach for image inpainting. The comparison of our approach with other ways is presented in Sect. 4. Section 5 discusses the findings and the work that has to be done moving forward.

2 Related Work

The authors of [3] suggest a method using deep convolutional generative adversarial network. The fundamental principle behind employing DCGANs for image inpainting and object removal is to train a generator network to create new pixels that perfectly blend in with the surrounding pixels and a discriminator network to identify the created pixels from the original pixels. The generator is trained to reduce the discriminator's ability to tell the difference between created and real pixels, while the discriminator is trained to increase that ability. Here, the author compares his approach with "Navier–Stokes" algorithm and Telea's inpainting algorithm; based on PSNR value, the performance of DCGAN was average, but visually the output is sharper and more natural [2]. Overall, DCGANs have been utilized successfully in a range of applications, including image restoration, image editing, and video processing; this is an effective tool for object removal and picture inpainting.

The research [12] deals with the problem of low-quality image inpainting at high resolutions caused by the constrained receptive fields of neural networks. The authors suggest a multi-scale consistency loss and an iterative refinement technique to enhance the inpainting outcomes at higher resolutions and optimize the network's intermediate feature maps. With this method, the color and structure from lower resolution forecasts are preserved without the need for additional inpainting network training.

The authors of [16] suggest a method using partial convolution. The encoder and decoder networks make up the UNet architecture. The decoder network is used to boost spatial resolution and produce a segmentation mask, while the encoder network is used to extract spatial information from the input picture and reduce its spatial resolution. The architecture is known as "U-Net" because of its U-shaped structure, which includes an expanding path (the decoder network) and a contracting path (the encoder network). The encoder network's contracting path is made up of a number of convolutional layers interspersed with max pooling layers. The input image's spatial resolution is decreased while the number of feature channels is increased. The decoder network's expanding route is made up of several convolutional layers interspersed with up sampling layers. The number of feature channels is reduced, while the spatial resolution of the feature map is increased. The UNet architecture incorporates skip links in addition to the encoder and decoder networks. Due to [1] the direct communication between the expanding and contracting paths made possible by these linkages, spatial information is preserved and segmentation results are more precise.

3 Proposed Methodology

This section discusses the methodology for our proposed model, system architecture, as well as techniques like the Generator UNet, Instance Normalization, VGG Extractor, and partial convolution layers applied on an archaeological dataset for image inpainting.

3.1 Partial Convolution Layers

Partial convolution also helps to maintain spatial consistency between the input and output data [16]. When convolving with incomplete data, traditional convolution operations can introduce artifacts or distortions into the output image. Partial convolution, on the other hand, only convolves with the valid pixels of the input data, which helps to maintain the spatial consistency between the input and output data. In our model shown in Fig. 1, the partial convolution is a custom layer that implements partial convolution operation, which only applies convolution to the unmasked parts of the input image. It takes an image and a binary mask as inputs and outputs a tuple of the convolved image and mask. This layer adds two new parameters, n channels and mono, in the init method [15]. The build method initializes the kernel weights and bias terms and calculates the required padding and window size for normalization. The call method applies partial convolution by padding the inputs, applying convolutions, calculating mask ratios, and normalizing image outputs. The compute output shape method computes the output shape based on the input shape.

3.2 Instance Normalization

Instance Normalization is a normalization technique that operates on individual samples in a batch before passing to the model as shown Fig. 1, and it can be useful for improving the stability and speed of training by reducing internal covariate shift. In the context of our inpainting model, we have implemented an Instance Normalization layer as a custom layer that has an epsilon attribute added to the variance to avoid division by zero. The layer creates two trainable variables scale and offset [18] which are used to scale and shift the normalized output. The layer first calculates the mean and variance of each sample along the height and width dimensions, but not the channel dimension using the built-in function.

3.3 Generator UNet

Our model makes use of partial convolutional layers in a UNet architecture. The binary mask used as the function's second input identifies the image's missing portions, in addition to the original image. An encoder network and a decoder network connected by skip connections make up the UNet architecture. In order to create the output image, the decoder network up samples the feature maps, while the encoder network down samples the input image [20]. A similar layer in the decoder network follows each layer in the encoder network, concatenating the encoder's feature maps with the decoder's up-sampled feature maps [13].

The partial convolutional layer is a modification of the standard convolutional layer that only convolves over valid pixels in the input image, as defined by the binary mask. The partial convolutional layer outputs a mask indicating which pixels in the output image were computed from valid input pixels.

The generator function allows the user to specify the normalization type to be used in the network, either Instance Normalization (In) or Batch Normalization (Bn), as well as whether the model is being trained or evaluated, respectively). The output of the generator function is a model that takes two inputs inputs image and input mask and outputs the inpainted image. The generator has been trained using a combination of pixel-wise mean squared error loss and adversarial loss.

3.4 VGG Extractor

VGG16 extractor is its pre-trained weights on the ImageNet dataset, which allows for transfer learning to be used in various tasks. This means that the weights learned from the ImageNet dataset can be used as a starting point for training on other datasets, which can reduce training time and improve performance. The extracted feature which helps in two types of losses:

1. Pixel-wise mean squared error (MSE) loss.
2. Perceptual loss.

3.5 Loss Function

These are some representation used to explain loss function.

I_{gt}	Ground truth image.
I_{out}	Output restored image generated by the network.
I_{comp}	Output restored image with non-hole regions replaced with the ground truth.
L_{per}	Perceptual loss.
L_{tv}	Total variation loss.
L_{style}	Style loss.
$L_{hole} L_{valid}$	Pixel losses.
L_{MSSSIM}	MSSSIM loss.
L_{total}	Total loss.

Perceptual loss: The perceptual loss [16], or feature reconstruction loss, is a technique that encourages restored images to have similar feature representations to the original image, rather than an exact pixel-by-pixel match. In our experiments, we utilized the VGG-16 CNN to extract feature representations from the image through several convolutional and fully connected layers. This loss function operates by minimizing the semantic gaps between the activated features in the input image and those in the reconstructed image, across multiple network layers.

$$L_{per} = \sum_{i=0}^{N-1} \left|\psi_i(I_{out}) - \psi_i(I_{gt})\right| + \sum_{i=0}^{N-1} \left|\psi_i(I_{comp}) - \psi_i(I_{gt})\right|. \tag{1}$$

Total variation loss: The nature of regularization losses and the total variation loss is relatively similar. It is in charge of preserving smoothness and spatial continuity in the restored image. Other computer vision applications like style transfer and digital signal processing for noise reduction also heavily rely on the same technique.

$$L_{tv} = \sum_{(i,j),(i,j+1)\in I} \left|I_{comp(i,j+1)}\right| + \sum_{(i,j),(i+1,j)\in I} \left|I_{comp(i+1,j)}\right|. \tag{2}$$

The Huber loss is a combination of the L1 and L2 norms for better regression. The Huber loss function can be written as:

$$L_{huber} = \frac{1}{2}\left(L_1\left(\left|\psi_i(I_{out}) - \psi_i(I_{gt})\right|\right) + L_2\left(\left|\psi_i(I_{out}) - \psi_i(I_{gt})\right|\right)\right). \tag{3}$$

Style loss: Style loss is a technique for incorporating texture-like elements into a recovered image rather than recreating the exact structure of the original image. It captures the image's overall look in terms of color tones and local compositions [19]. The feature maps are translated into Gram matrices, which are created by analyzing the correlations among these feature maps. The loss function aims to minimize the $L1$ distance between the entries of the Gram matrices derived from the style of the input picture and those from the reconstructed image. Furthermore, the I_{comp} created after each cycle is reduced, resulting in a more aesthetically pleasing output.

$$L_{\text{style}} = \sum_{i=0}^{N-1} \left|\xi(\psi_i(I_{\text{out}})) - \xi\left(\psi_i\left(I_{\text{gt}}\right)\right)\right| + \sum_{i=0}^{N-1} \left|\xi\left(\psi_i\left(I_{\text{comp}}\right)\right) - \xi\left(\psi_i\left(I_{\text{gt}}\right)\right)\right|. \quad (4)$$

Pixel losses: The goal of the pixel losses is to increase the accuracy of the per-pixel reconstruction for both regions with and without holes.

$$L_{\text{hole}} = \left|(1 - M) \cdot \left(I_{\text{out}} - I_{\text{gt}}\right)\right|, \quad (5)$$

$$L_{\text{valid}} = \left|M \cdot \left(I_{\text{out}} - I_{\text{gt}}\right)\right|. \quad (6)$$

The equation above displays the aggregate loss function. We utilize the hyper-parameter values (s) given in the following equation for our initial experiments.

Multi-scale Structural Similarity (MSSSIM) loss: The MS-SSIM loss function better preserves structural information during training because it considers the structural similarity of pictures at various scales. By doing this, the structural details of the images provided will be preserved. This may result in better performance and image quality for some jobs.

$$\text{MSSSIM}(x, y) = \frac{1}{n} \sum_{i=1}^{n} \text{SSIM}(x_i, y_i), \quad (7)$$

$$\text{filter size}(f) = 11, 7, 5, \quad (8)$$

$$\begin{aligned} L_{\text{MSSSIM}} &= \alpha \cdot \text{MSSSIM}(x, y), (f = 11) \\ &+ \beta \cdot \text{MSSSIM}(x, y), (f = 7) \\ &+ \gamma \cdot \text{MSSSIM}(x, y), (f = 5), \end{aligned} \quad (9)$$

$$\begin{aligned} L_{\text{total}} &= \lambda_1 L_{\text{tv}} + \lambda_2 L_{\text{valid}} + \lambda_3 L_{\text{hole}} \\ &+ \lambda_4 L_{\text{per}} + \lambda_5 L_{\text{style}} + \lambda_6 L_{\text{MSSSIM}}. \end{aligned} \quad (10)$$

Fig. 2 Different temples images from the dataset

4 Experimental Setup

For our experimental setup for the suggested image inpainting technique, Google Collab was used. Google Collab is a free cloud-based platform that offers a Jupyter Notebook environment and enables users to run Python code on potent GPU hardware, making it an excellent choice for training deep learning models. We were able to train and test our suggested model on archaeological datasets using Google Collab without any technology limitations and the GPU utilized is 14.09 GiB. To make a mobile application, we have used Android Studio.

4.1 Dataset

Images The dataset contains a collection of 3200 photographs showing various temples in different parts of India. Examples are shown in Fig. 2. These images were collected from photographs taken by the students of KLE Technological University, and all the images have been standardized to a resolution of 256 × 256 pixels. The dataset is divided into train, validation, and test sets in the proportion of 75%, 15% and 10%, respectively.

Masks A total of 3200 images were randomly selected from the palette of irregular MST masks to create a mask set (Fig. 3). The mask set has a mask rate ranging from 5 to 90%. The images were divided into training, validation, and test images in the ratio of 75%, 15%, and 10%, respectively, with all images having a size of 256 × 256.

4.2 Training the Model

The model was trained on an archaeology dataset for 100 epochs, utilizing a batch size of 16. The model's input size was set to 256 × 256 × 3, and a learning rate of 0.001 was used during training. To address specific characteristics, present in archaeological images such as occlusion and structural features, a new MS-SSIM

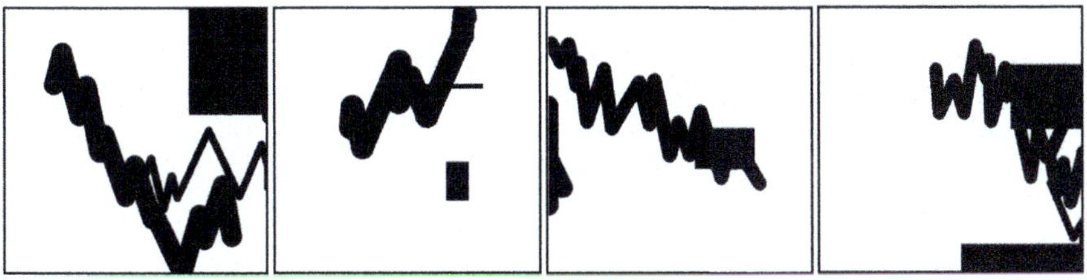

Fig. 3 Different mask samples used for training

loss function was employed to enhance model performance. During the training phase, the VGG16 model, pre-trained on the ImageNet dataset, was utilized for effective feature extraction. This extrapolated feature information played a crucial role in calculating both the Pixel-wise Mean Squared Error (MSE) loss and the Perceptual loss, contributing to the overall training process.

5 Results and Comparison

In this section, we discuss a complete performance analysis of the models with and without MSSSIM. Various metrics, including the pixel signal-to-noise ratio (PSNR), mean square error (MSE), and structural similarity index measure (SSIM) are evaluated. When a model with MSSSIM is employed, which internally provides a high SSIM value, the model is intended to demonstrate that structural information is preserved.

5.1 Comparing the Results

In this study, we used the UNet with MSSIM and Huber loss which will preserve the structure information in the image. This technique allows the model to leverage knowledge learned from a training. The model without MSSIM and Huber loss is compared in case to compare that the structure information is preserved stages (Fig. 4). This approach has resulted in a significant improvement in performance compared to training the model from scratch.

According to the Table 1, our model's SSIM value is better due to the use of MSSSIM, which helps to maintain the structure information that internally raises the SSIM value. In our model, MSE loss is decreased. The PSNR value is slightly lower than that of the other model as a result of the structure's softening and the use of the Huber loss in tern in addition to the l2-norm.

Fig. 4 Result comparison with ground truth: **a** ground truth, **b** masked image, **c** model without MSSIM, **d** model with MSSIM

Table 1 Quantitative result for model with MSSSIM and without

	PSNR ↑	MSE ↓	SSIM ↑
Model without MSSSIM	**32.564656**	0.00579	0.9026567
Model with MSSSIM (Ours)	31.804071	**0.00446**	**0.9590165**

5.2 *Post-training Quantization*

We used post-training quantization [8] to shrink our model from 260 to 63 MB after Google Collab training and attaining state-of-the-art performance. Analytical Clipping for Integer Quantization, or ACIQ, is a method for post-training weight quantization that restricts the value of weights to a specific range while minimizing the impact on the model's accuracy. The ACIQ technique works by examining the model's weight distribution and establishing a clipping range that restricts the weights maximum and minimum [11] values to a range that can be precisely represented by the target integer precision. Post-training quantization is a method for reducing the size of a trained model by reducing the precision of the model's weights and activations, usually to 8 bits or less [7, 9]. We were able to apply post-training quantization to our trained model using the TensorFlow. Lite package, which is easily accessible on Google Collab [10]. The trained model was first converted to a TensorFlow Lite format, then quantization was applied, and finally, the quantized model was converted back to a TensorFlow Lite format for deployment.

5.3 Deploying the Model on Edge Device

After quantization, we created an image-painting app using Android Studio and deployed the model there. As shown in Fig 5, the application has image view for mask, input image and output image, where two buttons are present one to access image from gallery and another from camera (Table 2).

Fig. 5 Application of inpainting in Android app

Table 2 Analysis of model before and after quantization

	Model size (MB) ↓	SSIM ↑
Model before quantization	260	**0.9590165**
Model after quantization	**63**	0.611023

6 Conclusion

In conclusion, our proposed image inpainting technique, leveraging the UNet architecture and MS-SSIM loss function, outperforms alternative approaches in restoring archaeological images. It offers higher accuracy and quality in both quantitative and qualitative evaluations. We also achieved a significant reduction in model size from 260 to 63 MB through post-training quantization, enabling mobile device deployment for on-site restoration of archaeological photos. The versatility of our method extends to image restoration, photo editing, and digital forensics, making it a valuable asset for professionals in various fields. Overall, our approach combines efficacy, practicality, and accessibility, contributing to the preservation and analysis of visual records in archaeology and other domains. Additionally, our model can be further refined and trained to excel in scenarios involving low structural and low-texture images, expanding its potential impact in diverse domains.

References

1. Dogan Y, Keles HY (2022) Iterative facial image inpainting based on an encoder-generator architecture. Neural Comput Appl 34(12):10001–10021
2. Elharrouss O, Almaadeed N, Al-Maadeed S, Akbari Y (2020) Image inpainting: a review. Neural Process Lett 51:2007–2028
3. Fu Q, Guan Y, Yang Y (2018) Image inpainting and object removal with deep convolutional gan. Technical report. http://stanford.edu/class/ee367/Winter2018
4. Guillemot C, Le Meur O (2013) Image inpainting: overview and recent advances. IEEE Signal Process Mag 31(1):127–144
5. Guo X, Yang H, Huang D (2021) Image inpainting via conditional texture and structure dual generation. In: Proceedings of the IEEE/CVF international conference on computer vision, pp 14134–14143
6. Kaur A, Raj A, Jayanthi N, Indu S (2020) Inpainting of irregular holes in a manuscript using unet and partial convolution. In: 2020 second international conference on inventive research in computing applications (ICIRCA). IEEE, pp 778–784
7. Kulkarni N, Singh N, Joshi Y, Hasabi N, Meena SM, Kulkarni U, Gurlahosur SV (2022) Hybrid optimization for dnn model compression and inference acceleration. In: 2022 2nd international conference on intelligent technologies (CONIT). IEEE, pp 1–8
8. Kulkarni U, Hosamani AS, Masur AS, Hegde S, Vernekar GR, Siri Chandana K (2022) A survey on quantization methods for optimization of deep neural networks. In: 2022 international conference on automation, computing and renewable systems (ICACRS). IEEE, pp 827–834
9. Kulkarni U, Meena SM, Gurlahosur SV, Benagi P, Kashyap A, Ansari A, Karnam V (2021) AI model compression for edge devices using optimization techniques. Mod Appr Mach Learn Cogn Sci Walkthrough Latest Trends AI 2:227–240

10. Kulkarni U, Meena SM, Gurlahosur SV, Bhogar G (2021) Quantization friendly MobileNet (QF-MobileNet) architecture for vision based applications on embedded platforms. Neural Netw 136:28–39
11. Kulkarni U, Meena SM, Joshua P, Kruthika K, Platel G, Jadhav K, Singh A (2021) Performance improvements in quantization aware training and appreciation of low precision computation in deep learning. In: Advances in signal processing and intelligent recognition systems: 6th international symposium, SIRS 2020, Chennai, India, 14–17 Oct 2020, revised selected papers 6, Springer, pp 90–107
12. Kulshreshtha P, Pugh B, Jiddi S (2022) Feature refinement to improve high resolution image inpainting. arXiv preprint arXiv:2206.13644
13. Li G, Li L, Pu Y, Wang N, Zhang X (2022) Semantic image inpainting with multi-stage feature reasoning generative adversarial network. Sensors 22(8):2854
14. Li W, Lin Z, Zhou K, Qi L, Wang Y, Jia J (2022) MAT: mask-aware transformer for large hole image inpainting. In: Proceedings of the IEEE/CVF conference on computer vision and pattern recognition, pp 10758–10768
15. Liu G, Shih K, Wang T-C, Tao A, Catanzaro B et al (2019) Image in-painting for irregular holes using partial convolutions, 26 Sept 2019. US Patent App. 16/360,895
16. Patel H, Kulkarni A, Sahni S, Vyas U (2021) Image inpainting using partial convolution. arXiv preprint arXiv:2108.08791
17. Suvorov R, Logacheva E, Mashikhin A, Remizova A, Ashukha A, Silvestrov A, Kong N, Goka H, Park K, Lempitsky V (2022) Resolution-robust large mask inpainting with fourier convolutions. In: Proceedings of the IEEE/CVF winter conference on applications of computer vision, pp 2149–2159
18. Wang C, Zhu Y, Yuan C (2022) Diverse image inpainting with normalizing flow. In: Computer vision—ECCV 2022: 17th European conference, Tel Aviv, Israel, 23–27 Oct 2022, proceedings, part XXIII. Springer, pp 53–69
19. Wang N, Ma S, Li J, Zhang Y, Zhang L (2020) Multistage attention network for image inpainting. Pattern Recogn 106:107448
20. Yang C, Lu X, Lin Z, Shechtman E, Wang O, Li H (2017) High-resolution image inpainting using multi-scale neural patch synthesis. In: Proceedings of the IEEE conference on computer vision and pattern recognition, pp 6721–6729

Analysis of Mispronunciation Detection and Diagnosis Based on Conventional Deep Learning Techniques

M. Soundarya and S. Anusuya

Abstract Mispronunciation Detection and Diagnosis (MDD) constitutes detecting the mispronounced chunk and diagnosing it by providing feedback about the mispronounced chunk to the model. MDD is a vital function of CAPT—Computer-Assisted Pronunciation Training, the subcategory in CALL—Computer-Assisted Language Learning. The ability of self-learning, handling huge unstructured datasets, having better evaluation metrics, and being cost-effective make deep learning models to outperform traditional models. In this work the recent deep learning methods utilizing i) input with Anti phones, Accented content, Non-Segmental error, Raw input waveform, Acoustic data – Phonetic data – Linguistic data in the part of training input ii) Model such as Transformer, Attention model, CTC-Connectionist – Temporal – Classification that are implemented on the L2-ARCTIC English dataset are analyzed. The evaluation metrics analyzed include accuracy of diagnosis and detection, F-measure. The transformer-based model shows a high F-measure value of 60.5% and a detection accuracy value of 88.2%. The hybrid CTC-ATT (CTC-Attention) model gives the second highest value of F-measure 56.08% and good performance in terms of both the diagnostic and detection accuracies of 74.96% and 75.45%, respectively. This shows that we can prefer the former method for detection and the latter method for both detection and diagnosis. Thus, based on the need for detection and diagnosis in MDD, these two models show the optimum overall performance.

Keywords Mispronunciation · Language · Features · Acoustic · Diagnosis · Detection · Phonemes · L2-ARCTIC English dataset · Encoder · Decoder

---

M. Soundarya (✉) · S. Anusuya
Saveetha School of Engineering, SIMATS, Chennai, India
e-mail: soundaryam2009.sse@saveetha.com

S. Anusuya
e-mail: anusuyas.sse@saveetha.com

V. K. Gunjan and J. M. Zurada (eds.), *Proceedings of 4th International Conference on Recent Trends in Machine Learning, IoT, Smart Cities and Applications*, Lecture Notes in Networks and Systems 873, https://doi.org/10.1007/978-981-99-9442-7_31

1 Introduction

Communication is an important part of life that helps us express our thoughts, emotions, and decisions in essential fields. Speech is one form of communication that can provide communication with more speed and clarity and can also express the emotions of human beings. Pronunciation is nothing but the way we use a word or syllable in a particular language, and it is an important feature in verbal communication that provides an efficient means of communication. In contrast to pronunciation, mispronunciation is the obviously incorrect way of pronouncing a syllable, word, or sound. The measure of the error in mispronunciation constitutes the term pronunciation error. The main linguistic causes of these pronunciation errors were mother-tongue interference [1], poor knowledge of the unknown language, and medical conditions. The various forms of the pronunciation error are portrayed in Fig. 1. Phonemes are the smallest part of the word that can take meaning. The phonemes may be consonants, vowels, diphthongs, triphthongs. If the sound produced by the speaker is not the sound expected by the listener, it constitutes an error in phonemes. For example, the word "smile" is mispronounced by the way of "smole." Segmental phonemes somewhat differ in sounds, and they are similar within a language. Examples are the sounds produced by "a," "e," "i," "o," and "u." Segments constitute consonants and vowels and are simple to determine in seclusion [2]. The supra-segmental feature, also known as the Prosodic feature or Prosody, is not restricted to a single consonant or vowel sound and extends over words, syllables, and phrases. The prime parameters include intensity, pitch, and duration. The secondary language characteristics include rhythm, tone, stress, meter, timing, nasalization, and the harmony involved over vowels or consonants. Both voiced and non-voiced consonants use the upper palate, teeth, lips, and breath (articulation) [3], but only the former use the vocal cord, which produces a vibration when the sounds are delivered, and the latter does not. For example, the sounds produced by "P" and "B" are unvoiced and voiced, respectively.

In long vowels and short vowels, the same words are pronounced, but both differ by the duration or length of the sounds. For example, in "sin-seen," "sin" represents short vowels and "seen" represents long vowels. The sounds of either two, three vowels combined in a single syllable give a diphthong, triphthong, respectively. An example of the diphthong is "cow" and the triphthong is "year." The variation of the pitch in sound helps to give all sorts of emotions to our speech communication, and this is nothing but intonation [4]. For example, while terminating any question, our voice raises. There are four forms of intonation: rising pitch, falling pitch, non-final pitch, and wavering pitch. Rising intonation is when we raise the pitch of our voice while we terminate a phrase or a sentence. Example yes or no question, will you give me your pen? Falling intonation occurs when our voice is lowered at the end of a sentence. The questions such as "Glad to meet you" and "How dare you do that?" that constitutes words such as where, when, what, how, and who are examples for Falling intonation

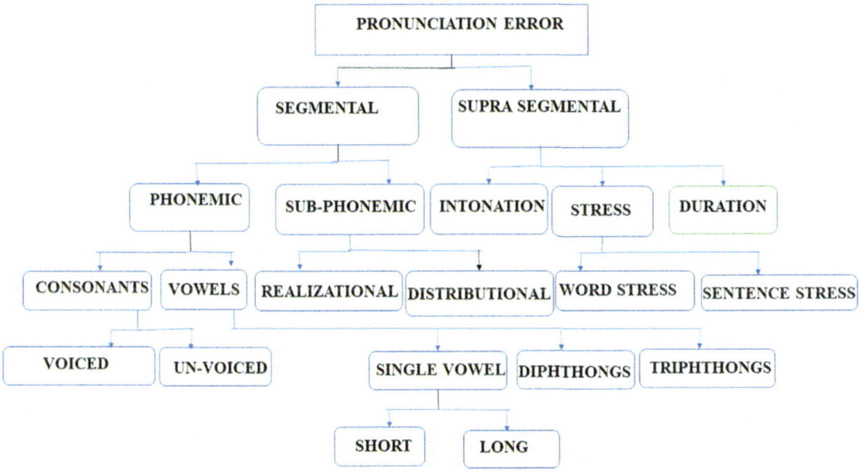

Fig. 1 Classification of pronunciation error

In non-final intonation, within the sentence, the pitch rises and then falls. For example, if I work smart, I will earn more money. Sentence that includes Introduction Phrases, Expressing Choices, and a Series of Words also falls under this category. Examples of the former are "In my opinion, this watch is very expensive" and "Do you want to go to the temple or movie?" and the latter is "I like eating pizza, burgers, sandwiches, and juice," respectively. Wavering intonation gives emotions within a used word, and for example, "You did?" can take several emotions such as astonished, eager, annoyed, and in the contract. In linguistics stress or accent [5], the importance given to a particular word or syllable in a sentence produce acoustics that varies in terms of loudness, length, and robustness compared to the other acoustics. People pronouncing the same words in different modulations in two different states or countries constitute an accent. Need for mispronunciation detection: Mispronunciation detection helps to offer efficient and fast communication with clarity which is mandatory for humans involved in various fields such as health care, business, foreign agreements, trading with foreign agencies, education, research fields, marketing. Need for machines for mispronunciation detection: Training humans who are non-native or learning a new language is a complex process in terms of time, number, availability of trainers, clarity, and performance, so we prefer the use of machines. Many types of machine learning procedures [6, 7] aid in providing training, and nowadays they have been substituted by deep learning [8, 9]. All these technologies support all sorts of people who belong to various groups of ages and genders, especially people affected by speaking disabilities [10] or any medical reasons.

2 Conventional Methodologies

Analysis of previous research includes the traditional HMM—Hidden Markov Model—which is a well-known and frequent technique used by speech recognizers for detecting pronunciation faults. The GoP—Goodness of Pronunciation—method is nothing but an upgraded variation of the Posterior Probability Method, which measures the pronunciation score at the phonetic range, which is useful to identify the mispronounced word or phonemes. The approaches discussed to detect errors in Categorical Pronunciation in which the phonemes are substituted, deleted, and inserted also provide feedback on mispronounced content. There are five methodologies, of which three constitute GoP and two constitute classifier-based techniques. The former includes structures such as HMM using Gaussian Mixture and DNN—Deep Neural Network—which are used to predict and model features of tone, pitch, and stress for each discrete term effectively. The revised GOP calculation gives pronunciation distinctions for learning a non-native language. Despite a smaller number of samples, a robust classifier like a Neural Network or Support Vector Machine can be constructed utilizing a common hidden layer network topology so that the generalization of specific phones is improved. It also aids in the extraction of additional features to distinguish between incorrect and correct phones.

Transition probabilities are mostly disregarded in almost all existing methods built on DNN and HMM. The Senone—Posterior probability and the probability for the change for developing a novel project using GOP in DNN-HMM also provide a reliant architecture. To incorporate this in Kaldi, which is a state-of-the-art kit useful in the domain of ASR—Automatic Speech Recognition. This research also provides results when the Senone is distributed across multiple states. In contrast to a method that aims to detect the errors in Categorized Pronunciations, including Phoneme Substitution, Phoneme Insertion, and Phoneme Deletion, a greater emphasis is placed on spotting mispronunciations that are classified as non-errors. The ASR system model uses a modified E2E—End-to-End—type to separate and analyze mispronunciation, with the hybrid combination CTC-Attention structure characterized in an E2E type model tangled. An anti-phone group is adopted in conjunction with a label-shuffling technique to create additional language training information for an innovative data-augmentation approach. For every word in the unique voice training datasets, the phoneme's label at each point of its located transcription is maintained as it is or arbitrarily changed with a random anti-phone label.

2.1 *Anti-phone Modeling*

The CAPT includes a fundamental component, MDD which performs detection of mispronounced phonemes and diagnosis by providing feedback to the users. Instead of the usual method of focusing only on the Categorical Error [11, 12] or Distortion Error, also called a non-Categorical Error, there is a need to concentrate on both errors. The non-Categorical Error occurs when attempting to resemble L1 language phonetics by L2 language speakers or by means of incorrect pronunciations among L1 and L2 [13, 14]. A vast variety of techniques were implemented to perform the MDD tasks for diagnosing both errors, but all these techniques have limitations in terms of phonological rules [15]. To overcome all the problems, an innovative method in ASR is proposed that is end-to-end based. The methodology is nothing but using anti-phones to intensify the original phones. The calculation of the original module is done with protocols in phonological rule books. A novel module, namely the transfer-learning pattern, is proposed, which breaks all the limitations of the original module.

A hybrid CTC-ATT architecture in E2E mode: This performs orientation between the audio vector sequence for input and the vector sequence for output, which can be updated. A combination of CTC and an encoder uses Markov assumptions to alleviate any orientation mismatch. The sequence of audio vector input is given by $I = I_1 \ldots I_a \ldots I_A$ and the length A. Similarly, the sequence for the sequence vector output is given by $W = W_1 \ldots W_b \ldots W_B$ and the length B. For analyzing the mathematical relation, we need to calculate the conditional probability distribution, which is given by

$$P_{\mathrm{ATT}}(W|I) = \Pi_{i=1}^{B} . P_{\mathrm{ATT}}(W_N|W_N{:}\,l-1, I), \tag{1}$$

where is the conditional probability value of W_N multiplied by the past state W_N: $l - 1$ in the above equation.

Then, $P_{\mathrm{ATT}}(W_B|I, W_1{:}\,l-1)$ is calculated using the joint network of the decoder and encoder. The sequence of hidden audio vector $H_E = (H_{E1} \ldots H_{ES})$ has been extracted from sequence audio vector input I by the encoder part using BLSTM—Bidirectional Long Short-Term Memory—and is given by

$$H_E = \mathrm{BLSTM}(I). \tag{2}$$

The length of the sequence of a hidden vector is denoted by S and ($S<A$) as we do a process of sub-sampling. In contrast, the decoder part uses ULSTM—Unidirectional Long Short-Term Memory. The output state, O_L for predicting the label of phonemes is given by

$$O_L = \mathrm{SoftMax}(\mathrm{Lin}\,L(\mathrm{LSTM}(O_{L-1}, C_L, h_L D))), \tag{3}$$

$$h_L D = \mathrm{ULSTM}(\mathrm{cat}(Y_{L-1}, C_L), h_{L-1} D). \tag{4}$$

O_{L-1}	the past output state.
$h_L D$	the present state of the decoder.
$h_{L-1} D$	past decoder state.
C_L	the context vector.

Lin $L(.)$ Linear Transformation.

The context vector is given by

$$\mathrm{CL} = \sum_{s=1}^{S} C_{L,s} h_{ES}. \tag{5}$$

The attention part uses soft alignment between the hidden audio vector and the state vector of the decoder to calculate the vector showing context and is given by

$$C_{L,S} = \mathrm{align}(h_L D, h_{ES}), \tag{6}$$

$$C_{L,S} = \frac{\exp(\mathrm{score}(h_L D, h_{ES}))}{\sum_{s=1}^{S} \exp(\mathrm{score}(h_L D, h_{ES}))}, \tag{7}$$

$$P_{\mathrm{CTC}}(W|I\,|) = \sum_{F} \cdot \prod_{S} \cdot P(T_S|T_{S-1}, W)\, P(T_S|I) P(W), \tag{8}$$

$$P(T_S|T_{S-1}, W). \tag{9}$$

The probability of state transitions is calculated using the score value, and the value of probability aids in MDD tasks by identifying transitions of mispronounced data from correctly pronounced data. Despite the state level, we also need to calculate probability at the frame level, which is given by

$$P\,(T_S|I) = \mathrm{SoftMax}(\mathrm{Linear}\, B(h_{ES})). \tag{10}$$

The condition for probability is that it should satisfy monotonic configurations, and for that, the losses of an ATT-based model—L_{ATT}—and a CTC—L_{CTC}—are combined with the weight of interpolation $\lambda \varepsilon$ [1, 1].

$$L_{\mathrm{HYB}} = L_{\mathrm{CTC}} + (1 - \lambda) L_{\mathrm{ATT}}. \tag{11}$$

L_{HYB} Hybrid architecture model is also considered in the test phase.

The monotonic configurations in this proposed attention model aid in acquiring more accurate and fast interpretations.

Generating a Set of Anti-phones

$$Ş = Ş_{\mathrm{CAN}} \cup Ş_{\mathrm{ANTI}} \tag{12}$$

$Ş$ set of anti-phones.
$Ş_{\mathrm{CAN}}$ set of canonical phonemes.

$\S_{ANTI}$ set of anti-phonemes created by appending any symbol, such as #, at the beginning of canonical phonemes.

Augmenting data and shuffling the label: The novelty in the augmentation process in this paper is a label-shuffle system in which a few of the training set data points are duplicated with labels of anti-phones.

Training phase: The non–accented unit is trained on LibriSpeech and the TIMIT Texas Instruments/Massachusetts Institute of Technology corpus, and secondarily, the accented model is trained using transfer learning. The dataset that is augmented is used in the decoder unit and encoder unit. The accented unit is fine-tuned and includes mispronunciations that are used to identify both categorical and non-categorical errors, which is the advantage of this work over existing work. The outcomes of the two methodologies were analyzed using the English dataset L2-ARCTIC [16], and the overall performance reached greater heights. The $F1$-score is improved by 11.05% with the baseline system using the end-to-end approach and by 27.71% with GOP methods using the score value of pronunciation.

2.2 Accent Modulation Method

The MDD techniques include features such as audio input, and language input, but still, the methods related to accents are also important, as people from non-native countries have various varieties of accents. We need to learn the L2 language for purposes of business, education, health care, research, etc. in foreign countries, and there are several methods focused on the L2 language [17]. We also add accent features to the network as accents also significantly influence the native language, which are modeled and enable end-to-end fashion [18–20]. The discriminating ability is proven to be worse, and this is improved by this accent aware model and further improved by adopting a finer-grained method.

Because of accented speech, there is a discrepancy between native language speakers and non-native language speakers. Considering this discrepancy, we propose a novel technique of introducing various accents in our audio features as an intermediate vector in our proposed system. The various accented data are fed a priori in both the training phase and the testing phase. The input vector, along with accented data, is fed into the encoder unit to produce the intermediate vectors. To combine the audio and accented data, we use two methods: Accent Modulation using Concatenation and Accent Modulation using Gating. In the former method, the next state acoustic transitional vector H′s is obtained by performing the operation of a concatenation of sentence level entrenched vector AE (one hot embedded) and present-state acoustic transitional vector Hs and is given by

$$H_S' = \text{Con}(Hs, AE)U_1 + B_0, \tag{13}$$

where U_1 and B_0 are weights and biases for the transformation matrix that aids in performing dimensionality reduction. In the latter method, the ability to discriminate is improved by fine-grained modulation, and we use the function sigmoid to produce gate vectors *Gs* that model similar and dissimilar vectors. The function ReLU performs further finer modulation using training parameters.

$$Gs = \text{Sig}([Hs, AE])U_2 + B_1, \tag{14}$$

$$H's = ReLu(Hs + [H_S \odot G_S]U_3 + B_2. \tag{15}$$

$\odot$ element dot product, and U_2, U_3, B_1, B_2 are parameters that can be trained.

$$L = (1 - B)\, L_{\text{ATT - CTC}} + BL_{\text{AC}}. \tag{16}$$

L is the parameter for the loss function, and *B* is the weight for interpolation, which shows the degree of dependence on backpropagated gradients.

The experimental outcome shows improved performance with the L2-ARCTIC dataset on hybrid when compared with the baseline method and the method using GOP.

2.3 End-to-End Model Using Raw Waveforms

MDD is a well-known process that performs detection and provides feedback for any type of dataset, including native and non-native. But there are major hindrances that exist to accessing to a large variety of non-native datasets, and labeling those datasets is also a cumbersome process. To overcome these issues, we prefer a neural model that works in an end-to-end fashion. This network accepts raw waveforms directly [21, 22] instead of any type of acoustic input features, and then, the raw waveforms are converted into sequences of vector representations. The SincNet module [23] is user-friendly as it has very few parameters and implements a bandpass filter in the core network using the trigonometric Sinc function.

Encoder Unit: The input audio sequences of length *T*,

$$I = (I_1, \ldots, I_T), \tag{17}$$

is given to the first unit encoder having a recurrent or convolutional network. The high-level state output sequences of length *S* are extracted and given by

$$H = (H_1, \ldots, H_S). \tag{18}$$

Here due to the process of down-sampling, the S is very less than T ($S<T$).

Attention Unit: The output of the encoder unit is potted for each step l ∈ [1, …, L] and the context vector C_l having a static length is calculated. The output sequence $O = (O_1, \dots O_L)$ is a set of recognized phonemes.

Decoder Unit: The hidden state Q_L is restructured in an autoregressive manner, and the next phoneme symbol Y_L is calculated by using two parameters: the context vector C_l and the past output sequence $O_1 : l - 1$.

Diagnostic results of the CTC Unit: The sequences of I and O are aligned at frame level using a token $\prec$blank$\succ$ to overcome imprudence between the training phase and testing phase. The goal of this model is to make the best use of the linear combination of $P_{\text{ATT}}(O|I)$ and $P_{\text{CTC}}(O|I)$, which is given by

$$L = \alpha \log P_{\text{CTC}}(O|I) + (1 - \alpha) \log P_{\text{ATT}}(O|I). \tag{19}$$

$P_{\text{ATT}}(O|I)$ posterior probability for the Attention model.
$P_{\text{CTC}}(O|I)$ posterior probability for the CTC model.

$$P_{\text{CTC}}(O|I) = \sum_{Z} p(O|Z)p(Z|I), \tag{20}$$

$$P_{\text{ATT}}(O|I) = \prod_{l=1}^{L} p(Ol|O1 : l - 1, I), \tag{21}$$

$$H = \text{encode}(I), \tag{22}$$

$$\alpha_{l,S} = \text{ATT}(QL, HS). \tag{23}$$

$\alpha_{l,S}$ shows the attention of hidden states in an autoregressive manner in Q_L and high-level state output sequences HS.

The summation of output sequence of high-level states gives the context vector

$$C_l - \sum_{\alpha_{l,S}} HS. \tag{24}$$

For range of attention operations,

$$P(O_l|O_{l-1}, H) = \text{decod}\left(Q_{l-1}, \left[O_{l-1}, C_{l-1}\right]\right). \tag{25}$$

The convolution performed by SincNet is given by

$$O(t) = I(t) * HI(t) = \sum_{l=0}^{L-1} I[l].HI[t - l], \tag{26}$$

where $I(t)$ is the input signal, $HI(t)$ is the filter signal of length L, and $O(t)$ is the output signal.

All the elements of the $h[\cdot]$ constraint can be learned. Using SincNet, the function S replaces the filter function H so that the dependency is restricted only to high and low cut-off frequencies. The impulse response and frequency response of SincNet are represented by $s[T, F_1, F_2]$ and $S[F, F_1, F_2]$ and are given by

$$s[T, F_1, F_2] = 2F_2\text{Sinc}(2\Pi F_2 t) - 2F_1\text{Sinc}(2\Pi F_1 t), \tag{27}$$

$$S[F, F_1, F_2] = \Pi\left(\frac{F}{2}.F_2\right) - \Pi\left(\frac{F}{2}.F_1\right). \tag{28}$$

The experimental results on the L2-Arctic dataset and TIMIT corpus for PER—Phoneme or Phone Error Rate, Recall, Precision, and $F1$-score using a hybrid ATT-CTC + SincNet module give the highest values of 5.50%, 50.09%, 55.31%, and 52.57%, respectively. This technique offers better experimental results in terms of end-to-end fashion [9].

2.4 Transformer-Based Model

The mispronunciation detection technique is used to identify errors in audio or speech signals when an unknown or non-native person is learning the L2 language. Depending on the configuration between the typical and predicted phonemes, the ASR can recognize the pronunciation when compared to the traditional score-based methods. In recent times, employing the transformer-dependent structure has performed better for modelling in a Sequence-to-Sequence manner [24]. The purpose of the target text in this method is to recognize the value of an error. With target text, the Transformer enters the feed-forward mode of operation [25], which supports faster speed and end-to-end transfer mode. In the absence of the target text, the Transformer operates in an Autoregressive mode [26, 27]. The Transformer model estimates the conditional probability for existing words or phonemes from the formerly decoded output target word or phoneme. In the training phase, if the target word or phoneme is trained, it is presumed to be decoded output, and thus the parallel means of operation are accomplished. But still, the Transformer functions in a sequential manner in the inference phase as no prior text is available. In order to solve all these issues, a model should satisfy prior text, unite the ASR, and be aligned.

The dataset for training includes input audio X, text target P to read, and standard text by the manual annotator. The proposed workflow is shown in Fig. 2.

$$X = (X_1, X_2, \ldots, X_m), \tag{29}$$

$$P = (P_1, P_2, \ldots, P_n, P_{n+1}). \tag{30}$$

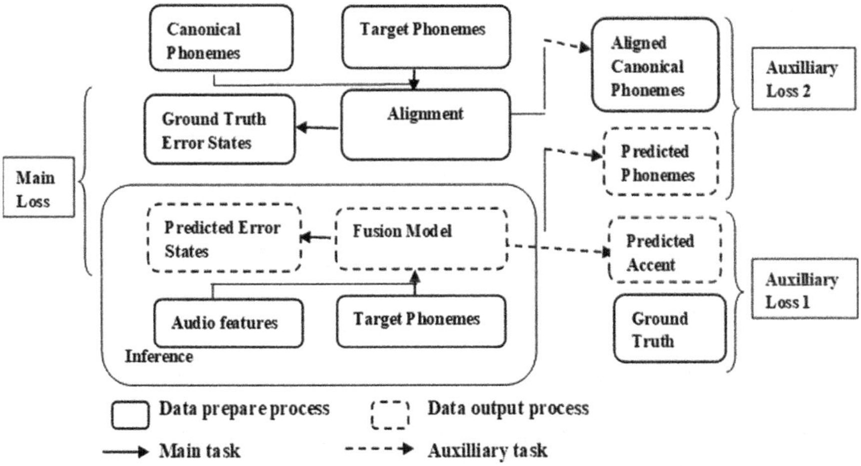

Fig. 2 Working-proposed model of transformer-based model

In our Transformer-based model, we focus on attention-dependent models and inference models. The former model prefers loss function cross-entropy between standard phonemes P and predicted phonemes $\hat{P}\,\hat{P}$

$$L = \text{Cross-Entropy}\left(P, \hat{P}\right). \tag{31}$$

The latter model enters autoregressive mode to identify the phonemes in a sequential manner. The identified phonemes are represented with the term EOS—End Of Sentence. The identified sequence $\hat{P}$ is aligned with the text target $T = (T_1, T_2, \ldots, T_k)$ using the algorithm Needleman–Wulscha. Once the alignment process is complete, the error can be identified as $E = (E_1, E_2, \ldots, E_k)$. In our model, we include both substitution and deletion errors, and their alignment is shown in Table 1.

The input for the encoder unit comprises 39-dimensional MFCC—Mel frequency Cepstral Coefficients. The encoder unit performs concatenation of the SoS tag—Start of Sentence—and one-dimensional standard canonical right-shifted phoneme, and this serves as the input for further decoding units. In the stage of inference, the sequence is replaced by a combination of SoS tags, and the phonemes are decoded in a regressive manner. The decoder unit predicts the probability value for the succeeding phonemes and EoS. So, the two tags SoS and EoS and 39 phonemes, along with the padding tag PAD, make a total of 42 tags that are considered for further classification.

Table 1 Orientation showing substitution and deletion error

	O N L Y
Text target	O W N L I Y
Pronounced sequences	A O N L I Y
States showing Error	1 0 0 0

The length of acoustic features in the attention part, phonemes recognized, and aligned text target sequences are $M, N + 1$, and K, respectively. To perform a complete E2E mode, we must perform an alignment process to identify an error in advance that includes a combination of standard phonemes with text target and accented information. The accented features are one-dimensional global features. The output sequence from this encoder unit is extracted and fed into a supplementary classifier for comparing the predicted accent A' with the accent in ground-truth A,

$$L_A = \text{Cross-Entropy}\left(A', A\right). \tag{32}$$

A further process of training is done to obtain the next state loss function of ASR using weight α.

$$L'_{\text{ASR}} = L_{\text{ASR}} + \alpha\, L_A. \tag{33}$$

A is the ground truth feature, and A' is the accent predicted feature.
L_{ASR} is the Loss function of A, and L'_{ASR} is the Loss function of A'.

Gated Recurrent Units, or preferably Global mean values, are utilized for processing the sequences. The further process of including the additional target text is included, the error sequences are identified using the binary loss function, and we use weights α, β to overcome the lost phoneme sequences. The training and inference processes are summarized in Table 2.

$$L_{\text{BIASR}} = \text{BI}\left(E', E\right). \tag{34}$$

E is the ground truth error, and E' is the accent predicted error.

$$L = L_{\text{EVAL}} + \beta L_{\text{ASR}} + \alpha L_A. \tag{35}$$

Table 2 Summary for the training stage and inference stage in transformer model

	Input—encoder unit	Input—decoder unit	Output—decoder unit
Phase1—training			
Data	Features of input	SoS-shifted canonical (standard) phonemes	Canonical (standard) phonemes—EoS
Loss	–	–	I ASR
Length	"m"	$1 + n$	$n + 1$
Dimension	Thirty-nine	One	Forty-two
Phase2—inference			
Data	Features of input	SoS—phonemes which are recognized	Phonemes which are recognized next—EoS
Length	"m"	END + EoS	END + EoS
Dimension	Thirty-nine	One	Forty-two

L_{BIASR} is the binary loss function, and L_{EVAL} is the evaluation loss function.

We conducted experiments using six blocks of the encoder and decoder, Toolkit Kaldi, L2 Arctic, and LibriSpeech datasets. The proposed method is implemented using the binary loss function, the $F1$ loss function, and the Focal loss function. The experimental results using the L2-Arctic dataset in this model with a Focal loss function gives a False Acceptance Rate, False Rejection Rate, Accuracy, Precision value, Recall value, and high $F1$-score value of 0.423%, 0.055%, 0.882%, 0.636%, 0.577%, and 0.605%, respectively, where the value of the $F1$-score is improved by 8.4%.

2.5 Non-autoregressive End-to-End Neural Modeling for MDD

Although analyzing the performance of the traditional techniques in MDD utilizing the score value is replaced with Transformer methods supporting an end-to-end fashion, but suffers from two significant challenges, which are the speed and availability of a poor number and quality of both non-native and native datasets. In Autoregressive mode, the pronunciation is dictated using the beam search-left to right algorithm, which is proven to fail to meet the constraints of latency as the speed of inference is too slow [28]. Also, this model does not perform better for large quantities of datasets. So, non-Autoregressive mode suits these categories of applications. After the free-phone recognition model that dictates a possible sequence of phones, we preferred the attention-based encoder–encoder model in conjunction with Connectionist temporal classification—CTC, referred to as ATT-CTC in short [29]. Then ATT-CTC is extended to mask CTC with the advantage of parallelism and the increased synergistic power of CTC [30, 31]. To solve these issues, the Transformer model that operates in non-Autoregressive mode [32, 33], which drastically improves the speed of inference, was designed, and it leads to better overall performance when stacked on top of the neural model. But all these models using this non-autoregressive mode do not accomplish CAPT tasks. So, the Transformer model with non-Autoregressive mode and masked CTC is preferred. In our proposed technique, the main difference from existing Autoregressive techniques is the use of a Dictation Unit. The techniques in this model use Transformer, where the high-end phonemes are decoded. In our proposed technique, we use confident scores, where the phonemes with low scores are masked by comparing them with phonemes with high scores using a language model for masking. This provides the dual advantage of parallelism and efficiency in terms of speed. The input audio vector is in a frame-wise manner and is given to the block of the encoder to get the phonemes in an intermediate form, which is further aligned in a latent manner to get the final phoneme sequence.

The function of the Pronunciation Unit is to compare the output phoneme sequences of the Dictation Unit with recognized phoneme sequences to detect errors and provide feedback. The labeled data lacks coverage of the varieties of accents, which may lead to the problem of overfitting. The output sequences of phonemes from the Dictation Unit are made into entrenched sequences, on which we get augmented sequences of phonemes by performing concatenation with their corresponding score values. This augmented sequence is given to the encoder (Transformer dependent) to get transitional entrenched sequences, followed by a Gated Recurrent Model comprising single layer or multi-layer, which provides the vector sequences at the utterance level, and the mathematical formulations are given as:

$$Fs = \mathrm{Sig}([Js; sE]U_1 + B_1), \tag{36}$$

$$J's = \mathrm{ReLu}(Js + [Js \odot Fs]U_2 + B_2, \tag{37}$$

where

$\odot$	dot product is the elementwise.
U_1, B_1, U_2, and B_2	are the parameters to be trained.
Fs	is the transitional entrenched sequence at timestep s.
sE	is the entrenched sequences at the utterance level.
$J's$	is the augmented transitional entrenched sequences.

The gating model stacked with the Neural Network, which is feed forward and normalized with Softmax, makes the final MDD decision. The experimental results on the L2-Arctic dataset show the result of Precision, Recall, and $F1$-score for Correct Detection of pronunciation, followed by Mispronunciation Detection at 91.70%, 90.80%, 91.25%, 45.66%, 48.46%, and 47.02%, respectively. The dataset used for this technology is the L2-ARCTIC English dataset, and the results outperform the traditional DNN-HMM method.

2.6 APL Embeddings' Method

The traditional approaches for calculating the probability of the log-posterior of the model can give satisfactory performance for detection but not for correction of pronunciation [34]. To obtain the diagnosis, the Extended Recognition model is used, where the decoding phase of ASR is modeled with context reliant phonological directions, but still the accuracy is affected, and it requires multiple training phases [35, 36]. Then End-to-End models such as CNN-RNN-CTC [37] (RNN: Recurrent Neural Network) are preferred, which need well-labeled large quantities of data and are hard to collect. In [38, 39], the canonical character order and canonical phoneme order are aligned along with acoustic data by means of attention mechanisms, respectively, but still, this model suffers from an insufficiency of data. Unlike the acoustic data, the phonetic and linguistic embeddings, which are noise-resistant, are extracted from an already trained ASR unit and given as inputs along with the acoustic data. The performance of the network is further enriched using the features of acoustic, phonetic, and linguistic (APL) together, and it is proven using the technique in the L2-ARCTIC dataset.

In our proposed system, we use three units of an encoder for acoustic embeddings, phonetic embeddings, linguistic embeddings and a common decoder unit. The Encoder Unit is constructed using two-layer stacks of CNN and four-layer stacks of RNN. The part of CNN that includes the first set of two-dimensional layers performing the convolution function is followed by a layer to normalize in a batch-wise manner and a layer performing dropout. The acoustic embeddings' encoder, ENA, includes input $A = [A_1 \ldots A_a, A_T]$ of length T having fbanks of dimension 80 and energy of dimension 1. The function of ENA showing input features encoded is

$$H_A = EA(X), \tag{38}$$

$$H_A = [h_{A1}, \ldots, h_{At}, \ldots, h_{AT}]. \tag{39}$$

The phonetic embeddings' encoder ENP includes input $P = [P1, \ldots, Pt, \ldots, PT]$ of same length and dimension, and its function showing encoded phonetic features is

$$H_P = EP(X), \tag{40}$$

$$H_P = [h_{P1}, \ldots, h_{Pt}, \ldots, h_{PT}]. \tag{41}$$

The linguistic embeddings encoder: ENL includes a canonical input sequence of phonemes $U = [U_1, \ldots, U_n, \ldots, U_N]$ of length N, and its function showing encoded phonetic features is

$$H_K, H_V = EL(U), \tag{42}$$

$$H_K = [H_{K1}, \ldots, H_{Km}, \ldots, H_{Kn}] \,\&\, H_V = [H_{V1}, \ldots, H_{VM}, \ldots, H_{VN}]. \tag{43}$$

N length.
K keys.
V values.
T length.

The Decoder Unit using the attention process integrates all the output vectors from the three Encoder Units. The sequence of ENA and ENP undergoes the operation of concatenation to form $HCt^{\wedge}$

$$HCt^{\wedge} = [HAt^{\wedge}, HPt^{\wedge}]. \tag{44}$$

The attention weights are normalized and given by

$$At^{\wedge m} = \frac{\exp(HCt^{\wedge}, HKTm)}{\sum_{m-1}^{N} \exp(HCt^{\wedge}, HKTm)}. \tag{45}$$

The sequences of ENA and ENL form a context vector by performing a weighted average.

$$Ct^{\wedge} = \sum_{m}^{N} \exp(\alpha t^{\wedge}, H_{VM}). \tag{46}$$

In order to overcome the presence of phonemes that are not in the given canonical set, we need to calculate the probability value Ot' using $Ct^{\wedge}$ and $HCt^{\wedge}$ in a frame-wise manner represented as

$$Ot^{\wedge} = \text{Softmax}\,(V[Ct^{\wedge}, HCt^{\wedge}] + B). \tag{47}$$

V is the matrix weight; B is the bias value in a layer of output.

The ultimate sequence for output phonemes is extracted using a Beam-Search algorithm over Ot^. Compared with the baseline system, our model shows an improvement of 9.93% for Detection Accuracy, 10.13% for DER—Diagnosis Error Rate, and 6.17% for the F-measure.

2.7 Hybrid CTC-ATT-Based Approach

The End-to-End model in ASR overcomes the need to train multiple modules with improved performance [39, 40]. Zhang et al. [41] offer MDD with a specifically customized End-to-End model, where a hybrid combination of attention-CTC is implemented [42]. In the model, the data is customized with additional acoustic data, and its label is shuffled. For every utterance, the label of the phones is replaced in a random manner, and the performance outperforms the traditional methods. Studies done by [37, 38, 43] also exhibited the hybrid combination and improved results for the detection of mispronunciation in End-to-End mode. The model discussed below is analogous to [39], except that it uses characters in sequence in order to create utterance embedding. Also, in this model, we prefer loss function CTC to time label data. The architecture of the hybrid CTC-ATT model is shown in Fig. 3, and the steps involved are as follows:

Audio data preparation using Kaldi Style: To extract acoustic features from audio data, here we use more independent transcript records; the acoustic features are extracted from audio data. For preparing data, many sample scripts like LibriSpeech, TIMIT, Switchboard, and WSJ: Wall Street Journal are also available in Kaldi.

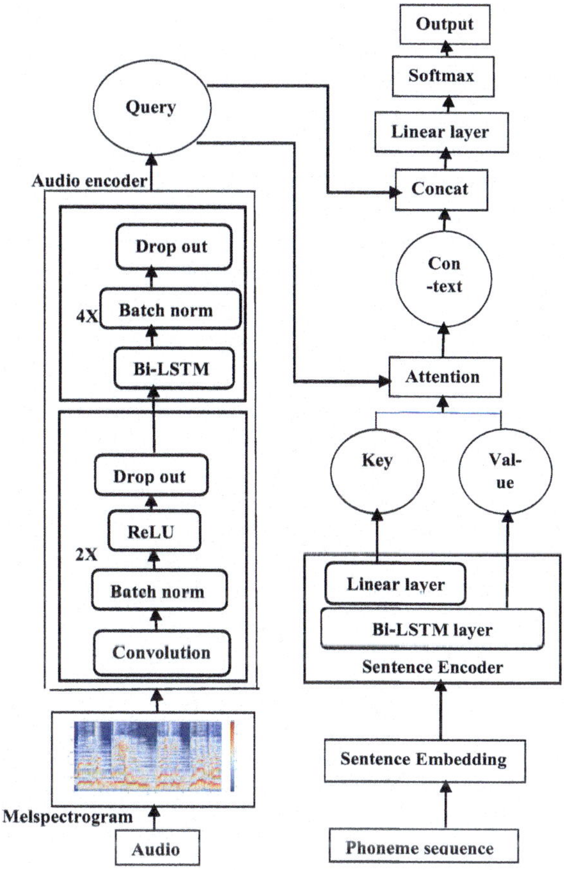

Fig. 3 Block diagram for hybrid CTC-ATT unit

Extracting the features: The characteristics are retrieved using *f*-filter bank features. Filter banks outperform MFCC in deep learning techniques because their characteristics are very closely correlated.

Training using our model: The approach proposed is like [39], but it does not employ sequence; it utilizes the sequence, which consists of phonemes, and the segmentation of the phrase is constructed. When no time label data is available, the CTC loss function is preferred.

CTC abbreviated as Connectionist Temporal Classification: CTC is the type of neural system that uses score values for solving sequence challenges with dynamic scheduling. It is trained using an RNN system like LSTM—Long Short-Term Memory models. A backpropagation technique is often used to alter the weights of NN using CTC scores. At the level of letters in phonemes or words, the CTC process typically crosses two stages: The first is an acoustic model, and the second is a translation model, which is nothing but an encoder and decoder, respectively.

The sequenced output is generated using alignment ordered from left to right to consider the Markov assumption between the series of inputs and outputs. The CTC adds a novel sign termed the blank symbol to the previous set of words as well as phonemes. Its probability distribution of phonemes or words along with the blank token is formed at each sampling interval. This algorithm chooses the word or phoneme with the highest likelihood. Language modeling includes a linguistic system model that can be constructed with the text available in this corpus. This model increases the CTC model's forecasted output much more. The CTC system's resultant transcript may be comparable to the input sounds, but it differs from the expected transcript. When the longer sequence is evaluated, the RNN faces vanishing and exploding gradients. An attention structure inside a Neural Network considers just a portion of a sequence belonging to input data when constructing an output data sequence. The encoder's output is sent to the decoding device, which uses the Beam-Search method for selecting the finest probable phonemic arrangement of sequences.

Our proposed hybrid set of models accepts two sets of inputs: acoustic inputs represented as

$$I = (I_1, I_2, \ldots, I_t, \ldots, I_M) \tag{48}$$

and input sequence of standard phonemes represented as

$$S = (S_1, S_2, \ldots, S_n, \ldots, S_{N+1}, S_N). \tag{49}$$

Our model includes three units: an encoder for audio, an encoder for sentences, and a decoder with attention.

$$\text{Encoder for audio}(I) = H_O, \tag{50}$$

$$\text{Encoder for Sentence}(S) = H_K, H_V, \tag{51}$$

$$\text{Decoder with Att}(H_O, H_K, H_V) = CV. \tag{52}$$

In the encoder for sentence unit, the input sequence of phoneme entrenched with vector is fed into the BLSTM—Bidirectional Recurrent Neural Network—to get the output sequence as

$$H_{V1}, H_{V2}, h_{V3}, \ldots, H_{VN} \quad \text{that denotes value,} \tag{53}$$

$$H_{K1}, H_{K2}, H_{K3}, \ldots, H_K \quad \text{that denotes the key.} \tag{54}$$

These sequences are fed into the Decoder Unit. *En* represents Entrenched sentence (*Sn*).

$$H_{VN} = f1(En)f2(.), \tag{55}$$

$$H_{KN} = f2(H_{VN}). \tag{56}$$

The encoder for the audio unit comprises the layers of CNN with RNN that accepts the input audio feature having dimension 243, fbank of dimension 80, energy of dimension one, and the sequences of output named query and is denoted as

$$(H_{O1}, H_{O2}, H_{O3}, \ldots H_{Ot-1}, H_{Ot}, \ldots H_{OT-1}, H_{OT}), \tag{57}$$

$$\mathrm{Hot}' = f3(It). \tag{58}$$

$f1(.)$ is the function of Bidirectional RNN, $f2(.)$ is the fully connected layer, and $f3$ is the function in CNN with RNN. Due to down-sampling, the length t is updated to t'. In the decoder with attention unit, the decoder part produces the context vector CV that contains information about the key and value. This unit utilizes both the past and next frame states of time, so after normalization using H_{OT}, H_{KN}

$$\alpha t'n = \frac{\exp(\mathrm{score}(H_{KN}, H_{OT'}))}{\sum_{n=1}^{N} \exp(\mathrm{score}(H_{KN}, H_{OT'}))}, \tag{59}$$

where α is the weight for attention

The updated context vector CVt' is calculated by taking the average using weight α in H_{VN} and is denoted as

$$CVt' = \sum_{t=1}^{T} \alpha t'nH_{VN}. \tag{60}$$

The ultimate sequence of output phonemes is extracted using the activation function Softmax and refined with the Beam-Search algorithm

$$Pt' = \mathrm{Softmax}\left(U\left[CVt'H_{OT'}\right] + B_0\right). \tag{61}$$

U is the matrix weight, and B_0 is the bias in a layer of output.

The evaluation metrics are Recall, Precision, and $F1$ value, and the dataset is L2-ARCTIC English datasets. The order of the evaluating parameters is shown in Fig. 4. The $F1$-measure range obtained is 52.51%, which is a 3.02% rise compared with [38]. The $F1$-measure value for the proposed system is 10% vowel–consonant augmentation, which is achieved with various augmentation procedures.

The $F1$-measure value is 56.08%, which is 6.79% improved in comparison with the reference model of the same. The measure of True Rejection and Correct diagnosis is 93.06% and 75.45%, respectively, for an implemented model that assumes a 10% confusing pair utilizing various techniques for augmentation. Our proposed technique

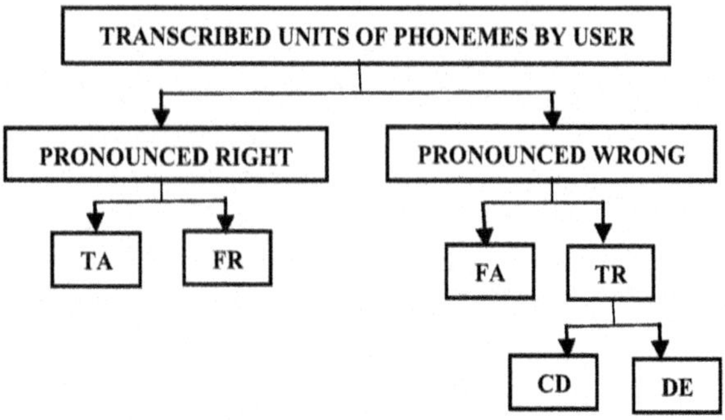

Fig. 4 Ordered hierarchical construction of evaluation parameter in MDD

renders a PER of 15.48% with 10% confusion pair augmentation. The limitations are the availability of very few speech datasets enabling E2E operation, which is unbalanced too. If large quantities of datasets are considered, there is less computation power and poor timing constraints.

The formulas to calculate performance metrics are,

$$\text{Precision} = \frac{\text{TR}}{\text{TR} + \text{FR}}, \tag{62}$$

$$\text{Recall} = \frac{\text{TR}}{\text{TR} + \text{FA}}, \tag{63}$$

$$F\text{-measure} = 2*\frac{\text{Precision*Recall}}{\text{Precision} + \text{Recall}}, \tag{64}$$

$$\text{Detection Accuracy} = \frac{\text{TA} + \text{TR}}{\text{TA} + \text{FR} + \text{FA} + \text{TR}}, \tag{65}$$

$$\text{Diagnosis Accuracy} = \frac{cD}{cD + DE}, \tag{66}$$

$$\text{Correct Diagnosis} = \frac{\text{No. of Correct Predictions}}{\text{Total no. of Predictions}}. \tag{67}$$

$$\text{Diagnosis Error} = \frac{\text{No. of Correct Predictions}}{\text{Total no. of Predictions}}. \tag{68}$$

TR True Rejection value (predicted negative and actually negative).
FR False Rejection value (predicted negative and actually positive).
TA True Acceptance value (predicted positive and actually positive).
FA False Acceptance value (predicted positive and actually negative).
CD Correct Diagnosis value.
DE Diagnosis Error value.

Table 3 Confusion matrix showing performance metrics

Predicted value		Actual value	
		Acceptance (1)	Rejection (0)
	Acceptance (1)	TA	FA
	Rejection (0)	FR	TR

Table 3 shows the confusion matrix for the performance metrics evaluated. If we obtain the TR, FR, TA, FA, CD, and DE using the confusion matrix, the metrics can be calculated using Eqs. (62–68).

3 Results and Discussions

The values of evaluation metrics for all methods discussed are compared in Table 4, and the values of metrics are depicted in the form of a bar chart in Fig. 5. The *F*-measure, which reflects the stability of both Precision and Recall, is compared. The transformer-based model and hybrid CTC-ATT model show *F*-measure values of 60.5% and 56.04%, respectively. The Diagnostic and Detection Accuracies are compared for further clarification regarding detection and diagnosis. The transformer-based model shows a detection accuracy of 88.2%, and the hybrid CTC-ATT model shows diagnostic and detection accuracy of 75.45% and 74.96%, respectively.

Table 4 Compared values of the metrics for evaluation

Methods	Precision value (%)	Recall value (%)	*F*-measure (%)	Diagnostic Accuracy (%)	Detection Accuracy (%)
A. Anti-phone modeling [44]	46.57	70.28	56.02	40.66	–
B. Accent modulation [45]	49.57	50.68	50.12	73.75	–
C. APL embeddings [46]	52.79	54.49	53.62	33.94	85.59
D. E2E MDD using raw waveforms [47]	55.31	50.09	52.57	60.96	70.17
E. Transformer-based model [48]	63.6	57.7	60.5	–	88.2
F. Non-autoregressive MDD [49]	45.66	48.46	47.02	73.22	–
G. Hybrid CTC-ATT [50]	56.04	56.12	56.08	74.96	75.45

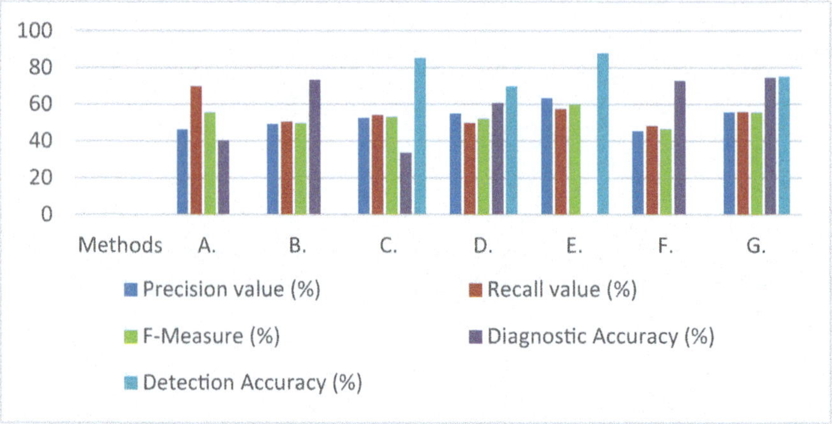

Fig. 5 Comparison chart of the evaluation metrics

4 Conclusion

The various conventional deep learning methods for MDD using the L2-ARCTIC English dataset are studied. The transformer-based model shows the highest *F*-measure value of 60.5% and detection accuracy of 88.2%. The hybrid CTC-ATT model shows the second highest value of *F*-measure (56.08%) and diagnostic and detection accuracy of 74.96% and 75.45%, respectively, which shows that we can prefer this method for both detection and diagnosis. Thus, based on the need for diagnosis in MDD, these two models can be preferred for better overall performance. Thus, future work must include implementing a novel technique for using the transformer-based model not only for detection but also for diagnosis purposes. The F-measure, diagnostic, and detection accuracies of the Hybrid CTC-ATT model might be improved by adopting a novel technique.

References

1. Li K, Qian X, Meng H (2017) Mispronunciation detection and diagnosis in L2 English speech using multi-distribution deep neural networks. IEEE/ACM Trans Audio Speech Lang Process 25(1). https://doi.org/10.1109/TASLP.2016.2621675
2. Qian X-J, Meng H, Soong F (2010) Capturing L2 segmental mispronunciations with joint-sequence models in computer-aided pronunciation training (CAPT). In: Proceedings of 7th international symposium conference on spoken language processing, pp 84–88. https://doi.org/10.1109/ISCSLP.2010.5684845
3. Mao S, Wu Z, Li X, Li R, Wu X, Meng H (2018) Integrating articulatory features into acoustic-phonemic model for mispronunciation detection and diagnosis in L2 English speech. https://doi.org/10.1109/ICME.2018.8486462

4. Li K, Zhang S, Li M, Lo W-K, Meng H (2010) Detection of intonation in L2 English speech of native Mandarin learners. In: Proceedings of 7th international symposium on conference on spoken language processing, pp 69–74. https://doi.org/10.1109/ISCSLP.2010.5684846
5. Yang X, Audhkhasi K, Rosenberg A, Thomas S, Ramabhadran B, Hasegawa-Johnson M (2018) Joint modeling of accents and acoustics for multi-accent speech recognition. Proc ICASSP: 1–5. https://doi.org/10.1109/ICASSP.2018.8462557
6. Chen Y, Huang C, Soong F (2009) Improving mispronunciation detection using machine learning. In: IEEE international conference on acoustics, speech, and signal processing. https://doi.org/10.1109/ICASSP.2009.4960721
7. Wei S, Hu G, Hu Y, Wang RH (2009) A new method for mispronunciation detection using support vector machine based on pronunciation space models. Speech Commun 51:896–905. https://doi.org/10.1016/j.specom.2009.03.004
8. Hu W, Qian Y, Soong FK, Wang Y (2015) Improved mispronunciation detection with deep neural network trained acoustic models and transfer learning based logistic regression classifiers. Speech Commun 67:154–166. https://doi.org/10.1016/j.specom.2014.12.008
9. Hu W, Qian Y, Soong FK (2015) An improved DNN-based approach to mispronunciation detection and diagnosis of L2 learner's speech. In: Proceedings of the SLaTE. Leipzig, Germany, pp 71–76. https://doi.org/10.1016/j.specom.2014.12.008
10. Wren Y, Roulstone S, Williams AL (2010) Computer-based interventions in interventions for speech sound disorders in children. Brookes Publishing, Baltimore. MD. USA, pp 275–294
11. Li W, Siniscalchi SM, Chen NF, Lee C-H (2016) Improving non-native mispronunciation detection and enriching diagnostic feedback with DNN-based speech attribute modeling. In: Proceedings of the ICASSP, pp 6135–6139. https://doi.org/10.1109/ICASSP.2016.7472856
12. Lo W, Zhang S, Meng H (2010) Automatic derivation of phonological rules for mispronunciation detection in a computer assisted pronunciation training system. In: Proceedings of the INTERSPEECH, pp 765–768. https://doi.org/10.21437/Interspeech.2010-280
13. Mao S, Li X, Li K, Wu Z, Liu X, Meng H (2018) Unsupervised discovery of an extended phoneme set in L2 English speech for mispronunciation detection and diagnosis. In: Proceedings of the ICASSP, pp 6244–6248. https://doi.org/10.1109/ICASSP.2018.8462635
14. Li X, Mao S, Wu X, Li K, Liu X, Meng H (2018) Unsupervised discovery of non-native phonetic patterns in L2 English speech for mispronunciation detection and diagnosis. In: Proceedings of the INTERSPEECH, pp 2554–2558. https://doi.org/10.21437/Interspeech.2018-2027
15. Ronen O, Neumeyer L, Franco H (1997) Automatic detection of mispronunciation for language instruction. In: Proceedings of speech communication and technology, pp 649–652. https://doi.org/10.21437/eurospeech.1997-231
16. Zhao G et al (2018) L2-ARCTIC: A non-native English speech corpus. In: Proceedings of INTERSPEECH, pp 2783–2787. https://psi.engr.tamu.edu/l2-arctic-corpus/
17. Li K, Mao S, Li X, Wu Z, Meng H (2018) Automatic lexical stress, and pitch accent detection for L2 English speech using multi-distribution deep neural networks. Speech Commun 96:28–36. https://doi.org/10.1016/j.specom.2017.11.003
18. Pratap V, Sriram A, Tomasello P, Hannun A, Liptchinsky V, Synnaeve G, Collobert R (2020) Massively multilingual ASR: 50 languages. 1 model. 1 billion parameters. In: Proceedings of INTERSPEECH, pp 4751–4755. https://doi.org/10.21437/interspeech.2020-2831
19. Vergyri D, Lamel L, Gauvain J-L (2010) Automatic speech recognition of multiple accented English data. In: Proceedings of INTERSPEECH. https://doi.org/10.21437/Interspeech.2010-477
20. Tong S, Garner PN, Bourlard H (2018) Cross-lingual adaptation of a CTC-based multilingual acoustic model. Speech Commun 104:39–46. https://doi.org/10.1016/j.specom.2018.09.001
21. Palaz D, Collobert R, Magimai-Doss M (2013) Estimating phoneme class conditional probabilities from raw speech signal using convolutional neural networks. In: Proceedings of INTERSPEECH, pp 1766–1770. https://doi.org/10.48550/arXiv.1304.1018
22. Watanabe S, Hori T, Kim S, Hershey JR, Hayashi T (2017) Hybrid CTC/attention architecture for end-to-end speech recognition. IEEE J Sel Top Signal Process 11(8):1240–1253. https://doi.org/10.1109/JSTSP.2017.2763455

23. Ravanelli M, Bengio Y (2018) Interpretable convolutional filters with sincnet. arXiv preprint arXiv:1811.09725. https://doi.org/10.48550/arXiv.1811.09725
24. Moritz N, Hori T, Le J (2020) Streaming automatic speech recognition with the transformer model. IEEE In: International conference on acoustics. Speech and signal processing (ICASSP). IEEE, pp 6074–6078. https://doi.org/10.1109/icassp40776.2020.9054476
25. Gu J, Bradbury J, Xiong C, Li VO, Socher R (2018) Non-autoregressive neural machine translation. arXiv preprint arXiv:1711.02281. https://doi.org/10.48550/arXiv.1711.02281
26. Zhang Q, Lu H, Sak H, Tripathi A, McDermott E, Koo S, Kumar S (2020) Transformer transducer: a streamable speech recognition model with transformer encoders and RNN-t loss. In: IEEE international conference on acoustics, speech, and signal processing, pp 7829–7833. https://doi.org/10.1109/icassp40776.2020.9053896
27. Dong L, Xu S, Xu B (2018) Speech-Transformer: a no-recurrence sequence-to-sequence model for speech recognition. In: IEEE international conference on acoustics, speech and signal processing, pp 5884–5888. https://doi.org/10.1109/icassp.2018.8462506
28. Vaswani A, Shazeer N, Parmar N, Uszkoreit J, Jones L, Gomez AN, Kaiser L, Polosukhin I (2017) Attention is all you need. In: Proceedings of NIPS, pp 5998–6008
29. Lo T-H, Weng S-Y, Chang H-J, Chen B (2020) An effective end-to-end modeling approach for mispronunciation detection. In: Proceedings of INTERSPEECH, pp 3027–3031. https://doi.org/10.48550/arXiv.2005.08440
30. Kim S, Hori T, Watanabe S (2017) Joint CTC-attention based end-to-end speech recognition using multi-task learning. In: Proceedings of ICASSP, pp 4835–4839. https://doi.org/10.48550/arXiv.1609.06773
31. Higuchi Y, Watanabe S, Chen N, Ogawa T, Kobayashi T (2020) Mask CTC: non-autoregressive end-to-end ASR with CTC and mask predict. In: Proceedings of INTERSPEECH, pp 3655–3659. https://doi.org/10.21437/Interspeech.2020-2404
32. Bai Y, Yi J, Tao J, Tian Z, Wen Z, Zhang S (2020) Listen attentively, and spell once: whole sentence generation via a non-autoregressive architecture for low-latency speech recognition. In: Proceedings of INTERSPEECH, pp 3381–3385. https://doi.org/10.48550/arXiv.2005.04862
33. Chi EA, Salazar J, Kirchhoff K (2021) Align-refine: non-autoregressive speech recognition via iterative realignment. In: Proceedings of NAACL, pp 1920–1927. https://doi.org/10.18653/v1/2021.naacl-main.154
34. Witt SM, Young SJ (2000) Phone-level pronunciation scoring and assessment for interactive language learning. Speech Commun 30(2–3):95–108. https://doi.org/10.1016/S0167-6393(99)00044-8
35. Lo A-K, Zhang S, Meng H (2010) Automatic derivation of phonological rules for mispronunciation detection in a computer-assisted pronunciation training system. In: Proceedings of INTERSPEECH, pp 765–768. https://doi.org/10.21437/Interspeech.2010-280
36. Qian X, Soong F, Meng H (2010) Discriminatively-trained acoustic models for improving mispronunciation detection and diagnosis in computer aided pronunciation training (CAPT). In: Proceedings of INTERSPEECH. https://doi.org/10.21437/Interspeech.2010-278
37. Leung W-K, Liu X, Meng H (2019) CNN-RNN-CTC based end-to-end mispronunciation detection and diagnosis. In: IEEE international conference on acoustics. Speech and signal processing (ICASSP), pp 8132–8136. https://doi.org/10.1109/ICASSP.2019.8682654
38. Feng Y, Fu G, Chen Q, Chen K (2020) SEDMDD: towards sentence dependent end-to-end mispronunciation detection and diagnosis. In: IEEE international conference on acoustics, speech and signal processing, pp 3492–3496. https://doi.org/10.1109/ICASSP40776.2020.9052975
39. Fu K, Lin J, Ke D, Xie Y, Zhang J, Lin B (2021) A full text-dependent end to end mispronunciation detection and diagnosis with easy data augmentation techniques. arXiv preprint arXiv:2104. https://doi.org/10.48550/arXiv.2104.08428
40. Zhang Y, Pezeshki M, Brakel P, Zhang S, Laurent C, Bengio Y, Courville A (2016) Towards end-to-end speech recognition with deep convolutional neural networks. In: Proceedings of the annual conference of the international speech communication association. INTERSPEECH [online], pp 410–414. https://doi.org/10.48550/arXiv.1701.02720

41. Amodei D, Anubhai R, Battenberg E, Case C, Casper J, Catanzaro B, Chen J, Chrzanowski M, Coates A, Diamos G, Elsen E, Engel J, Fan L, Fougner, Han T, Hannun A, Jun B, LeGresley P, Lin L, Narang S, Ng A, Ozair S, Prenger R, Raiman J, Satheesh S, Seetapun D, Sengupta S, Wang Y, Wang Z, Wang C, Xiao B, Yogatama D, Zhan J, Zhu Z (2016) Deep speech 2: end-to-end speech recognition in English and Mandarin. In: 33rd international conference on machine learning, pp 312–321. Available at http://arxiv.org/abs/1512.02595
42. Zhang L, Zhao Z, Ma C, Shan L, Sun H, Jiang L, Deng S, Gao C (2020) End-to-end automatic pronunciation error detection based on improved hybrid CTC/attention architecture. Sensors 207:1809 [online]. Available at https://www.mdpi.com/1424-8220/20/7/1809
43. Yan B-C, Wu M-C, Hung H-T, Chen B (2020) An end-to-end mispronunciation detection system for L2 English speech leveraging novel anti-phone modeling. In: INTERSPEECH. ISCA, pp 3032–3036 [online]. https://doi.org/10.48550/arXiv.2005.11950
44. Yan B-C, Wu M-C, Hung H-T, Chen B (2020) An end-to-end mispronunciation detection system for L2 English speech leveraging novel anti-phone modeling, audio and speech processing (EESS.AS). In: Computation and language (CS.CL). Sound (CS.SD). Accepted by INTERSPEECH (2020). https://doi.org/10.48550/arXiv.2005.11950
45. Jiang S-WF, Yan B-C, Lo T-H, Chao F-A, Chen B (2021) Towards robust mispronunciation detection and diagnosis for L2 English learners with accent-modulating methods. In: IEEE automatic speech recognition and understanding workshop (ASRU). https://doi.org/10.48550/arXiv.2108.11627
46. Ye W, Mao S, Soong F, Wu W, Xia Y, Tien J, Wu Z (2022) An approach to mispronunciation detection and diagnosis with acoustic, phonetic and linguistic (APL) embeddings. In: IEEE international conference on acoustics, speech and signal processing (ICASSP). https://doi.org/10.48550/arXiv.2110.07274
47. Yan B-C, Chen B (2021) End-to-end mispronunciation detection and diagnosis from raw waveforms. In: IEEE 29th European signal processing conference (EUSIPCO). https://doi.org/10.23919/EUSIPCO54536.2021.9615987
48. Zhang Z, Wang Y, Yang J (2021) Text-conditioned transformer for automatic pronunciation error detection. Speech Commun J Audio Speech Process (EESS.AS) Speech Commun 130. https://doi.org/10.48550/arXiv.2008.12424
49. Wang H-W, Yan B-C, Chiu H-S, Hsu Y-C, Chen B (2022) Exploring non-autoregressive E2E neural modeling foe English mispronunciation detection and diagnosis. In: IEEE international conference on acoustics, speech and signal processing (ICASSP). https://doi.org/10.48550/arXiv.2111.00844
50. Baranwal N, Chilaka S (2022) Improved mispronunciation detection system using a hybrid CTC-ATT based approach for L2 English speakers. In: Sound (CS.SD) A (EESS.AS), audio and speech processing. https://doi.org/10.48550/arXiv.2201.10198

A Deep Learning Model-Based Approach for Brain Tumor Detection in Low-Brightness and Low-Contrast MRI

S. Komal Kour, Aashrith Sagar Ramachandrula, Prerna Patnaik, and T. Adilakshmi

Abstract Magnetic resonance imaging (MRI) has become a valuable diagnostic assessment means for the detection, segmentation, and characterization of brain tumors. However, low brightness and low contrast in MRI images pose a significant challenge for accurate tumor detection, especially in the early stages. Several approaches have been proposed to address this challenge, including image enhancement and filtering techniques. However, these methods often result in loss of image details, making it difficult to discern the tumor regions from the non-tumor ones. To overcome these limitations, deep learning-based approaches have gathered attention in recent years for their capability to automatically learn features from the input images and achieve high accuracy in various medical imaging tasks. The aim of our research is to present a deep learning-based methodology for detecting brain tumors in low-brightness and low-contrast MRI images. We employ a neural network with convolutions' (CNN) architecture, which has been proven to be effective in acquiring complex image features. Previous studies have used deep learning techniques for brain tumor segmentation and detection (Ramin Ranjbarzadeh et al. in Brain tumor segmentation based on deep learning and an attention mechanism using MRI multi-modalities brain images [1]). However, these studies did not specifically address the problem of low brightness and low contrast in MRI images. In contrast, our proposed method is designed to capture the subtle differences between tumor regions and non-tumor regions in such MRI images. Our CNN model has been trained and validated on a larger dataset of MRI images, including both normal and tumor-containing images. Our results demonstrate that our proposed method achieves high accuracy and specificity in detecting brain tumors, even in low-brightness and low-contrast MRI images. Additionally, our method has the potential to aid healthcare professionals in precisely and promptly pronouncing tumors in brain, resulting in better

S. Komal Kour (✉) · A. S. Ramachandrula · P. Patnaik · T. Adilakshmi
Department of Computer Science and Engineering, Vasavi College of Engineering, Hyderabad, India
e-mail: komalkour@staff.vce.ac.in

T. Adilakshmi
e-mail: hodcse@staff.vce.ac.in

V. K. Gunjan and J. M. Zurada (eds.), *Proceedings of 4th International Conference on Recent Trends in Machine Learning, IoT, Smart Cities and Applications*, Lecture Notes in Networks and Systems 873, https://doi.org/10.1007/978-981-99-9442-7_32

patient results. To sum up, our proposed approach, which is based on learning from images more deeply, has the potential to identify brain tumors in MRI images with low brightness and contrast with a promise of high accuracy. This approach could be incorporated into clinical processes, improving the detection accuracy thereby saving patient lives.

Keywords Brain tumor detection · Low brightness · Low contrast · Health care · Medical imaging

1 Introduction

Brain tumors are not only a significant global health challenge, but also result in a high morbidity and mortality rate. The American Brain Tumor Association reported that more than 7 lakh individuals in the USA are presently suffering with brain tumors, and an estimated 75,000+ new cases are being diagnosed every year [2]. The way to detect the tumors that are present in brain and treatment of brain tumors depends on accurate imaging techniques, including magnetic resonance imaging (MRI). However, the low brightness and low contrast of MRI images can pose a significant challenge to the detection of brain tumors. Traditionally, brain tumors were detected through **in**vasive methods, such as biopsy or surgical resection. These methods were associated with significant risks, including neurologic deficits and infection [3]. With the advent of non-invasive imaging techniques such as MRI, the detection of brain tumors has become less invasive and more accurate [4]. However, the low brightness and low contrast of MRI images can still pose a significant challenge to the accurate detection of brain tumors.

Substantially, convolutional neural networks (CNNs) have proved themselves to be successful in automatically acquiring characteristics from medical images, encompassing MRI images that possess low contrast and brightness. The purpose of this paper is to suggest a method based on deep learning for detecting brain tumors in low-brightness and low-contrast MRI images by utilizing CNNs.

Our approach utilizes a novel architecture that includes multiple convolutional layers for the extraction of features from the input images. The model is trained with a broad dataset of Magnetic Resonance Images and their corresponding labels, enabling it to discern the subtle disparities between tumors and the ones that are non-tumor regions in MRI images with low contrast and brightness. We evaluate our approach on a brain tumor dataset which is publicly available and compare our results to existing intricate methods.

Our investigation showcases the power that we can achieve by utilizing the merits of deep learning-based methodologies for detecting brain tumors in MRI images with low brightness and contrast. By accurately detecting the brain tumors that are in MRI images, our approach has the possibility to improve the patient outcomes by enabling earlier detection and treatment. Additionally, our approach may be applied to other medical imaging tasks that require accurate detection of subtle features in low-contrast and low-brightness images.

2 Literature Review

Globally, brain tumors are still on top of the list in the causes of cancer-related deaths across the globe. As stated by the World Health Organization (WHO), approximately 299,000+ patients of primary brain tumors are diagnosed worldwide each year [5]. Brain tumors can cause a wide range of symptoms, including headaches, seizures, and neurological deficits, and the detection and process to treat the brain tumors depend severely on imaging techniques, particularly MRI. Approaches based on deep learning have demonstrated encouraging outcomes in enhancing the precision of detecting brain tumors in MRI images. CNNs have shown their capability in learning features from medical images automatically, including low-contrast and low-brightness MRI images [6]. CNNs have been used in a variety of medical imaging applications, including brain tumor detection, and have shown significant improvements in accuracy over traditional methods of machine learning.

Several studies have proposed approaches that are based on either machine or deep learning for brain tumor detection in low-brightness and low-contrast MRI images. For example, Wang et al. [7] proposed an approach based on CNNs for brain tumor detection in MRI images. The strategy utilized a deep residual network architecture and achieved an overall sensitivity of 0.85 and a specificity of 0.96 in detecting the brain tumors. Another study by Menze et al. [8] proposed an approach based on deep learning for the segmentation of brain tumor using MRI images. Here also, in the method they utilized a CNN-based architecture and achieved brilliant results in segmenting particularly brain tumors in MRI images. The authors also showed that their method was more robust to noise and artifacts than traditional segmentation methods.

Some other approaches have also been used for the classification of brain tumor subtypes. For example, Abd-Ellah et al. [9] presented a CNN-based approach for the classification of glioma subtypes in MRI images, demonstrating the potential of DL-based approaches for the classification of brain tumor subtypes. While deep learning-based approaches have shown promising results in brain tumor detection and classification, despite the staggering results of deep learning-based approaches in brain tumor detection, there still exist some significant challenges that need to be addressed. One critical obstacle is the limited availability of labeled data.

To address this issue, several studies have proposed the use of transfer learning, where pre-trained CNNs are fine-tuned for medical imaging tasks. For example, Mathew et al. [10] proposed a transfer learning-based approach for brain tumor segmentation in MRI images. Their method utilized a pre-trained CNN and achieved exceptional results in brain tumor segmentation, demonstrating the potential of transfer learning for medical imaging tasks. Another challenge in deep learning-based approaches for brain tumor detection is the need for interpretability. Deep learning-based approaches are often viewed as "black boxes" due to their complexity, making it difficult to understand how they make decisions. Several studies have proposed various methods in order to improve the interpretability of approaches based on deep learning for medical imaging tasks. Gudhe et al. [11] proposed a method for generating saliency maps to visualize the regions of an MRI image that contributed the most to a CNN's decision for brain tumor detection.

In conclusion, deep learning-based approaches have been showing promise in improving the accuracy of brain tumor detection in low-brightness and low-contrast MRI images. CNN-based architectures have been shown to be effective in automatically learning features from medical images.

3 Dataset

The brain tumor MRI dataset contains a total of 7022 images, which are used for training and testing machine learning models for brain tumor classification. Out of these, 5712 images are part of the training dataset, and 1311 images are used for testing the models. The dataset includes four types of brain tumors, glioma, meningioma, pituitary, and no tumor. Specifically, the dataset contains 1612 glioma images, 1645 meningioma images, 1757 pituitary images, and 2000 images that show no tumor. Each image in the dataset is resized to have a height of 256 pixels and a width of 256 pixels, and each batch contains 256 images. The images in the dataset are randomly shuffled during training to prevent the model from learning the order of the images and overfitting to the data and the shuffle is uniformly maintained across the dataset.

This dataset is an important resource for developing and testing machine learning and deep learning algorithms for accurate brain tumor detection and classification. The dataset is balanced, with an equal number of images in each of the four tumor categories, which ensures that the machine learning models are not biased toward any particular type of tumor. Additionally, the dataset is annotated, with each image having a corresponding label indicating whether the tumor is present or not. The dataset has many potential applications in the field of medical imaging and machine learning, including developing predictive models for the early detection of brain tumors and assisting healthcare professionals in providing better patient care. The researchers who created the dataset used a neural network with multiple convolutions to classify the MRI images of brain into roughly four different categories. The CNN model achieved high accuracy on the test set, demonstrating the potential of the dataset for developing accurate machine learning models.

In short, the brain tumor MRI dataset containing 7022 images is a valuable resource for developing and testing machine learning algorithms for accurate brain tumor classification. The dataset's balance and annotation make it an excellent resource for developing predictive models and assisting healthcare professionals in providing better patient care.

4 Proposed Model

CNNs have demonstrated encouraging outcomes in various image classification tasks, including the detection of brain tumors. Our proposed model for brain tumor detection utilizes a CNN architecture that combines convolutional and fully connected neural networks to effectively detect brain tumors in low-brightness and low-contrast MRI images. The model begins with an input layer of shape (256, 256, 3) to match the size of our images, followed by a series of convolutional layers with increasing filter sizes (32, 64, 128) and kernel sizes of 3. After every layer of convolution, a max pooling layer is utilized to decrease the spatial dimensions of the output. This permits the model to capture the most important characteristics of the image while lessening the computational intricacy of the model. For each convolutional layer, the activation function employed is the rectified linear unit (ReLU), which has been demonstrated to be successful in tasks related to image classification. The ReLU activation function is used in the convolutional and dense layers as it has been shown to be effective in deep learning tasks and helps to prevent the vanishing gradient problem. After the convolutional layers, we flatten the output and feed it into a fully connected neural network consisting of two dense layers with 64 and 32 units, respectively. To improve the stability of the network and prevent overfitting, we incorporate batch normalization and dropout layers between the dense layers. Batch normalization is a technique that standardizes the output of a layer to possess a mean of zero and a variance of one, which can hasten the training procedure and enhance generalization. During training, dropout is applied to randomly drop out a portion of the nodes in the layer to avoid the network relying too much on a specific feature. For binary classification of tumor or non-tumor regions, the output layer consists of a single unit with a sigmoid activation function. The Adam optimizer, an extended version of optimizer stochastic gradient descent (SGD), is used to improve the convergence of model during training by making use of adaptive learning rates and momentum. It maintains separate learning rates for each weight in the network, and the learning rates are adapted based on the moving average of the gradients. This allows the optimizer to converge faster and more reliably than traditional SGD. In our proposed model, we use a learning rate of 0.01 for Adam, which has been observed to be effective for a wide variety of deep learning tasks. The binary cross-entropy loss function is used as it is usually preferred for tasks that involve binary classification, and the difference between the predicted and actual output is measured.

Overall, this proposed model architecture (see Fig. 1) with the Adam optimizer and binary cross-entropy loss function proves to be a powerful tool for accurate brain tumor detection in low-brightness and low-contrast MRI images. In addition to the proposed model architecture, several key techniques are used to enhance its performance. Batch normalization is applied after each dense layer to improve the stability of the network during training. By normalizing the output of a layer to have unit variance along with zero mean, this technique can make the training process complete quickly and improve generalization. Dropout is another technique used to prevent overfitting of the model. This technique randomly drops out a fraction of the

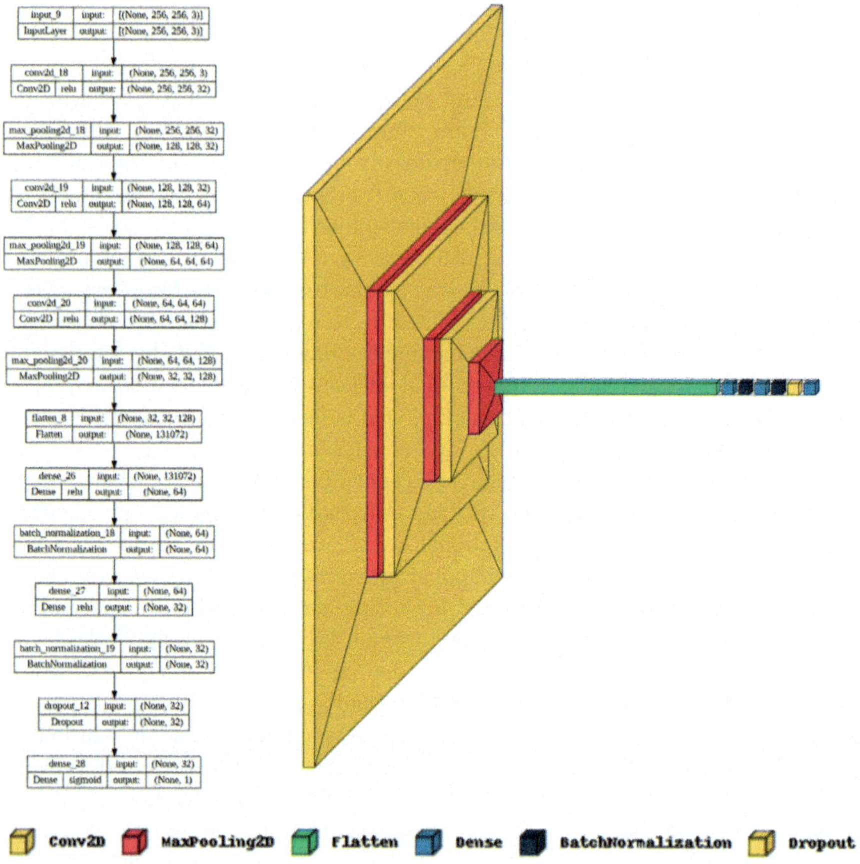

Fig. 1 Proposed model of brain tumor detection in low-brightness and low-contrast MRI

nodes in the layer during training, preventing the network from relying too heavily on any one feature. Finally, the use of the ReLU activation function plays a significant role in the convolutional and dense layers and prevents the vanishing gradient problem, thereby improving the model's performance. This function effectively sets negative values to zero, preventing the gradient from becoming too small, and is computationally efficient, thereby speeding up training time.

5 Results

The model which was proposed for detecting brain tumors was trained for 10 epochs utilizing an MRI image dataset. We were able to achieve, in the context of training, an accuracy of 98.78%, and during the validation, the results were 96.41% accurate.

As each epoch passed by, the model's performance improved, as evidenced by the increasing validation accuracy, which started at 71.27% and reached 96.41% by the end. A graph showing the variation of accuracy and validation accuracy with each epoch has been displayed (see Fig. 2). During training, the loss function was employed to assess the model's performance. The loss value decreased from 0.6332 at the beginning to 0.0412 by the end, indicating that the trained model had the capability and learned the necessary features to correctly and precisely detect brain tumors from MRI reports with low brightness and contrast. A sample of images obtained by varying the brightness has been shown (see Fig. 3). There is also a significant observation in the context of validation loss, which progressively reduced from its initial value in the first-ever epoch to a lower value in the final one, reaching a lowest value of 0.1199.

These outcomes, as suggested by the above figures, indicate that there was effective learning by the model about the essential features for accurately detecting brain tumors in MRI images with low brightness and contrast, with a high validation

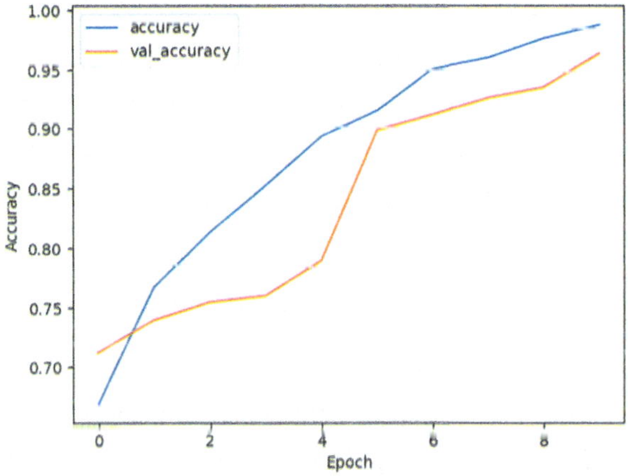

Fig. 2 Accuracy versus validation accuracy graph for the model's history

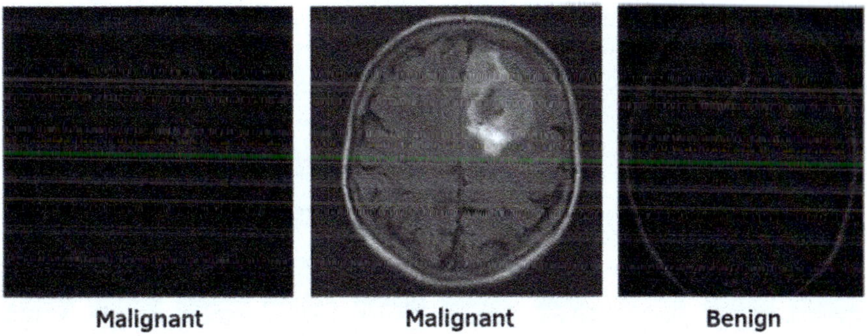

Fig. 3 Images with varied brightnesses

accuracy of 96.41%. Overall, the proposed model has the potential to aid healthcare professionals in precisely and promptly diagnosing brain tumors, resulting in improved patient outcomes.

In order to assess the overall effectiveness of the trained model that we have proposed, various performance metrics are employed. These include accuracy, measure of recall along with precision, and finally $F1$-score, which have been used to evaluate the accuracy of both the true and predicted classes. The equations for each of these metrics (1, 2, 3, and 4) are shown below in mathematical notation. These metrics allow for a quantitative evaluation of the model's performance in detecting brain tumors in low-brightness and low-contrast MRI images. Each of these metrics has been evaluated on a sample of 500 images for the proposed model and presented in tabular form

$$\text{Accuracy} = \frac{\text{TP} + \text{TN}}{\text{TP} + \text{TN} + \text{FP} + \text{FN}}, \quad (1)$$

$$\text{Recall} = \frac{\text{TP}}{\text{FN} + \text{TP}}, \quad (2)$$

$$\text{Precision} = \frac{\text{TP}}{\text{FP} + \text{TP}}, \quad (3)$$

$$F1\ \text{Score} = \frac{2 \times \text{Recall} \times \text{Precision}}{\text{Recall} + \text{Precision}}. \quad (4)$$

An evaluation of the CNN model was conducted using a sample of 500 brain tumor images, some of which were obtained by making brightness and contrast variations (see Fig. 4). Table 1 displays the performance evaluation metrics, which include accuracy, along with recall and precision, finally $F1$-score. Accuracy assesses the overall authenticity of the predictions made by the model and yielded a value of 0.96. Precision measures the model's ability to correctly identify positive instances and resulted in a value of 0.97. Recall, also referred to as sensitivity, is primarily used to indicate and quantify the model's capability to identify all actual positive instances, and it achieved a value of 0.98. Finally, the $F1$-score combines precision and recall into a single metric, which obtained a value of 0.97, indicating the overall performance of the model.

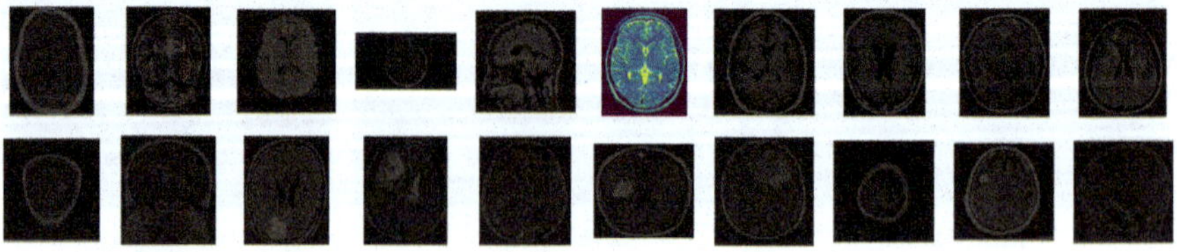

Fig. 4 Randomly selected images with brightness and contrast adjustments

Table 1 Performance evaluation metrics of the CNN model on a sample of 500 images

Accuracy	Recall	Precision	$F1$-score
0.96	0.98	0.97	0.97

Table 2 Comparison with other existing models

Related work	Features	Model	Accuracy (%)
Foeady [12]	Preprocessing, morphological operations, GLCM features, SVM classifier	Support vector machine (SVM)	82.35
Wasule [13]	K-nearest neighbors (KNN) classifier, GLCM	K-nearest neighbors (KNNs)	86
Proposed model	Deep learning-based approach, convolutional neural network (CNN)	CNN	95.2

The proposed model for brain tumor detection in low-brightness and low-contrast MRI images using a convolutional neural network (CNN) with multiple convolutional layers appears to be a promising approach. Comparing this model with other existing models in the literature (see Table 2), it can be seen that this approach performs well in terms of both the specificity along with a measure of the accuracy. One recent study used a deep learning-based approach to detect brain tumors from MRI images using a combination of two CNNs, with one CNN focusing on tumor detection and the other CNN identifying tumor subregions. While this model achieved high accuracy, it required a large number of MRI images to train the CNNs, which can be challenging for institutions with limited access to MRI data. Another study employed a support vector machine (SVM) for the detection of the tumors in the brain with MRI images. Although SVMs have been widely used for medical image analysis, the model's performance depended substantially on the extraction of features, which in most of the times can become time-consuming and subjective. In contrast, the proposed model uses a relatively simple architecture with multiple convolutional layers and a small number of dense layers. This model has been trained using a large dataset of MRI images, and its performance is measured using common evaluation metrics, including accuracy, recall, precision, and $F1$-score. On the whole, this model shows great potential for assisting clinicians in accurately detecting brain tumors from low-brightness and low-contrast MRI images.

6 Conclusion

To summarize, we introduced a methodology based on deep learning for detecting tumors in the brain using MRI images with low brightness and contrast. Our approach utilizes a convolutional neural network (CNN) with multiple layers of convolution that can automatically take the preprocessed image as input in order to learn as many features as possible from the input images. The CNN architecture is designed to capture the subtle differences between regions with tumor and the ones without tumor in low-brightness and low-contrast images. Our trained CNN model was verified using a substantial collection of MRI images, which included both normal images and images with tumors. The outcomes we obtained indicate that our suggested technique can accurately and specifically identify brain tumors, even in MRI images with low brightness and contrast. Our proposed model makes use of the ramp function which is also known as the rectified linear unit (ReLU) which is the main activation function in the convolutional and dense layers, which has been shown to be effective in deep learning tasks and helps to prevent the vanishing gradient problem. We utilized standard techniques in deep learning for binary classification problems, specifically the binary cross-entropy, which is the loss function in our method and the Adam optimizer. Our model achieved an $F1$-score of 0.97, a recall of 0.98, and a precision of 0.96 when evaluated on a test set of MRI images. These results demonstrate the effectiveness of our approach for accurately detecting brain tumors in low-brightness and low-contrast MRI images. The method we proposed can used to assist the clinicians in the proper and timely detection of brain tumors, leading to improved patient outcomes. Further research could explore the use of our model in a clinical setting and its integration into existing medical imaging software.

References

1. Ranjbarzadeh R, Kasgari AB, Ghoushchi SJ, Anari S, Naseri M, Bendechache M. Brain tumor segmentation based on deep learning and an attention mechanism using MRI multi-modalities brain images
2. https://braintumor.org/brain-tumors/
3. Kettritz U. Minimally invasive biopsy methods—diagnostics or therapy? Personal opinion and review of the literature
4. Harel NY, Strittmatter S-M. Functional MRI and other non-invasive imaging technologies: providing visual biomarkers for spinal cord structure and function after injury author links open overlay panel
5. Mitusova K, Peltek OO, Karpov TE, Muslimov AR, Zyuzin MV, Timin AS. Overcoming the blood–brain barrier for the therapy of malignant brain tumor: current status and prospects of drug delivery approaches
6. LaViolette PS, Daun MK, Paulson ES, Schmainda KM. Effect of contrast leakage on the detection of abnormal brain tumor vasculature in high-grade glioma
7. Wang W, Wu X, Yuan X, Gao Z (2020) An experiment-based review of low-light image enhancement methods. IEEE Access 8:87884–87917. https://doi.org/10.1109/ACCESS.2020.2992749
8. Menze BH et al (2015) The multimodal brain tumor image segmentation benchmark (BRATS). IEEE Trans Med Imaging 34(10):1993–2024. https://doi.org/10.1109/TMI.2014.2377694
9. Abd-Ellah MK, Awad AI, Hamed HFA, Khalaf AAM (2019) Parallel deep CNN structure for glioma detection and classification via brain MRI images. In: 2019 31st international conference on microelectronics (ICM), Cairo, Egypt, pp 304–307. https://doi.org/10.1109/ICM48031.2019.9021872

10. Deepak S, Ameer PM (2019) Brain tumor classification using deep CNN features via transfer learning. Comput Biol Med 111:103345. ISSN 0010-4825. https://doi.org/10.1016/j.compbiomed.2019.103345
11. Gudhe N, Behravan H, Sudah M, Okuma H, Vanninen R, Kosma V-M, Mannermaa A (2021) Multi-level dilated residual network for biomedical image segmentation. Sci Rep 11. https://doi.org/10.1038/s41598-021-93169-w
12. Foeady AZ, Novitasari DCR, Asyhar AH, Firmansjah M (2018) Automated diagnosis system of diabetic retinopathy using GLCM method and SVM classifier. In: 2018 5th international conference on electrical engineering, computer science and informatics (EECSI), Malang, Indonesia, pp 154–160. https://doi.org/10.1109/EECSI.2018.8752726
13. Wasule V, Sonar P (2017) Classification of brain MRI using SVM and KNN classifier. In: 2017 third international conference on sensing, signal processing and security (ICSSS), Chennai, India, pp 218–223. https://doi.org/10.1109/SSPS.2017.8071594

Implementation of IoT and Data Analytics in Smart Agricultural System

K. T. Rashmi, Sanjeev S. Tigadi, K. S. Nikhil, Dharnesh Narayana Swamy, and S. Darshan Nagesh

Abstract Headway in innovation is impacting the world at extraordinary speed. New techniques and frameworks are being created all through the world in various application regions. This headway can demonstrate as a help to agriculture industry to fulfill the rising needs of food and feed all over the planets. This paper proposes a non-contact plant development observing framework utilizing Machine Learning and IOT. A mechanized water system framework was created for improved utilization of water in farming. The framework has a conveyed remote organization of soil wetness, mugginess, and temperature sensors set at the root zone of plants. Arduino microcontroller gets the sensor data and sets off the result gadgets. An instrument was created relying upon the edge upsides of the dirt dampness, mugginess and temperature sensor, and sort of plant. Customized into Arduino to control the temperature of water took care of to the water system field. Sensors and Arduino are associated with one another with wire. As a result of low-energy utilization and cost, the framework can possibly be helpful in water-restricted geologically disconnected regions.

Keywords Machine Learning · Internet of Things · Smart cultivating · Data analytics · ThingSpeak · Arduino · Pump

1 Introduction

In India, agriculture has its own set of experiences. Assuming we check the measurements of nations, we can say that India positions second in ranch yield. About 13.7% of the Gross Domestic Product (GDP) was represented agriculture and associated areas in 2013; around half of the labor force relies upon cultivating [1]. The commitment of horticulture to India's GDP is consistently diminishing with the nations less consideration over ranchers. Still farming is the broadest area and assumes a huge part in the general social texture of India. Today auto-portable businesses, plants,

K. T. Rashmi (✉) · S. S. Tigadi · K. S. Nikhil · D. N. Swamy · S. Darshan Nagesh
Department of ECE, Don Bosco Institute of Technology, Bengaluru, India
e-mail: rashmikt@dbit.co.in

V. K. Gunjan and J. M. Zurada (eds.), *Proceedings of 4th International Conference on Recent Trends in Machine Learning, IoT, Smart Cities and Applications*, Lecture Notes in Networks and Systems 873, https://doi.org/10.1007/978-981-99-9442-7_33

and production lines are growing their regions in India. One side these are gainful as far as we are concerned as it gives work and items, and at the opposite side, these are contaminating our current circumstance. Furthermore, has turned into a serious test for an unnatural weather change [2]. An Earth-wide temperature boost is influencing our environment and we face questionable change in our current circumstance which harms our agrarian yield. As told in above section, in India around half of individuals relies upon agribusiness. Also, an Earth-wide temperature boost is one of the serious issues for them. As of late, India confronted a major dry spell in Vidarbha area of Maharashtra. This declared by the government following one year was no more. A portion of the ranchers was endured a lot by this dry spell even some did self-destruction [3]. A portion of the harvests of explicit season cannot get expected climatic circumstances because of questionable change in our current circumstance because of nursery impact. A portion of the Industries is contaminating waterways by catapulting hurtful synthetics in it. Green plants do not develop because of lacking watering and some because of deficient temperature and so forth. Green house is one and only arrangement against questionable environmental change or an Earth-wide temperature boost. Green house assists our ranchers with establishing a fake climate for keeping up with required temperature, mugginess, and light. Today, greenhouse frameworks are accessible in market as client wants it to work like. In any case, the issue is client needs to notice it persistently to check regardless of whether it is working appropriately [4]. There is no successful observing of greenhouse frameworks to screen the greenhouse environment conditions at unbounded distance.

As our nation is a farming focused country and the rate at which water assets are exhausting is a perilous danger, thus there is a need of a savvy and productive method of water system [5]. This task is intended to foster a programmed water system framework which controls the siphon on detecting the dampness content of the soil. In the field of agribusiness, utilization of the legitimate strategy for water system is significant. The principal benefit of this task is to decrease human intercession yet guarantee appropriate water system. This component contains an Arduino that gets the sign of fluctuating dampness content in the dirt through the sensor game plan. When the regulator gets this sign, it creates a result that drives a hand-off for working the water siphon. The situation with the dirt is sent to Arduino. The detecting course of action is made by utilizing soil dampness sensor. The framework has an organization of soil dampness sensor and a temperature sensor set in the root zone of the plants.

The framework has an organization of soil wetness sensor and a temperature sensor put in the root zone of the plants. A microcontroller handles sensor data, triggers siphon, and communicates information to the Arduino [6]. A calculation was created with edge upsides of temperature and soil wetness that was modified into a microcontroller to control water amount.

2 Literature Analysis

Rice yield assessment a pixel scale utilizing vegetation lists from automated elevated frameworks [7]. Ideal and precise forecast of rice yield data is firmly connected with individuals' occupation, which has been appended incredible significance by all degrees of government. Satellite remote detecting gives the likelihood to huge scope crop yield assessment, yet they are generally restricted by spatial and phantom goal. Automated Aerial Vehicles' (UAVs) remote detecting with hyper unearthly sensors can get high spatial fleeting goal and hyper otherworldly pictures on request. By and large, time-series Vegetation Indices (VIs) are utilized for assessing grain yield.

Utilization Deep Neural Network Prediction of Crop Yield: A Case Study of Soybean Yield in Lauderdale County, Alabama, USA [8]. Total populace is continually expanding, and having adequate yield production is fundamental. Observing harvest development and yield assessment is vital for the financial improvement of a country. The expectation of harvest yield straightforwardly affects public and global economies and assume significant part in the food the board and food security. Profound learning acquires significance on crop checking, crop type grouping, and harvest yield assessment applications with the new advances in picture arrangement utilizing profound Convolutional Neural Networks.

Crop Yield Prediction and Fertilizer Utilization Efficiency [9]. As a horticulture country, India's economy is heavily reliant on agribusiness yield development and agro industry products. Information mining is a new research area in crop yield analysis. In horticulture, yield expectation is critical. Any rancher is interested in knowing how much yield he can expect. Examine the various related attributes such as area, pH value, and alkalinity of the still up in the air. Along with it, the level of supplements such as Nitrogen (N), Phosphorus (P), and Potassium (K).

Forecast of significant harvest yields of Tamil Nadu utilizing K-implies and Modified KNN [10]. Horticulture is the chief wellspring of job for in excess of 40% of the number of inhabitants in this state. As per the Food and Agricultural Organization (FAO) researchers, somewhere in the range of 2010 and 2050, the total populace will increment by 33%. The interest for crop creation will increment by 60% higher than the ongoing creation. Thus, expectation assumes a significant part to figure out the interest of harvest creation for boosting the yield. For that in this paper, we propose an expectation technique for the significant yields of Tamil Nadu utilizing K-implies and Modified K Nearest Neighbor (KNN) [11]. MATLAB and WEKA are utilized as the instrument for bunching and order separately. The number outcome shows that our technique is superior to customary information mining approach.

Rice crop yield forecast in India is utilizing support vector machines [12]. Food creation in India is to a great extent reliant upon oat crops including rice, wheat, and different heartbeats. The maintainability and efficiency of rice developing regions are reliant upon reasonable climatic circumstances. Fluctuation in occasional environment conditions can make hindering difference, with episodes of dry season diminishing creation. Growing improved strategies to anticipate crop efficiency in various climatic circumstances can help rancher and different partners in better dynamic with regard to agronomy and yield decision [13]. AI procedures can be utilized to further develop forecast of harvest yield under various climatic situations.

Crop System, a Simulation Model of Cropping Systems: Water/Nitrogen Budgets and Crop Yield [14]. Water and nitrogen are two essential resources for growing a harvest in horticulture. However, their administration cannot be investigated independently of climate, soil qualities, field hydrology, crop attributes, crop turn, and other factors. This paper depicts the Crop System's water, nitrogen, and yield development components, as well as a detailed trimming frameworks reproduction model, and provides initial confirmation of these components.

3 Proposed System

This framework is made up of a variety of different equipment and programming parts. The equipment includes various sensors to measure soil moisture, light levels, and other factors. The product includes an Android-based application that interacts with the Arduino board and other equipment. The application includes signs and a dataset with readings from sensors, which are embedded using the equipment. The improvement of the water system framework using remote organization is an answer for improving water protection and progress in water system process. This study tries to improve water management on farmland by checking the water level in the dirt compared to the plants being grown and the amount of water adapted to reproduce the effect of rainfall. The primary purpose of the Analysis Phase is to gather necessary information as shown in Fig. 1.

The method involved with get-together prerequisites is generally more than essentially asking the clients what they need and recording their responses on paper. Contingent upon the intricacy of the application, the interaction for social occasion prerequisites has an obviously characterized course of its own. This cycle comprises a gathering of repeatable cycles that use specific strategies to catch, record, convey, and oversee necessities. Framework configuration is the most common method of characterizing the engineering, parts, modules, connection points, and data for a framework to meet specified requirements. Frameworks could be used to improve items. There is a crossover with the fields of framework review, framework design, and framework design. If the more comprehensive topic of item improvement "confuses the viewpoint of presentation, planning, and assembling into a single way of dealing with the item's progress", then the plan at this point is to demonstrate that the promotion has taken the data and made a plan for the item to be conducted. The Framework Configuration process is used to create frameworks that meet the specific requirements of a client. Until the 1990s, frameworks' configuration had a key role in the information handling industry. The 1990s saw a gradual normalization of equipment and programming, which made it possible to build reliable measurement frameworks.

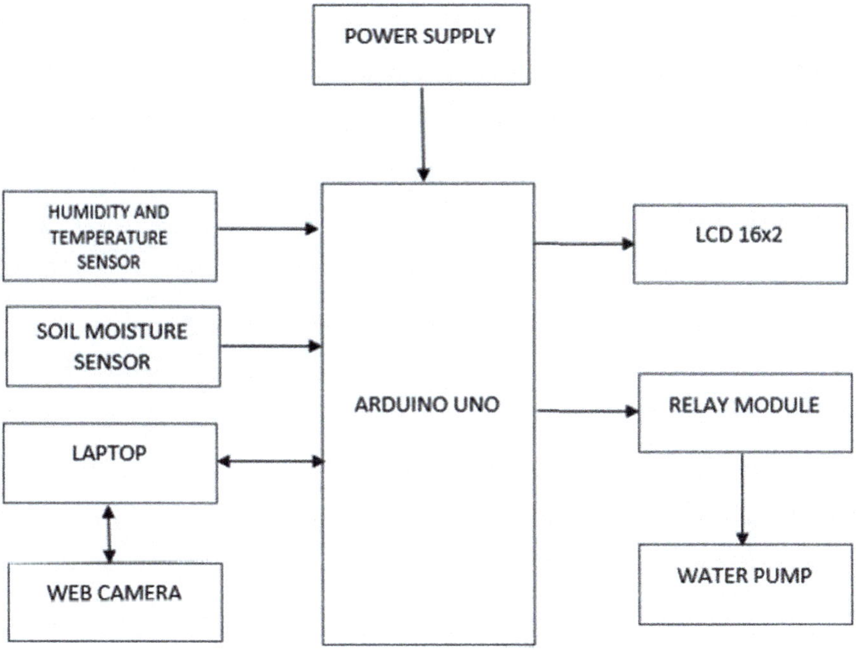

Fig. 1 Methodology flow diagram

The growing importance of programming on traditional stages has improved computer programming discipline. Object-based investigation and planning techniques are quickly becoming the most widely used strategies for the planning of PC frameworks The UML has evolved into the standard language for object-oriented analysis and planning. It is commonly used for demonstrating programming frameworks, but it is also being used for high-level non-programming frameworks and associations. Framework configuration is a critical stage in the programming improvement process. The plan's motivation is to design the solution to a problem based on the required documentation. As a result, the plan of the undertaking is the first stage in the solution to the problem. The framework's design is perhaps the most fundamental variable influencing the nature of the product. The plan stage's goal is to deliver a general product plan. It expects to sort out the modules that should be in the framework in order to effectively meet all of the framework prerequisites. The plan will detail this plethora of modules, their relationships with other modules, and the ideal outcome from each module as shown in Fig. 1.

Fig. 2 Arduino Uno

4 Methodology

4.1 Arduino Uno

The ATmega328-based Arduino Uno is a microcontroller board. A 16 MHz resonator, a USB connection, a power jack, and a reset button are among the 14 input/output pins on the device. This is a kit that includes everything needed to connect a microcontroller to a computer via USB or power it via an AC-to-DC connector or battery, as shown in Fig. 2.

4.2 Moisture Sensor

In this sensor, we use two probes that are inserted into the soil based on moisture levels. Analog output voltages will range from 0.60 to 12 V. Soil dampness sensors measure the amount of water in the soil. A dirt dampness test is made up of several dirt dampness sensors. A frequency space sensor, similar to a capacitance sensor, is a common type of soil dampness sensor used in business. Another sensor, the neutron dampness check, makes use of water's arbitrator properties for neutrons as shown in Fig. 3.

4.3 DHT 11 Sensor

A mugginess sensor (also known as a hygrometer) detects, measures, and reports humidity as well as air temperature. Relative stickiness is defined as the proportion

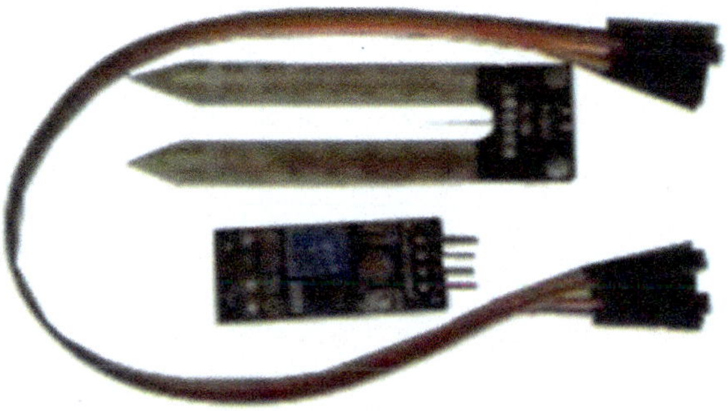

Fig. 3 Soil moisture sensor

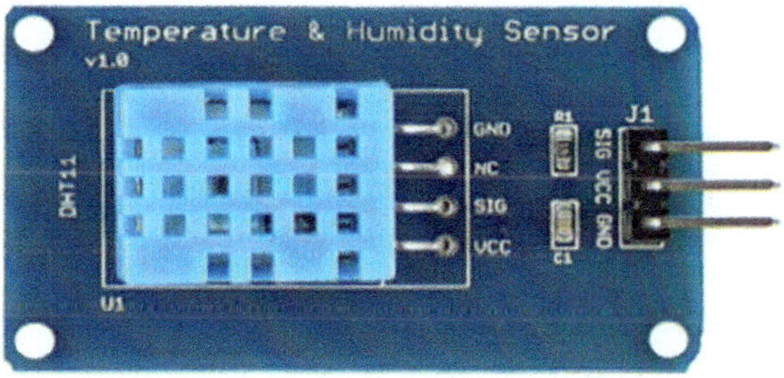

Fig. 4 DHT 11 sensor

of dampness in the air to the most elevated measure of dampness at a given air temperature. When looking for solace, relative mugginess becomes an important factor as shown in Fig. 4.

4.4 Light-Dependent Resistor

Light-dependent resistors, LDRs, or photograph resistors are electronic parts that are many times utilized in electronic circuit plans where recognizing the presence or the degree of light is vital. LDRs are altogether different to different types of resistor like the carbon film resistors, metal oxide film resistors, and metal film resistors, and similar components are commonly used in other electronic plans. They are expressly designed for their light responsiveness and the resulting obstruction adjustment. These electronic components are known by a variety of names, including light ward resistor, LDR, photograph resistor, and photograph cell, photocell, or photoconductor as shown in Fig. 5.

Fig. 5 LDR

4.5 pH Meter

Soil pH meters are gadgets used to gauge the causticity or alkalinity of a dirt. They work by estimating the hydrogen particle movement and this is communicated through the potential for hydrogen or 'pH'. The pH scale ranges from 0 to 14, with 0 being extremely acidic, 7 being neutral, and 14 being neutral as shown in Fig. 6.

4.6 Liquid Crystal Display

A liquid crystal display units are the most important output devices in embedded projects. It gives the user a lot of versatility because he can display whatever data he wants on it. However, there should be a appropriate approach to use LCD interface. In LCD, a driver/controller is employed. It is stupid that it does not know how to communicate with a microcontroller as shown in Fig. 7.

4.7 Relay

This is a little and simple to utilize 1 channel transfer 12 V power supply board. It can be used to control one 240 V power device directly from Arduino, Raspberry Pi, and

Fig. 6 pH meter

Fig. 7 LCD panel

other microcontrollers or low voltage circuits. The board utilizes a great hand-off, which can deal with a limit of 10 A/250 V AC or 15 A/125 V AC as shown in Fig. 8.

4.8 Water Pump

Micro Submersible Pump DC 3–6 V Miniature DIY project for a mini water pump for a fountain garden or a mini water flow system. This is a low-cost, small-size Submersible Pump Motor that runs on a 3–6 V power supply. With an extremely low current usage of 220 mA, it can take up to 120 L per hour. Connect the cylinder line to the engine outlet, submerge it in water, and turn it on. Check to see if the water level is consistently higher than the engine. Running through may cause the engine to overheat, as well as causing disturbance as shown in Fig. 9.

Fig. 8 Relay module

Fig. 9 DC pump

4.9 *Software Design*

In the high-level plan, the proposed practical and non-practical prerequisites of the software are portrayed. By and large answer for the design is created which can deal with those necessities. This section includes the accompanying thought. Plan and thought Data stream chart DESIGN CONSIDERATION. There are a few plan contemplations gives that should be tended to or settled prior to getting down planning a total answer for the framework. The information stream outline in a flow diagram is a graphical representation of the progression of information through a data framework. DFD is exceptionally helpful in figuring out a framework and can be effectively utilized during examination. A DFD shows the progression of information via a framework. It considers a framework to be a capability that converts contributions to desired yields. Any mind-boggling framework will not play out this change in a single advance, and an information will frequently go through a progression of changes before it becomes the result. With an information flow sketch, customers can imagine how the framework will work, the framework will achieve and how the framework will run, old framework information flow diagrams can be created and contrasted and another framework information flow diagram to make correlations with the execution a more effective framework. Information stream outlines can be used as a visual representation of how information entered into a project will ultimately impact the construction of the entire framework. Area beneath each module's information stream chart makes sense of the overall module information.

Information Collection: In this module, datasets are gathered from different sources like GKVK, DES, agmarknet.gov.in, and krishimaratavahini. The datasets incorporate data like temperature, precipitation, value, region, creation, and yield of the past 5 years (2013–2018). Figure 10 shows the data collection flow diagram.

Information Preprocessing: Data pre-handling incorporates eliminating of the undesirable credits from our datasets. Highlight extraction is finished to separate just the characteristics that influence the development of the plant as shown in Fig. 11.

Plant Growth Monitoring Using Yolo

A plant was heedlessly picked and the proposed framework was executed to screen the development of the plant under normal regular circumstances as showed up in the figures. The banana plant was set before the framework a distance away of 15 cm

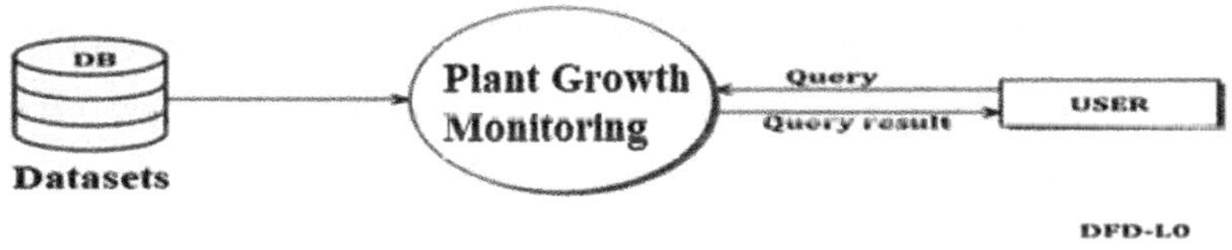

Fig. 10 Data collection flow diagram

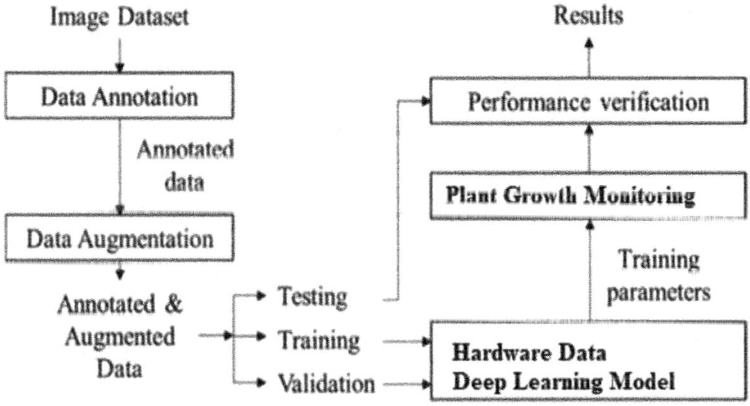

Fig. 11 Data flow diagram

from the foundation of the framework. The framework was changed to parcel the space in a system of 50 × 40 with a phase size of 2 cm. The framework was reserved to require two discernments in a day at a break of 5 h and to store the data in the internal memory of the framework.

YOLO, in a solitary look, takes the whole picture and predicts for these containers the jumping box facilitates and class probabilities. YOLO's most prominent benefit is its extraordinary speed, it is incredibly quick, and it can deal with 45 casings each second. Among the three variants of YOLO, variants 3 and 5 are the quickest and more exact regarding recognizing little items. In the proposed calculation, YOLO comprises absolute 106 layers. The engineering is comprised three unmistakable layer structures. The lingering layer which, first and foremost, is shaped when enactment is effectively sent to a more profound layer in the brain organization. In a remaining arrangement, results of layer 1 are added to the results of layer 2. Second is the recognition layer which performs discovery at three distinct scales or stages. Size of the lattices is expanded for recognition. Third is the up-inspecting layer which expands the spatial goal of a picture. Here picture is up inspected before it is scaled. Likewise, connection activity is utilized, to link the results of past layer to the current layer. Option activity is utilized to add past layers. In the figure, the pink hued blocks are the lingering layers, orange ones are the identification layers, and the green are the up-testing layers. Identification at three distinct scales is as displayed in Fig. 12.

YOLO predicts three distinct sizes of expectation. The location layer is utilized to recognize include guides of three distinct sizes, with steps 32, 16, 8 separately. This implies that identifications are made on sizes of 13 × 13, 26 × 26, and 52 × 52 with a contribution of 416 × 416.

The working of Yolo is referenced underneath.

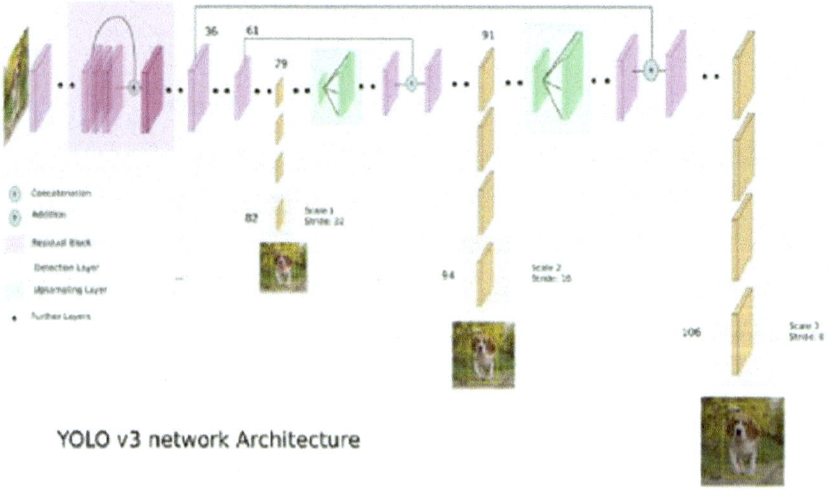

Fig. 12 YOLO v3 system

Darknet: This calculation is executed utilizing an open-source brain network structure, i.e., Darknet which was created in C Language and CUDA innovation to deliver expedient estimations on a GPU important for continuous expectations.

- DNModel.py: Darknet Model document is a PC vision code utilized for building the model utilizing the setup record and it attaches each layer.
- Util.py: Contains every one of the equations utilized.
- imageprocees.py: Required to play out the picture handling task. It takes every one of the info pictures to resize them and perform Up-inspecting additionally performs translate work.
- detect.py: The primary code which is rushed to perform object recognition. This code utilizes every one of the previously mentioned documents to perform object location and carries out every one of the roles as per the YOLO idea.

5 Results and Discussion

The sensor readings are uploaded in the ThingSpeak cloud platform and data analysis is performed by comparing the Comma Specified Values' (CSVs) data with the target plant's height and width. The final output shows us which crop can be grown in that particular environment. The height and width are measured using a webcam which works on YOLO principle. In the ThingSpeak cloud platform, API key is generated, which is called as write API key. This API key is fed in the Arduino UNO board by which we receive the sensor data in the cloud as shown in Fig. 13.

In ThingSpeak platform, we can see the real-time data of all the sensors used in this project for example as shown in Figs. 14 and 15.

Fig. 13 Web camera detection

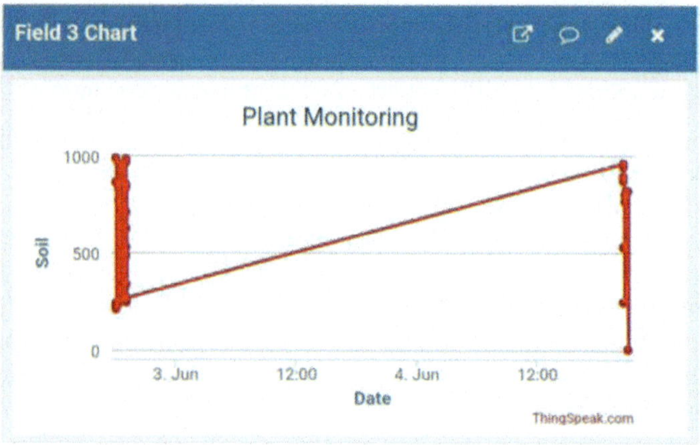

Fig. 14 Soil moisture readings

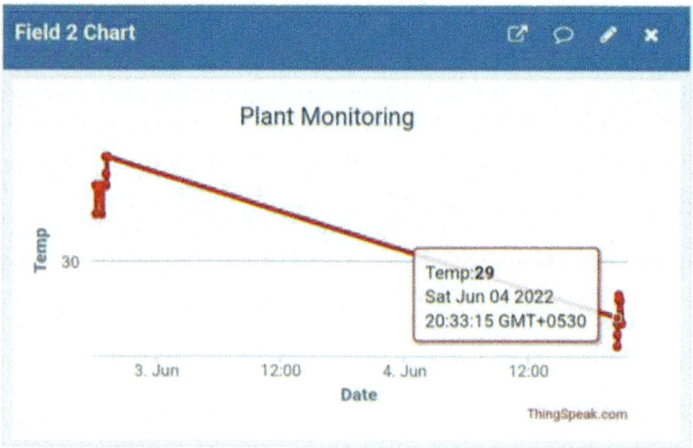

Fig. 15 Temperature readings

6 Conclusion

Image processing is the use of a great deal of methods and estimations to a mechanized picture to research, improve, or smooth out picture credits. Along these lines, quite possibly the main trouble is the shortfall of suitable checking and control frameworks for capable developing. The standard objective of this proposed work is to develop a plant observing system using picture handling. Consequences can be assumptions which gives an improved result as the difference with various computations. As the primary point of convergence of this application is straightforward, this application is arranged so it upholds the Multi-Lingual thought. This application is valuable for farmer and the office where they are and can without any remarkable stretch secure their Plants and there will be an addition in the development of creation.

References

1. Gandhi N et al (2016) Rice crop yield forecasting of tropical wet and dry climatic zone of India using data mining techniques. In: IEEE international conference on advances in computer applications (ICACA)
2. Eswari KE, Vinitha L. Crop yield prediction in Tamil Nadu using Bayesian network. Int J Intell Adv Res Eng Comput 6(2). ISSN: 2348-2079
3. Mishra S, Paygude P, Chaudhary S, Idate S. Use of data mining in crop yield prediction. In: IEEE Xplore compliant-part number: CFP18J06-ART. ISBN: 978-1-5386-0807-4; DVD part number: CFP18J06DVD. ISBN: 978-1-5386-0806-7
4. Chlingaryana A, Sukkarieha S, Whelanb B (2018) Machine learning approaches for crop yield prediction and nitrogen status estimation in precision agriculture: a review. Comput Electron Agric 151:61–69
5. Patil D et al (2017) Rice crop yield prediction using data mining techniques: an overview. Int J Adv Res Comput Sci Softw Eng 7(5)
6. Shakil Ahamed ATM, Mahmood NT, Hossain N, Kabir MT, Das K, Rahman F, Rahman RM. Applying data mining
7. Techniques to predict annual yield of major crops and recommend planting different crops in different districts in Bangladesh. 978-1-4799-8676-7/15/$31.00 copyright 2015 IEEE SNPD 2015, 1–3 June 2015
8. Menaka K, Yuvaraj N (2016) Indian J Innov Dev 5(12), Oct 2016. ISSN (online): 2277-5390, ISSN (Print): 22775382
9. Dahikar SS, Rode SV (2014) Machine learning by Tom M. Mitchell. Agricultural crop yield prediction using artificial neural network approach. Int J Innov Res Electr Instrum Control Eng 2(2). ISBN: 0070428077
10. Ramesh D, Vishnu Vardhan B. Analysis of crop yield prediction using data mining techniques. Int J Res Eng Technol. EISSN: 2319-1163
11. Medar RA (2014) A survey on data mining techniques for crop yield prediction. Int J Adv Comput Sci Manage Stud 2(9). ISSN: 2231-7782
12. Pooja MC et al (2018) Implementation of crop yield forecasting using data mining. Int Res J Eng Technol (IRJET), p-ISSN: 2395-0072
13. Jeong JH et al (2016) Random forests for global and regional crop yield predictions. PLOS-One. https://doi.org/10.1371/journal.pone.0156571, 3 June 2016
14. Majumdar J, Naraseeyappa S, Ankalaki S (2017) Analysis of agriculture data using data mining techniques: application of big data. Journal

IoT-Based Healthcare System: A Blockchain Approach

S. Komal Kour, Guradi Vinay Reddy, N. V. S. Sravani, and T. Adilakshmi

Abstract This study introduces a blockchain-based Internet of Things (IoT)-based healthcare system. By utilizing the immutable and decentralized characteristics of blockchain technology, the suggested system addresses the issues of data security, privacy, and interoperability in health care. By integrating IoT devices with blockchain, the system ensures secure and trusted management of healthcare data, enabling efficient data exchange among stakeholders. The paper presents the architecture, components, and mechanisms of the blockchain-based IoT healthcare system, highlighting its potential benefits and future directions.

Keywords Internet of Things · Health care · Blockchain · Data security

1 Introduction

The integration of Internet of Things (IoT) devices in health care has ushered in a new era of patient care and data management. These interconnected devices, ranging from wearable health trackers to remote monitoring systems, generate vast amounts of valuable healthcare data. However, this proliferation of data brings forth significant challenges in terms of data security, privacy, and interoperability. Traditional centralized systems have proven inadequate in addressing these challenges effectively, leading to a growing interest in leveraging blockchain technology to enhance the security and efficiency of IoT-based healthcare systems. Blockchain, the distributed ledger technology underlying cryptocurrencies like Bitcoin, offers a decentralized and tamper-proof platform for securely managing data and transactions. By design, blockchain provides immutability, transparency, and data integrity, making it an

S. Komal Kour (✉) · G. V. Reddy · N. V. S. Sravani · T. Adilakshmi
Department of Computer Science and Engineering, Vasavi College of Engineering, Hyderabad, India
e-mail: komalkour@staff.vce.ac.in

T. Adilakshmi
e-mail: hodcse@staff.vce.ac.in

V. K. Gunjan and J. M. Zurada (eds.), *Proceedings of 4th International Conference on Recent Trends in Machine Learning, IoT, Smart Cities and Applications*, Lecture Notes in Networks and Systems 873, https://doi.org/10.1007/978-981-99-9442-7_34

ideal solution for healthcare systems that handle sensitive patient information. Blockchain enables secure data sharing and collaboration among various stakeholders, including healthcare providers, researchers, and patients, while ensuring privacy and maintaining control over personal health.

In the proposed blockchain-enabled IoT healthcare system, IoT devices collect and transmit real-time patient health data, such as vital signs, activity levels, and medication adherence, to the blockchain network. In a transaction record (Fig. 1), the "address" and "balance" columns serve to display the addresses of the various blocks generated during the transactions. The "address" column shows the specific addresses involved in each transaction, while the "balance" column indicates the remaining balance after each transaction has been completed. This network, comprised interconnected nodes, records and validates the data in a transparent and tamper-proof manner. The decentralized nature of the blockchain ensures that no single entity has complete control over the data, reducing the risk of data breaches and unauthorized access. To ensure data privacy, encryption techniques can be employed, allowing only authorized parties to decrypt and access sensitive healthcare information. Access control mechanisms, implemented through smart contracts, govern the permissions and privileges of different participants within the blockchain network. These mechanisms enable patients to maintain control over their own health data and grant access to healthcare providers and researchers on a need-to-know basis. Interoperability is another critical aspect of the blockchain-enabled IoT healthcare system. Healthcare data from various sources, such as electronic health records (EHRs) and IoT devices, often reside in disparate systems and formats. Blockchain can act as a unifying platform that enables smooth data exchange and interoperability between various healthcare systems, overcoming the existing challenges of data silos and incompatible data formats. In this paper, we propose a comprehensive exploration of the blockchain-enabled IoT healthcare system, discussing its architecture, components, and mechanisms. We will delve into the technical aspects of blockchain integration, including smart contract development, data encryption, access control, and interoperability protocols. Furthermore, we will address security considerations, ethical implications, and potential use cases of this innovative solution. By leveraging blockchain technology within the IoT healthcare domain, we aim to enhance data security, privacy, and interoperability, ultimately improving patient care, enabling groundbreaking research, and revolutionizing the healthcare industry as a whole.

2 Literature Review

Blockchain-enabled healthcare systems, integrated with the Internet of Things (IoT), have emerged as a promising solution to address critical challenges in the healthcare industry such as data security, privacy, and interoperability. Several research papers and publications have explored the potential of this integration to revolutionize healthcare data management.

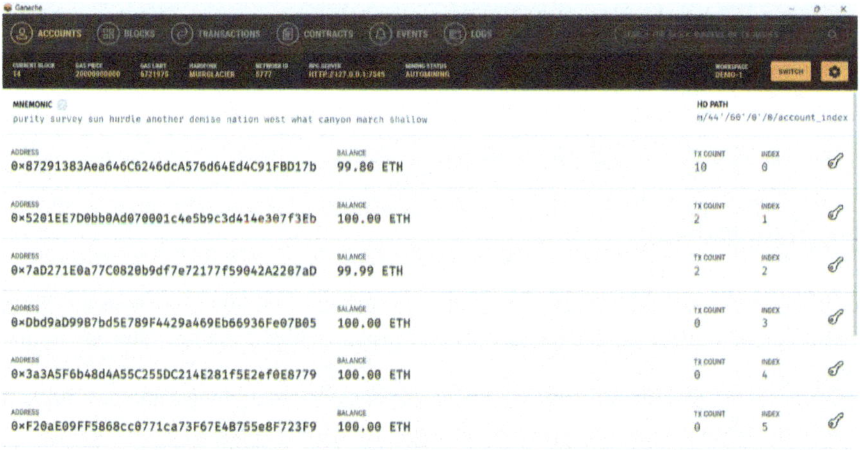

Fig. 1 Pictorial representation of transaction records

Providing a comprehensive review of blockchain technology in healthcare, Johnson [1] emphasizing its benefits in enhancing Secure data, streamlining patient record management, and improving interoperability. Gupta and Sharma [2] focus on the vital role of blockchain in securing IoT-enabled healthcare systems, proposing a solution that enhances security and data integrity, particularly in protecting sensitive medical data. Chen et al. [3] address the critical issue of privacy-preserving data sharing in blockchain-based healthcare systems. They suggest a structure that ensures private and safe data sharing across healthcare stakeholders, with a big focus on protecting patient privacy. The potential of blockchain in health care is still untapped, according to Schatsky et al. [4], discussing its applications in supply chain management, patient data sharing, and clinical trials. They advocate for broader adoption of blockchain to fully realize its benefits in the healthcare domain.

Dhillon [5] provides a non-technical introduction to blockchain in health care, outlining its potential impact on enhancing data security, improving interoperability, and empowering patients. Shrestha et al. [6] delve into various blockchain applications in health care, including patient data management, the medical supply chain, and healthcare research. They discuss both the benefits and limitations of implementing blockchain in healthcare contexts. Xu et al. [7] focus on health data management and privacy preservation using blockchain technology. They propose a framework that ensures secure and private healthcare data sharing while addressing the associated challenges and potential solutions. Mirsky et al. [8] discuss the revolutionary potential of blockchain in healthcare information sharing. The paper highlights benefits like decentralized data storage, secure data exchange, and data integrity, while also addressing potential challenges and regulatory considerations for implementing blockchain in health care.

In summary, the reviewed literature underscores the promising potential of blockchain-enabled healthcare systems integrated with IoT. These systems aim to overcome crucial data-related challenges and ensure secure and trusted data exchange among healthcare stakeholders. However, it is evident that further research and widespread adoption are essential to fully unlock the potential of this transformative technology in the healthcare sector.

3 Methodology

In this section, we will cover system design and requirements, blockchain implementation and integration, data security and privacy enhancements, interoperability, and performance evaluation. We will analyze the existing healthcare infrastructure and IoT devices to identify requirements and design a system architecture. We will deploy a suitable blockchain platform, develop smart contracts, and integrate IoT devices with the network. Encryption techniques and access control policies will be implemented for data security. Interoperability standards and data exchange mechanisms will be established, and the system's performance will be evaluated through tests and validations.

3.1 System Design and Requirements

In this initial phase, we identify the specific requirements and objectives of the blockchain-enabled IoT healthcare system. We conduct a thorough analysis of the existing healthcare infrastructure, IoT devices, and data sources to understand the technical and functional aspects. This analysis helps us to design a system architecture that aligns with the unique needs of the healthcare domain. To enable seamless integration and interoperability across various healthcare systems and Internet of Things devices, we develop the data models, schemas, and communication protocols.

3.2 Blockchain Implementation and Integration

In this stage, we choose a blockchain platform that is appropriate for the needs of the healthcare system, such as Ethereum or Hyperledger. We take into account aspects like scalability, security, and consensus mechanisms when designing and deploying the blockchain network. Within the blockchain network, smart contracts are created and implemented to manage activities like access control, data validation, and secure transactions. We also implement integration mechanisms to connect IoT devices with the blockchain network, allowing secure and reliable transmission of healthcare data. The provided figure (Fig. 2) portrays the transaction history. The leftmost column presents the block number, followed by the date and time at which the transaction

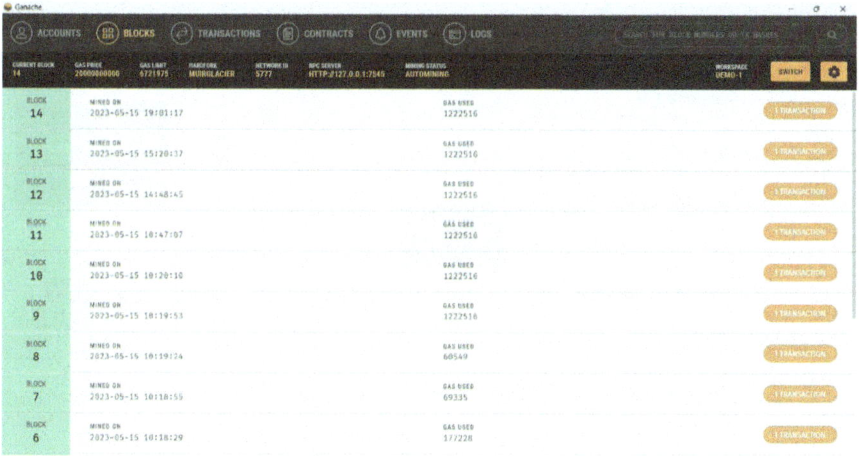

Fig. 2 Transaction and block details

was mined. The gas units consumed by each transaction are displayed, and on the rightmost side, the number of transactions is indicated.

3.3 *Data Security and Privacy Enhancements*

Data security and privacy are of paramount importance in healthcare systems. To address these concerns, we implement encryption techniques to ensure the confidentiality of sensitive healthcare data. Access control policies and permissions are defined within the blockchain network to restrict data access and maintain data integrity. To strengthen privacy protection, we investigate privacy-preserving methods including differential privacy and zero-knowledge proofs. In order to provide secure and permitted access to healthcare data, user authentication and identity management procedures are included into the system.

3.4 *Interoperability and Data Exchange*

One of the major issues facing healthcare systems is interoperability, given the diverse range of IoT devices and healthcare systems in use. To achieve seamless data exchange and interoperability, we establish standards and protocols for data exchange, such as HL7 FHIR. We develop data mapping and transformation mechanisms to enable compatibility and integration among different data sources. Data validation and verification processes are implemented to ensure the accuracy and

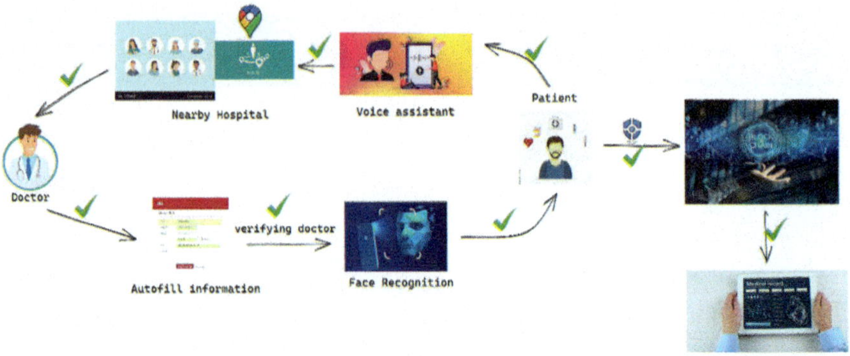

Fig. 3 Proposed model of IoT health care using blockchain

reliability of shared healthcare data. We explore the use of interoperability frameworks and protocols to facilitate efficient and secure data exchange between different healthcare systems.

3.5 Performance Evaluation and Validation

In this final phase, we evaluate the performance of the blockchain-enabled IoT healthcare system. We define performance metrics, such as response time, throughput, and resource utilization, to assess the system's efficiency and scalability. Experiments and tests are conducted under various load conditions and data volumes to measure the system's performance. The obtained results are analyzed to identify any bottlenecks or areas for improvement. We also validate the system's functionality and performance through pilot studies or real-world deployments in collaboration with healthcare providers. To assess the system's usability, efficacy, and potential to enhance healthcare outcomes, user feedback is gathered. The proposed model used for IOT in health care using blockchain is shown in the picture below (Fig. 3).

4 Implementation

The implementation of a blockchain-enabled IoT healthcare system involves several key steps. First, we start with system design and requirements analysis. This phase entails a comprehensive assessment of existing healthcare infrastructure and IoT devices to identify specific system requirements. This analysis forms the basis for designing a robust architecture that ensures smooth integration and interoperability among various healthcare systems and IoT devices. It also includes defining data models, schemas, and communication protocols to enable efficient data exchange and management.

Next, we move on to platform selection, where we evaluate blockchain platforms such as Ethereum based on factors like scalability, consensus mechanisms, and smart contract support. Once the platform is chosen, we configure it to align with project-specific needs. For local development and testing, tools like Ganache CLI come in handy, allowing us to set up a private blockchain network with custom parameters.

Developing smart contracts is a critical aspect of the implementation. These contracts encapsulate the business logic and rules governing interactions within the blockchain network. Using languages like Solidity and development environments like Remix, we create smart contracts to handle tasks like data validation, access control, and secure transactions. Deployment of the blockchain network follows, involving the deployment of smart contracts onto the chosen platform, making them accessible to system participants. Ganache CLI remains a useful tool for configuring the blockchain network during local development and testing.

Integration with IoT devices is another essential step, ensuring compatibility between these devices and the blockchain-enabled system. We establish secure communication channels between IoT devices and the blockchain network for the secure transmission of healthcare data. Robust encryption techniques are employed to protect sensitive data during transmission.

To maintain system integrity and security, we implement access control mechanisms. Smart contracts define access control policies and permissions, with cryptographic techniques used to verify participant identities and grant authorized access to the blockchain network. Wallets like MetaMask serve as secure means for managing user identities and interacting with the blockchain network.

Enhancing data security and privacy is paramount. We employ encryption algorithms like AES or RSA to encrypt sensitive healthcare data stored within the blockchain network. Secure storage mechanisms further bolster the protection of encrypted data. Approaches for protecting privacy, such as zero-knowledge proofs, may be applied to enhance patient data privacy while allowing for selective disclosure of information.

Interoperability remains a focus, ensuring seamless integration with existing healthcare systems. We define data exchange formats and protocols to facilitate compatibility and smooth data flow, often leveraging industry standards like HL7 FHIR. Additionally, we develop data mapping and transformation mechanisms to promote interoperability among diverse healthcare systems. To ensure system functionality and reliability, rigorous testing and validation processes are conducted. Unit tests scrutinize smart contracts for correctness and security, while integration tests assess the interaction between IoT devices and the blockchain network. Real-world deployments and pilot studies, often in collaboration with healthcare providers, validate the system's performance and functionality (Fig. 4).

Finally, performance optimization ensures the efficient operation of the blockchain-enabled IoT healthcare system in real-world scenarios. This comprehensive approach to implementation ensures that the system is technically robust and tailored to the specific needs of the healthcare industry.

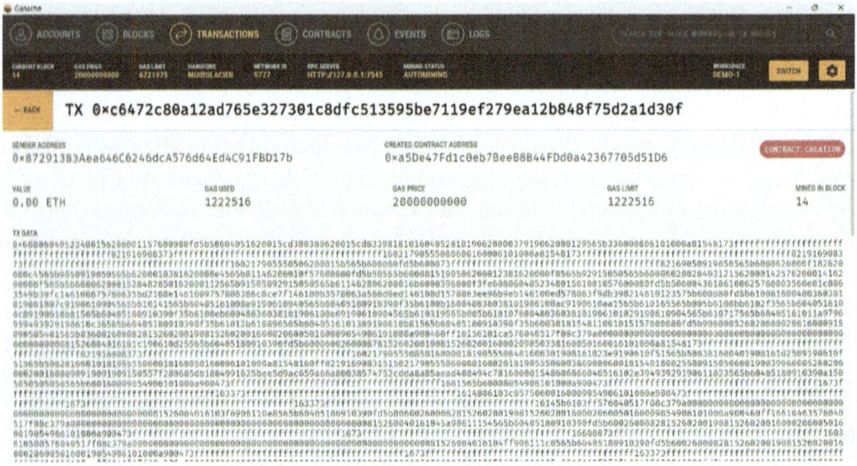

Fig. 4 Illustration of block details

5 Results

The successful implementation of the blockchain-enabled IoT healthcare system has ushered in a new era of healthcare data management. By seamlessly integrating various healthcare systems and IoT devices, the system has overcome historical barriers to data exchange and interoperability within the healthcare industry. Smart contracts have emerged as a critical tool in ensuring the integrity of healthcare data, enforcing access control, and facilitating secure transactions. This innovation not only enhances the trustworthiness of healthcare records but also streamlines administrative processes and reduces the risk of errors. Moreover, the robust data security and privacy measures, including advanced encryption techniques and access control mechanisms, provide a strong shield for sensitive patient information, instilling confidence in both patients and healthcare providers.

In addition to its technical prowess, the system has successfully established interoperability with existing healthcare systems through the adoption of standards like HL7 FHIR. This seamless integration allows for efficient data flow and communication between different healthcare providers and systems, ultimately leading to better-informed decisions and improved patient care. Rigorous testing and validation processes have further underscored the system's reliability and functionality, instilling confidence in its potential to revolutionize healthcare data management. Overall, the results of this implementation highlight the system's capacity to enhance the security, privacy, interoperability, and overall performance of healthcare systems, setting the stage for a future where healthcare delivery is more efficient, accurate, and patient-centered (Fig. 5).

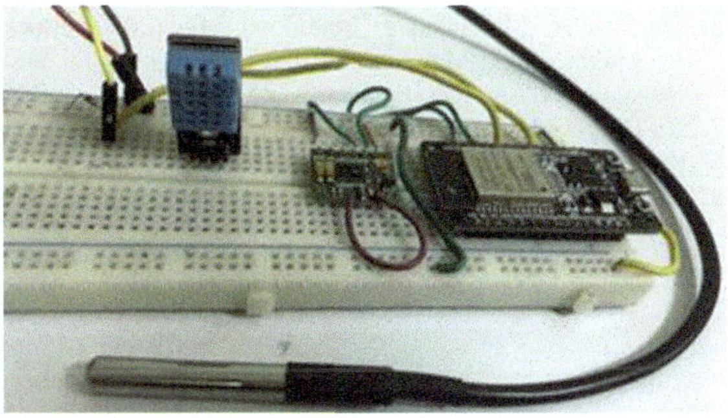

Fig. 5 IoT-designed healthcare system

6 Conclusion

The successful implementation of a blockchain-enabled IoT healthcare system holds tremendous promise for transforming healthcare data management and elevating the quality of healthcare services. By seamlessly integrating healthcare systems with IoT devices, we achieved smooth data exchange and enhanced interoperability. The use of smart contracts ensured data validation, access control, and secure transactions within the blockchain network. The system also demonstrated robust data security and privacy measures, including encryption techniques and access control mechanisms. We effectively established interoperability with existing healthcare systems, facilitating efficient data flow. Rigorous testing and validation processes confirmed the system's functionality and reliability. In summary, these results underscore the significant potential of blockchain and IoT technologies to enhance the security, privacy, interoperability, and performance of healthcare systems. The successful implementation of this system sets the stage for improved healthcare data management, streamlined processes, and enhanced patient care. As technology continues to advance, blockchain-enabled IoT healthcare systems are poised to play a pivotal role in shaping the future of healthcare delivery.

References

1. Johnson S (2018) Blockchain technology in healthcare: a comprehensive review. IEEE Trans Emerg Top Comput 6(2):245–256
2. Gupta A, Sharma B (2018) Securing IoT-enabled healthcare systems using blockchain technology. IEEE Internet Things J 5(3):2134–2141
3. Chenetal C (2019) Privacy-preserving data sharing in blockchain-based healthcare systems. IEEE Trans Dependable Secure Comput 16(4):671–683
4. Schatsky D, Muraskin J, Berk P (2017) Blockchain: the untapped potential for healthcare. Deloitte University Press
5. Dhillon R (2017) Blockchain in healthcare. In: Drescher D (ed) Blockchain basics: a non-technical introduction in 25 steps. Apress, pp 203–208

6. Shrestha A, Mahmood P, Mahmood MA (2020) Blockchain for healthcare applications. In: Mahmood A, Mahmood P (eds) Blockchain for enterprise applications and healthcare management. Springer, Berlin, pp 135–162
7. Xu D, Zhu M, Wei R (2020) Blockchain for health data management and privacy preservation. Springer, Berlin
8. Mirsky SH, Kashani N, Ransbotham SM (2018) Blockchain in healthcare: the next revolution in information sharing? In: Wüst K, Gervais A (eds) The blockchain and the new architecture of trust. Springer, Berlin, pp 225–250

Range-Free Anchor-Based Localization in WSN Using Optimized Support Vector Machine Derived by Teamwork Optimization Algorithm

V. Ch. Sekhar Rao Rayavarapu, Arunanshu Mahapatro, and R. Divya Kanti

Abstract In recent days, node localization became complex in Wireless Sensor Networks (WSNs). Localization is differentiated into two types, namely range-free as well as range-based models. Localization related to topological information and nodes is referred to as range-free approaches. In range-based approaches, distances among the nodes are validated. The scope of this paper is on range-free anchor-based localization. It is performed using the Teamwork Optimization Algorithm (TOA) based on anchor nodes. Through the range-free concept, the nodes are assigned randomly. The unknown node locations are formulated using the Optimized Support Vector Machine (SVM) with TOA. This SVM-based TOA technique is utilized to compute the unknown node coordination to guarantee localization accuracy via the multi-objective problem. The validation outcome displayed that the recommended approach attained an enhanced performance rate in Distance Vector Hop (DV-Hop) in terms of computational complexity, accuracy, and communication cost.

Keywords Range-free anchor-based localization · Wireless Sensor Network · Support vector machine · Teamwork optimization algorithm · Communication system

1 Introduction

WSN is a network that is developed by utilizing a huge number of sensor nodes and a self-arranged network. Each node presented in WSN has several capabilities to perform communication, calculation, storing, and sensing [1]. Most of the sensor nodes are generated by utilizing batteries, and also they are utilized in several

V. Ch. Sekhar Rao Rayavarapu (✉) · R. Divya Kanti
Department of ECE, LENDI Institute of Engineering & Technology, Vizianagaram, Andhra Pradesh, India
e-mail: sekharrao.81@gmail.com

A. Mahapatro
ETC Department, VSSUT Burla, Sambalpur, Odisha, India

V. K. Gunjan and J. M. Zurada (eds.), *Proceedings of 4th International Conference on Recent Trends in Machine Learning, IoT, Smart Cities and Applications*, Lecture Notes in Networks and Systems 873, https://doi.org/10.1007/978-981-99-9442-7_35

areas for monitoring the physical quantities. WSN applications are widely utilized in different areas like monitoring the environment, protecting the crop, military, health, underwater surveillance, and so on [2]. In monitoring the physical quantities, it is highly required to monitor the geographical position, because sensed data without the information of location is considered useless [3].

In WSN, attaining the node position is referred to as localization and also it became an essential key element [4]. In order to know when the localization is performed, accurate range and angle among the nodes information are essential. Node localization achieved in WSN approaches is classified as range-free and range-based localization [5]. In range-based localization approaches, it is important to recognize the distance accurately among the sensor nodes. The major feature of range-based approaches is a superior precision rate still, it always needs additional hardware for the implementation so it became so costly [6]. In range-free approaches, it only needs to understand the connection among the node and also they did not need any additional hardware. Still, the accuracy of localization in the range-based approach is minimal [7].

In recent days, DV-Hop is termed as a widely used range-free localization model, and it is easy to implement, but its precision rate is not enough for some applications [8]. Then, the anchor-free approaches are used to locate the unknown sensor node by connectivity data in the network regarding the location of the anchor node [9]. The mobile anchor node-based localization models are classified into two types such as usage of path-planning techniques and usage of mobility approaches. Most of the existing analyses are mainly focused to resolve the localization issues with single optimization approaches. But, later multi-objective techniques are developed and utilized to resolve complex problems [10]. So, it is highly essential for developing a new range-free anchor-based node localization model by considering the complexities presented in existing models. Several contributions of the developed range-free anchor-based node localization model are listed below.

- To develop a range-free localization model with anchor nodes based on machine learning and an efficient heuristic algorithm for enhancing the node localization performance in the WSN communication system.
- To integrate the localization approach named SVM-TOA to localize the unknown nodes with parameter optimization in SVM using suggested TOA for maximizing the localization accuracy and also lowers the communication cost and computational cost.
- To analyze the efficiency of the suggested range-free localization model over a different conventional framework.

The rest of the parts of the suggested model is elaborated as follows. The literature review of existing localization approaches is displayed in Phase II. The range-free localization using optimization approaches is represented in Phase III. The localization approaches based on optimized SVM are attained in Phase IV. The result and discussion as well as the conclusion part of the suggested model are presented in Phases V and VI, respectively.

2 Literature Work

2.1 Related Works

In 2013, Ou and He [11] have developed a path-planning technique, that effectively assured the path in mobile anchor nodes and effectively minimized the error of localization and their location. The approach named obstacle-resistant trajectory was designed for handling the obstacles presented in the sensing region. The efficacy rate of the recommended approach was observed over a sequence of validation with a network simulator. Thus, the suggested framework scored a better performance rate than the traditional framework in localizing the sensor node. In 2009, Mourad et al. [12] have initiated an approach for mobile ad hoc network-based self-localization. The major goal of the approach was to attain minimal memory and computation sources for the analysis. Here, error estimation for propagation was performed based on interval formulation in the state space model. Waltz approaches were applied for contracting the solution and also enhanced the position of the mobile node when validation was performed. Validation outcomes of the mobile node trajectory approach were fused effectively and achieved effectively better outcome rate.

In 2018, Alomari et al. [13] have recommended a novel path-planning scheme for localization in WSN. The developed approach assured that entire nodes can obtain the localization data and validate their position with a superior accuracy rate. The developed approach also has overcome the issues related to collinearity in terms of precision, a ratio of localization, and the path length in mobile anchors. In 2022, Sabal and Min [14] have introduced a static trajectory for the localization of WSN in mobile beacon nodes. The trajectory assured that the success ratio of localization was enhanced effectively for a minimal range of communication in the anchor node along with an improved localization precision rate. The developed approaches effectively have overcome the collinearity issue and also checked the unknown sensor node for receiving higher quality beacon position.

In 2023, Cao and Xu [15] have suggested an Optimum Anchor Nodes Subsets (OANS DV-Hop) approach. In the developed approach, every anchor node is localized with another anchor node. Then, an optimization approach was developed for generating an optimum subset combining other anchors. Later, revalidation was performed by applying an anchor node in OANS and broadcasting a new hop size to the unknown node. The fitness function for OANS was developed, and also optimization approach was applied to enhance the accuracy of localization. The experimental outcome displayed that recommended framework secured a better performance rate than the traditional framework in a different network environment

2.2 Problem Statement

Localization is considered a highly essential problem in WSN due to its need over a huge number of applications and also provides a superior precision rate to perform the analysis effectively, and also it required more memory resources for the mobile node. Path-planning algorithm [11] attains minimal localization error and also a superior percentage rate in the localized sensor node. But, it requires more computational time and also superior memory overhead. Waltz algorithm [12] describes about how to achieve robustness rate effectively in various path planning algorithms for estimating the mobile node location. Still, it did not resolve the environmental imperfection rate for attaining accurate outcomes. Static path-planning model [13] entire node has the capability for receiving information about localization, and also they validate their location with superior localization accuracy. Yet, it suffers from local optima issues and redundant points. Static trajectory [14] has minimal localization time than conventional models and effectively overcome the collinearity issue. But, it is highly complex to implement. OANS DV-Hop [15] can implement simply. But, it easily falls into overlapping issues. Several complexities attained in the existing localization model are needed to be resolved, and so, a new anchor-based localization approach in WSN is highly essential with several approaches.

3 Range-Free Anchor-Based Node Localization in WSN Using an Optimization Algorithm

3.1 Node Optimization and Multi-objective Function

To reduce the computational overhead effectively, an optimized node localization model is utilized. Initially, unknown nodes are considered for localization and also their Euclidean distances are calculated. Then, the unknown node location is analyzed and attained coordinates of an unknown node (x, y, z) as the outcome. Further, the coordinates of an unknown node (x, y, z) are tuned by utilizing TOA and gained optimized coordinates of the unknown node $\text{opt}(x, y, z)$. The objective function for the developed TOA-based node optimization model is provided in Eq. (1).

$$\text{ft}_1 = \underset{\{\text{opt}(x,y,z)\}}{\arg\min}\left(\sum \text{Eudis}(\text{HOP}, (x, y, z))\right), \tag{1}$$

$$\text{HOP} = \text{avg hop size} + \text{hop count}. \tag{2}$$

The variable Eudis represents the Euclidean distance and avg hop size displays the average hob size. Euclidean distance is referred to as the span between the hop and the coordinates.

$$H(\text{HOP}, (x, y, z)) = \sqrt{\sum_{j=1}^{y} ((x, y, z)_j - \text{HOP}_j)^2}. \quad (3)$$

Here, the term HOP and (x, y, z) represent the Euclidean attained in y space, HOP_j and $(x, y, z)_j$ are the Euclidean vector, and y is the space.

3.2 *Teamwork Optimization Algorithm*

TOA is utilized in the developed node localization model to tune the unknown node coordinates (x, y, z) and also to optimize the parameters of SVM like epoch count for enhancing the localization performance. TOA [16] is a population-related framework and it is structured by several characteristics of team members to satisfy the goal allocated for the team. The team members are considered the search agent and this approach is mainly designed to overcome the optimization problem. In TOA, a supervisor is chosen based on Eq. (4).

$$G = A_g. \quad (4)$$

Here, the number of a team member is given as g, and the supervisor is presented as G. Different phases of TOA are explained as follows:

Phase 1—guidance from the supervisor: The positions of team members are renewed with the help of the supervisor and also share different data to achieve the goal. These phases are resumed by Eqs. (5), (6), and (7).

$$Y_q^{G1} : y_{q,z}^{G1} = y_{q,z} + v \times (G_z - L \times y_{q,z}), \quad (5)$$

$$Y_q = \{Y_q^{G1}, K_q^{G1} < K_q, Y_q \text{, else}, \quad (6)$$

$$L = \text{round}(1 + p). \quad (7)$$

Here, the new status is given Y_q^{G1} and the objective function is presented K_q^{G1}. The updated index L and the random variable v are presented in the interval [0,1].

Phase 2—information sharing: Here, the whole team member uses the data of other members presented in the team, who achieved the job effectively for upgrading the overall efficacy rate. The team member updating phase is presented in Eqs. (8), (9), and (10).

$$Y^{n,q} : y_z^{n,q} = \frac{\sum_{b=1}^{V_q} h_{b,z}^{d,q}}{V_q}, \quad (8)$$

$$Y_q^{G2} : y_{q,z}^{g2} = y_{q,z} + v \times \left(y_z^{n,q} - L \times y_{q,z}\right) \times \text{sign}\left(K_q - K^{n,q}\right), \tag{9}$$

$$Y_q = \left\{Y_q^{G2}, K_q^{G2} < K_q, Y_q \text{, else.}\right. \tag{10}$$

The team member average is showcased $Y^{n,q}$ and the fitness function is termed as $K^{n,q}$. The value of an objective function is presented Y_q^{G2} and K_q^{G2} in the above equation.

Phase 3—individual activity: Here, all the co-workers provide an effort to improve the efficacy rate based on the latest condition and they are provided in Eq. (11) and Eq. (12).

$$Y_q^{G3} : y_{q,z}^{G3} = y_{q,z} + (-0.01 + v \times 0.02) \times y_{q,z}, \tag{11}$$

$$Y_q = \left\{Y_q^{G3}, Y_q^{G3} < K_q, Y_q \text{, else.}\right. \tag{12}$$

The member's latest status is given Y_q^{G3}, and the objective function is presented K_q^{G3}. This approach updates their position which until satisfies the end condition. The pseudo-code for TOA is offered in Algorithm 1.

Algorithm 1 Developed TOA

```
Initialize TOA
  Initiate the information related to issues
  Estimate the team member V and iteration M
  Initiate the population matrix randomly
  Validate the fitness function
    For m = 1 : M
      Supervisor solution renewed by Eq. (4)
        For q = 1 : V
          Stage 1 (supervisor guidance):
            Update Y_q by Eqs. (5) and (7) at the initial phase
          Stage 2 (information sharing):
            Determine the best team member by Eq. (8)
            Renew Y_q using Eqs. (9) and (10) in the next phase
          Stage 3 (individual activity):
            Update Y_q by Eqs. (11) and (12) in the final phase
        End For q = 1 : V
      Save the best solution
    End For m = 1 : M
  The best solution is obtained
End TOA
```

4 Range-Free Anchor-Based Node Localization in WSN Using Optimized Support Vector Machine

4.1 Training and Testing of SVM

Traditionally, most of the research work has utilized the optimization algorithm as the node localization strategy, but in the proposed model, the node localization is performed automatically by utilizing the machine learning mechanism named SVM, where the target is not given randomly and so, initially the node location has been optimized with TOA that is fixed as the target for the SVM network. This makes the network automatically generate the optimized node coordinates. SVM is a widely utilized supervised learning approach. Most of the supervised models has utilized the dataset to perform effective analysis in the form of a row and also train the parameters of the approach for attaining fit dataset outcomes. The dataset is divided into two phases as training phase and the testing phase for attaining an effective performance rate. In the training phase, a huge amount of data from the optimized solution is utilized, and for the testing phase, a minimal number of data is utilized for the analysis to check the outcome of the trained model. In the SVM training phase, the input like position, RSS, packet size, and hob count of the unknown node is offered to the training network. Here, the optimized unknown coordinate is assumed as the target that is presented $\text{opt}(x, y, z)$. In the SVM testing phase, unknown node information is provided randomly as the input to the trained SVM network for node localization and it provides the accurate localization result as optimized unknown node coordinates $\text{opt}(x, y, z)$.

4.2 Support Vector Machine

SVM classifier is a commonly utilized learning approach. SVM is considered a supervised learning approach along with several learning validation data utilized for classification as well as regression. The base spacing presented in an isolated hyperplane for the nearest instance is enhanced in SVM. SVM decision function to classify the unknown point z is referred to as in Eq. (13).

$$f(z) = qz + l = \sum_{u-1}^{F_r} \alpha_u p_u j_u . z + l. \tag{13}$$

Here, the term F_r indicates the number of support vectors. The solution of mapping in input space to superior dimension feature space is presented as given in Eq. (14).

$$\varphi(z_u).\varphi(j_u) = k(z_u . j_u). \tag{14}$$

Here, the kernel function is given $k(z_u.j_u)$ for the dot product in the feature space. Finally, the SVM decision function is presented in Eq. (15).

$$f(z) = \sum_{u=1}^{F_r} \alpha_u p_u k(j_u.z) + l. \quad (15)$$

The decision function that fulfills Mercer's theorem is utilized as kernels. Different kinds of kernels are the polynomial kernel, Gaussian kernel, and linear kernel.

4.3 Optimized SVM

The main goal of the optimized SVM is to provide enriched node localization in the WSN network. SVM performs efficiently with clear margin separation and it is efficient in the high dimensional area. But, it lagged when enormous numbers of data are utilized for the analysis and it needs more time for the analysis. So, an optimized SVM approach is utilized for the analysis. Here, the epoch counts in SVM are tuned in the limit of [50,100] with TOA approach. The fitness function for the suggested localization model is presented in Eq. (16).

$$\mathrm{Ft}_2 = \underset{\{\mathrm{EP}_p^{\mathrm{SVM}}\}}{\arg\min}\left(\frac{1}{\mathrm{Acy}} + \mathrm{Cmc} + \mathrm{Cpc}\right). \quad (16)$$

Here, the term $\mathrm{EP}_p^{\mathrm{SVM}}$ indicates the SVM epoch count, Acy denotes the accuracy, Cmc is referred to as the communication cost, and Cpc represents the computational cost. Accuracy is defined as the validation of closeness for an appropriate rate, and it is presented in Eq. (17).

$$\mathrm{Acy} = \frac{(\mathrm{rb} + \mathrm{rc})}{(\mathrm{rb} + \mathrm{rc} + \mathrm{rd} + \mathrm{re})}. \quad (17)$$

Here, the false negative is termed as re, false positive is referred to as rd, a true negative value is presented as rc, and a true positive value is presented as rb, respectively. Communication cost is defined as the cost related to reception and wireless transmission. It combines the whole radio part, analogy, and digital circuit for processing and also receives and transmits the information in terms of bits. Computational cost is referred to as the execution of the node localization algorithm in the embedded platform. The structural view of optimized SVM is presented in Fig. 1.

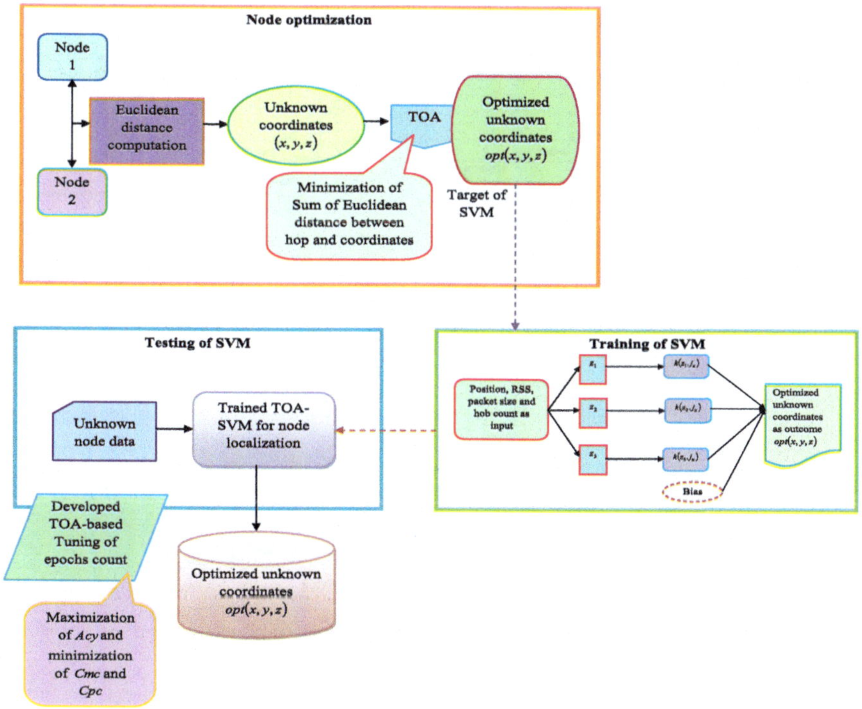

Fig. 1 Structural view of developed TOA-SVM-based node localization

5 Result and Discussions

5.1 Simulation Setup

The latest anchor-free range-based localization model in WSN was implemented in MATLAB 2020a and also their outcome was validated over several approaches. The anchor nodes were presented in different ranges such as 50, 100, and 150, and also the numbers 200, 250, and 300 unknown nodes were utilized to perform effective analysis. The developed model was contracted over heuristic approaches like PSO [17] and GWO [18] and also with existing localization approaches like SVM-PSO [19] and SVM-GWO [20].

5.2 Convergence Analysis

The convergence analysis employed in the recommended framework was contrasted with multiple frameworks presented in Fig. 2. The developed model achieved 21.1

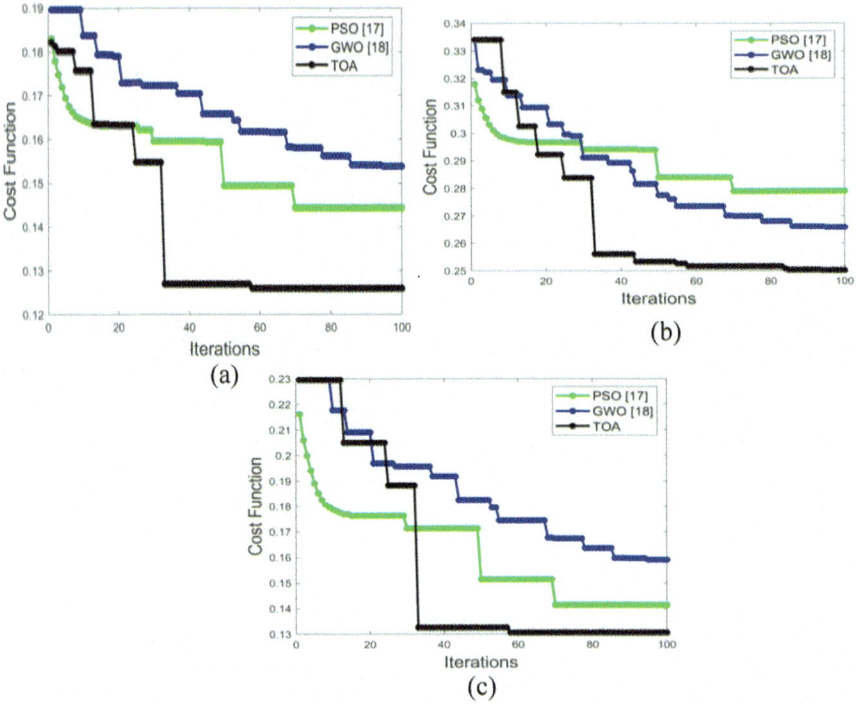

Fig. 2 Convergence analysis on suggested range-free anchor-based localization model with heuristic approaches over **a** node range at 200, **b** node range at 250, and **c** node range at 300

and 26.58% superior to traditional frameworks like PSO and GWO, respectively, in Network 1 at the 40th iteration. Similarly, several analyses employed in the recommended framework performed scored effectively higher efficacy rates than the existing localization framework. Thus, the developed model has a minimal convergence rate than a traditional model.

5.3 *Analysis on Developed Node Localization Model Based on Average Localization Error*

In Fig. 3, the average localization analysis for the developed model based on range-free anchor-based localization in WSN is displayed. Several observations achieved in the recommended model achieved effectively great performance rates than the conventional approaches. The suggested TOA model attained 19.23% higher than PSO and 8.69% enhanced than GWO in the unknown node count 300. Similarly, the analysis based on average localization error in the developed node localization

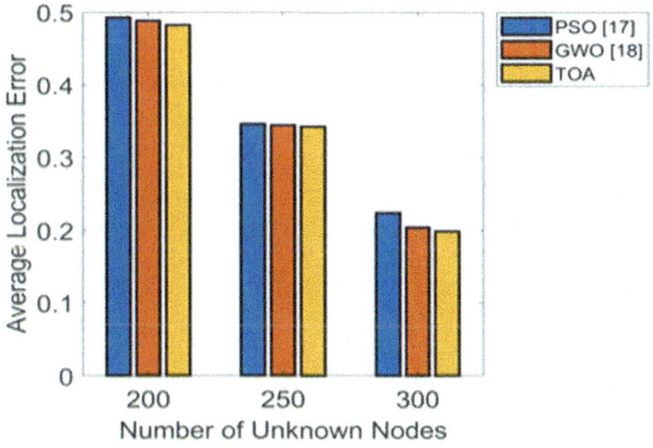

Fig. 3 Analysis based on average localization error on developed range-free anchor-based localization model in WSN with heuristic approaches

model is performed under a different number of unknown nodes and also attained better performance than conventional models.

5.4 Analysis on the Developed Node Localization Model Based on Localization Error Variance

The analyses on localization error variance over several traditional models are showcased in Fig. 4. Localization error analysis performed in the suggested model achieved 29.16% superior to PSO and 5.55% enhanced than GWO in iteration 250. Here, the developed model achieved a minimal rate than the existing model. Thus, the developed model efficiently enhances the efficacy rate in the suggested model as well as offered a superior performance rate in a communication system.

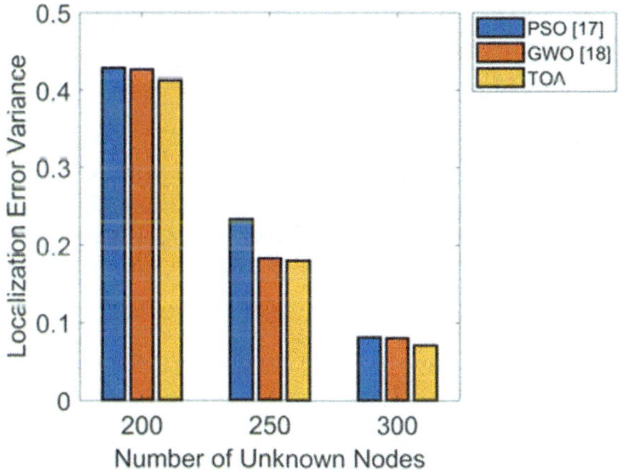

Fig. 4 Analysis based on localization error variance on developed range-free anchor-based localization model in WSN over several heuristics approaches

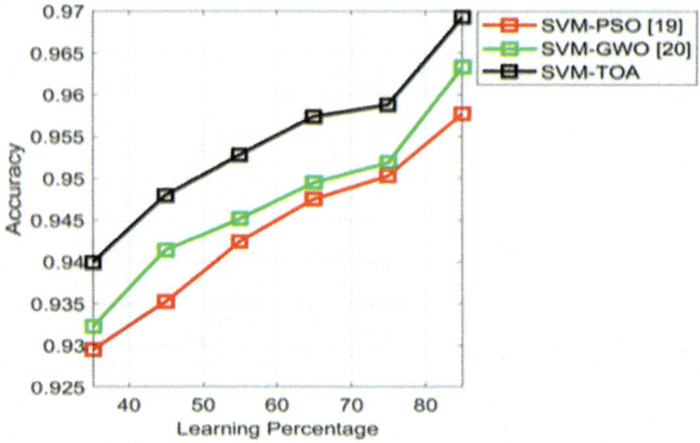

Fig. 5 Accuracy analysis on developed range-free anchor-based localization model in WSN over several heuristics approaches

5.5 Accuracy Analysis on Developed Node Localization Model

Accuracy observation performed on the suggested model is presented in Fig. 5 along with several existing approaches. The recommended framework yields effectively better efficacy rate than the existing approaches. The suggested range-free anchor-based localization approach attained 1.05% better than SVM-PSO and 0.844% enhanced than SVM-GWO. Thus, the suggested framework scored better outcomes than the conventional approaches.

5.6 Overall Analysis on the Suggested Node Localization Approach

The overall research performed in the suggested node localization model by comparing it with conventional models is displayed in Table 1. Accuracy analysis performed in the suggested range-free anchor-based localization model in WSN attains 1.39% better than SVM-PSO and 0.58% superior to the SVM-GWO model. Thus, the suggested framework showcased a better efficacy rate than the traditional localization approach in the WSN communication system.

Table 1 Overall efficacy analysis on developed range-free anchor-based localization models in WSN with conventional approaches

Measures	SVM-PSO [19]	SVM-GWO [20]	SVM-TOA
Accuracy	0.93234	0.93986	0.94532
Sensitivity	0.91505	0.92307	0.93202
Specificity	0.96745	0.97396	0.97234
Precision	0.98279	0.9863	0.98559
FPR	0.032547	0.026037	0.027665
FNR	0.084947	0.076933	0.067985
NPV	0.96745	0.97396	0.97234
FDR	0.017214	0.013701	0.014407
F_1-score	0.94771	0.95364	0.95806
MCC	0.85661	0.87221	0.88254

6 Conclusion

A novel heuristic approach was developed for range-free anchor-based localization in WSN to enhance the localization performance rate effectively. The unknown node localization was determined based on TOA over range-free concept. TOA was mainly utilized to validate the coordinates of unknown nodes and also to assure the accuracy of localization by resolving the multi-objective issue. Localization error analysis employed in the recommended model secured 29.16% superior to PSO and 5.55% enhanced than GWO in iteration 250. Likewise, cost function observation employing recommended model scored a high-performance rate of 21.1% and 26.58% superior to traditional models such as PSO and GWO, respectively. Hence, the recommended framework scored an effectively higher efficacy rate than the traditional models in range-free anchor-based localization in the WSN communication system.

References

1. Han G, Yang X, Liu L, Zhang W, Guizani M (2020) A disaster management-oriented path planning for mobile anchor node-based localization in wireless sensor networks. IEEE Trans Emerg Top Comput 8(1):115–125
2. Lin Y, Tao H, Tu Y, Liu T (2019) A node self-localization algorithm with a mobile anchor node in underwater acoustic sensor networks. IEEE Access 7:43773–43780
3. Sabale FK, Mini S (2021) Transmission power control for anchor-assisted localization in wireless sensor networks. IEEE Sens J 21(8):10102–10111
4. Ou C-H (2011) A localization scheme for wireless sensor networks using mobile anchors with directional antennas. IEEE Sens J 11(7):1607–1616
5. Ssu K-F, Ou C-H, Jiau HC (2005) Localization with mobile anchor points in wireless sensor networks. IEEE Trans Veh Technol 54(3):1187–1197
6. Xiao B, Chen L, Xiao Q, Li M (2010) Reliable anchor-based sensor localization in irregular areas. IEEE Trans Mob Comput 9(1):60–72
7. Liu K, Liu X, Li X (2016) Guoguo: enabling fine-grained smartphone localization via acoustic anchors. IEEE Trans Mob Comput 15(5):1144–1156

8. Wang X, Qian Z, Wang X, Huang L (2019) Robust localization for cognitive IoT via the mobile anchor node based on the diameter-varying spiral line. IEEE Access 7:28487–28497
9. Huang J-F, Chang G-Y, Chen G-H (2015) A historical-beacon-aided localization algorithm for mobile sensor networks. IEEE Trans Mob Comput 14(6):1109–1122
10. Sabale K, Sapre S, Mini S (2021) Obstacle handling mechanism for mobile anchor assisted localization in wireless sensor networks. IEEE Sens J 21(19):21999–22010
11. Ou C-H, He W-L (2013) Path planning algorithm for mobile anchor-based localization in wireless sensor networks. IEEE Sens J 13(2):466–475
12. Mourad F, Snoussi H, Abdallah F, Richard C (2009) Anchor-based localization via interval analysis for mobile ad-hoc sensor networks. IEEE Trans Signal Process 57(8):3226–3239
13. Alomari A, Comeau F, Phillips W, Aslam N (2018) New path planning model for mobile anchor-assisted localization in wireless sensor networks. Wireless Netw 24:2589–2607
14. Sabal K, Mini S (2022) Path planning mechanism for mobile anchor-assisted localization in wireless sensor networks. J Parallel Distrib Comput 165:52–65
15. Cao Y, Xu J (2023) DV-Hop-based localization algorithm using optimum anchor nodes subsets for wireless sensor network. Ad Hoc Netw 139:103035
16. Dehghani M, Trojovský P (2021) Teamwork optimization algorithm: a new optimization approach for function minimization/maximization. Sensors 21(13)
17. Iqbal A, Singh GK (2021) PSO-based controlled six-phase grid connected induction generator for wind energy generation. CES Trans Electr Mach Syst 5(1):41–49
18. Anbumani K, Ranihemamalini R, Pechinathan G (2017) GWO based tuning of PID controller for a heat exchanger process. In: 2017 Third international conference on sensing, signal processing and security (ICSSS), pp 417–421
19. Azimirad V, Hajibabzadeh M, Shahabi P (2017) A new brain-robot interface system based on SVM-PSO classifier. In: 2017 Artificial intelligence and signal processing conference (AISP), pp 124–128
20. Li B, Luo C, Wang Z (2020) Application of GWO-SVM algorithm in arc detection of pantograph. IEEE Access 8:173865–173873

Deep Learning Model for ECG Analysis and Prediction Using mHealth Datasets

K. M. Yogesh, S. Arpitha, and D. H. Manjaiah

Abstract The popularity of Human Activity Recognition (HAR) as a research topic has increased because of its many applications. New approaches to HAR concerns have emerged as a result of deep learning. Here, a deep network architecture based on residual bidirectional Long Short-Term Memory (LSTM) is recommended. The ability of a bidirectional link to integrate the forward state of positive time and the backward state of negative time is one advantage of the new network. Second, the gradient vanishing problem is successfully solved because remaining connections between stacked cells allow gradients to function as a shortcut. In order to raise the recognition rate, the proposed network generally shows improvements on the spatial (stacked residual connections) and theoretical (using bidirectional cells) dimensions. Testing with the opportunity dataset and the public domain UCI dataset greatly increases accuracy in comparison to earlier findings.

Keywords Deep learning · Electrocardiogram · LSTM · mHealth

1 Introduction

Nearly everyone in this century, the twenty-first, is dependent on technology. This amazing technological advancement has improved lives in a variety of ways. The number of individuals suffering from heart disease is rising daily, mostly due to the unhealthy lifestyles that people are leading. As a result, systems for precisely sensing, gathering, recording, and analyzing the patients' physical status have been

S. Arpitha and D. H. Manjaiah: These authors contributed equally to this work.

K. M. Yogesh (✉)
Computer Science and Engineering, Ramaiah University of Applied Sciences, Bangalore, Karnataka 560058, India
e-mail: yogeshkm.mrj@gmail.com

S. Arpitha · D. H. Manjaiah
Computer Science, Mangalore University, Mangalore, Karnataka 574199, India

V. K. Gunjan and J. M. Zurada (eds.), *Proceedings of 4th International Conference on Recent Trends in Machine Learning, IoT, Smart Cities and Applications*, Lecture Notes in Networks and Systems 873, https://doi.org/10.1007/978-981-99-9442-7_36

developed over the past few decades [1]. Systems may now automatically learn from experience and develop without explicit programming thanks to machine learning. Artificial intelligence is being used in this situation.

Machine learning approaches often fall into the following categories: supervised learning, unsupervised learning, semi-supervised learning, and reinforcement learning [2]. The links and dependencies between the input characteristics and the goal prediction output are determined using supervised learning algorithms [3]. Because the model is trained using unlabeled data for unsupervised learning, we may estimate the output values for new data based on the associations that it learnt from the prior datasets. Semi-supervised learning comes between supervised and unsupervised learnings since it employs both labeled and unlabeled data [4]. Machine learning algorithms that use reinforcement interact with their surroundings by taking actions and then identifying successes or failures. In order to maximize performance, machines and software agents are permitted to autonomously identify the best behavior within a given situation. We used supervised machine learning to train our model.

Supervised learning, unsupervised learning, semi-supervised learning, and reinforcement learning are the four main categories that machine learning techniques fall under [2]. Examples include the discovery of model dependencies and links between the characteristics of the input and the desired prediction output using supervised learning algorithms [3]. Given that the model is trained using unlabeled data for unsupervised learning, we may estimate the output values for fresh data based on the associations that the model learned from the prior datasets. Semi-supervised learning lies between supervised and unsupervised learnings since it employs both labeled and unlabeled data. Algorithms for reinforcement-based machine learning engage with their surroundings by acting and then assessing whether their actions were successful or unsuccessful.

ECG gives us a sequence of sinus rhythms that describe the state of the heart [5]. Notes that the ECG is the most used test for diagnosing cardiac disorders since it is the least expensive and effective at detecting specific illnesses. An ECG telemonitoring device assists patients in avoiding frequent hospital visits and enables ECG patient monitoring. Patients' ECG readings can be used to detect various cardiac conditions. Occasionally, patients can receive remote monitoring diagnoses from wireless healthcare providers. These services have advanced significantly in the 2 Introduction present time, and they can now provide high-quality services for infants and young children as well as patients with cognitive and physical disabilities [1, 2]. ECG pattern identification is also frequently helpful as a heart disease early warning method. As a consequence, we intend to identify unusual heart rate patterns in ECG readings and determine which algorithm produces more accurate results for a given cardiac condition.

2 Related Work

The general target of an ECG is to perceive the condition and capacity of the heart. The ECG has been found as the most utilized and uncomplicated route for exact and fast finding of various cardiovascular episodes [5]. In the United States (US), the United States Preventive Services Task Force (USPSTF) does not recommend an ECG for routine screening in patients with no signs and the individuals who have lower opportunities to have coronary illness [6]. This is on the grounds that an ECG may demonstrate a wrong nearness of an issue, inciting misdiagnosis, the proposition of prominent procedure, or over treatment. To keep away from such difficulty, machine learning (ML) can be utilized. Throughout the years, a lot of crude information that has been assembled that can be useful. Information obtaining techniques have been created for as long as decades to precisely detect, gather, record, and investigate the patients' physical condition [1]. To a consistently expanding degree, data is being assembled and taken care of in databases or of course spreadsheets [7]. Machine learning is the ideal way to deal with find in advance darkens regularities and examples from various informational collections, with the desire that machines can help in the much of the time dull and botch slanted methodology of getting data from test data, and help people to explain and mastermind their discernment [7]. Recently, a variety of techniques have been used for successful arrhythmia classification, including genetic algorithms, fuzzy logic, self-organizing maps, hidden Markov models, and SVMs [8]. Because it takes a lot of time and effort to examine ECG signal patterns for diagnosis, it is necessary to use computerized methods for diagnosing cardiac illnesses [9].

3 mHealth Data Description

The collected mHealth dataset comprises Human Activities (HAs) and vital signs recorded for ten people performed by 12 different physical activities shown in Table 1. Shimmer2 wearable sensors were used for the recordings and stored in UCI machine learning repository which is accessible publicly available in machine learning repository. The sensors were placed respectively at three different positions of human body (1) chest, (2) right lower arms (RLAs), and (3) left ankle (LA) shown in Fig. 1 and attached by using elastic straps [10]. By integrating many sensors, we can quantify the motion experienced by various body sections, including the acceleration, rate of rotation, and magnetic field direction, better capturing the dynamics of the body. Each person was done all 12 various activities; for each activity, they took one min time duration; and only the following attribute was considered: *XYZ* axis of accelerometer, gyro meter, and magnetometer sensor. The primary steps of our work evoked by downloading and reprocessing from machine repository. Table 1 shows sample physical activity datasets. Because different attributes have different values and scales, it is essential to normalize before applying deep learning model.

Table 1 Data sample of mHealth dataset

Human physical activity (HPA)	Actual time taken in seconds	Instances	Time
Standing still	61.44	3072	1 min
Sitting and relaxing	61.44	3072	1 min
Lying down	61.44	3072	1 min
Walking	61.44	3072	1 min
Climbing stairs	61.44	3072	1 min
Waist bends forward	**20 s**	**1024**	**20 s**
Front elevation of arms	**20 s**	**1024**	**20 s**
Knees bending	61.44	3072	1 min
Cycling	61.44	3072	1 min
Jogging	61.44	3072	1 min
Running	61.44	3072	1 min
Jump front and back	61.44	3072	1 min

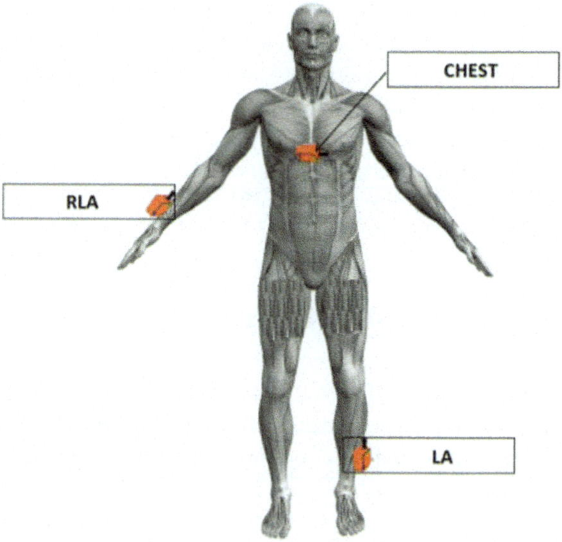

Fig. 1 Graphical representation of sensor deployment on human body

4 Data Acquisition and Organization

The Mobile Health dataset, which is primarily the work of [10], is the data source for the proposed research study. It is hosted in the machine learning archive of the center for machine learning and intelligent systems at the University of California, Irvine (UCI), and is freely available to the research community at https://archive.ics.uci.edu/ml/datasets/MHEALTH+Dataset. This mHealh data is utilized as secondary information to learn more about how different human activities affect heart rate activity. The Mobile Health dataset contains recordings of the bodily motions and vital signs of ten participants while they underwent a range of physical activities. Sensors are attached to the participants' chest, right lower arm (RLA), and left ankle (LA). These sensors track the body parts' acceleration, rate of rotation, and magnetic field direction 1. The sensor put on the chest enables for basic cardiac monitoring, testing for various arrhythmias, and analyzing the impact of activity on the ECG in addition to giving two-lead ECG values.

5 Deep Neural Network

The Artificial Neural Networks (ANNs) simply consist of three layers, namely the input, hidden, and output layers [11, 12]. Moreover, each layer has a number of neurons to extract the features such that the data from the input layer is processed and transferred into the next hidden layers. ANN performs a series of mathematical operations through the activation functions and the computational cost is reduced by using the dropout technique to downsize the number of parameters. Generally, different types of ANN have been developed, and the fully connected layers are commonly used in which each neuron in one layer is connected to every neuron in the adjacent layers. The final output layer generate a list of probabilities such that assigns for each sample in the dataset a corresponding probability Fig. 2 shows the general steps involved to train and test the machine learning models.

The deep learning model is a common part of machine learning algorithms. Recently, due to the quick advances in computing resources and technologies, so that complex, huge amounts of datasets are generated which can be efficiently analyzed using deep learning models. In addition, it is clear that a large number of applications in which deep learning architectures have been developed and applied based on the type and complexity of data such as bioinformatics, speech recognition, drug design, medical image interpretation, and social network filtering [13].

The Convolutional Neural Networks (CNN) are widely used in image processing [14]. CNN consists of many layers in which every layer has a set of neurons. The main components of CNN are three layers, namely convolutional layer, activation layer, and polling layer [15]. Finally, there is a last fully connected layer or the output layer that generates the network predictions as shown in Fig. 3.

6 Architecture of the Proposed CNN Network

In spite of the fact that the residual connections for CNN have been used in many applications and various research domains [16]. In this section, the proposed Res-CNN model to deal with the Human Activity Recognition (HAR) dataset is described. Moreover, the Res-CNN model consists of many similar blocks where we select modules and combine them to build the final model.

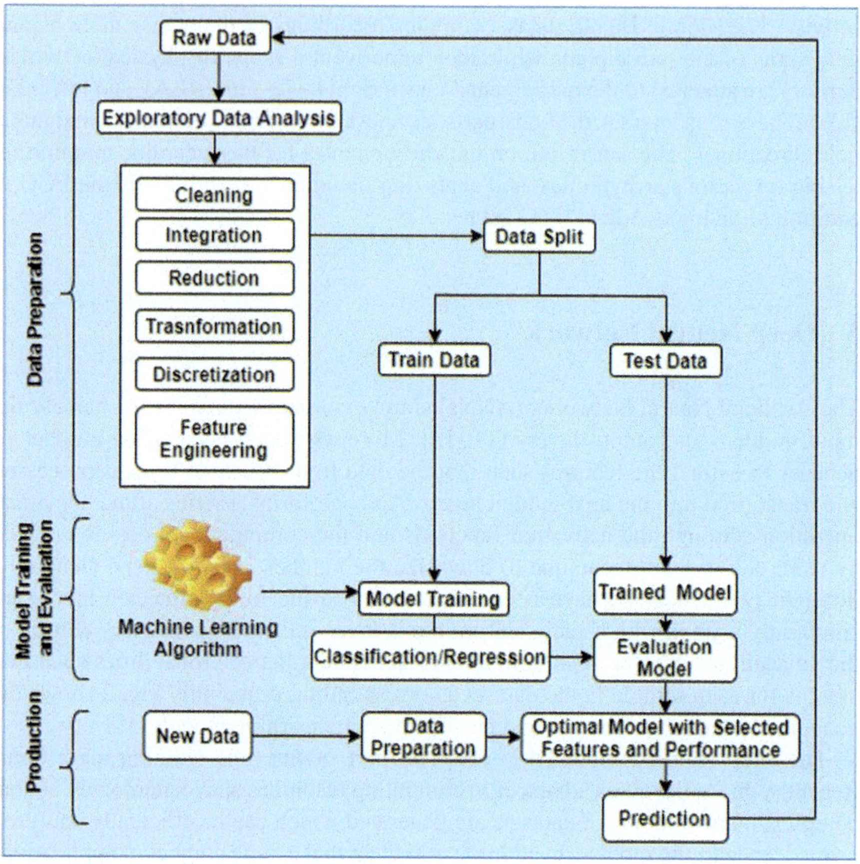

Fig. 2 General steps involved to train and test the machine learning models

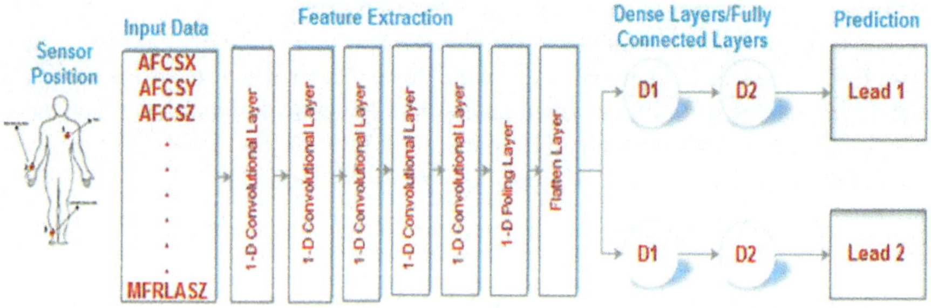

Fig. 3 Architecture of deep residual CNN model

Deep neural networks frequently experience gradient vanishing, which is an issue in general. As a result of the residual, bidirectional, and stacked layers—hence the term "Deep Residual Bidir-LSTM" some bottom layers that would normally be too difficult to optimize via back propagation are now addressed. Residual connections serve as gradient shortcuts when batch normalization is applied on top of each residual layer. It guards against overfitting by keeping constraints in the hidden layer feature space from becoming overly complicated and by avoiding outlier values during testing.

In Fig. 3, the information flows in the direction from the input layer into the output layer. Moreover, there are many similar blocks of hidden layers called "residual layer" that have residual connections inside. Moreover, each residual layer contains two CNN layers. The detailed description of the residual layer is presented in Table 2, which contains four CNN cells in sum. In addition, the activation functions are unified with ReLU, because it always outperforms with deep networks to counter gradient vanishing.

7 Experimental Results

The experiments are carried out using the proposed model with the public HAR dataset. The computer that used to accomplish these experiments is described as follows: CPU core i7, 8 GB RAM as well as GPU card. The following subsections present the different experiments that are carried out using various sensors.

7.1 Arm Position

This section presents the experimental results for the attributes generated from arms' sensors. Figure 4 shows the plot of LEAD1 and LEAD2 values using arm sensors. From this figure, the values of LEAD1 and LEAD2 are almost the same, while Fig. 5 presents LEAD1 prediction for the test dataset using arm sensors. Because of the value of R^2 is a small value 0.0614667, So that there exists a clear difference between the predicted and the test values. This means that the model is not forecast LEAD1 properly. Similarly, Fig. 6 presents LEAD2 prediction for the test dataset using arm sensors. It is clear that the behavior for the model is similar to the LEAD1 case. Where the R^2 in Tables 3 and 4, the proposed model achieves the following scores, 0.0727415, 0.00231712, 0.0481365, and 0.0236181 for R^2, MSE, RMSE, and MAE, respectively. The MSE value for LEAD1 and LEAD2 is almost the same.

Table 2 Layers of deep residual CNN models

Layer type	Output and shape	Param#	Connected to
input_1 (InputLayer) conv1d_1 (Conv1d)	(None, 4, 1) (None, 4, 128)	0 256	input_1[0][0]
batch_normalization_1	(None, 4, 128)	512	conv1d_1[0][0]
activation_1 (Activation)	(None, 4, 128)	0	batch_normalization_1[0][0]
conv1d_2 (Conv1d)	(None, 4, 64)	8256	activation_1[0][0]
batch_normalization_2	(None, 4, 64)	256	conv1d_2[0][0]
activation_2 (Activation)	(None, 4, 64)	0	batch_normalization_2[0][0]
dropout_1 (Dropout)	(None, 4, 64)	0	activation_2[0][0]
conv1d_3 (Conv1d)	(None, 4, 128)	24,704	dropout_1[0][0]
max_pooling1d_1 (MaxPooling1D)	(None, 4, 128)	0	activation_1[0][0]
add_1 (Add)	(None, 4, 128)	0	conv1d_3[0][0]
max_pooling1d_1[0][0] dropout_2 (Dropout)	(None, 4, 128)	0	add_1[0][0]
batch_normalization_3	(None, 4, 128)	512	dropout_2[0][0]
activation_3 (Activation)	(None, 4, 128)	0	batch_normalization_3[0][0]
conv1d _4 (Conv1d)	(None, 2, 64)	24,640	activation_3[0][0]
batch_normalization_4	(None, 2, 64)	256	conv1d_4[0][0]
activation_4 (Activation)	(None, 2, 64)	0	batch_normalization_4[0][0]
dropout_3 (Dropout)	(None, 2, 64)	0	activation_4[0][0]
conv1d_5 (Conv1d)	(None, 1, 128)	24,704	dropout_3[0][0]
max_pooling1d_2 (MaxPooling1D)	(None, 1, 128)	0	dropout_2[0][0]
add_2 (Add)	(None, 1, 128)	0	conv1d_5[0][0]
max_pooling1d_2[0][0] batch_normalization_5	(None, 1, 128)	512	add_2[0][0]
activation_5 (Activation)	(None, 1, 128)	0	batch_normalization_5[0][0]
conv1d_6 (Conv1d)	(None, 1, 64)	24,640	activation_5[0][0]

(continued)

Table 2 (continued)

Layer type	Output and shape	Param#	Connected to
batch_normalization_6	(None, 1, 64)	256	conv1d_6[0][0]
activation_6 (Activation)	(None, 1, 64)	0	batch_normalization_6[0][0]
dropout_4 (Dropout)	(None, 1, 64)	0	activation_6[0][0]
conv1d_7 (Conv1d)	(None, 1, 128)	24,704	dropout_4[0][0]
max_pooling1d_3 (MaxPooling1D)	(None, 1, 128)	0	add_2[0][0]
add_3 (Add)	(None, 1, 128)	0	conv1d_7[0][0]
max_pooling1d_3[0][0] batch_normalization_7	(None, 1, 128)	512	add_3[0][0]
activation_7 (Activation)	(None, 1, 128)	0	batch_normalization_7[0][0]
conv1d_8 (Conv1d)	(None, 1, 64)	24,640	activation_7[0][0]
batch_normalization_8	(None, 1, 64)	256	conv1d_8[0][0]
activation_8 (Activation)	(None, 1, 64)	0	batch_normalization_8[0][0]
dropout_5 (Dropout)	(None, 1, 64)	0	activation_8[0][0]
conv1d_9 (Conv1d)	(None, 1, 128)	24,704	dropout_5[0][0]
max_pooling1d_4 (MaxPooling1D)	(None, 1, 128)	0	add_3[0][0]
add_4 (Add)	(None, 1, 128)	0	conv1d_9[0][0]
max_pooling1d_4[0][0] flatten_1 (Flatten)	(None, 128)	0	add_4[0][0]
dense_1 (Dense)	(None, 128)	16,512	flatten_1[0][0]
dense_2 (Dense)	(None, 128)	16,512	flatten_1[0][0]
LEAD1_output (Dense)	(None, 1)	129	dense_1[0][0]
LEAD2_output (Dense)	(None, 1)	129	dense_2[0][0]
Total params	217,602		
Trainable params	216,066		
Non-trainable params	1536		

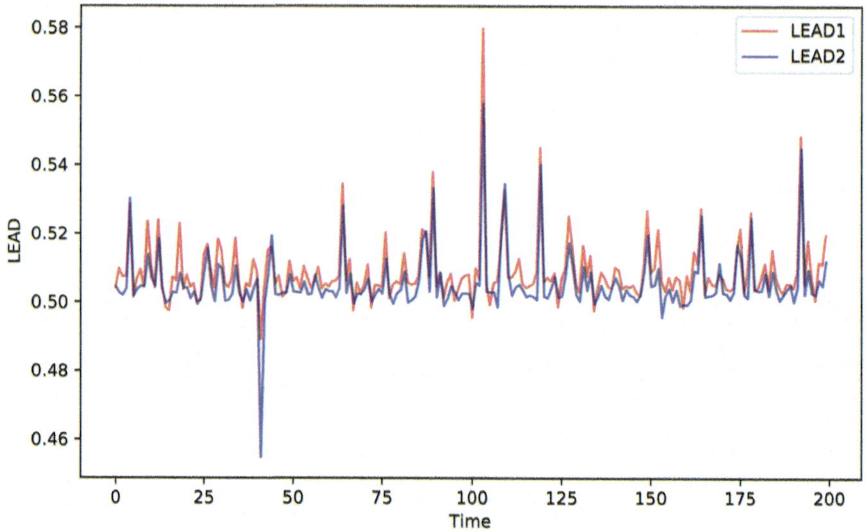

Fig. 4 Plot of LEAD1 and LEAD2 values using arm sensors

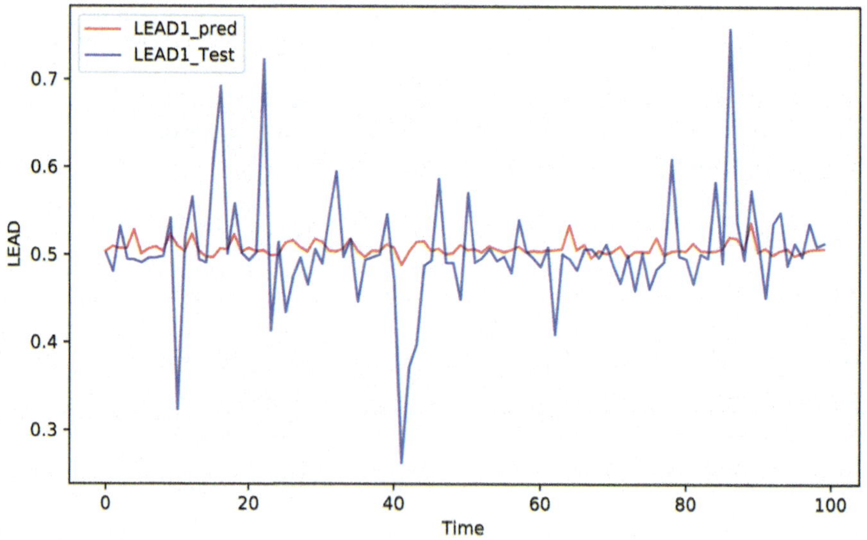

Fig. 5 LEAD1 prediction for test dataset using arm sensors

7.2 Chest

The experimental results for the attributes generated from chest sensors are presented in this section. Figure 7 shows the plot of LEAD1 and LEAD2 values generated using

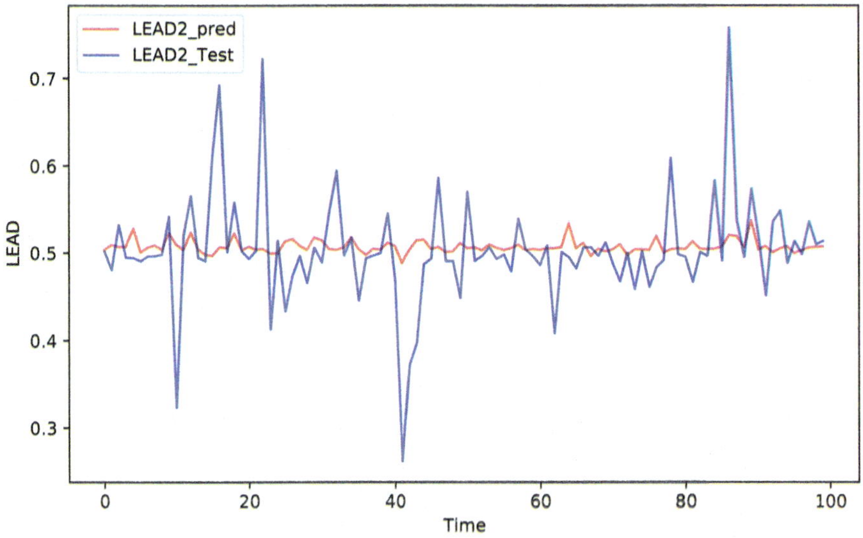

Fig. 6 LEAD2 prediction for test dataset using arm sensors

Table 3 Summary of score results for LEAD1 using ECC dataset

LEAD_1 (L1)					
Position	Time	R^2_L1	MSE_L1	RMSE_L1	MAE_L1
Time	21,404.7	0.00725454	0.00235147	0.048492	0.0271749
Chest	30,746	0.0108866	0.00240494	0.0490402	0.0273423
Left ankle	53,750.1	0.0407855	0.00230492	0.0480096	0.0266036
Right lower arm	51,296.5	0.0614667	0.00225269	0.0474625	0.0271988
All	202,930	0.11415	0.00208401	0.045651	0.0256458

Table 4 Summary of score results for LEAD2 using ECC dataset

LEAD_2 (L2)					
Position	Time	R^2_L2	MSE_L2	RMSE_L2	MAE_L2
Time	21,404.7	0.0134076	0.00251157	0.0501155	0.0240464
Chest	30,746	− 0.014871	0.00255617	0.0505586	0.0255641
Left ankle	53,750.1	0.0319458	0.00242746	0.0492692	0.0236175
Right lower arm	51,296.5	0.0727415	0.00231712	0.0481365	0.0236181
All	202,930	0.0760295	0.00228727	0.4782554	0.0237987

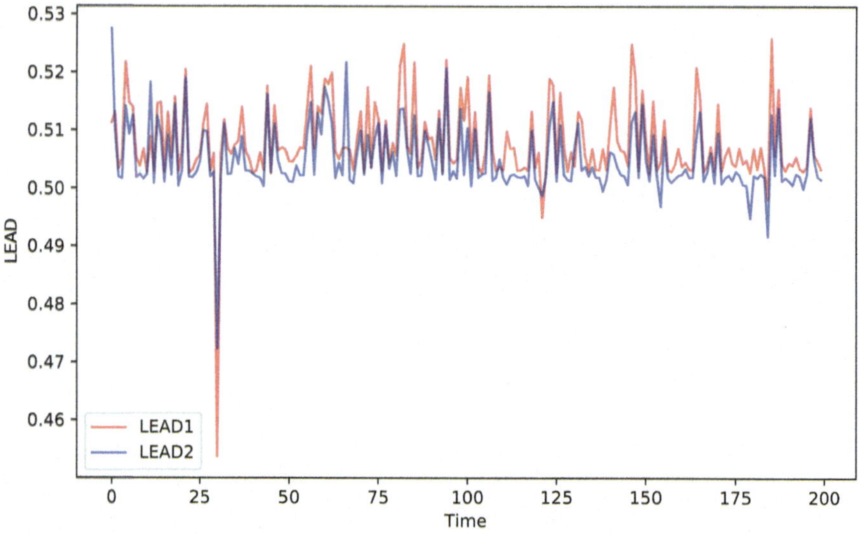

Fig. 7 Plot of LEAD1 and LEAD2 values using chest sensors

arm sensors, while Fig. 8 presents LEAD1 prediction for test dataset using chest sensors. Similarly, Fig. 9 presents LEAD2 prediction for the same dataset using chest sensors. It is clear that the behavior of the model is the same for the LEAD1 and LEAD2 cases since the difference between MSE for LEAD1 and LEAD2 is very small. The proposed model achieves the following MSE scores for LEAD1 and LEAD2, 0.00240494 and 0.00255617, as shown in Tables 3 and 4. The MSE value for LEAD1 and LEAD2 is almost the same.

7.3 *Left Ankle*

This section presents the experimental results for the attributes generated from left ankle sensors. Figure 10 shows the plot of LEAD1 and LEAD2 values generated using left ankle sensors. From this figure, the values of LEAD1 and LEAD2 are very nearly to each other, while Figs. 11 and 12 present LEAD1 and LEAD2 prediction for the test dataset using left ankle sensors. From these figures, there exists a clear difference between the predicted and the test values. This means that the model does not forecast LEAD1 and LEAD2 properly. However, for LEAD1 prediction, the proposed model scores the following values 0.0407855, 0.00230492, 0.0480096, and 0.0266036 for R^2, MSE, RMSE, and MAE, respectively. Generally, the metrics values for LEAD1 and LEAD2 are almost the same as shown in Tables 3 and 4.

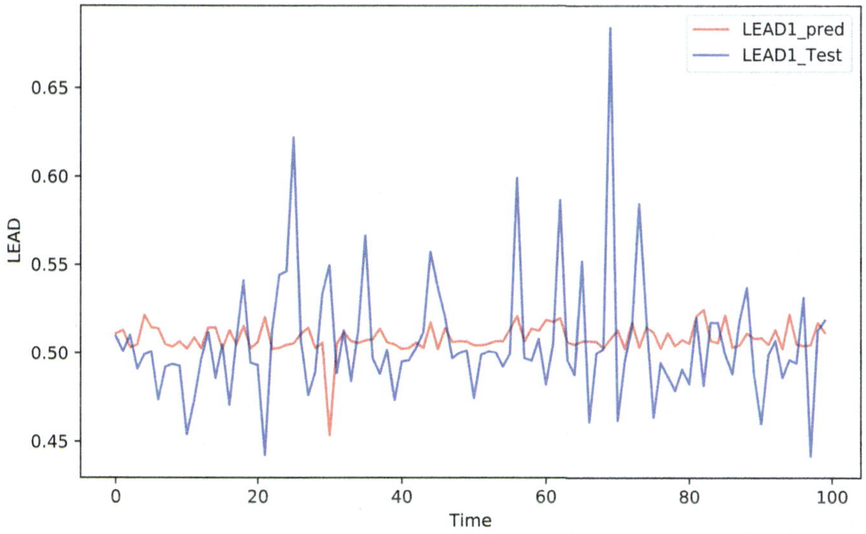

Fig. 8 LEAD1 prediction for test dataset using chest sensors

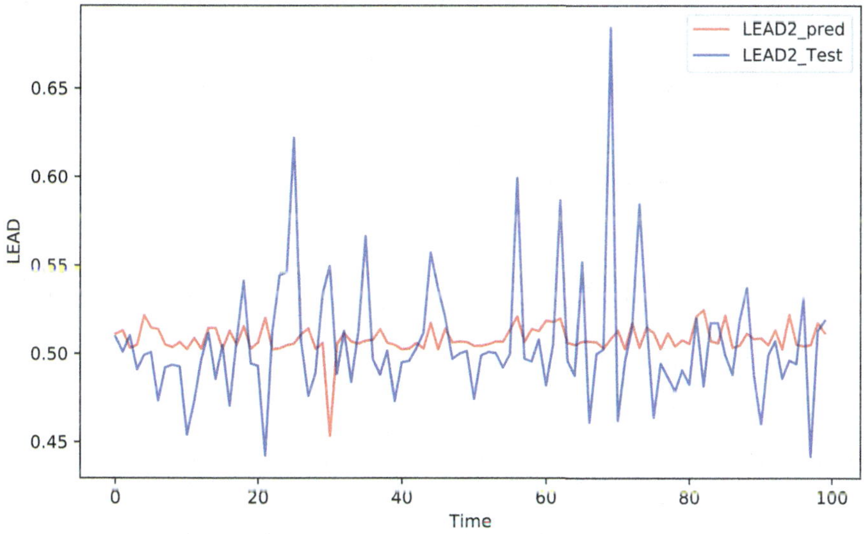

Fig. 9 LEAD2 prediction for test dataset using chest sensors

7.4 Time

This section presents the experiment that carried out using the time attribute. Figure 13 shows the plot of LEAD1 and LEAD2 values based on the time column. From this figure, the values of LEAD1 and LEAD2 are very nearly to each other such

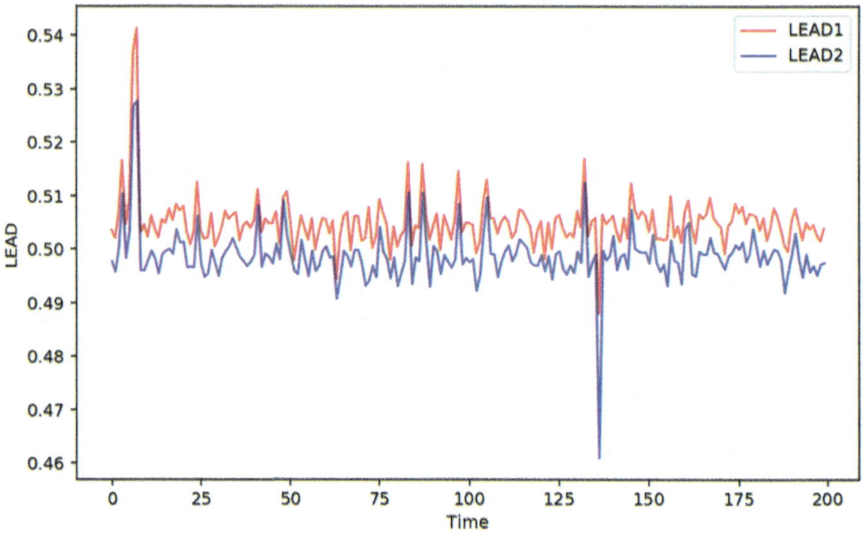

Fig. 10 Plot of LEAD1 and LEAD2 values using left ankle sensors

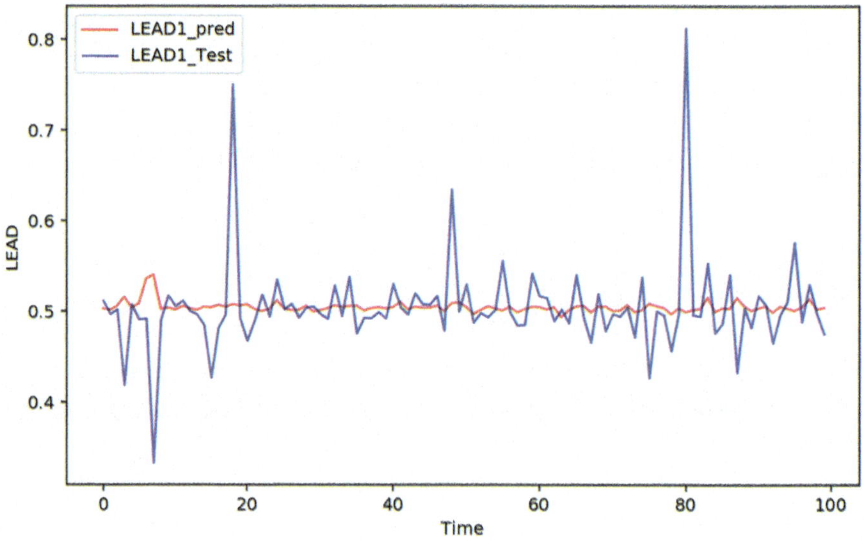

Fig. 11 LEAD1 prediction for test dataset using left ankle sensors

that the values are rising and down together. LEAD1 and LEAD2 predictions for the test dataset using time column are shown in Figs. 14 and 15. Moreover, there exists a big difference between the predicted and the test values. Thus, the model is not performed well to predict the LEAD1 and LEAD2 properly. Generally, the metrics values for LEAD1 and LEAD2 are almost the same as shown in Tables 3 and 4.

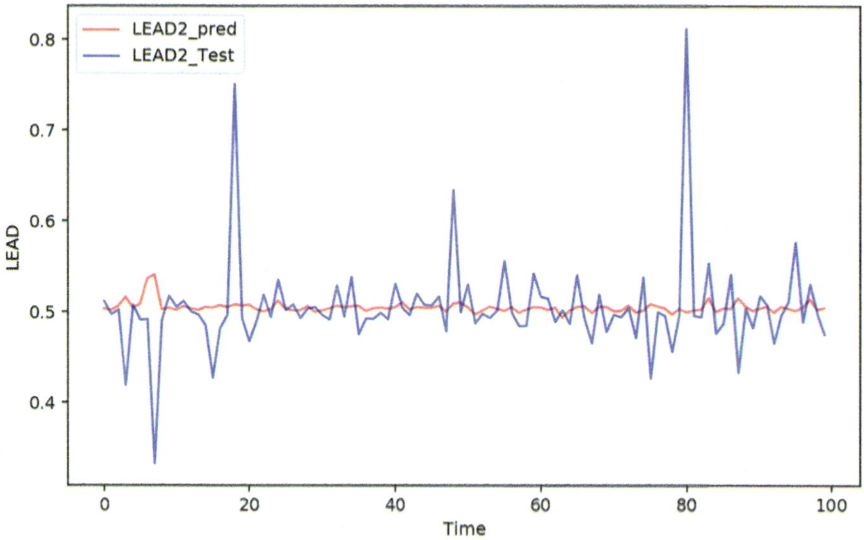

Fig. 12 LEAD2 prediction for test dataset using chest sensors

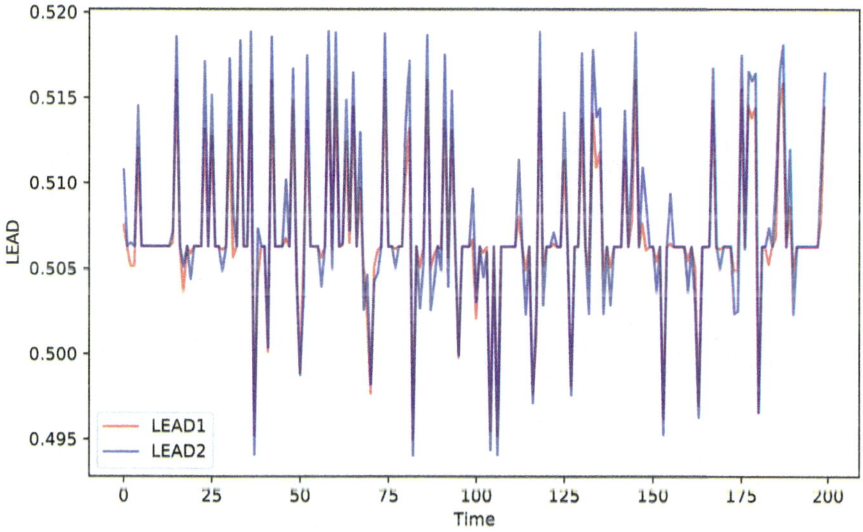

Fig. 13 Plot of LEAD1 and LEAD2 values using time

7.5 All

Finally, the whole attributes generated from different sensors are used to perform the last experiment. Similarly, Fig. 15 shows the plot of LEAD1 and LEAD2 values generated using the entire sensors. It is obvious that the values of LEAD1 and LEAD2

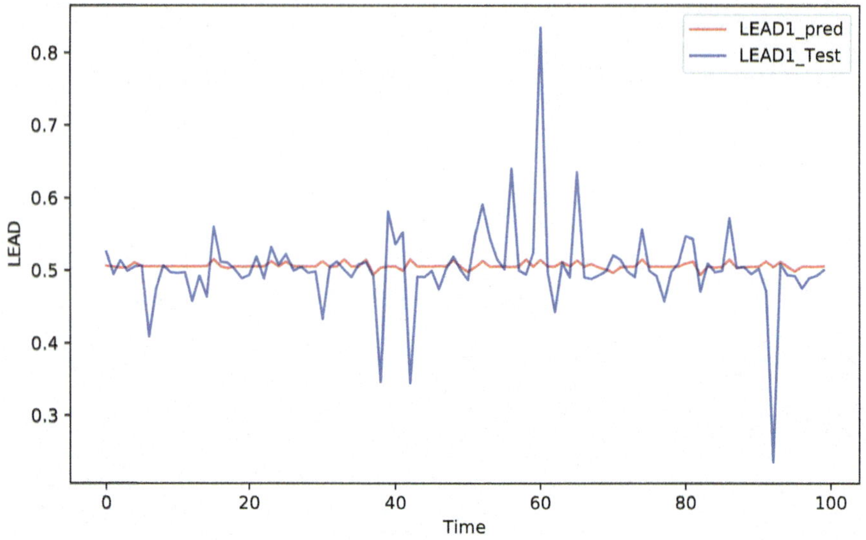

Fig. 14 LEAD1 prediction for test dataset using time

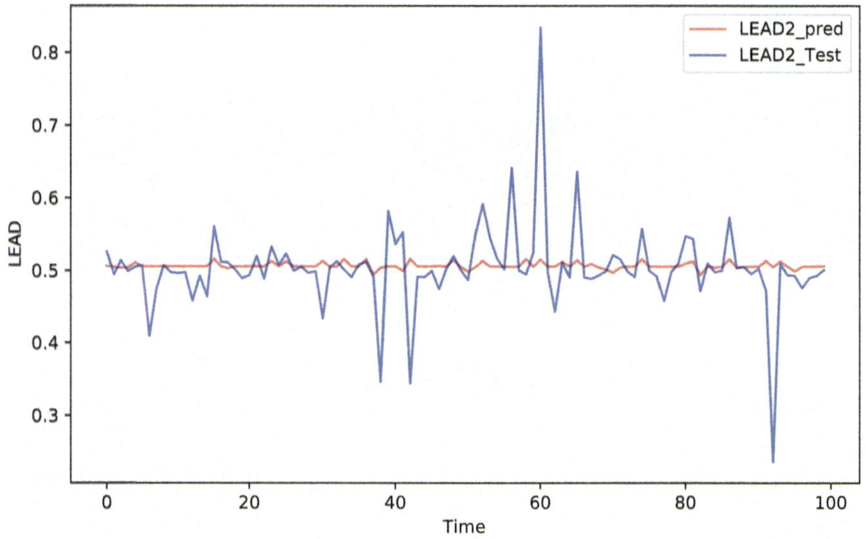

Fig. 15 LEAD2 prediction for test dataset using time

are very close. Figures 16 and 17 present LEAD1 and LEAD2 prediction using test data. From these figures, the performance of the model is better than the previous cases to forecast LEAD1 and LEAD2 properly. For instance, to predict LEAD1, the proposed model scores the following values 0.11415, 0.00208401, 0.045651, and 0.0256458 for R^2, MSE, RMSE, and MAE, respectively. Generally, the metrics values for this model outperform the other models as shown in Table 3.

8 Conclusion

In this study, the description of the HAR dataset is done, and an overview of different models for analyzing this dataset is presented. The proposed model is consisted of CNN networks that have been used in many domains such as image processing and weather prediction. The proposed framework of the deep CNN network is used to enhance the prediction ability for the HAR dataset. In the experimental results, the proposed model is able to reduce the MSE error, by 0.00240494, for the chest sensors dataset and all sensors, by 0.00208401. In future work, the proposed model can be more efficient by using the grid search to tune parameters of the model such that the range of values for each parameter must be changed manually. Moreover, also the model's architecture evolves by adjusting the shape, or adding, and removing many layers. Finally, the proposed model can be applied to other areas such as weather forecasting, stock prediction, and trajectory prediction.

References

1. Glavic M (2019) (Deep) reinforcement learning for electric power system control and related problems: a short review and perspectives. Annu Rev Control 48:22–35
2. Asfaqul Islam A (2015) A real-time ECG warning system on myocardial infarction, hyperkalemia and atrioventricular block
3. Kalantari LS, Bakr MH (2016) Wideband cloaking of objects with arbitrary shapes exploiting adjoint sensitivities. IEEE Trans Antennas Propag 64(5):1963–1968
4. Patro KK, Reddi SPR, Khalelulla SE, Rajesh Kumar P, Shankar K (2020) ECG data optimization for biometric human recognition using statistical distributed machine learning algorithm. J Supercomput 76:858–875
5. Neophytou N (2012) ECG event detection & recognition using time-frequency analysis. PhD thesis, University of Cyprus
6. Fan T, Smallman DP (2014) Screening for coronary heart disease with electrocardiography. Am Fam Phys 89(2):79–80
7. Tithi SR, Aktar A, Aleem F (2018) Machine learning approach for ECG analysis and predicting different heart diseases. PhD thesis, BRAC University
8. Gupta K, Chatur P (2012) ECG signal analysis and classification using data mining and artificial neural networks. Int J Emerg Technol Adv Eng 2(1):56–60
9. Sao P, Hegadi R, Karmakar S (2015) ECG signal analysis using artificial neural network. Int J Sci Res Natl Conf Knowl Innov Technol Eng 82–86
10. Banos O, Garcia R, Holgado-Terriza JA, Damas M, Pomares H, Rojas I, Saez A, Villalonga C (2014) mhealthdroid: a novel framework for agile development of mobile health applications. In: Ambient assisted living and daily activities: 6th international work—conference, IWAAL 2014, Belfast, UK, 2–5 Dec 2014. Proceedings 6. Springer, Berlin, pp 91–98
11. Hassoun MH et al (1995) Fundamentals of artificial neural networks. MIT Press, Cambridge
12. Arabshahi P (1996) Fundamentals of artificial neural networks [book reviews]. IEEE Trans Neural Networks 7(3):793

13. Zhao Y, Yang R, Chevalier G, Xu X, Zhang Z (2018) Deep residual BIDIR-LSTM for human activity recognition using wearable sensors. Math Probl Eng 2018:1–13
14. Aghdam HH, Heravi EJ (2017) Guide to convolutional neural networks. Springer, New York, 10(978-973), 51
15. Milosevic N, Corchero M, Gad AF, Michelucci U (2020) Introduction to convolutional neural networks: with image classification using PyTorch. Springer, Berlin
16. Khan A, Sohail A, Zahoora U, Qureshi AS (2020) A survey of the recent architectures of deep convolutional neural networks. Artif Intell Rev 53:5455–5516

Implementing Service-Oriented Game-Theoretic Security Scheme for IoV Networks in Self-driving Cars

Usha Divakarla and K. Chandrasekaran

Abstract The security, privacy, and ethical issues surrounding the implementation of connected vehicles (CVs) are numerous. This paper provides a review of recent studies that investigate various IoT-related topics in self-driving cars, such as network architectures, security, and routing. It also suggests the implementation of Service-Oriented Game-Theoretic Security (SOS), which protects against phishing, DoS, and Sybil attacks. Here, we have combined our plan with Named the Networking (NDN), a method that focuses more on the substance and verifies the accuracy of the received data rather than examining the reliability of the sender. In order to accommodate our implementation and simulate the suggested method in a smaller-scale setting, we additionally change the current alert system model.

Keywords Internet of vehicles · Connected vehicles · Autonomous vehicles · Named data networking · Forwarding information base · Service-oriented security · Neighbor hop count

1 Introduction

The car industry has produced enormous benefits, such as minimizing the requirement for human attention while driving and making efficient use of travel time, thanks to its unrelenting improvement and automation advancements around-the-clock. Better help and communicative systems reduce the likelihood of accidents. A "self-driving car" is one that, as the name implies, operates by using sensors, cutting-edge technology, such as Global Positioning System (GPS), Light Detection and Ranging (Lidar), Sonar (Sound Navigation and Ranging), and other elements of

U. Divakarla (✉)
N M A M Institute of Technology, Nitte, Karkala, India
e-mail: ushachavali@gmail.com

K. Chandrasekaran
National Institute of Technology Karnataka, Surathkal, India
e-mail: kch@nitk.edu.in

V. K. Gunjan and J. M. Zurada (eds.), *Proceedings of 4th International Conference on Recent Trends in Machine Learning, IoT, Smart Cities and Applications*, Lecture Notes in Networks and Systems 873, https://doi.org/10.1007/978-981-99-9442-7_37

its surrounding environment to estimate and control the vehicle's speed, acceleration, and distance from other obstacles in order to maneuver safely. Technically speaking, autonomous vehicles (AVs) are self-driving automobiles that require minimum driver input [1].

Manufacturers are, however, currently gradually phasing in increasing levels of autonomy until full autonomy is attained. Atkins provided succinct definitions for connected vehicle (CV) and autonomous vehicle (AV), defining that CV can be described as "an automobile fitted with technology that allows it to establish connections and exchange data with other in-car devices, nearby vehicles, and external networks." AV can be described as "a vehicle capable of performing the typical operational tasks of a conventional car, such as driving, changing lanes, parking, etc., without the need for human intervention" [2].

AVs rely on their sophisticated sensor technology to gather information from their surroundings, such as images of other vehicles, obstacles, road signs, traffic lights, pedestrians, cyclists, infrastructure, and other traffic elements on their path [3]. The GPS system is used to determine geolocation coordinates. It also considers sensor inputs from sensors that measure things like wind speed, moisture content, and sunlight intensity. Once the inputs are obtained in raw form, they must be processed repeatedly to provide valuable outputs that may be used to compute the dynamic variables, such as velocity, acceleration, direction, and to update them to ensure safe driving. Unlike AVs, CVs mainly rely on wireless connections with their nearby vehicles (V2V communications) and infrastructures (V2I communications) and function as a result with improved mobility [4]. With the use of safe intra- and inter-vehicular communications, connected and autonomous vehicles (CAV) are capable of safely navigating a regular car. CAV considers the best of both the worlds of CV and AV.

2 Literature Review

Numerous academics and writers have already examined the difficulties linked automobiles in IoV networks confront and suggested solutions to those difficulties. This work unifies many ideas put out in other studies and discusses the unique difficulties and constraints of each paper.

The benefit of the redundancy already present in heterogeneous sensors is utilized by the authors in [5] to identify malicious data injection attempts. The method for detecting attacks is based on the analysis of pairwise correlation of vehicle sensors. Vehicle data contexts are determined using cluster analysis.

It is necessary to determine the context and collect the relevant cluster data for each time interval. Pairwise cross-correlation is performed for each pair after obtaining the cluster data. On the basis of the correlation pairs, the mean and standard deviation are then determined. If the correlation varies significantly, or above a certain threshold, it is recognized as an anomaly. This work has two drawbacks.

1. It tries to detect an attack rather than preventing it.

2. It is based on the internal environment (based on sensors) or in other words concerned only with respect to a single vehicle in the IoV.

The authors of [6] provide an overview of several potential attack variants and categorize attacks based on various methods used by other academics to identify them. The majority of potential attacks consist of the following:

1. Denial-of-service (DoS) attack: A DoS attack involves flooding the target device with traffic at several network tiers in an effort to bring down the computer or network because it is unable to handle heavy loads. Making it inaccessible to its intended users is the goal of the attacker.
2. Sybil attacks: One of the most well-known destructive attacks, this one involves a hostile vehicle using several certified identifiers concurrently or sequentially. Attackers using Sybil could also disrupt networks, squander resources, and endanger security. The goal of these attacks is to weaken the influence of a reliable system and seize control of most of the network.
3. Security approaches to V2X communications could be broadly categorized as proactive and reactive mechanisms. Proactive approach prevents major attacks/ incidents before they happen by enforcing security policies like Public Key Infrastructure (PKI), digital signatures and certificates, tamper-proof hardware, etc.

Named data networking (NDN) architecture uses named data labels instead of data containers as IP does today, securing the data directly [7]. Segments transmitted in this architecture are called as "Interest Packets" which contains the "name" of the data being requested. This is used to access the corresponding repository and fetch the content. Important differences to be considered between routing in IP and NDN networks are

1. IP routing with robust security is still under research. However, a NDN routing protocol in an NDN application is named and secured NDN data packets.
2. IP Forward Information Basing (FIB) has only one next hop to forward its incoming data packets. NDN allows FIB to support multiple next hops for forwarding incoming packets without worrying about interest looping (also known as multipath forwarding).
3. NDN is capable of self-learning to discover data availability without the need of any routing protocols in a network of small size.
4. In NDN, FIB is just one, but not the sole input factor in forwarding decisions. The stateful forwarding plane changes the requirements and routing protocols accordingly.

For communication across insecure channels, authors of [8] created a cloud-aided, lightweight, certificate-less authentication protocol with anonymity. The proposed protocol complies with the vast majority of security requirements.

1. Anonymity—Hash function's irreversibility property is used to encrypt the index values, so that the adversary cannot get user's true identity.

2. Forward Secrecy—An adversary cannot generate the forward session key as the timestamp decrypted by him/her is not the same as the original one.
3. Non-repudiation—No information can be forged, as no one can get an user's private key.
4. Mutual Authentication—Both the parties in a transaction verify each other's signature and mutually validate each other's authenticity.
5. Session Key Management—The session key is generated by user's public key and Tc; in the insecure channel, the adversary cannot obtain, as it is generated by their own that does not pass through the channel.

Authors in [9] propose a new conditional privacy-preserving authentication scheme which is based on the signature scheme developed by them for a vehicular sensor network. The proposed scheme consists of five algorithms—(1) setup, (2) extract, (3) sign, (4) verification, and (5) batch-verification. It uses a single-way hash function called as Map-to-Point. There is no need for pairing operations in this scheme. It achieves the following security requirements—(1) Message Authentication and Integrity, (2) Identity privacy preserving, and (3) Traceability. Elliptic Curve Digital Signature Algorithm (ECDSA) is used in its core algorithm. Even though it is computationally more secure and efficient than the other existing digital signature algorithms, implementing it in real-world applications is not so easy, as the support for the ECC, especially modern curves, is lacking from many libraries.

Authors in [5] proposed a technique called "message ferrying" to enhance the message propagation in low-density areas. In this technique, public transport vehicles are utilized to collect as much traffic data as possible while transiting the densest areas of the city. These data could be fetched from the neighboring vehicles in the least dense areas in case of a Sybil attack as a backup and proceed forward without any deviations from the path followed.

Authors in [10] propose a blockchain-based secure authentication design for Vehicle-to-Roadside Unit (V2R) data exchange. For each vehicle in IoV, an independent session key is generated. The corresponding V2V group key management scheme is developed for data sharing among neighboring vehicles.

Authors in [11] create a public key from hop count value, reducing the number of steps (three steps) in traditional cryptography to two-way cryptanalysis and provide robust security to the network. IoT network is modeled into Stackelberg Game Theory to understand the approach in terms of attackers and defenders, where each party tries to maximize its payoff using optimum strategies. If either party deviates from the strategy, they would incur losses. This point is called the Nash Equilibrium point. Amalgamating this scheme with NDN makes gives it an additional intrinsic layer of security by implementing a data centric approach on the Neighboring Hop Count (NHC) table.

To prevent accidents, the authors of [12] suggest three different forms of alert messages. The fundamental components of CV technology are two On-Board Units (OBUs) and a Roadside Unit (RSU). The OBU is located within an automobile, while the RSU is frequently situated at a junction or along a route. These inbound messages, also known as BSMs, are designed to analyze the trajectory of the currently driving vehicle and any nearby autos in order to identify any potential impending incidents. Three V2V applications—Front Collision Warning (FCW), Emergency Electronic Brake Light (EEBL), and Blind Spot Warning (BSW)—are used to achieve this. The primary goal is to determine the relative velocity using the position vectors of the host vehicle (HV) and reference vehicle (RV). Then, utilizing this viewpoint, three algorithms are created to produce precise predictions about future events before they occur and issue pertinent cautions.

1. First Collision Warning (FCW): This feature warns the driver if the vehicle in front of them stops suddenly.
2. The Emergency Electronic Brake Light (EEBL) program alerts the driver when the vehicle in front of it abruptly stops and is hidden from the driver's vision by another vehicle.
3. When a car is traveling diagonally in the next right or left lane, the Blind Spot Warning (BSW) app warns the driver. This warning notifies the driver if there is a vehicle in their blind zone when attempting a lane change.

Advantages and drawbacks of these architectures are summarized in Tables 1 and 2. Authors in [14] use EdgeArc architecture, which is a hybrid architecture (consisting of both centralized and decentralized). Centralized architecture is implemented at higher levels of the network, and decentralized architecture is implemented at lower levels of the network.

Table 1 Summarized table of existing works with drawbacks

Proposed work	Novelty	Drawbacks
Hasan et al. [6]	Taxonomy on the misbehavior detection of most potential attacks on V2X. Classification into entity centric and data centric approach	Discuss on detection approaches but not on preventive measures
Ganeshan et al. [13]	Cluster analysis to detect attacks based on redundancy in heterogeneous sensors	Not a preventive measure but a recovery mechanism
Lo et al. [9]	EDCSA-based privacy-preserving authentication scheme for VANET	Complexity and scalability issues
Kitani et al. [5]	Message ferrying technique	Lacks authentication and not organized structure
Tan et al. [10]	Blockchain-based secure authentication for V2R	Implementation is expensive, scalability issue and poor performance for dynamic network
Dokur et al. [12]	Proposed three alert systems—FCW, EEBL, BSW	Not suitable for this work. Modified as per requirement

Table 2 Comparison of different possible architectures for IoV networks

Architecture	Advantages	Drawbacks
Centralized architecture	Efficient, quicker message exchanges Unity of command-single/fewer admin Easy to trace	Single point of failure Limited scalability Bandwidth bottleneck
Decentralized architecture	Lesser probability for DDOS attack Efficient management of data Easy scalability	Not suitable for real world due to lack of organized structure
Publish/subscribe architecture	Asynchronous-adding or removing subscribers accordingly Greater scalability and flexibility	Requires precise policy Testing could be difficult
Broadcast architecture	Efficient communication Lower traffic loads Low cost deployment	Requires high bandwidth Personal communication is not possible

3 Proposed Methodology

In this section, we discuss the architecture, vehicular interaction method, security, and implementation features of the proposed system. These are the main points necessary for a fully realized system which can be implemented in the real world.

3.1 Architecture

Self-driving cars are constantly worked on field due to the multitude of complications they face. One major such component that we focus in this paper is the IoT system where the vehicles get to interact with each other and with other systems such as traffic lights. In the IoT system proposed, we have all the vehicles connected to a central cloud server based on the edge-based architecture for CVs, called EdgeArC which enables the vehicle to collaborate with surrounding vehicles, offloading workloads to edge servers and connected devices/things (denoted as XEdge), and remote cloud [14].

The security and integrity of such a system are vital to make sure that there is no malicious party trying to give bad information to the vehicles, which can lead to a fatal accident.

Hence, we first use NDN, networking for communications within the IoT network. This gives us an initial intrinsic security within the communications of the system, which can be further added upon by using other forms of security. We also use the Service-Oriented Game-Theoretic Security (SOS) architecture [11] as shown in Fig. 2 for encrypting and decrypting the data for higher security. In the SOS system, we store the hop counts between all nodes of the network in a neighboring hop count (NHC) table in each individual node (Fig. 1).

The hop count from this NHC table is used as the public key while encrypting and decrypting data in a dynamic formula of $P_{ij} = h_{ij}$ to give a higher order of security.

Fig. 1 Proposed algorithm to generate alert message

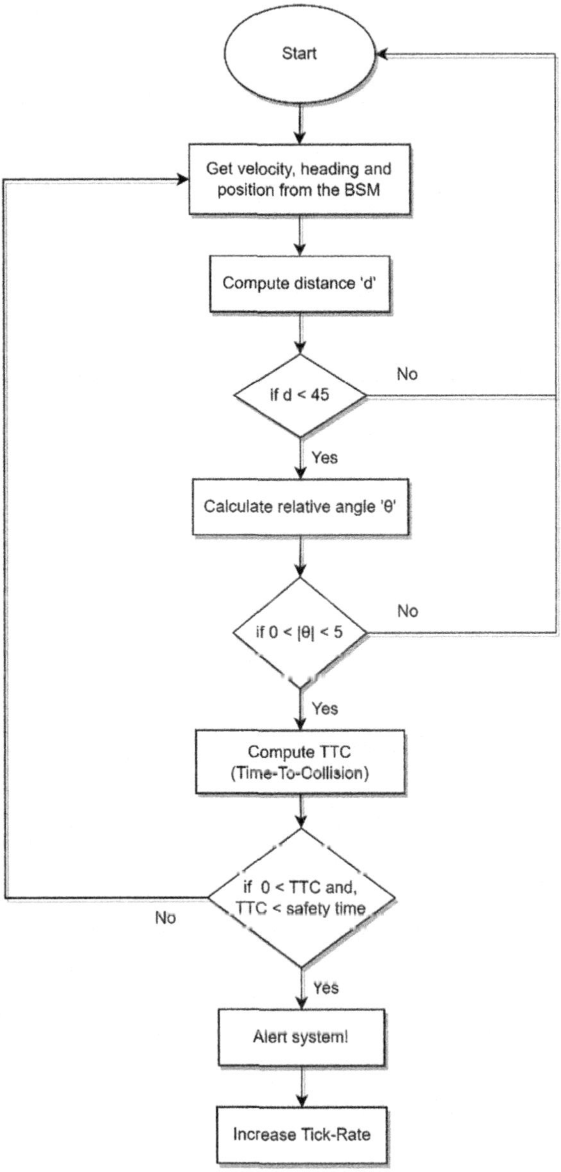

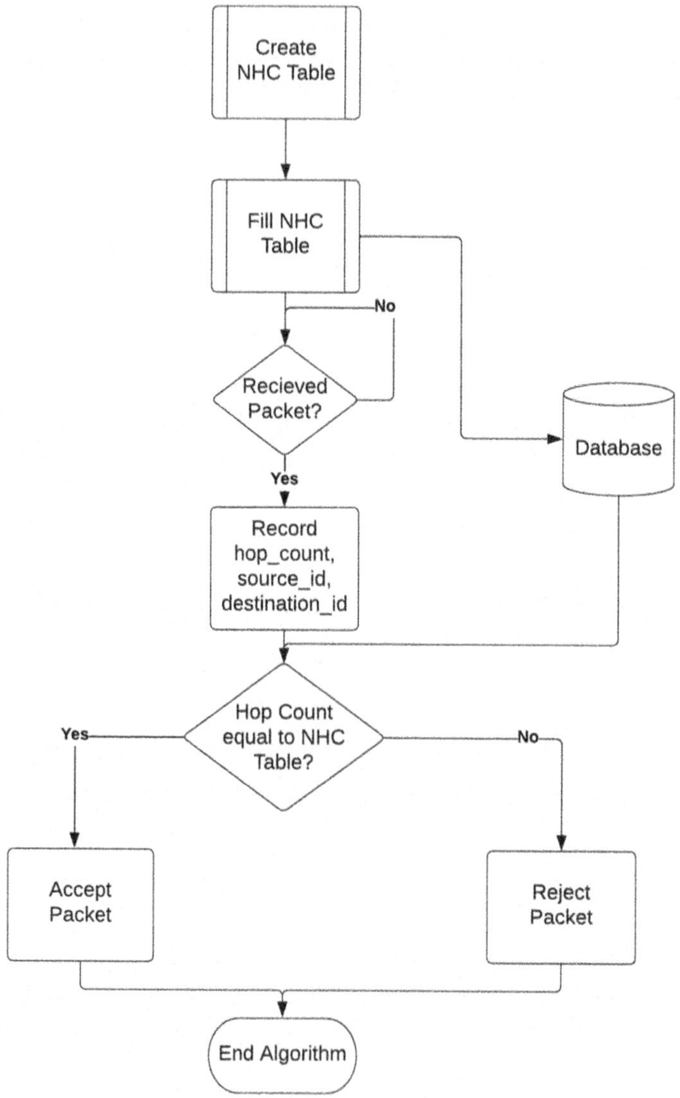

Fig. 2 SOS algorithm

To decrypt the data, we use the decryption function $F_{\text{decry}} = \log_n P_{ij}$ to give us the hop count again. This method of encryption and decryption is done with ease using the NHC table. Also, using the NHC table, the public key is known on both nodes before-hand, hence eliminating the first step of the three way handshake where the public key needs to be communicated to both nodes, making the connection 33% faster compared to regular encryption methods.

When we share a data packet, we first check the hop count associated with it and the source node. If the hop count does not match our NHC table, we immediately drop the packet, preventing malicious nodes from accessing the data. If it is a match, we use the formula to get the public key, which is used to decrypt the final data communicated as shown in Fig. 1.

4 Implementation

We can now implement a system using ML for the self-driving cars and we can analyze the communications between the IoT systems with the goal of checking the feasibility of the proposed SOS system with the IoV approach. We do the simulation as shown in Fig. 3 with constraints set to simulate a scenario of congested traffic, which is done to achieve highest number of interactions of the autonomous vehicles with each other.

The important points of the given approach are the security against attack, the speed of communication to prevent any accidents, and also the overhead per message. These metrics tell us the usability of SOS system in a realistic scenario and we can be sure that it is geared to the needs of speed and security of networks of the future.

We have built a machine learning model using the evolutionary technique known as NEAT, which stands for NeuroEvolution of Augmenting Topologies, within the simulation. In reinforcement learning scenarios, neuroevolution, which entails evolving artificial neural networks through evolutionary algorithms, has

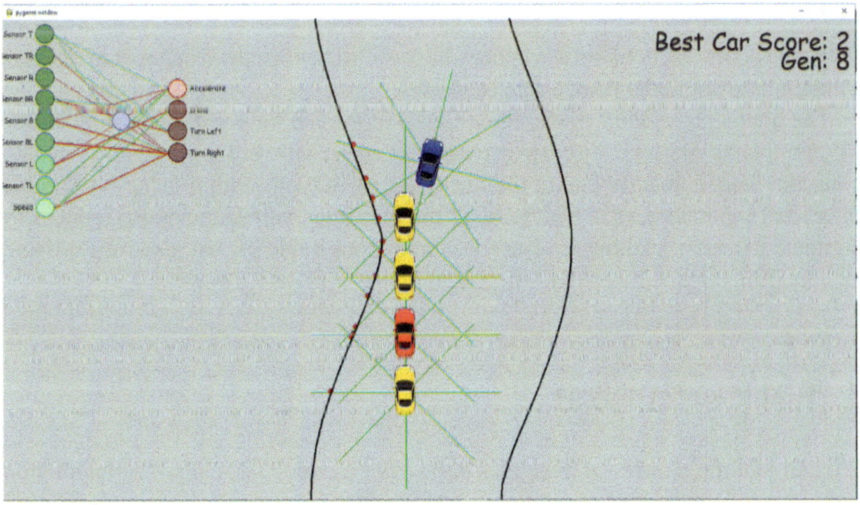

Fig. 3 Simulation of the NEAT model on five cars

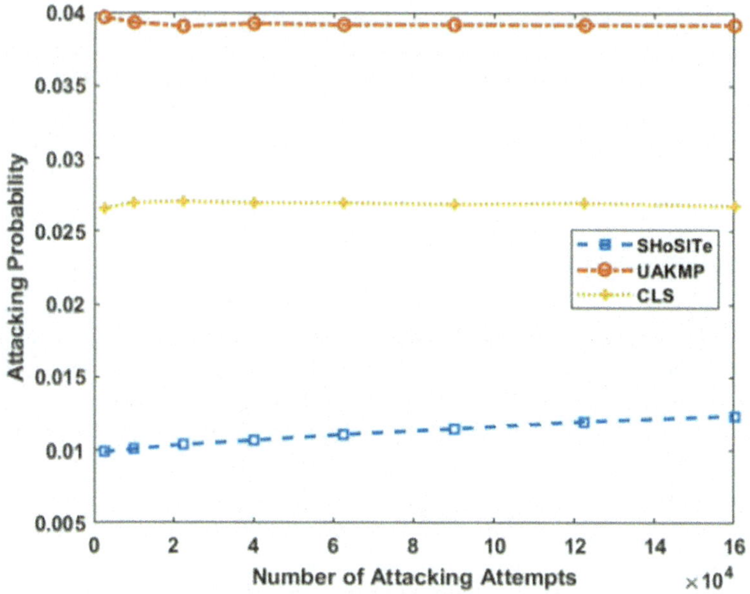

Fig. 4 Attack probability

demonstrated remarkable success, particularly when dealing with concealed state information.

The message overhead in an SOS system increases along with the total nodes in the network, which resembles the entire IoV system. The total number of autonomous vehicles connected in this manner can vary greatly, and so the central servers resolving all interactions between the vehicles must have high throughput to maintain availability even in the peak usage scenarios.

The attack probability as shown in Fig. 4 of the SOS algorithm goes up number of attempts as opposed to other security schemes, where the probability remains constant. This is mainly due to the drawback of SOS algorithm using NHC tables instead of separate keys, which reduces the message overhead and also reduces the computation time, making it ideal for autonomous vehicles, but also runs a risk of attacks from a malicious party with much higher processing power.

5 Result and Analysis

We first plot the number of messages passed per second, also called the interactions per second, between the autonomous vehicles. This is done to show the high needs of message passing in an IoV autonomous vehicle setup.

We see from Fig. 5 that the number of messages passed between vehicles in this scenario can go from an average to its peak of more than double based on the

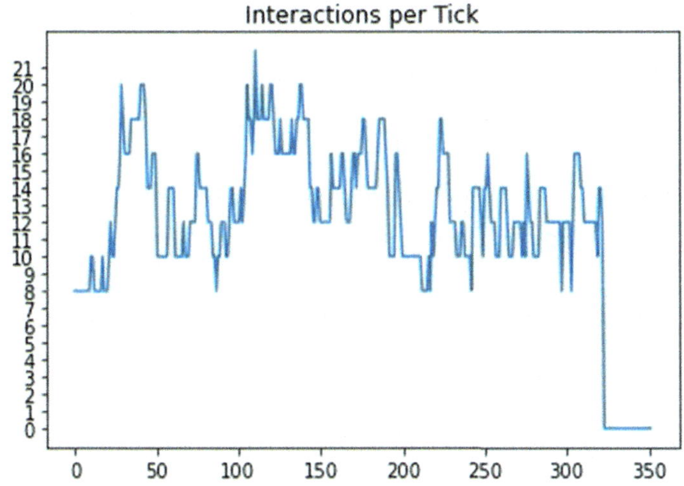

Fig. 5 Number of messages passed per tick

congestion of traffic. This gives us a basis of needs to ensure that the server must be able to attend loads more than double the average load, with good response time and availability. The central server resources must be allocated based on these needs.

We also plot the right sensor data and simulate a few random attacks, which are promptly ignored and reset by the system based on the comparison on spike method we have explained before.

6 Conclusion

Vehicles are rapidly evolving into the most advanced mobile devices within the realm of the Internet of Things (IoT). They are equipped with extensive arrays of integrated sensors and onboard computers that connect to the cloud through wireless technologies. Nevertheless, establishing a cloud-based IoT infrastructure for Connected Cars is a complex endeavor, not without its challenges. Papers discussed in the introduction provide insights into the challenges and issues in the development of self-driving cars and the solutions proposed to address them. We proposed a four-layer architecture combined with NDN, using the SOS security approach to give us the best speed and security. We also looked into the implementation features that ensure the feasibility of the proposed system in a real-time environment under various aspects. The integration of IoT, autonomous systems, and advanced network architectures is essential for the successful deployment of autonomous vehicles, which is expected to transform the way we live and travel.

In the future, we plan to simulate our scheme on CARLA simulation software to test it in virtual road traffic and later on validating, implement it on a real testbed securing all the vehicles from potential attacks like DoS, Sybil, and False Injection attacks in an IoV network.

References

1. Litman T (2017) Autonomous vehicle implementation predictions. Victoria Transport Policy Institute, Victoria, BC, Canada
2. ATKINS (2016) Research on the impacts of connected and autonomous vehicles (CAVs) on traffic flow. Stage 2: Traffic modelling and analysis technical report, Department for Transport. ATKINS, London, UK
3. Gibbs S (2023) Google's self-driving car: how does it work and when can we drive one? Guardian 2014, 5–29. Available online: https://www.theguardian.com/technology/2014/may/28/google-self-driving-car-how-does-it-work. Accessed on 01 Mar 2023
4. Jadaan KZ, Abukhalil S, Yazan (2017) Connected vehicles: an innovative transport technology. Procedia Eng 187:641–648. https://doi.org/10.1016/j.proeng.2017.04.425
5. Kitani T, Shinkawa T, Shibata N et al (2008) Efficient VANET-based traffic information sharing using buses on regular routes. In: Vehicular technology conference, VTC Spring. IEEE
6. Hasan M, Mohan S, Shimizu T, Lu H (2020) Securing vehicle-to-everything (V2X) communication platforms. IEEE Trans Intell Veh 5(4):693–713. https://doi.org/10.1109/TIV.2020.2987430
7. Afanasyev A, Burke J, Refaei T, Wang L, Zhang B, Zhang L. A brief introduction to named data networking
8. Shen J, Gui Z, Ji S, Shen J, Tan H, Tang Y (2018) Cloud-aided lightweight certificateless authentication protocol with anonymity for wireless body area networks. J Netw Comput Appl 106:117–123
9. Lo N-W, Tsai J-L (2016) An efficient conditional privacy-preserving authentication scheme for vehicular sensor networks without pairings. IEEE Trans Intell Transp Syst 17(5):1319–1328. https://doi.org/10.1109/TITS.2015.2502322
10. Tan H, Chung I (2020) Secure authentication and key management with blockchain in VANETs. IEEE Access 8:2482–2498. https://doi.org/10.1109/ACCESS.2019.2962387
11. Kar P, Misra S, Mandal AK, Wang H. SOS: NDN based service-oriented game-theoretic efficient security scheme for IoT networks
12. Dokur O, Katkoori S (2022) Three connected V2V applications based on DSRC basic safety messages. In: 2022 International conference on connected vehicle and expo (ICCVE), Lakeland, FL, USA, pp 1–6. https://doi.org/10.1109/ICCVE52871.2022.9743088
13. Ganesan A, Rao J, Shin K (2017) Exploiting consistency among heterogeneous sensors for vehicle anomaly detection, SAE technical paper 2017-01-1654. https://doi.org/10.4271/2017-01-1654
14. Lu S, Ammar N, Ganlath A, Wang H, Shi A (2022) A comparison of end-to-end architectures for connected vehicles. In: 2022 Fifth international conference on connected and autonomous driving (MetroCAD), Detroit, MI, USA, pp 72–80. https://doi.org/10.1109/Metro-CAD56305.2022.00015

A Comprehensive Review of the Potential Use of Genetic Analysis in Precision Medicine for Rheumatoid Arthritis

A. Ezhil Grace and R. Thandaiah Prabu

Abstract Rheumatoid Arthritis (RA) is a joint-related inflammatory condition, that causes pain, swelling, stiffness, fatigue, and deformities. Recent advances in genetic technologies have identified various genetic risk factors, contributing to the complex nature of RA with a polygenic inheritance pattern. Precision medicine has the potential to improve RA patient outcomes by utilizing a patient's genetic makeup to customize medical therapies. By assessing a patient's genetic risk factors, clinicians can predict susceptibility to RA and select personalized treatment strategies, potentially identifying new drug targets and developing more effective therapies with fewer side effects. As precision medicine continues to evolve, the integration of genetic technologies and our expanding understanding of the pathophysiology of RA present exciting opportunities to provide a more targeted and personalized approach to the management of RA, leading to better patient outcomes. This review paper emphasizes the potential of genetic analysis in precision medicine to revolutionize the treatment of RA.

Keywords Rheumatoid Arthritis · Autoimmune diseases · Genetics · Pathophysiology · Precision medicine · Treatment

1 Introduction

RA is a persistent autoimmune disorder resulting in pain, stiffness, and reduced mobility. Genetics may contribute, but medication and lifestyle alleviate symptoms and slow progression. Family history increases risk, with factors accounting for two-thirds of variation. Genes impact on human health and disease, including disease development and progression. RA development is linked to genetics, with specific

A. Ezhil Grace (✉) · R. Thandaiah Prabu
Saveetha School of Engineering, SIMATS, Chennai, Tamil Nadu, India
e-mail: ezhilgracea9072.sse@saveetha.com

R. Thandaiah Prabu
e-mail: thandaiahprabur.sse@saveetha.com

V. K. Gunjan and J. M. Zurada (eds.), *Proceedings of 4th International Conference on Recent Trends in Machine Learning, IoT, Smart Cities and Applications*, Lecture Notes in Networks and Systems 873, https://doi.org/10.1007/978-981-99-9442-7_38

involvement of the HLA gene complex and immune or inflammation-related genes. Environmental factors also play a role, so having genetic variations does not guarantee to the development of RA. Next-generation sequencing and CRISPR/Cas9 gene editing have led to new genetic variant discoveries and offer the potential for gene therapies. These advances expand our understanding of genetics' role in health and disease and could revolutionize medicine [1].

2 Pathophysiology of RA

RA's pathophysiology involves genetic and environmental factors that trigger an abnormal immune response, causing inflammation and joint damage. Autoantibodies and cytokines perpetuate inflammation, leading to deformities and organ damage. Treatment aims to reduce inflammation and prevent joint damage. In RA, the immune system mistakenly targets and harms the healthy tissue in the joints, resulting in inflammation and damage [2].

RA inflammation attacks synovial tissue causing inflammation, immune cell infiltration, and cytokine production leading to joint damage. Genes and immune pathology are predictors of disease activity and joint injury progression.

Figure 1 shows unique DNA methylation regions associated with arthritis-related genes in four groups. The groups include Caucasians and African Americans with and without RA, and the data present the number of unique methylation regions identified in buccal cells and blood monocytes for each group.

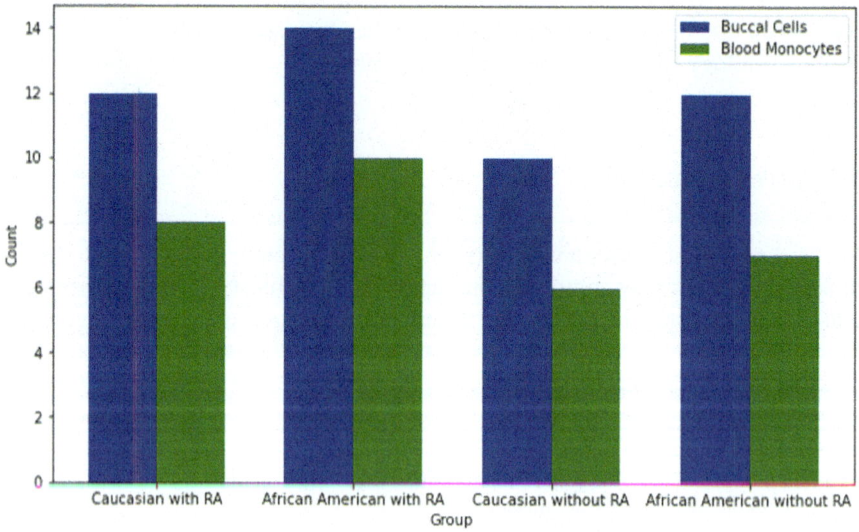

Fig. 1 Unique DNA methylation regions linked to arthritis-related genes [3]

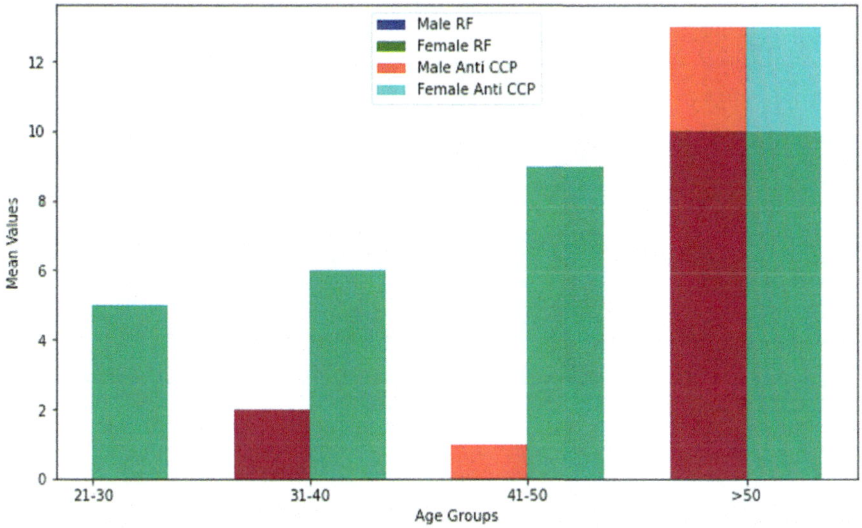

Fig. 2 Sex-specific mean values of RF and anti-CCP antibodies [4]

3 Genetics of RA

3.1 Genetic Factors Linked to RA Risk

Numerous genetic risk factors are associated with RA. The Rheumatoid Factor (RF) and anti-CCP antibody mean values for different age groups and genders are shown in a grouped bar plot in Fig. 2. The study emphasizes the importance of understanding genetic markers in RA, given that the condition is inherited.

The evaluation of gene classifiers is crucial for accurately identifying and classifying RA patients. In order to evaluate the performance of gene classifiers in RA classification, various metrics were utilized. Figure 3 presents a bar chart depicting the performance metrics for different gene classifiers, including precision, F_1-score, accuracy, and recall. The classifiers, such as SVM, XGBoost, KNN, Decision Tree, Logistic Regression, and Random Forest, were assessed based on these metrics.

3.2 The Role of Rare and Prevalent Genetic Variants in the Development of RA

Certain genetic markers have been shown to increase the likelihood of developing RA, especially in seropositive patients. HLA-DRB1 alleles have been identified as significant risk factors for seropositive RA. Integrating genetic association research with other fields, such as epigenetics and proteomics, is crucial for personalized

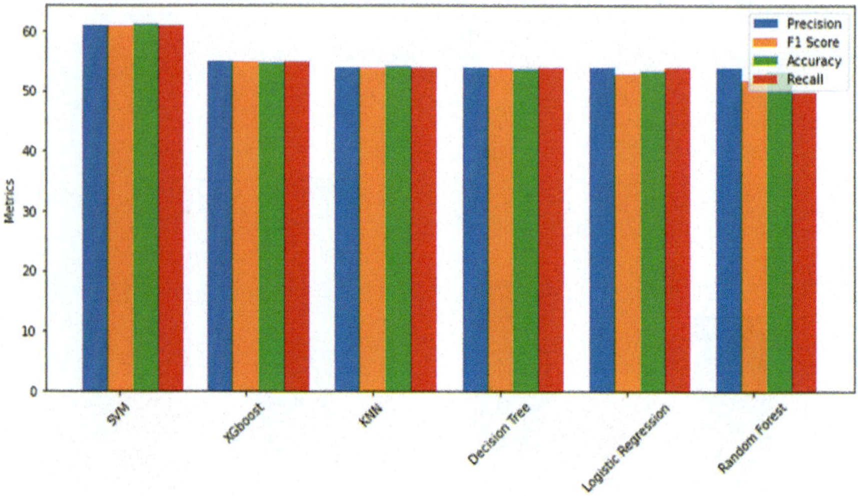

Fig. 3 Performance metrics of RA gene classifier [5]

therapy and prevention of chronic RA. Both common and rare genetic variations contribute to RA, with rare variants having a more pronounced impact [1].

4 Precision Medicine in RA

4.1 The Potential of Precision Medicine in RA Patient Care

Precision medicine considers genetic, environmental, and lifestyle factors to provide personalized treatment plans. Identifying and quantifying genetic variation are crucial to find and ranking therapeutic targets based on human genetics. A deeper understanding of genetics can enhance drug development by identifying genes that regulate drug action and metabolism and their variations. Large-scale sequencing can be useful in identifying such genes and their variations. Table 1 provides a thorough summary of the genetic risk factors for the development of RA, including family history and research on identical twins.

Table 1 Risk factors for RA [6]

Risk factor	Risk percentage
Family history of RA	0.8%
No family history of RA	0.5%
The identical twin of a person with RA	12–15% chance of developing RA

4.2 How Clinicians Can Use Genetic Risk Factors to Predict RA Susceptibility

Clinicians can use genetic testing to pinpoint RA risk factors and develop individualized preventative and therapy plans. In addition to genetic testing, clinicians may take into account other variables such as family history, smoking, and diet. Increased awareness of RA and its genetic markers can lead to earlier diagnosis and personalized treatment plans. The HLA genes are responsible for producing proteins that assist the immune system in distinguishing between pathogenic and self-cellular components. Moreover, many other genes associated with RA also impact the immune system [1].

4.3 Personalized Treatment Strategies Based on Genetic Makeup

Advanced genotyping technologies like next-generation sequencing and SNP microarray have facilitated studies investigating the relationship between various traits and the genome. Integrating these findings into clinical practice involves techniques such as evaluating target genes, repurposing drugs, and utilizing MR-based causal inference. Genetic data offer valuable insights into the causal relationships between traits, which can lead to meaningful associations in observational studies [7].

Machine learning approaches have been used to predict how RA patients would react to anti-TNF treatment. By conducting a microarray experiment, researchers identified the most significantly annotated outcome genes and discovered enriched gene sets with interconnected functional relevance, using the concept of existence dependency [8].

5 Current Therapies for RA

Genetic disorders occur in every cell due to the shared set of inherited genes, making them incurable but manageable with ongoing gene therapy research. Mucosal inflammation and microbiological factors are under investigation as potential causes of RA. A preclinical stage of RA has been identified, characterized by autoantibodies, and categorized as local or systemic autoimmunity [1].

Current RA therapies have limitations, including side effects, incomplete response, high cost, lifelong treatment, and limited impact on joint damage. Treatment efficacy varies among individuals, and long-term use presents challenges [2].

Figure 4 presents a graph comparing the performance of the two-stage technique and direct classification in predicting drug response for RA patients. The

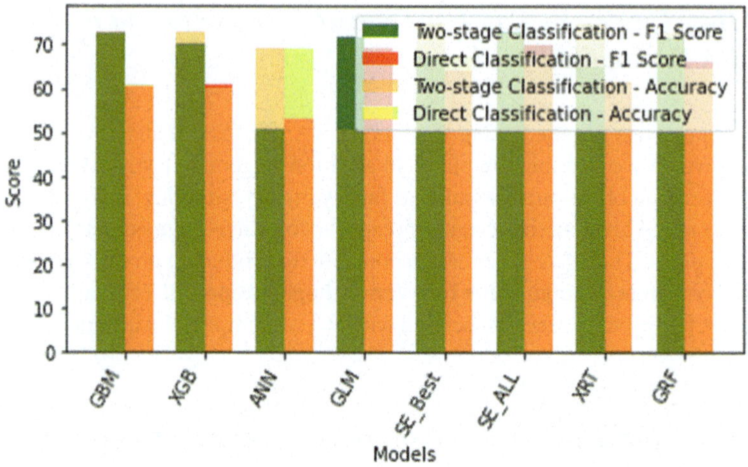

Fig. 4 Performance comparison of two-stage technique and direct classification for drug response prediction in RA patients [9]

graph demonstrates the accuracy and F_1-score metrics for various models, including Stacked Ensemble, Boosting Ensemble, Bagging Ensemble, and Deep learning models. It provides insights into the relative performance of these models on the test set.

Based on genetic findings, a study identified 87 high-confidence risk genes for RA and suggested potential new drug targets and treatments. These studies provide an important new understanding of the genetic basis of RA and open up interesting new treatment approaches [10].

6 Conclusion and Future Directions

Precision medicine in Rheumatoid Arthritis (RA) offers personalized treatment, monitoring, and early intervention, utilizing biomarkers, genomic profiling, and wearable technology. Genetic findings have identified high-confidence risk genes for RA, presenting potential drug targets.

Efforts in precision medicine aim to develop effective therapies with fewer side effects, addressing patient stratification, biomarker identification, and data analysis challenges. Personalized treatment, disease monitoring, and early intervention strategies are crucial for better outcomes. Continued research in precision medicine is vital, focusing on biomarker identification, data analysis, and novel treatments. Collaboration among researchers, clinicians, and patients drives progress and implementation. Precision medicine's development in RA improves quality of life, lowers disease activity, and opens up new therapeutic possibilities.

References

1. Padyukov L (2022) Genetics of rheumatoid arthritis. In: Seminars in immunopathology, vol 44, no 1. Springer Science and Business Media Deutschland GmbH, pp 47–62. https://doi.org/10.1007/s00281-022-00912-0
2. Bharadwaj S, Sarma P (2020) Symptomization of rheumatoid arthritis in patients on pathological examination: a case study. In: 2020 IEEE international students' conference on electrical, electronics and computer science, SCEECS 2020. Institute of Electrical and Electronics Engineers Inc., Feb. https://doi.org/10.1109/SCEECS48394.2020.29
3. Craig G, Kenney H, Nilsson EE, Sadler-Riggleman I, Beck D, Skinner MK (2021) Epigenome association study for DNA methylation biomarkers in buccal and monocyte cells for female rheumatoid arthritis. Sci Rep 11(1). https://doi.org/10.1038/s41598-021-03170-6
4. Narasingam A, Christina E, Samuel J, Vijaykanth N (2012) Studies on rheumatoid arthritis cases in Chennai, Tamil Nadu using anti-cyclic citrullinated peptide (anti-CCP) assay
5. Sana A, Khawaja SA, Farooq M (2021) Rheumatoid arthritis genes classification using ML classifiers. In: 4th International conference on innovative computing, ICIC 2021. Institute of Electrical and Electronics Engineers Inc. https://doi.org/10.1109/ICIC53490.2021.9693009
6. Kuo CF et al (2017) Familial aggregation of rheumatoid arthritis and co-aggregation of autoimmune diseases in affected families: a nationwide population-based study. Rheumatology (U K) 56(6):928–933. https://doi.org/10.1093/rheumatology/kew500
7. Sonehara K, Okada Y (2021) Genomics-driven drug discovery based on disease-susceptibility genes. Inflamm Regeneration 41(1). https://doi.org/10.1186/s41232-021-00158-7
8. Sieberts S et al (2016) Crowdsourced assessment of common genetic contribution to predicting anti-TNF treatment response in rheumatoid arthritis. Nat Commun 7. 10.1038/ncomms12460
9. Chen S, Gupta N, Galbraith WB, Shah V, Cirrone J (2022) Prediction of drug effectiveness in rheumatoid arthritis patients based on machine learning algorithms, Oct 2022. [Online]. Available: http://arxiv.org/abs/2210.08016
10. Birga AM, Ren L, Luo H, Zhang Y, Huang J (2022) Prediction of new risk genes and potential drugs for rheumatoid arthritis from multiomics data. Comput Math Methods Med 2022. https://doi.org/10.1155/2022/6783659

Clean Sustainable Society Using CNN

Shivani Sharma, Taniya Goel, Radhika Piplani, and Bipin Kumar Rai

Abstract Waste disposal is a significant challenge in India. The rapid increase in the global human population is leading to a surge in mass production. The prevailing model of mass production and disposal follows a linear approach, wherein waste is primarily lost and not recycled or reused. This not only exacerbates environmental problems by increasing pollution but also results in economic losses due to the wasted potential of these materials. This paper basically focuses on an intelligent waste management system for redirecting the various kinds of trash materials to their correct positions by uploading the image of waste in the web application and it will classify images into following categories: reuse, recyclable, and organic waste. To identify the category of waste items, a machine learning model is used, followed by the implementation of a Convolutional Neural Network model to direct the item to the appropriate trash bin.

Keywords Waste management · Mass production · Mass disposal · Pollution · Economic loss · Reuse · Recyclable · Organic waste · Convolutional Neural Network

1 Introduction

In India day by day the rapid population is growing, and the condition is getting worsen by the continuously increasing waste in metric of ton per day. If we see, the cities are getting smarter nowadays, but waste is not as well, and sometimes, the garbage in the bins gets to above the point that causes smell, dirt, and various types

S. Sharma (✉) · T. Goel · R. Piplani
Department of Information Technology, ABES Institute of Technology, Ghaziabad, Uttar Pradesh, India
e-mail: shivani.shivanisharma@gmail.com

B. K. Rai
Department of Computer Science and Engineering, Dayananda Sagar University, Bengaluru, India
e-mail: bipinkrai@gmail.com

V. K. Gunjan and J. M. Zurada (eds.), *Proceedings of 4th International Conference on Recent Trends in Machine Learning, IoT, Smart Cities and Applications*, Lecture Notes in Networks and Systems 873, https://doi.org/10.1007/978-981-99-9442-7_39

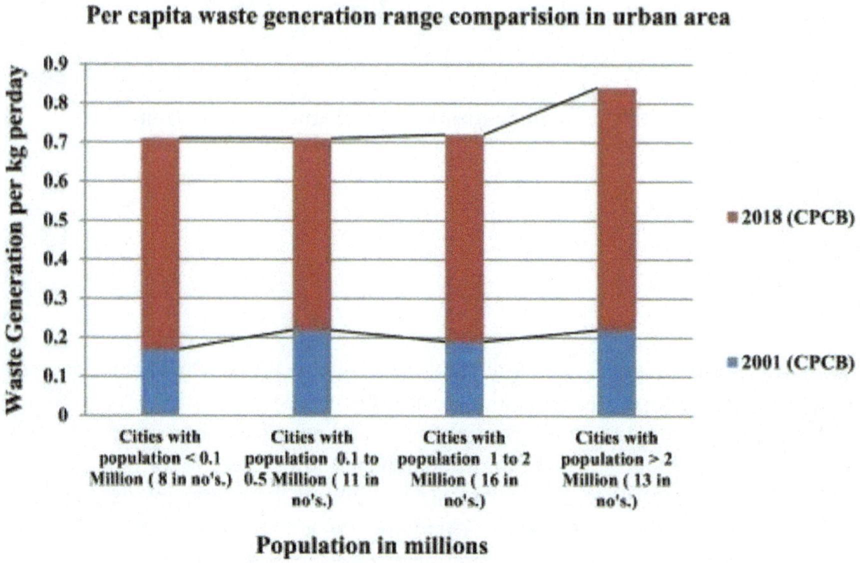

Fig. 1 India's waste generation and its future challenges [5]

of flu and diseases. Everyday country produces 1.50 lac metric tons of solid waste. However, only 20–30% (27,000 metric ton per day) of the total waste generated is being processed while remaining 70–80% (108,000 metric ton per day) is dumped into landfills sites [1]. Whereas in urbanized areas approximately 170,000 tons of waste are generated nearly every day, the waste generation is approximately 0.17 kg per person per day in smaller cities. In contrast, larger cities generate 0.62 kg of waste per person per day [2]. In India 0.57 kg of waste is generating per person per day, and if we see globally, that is 0.74 kg of waste is generating per person every day [3]. India is producing 277.1 million ton of waste every year, which is likely to be treated 387.8 million ton in 2030 if it does not find a right way of disposing the waste and the landfilling, and moreover, 543.3 million ton of waste will be touched by 2050 as a result of the rising urbanization of the population and the expansion of the economy. India seems to be leading third position of solid waste generation. However, India is figured to be projected to increase by 900 gm of waste everyday by 2050 [4]. The following graph shown in Fig. 1 describes the India's waste generation and its future challenges.

1.1 Research Objective

- To design and implement a web portal for users to interact with the waste management system.

- To integrate Google Maps API for location-based services, such as finding nearby NGO's, industries, and slum areas.
- To develop a real-time waste management system using CNN for accurate waste classification.
- To explore future possibilities for enhancing the waste management system, such as incorporating machine learning algorithms for predictive analytics and optimizing waste collection routes.

2 Literature Review

Malik et al. [6] proposed "Waste Classification for Sustainable Development by Image Recognition" which provides an overview of current deep learning techniques that are utilized for waste classification. The paper emphasizes the use of Convolution Neural Networks (CNNs) to classify waste effectively [6]. Chaturvedi et al. [7] describe an approach for detecting and sorting municipal solid waste through the use of the ResNet-50 model. The classification model considers waste materials such as cardboard, glass, metal, paper, and plastic [7]. Pardini et al. [8] put forward this innovation into an IoT-based waste management model. This effective model works by using sensors and communication technologies in which data will be collected from smart bins and transmitted over this online platform, where the user is having access to check availability of their perspective compartments disperse around a city. It is the Web Version and mobile application of a smart container that facilitates the implementation of the new waste management system [8]. Imran et al. [9] bring forward the "Data Analysis" approach which utilizes predictive analysis with the aid of Quantum Geographic Information System (QGIS). This system employs virtual analytical tools to accurately assign waste bins by forecasting the most efficient route for the garbage collector truck [9]. Vishnu et al. [10] have enhanced the waste management systems by creating and validating a "Hybrid Network" that enables efficient management of trash bins in public and residential areas. This system is developed on an IoT platform, utilizing PBLMU and HBLMU, and can perform long-distance data transmission with the Lora network. The system also includes a Wi-Fi functionality for homes, which enables users to visualize the trash bin levels through an intelligent GUI-monitored system [10]. Sharma et al. provide real-time data about the garbage level in various locations throughout the city/town. The system enables garbage analysts to monitor each dustbin's fill level accurately, allowing for better waste management throughout the city in real-time [11]. In Bano et al. [12], proposed a Smart Bin Mechanism (SBM) for real-time tracking of trashcans in smart cities. The SBM operates on the principles of Artificial Intelligence of Things (AIoT) and aims to implement concept of reduce, recycle, and reuse in waste management. With the SBM, users can obtain information about each bin at real-time, preventing overloading and ensuring efficient waste management. The system employs fuzzy logic for decision-making to identify the most appropriate locations in cities for installing trash bins, optimizing the waste collection process [12].

3 Methodology

The proposed solution of clean sustainable society using CNN focuses on a real-time waste management system (WMS) to make proper waste disposal, reuse, and recycle of waste materials. With this proposed model, identification and classification of the waste material are done, and then, it is redirected to the right category which will be picked up by the nearby services. The service of Google Maps is used to detect the nearby slum area and NGO's. A Google Map can either show a single location's position or the route between locations. These waste materials are collected to NGOs, industries, and disposable sites. Reusable materials will be collected by NGOs, recyclable materials will be collected by tied-up industries, and non-recyclable (i.e., organic waste) will be sent to disposable sites for incineration, vermicomposting, and landfilling.

3.1 Process Flow Diagram

Process flow diagram (PFD) shows the series of operation, including its components and workpieces. It is an essential technique for understanding and analyzing the process design. The action and protocol being used in our web portal are shown in Fig. 2.

Processes:

Process 1: In this process, the pick of any waste material from various sources, such as households, businesses, or public places, acts as an input for the system.
Process 2: Now, the user can view nearby slum areas.
Process 3: There will be an option of adding the waste material and classifying its category.
Process 4: This process is divided into three sub-processes:

Process 4.1: The waste material which is recyclable materials will be send to tied-up industries.
Process 4.2: The waste material which is reusable materials will be send to NGO.
Process 4.3: The waste material which are non-recyclable (i.e., organic waste) will be sent to disposable sites.

Process 5: Reward system to involve people to get more interested toward our waste management web application.

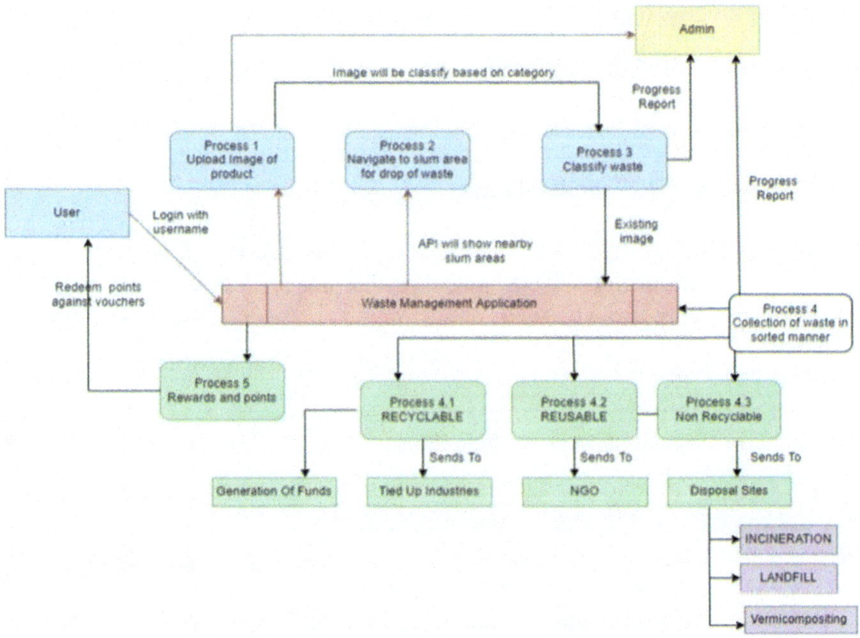

Fig. 2 Process flow diagram showing the relationship between the components of the WMS

4 Implementation

4.1 Working Approach Represented in the Form of Pseudo Code Described

The process of waste classification and routing in the proposed waste management system can be summarized by the following pseudo code:

4.1.1 image ← INPUT

```
CNN(image)
CNN_model = generate_CNN_model
Waste_category = classify(CNN_model, image)
IF(Waste_category == Reusable)
  Send to NGO
IF(Waste_category == Recyclable)
  Send to Industry
```

```
IF(Waste_category == Organic)
  Send to Disposable_site
```

4.1.2 Classify (CNN_model, Image)

ResultSet = CNN_model.predict(image)
IF(ResultSet == 1)
  prediction = Recyclable Waste or Reusable waste
ELSE
  prediction = Organic Waste

The parameters used in above pseudo code are:

image = image or waste via camera or by upload
CNN_model = CNN Model which is trained
Waste_category = category of waste detected
ResultSet = holds 0 or 1

This pseudo code demonstrates the sequential steps involved in processing waste images using a Convolutional Neural Network (CNN) model. The input image is passed through the CNN model to classify the waste into specific categories. Depending on the determined waste category, appropriate actions are taken to route the waste to the designated destinations. The utilization of this pseudo code ensures efficient waste management and facilitates the proper disposal, recycling, and reusability of waste materials.

4.2 User Dashboard Layout

After the authentication, the user will access the working modules such as: Classify waste category, Add waste Product after classifying the category, View Wastes Upload which has been uploaded by an particular user, View nearby NGO's, View Nearby Slum Area's, View Society and View Industries by using integrated Google Map API, View pickup schedule which has been recently scheduled by user, and the demonstration is shown in Fig. 3.

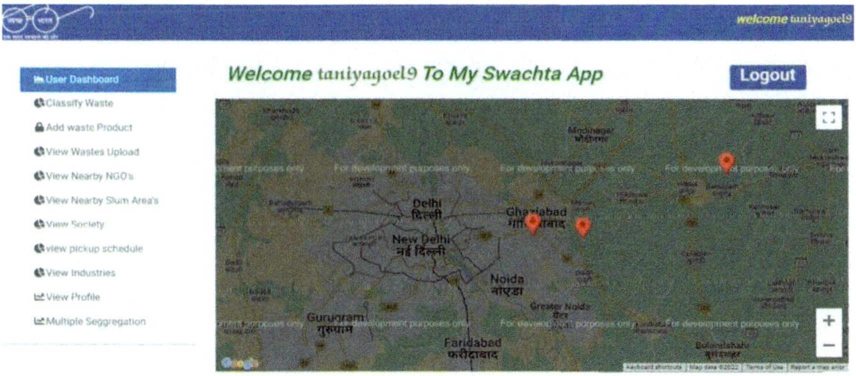

Fig. 3 User dashboard

4.3 CNN Model Layout

In today's world, the concept of clean and sustainable society has become more important than ever before. The waste prediction uses CNN model and binary classification technique to categorize the waste in recyclable or non-recyclable (i.e., organic waste), according to the category which found that the waste pickup will be scheduled. Once the waste is classified, appropriate waste pickup schedules can be established. For recyclable waste, pickups can be scheduled as per the regular waste collection routines. This ensures that recyclable materials are efficiently collected and directed to the appropriate recycling processes, thereby reducing the strain on landfills and promoting the reuse of valuable resources. By implementing the waste prediction system using CNN models and binary classification, we can accurately determine whether waste items are recyclable or non-recyclable. This approach allows us to analyze visual data, such as images of waste items, and classify them with a high degree of accuracy. This process is shown in Fig. 4.

5 Results and Discussion

5.1 Real-Time Image Procession of CNN Model

A proposed solution for the sustainable development of smart cities is to implement a Convolutional Neural Network (CNN) model for classifying real-time waste images based on their features. This model can be trained using large datasets of waste images to accurately distinguish between categories of waste such as recyclable, reusable, and organic waste. The dataset which is used in waste classification for training contains the 22,500 images of organic and recyclable objects. The analysis has been made in three steps: the first includes data preparation and builds model

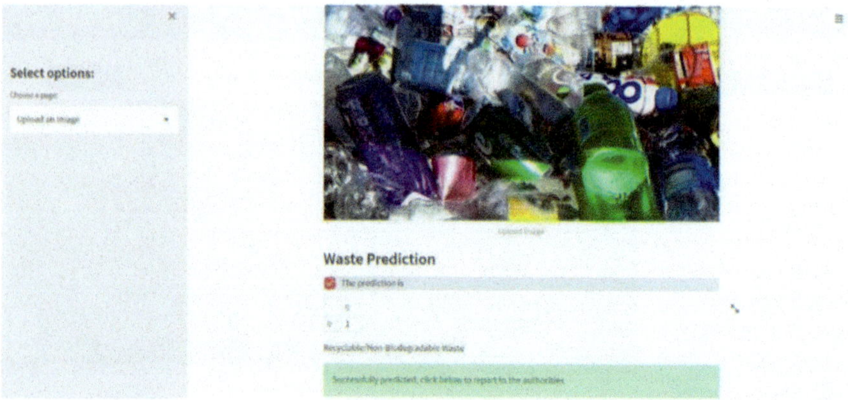

Fig. 4 Categorization of waste

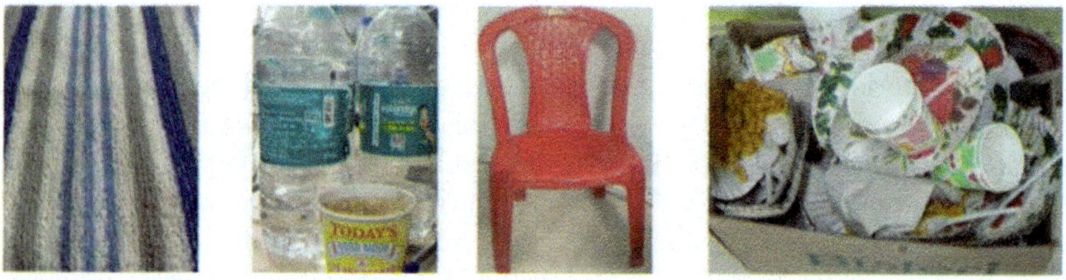

Fig. 5 Real-time images used for testing the CNN model working

CNN to recognize waste, the second includes CNN model with data augmentation, and the third used a transfer learning with pre-trained MobileNet V2 model to achieve the 88.93% accuracy. The classification algorithm is used in classifying the waste material which is binary. Figure 5 shows real-time images on which the model is being tested.

6 Conclusion

In conclusion, the proposed solution in the form of web application provides an effective solution for proper waste disposal, recycling, and reusing of waste materials. The application incorporates various modules such as waste classification, waste product addition, waste pickup scheduling, nearby NGO's and slum area's identification, and society and industry identification using Google Maps API. The waste classification module uses a CNN model and binary classification technique to categorize the waste into recyclable and non-recyclable categories. This helps in the efficient and effective scheduling of waste pickups by the nearby services. With this application, users can easily identify and classify the waste they produce, schedule pickups for proper disposal, and even donate to nearby NGO's or give waste to industries or slum areas according to the category found. Overall, this methodology has the potential to make a significant impact on the environment by reducing the amount of waste that goes to landfills and promoting the reuse and recycling of waste materials. By using this application, individuals and communities can contribute to a sustainable and healthier environment.

References

1. Nwofe PA (2015) Management and disposal of municipal solid wastes in Abakaliki metropolis, Ebonyi state, Nigeria. Int J Sci Res Environ Sci 3(3):107–118. https://doi.org/10.12983/ijsres-2015-p0107-0118
2. Connett P (2007) Zero waste: a key move towards a sustainable society. Res Gate 91(2):399–404
3. Choudhary S, Choudhary S (2019) A research paper on solid waste management. Jetir 6(3):657–662. [Online]. Available: https://www.jetir.org/view?paper=JETIREO06150
4. Zaman AU (2017) A strategic framework for working toward zero waste societies based on perceptions surveys. Recycling 2(1):1–15. https://doi.org/10.3390/recycling2010001
5. Akhilesh Kumar AA. Recent trends in solid waste management status, challenges, and potential for the future Indian cities. [Online]. Available: https://www.sciencedirect.com/science/article/pii/S2666049020300244
6. Chen C et al (2022) Waste classification for sustainable development using image recognition with deep learning neural network models sustainability waste classification for sustainable development using image recognition with deep learning neural network models, no June. https://doi.org/10.3390/su14127222
7. Chaturvedi S, Yadav BP, Siddiqui NA (2021) An assessment of machine learning integrated autonomous waste detection and sorting of municipal solid waste, no Rathi 2006
8. Pardini K, Rodrigues JJPC, Diallo O, Das AK, de Albuquerque VHC, Kozlov SA (2020) A smart waste management solution geared towards citizens. Sensors (Switzerland) 20(8):1–15. https://doi.org/10.3390/s20082380
9. Imran S. Ahmad, Kim DH (2020) Quantum GIS based descriptive and predictive data analysis for effective planning of waste management. IEEE Access 8:46193–46205. https://doi.org/10.1109/ACCESS.2020.2979015
10. Vishnu S et al (2021) IoT-enabled solid waste management in smart cities. Smart Cities 4(3):1004–1017. https://doi.org/10.3390/smartcities4030053
11. Sharma S, Rai BK, Gupta M, Dinkar M (2023) DDPIS: Diabetes disease prediction by improvising SVM. International Journal of Reliable and Quality E-Healthcare (IJRQEH) 12(2):1–11. https://doi.org/10.4018/IJRQEH.318090
12. Bano A, Ud Din I, Al-Huqail AA (2020) AIoT-based smart bin for real-time monitoring and management of solid waste. Sci Program 2020. https://doi.org/10.1155/2020/6613263

Framework for 2D MRI Brain Tumour Segmentation Employing Znet-Based Deep Learning

P. S. Tejashwini, J. Thriveni, and K. R. Venugopal

Abstract The spotting and segmentation of brain tumours using MR images presents substantial problems to the medical community. For potentially life-saving actions and the ability for doctors to select efficient treatment modalities, early detection and accurate localization of brain tumours are essential. This research presents a unique structure called Znet that makes use of deep neural networks (DNN) and data augmentation methods with the goal to forecast and segment brain tumours in medical imaging. The architecture uses skip-connections, encoder–decoder models, and data amplification to create synthetic situations using the knowledge from a small number of expertly described tumours. Experimental results, which generated a high mean dice similarity coefficient (DSC) of about 0.98 as part of model training and 0.94 for the independent testing dataset, show the effectiveness of the Znet model. These results show how correct data estimation in the area of healthcare data can be improved by deep learning (DL).

Keywords Region segmentation · Augmentation · Neural networks · Up sampling · Down sampling residual blocks

P. S. Tejashwini (✉) · J. Thriveni
Research Scholar, UVCE, Assistant Professor, Department of CSE (Data Science), Bangalore Institute of Technology, Bengaluru 560004, India
e-mail: tejashwinirnk@gmail.com

P. S. Tejashwini
Professor, Department of Computer Science and Engineering, University of Visveshwaraya College of Engineering, Bengaluru 560001, India

K. R. Venugopal
Former Vice Chancellor, Bangalore University, Bengaluru, Karnataka, India

V. K. Gunjan and J. M. Zurada (eds.), *Proceedings of 4th International Conference on Recent Trends in Machine Learning, IoT, Smart Cities and Applications*, Lecture Notes in Networks and Systems 873, https://doi.org/10.1007/978-981-99-9442-7_40

1 Introduction

The segmentation of anatomical structures and physiologically important regions is a crucial task in medical imaging. Segmented brain maps are an essential component of computer-aided diagnosis and play a crucial role in clinical decision-making. These maps offer crucial data that helps medical practitioners make well-informed decisions regarding patient care. DL methods have promulgated a lot of attention for health-care image segmentation because of their superior capability in precisely detecting and predicting target structures inside of images. Among the many segmentation techniques, the fully convolutional network (FCN) [1] and multilayer autoencoder approaches have stood out as being particularly effective for the segmentation of medical pictures. These techniques make use of deep learning to produce accurate and trustworthy segmentation results in the medical industry. The FCN setup for healthcare data segmentation is shown in Fig. 1.

The Unet architecture, depicted in Fig. 2, serves as an enhanced variant of the FCN framework. It has been specifically designed to handle situations with a restricted amount of training data, while also generating more precise segmentation results [2]. Unet is structurally made up of a path that contracts on the left and expands on the right. The Unet design is characterized by its extensive usage of feature channels and the presence of skip connections. The integration of skip connections within convolutional networks serves the purpose of augmenting gradient flow across numerous layers, thereby emphasizing a broader range of features within the image and producing more refined segmentation results [3].

1.1 Organization

The organization of a report on a DL architectures for brain tumour prediction and analysis using MRI images could include several sections. The first section explains

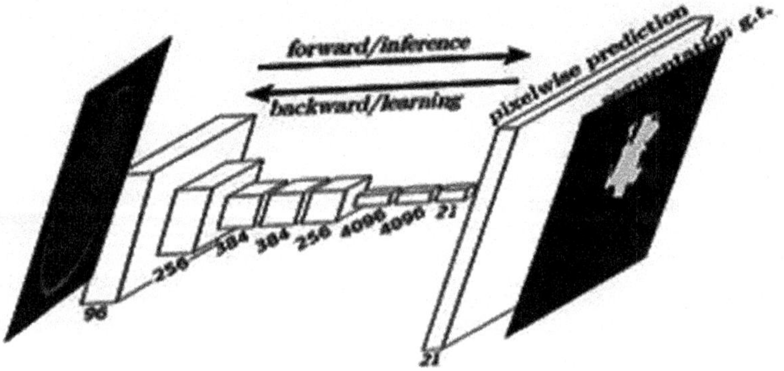

Fig. 1 Fully convolutional network

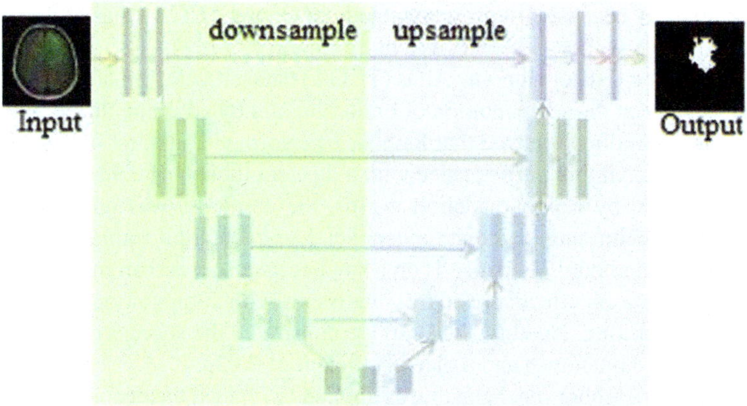

Fig. 2 Unet architecture [3], an expanded and modified version of the FCN, takes MRI image as input and generates a corresponding mask that accurately traces the contours of the detected tumour

the introductory material, which gives some background on the significance of accurate brain tumour analysis and early diagnosis in the medical community. The second section refers to a literature review, which would summarize current research and identify gaps that the current model aims to fill. The third section would discuss the proposed model in detail, including the architecture. This section could also consist of figures or diagrams to help visualize the approaches. In the fourth section, we will present the conclusion of the performed experiments utilizing the model, which will encompass data augmentation techniques, evaluation metrics, accuracy, and performance analysis. Additionally, we will provide visualizations of the tumour segmentations generated during the experiments. The conclusion, which would summarize the key findings of the study and underline its importance for the medical community, would be the last part of the report.

2 Related Work

Anaraki et al. [4] offer a unique approach for the noninvasive categorization of various grades of glioma, the most prevalent kind of significant adult brain tumours. CNN and a genetic algorithm (GA) are in the procedure, as opposed to conventional methodologies that depend on trial-and-error or predetermined system to generate the architecture of the CNN.

Bagging is employed in the suggested approach as an ensemble process to lower prediction error variance. The findings show that MRI scans can precisely categorize glioma, grades, and various tumour types, making them a viable tool for the detection of early stage brain tumours. The approach may be used in practice to help doctors classify tumours because of how adaptable it is. On the TCGA dataset, this method builds several networks and achieves a high accuracy of 0.94.

Using the The Cancer Genome Atlas (TCGA) dataset, Ghosh et al. [5] carried out a series of tests on the segmentation of brain tumours by adopting different models. The validation findings showed that RESNET performed at its best in this situation, achieving the maximum performance with a dice coefficient of 0.93.

In this article by Ranjbarzadeh et al. [6], the emphasis was on implementing deep learning techniques, to create a beneficial segmentation framework for brain tumours. This methodology depends on a smaller data subset than methods that use the complete dataset, effectively addressing issues with computation time, resource usage, and overfitting. Notably, the study demonstrated the success of their strategy with competitive outcomes and a dice coefficient of 0.92.

The nearest neighbor mechanism was applied to replace the deconvolution block in the Unet framework, and two more convolutional layers were included in [7]. With a dice coefficient (*D*) of 0.88 and an intersection over union (IoU) of 0.86.

Goetz et al. [8] employed the novel algorithm for the segmentation of brain tumours. The utilization of ExtraTrees, which introduces increased randomness as part of the training phase, led to favorable results, as evidenced by their reported dice coefficient (*D*) of 0.83.

3 Proposed Methodology

Znet is a CNN architecture designed for efficient and accurate brain tumour diagnosing from MRI scans. The model involves of several layers, comprising convolutional, pooling, and fully connected layers, allowing it to learn hierarchical features from the input images. The architecture includes skip connections and residual blocks, which help to mitigate the vanishing gradient difficulties and promote more deeply networks to be trained effectively. The model was also shown to be effective in detecting different category of brain tumours, comprising gliomas, meningiomas and pituitary adenomas, suggesting its potential applicability across different types of clinical scenarios (Fig. 3).

$$\mathrm{ReL}(z) = \begin{cases} 0, & z \leq 0 \\ z, & \text{otherwise} \end{cases} \tag{1}$$

$$\sum(z) = \frac{1}{1 + \mathrm{e}^{-z}} \tag{2}$$

$$l_n = -w\left[y_n.\log x_n + (1 - y_n).\log(1 - x_n)\right] \tag{3}$$

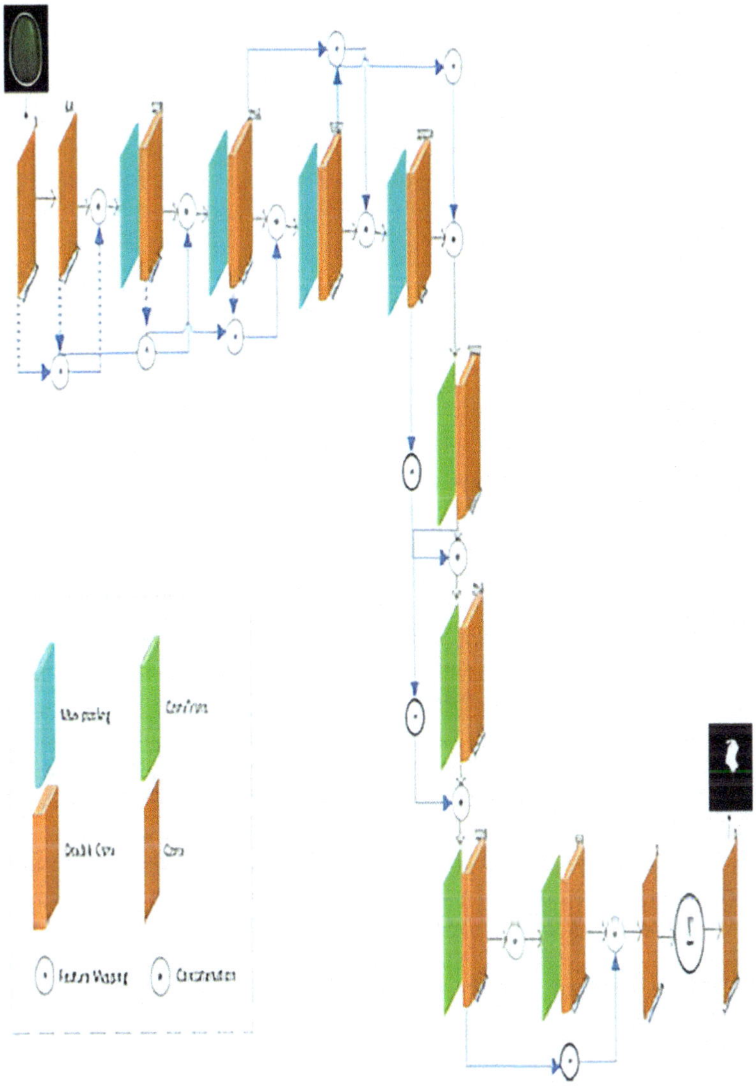

Fig. 3 ZNet architecture

The encoder network consists of layers of convolution with batch normalization and ReLU activation algorithms, along with layers that deploy max-pooling to down-sample the feature maps. By adding more filters, this architecture eventually cut down the spatial resolution of the feature of the maps.

While the decoder network uses batch normalization and ReLU techniques together with transposed convolutional layers. The encoder and decoder networks' associated feature of the maps are combined at the identical spatial resolution using skip connections, which are also included. These skip connections allow the model to enhance segmentation in the decoder network using high-level properties discovered in the encoder network. This design efficiently captures both low-level and high-level information by integrating the encoding and decoding operations, enabling precise and thorough segmentation outcomes. The decoder network's last set of 1×1 convolutional multiple layers with sigmoid activation functions, as a component of the Znet architecture, produces the input image's final segmentation mask.

4 Results and Discussion

4.1 Dataset

The Cancer Genome Atlas (TCGA) research network created the publicly accessible TCGA LGG dataset, which contains genetic and clinical evidence on individuals with low-grade gliomas (LGG). The collection comprises clinical evidence including age, sex, and survival status along with data from 530 individuals with 483 initial tumours and 47 recurring tumours.

The FLAIR MRI pictures, which also include LGG enhanced tumour images, are the subject of our study. Neuroradiologists with the necessary training carefully examined, annotated, and labeled the FLAIR pictures of 110 patients. When pathology was present, the experts determined the limits of each FLAIR abnormality (a tumour). The real MRI images and the manually annotated masks made the collection of training database.

As an essential preliminary process in the discipline of data science and artificial intelligence applications, data augmentation for visuals has experienced a fast hit in popularity. It accomplishes very well when there is a dearth of training data or when the volume of data is necessary for strengthening computer models. Additionally, data augmentation improves model performance and helps resolve overfitting difficulties [9]. The picture augmentation processes necessary for our investigation were carried out using Albumentations, an open-source, free Python tool [10].

The flexible Albumentations library offered a wide range of data augmentation methods. In contrast to HorizontalFlip, which created unique images by turning the input data horizontally around the y-axis. VerticalFlip created new images by vertically flipping the input data around the x-axis.

To add unpredictability, RandomRotate90 rotated the data input images by 90 degrees. By applying shifts, scaling, and rotations to the input images, ShiftScaleRotate was utilized to create upgraded samples that included more variants and diversified the dataset. The input images' pixels were all transformed to floating-point numbers between 0 and 1 in the final stage. Additionally, Fig. 4 illustrates a subset of the augmented data samples obtained through the utilization of the Albumentations Python library.

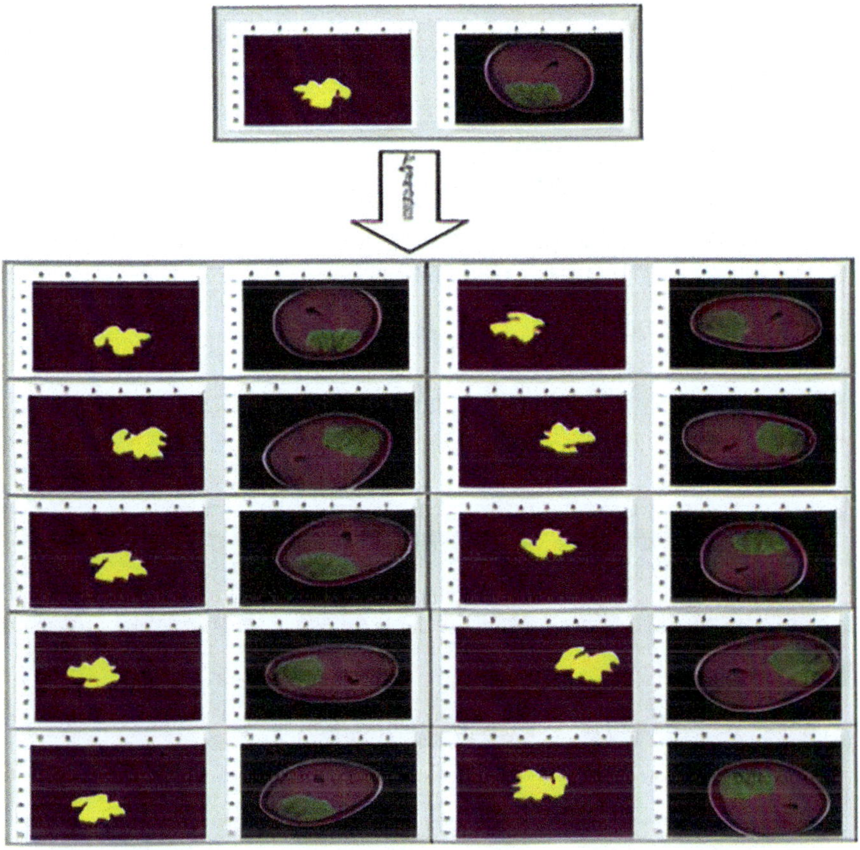

Fig. 4 Data augmentation samples

4.2 Evaluation Metrics

The Znet architecture obtained a DSC of 0.90 ± 0.02, sensitivity of 0.86 ± 0.03, and specificity of 0.99 ± 0.01 on the TCGA LGG dataset, outperforming other state-of-the-art methods in terms of accuracy. The processing time of the Znet approach was reported to be 5 s per MRI slice, which is faster compared to other methods (Table 1).

4.3 Results

Comparing the suggested system's performance to that of the well-known Unet segmentation technique. The evaluation included 432 MR pictures from the collection that were randomly selected. The results clearly demonstrate that Znet is superior

Table 1 Suggested model training parameters

Parameter	Values
Epoch	250
Optimizer	Adam
Batch Size	50
Training rate	tr = 5 * e*−4
Input size	128 * 128 * 3

to Unet in regarding of assessment metrics including dice coefficient, dice loss, pixel accuracy, F_1 score, and Matthews correlation coefficient (MCC). Figure 5 shows how the proposed architecture performs better than Unet.

The training procedure for detecting brain tumours on MR images took approximately 5–6 min per epoch. The training dataset and the testing dataset are the two

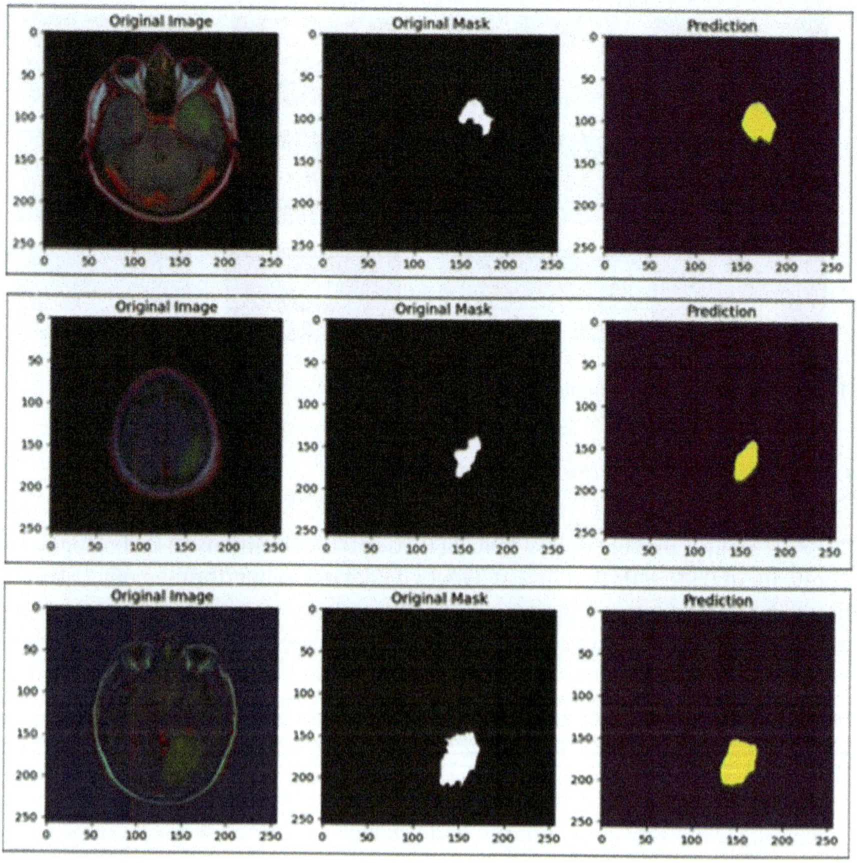

Fig. 5 Data segmentation results

main data groups that are commonly separated in applications for data science. The training data is further divided into the framework training dataset and the framework validation set. This department oversees model evaluation on unknowable data and guards against overfitting during training.

4.4 Performance Analysis

A loss graph indicates the discrepancy between a model's anticipated output and the actual target output. The *y*-axis in a loss graph reflects the loss value function, and the *x*-axis the number of learning iterations.

The IoU graph determines how much masking for segmentation and ground truth overlap. To calculate the IoU, the expected and ground truth areas' intersection area is subdivided by the dimension of their union. A score of 1, which ranges from 0 to 1, indicates an exact match between the predicted and ground truth regions.

The dice score is a tool used in image segmentation and medical imaging to determine whether two sets of binary or category data are equivalent or overlap (Fig. 6).

5 Conclusion

In this paper, we provide a unique approach for MR image segmentation that uses tagged images and deep learning ideas, particularly convolutional neural networks and data augmentation techniques. The proposed architecture includes an auto encoder–decoder structure, skip connections, and residual neural networks. Skip connections make learning easier and lessen the issue of fading gradients by allowing the merging of output from the previous layer with the subsequent layer [40]. Feature maps are used to map the dimensions between the input and output layers. The model was trained on a server with two NVidia Titan 12 GB GPUs over the course of 250 epochs. On the testing dataset, the dice accuracy was around 0.96, whereas during training it achieved a die accuracy of 0.98. A pixel accuracy of 0.998, an F_1 score of 0.83, and an MCC of 0.83 were also included as assessment criteria. Comparatively, the benchmark Unet model earned an MCC of 0.79, an F_1 score of 0.79, and dice accuracy of 0.85

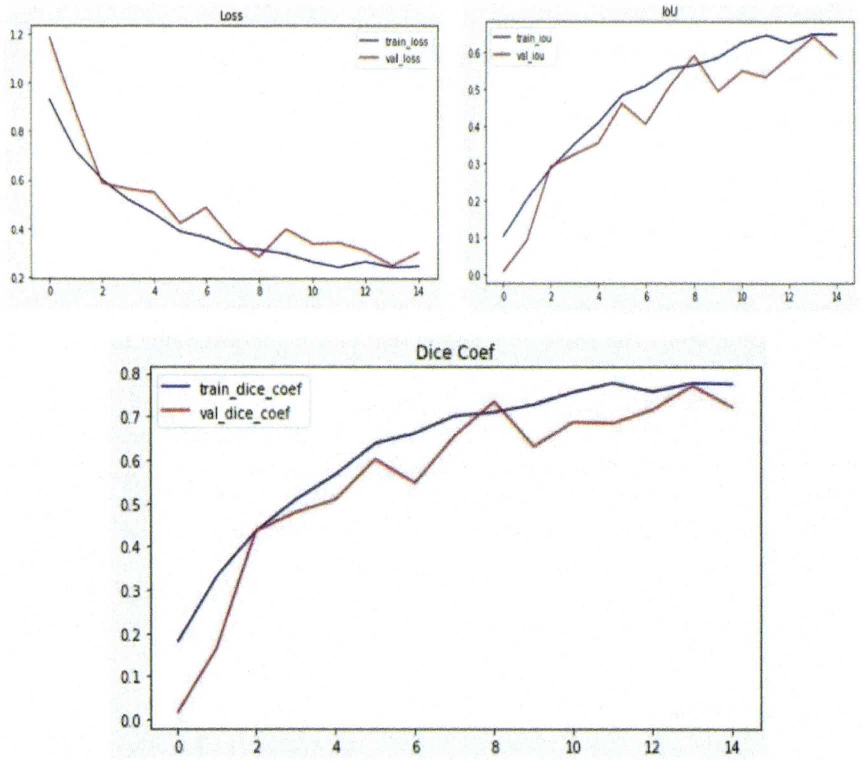

Fig. 6 Suggested model's validation and training performance

References

1. Rehman MU et al (2020) Bu-net: brain tumor segmentation using modified u-net architecture. Electronics 9(12):2203
2. Zhang J (2019) UNet-line by line explanation. Towards data science. Technical report [Online]. Available: https://towardsdatascience.com/
3. Ronneberger O, Fischer P, Brox T (2015) U-net: convolutional networks for biomedical image segmentation. Medical image computing and computer-assisted intervention—MICCAI 2015: 18th international conference, Munich, Germany, 5–9 Oct 2015, Proceedings, Part III 18. Springer International Publishing
4. Anaraki AK, Ayati M, Kazemi F (2019) Magnetic resonance imaging-based brain tumor grades classification and grading via convolutional neural networks and genetic algorithms. Biocybern Biomed Eng 39(1):63–74
5. Ghosh S, Santosh KC (2021) Tumor segmentation in brain MRI: U-Nets versus feature pyramid network. In: 2021 IEEE 34th international symposium on computer-based medical systems (CBMS). IEEE
6. Ranjbarzadeh R et al (2021) Brain tumor segmentation based on deep learning and an attention mechanism using MRI multi-modalities brain images. Sci Rep 11(1):1–17
7. Kamrul Hasan, SM, Linte CA (2018) A modified U-Net convolutional network featuring a nearest-neighbor re-sampling-based elastic-transformation for brain tissue characterization

and segmentation. In: 2018 IEEE Western New York image and signal processing workshop (WNYISPW). IEEE

8. GoetzM, Weber C, Bloecher J, Stieltjes B, Meinzer HP, Maier-Hein K (2014) Extremely randomized trees based brain tumor segmentation. In: Proceeding of BRATS challenge—MICCAI, vol 14, pp 6–11, 14 Sept 2014
9. Khalifa NE, Loey M, Mirjalili S (2022) A comprehensive survey of recent trends in deep learning for digital images augmentation. Artif Intell Rev 1–27
10. Buslaev A et al (2020) Albumentations: fast and flexible image augmentations. Information 11(2):125

Multiple-Emotion Recognition in Gujarati Language Using Natural Language Processing

Pushkar Ambatkar, Aishawariya Athawale, Jigna Sherasiya, Ankan Majumdar, and Savita Gandhi

Abstract Understanding emotions when we talk to computers is essential and is an important aspect of human–computer interaction, and it has been widely researched for the English language. However, the problem of emotion recognition in native languages other than English has received less attention. This paper aims to develop and test a robust multiple-emotion ('joy', 'surprise', 'love', 'sadness', 'fear', and 'anger') detection model for the Gujarati language using natural language processing techniques. A dataset of labelled English tweets with emotions was translated into Gujarati using machine translation, and then classification algorithms were applied. The results of this study suggest that it is possible to build a robust emotion recognition model for Gujarati language and provide a foundation for future research in this area.

Keywords Multiple-emotion recognition · Gujarati language · Supervised learning · LSTM · Natural language processing

1 Introduction

Emotion recognition plays a pivotal role in human–computer interaction, representing a focal point of investigation within the domain of natural language processing. The capacity to discern emotions within written text holds immense potential for diverse applications spanning domains like social media [1], customer service [2], and mental health [3, 4]. Despite the paramount importance of emotion recognition, a majority of prior research has predominantly concentrated on the English language, leading to a scarcity of studies examining other native languages. This paper seeks to bridge this void by embarking on the development and assessment of a resilient emotion detection model, extending beyond merely positive and

P. Ambatkar · A. Athawale · J. Sherasiya · A. Majumdar (✉) · S. Gandhi
Department of Computer Science, Gujarat University, Ahmedabad, India
e-mail: majumdarankan7@gmail.com

S. Gandhi
e-mail: savita.gandhi@glsuniversity.ac.in

V. K. Gunjan and J. M. Zurada (eds.), *Proceedings of 4th International Conference on Recent Trends in Machine Learning, IoT, Smart Cities and Applications*, Lecture Notes in Networks and Systems 873, https://doi.org/10.1007/978-981-99-9442-7_41

negative classes, for the Gujarati language, employing advanced natural language processing techniques.

The importance of emotion detection in online tweets and comments lies in its ability to provide insights into the mental state of a user. By analysing the emotions like 'joy', 'surprise', 'love', 'sadness', 'fear', and 'anger' expressed in a user's online communication, we can gain an understanding of their emotional state, which may not be visible in face-to-face interactions. For example, social media platforms are often used as a means of expressing one's thoughts and feelings, and users may share their personal experiences, opinions, and emotions through comments and tweets. Emotion detection in these online interactions can provide valuable information to mental health professionals, social workers, and other stakeholders who work with vulnerable populations. Additionally, businesses can also benefit from emotion detection in online communication [5], as it can provide insights into customer sentiment, which can be used to improve customer service, product development, and marketing strategies [6]. Therefore, the importance of above-mentioned emotion detection in online tweets and comments lies in its potential to help us understand the mental state of a user, which can inform decision-making and allow for necessary actions to be taken if needed.

The Gujarati language boasts its distinctive traits, demanding a model proficient in capturing these subtleties to excel in emotion recognition. In this research, we delve into the effectiveness of diverse supervised machine learning algorithms to discern multiple emotions, including 'joy,' 'surprise,' 'love,' 'sadness,' 'fear,' and 'anger,' within the Gujarati language. To embark on this journey, we started with a dataset of 20,000 meticulously labelled sentences in English. Using advanced machine translation techniques, we adeptly translated this dataset into Gujarati. Subsequently, we diligently undertook manual data refinement, resulting in a pristine dataset comprising 19,600 labelled sentences in the Gujarati language.

In conclusion, the study aims to develop and test a robust emotion detection model for the Gujarati language, which is a widely spoken language in the western part of India. We explore the performance of different supervised machine learning algorithms on both English and Gujarati datasets for emotion classification. The performance of the models improved after the removal of stop words for the English language and reduced for the Gujarati language. The long short-term memory (LSTM) algorithm performed better for emotion classification in both languages. The results of this study will be helpful for researchers working on emotion detection and classification in native languages other than English.

2 Related Work

There have been many types of research done in the past for sentiment analysis in the Gujarati language. But most of the work is done for classifying a sentence as positive, negative, or neutral.

In their research paper, "A sentiment analysis of Gujarati Text using Gujarati Senti Word Net" Patel and Gohil [7] presented a methodology for constructing the Gujarati Senti Word Net using Hindi and Indo WordNet. The resulting G-SWN contains synsets with corresponding polarity scores and part-of-speech tags. The researchers evaluated the accuracy of G-SWN on Twitter data using the unigram presence method and simple scoring method. The classifiers' accuracy using G-SWN was found to be approximately 52%.

In the article titled "An Approach to Sentiment Analysis on Gujarati Tweets" Joshi and Vekariya [8] have presented a method for categorizing Gujarati tweets into positive, negative, and neutral sentiment classes. The authors employed POS tagging along with support vector machine (SVM) for classification and achieved an impressive accuracy of 92%. This study offers insights into the potential for sentiment analysis on Gujarati language text data and demonstrates the efficacy of using machine learning algorithms for the task.

In their work titled "GoEmotion: A Dataset of Fine-Grained Emotions," Demszky et al. [9] have curated the most extensive manually annotated dataset known to date, specifically designed for fine-grained emotion detection. This dataset encompasses a staggering 58,000 English comments sourced from Reddit, with each comment being meticulously categorized into one of 27 distinct emotion categories or marked as neutral. To conduct their experiments, the authors harnessed the power of a BERT-based model, which yielded an average F_1 score of 0.46. The GoEmotion dataset, along with its experimental outcomes, effectively underscores the potential and viability of identifying nuanced emotions within vast text repositories, including online comments and social media posts."

3 Methodology

As illustrated in Fig. 1, our methodology for this study encompassed several key stages, which included dataset preparation, stop word elimination, tokenization, model training, model testing, and classification. In the initial step, we gathered a dataset comprising labelled English tweets with emotions such as 'joy,' 'surprise,' 'love,' 'sadness,' 'fear,' and 'anger.' Subsequently, we translated this dataset into Gujarati using machine translation techniques. Following this, we carried out stop-word removal to filter out common words that hold no significance in emotion classification. Tokenization was then implemented to break down the text into individual words or tokens. This processed dataset served as the foundation for training the emotion recognition model, employing various classification algorithms including logistic regression, support vector machines (SVM), multinomial Naïve Bayes, K-nearest neighbour, decision tree classifier, random forest classifier, long short-term memory, and bi-directional long short-term memory. We rigorously assessed the model's performance by subjecting it to testing on a separate dataset. Ultimately, we identified the most effective algorithm, which we then utilized for emotion classification in the translated Gujarati dataset.

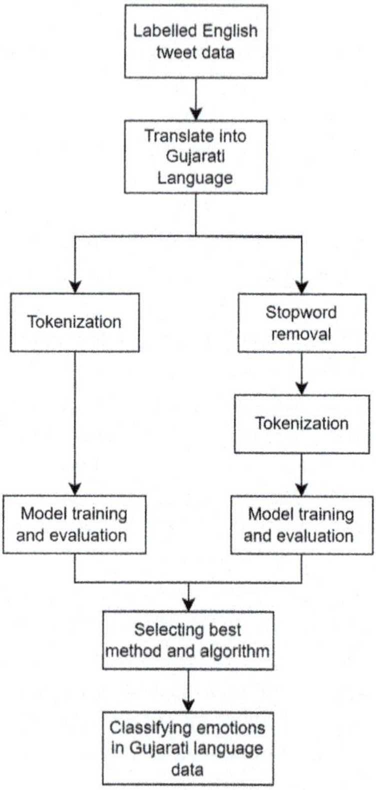

Fig. 1 Graphical illustration of proposed approach

4 Data

For this project, we leveraged the Emotions dataset for Natural Language Processing (NLP) sourced from Kaggle [10], a rich collection featuring 20,000 labelled English sentences. This dataset encompassed six distinct emotions: anger, fear, joy, love, sadness, and surprise, as aptly illustrated in Fig. 2. Given our objective of extending emotion recognition to the Gujarati language, we employed the services of Amazon Web Services' translate API to skilfully translate the dataset, as depicted in Fig. 3. In this endeavour, we adopted the batching method, allowing us to aggregate emotions and efficiently utilize cloud resources. Following the translation process, we meticulously carried out manual cleaning to ensure data accuracy. This step was imperative as certain words did not translate accurately, and the dataset contained instances of inappropriate language. Ultimately, our diligent efforts resulted in a final Gujarati dataset comprising 19,600 labelled sentences, a robust sample size well-suited for emotion recognition. We made the reasonable assumption that emotions remained consistent during the translation process, making the resultant dataset a suitable foundation for constructing a resilient emotion detection model for the Gujarati language.

	sentence	emotion
0	i am i feel like it s important to keep on tak...	joy
1	i can say is that despite my occasional jokes ...	joy
2	i do like hearing about ministries that reach ...	fear
3	i know i cannot rest of my laurels and its a a...	joy
4	im not saying cut everyone out of your life bu...	joy
...	...	...
19995	i cant help feeling this way	sadness
19996	im a bit afraid the cookie is what is making m...	joy
19997	i started to feel melancholy and uncertain and...	sadness
19998	i believe that if i by myself make a person fe...	fear
19999	i feel paranoid but atleast now i get some com...	fear

20000 rows × 2 columns

Fig. 2 Glimpse of dataset in English language

	sentence	emotion
0	મને વાર્તા પર હેન્ડલ મળ્યું અને તે ખરેખર લાગણી...	love
1	હું અચાનક નારાજ થઈ રહ્યો હતો	anger
2	હું આ ટાઇપ કરું છું હું મારી અસ્વીકૃત વિશાળ મક...	anger
3	હું ઉન્મત્ત થઈ રહ્યો હતો ભગવાનનો આભાર મારી પાસ...	anger
4	મારી પાસે ખરેખર વાત કરવા માટે કંઈ નથી કે હું ક...	sadness
...	...	...
19595	[illegible]	sadness
19596	મને યાદ છે કે સૌંદર્ય ખરેખર જોનારાની આંખમાં છે...	fear
19597	હું ગુસ્સો ન અનુભવું છું કારણ કે મને તે રીતે અ...	anger
19598	મને આવું કરવા માટે મૂર્ખ મૂર્ખ લાગે છે	sadness
19599	મને લાગે છે કે સ્ટેજ વચ્ચેના આ વિચિત્ર છે	surprise

19600 rows × 2 columns

Fig. 3 Glimpse of dataset in Gujarati language

4.1 Pre-processing

Prior to the implementation of any algorithm, text pre-processing, as highlighted in Ref. [11], plays a pivotal role in minimizing noise and enhancing accuracy. This critical procedure involves cleaning the text data. In the case of the English dataset utilized in our study, the sole pre-processing required was the removal of stop words—those common words that contribute little to emotion classification. However, the Gujarati dataset posed a unique challenge due to its translation from English, resulting in several English words lacking corresponding translations in Gujarati. Consequently, a manual cleaning process was undertaken to ensure the dataset remained free from non-Gujarati words and any undesirable language. Once this cleaning was accomplished, the subsequent step involved tokenization, which entailed the segmentation of text into individual words or tokens. This tokenization step proved essential to facilitate the smooth operation of our classification algorithms.

The dataset, after undergoing preprocessing and tokenization, was subsequently divided into a 70:30 ratio, with the larger portion allocated for training and the remaining for testing. The training dataset was employed for instructing the classification algorithms, while the testing dataset was used in assessing the proficiency of the trained models.

5 Classification Algorithms

Multiple classification algorithms have been applied for recognizing emotions in the Gujarati language. In this section, we will discuss the algorithms used in our study.

5.1 Multinomial Naïve Bayes

The Multinomial Naïve Bayes algorithm is a renowned choice for text classification, as referenced in [12]. It operates on the foundational principles of the Bayes theorem, which posits that the occurrence of a specific feature within a class remains independent of the presence of other features.

5.2 Logistic Regression

Logistic regression is another popular algorithm for text classification as cited in [13, 14]. This statistical technique effectively models the likelihood of a binary result by harnessing the power of a logistic function.

5.3 K-Nearest Neighbour

K-Nearest neighbour (KNN) finds its place as a nonparametric algorithm specially employed for classification tasks, as referenced in [15]. It draws its strength from the notion of similarity between data instances. This algorithm classifies an instance by examining the classes of its k closest neighbours.

5.4 Support Vector Machines (SVM)

Support vector machines (SVM) represent a class of supervised learning algorithms capable of handling both classification and regression tasks. Within the realm of text classification, SVM stands as a potent tool, as noted in [16]. This algorithm operates on the principle of identifying the optimal hyperplane to effectively segregate data into distinct classes. In our research, we employed SVM for recognizing emotions in the Gujarati language.

5.5 Decision Tree Classifier

The decision tree classifier serves as a nonparametric technique in supervised learning. It follows a two-step process: first, it learns a model from the training data, and then it employs this model to make predictions. This involves the creation of a decision tree that takes into account the features and their corresponding values, subsequently utilizing this tree to classify incoming data [17].

5.6 Random Forest Classifier

Random Forest Classifier, an ensemble learning technique versatile in handling classification, regression, and various other tasks, operates by assembling a diverse array of decision trees. Its output is determined by either selecting the most frequent class (for classification) or calculating the mean prediction (in the case of regression) from the collective wisdom of these individual trees [18].

5.7 Long Short-Term Memory (LSTM)

LSTM, or Long Short-Term Memory, represents a specialized variant of recurrent neural networks (RNNs) adept at capturing long-term dependencies in data.

It possesses the remarkable ability to retain knowledge of past input values over an extended duration and employ this memory to generate predictions. LSTM finds particular utility in handling and classifying sequential data, thanks to its innate capability to selectively retain or discard information, rendering it highly suitable for tasks like emotion recognition [19, 20].

5.8 Bi-directional Long Short-Term Memory (BiLSTM)

Bi-directional Long Short-Term Memory (BiLSTM) represents a specialized iteration of LSTM that evaluates data in both the forward and backward directions. This unique approach enhances its applicability to tasks where the contextual understanding of the data plays a pivotal role [21, 22].

Each of the algorithms was applied to the pre-processed dataset to determine the best algorithm for emotion detection in the Gujarati language.

6 Experiments, Results, and Discussion

Assessment criteria play a pivotal role in evaluating the effectiveness of machine learning models. In the realm of classification models, multiple metrics are employed to gauge their performance, encompassing accuracy, precision, recall, and the F_1 score.

Accuracy stands as a widely used assessment measure, gauging the proportion of accurate forecasts amidst the entire prediction count. Its computation involves dividing the count of accurate predictions by the total number of predictions generated. A heightened accuracy score signifies that the model excels in generating correct predictions.

$$\text{Accuracy} = \frac{\text{True Positives} + \text{True Negatives}}{\text{True Positives} + \text{False Positives} + \text{True Negatives} + \text{False Negatives}} \tag{1}$$

Precision is expressed as the ratio of true positives (positively predicted instances correctly identified) to the overall number of positive predictions. It serves as a gauge of the model's ability to minimize false positives. A heightened precision score denotes that the model produces fewer false positive predictions, highlighting its accuracy in positive identification.

$$\text{Precision} = \frac{\text{True Positives}}{\text{True Positives} + \text{False Positives}} \tag{2}$$

Recall, sometimes referred to as sensitivity or the true positive rate, quantifies the proportion of true positives among all the actual positive instances in the dataset. This metric assesses the model's proficiency in recognizing true positives. A robust recall score signifies that the model is minimizing the occurrence of false negative predictions.

$$\text{Recall} = \frac{\text{True Positives}}{\text{True Positives} + \text{False Negatives}} \quad (3)$$

The F_1 score is like the sweet spot, finding the harmonious balance between precision and recall. It is a well-rounded measure that considers both precision and recall when assessing the model's overall performance. This score effectively merges precision and recall into one single metric, with a scale from 0 to 1, where 1 marks the pinnacle of performance excellence.

$$F_1 \text{ Score} = \frac{2 * \text{Precision} * \text{Recall}}{\text{Precision} + \text{Recall}} \quad (4)$$

In classification tasks, it is vital to look at all the evaluation measures as a unified score. Take, for instance, a model with strong precision but weak recall—it might spot fewer positive instances, not always suitable for certain uses. Conversely, a model with high recall but less precision could generate numerous false positives, potentially resulting in incorrect predictions. That is why finding the right balance between precision and recall is key to achieving a solid F_1 score, ensuring that both aspects of model performance are taken into account.

Our research centred on applying a range of supervised machine learning algorithms to forecast multi-label emotions in both English and Gujarati languages. We executed these algorithms using Python 3.8, making good use of various libraries such as scikit-learn, Keras, pandas, matplotlib, plotly, and seaborn. The dataset employed in our study is detailed in the Dataset section, and we split it into a 70:30 ratio for the purposes of training and testing. We trained all our models using datasets in both English and Gujarati. To gain deeper insights, we assessed the algorithms' performance with and without the removal of stop words. This allowed us to gauge the influence of stop words on our model's performance.

Tables 1, 2, 3 and 4 and Figs. 4, 5, 6 and 7 present a comprehensive view of the performance metrics for all the algorithms on both English and Gujarati language datasets, both before and after removing stop words. It is evident from the tables that the algorithms exhibited enhanced performance following the removal of stop words in both the English and Gujarati datasets. Notably, the Multinomial Naïve Bayes algorithm demonstrated decent performance in the English dataset, achieving an F_1 score of 0.59. However, its performance declined when applied to the Gujarati dataset after stop word removal, resulting in an F1 score of 0.49. Conversely, the Logistic Regression algorithm excelled with an F_1 score of 0.85 on the English dataset but displayed average performance on the Gujarati dataset, achieving an F_1 score of 0.55 post-stop word removal. The K-Nearest Neighbours algorithm performed admirably with an F_1 score of 0.79 on the English dataset but faced challenges on the Gujarati dataset, recording an F_1 score of 0.38 after stop word removal.

The Support Vector Machine (SVM) classifier exhibited significant variations in performance between the English and Gujarati language datasets, recording F_1 scores of 0.83 and 0.53, respectively, post the stop word removal. On the other hand, the Decision Tree Classifier algorithm displayed slightly superior performance on the English language dataset, achieving an F_1 score of 0.86 after stop word removal, whereas on the Gujarati language dataset, it achieved a lower F_1 score of 0.46. Similarly, the Random Forest Classifier algorithm showcased commendable performance on the English language dataset, boasting an F_1 score of 0.88 following stop word elimination. However, its performance dipped on the Gujarati language dataset, yielding an F_1 score of 0.54.

The LSTM algorithm and its counterpart, the BiLSTM algorithm, were part of our implementation, and their performance displayed promising outcomes. In particular, the LSTM algorithm achieved an F_1 score of 0.68 on the Gujarati language dataset post-stop word removal, while the BiLSTM algorithm obtained an F_1 score of 0.67 on the same dataset. These results strongly indicate that the implemented algorithms hold the potential to accurately forecast multi-label emotions in both English and Gujarati languages. Furthermore, the findings emphasize that the removal of stop words from the datasets can serve as a valuable technique to further enhance the models' performance.

Table 1 Performance of different algorithms on English dataset before removal of stop words

Algorithms	Test accuracy	Precision	Recall	F_1-score
Logistic regression	0.88	0.85	0.85	0.84
K-nearest neighbours	0.73	0.73	0.73	0.72
Support vector machine	0.84	0.85	0.84	0.83
Decision tree	0.86	0.86	0.86	0.86
Random forest	0.86	0.86	0.86	0.85
Multinomial Naïve Bayes	0.64	0.72	0.64	0.54
LSTM	0.86	0.82	0.81	0.81

Table 2 Performance of different algorithms on English dataset after removal of stop words

Algorithms	Test accuracy	Precision	Recall	F_1-score
Logistic regression	0.85	0.86	0.85	0.85
K-nearest neighbours	0.8	0.8	0.8	0.8
Support vector machine	0.84	0.85	0.84	0.84
Decision tree	0.87	0.87	0.87	0.87
Random forest	0.88	0.89	0.89	0.89
Multinomial Naïve Bayes	0.67	0.73	0.67	0.6
LSTM	0.86	0.82	0.81	0.82

Table 3 Performance of different algorithms on Gujarati dataset before removal of stop words

Algorithms	Test accuracy	Precision	Recall	F_1-score
Logistic regression	0.58	0.59	0.58	0.56
K-nearest neighbours	0.48	0.4	0.42	0.4
Support vector machine	0.57	0.5931	0.57	0.55
Decision tree	0.47	0.46	0.47	0.47
Random forest	0.57	0.58	0.57	0.55
Multinomial Naïve Bayes	0.54	0.57	0.54	0.5
LSTM	0.78	0.73	0.73	0.73
BLSTM	0.78	0.73	0.72	0.73

Table 4 Performance of different algorithms on Gujarati dataset after removal of stop words

Algorithms	Test accuracy	Precision	Recall	F_1-score
Logistic regression	0.58	0.59	0.58	0.56
K-nearest neighbours	0.4	0.4	0.4	0.38
Support vector machine	0.56	0.59	0.56	0.53
Decision tree	0.47	0.46	0.47	0.46
Random forest	0.56	0.58	0.56	0.54
Multinomial Naïve Bayes	0.53	0.57	0.53	0.49
LSTM	0.74	0.7	0.68	0.69
BLSTM	0.73	0.69	0.66	0.67

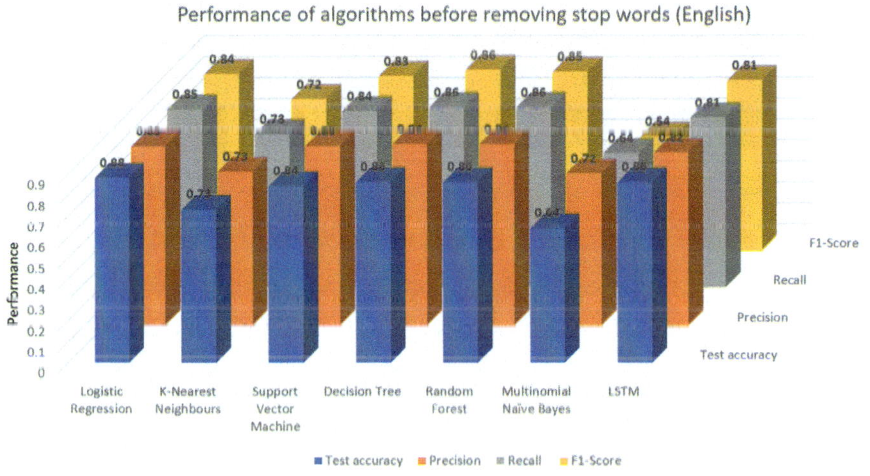

Fig. 4 Graphical representation of results on English dataset of algorithms before removal stop words

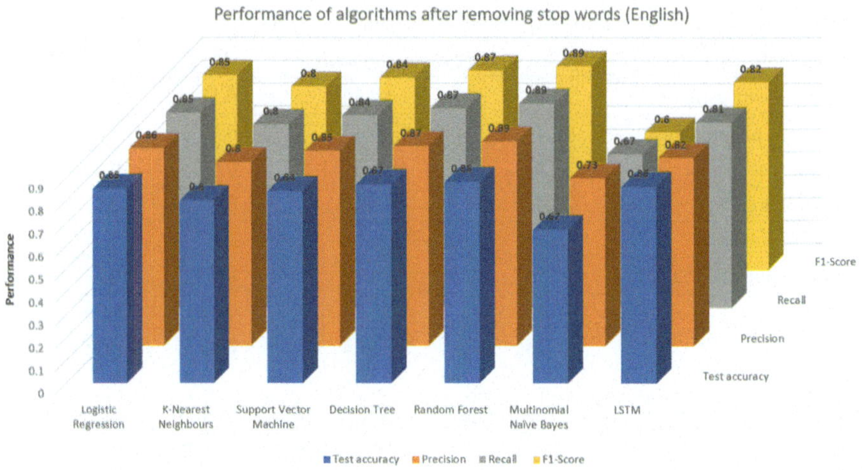

Fig. 5 Graphical representation of results on English dataset of algorithms after removal stop words

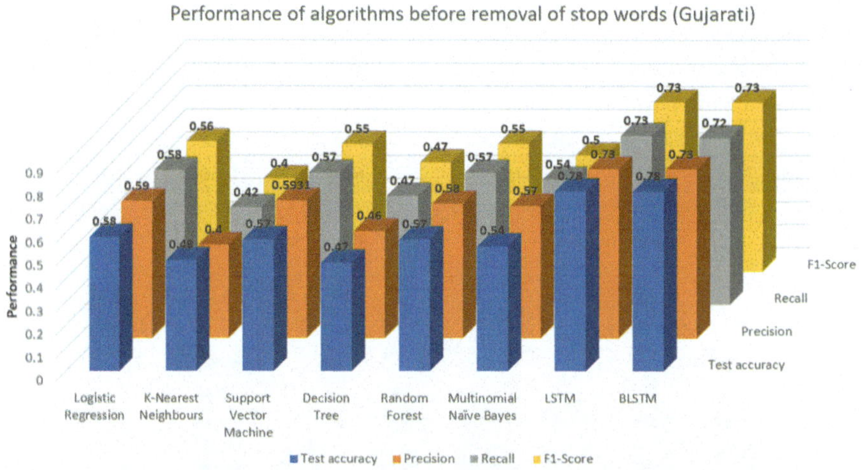

Fig. 6 Graphical representation of results on Gujarati dataset of algorithms before removal of stop words

Further, the LSTM model was used for classifying a Gujarati language text's emotion, and the output is shown in Fig. 8.

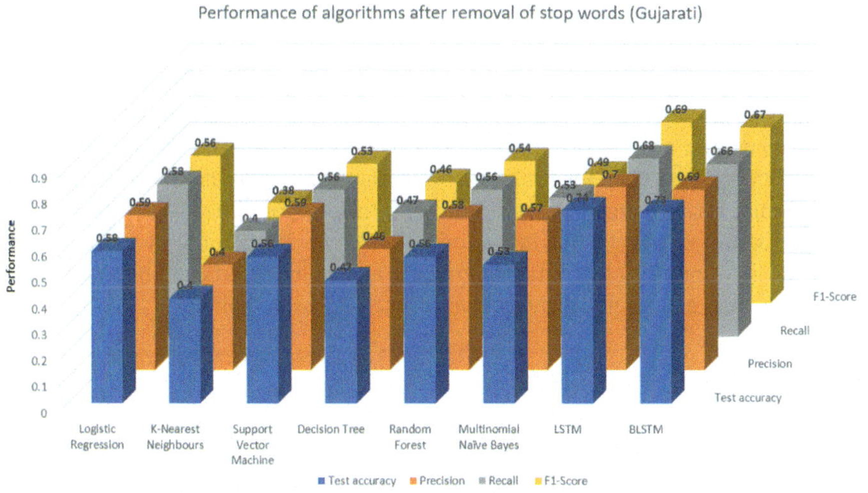

Fig. 7 Graphical representation of results on Gujarati dataset of algorithms after removal of stop words

```
{0: ('હું તરત જ બધું વિશે વિચિત્ર લાગણી સાથે સંબંધિત',
  ('anger', 0.63582534, 0.55160934)),
 1: ('મેં આશ્ચર્ય અનુભવવાનું બંધ કરી દીધું છે',
  ('anger', 0.7160984, 0.8529369)),
 2: ('હું જીવનથી એટલો ભરેલો છું કે મને લાગે છે',
  ('sadness', 0.99982125, 0.7889222)),
 3: ('મેં પવન કર્યો અને કહ્યું કે રમુજી નથી લાગતું',
  ('anger', 0.48366123, 0.725867)),
 4: ('મને સમજાયું કે જ્યારે હું મારા મનને રેસ કરું છું અને હું બેચેન અનુભવવાનું શરૂ કરું છું ત્યારે મને ધૂમ્રપાન કરવાની ઇચ્છા મળે છે',
  ('anger', 0.9999664, 0.718779)),
 5: ('હું અતાર્કિક રીતે ખૂબ ફેન્કી લાગે', ('anger', 0.7611769, 0.5784984)),
 6: ('હું વિચિત્ર અને ગભરાટ અનુભવું છું', ('fear', 0.9366019, 0.7049196)),
 7: ('મને લાગણી છે કે તે ખુશ અને આનંદિત હતી',
  ('sadness', 0.44991305, 0.65245396)),
 8: ('મને તે કહેતા થોડી વિચિત્ર લાગે છે', ('fear', 0.9469613, 0.59961))}
```

Fig. 8 Result of LSTM model on Gujarati language dataset

7 Conclusion

The outcomes of our experiment reveal that removing stop words from the text indeed has a substantial impact on how well our emotion classification models perform. For the English language, this tweak has led to enhanced model performance, but interestingly, for Gujarati, it has resulted in reduced accuracy. Furthermore, the *K*-Nearest Neighbours algorithm did not fare well in either language when it came to emotion classification. On the flip side, the LSTM algorithm outshone the others in both English and Gujarati, underscoring the superiority of deep learning over traditional machine learning methods. To further enhance our models, we can consider adding more data for training and testing, and exploring advanced deep learning models like transformer-based ones. This study underscores the importance of crafting emotion detection models for native languages like Gujarati, enabling us to classify emotions beyond the simplistic positive and negative labels. It also offers valuable insights into the challenges and potential remedies for building such models.

To sum up, this paper has showcased the feasibility of constructing a sturdy emotion recognition model for the Gujarati language through the application of natural language processing techniques. The findings from this investigation not only lay the groundwork for forthcoming research in this domain but also indicate the potential for crafting emotion recognition models for diverse native languages. This research underscores the importance of further exploration into multi-emotion recognition for non-English native languages, as it holds significant implications for enhancing human–computer interaction across various domains.

References

1. Barnaghi P, Ghaffari P, Breslin JG (2016) Opinion mining and sentiment polarity on twitter and correlation between events and sentiment. In: Proceedings of the 2016 IEEE 2nd international conference on big data computing service and applications, BigDataService 2016, pp 52–57, May 2016. https://doi.org/10.1109/BIGDATASERVICE.2016.36
2. Płaza M et al (2022) Emotion recognition method for call/contact centre systems. Appl Sci 12(21):10951. https://doi.org/10.3390/APP122110951
3. Qi W, Li Y (2022) Research on emotion recognition and mental state prediction of college students based on deep neural network. Mob Inf Syst 2022. https://doi.org/10.1155/2022/1607177
4. Simcock G et al (2020) Associations between facial emotion recognition and mental health in early adolescence. Int J Environ Res Public Health 17(1). https://doi.org/10.3390/IJERPH17010330
5. Isiaka F, Adamu Z (2022) Custom emoji based emotion recognition system for dynamic business webpages. Int J Intell Comput Cybern 15(4):497–509. https://doi.org/10.1108/IJICC-11-2021-0254/FULL/XML
6. Koc E, Boz H (2019) Development of hospitality and tourism employees' emotional intelligence through developing their emotion recognition abilities. J Hosp Market Manag 29(2):121–138. https://doi.org/10.1080/19368623.2019.1608885
7. Patel D, Gohil L (2019) A sentiment analysis of Gujarati text using Gujarati Sento Word Net article. Int J Innov Technol Explor Eng 9:2278–3075. https://doi.org/10.35940/ijitee.I8443.078919
8. Joshi VC, Vekariya VM (2023) An approach to sentiment analysis on Gujarati tweets. Adv Comput Sci Technol 10(5):1487–1493. Accessed: 06 Mar 2023. [Online]. Available: http://www.ripublication.com
9. Demszky D, Movshovitz-Attias D, Ko J, Cowen A, Nemade G, Ravi S (2020) GoEmotions: a dataset of fine-grained emotions, pp 4040–4054, May 2020. https://doi.org/10.48550/arxiv.2005.00547
10. Saravia E, Toby Liu HC, Huang YH, Wu J, Chen YS (2018) Carer: contextualized affect representations for emotion recognition. In: Proceedings of the 2018 conference on empirical methods in natural language processing, EMNLP 2018, pp 3687–3697. https://doi.org/10.18653/V1/D18-1404
11. Haddi E, Liu X, Shi Y (2013) The role of text pre-processing in sentiment analysis. Procedia Comput Sci 17:26–32. https://doi.org/10.1016/J.PROCS.2013.05.005
12. Xu S, Li Y, Wang Z (2017) Bayesian multinomial Naïve Bayes classifier to text classification. In: Lecture notes electrical engineering, vol 448, pp 347–352. https://doi.org/10.1007/978-981-10-5041-1_57/COVER

13. Pranckevicius T, Marcinkevicius V (2017) Application of logistic regression with part-of-the-speech tagging for multi-class text classification. In: 2016 IEEE 4th workshop on advances in information, electronic and electrical engineering, AIEEE 2016—Proceedings, Jan 2017. https://doi.org/10.1109/AIEEE.2016.7821805
14. Shah K, Patel H, Sanghvi D, Shah M (2020) A comparative analysis of logistic regression, random forest and KNN models for the text classification. Augment Hum Res 5(1):1–16. https://doi.org/10.1007/S41133-020-00032-0
15. Chen Z, Zhou LJ, Da Li X, Zhang JN, Huo WJ (2020) The Lao text classification method based on KNN. Procedia Comput Sci 166:523–528. https://doi.org/10.1016/J.PROCS.2020.02.053
16. Goudjil M, Koudil M, Bedda M, Ghoggali N (2018) A novel active learning method using SVM for text classification. Int J Autom Comput 15(3):290–298. https://doi.org/10.1007/S11633-015-0912-Z/METRICS
17. Shi L, Weng M, Ma X, Xi L (2023) Rough set based decision tree ensemble algorithm for text classification. J Comput Inf 6:89–95. Accessed: 06 Mar 2023. [Online]. Available: http://www.jofci.org1553-9105/
18. Pranckevičius T, Marcinkevičius V (2017) Comparison of Naïve Bayes, random forest, decision tree, support vector machines, and logistic regression classifiers for text reviews classification. Balt J Mod Comput 5(2):221–232. https://doi.org/10.22364/bjmc.2017.5.2.05
19. Luan Y, Lin S (2019) Research on text classification based on CNN and LSTM. In: Proceedings of the 2019 IEEE international conference on artificial intelligence and computer applications, ICAICA 2019, pp 352–355, Mar 2019. https://doi.org/10.1109/ICAICA.2019.8873454
20. Ma J, Zheng WL, Tang H, Lu BL (2019) Emotion recognition using multimodal residual LSTM network. In: MM 2019—Proceedings of the 27th ACM international conference on multimedia, pp 176–183, Oct 2019. https://doi.org/10.1145/3343031.3350871
21. Li C, Bao Z, Li L, Zhao Z (2020) Exploring temporal representations by leveraging attention-based bidirectional LSTM-RNNs for multi-modal emotion recognition. Inf Process Manag 57(3):102185. https://doi.org/10.1016/J.IPM.2019.102185
22. Senthilkumar N, Karpakam S, Gayathri Devi M, Balakumaresan R, Dhilipkumar P (2022) Speech emotion recognition based on bi-directional LSTM architecture and deep belief networks. Mater Today Proc 57:2180–2184. https://doi.org/10.1016/J.MATPR.2021.12.246

Automation of Mechanical Ventilation for Optimal Pressure Predictions Using Machine Learning

Jayshree Ghorpade-Aher and Srushti Ijeri

Abstract The onset of Covid-19 has brought a lot of concerns in the new era. This has affected a majority of people across the globe and has resulted to a significant rise in the number of patients who are also suffering from acute respiratory distress syndrome (ARDS), severe acute respiratory syndrome (SARS), and other types of respiratory failures which calls for the need of mechanical ventilation. The accurate prediction of the settings for mechanical ventilation along with the response of the lungs of the patient toward the adjustments of pressure will be helpful in improving the efficiency of the medical care treatment. A particular kind of sequence prediction problem is the prediction of lung pressure. The most effective technique for solving the sequence prediction problem is called Long Short-Term Memory (LSTM). LSTM has an advantage over standard Recurrent Neural Network (RNN), since it can selectively recall patterns over time. Thus, the proposed algorithmic approach addresses the challenges of pressure prediction in mechanical ventilation and depicts healthcare assistance with the optimal pressure predictions.

Keywords Mechanical ventilation · Machine learning · LightGBM · COVID-19

1 Introduction

Mechanical ventilation is the life supporter that helps for breathing, when one cannot breathe naturally. It is given when a patient needs intensive care; it can be either during a surgery or when the patient is having severe problems while respiration. Lung failure occurs due to variety of illnesses, like pneumonia, heart disease, COVID-19, etc. A person who has trouble breathing or is unable to breathe on their own needs assistance with breathing from outside the body. The patient receives this assistance from a mechanical ventilator, and after stabilizing, the ventilator is taken off of

J. Ghorpade-Aher · S. Ijeri (✉)
School of Computer Engineering and Technology, Dr. Vishwanath Karad, MIT World Peace University, Pune, Maharashtra, India
e-mail: isrushti02@gmail.com

V. K. Gunjan and J. M. Zurada (eds.), *Proceedings of 4th International Conference on Recent Trends in Machine Learning, IoT, Smart Cities and Applications*, Lecture Notes in Networks and Systems 873, https://doi.org/10.1007/978-981-99-9442-7_42

them. The physician configures this device to supply air when the patient requires assistance in breathing and his condition is critical, according to his lung capacity needs. Ventilators use a PID control algorithm to control the oxygen concentration in the patient's body. The controller takes into account various body data such as frequency of breathing and oxygen levels to provide adequate oxygen and stabilize the patient. The ventilators are operated by trained medical staff and adjust the breathing frequency in response to the patient's changes. However, PID controllers have limitations, especially for integrated processes with large time delays. These limitations can be eliminated using deep learning, resulting in more efficient and budget-friendly technology that requires less manpower, especially during the pandemics. The Delta variant of COVID-19, caused pneumonia in both lungs of the infected people, and many infected patients also developed acute respiratory distress syndrome (ARDS) [18,deci20] so they required special medical care treatment, such as a ventilator - mechanical ventilation. COVID-19 variants had a huge impact on the immune system, ground-glass clouding, and various carcinogenic changes in the lungs due to SARS-CoV-2 and other variants are crucial areas of focus. According to the Johns Hopkins Medicine Centre, 2.2% of all COVID-19 cases worldwide, people are dying from respiratory diseases. The ML algorithms can help with the prediction of lung pressure prediction, where current recommendations call for efficient use of machine learning algorithms. These models have a great role in enhancing the accuracy of settings, minimizing treatment errors, and expediting the process of weaning patients off ventilation support.

2 Mechanical Ventilators

The main objectives of automated pressure prediction using smart technologies are to improve the accuracy and efficiency of pressure prediction in different fields, such as healthcare, manufacturing, and engineering. Specifically, in healthcare, pressure prediction using machine learning can help medical professionals to monitor patients more accurately and efficiently, leading to better diagnoses and treatments. In manufacturing and engineering, pressure prediction using machine learning can help optimize the performance of machines and equipment, improving productivity and reducing maintenance costs. Additionally, pressure prediction using machine learning can help predict and prevent system failures, reducing downtime and improving safety. The ventilators are used in various medical scenarios to aid with breathing for patients who are unable to breathe adequately on their own. Some common situations where the mechanical ventilators are used are depicted in Table 1 for respiratory healthcare assistance. It is important to remember that ventilation is a specialized medical intervention that requires careful monitoring and expertise to ensure the safety and well-being of the patient. The decision to use a ventilator is typically made by healthcare professionals based on the individual patient's reports. Table 2 represents the common effects on the lungs that are associated with mechanical ventilation, while necessary in many medical situations.

Table 1 Respiratory assistance of mechanical ventilators

Sr. No.	Parameters	Need for ventilation
1	Respiratory failure	When a person's respiratory system fails to function properly and cannot maintain adequate oxygen levels or remove carbon dioxide from the body. This can occur due to conditions such as pneumonia, ARDS, COPD, or asthma
2	Anesthesia	Anesthesia cause temporary suppression of natural breathing, so ventilators are used to provide controlled and assisted breathing to the patient while they are unconscious
3	Coma or unconsciousness	In a coma or unconscious state, a person can't breathe independently, so mechanical ventilation supports their respiratory function until they regain consciousness or recover
4	Neuromuscular disorders	Certain conditions like muscular dystrophy, amyotrophic lateral sclerosis can lead to paralyzed respiratory muscles
5	Respiratory distress in newborns	Premature infants or newborns with weak need mechanical ventilation for breathing until their lungs function properly

Table 2 Effect of mechanical ventilators on the lungs

Sr. No.	Parameters	Effect on lungs
1	Barotrauma	High levels of pressure can damage to the delicate lung tissues, leading to barotrauma. It can manifest as lung over inflation, pneumothorax, or air leakage into the surrounding tissues
2	Ventilator associated pneumonia (VAP)	The artificial airway and the presence of the ventilator provide an opportunity for bacteria to enter the lungs. VAP is a serious infection that requires treatment with antibiotics
3	Ventilator-induced lung injury (VILI)	Can be caused by factors such as high-pressure settings, excessive tidal volumes, or repeated alveoli opening and closing (atelectrauma). VILI can lead to inflammation, lung tissue damage, and impaired gas exchange
4	Oxygen toxicity	Prolonged exposure to high O_2 concentrations can cause oxidative stress and lung damage, known as oxygen toxicity, resulting in inflammation, injury & impaired lung function
5	Ventilator-induced diaphragmatic dysfunction	Prolonged mechanical ventilation causes faintness and atrophy of the diaphragm, the muscle responsible for breathing and it can result in difficulties weaning the patient off the ventilator

These effects are not universal and may differ on the basis of factors like the patient's health, ventilation duration, the settings used, and the care provided.

3 State-of-the-Art Study

The hidden Markov models (HMMs) algorithm is a fitting approach to predict ventilator pressure in sedated patients. The study [1] suggests that a particular threshold value of asynchrony events is expected to occur. In contrast to earlier studies that monitored patients with specific conditions for a limited period, the author of this paper analyzed the whole duration of ventilation in ICU patients with diverse critical illnesses. The paper [2] introduces an algorithm to forecast pressure in the respiratory circuit across many severe lung conditions. The most effective solution for this is the LSTM algorithm because of its capacity to selectively recognize patterns over a longer period. While the Recurrent Neural Network (RNN) is well-suited for short-term pattern recognition, LSTM is better suited for sequence prediction tasks. This paper addresses the challenges of predicting the pressure in mechanical ventilation and proposes an algorithm that predicts respiratory circuit pressure across various lung conditions, which is also a sequence prediction problem. Machine learning methods [3] calculate the duration of mechanical ventilation (MV) early on is a promising approach, but its accuracy is still limited. Previous studies have highlighted the limitations of intensivists in accurately predicting the duration of MV. It is challenging to compare the results of this study with other published machine learning predictions of MV duration as the study focused on predicting MV duration for a period longer than 48 h, while prior studies predicted for different outcomes. A study [4] that utilized a simple and effective machine learning algorithm that exhibits a high level of specificity and negative predictive value (NPV) in predicting challenging outcomes like mechanical ventilation is discussed. The algorithm relied on initial triage vitals and basic demographics and achieved a consistent accuracy rate of over 85% [5, 7]. The research further demonstrated that machine learning algorithms can be reliable even when dealing with missing data. Currently, most models depend on the analysis of lung mechanics [6, 8] data that is conducted offline. Future research could concentrate on creating real-time monitoring and prediction capabilities, which could be especially beneficial for adjusting mechanical ventilation settings during critical care. Among five machine learning methods [9–11], the random forest of regression trees was the most effective in predicting outcomes. However, because the study cohort was imbalanced, the sensitivity of the models was limited. It is worth noting that machine learning models do not necessarily outperform experienced clinicians [12] who can integrate multiple parameters with high accuracy based on their experience.

4 Proposed Algorithmic Approach

LightGBM is a gradient boosting framework designed to be efficient and effective. It is optimized for large datasets and can handle a large number of features and training samples efficiently. This makes it well-suited for problems with high-dimensional data like the one we are using for finding the results. Gradient boosting builds an ensemble of weak learners (decision trees) [13] to make accurate predictions. It iteratively improves the model by minimizing the loss function through gradient descent. This iterative process helps capture complex patterns and interactions in the data, leading to better predictive performance. *Recurrent Neural Networks* (RNNs) can be used in the context of mechanical ventilation and machine learning to enhance certain aspects of the ventilation process. RNNs are a type of artificial neural network that can capture sequential information and make predictions based on that context. By processing the time-series data of pressure, flow, and volume waveforms, RNNs can learn patterns and relationships that are indicative of lung conditions or ventilation settings. This analysis can assist in identifying abnormalities, detecting patient-ventilator asynchrony, and optimizing ventilator parameters [14] *Long Short-Term Memory Networks* (LSTMs) can be trained on historical patient data to predict optimal ventilation parameters for individual patients. By considering patient characteristics, clinical measurements, and desired outcomes, LSTMs can learn patterns that help predict appropriate tidal volumes, respiratory rates, or inspiratory and expiratory times. These predictions can assist clinicians in setting initial ventilator parameters or adjusting them dynamically based on patient condition. K-fold cross-validation and group k-fold cross-validation are techniques commonly used in machine learning, including in the context of mechanical ventilation, to evaluate and validate models. These techniques help assess the performance of machine learning models on different subsets of data and mitigate issues related to data imbalance or correlation. *K-fold Cross-Validation:* K-fold CV separates the existing data into K likewise sized folds. A robust estimate is deduced for the model's performance by averaging the evaluation results across the K iterations. *Group K-fold Cross-Validation:* Group k-fold cross-validation is particularly useful when the data has group or cluster structures, such as multiple patients or multiple ventilation sessions from the same patient. In group *K-fold* cross-validation, the data is split into K folds keeping the samples from the same group together within a fold. This is important to avoid scenarios where data from the same group is split between training and validation sets, which could lead to biased evaluation results. Group k-fold cross-validation can be applied in the mechanical ventilation domain to account for patient-specific variations or to evaluate models on specific cohorts, such as patients with a certain lung condition or undergoing a particular ventilator mode.

A basic regression equation for predicting the pressure using linear regression takes the form as revealed in Eq. (1).

$$\begin{aligned}\text{Pressure} = \beta_0 + (\beta_1 * \text{breath}_{\text{id}}) + (\beta_2 * R) + (\beta_3 * C) \\ + \left(\beta_4 * \text{time}_{\text{step}}\right) + (\beta_5 * u_{\text{in}}) + (\beta_6 * u_{\text{out}})\end{aligned} \quad (1)$$

As shown in Eq. (1), β_0, β_1, β_2, β_3, β_4, β_5, and β_6 represents the regression coefficients that will be estimated during the training process. The variables corresponding to the input features from the dataset are described below where *breath_id*: unique id for each breath, *R*: lung attribute indicating how restricted the airway is, *C*: lung attribute indicating how compliant the lung is, *Time step*: time_stamp, u_{in}: control input for inspiratory valve. Range [0–100%], u_{out}: control input for expiratory valve Range [0 or 1] closed or open, *Pressure*: pressure in the respiratory circuit. The proposed model as in Eq. (2) implements the *LightGBM function*() which is a composition of multiple decision trees, with the algorithmic equational approach as depicted in Eq. (3) for determining the estimates of 'y'.

$$y = f(x) \tag{2}$$

where 'y' ← Optimal pressure predictions (*Op_P*) and 'x' ← Extraneous input features

$$y = \{y \in Op\_p | 3 < Op\_p < 25 \text{ cm H}_2\text{O}\} \tag{3}$$

The cross-validation (CV) mean score, also known as the coefficient of variation of the mean, can be calculated to assess the variability of the mean values in a dataset. Equation (4) defines the ratio of the standard deviation (std) to the mean, expressed as a percentage, where 'σ' represents the standard deviation and 'μ' represents the mean value. The obtained CV mean score of 0.4716 and (std) of 0.0051 demonstrate the strong correlation among the scores.

$$\text{CV Mean} = \frac{\sigma}{\mu} \times 100 \tag{4}$$

5 Experimental Analysis

Feature engineering processes the knowledge to extract the relevant data. It improves performance of machine learning (ML) and deep learning (DL) algorithms [15]. The experimental research study and analysis aims to improve the mechanical ventilation control through machine learning (ML) and deep learning (DL) algorithmic optimization techniques. Figure 1 illustrates the experimental flow of the proposed model.

It is believed that neural networks and deep learning can provide better generalization across lungs with different characteristics compared to the current industry standard of PID controllers. Matplotlib, a robust Python tool, is extensively utilized for visualizing data by leveraging other libraries like Pandas, Sklearn, and NumPy. It enables the creation of graphical representations, including statistical inferences and 2D (and occasionally 3D) plotting of arrays. In our project, we employed Matplotlib

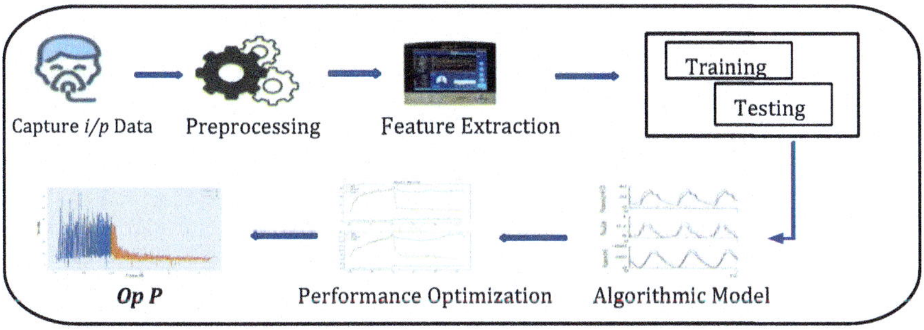

Fig. 1 Experimental process flow of the proposed optimal pressure prediction (*Op_p*) model

to generate a graph illustrating the u_{in}, u_{out}, and *pressure* variance of a specific patient at different time intervals with respect to density.

Figure 2 serves as an illustrative example of a graph produced using Matplotlib that depicts the relation of all the attributes with Density, so as to use the correct variables to find the expected target outcome. CV mean and mean absolute error (MAE) is a metric employed to evaluate and summarize the performance of a machine learning model, measuring the average absolute difference between predicted and actual values. First the login activity validates the user data. Then, enter patients Lung attribute Parameters (*R* and *C*). After that enter the *Time stamp* and u_{in}, u_{out} parameters. Later, raw data is sent to server application, and input data to trained model. Output data (predicted pressure) sent from server is displayed on the screen and plot graph for u_{in}, u_{out} and predicted pressure. The experimental analysis of the proposed algorithm on Google Brain—Ventilator Pressure Prediction [4] showed an improved performance and giving the least difference between the actual values and the predicted values.

As depicted in Fig. 3a, it is observed that all the values in the column pressure are below 25 cm H_2O which is the correct setting for patients suffering from ARDS. Thus, the proposed model is trained properly and gives accurate prediction that can be used for optimal pressure prediction for medical assistance.

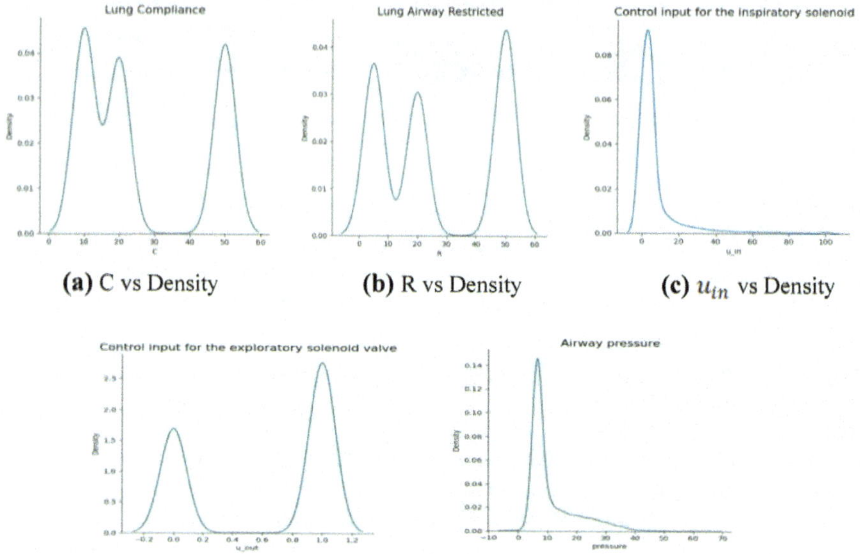

(a) C vs Density **(b)** R vs Density **(c)** u_{in} vs Density

(d) u_{out} vs Density **(e)** pressure vs Density

Fig. 2 Correlation of all attributes with density distribution

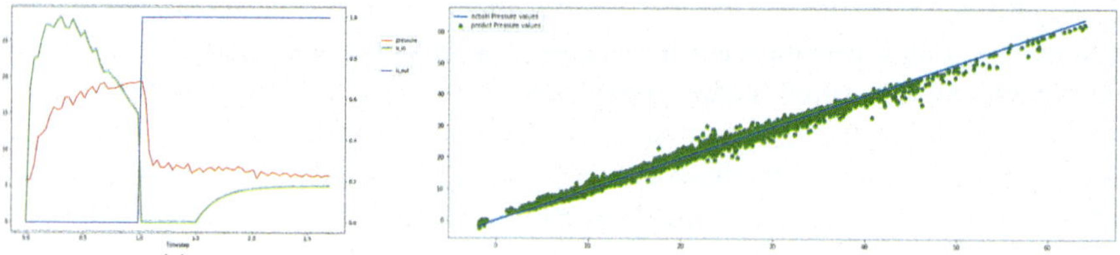

(a) (b)

Fig. 3 Proposed analysis for optimal pressure prediction (*Op_P*)

6 Conclusion

The purpose of this research is to develop a suitable algorithm using ML that can aid doctors in calculating lung pressure settings for patients with diverse conditions during mechanical ventilation. By achieving this, the algorithm will alleviate the burden on healthcare workers, especially during situations like pandemics. Predicting lung pressure is a type of problem that involves predicting sequences. Among various solutions, Long Short-Term Memory works well for sequence prediction. LightGBM enables forecasting of appropriate pressures by offering quicker training speed with good efficacy. In comparison to other approaches for predicting optimal pressures, this research study proposes the algorithmic technique for optimal prediction results.

References

1. Marchuk Y, Magrans R, Sales B, Montanya J, López Aguilar J, de Haro C, Gomà G, Subirà C, Fernández R, Kacmarek R, Blanch L (2018) Predicting patient ventilator asynchronies with hidden Markov models. Sci Rep 8
2. Sable NP, Wanve O, Singh A, Wable S, Hanabar Y (2023) Pressure prediction system in lung circuit using deep learning. ICT with intelligent applications. Smart innovation, systems and technologies, vol 311. Springer, Singapore
3. Wadne V et al (2022) Pressure prediction system in lung circuit using deep learning and machine learning. IRJET 09
4. Data set: Google brain—ventilator pressure prediction | Kaggle
5. Sayed M, Riaño D, Villar J (2021) Predicting duration of mechanical ventilation in acute respiratory distress syndrome using supervised machine learning. J Clin Med 10:3824
6. Holder-Pearson L, Geoffrey Chase J (2021) Physiologic-range flow and pressure sensor for respiratory systems. HardwareX. 10:e00227. https://doi.org/10.1016/j.ohx.2021.e00227
7. Roesthuis L, van den Berg M, van der Hoeven H (2021) Non-invasive method to detect high respiratory effort and transpulmonary driving pressures in COVID-19 patients during mechanical ventilation. Ann Intensive Care 11:26
8. Rahmanzade R, Rahmanzadeh R, Tabarsi P, Hashemian SM (2020) Noninvasive versus invasive ventilation in COVID-19: one size does not fit all! Anesth Analg
9. Botta M, Tsonas AM, Pillay J, Boers LS, Algera AG, Bos LD, Dongelmans DA, Hollmann MW, Horn J, Vlaar AP et al (2020) Ventilation management and clinical outcomes in invasively ventilated patients with COVID-19. Lancet Respir Med 9:139–148
10. Chang R, Elhusseiny KM, Yeh Y-C, Sun W-Z (2021) COVID-19 ICU and mechanical ventilation patient characteristics and outcomes—a systematic review and meta-analysis. PLoS ONE 16:e0246318
11. Ervin JN, Rentes VC, Dibble ER, Sjoding MW, Iwashyna TJ, Hough CL, Sales AE (2020) Evidence-based practices for acute respiratory failure and acute respiratory distress syndrome: a systematic review of reviews. Chest 158:2381–2393
12. Yu C, Liu J, Zhao H. Inverse reinforcement learning for intelligent mechanical ventilation and sedative dosing in intensive care units. BMC Med Inform Decis Making 19:57
13. Ghorpade-Aher J, Sonkamble B (2022) Effective feature selection using ensemble techniques and genetic algorithm. ICICT, LNNS, vol 236. Springer, Singapore
14. Campbell RS, Davis BR (2002) Pressure-controlled versus volume-controlled ventilation: does it matter? Respir Care 47:416–424
15. Ervin JN, Rentes VC, Dibble ER, Sjoding MW, Iwashyna TJ, Hough CL, Gong MN, Sales AE (2020) Evidence-based practices for acute respiratory failure and acute respiratory distress syndrome: a systematic review of reviews. Chest 158:2381–2393

Analysis of Various GNNs in Solving MaxCut Problem

Hiba Hameed and Ajeesh Ramanujan

Abstract Extraction of information from graph data is essential since most data in the real world are dynamic, large, and without fixed structure, unlike images. Graph neural networks (GNNs) harness the power of graphs, efficiently examining graph data and help make inferences from complex data structures, making them an invaluable tool in domains like social network analysis. GNNs have shown promise in solving combinatorial optimization problems which involve extracting a good candidate solution by scanning over the search space. The vastness of the search space often renders the optimal solution search a difficult task. Many previous works demonstrate the use of GNNs in solving problems like graph coloring and MaxCut problem. In this paper, we analyze the ability of different GNN architectures to tackle the MaxCut problem, posed as a Quadratic Unconstrained Binary Optimization (QUBO) problem, minimizing the loss function based on the QUBO objective.

Keywords Combinatorial optimization · Graph neural networks · Maximum cut · Quadratic unconstrained binary optimization

1 Introduction

Graph neural networks (GNNs) have garnered much attention in recent years as a powerful framework for dealing with graph-structured information. Their capacity to retrieve information from dynamic connectivity and relationships within graphs led to their successful application in various domains, including social network analysis, recommendation systems [1, 2], and molecular chemistry [3]. They lend the ability to understand complex systems and make informed decisions using the meaningful representation they extract, by leveraging the inherent structure and connectivity of graphs. The expressive power and scalability of GNNs make them well-suited for numerous graph-related tasks. Discovery of new antibiotics, such as Halicin, was

H. Hameed (✉) · A. Ramanujan
College of Engineering, Trivandrum, Kerala, India
e-mail: delta.hiba5@gmail.com

V. K. Gunjan and J. M. Zurada (eds.), *Proceedings of 4th International Conference on Recent Trends in Machine Learning, IoT, Smart Cities and Applications*, Lecture Notes in Networks and Systems 873, https://doi.org/10.1007/978-981-99-9442-7_43

facilitated by GNN's ability to predict molecular properties and screen potential candidates efficiently [4]. Additionally, GNNs have been instrumental in improving the accuracy of the estimated time of arrival (ETA) predictions in mapping applications like Google Maps, leveraging graph representations of road networks and historical traffic patterns [5]. Also, recommendation systems for Pinterest [2], Uber Eats [6], etc. have utilized GNN.

One particularly intriguing area where GNNs have shown promise is in solving combinatorial optimization problems. These problems involve extracting a good candidate solution by scanning over the search space. This vastness of the search space depends on the problem size, rendering the optimal solution search a difficult task. Some instances of these problems include the traveling salesman problem, set cover, integer programming, and graph partitioning. These problems are notoriously challenging to solve optimally due to their NP-hard nature, requiring exponential time in the worst case.

The MaxCut problem is a classic optimization task that finds practical use in network design, image segmentation, and clustering, to name a few. Given an undirected graph, the objective is to partition its vertices into two disjoint sets, in a way that the total edges or link weights crossing the sets are maximized. Finding the Maximum cut is rated NP-hard, due to the challenging nature to scan potential solutions over large-scale graphs. By posing the MaxCut as a QUBO problem [7], the task is transformed into finding the binary variable assignments that maximize the binary objective function.

In this paper, we focus on analyzing the effectiveness of different graph neural network architectures in tackling the MaxCut problem formulated as QUBO problem, based on the work in [7]. The architectures used are based on the works in [8–14].

2 Preliminaries

This section introduces mathematical notations and concepts used in this paper.

2.1 Graph Neural Networks

GNNs operate by repeatedly revising node representations based on information from neighboring nodes, allowing them to capture and leverage the inherent relational structure in graph.

The main essence of GNNs is the concept of message passing. During each iteration, nodes aggregate information from their neighbors, typically by summing or averaging their neighbor's representations. This aggregated information is then combined with the node's own representation, using a learnable update function that captures the node's local context within the graph. Through multiple iterations of message passing and updating, GNNs allow nodes to refine their representations by considering increasingly larger and more distant neighborhoods.

GNNs can also incorporate graph-level aggregation mechanisms to capture global information about the entire graph. GNN outputs can be in the form of predictions for centered on node-like node classification or link prediction or focusing on the entirety of the graph as in graph classification or generation.

2.2 *Maximum Cut Problem*

The Maximum Cut problem (MaxCut) comes under the graph combinatorial optimization problem that tries to segregate the node set into two disjoint sets, in a manner that maximizes edge count or sum of edge weights crossing the partition. MaxCut finds practical use in network design, image segmentation, data clustering, and VLSI circuit layout [15, 16].

Formally, an undirected graph can be notated as $G = (V, E)$. Here, V and E represent the node and edge set respectively. The goal of the MaxCut problem is to determine a partition $(S, V\backslash S)$ of V in a manner that makes the maximizes number of edges crossing the partition. Here, S represents the "cut" set, and $V\backslash S$ represents the "uncut" set.

2.3 *Quadratic Unconstrained Binary Optimization*

QUBO is a mathematical framework used to solve binary optimization problems. In QUBO, the objective function is quadratic, and the variables can take binary values, typically 0 or 1. The task is to determine the binary assignment that minimizes or maximizes the objective function. QUBO problems have applications in various fields, such as combinatorial optimization, machine learning, and quantum computing. Various combinatorial optimization tasks can be formulated as QUBO instances such as Maximum Clique (MC), Maximum Independent Set (MIS), determination of chromatic number, Quadratic Assignment Problems (QAP), Constraint Satisfaction Problems (CSP), and Satisfiability (SAT) problem [17].

To model the MaxCut problem using QUBO, a binary variable x_i for each vertex i is defined, with 1 if vertex i belongs to one set. The value 0 indicates membership on the other set. The QUBO objective is then defined as:

$$H_{\text{QUBO}} = x^{\text{T}} Q x = \sum_{i,j} x_i Q_{ij} x_j \tag{1}$$

This can also be expressed as the following Hamiltonian.

$$H_{\text{MAXCUT}} = \sum_{i<j} A_{ij}(x_i + x_j - 2x_i x_j) \tag{2}$$

In the equation, Q stands for QUBO matrix which is a $N * N$ matrix of constants that are equal to the linear and quadratic coefficients. The quantity $x_i + x_j - 2x_i x_j$ will be equal to 1 if the edge (i, j) has been cut, else 0. The aim is to maximize this quantity, and since QUBO prefers minimization, this can be also formulated as

$$\text{Minimize} - 1 * H_{\{\text{MAXCUT}\}} \tag{3}$$

3 Related Work

The experiments in this work are based on the architectures introduced in the following papers.

The analysis done in this work is primarily based on the research work "Combinatorial optimization with physics-inspired graph neural networks" [7], which obtains a differentiable loss function for unsupervised training on the node representations of a graph neural network (GNN), by applying a relaxation strategy by which the constraints on the decision variables are eliminated. The GNN is trained to generate soft probabilities which indicates the probability of each vertex's membership to one of two partitions. In the end, binary labels can be obtained using simple projection heuristics.

3.1 GNN Models

The paper "Semi-Supervised Classification with Graph Convolutional Networks" [11] introduces Graph Convolutional Networks (GCNs), a graph-specific variant of convolutional neural networks that facilitate capturing the local graph structure and node attributes using spectral approach. The convolution proposed is a first-order approximation [11] and scales linearly with the number of edges.

The model proposed eliminates the need for graph-based regularization that is found in traditional methods by accounting for the adjacency matrix of the graph while learning. This helps in diffusing the gradient obtained from supervised loss, allowing for obtaining a representation of unlabeled nodes [11]. The highlight of the work is the construction of a layer-wise propagation rule based on spectral convolutions on graphs. This involves signal multiplication with a filter in the Fourier domain [11]. To make the computation efficient, the authors approximate the filter using a limited expansion of Chebyshev polynomials [11], avoiding the costly eigendecomposition of the graph Laplacian.

The paper "Convolutional Neural Networks on Graphs with Fast Localized Spectral Filtering" [8] also based on spectral convolutions, introduces ChebNet which instead of using a fixed filter as in GCN, uses a kth order Chebyshev polynomials as a series of approximation filters.

The paper "Topology Adaptive Graph Convolutional Networks" [9] addresses the computational expensiveness and information loss that results from the spectral convolutions, approximations, and use of fixed-size filters. The authors introduce convolutions on the vertex domain and avoid the determination of the graph Laplacian spectrum. This is achieved as multiplication by polynomials of the adjacency matrix, and limits to a maximum degree of 2, thereby lending lower computational complexity than spectrum-based methods [9]. The authors demonstrate that using only one filter size can cause information loss and reduced accuracy in graph convolutional neural networks. By using a set of filters of varying sizes, their proposed TAGCN can better adapt to the topology of the graph and improve classification accuracy.

The paper "Dynamic Graph CNN for Learning on Point Clouds" [13] proposes a graph neural network module called EdgeConv that learns from point clouds. EdgeConv operates on graphs that are generated at each layer. EdgeConv takes as input a point cloud and a set of k-nearest neighbors for each point and outputs a point-wise feature vector by aggregating the features of its neighbors [13].

The paper "Simplifying Graph Convolutional Networks" [14] suggests a simpler version of a type of neural network called Graph Convolutional Networks (GCNs). This new approach, called SGC, reduces unnecessary complexity and repetitive calculations by removing nonlinearities between the GCN layers [14].

The paper "Attention-based Graph Neural Network for Semi-supervised Learning" [12] introduces AGNN which incorporates an attention mechanism for learning the local summary of the node neighborhood [12], to capture the relevance of different neighbors. It reduces the model complexity by discarding all the intermediate fully connected layers, resulting in a linear model with only a single scalar parameter at each intermediate layer [12].

4 Experiments and Results

The analysis is done on the models ChebNet [8], TAGGNN [9], GraphSAGE [10], GCN [11], AGNN [12], EdgeConv [13], SGCN [14], and is evaluated against two algorithms, Simulated Annealing (SA) by openjij library and the greedy method provided by cvxgraphalg library.

The dataset used is randomly generated regular graphs with 10^3, 10^4, 10^5 nodes and degrees ranging from 3 to 9. All GNN models are run with 2-, 4-, and 6-layered architectures. The node features are randomly initialized. Each model configuration is run for two shots for each graph. Early stopping techniques are employed to identify the best cost obtained, but let the model run for 1000 additional epochs to observe whether improvements are observed.

For larger graphs, memory constraints become an issue even while creating the Q matrix. To overcome the memory constraints, the matrix and loss calculation operations make use of sparse matrix calculation, resulting in a large difference in memory requirement.

Without dropout, the models quickly reach early stopping and the loss or cost stagnates. Whereas with a dropout of 0.5, the models are able to explore further and get better results as epochs progress for most models. However, this exploration is at the expense of large runtimes as observed from the scatterplot in Fig. 1. For graphs of lower orders, the effect of dropout is minimal for less number of layers and has a negative effect as layer size increases, except for GCN, as observed in the leftmost plots in Figs. 2, 3, 4, and 5. GCN gains a clear advantage from dropout in all cases as shown in Fig. 2. For ChebNet, dropout has negative effects across all cases, whereas for AGNN, dropout gives an advantage only for large graphs with less number of layers and as layers increase, performance falls as shown in Fig. 3. For graphs with 10^4, 10^5 nodes TAGNN, SGConv, EdgeConv, GraphSAGE benefits from dropout significantly as shown in the middle and rightmost plots in Figs. 4 and 5.

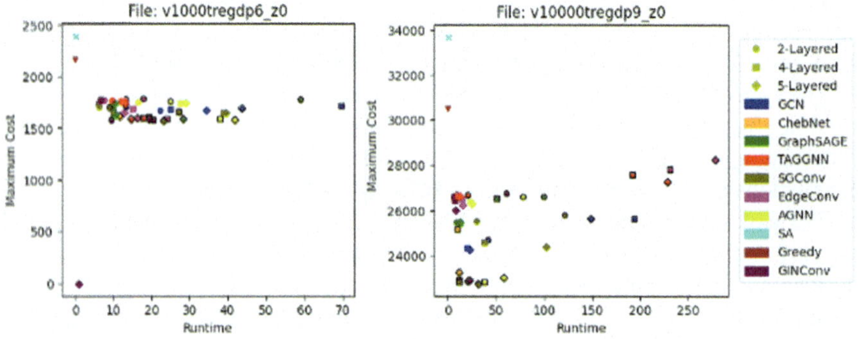

Fig. 1 Runtime versus cost for few graph instances. SA and Greedy deliver best cost in minimal runtime in all the cases. The edged and non-edged data points indicate models with and without dropout respectively. GNN models with dropout especially for 4 and 6 layers are found to have longer runtimes

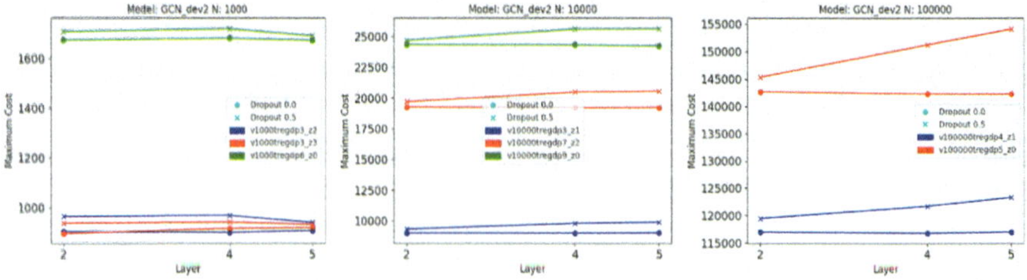

Fig. 2 Cost versus number of layers for GCN. From left to right, plots for graphs of size 10^3, 10^4, and 10^5 respectively. Each color represents one graph instance

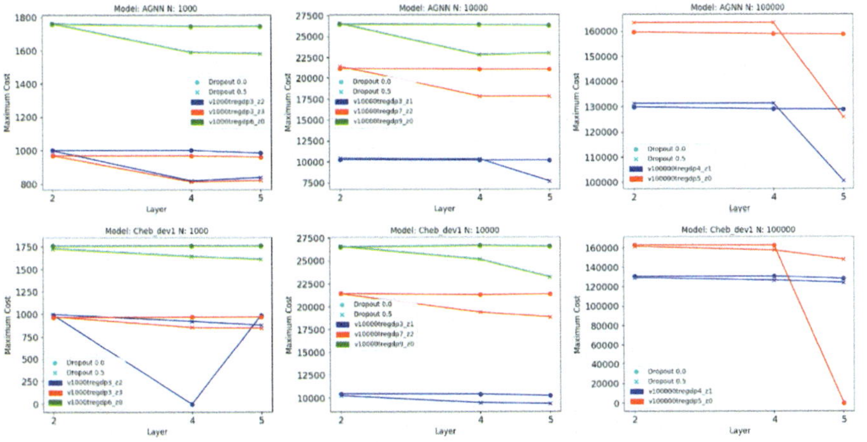

Fig. 3 Cost versus number of layers for AGNN and ChebNet. From left to right, plots for graphs of size 10^3, 10^4, and 10^5 respectively. Each color represents one graph instance

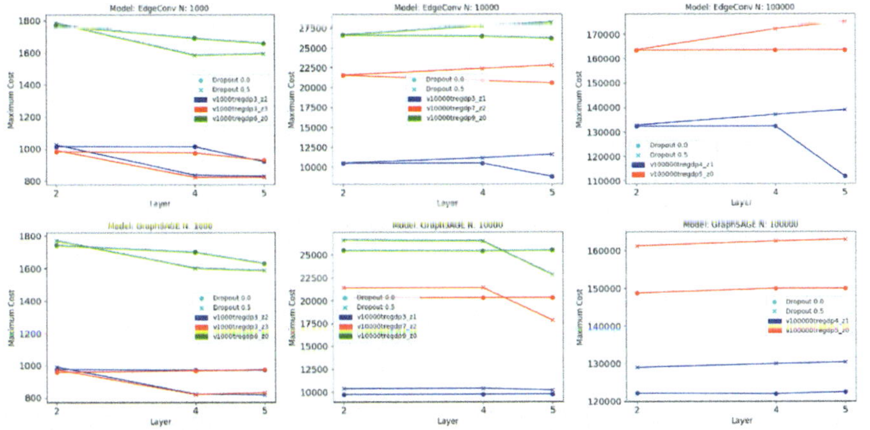

Fig. 4 Cost versus number of layers for EdgeConv, GraphSAGE. From left to right, plots for graphs of size 10^3, 10^4, and 10^5 respectively. Each color represents one graph instance

Without dropout, as layers increase, there are no significant performance changes or tend to have a negative effect. With dropout, as layers increase EdgeConv, GCN, TAGNN, GraphSAGE, SGConv performs better, whereas performance of ChebNet and AGNN deteriorates.

To illustrate the model which outputs better results comparatively across the experiments, a bar graph is provided in Fig. 6 from which it can be observed that among the GNNs experimented with, the EdgeConv and TAGNN model gives better output than others more frequently. The performance of other models varies for different graph scales.

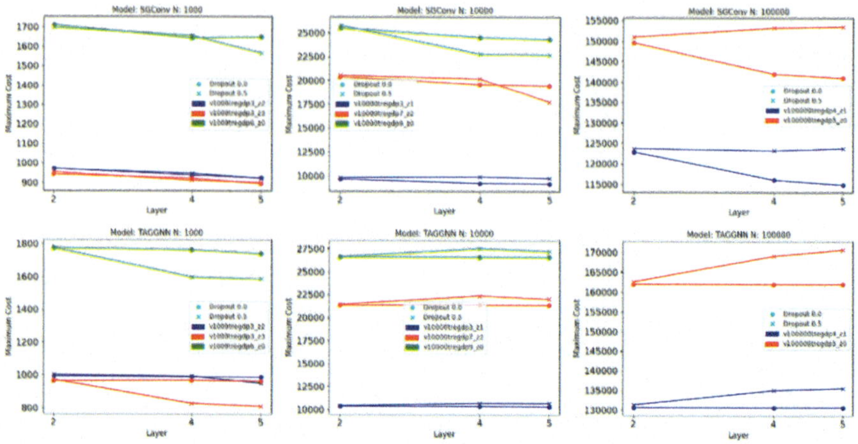

Fig. 5 Cost versus number of layers for SGConv, TAGNN. From left to right, plots for graphs of size 10^3, 10^4, and 10^5 respectively. Each color represents one graph instance

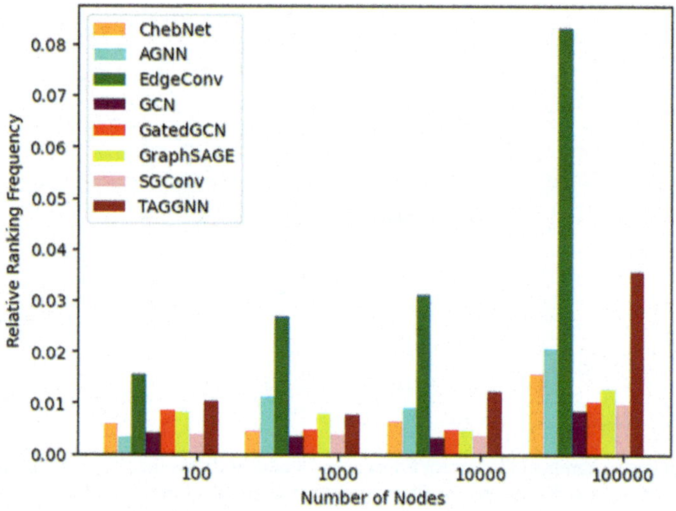

Fig. 6 Relative ranking frequency for graph neural network models

5 Conclusion

In this analysis, various GNN architectures were evaluated for solving the MaxCut problem, posed as a QUBO problem. The GNN models were compared against Simulated Annealing, and the greedy method. The results of the analysis indicate that GNN models generally achieve lower-quality solutions and have longer runtimes compared to SA and the greedy method. Among the GNN models tested, the Edge-Conv and TAGGNN models produced better results more frequently compared to other models, although the performance of other models varied across different graph scales. GCN, ChebNet, TAGNN, GraphSAGE, SGConv models do achieve improvement for large graphs with dropout. For small graphs, deeper models were found to reduce performance when using dropout.

For future work, it could be beneficial to explore the use of embeddings that capture specific graph properties instead of random initialization.

Overall, this analysis provides insights into the performance of different GNN architectures for solving combinatorial optimization problems, specifically the MaxCut problem. The findings can guide future research in improving GNN models for combinatorial optimization and exploring alternative techniques to enhance their performance.

References

1. Fan W, Ma Y, Li Q, He Y, Zhao E, Tang J, Yin D (2019) Graph neural networks for social recommendation. In: Proceedings of the World Wide Web conference (WWW) 2019
2. Ying R, He R, Chen K, Eksombatchai P, Hamilton WL, Leskovec J (2018) Graph convolutional neural networks for web-scale recommender systems.
3. Fout A, Aspuru-Guzik A, Radford A (2017) Protein interface prediction using graph convolutional networks. In: Advances in neural information processing systems (NeurIPS)
4. Stokes JM, Yang K, Swanson K, Jin W, Cubillos-Ruiz A, Donghia NM, MacNair CR, French S, Carfrae LA, Bloom-Ackermann Z, Tran VM, Chiappino-Pepe A, Badran AH, Andrews IW, Chory EJ, Church GM, Brown ED, Jaakkola TS, Barzilay R, Collins JJ. A deep learning approach to antibiotic discovery. In: Proceedings of the 30th ACM international conference on information & knowledge management
5. She J, Wong D, Lange O, Hester T, Perez L, Nunkesser M, Lee S, Guo X, Wiltshire B, Battaglia PW, Gupta V, Li A, Xu Z, Li Y, Veličković P (2021) ETA prediction with graph neural networks in Google maps. In: Proceedings of the 30th ACM international conference on information & knowledge management, Oct 2021
6. Jain A, Liu I, Sarda A, Molino P (2019) Food discovery with Uber eats: using graph learning to power recommendations
7. Schuetz MJ, Brubaker JK, Katzgraber HG (2022) Combinatorial optimization with physics-inspired graph neural networks. Nat Mach Intell 4:367–377
8. Defferrard M, Bresson X, Vandergheynst P (2016) Convolutional neural networks on graphs with fast localized spectral filtering. Proc Neural Inf Process Syst (NeurIPS) 2016:3844–3852
9. Du J, Zhang S, Wu G, Moura JM, Kar S (2017) Topology adaptive graph convolutional networks. ArXiv /abs/1710.1037
10. Hamilton WL, Ying R, Leskovec J (2017) Inductive representation learning on large graphs. In: Advances in neural information processing systems (NeurIPS), pp 1024–1033
11. Kipf T, Welling M (2016) Semi-supervised classification with graph convolutional networks. ArXiv /abs/1609.02907
12. Thekumparampil KK (2018) Attention-based graph neural network for semi-supervised learning. ArXiv abs/1803.03735
13. Wang Y, Sun Y, Liu Z, Sarma SE, Bronstein MM, Solomon JM (2018) Dynamic graph CNN for learning on point clouds. ACM Trans Graph 38(5)
14. Wu F, Zhang T, Fifty C, Yu T, Weinberger KQ (2019) Simplifying graph convolutional networks. In: International conference on machine learning (ICML)

15. Barahona F, Grötschel M, Jünger M, Reinelt G (1988) An application of combinatorial optimization to statistical physics and circuit layout design. Oper Res
16. Sun C, Sun Y, Dai R (2018) ADMM for combinatorial graph problems
17. Glover F, Kochenberger G, Du Y (2018) A tutorial on formulating and using QUBO models. ArXiv /abs/1811.11538

RythmNet: A Novel Convolutional Neural Network for Multiclass ECG Classification

E. Sathish, N. Keerthika, M. Vijayalakshmi, V. Kiruthika, and V. Shobhana

Abstract An effective non-invasive diagnostic tool for identifying heart problems is the electrocardiogram (ECG). It's essential to accurately categorize ECG data in order to monitor and diagnose a variety of heart conditions. This work suggests a method that effectively classifies arrhythmias using ECG data utilizing deep learning algorithms. The system takes into account the ECG signals that fall into the following four categories are supraventricular premature beat (S), premature ventricular contraction (V), fusion beats (F), and unclassifiable beat (Q). The system surpasses current state-of-the-art techniques by employing an intuitive convolutional neural network (CNN) architecture called "RythmNet," which results in an overall accuracy of 97.1%. The findings show that the suggested method has the ability to reliably and efficiently classify arrhythmias, which can be extremely important in the detection and follow-up of heart disorders.

Keywords Electrocardiogram · Heart beats · Convolution neural network · Deep learning

---

E. Sathish · N. Keerthika (✉) · M. Vijayalakshmi · V. Kiruthika · V. Shobhana
Vellore Institute of Technology, Chennai, Tamil Nadu, India
e-mail: keerthika.n2021a@vitstudent.ac.in

E. Sathish
e-mail: sathish.e@vit.ac.in

M. Vijayalakshmi
e-mail: vijayalakshmi.m2021@vitstudent.ac.in

V. Kiruthika
e-mail: kiruthika.v@vit.ac.in

V. Shobhana
e-mail: shobhana.v2021@vitstudent.ac.in

V. K. Gunjan and J. M. Zurada (eds.), *Proceedings of 4th International Conference on Recent Trends in Machine Learning, IoT, Smart Cities and Applications*, Lecture Notes in Networks and Systems 873, https://doi.org/10.1007/978-981-99-9442-7_44

1 Introduction

ECG is a medical test that measures the electric impulses generated by the heart. It is a non-invasive diagnostic tool used to detect and diagnose various heart conditions such as arrhythmias, heart attacks, and other cardiac diseases. The test records the electrical impulses generated by the heart as it beats, and produces a visual representation of these impulses in the form of a graph or waveform. ECGs are often performed in hospitals, clinics, or doctor's offices, and are usually administered by trained medical professionals. The results of an ECG can help healthcare providers to determine the health and functioning of the heart, and to identify any abnormalities or irregularities that may require further medical attention. In emergency situations, such as during a heart attack or other cardiac event, ECGs can quickly diagnose the problem and initiate life-saving interventions [1].

For several decades, researchers have been particularly interested in the ECG signal as a nonlinear signal for analysis. ECG is commonly used to model biometric systems, as traditional authentication methods relying on fingerprints and face recognition have become more susceptible to falsification. ECG has advantages such as being unique to every individual and difficult to forge as it can only be recorded in living things. Therefore, differentiating between ECG signals can be helpful, and ECG also plays a crucial role in diagnosing cardiovascular diseases. The standard ECG waveform consists of P-wave, Q-wave, R-wave, S-wave, and T-wave, as shown in Fig. 1. As ECG has numerous applications in various fields, such as medical, Internet of Things, cryptography, and wearable sensors, researchers have conducted significant research work on ECG, as evidenced in a comparative analysis on ECG [2].

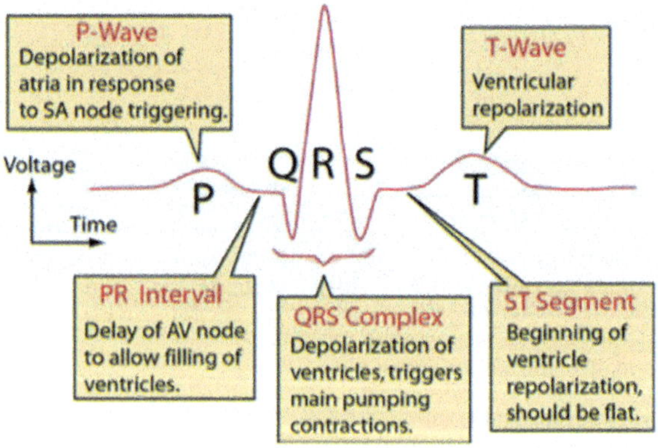

Fig. 1 PQRS wave form [5]

1.1 Frequency Range of ECG

The frequency range of an ECG refers to the range of electrical frequencies that can be detected and recorded by the ECG machine during a cardiac cycle. The frequency range of an ECG signal is between 0.05 and 150 Hz. The lower frequency limit of 0.05 Hz is associated with the baseline or DC component of the ECG signal, while the upper limit of 150 Hz is associated with the higher-frequency components of the signal, such as the QRS complex and *T* wave.

ECG machines filter signals outside the frequency range to eliminate noise and interference, ensuring clear and accurate recordings. Specialized recordings may be needed for specific arrhythmias or cardiac abnormalities at higher or lower frequencies [3].

1.2 ECG Heart Beat Analysis

The ECG heartbeats analysis includes the examination of the electric pattern of impulse signal produced by the activity of the heart which is captured by the ECG signal. The main purpose of this analysis is to identify the abnormalities, irregularities, or any deviation which is a sign of various heart diseases or heart conditions. There are various categories of analysis they are as follows Heart Rate Estimation: this is the distance between consecutive R waves (RR) and the calculation of heartbeats per minute. Supraventricular Premature Beat (S): This is a premature heartbeat that starts in the atria rather than the ventricles (lower chambers) of the heart. It can happen to healthy people and can be linked to ailments like electrolyte imbalances, coffee consumption, underlying heart issues, or anxiety [4]. Premature Ventricular Contraction (V): The ventricles are where aberrant heartbeats called PVCs begin. When the regular cardiac beat is interfered with by electrical impulses from the ventricles, they take place. PVCs may necessitate testing and treatment because they may be an indication of underlying heart problems [5] Fusion of ventricular and normal beats (F): Fusion beats are a combination of normal and pathological signs on an ECG that happens when a normal heartbeat and a premature ventricular contraction happen at the same time. They could be innocuous, but they could also be signs of more severe heart issues that need more testing and treatment [6]. Unclassifiable beat (Q): On an ECG, unclassifiable beats cannot be categorized as normal or pathological. They could be brought on by noise or artifacts in the recording or technological problems with the recording itself [6].

2 Related Study

The authors proposed an expert system that uses a decision tree algorithm to automatically detect and classify arrhythmias in ambulatory ECG recordings. The system consists of three main components: beat segmentation and feature extraction, beat classification, and rhythm analysis. The system achieves a high sensitivity of 98.6% and a positive predictive value of 98.5% for detecting arrhythmias, demonstrating its effectiveness in aiding the diagnosis and management of cardiac arrhythmias [7]. The authors evaluate the performance of ECG beat classifiers on the MIT-BIH Arrhythmia Database, comparing several algorithms in correctly identifying different types of beats. Most classifiers have similar performance in detecting normal beats, but their performance degrades significantly when detecting abnormal beats. The paper also discusses approaches to improving the performance of classifiers, such as using ensemble methods and feature selection techniques. The authors emphasize the importance of accurately detecting abnormal beats in ECG recordings, as these are often associated with cardiac abnormalities, and note that there is still room for improvement in detecting rare and complex arrhythmias [8]. The authors present a comparative study of different methods for ECG classification using dimensionality reduction. The authors explore the performance of several methods including Non-negative Matrix Factorization (NMF), Principal Component Analysis (PCA), Independent Component Analysis (ICA), support vector machine (SVM), and Linear Discriminant Analysis (LDA), Decision Trees (DT), Random Forest (RF), Artificial Neural Network (ANN) on a publicly available ECG dataset. In the dataset (MIT-BIH arrhythmia database), the highest accuracy achieved was 97.3% using a combination of SVM and PCA classification. In another dataset (the PTB diagnostic ECG database), the highest accuracy achieved was 91.5% using a combination of independent component analysis (ICA) and decision tree classification [9]. G. Andreao et al. compare the performance of different machine learning algorithms for ECG signal classification. The paper evaluates ANN, RF, DT, and SVM on a dataset of ECG recordings. The authors achieved high accuracy rates of up to 99% for some classes of arrhythmia using a neural network approach. The paper highlights the importance of accurate ECG classification for early diagnosis of cardiovascular diseases and for improving patient outcomes. The authors also emphasize the need for continued research in this field to improvise the efficiency and accuracy of ECG classification methods [10]. A study inferred that, "Deep Residual Learning for ECG Classification from a Single Lead ECG Recording" proposes a novel approach for ECG classification using deep residual learning. The proposed method uses a convolutional neural network with residual connections and achieves high accuracy in detecting myocardial infarction. The paper compares the proposed approach with other state-of-the-art approaches and shows its superior performance. The approach is computationally efficient and has potential applications in real-time ECG diagnosis. Overall, the paper presents an innovative method that can significantly improve the accuracy 97.2% of ECG diagnoses and assist healthcare professionals in making more informed decisions [1].

3 Dataset Description

The MIT-BIH Arrhythmia Database is a collection of ECG recordings that are commonly used to evaluate and test arrhythmia detection algorithms. It was created in the 1980s at the Massachusetts Institute of Technology (MIT) and includes over 3000 recordings from 47 different subjects. Each recording is approximately 30 min long and contains annotations that mark the onset and end of different cardiac arrhythmias. The MIT-BIH Arrhythmia Data has been widely utilized by researchers and developers to estimate the accuracy and performance metrics in order to detect the different types of arrhythmias using detection algorithms. It has also been used to study the characteristics of different types of arrhythmias and to develop new diagnostic tools for identifying and treating arrhythmias. The dataset is freely available online and can be downloaded from PhysioNet, a repository of biomedical data maintained by the MIT Laboratory for Computational Physiology. The dataset includes recordings in both raw and annotated formats, as well as software tools for processing and analyzing the data.

4 Methodology

Deep learning with convolutional neural network (CNN) is a popular method for ECG signal analysis. CNNs are a type of ANN that can learn to recognize patterns in ECG signals by using a series of convolutional layers to extract various features from data. These features are passed through fully connected layers to make a prediction. In Fig. 2 the entire framework for the proposed system is shown. In this study, the model can achieve good accuracy in classification for various tasks in the heartbeat such as counting the heart rate, S, V, F, and Q (Fig. 3).

To evaluate heartbeat categories, ECG signals were collected from 187 subjects. Data analysis commonly involves pre-processing techniques like sampling, filtering, and dimensionality reduction. In this study, the data was up-sampled to obtain an equal number of data samples for each category. The CNN architecture consists of three convolutional layers, with a filter size of 64 and kernel size of 5×5. Each layer is tracked by a corresponding max pooling layer with a unit of 2×2 matrix size. The final layer consists of three dense layers of filter size 64, and the last layer uses the softmax activation function for 5-class classification. The use of a filter size of 64 in the convolutional layers helps to retrieve features effectively from the signal. A kernel size of 5×5 is appropriate for capturing local spatial features in the input data. The corresponding max pooling layer with a unit size of 2×2 helps to reduce the spatial dimensions of the output of each convolutional layer.

The three dense layers with a filter size of 64 serve to capture high-level features from the output of the convolutional layers. The softmax activation function is used in the last layer to produce probabilities for the 5-class classification. The optimizer used in this architecture is Adam, which is an efficient optimization algorithm that adapts

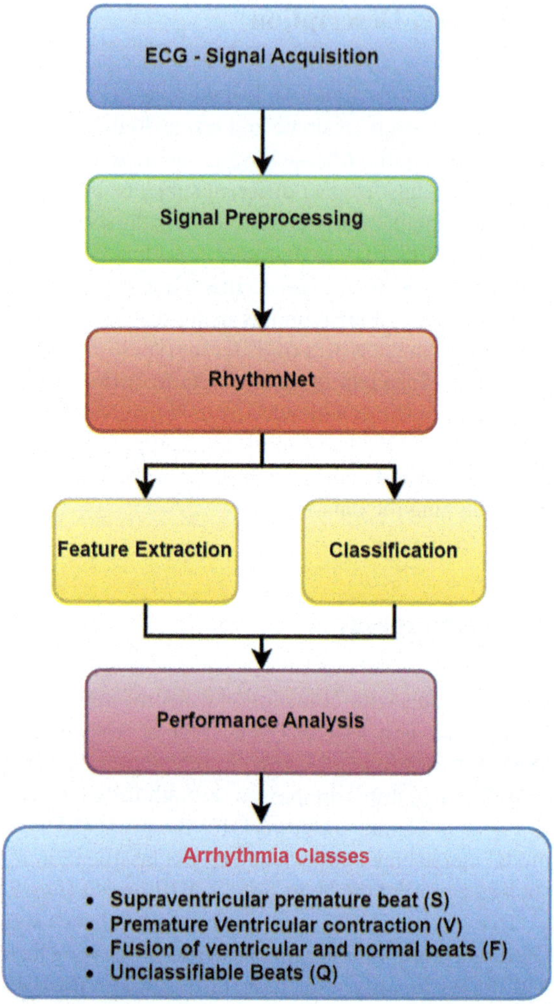

Fig. 2 Proposed system (Rhythm Net)

learning rates on a per-parameter basis. The categorical cross-entropy loss function is employed, which is well-suited for tackling multiclass classification problems. The Adam optimizer is a popular optimization algorithm that combines the advantages of both the AdaGrad and RMS Prop optimizers. It maintains exponentially decaying averages of past gradients and squared gradients and adapts the learning rate for each parameter based on the estimated second moment of the gradients. It is derived as

$$m_t = \beta_1 m_{t-1} + (1 - \beta_2) g_t \tag{1}$$

$$v_t = \beta_2 v_{t-1} + (1 - \beta_2) g_t^2 \tag{2}$$

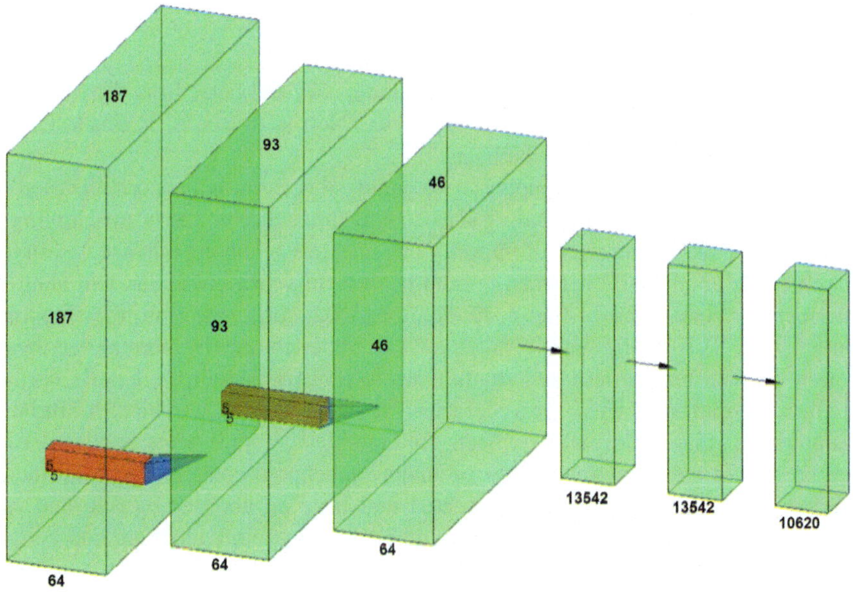

Fig. 3 Proposed model—RhythmNet

$$\widehat{m_t} = \frac{m_t}{\left(1 - \beta_1^t\right)} \tag{3}$$

$$\widehat{v_t} = \frac{v_t}{\left(1 - \beta_2^t\right)} \tag{4}$$

$$\theta_t = \theta_{t-1} - \frac{\alpha \widehat{m_t}}{\sqrt{\widehat{v_t}}} + \epsilon \tag{5}$$

where

g_t	gradient of the loss with respect to the parameters at time '*t*'
m_t	first moment (mean) of the gradients up to time '*t*'
v_t	second moment (uncertainty variance) of the gradients up to time '*t*'
$\widehat{m_t}$	bias corrected estimate of the first moment at time '*t*'
$\widehat{v_t}$	bias corrected estimate of the second moment at time '*t*'
α	learning rate
β_1 and β_2	exponential decay rates for the first moment and second moments, resp.
ϵ	small constant to avoid division by zero.

5 Results and Discussion

In this study, we analyzed a total of 1000 samples of ECG beats from the MIT-BIH dataset and classified them into five categories: Normal Beats, S, V, F, and Q. We used a CNN model to classify the beats.

The data preprocessing involves two main stages. Initially, normalization is applied to achieve consistent scaling across all features. Subsequently, upsampling is utilized to boost the presence of minority classes, thus achieving balance within the dataset. The comparison between normal heart beat, supraventricular, and ventricular ectopic beat is shown Fig. 4. Normal heartbeats originate from the sinoatrial node, while SVEBs originate from the atria or upper chambers. Normal heartbeats follow a consistent pattern and rhythm, while SVEBs disrupt the heart's normal rhythm. Normal heartbeats have a characteristic shape on an ECG, while SVEBs have an abnormal shape. Normal heartbeats are classified as sinus beats, while SVEBs are premature atrial contractions or premature junctional contractions. Ventricular ectopic beats occur earlier than expected and have an abnormal shape and can be caused by heart disease, electrolyte imbalances, caffeine, nicotine, and certain medications.

Our trained model achieved an overall classification accuracy of 97.1%. The accuracy for each class was: Normal Beats 99.2%, Q—91.0. The confusion matrix showed that the model performed well in correctly classifying Normal Beats, V, and S with high accuracy. However, it had difficulty in distinguishing between Q and FS, which were misclassified in some cases. In summary, our CNN model was able to accurately classify ECG beats into five categories with an overall accuracy of 97.1%. The model exhibits the potential for automating the analysis of ECG signals to detect and diagnose arrhythmias. However, further improvement may be necessary to improve the accuracy classification of Q and F.

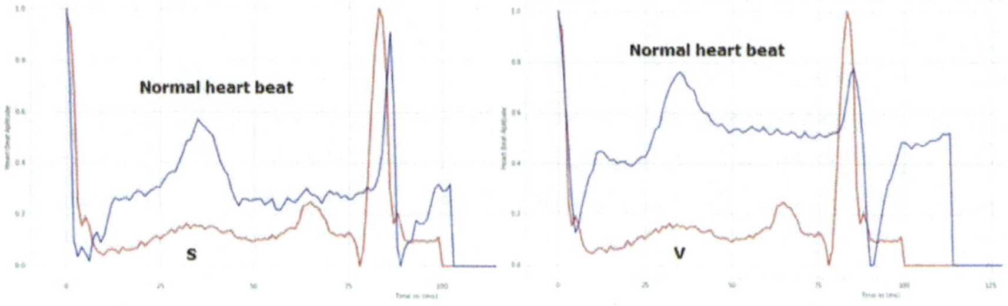

Fig. 4 Comparison of a normal heartbeat, supraventricular and ventricular ectopic beat

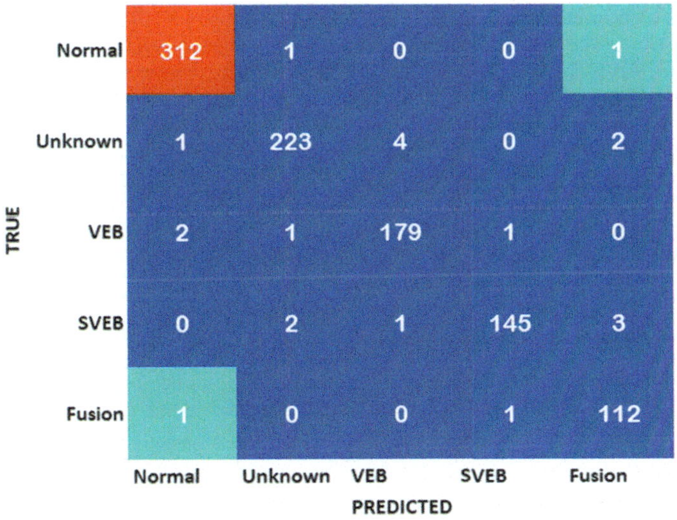

Fig. 5 Confusion matrix for the multiclass classification

6 Conclusion

The evaluation and categorization of 1000 ECG heartbeats using a CNN model encompassing five classes (Normal Beats, Q, V, S, and F) yielded an overall classification accuracy of 97.1%. The findings imply that employing a CNN-based model can prove efficacious in discerning and categorizing ECG beats across various classes. These outcomes hold significant implications for diagnosing heart diseases and monitoring patients with cardiac conditions. In order to evaluate the model's performance and therapeutic applicability, larger datasets and more ECG beat types are needed. These studies should contain many people and arrhythmias to evaluate the model's efficacy. The model can be expanded by making arrhythmia analysis more flexible and clinically relevant (Fig. 5).

References

1. Hannun AY, Rajpurkar P, Haghpanahi M, Tison GH, Bourn C, Turakhia MP, Ng AY (2019) Cardiologist-level arrhythmia detection and classification in ambulatory electrocardiograms using a deep neural network. Nat Med 25(1):65–69
2. Cheng J, Zou Q, Zhao Y (2021) ECG signal classification based on deep CNN and BiLSTM. BMC Med Inform Decis Mak 21(1):365
3. Kachuee M, Fazeli S, Sarrafzadeh M (2018) ECG heartbeat classification: a deep transferable representation. In: 2018 IEEE international conference on healthcare informatics (ICHI). IEEE, pp 443–444
4. Fox DJ, Tischenko A, Krahn AD, Skanes AC, Gula LJ, Yee RK, Klein GJ (2008) Supraventricular tachycardia: diagnosis and management. Mayo Clinic Proc 83(12):1400–1411

5. Ruberman W, Weinblatt E, Goldberg JD, Frank CW, Shapiro S (1977) Ventricular premature beats and mortality after myocardial infarction. N Engl J Med 297(14):750–757
6. Das MK, Ari S (2014) ECG beats classification using mixture of features. Int Sch Res Not 2014:178436. https://doi.org/10.1155/2014/178436
7. Nova: ECG signal delineation and compression. SlideServe. https://www.slideserve.com/nova/ecg-signal-delineation-and-compression
8. ECG (electrocardiogram)—hyperphysics. HyperPhysics. http://hyperphysics.phyastr.gsu.edu/hbase/Biology/ecg.html
9. Alcala-Ortega JR, López-López AE, Galván-Tejada JR, Gutiérrez-Rivas EJ, Saucedo-Ramírez JA (2019) A comparative study of different methods for ECG classification using dimensionality reduction. Comput Methods Programs Biomed 172:55–66
10. Andreao RG, Silva RGS, Gomes RS, Lima AF, Pereira LAM, Conci A (2020) A comparative study of machine learning approaches for ECG signal classification. IEEE Access 8:96106-96F119

IoT-Based Smart Campus

Saba Shaikh, Rupali Kuskar, Niraj Padalwar, Mohini More, and Radhika D. Joshi

Abstract A smart campus allows various devices and applications to connect so that students and staff can be connected to them which causes improved efficiencies and enhances the experience of everyone in campus. The proposed system describes the development of a smart campus by use of Internet of Things (IoT) technology. IoT is a platform where devices (sensors and actuators) can be connected, monitored, and controlled from any location throughout the network infrastructure. By development of smart campus, it is possible to connect different entities in campus as part of IoT-based system and make monitoring and control in real time possible. Smart campus idea is implemented by combining following features, automatic light, fan control, attendance, dustbin level, soil moisture, drinking water, and air quality monitoring. The system collects real-time data and takes corrective actions and displays it on the ThingSpeak server and on Blynk android application for user convenience.

Keywords IoT · Face recognition · ESP-32 · ThingSpeak · Blynk app · Sensors and actuators

S. Shaikh (✉) · R. Kuskar · N. Padalwar · M. More · R. D. Joshi
Department of Electronics and Telecommunication Engineering, College of Engineering, Pune, Pune, Maharashtra, India
e-mail: shaikhss20.extc@coeptech.ac.in

R. Kuskar
e-mail: kuskarrb20.extc@coeptech.ac.in

N. Padalwar
e-mail: padalwarnm18.extc@coeptech.ac.in

M. More
e-mail: Moremg19.extc@coeptech.ac.in

R. D. Joshi
e-mail: rdj.extc@coeptech.ac.in

V. K. Gunjan and J. M. Zurada (eds.), *Proceedings of 4th International Conference on Recent Trends in Machine Learning, IoT, Smart Cities and Applications*, Lecture Notes in Networks and Systems 873, https://doi.org/10.1007/978-981-99-9442-7_45

1 Introduction

IoT plays an important role in improving the quality of life, and it is the concept of connecting various day to day devices and allowing them to communicate and exchange data with each other using Internet. A smart campus refers to the use of technology (IoT) and data-driven solutions to enhance the experience and operations of a traditional campus which will ease communication between staff and device and have proper action on time and it also increases the safety, efficiency, sustainability, and campus experience both from learning and teaching side [1, 2]. In campus, it was observed that student and staff face many problems. Many times, it is observed that students use unfair means for attendance and even if no one is present in class the lights and fan are ON which causes wastage of electricity so to have proper resource management and regularity in classroom we can deploy automatic light and fan control, attendance system based on face recognition [3, 4]. Housekeeping staff need to keep watch on the level of dustbin by checking manually which utilizes extra time and efforts. Here, we can use IoT technology which enable you to have to proper alert when dustbin about to fill [5]. The health of plants depends on the level of moisture of soil. Traditional methods do not allow the availability of this parameter and affects plant. Proposed system is a sustainable watering system for the garden that will monitor and display the soil moisture level and regulate water consumption [6]. Drinking water quality system is responsible to monitor the quality of drinking water in the tank by measuring the Total Dissolved Solid (TDS) value and Air Quality Monitoring. This air quality monitoring system monitors the Air Quality Index (AQI) and Carbon Monoxide (CO) concentration in air and uploads both the data on ThingSpeak server [7–9].

2 Literature Review

There are several researches related to IoT-based smart campus, but many of them are confined to only some specific area of application such as classroom or lab control. Very few researches give a concept of implementing IoT in overall campus. Kantipudi et al. [10] studied different types of smart campuses using IoT and performance of different layers used on IoT modules and their applications for educational purposes. Enugala et al. [11] implemented IoT-based smart classroom environment by combining facilities like smart chairs, cloud storage, and smart attendance recording system. Liu et al. [12] analyzed students learning status based on monitored data. Roy [13] proposed microcontroller-based automated room light and fan controller using temperature sensor. Kolhatkar [14] proposed Smart E-Dustbin for handling the garbage in smart way by use of IoT. Hadi and Sholihul [15] designed garden irrigation system which remotely controls water pump and soil moisture in garden. Satheesh Kumar [16] monitors the quality of water is based on TDS/EC, i.e., hydrogen ion concentration in water, temperature of water, CO, and methane series present in it. Barot et al. [17] analyzed low cost air pollutant sensor, error sources, and calibration methods for air quality monitoring. From literature survey, it was inferred that there a system combining multiple facilities for converting traditional campus to a smart campus does not exist. There is a need for system that makes a smart campus by smart energy management, alerting system, and air and drinking water quality monitoring to enhance the environment and make real-time data of various physical quantities available remotely.

3 Methodology

In order to design energy efficient system on campus, IoT-based approach is useful where sensors and actuators placed in campus can be connected together for environmental parameters monitoring and control of various field devices. This smart campus system integrates different application area in campus and allows the data to be displayed on web and mobile phone. Smart campus is essential for improving efficiency and student experience and sustainability.

Flowchart. The flowchart of proposed system is illustrated in Fig. 1. In the beginning of process, the Wi-Fi connectivity is established. Once this step is done, physical parameters are sensed using field devices. These measured quantities are not always in the presentable form. Therefore, this data is converted to appropriate from for controller. In this system, ESP 32 is considered for this purpose. The two branches after this step are due to the use of two different circuits in this system. This division is done in accordance to the application area and IoT platform used for different facilities. The conditional blocks in the left are for energy management and alert system, and conditions for classroom occupancy, attendance, irrigation, and dustbin are mentioned. Depending on the condition, the process is executed and data is displayed on Blynk android application. The attendance is obtained in .csv format. The processes on the right side are for measuring and displaying drinking water and air quality on ThingSpeak server and LCD display.

Block diagram. The proposed system divides the facilities in two parts and different hardware is required for each. The first system is "smart energy management and alert system" and the other one is "drinking water and air quality monitoring." The block diagram for each part is illustrated in Figs. 2 and 3.

Smart energy management and alert system (Fig. 2) gets input from three different sensors placed in field. PIR sensor is responsible for occupancy detection in smart classroom. Soil moisture sensor is placed in garden soil in order to measure the moisture level of soil, whereas ultrasonic sensor is placed on the bottom of the lid to measure the trash level in smart dustbin. ESP 32 processes this data and sends these values on Blynk mobile application. Depending on the input from PIR sensor, the state of LED and fan is controlled. Similarly, the water pump is controlled as per the soil moisture level. As illustrated in Fig. 3, drinking water and air quality monitoring system includes TDS sensor for water quality monitoring, MQ135 sensor

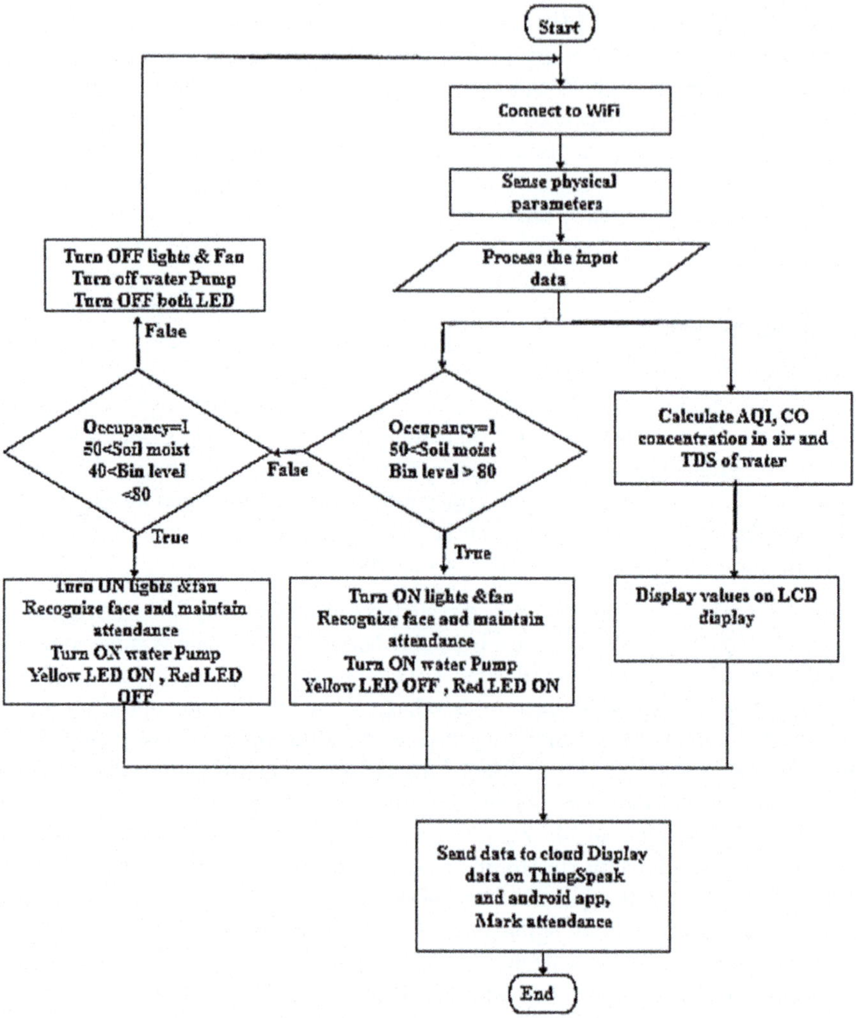

Fig. 1 Flowchart of proposed system

Fig. 2 Smart energy management and alert system

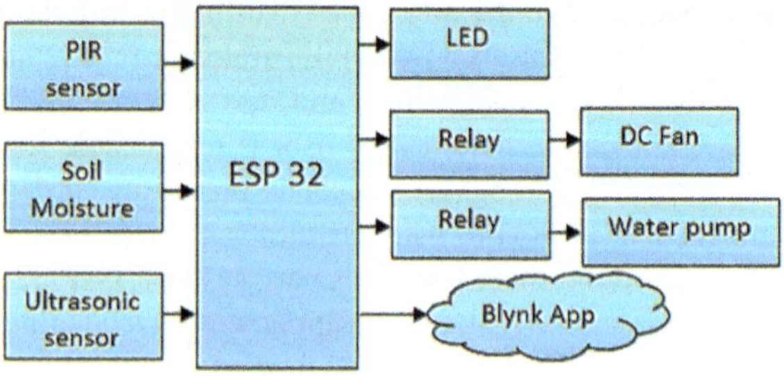

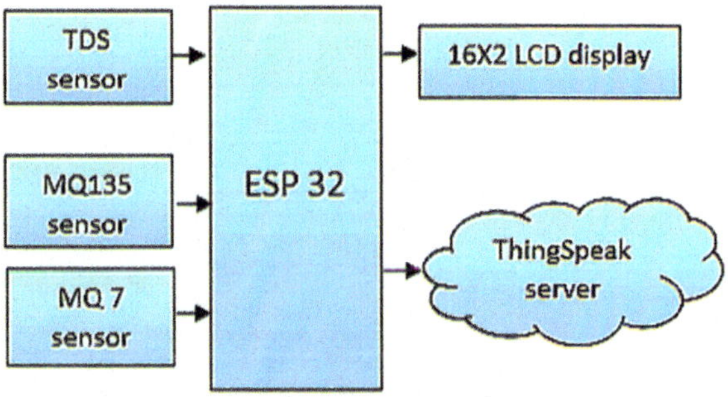

Fig. 3 Drinking water and air quality monitoring

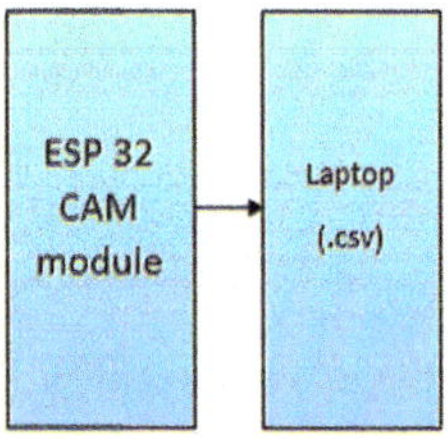

Fig. 4 Face recognition system

for calculating AQI, and MQ7 sensor to sense the CO concentration in air. The measured parameters are displayed on LCD display and sent on ThingSpeak server for monitoring online. Figure 4 illustrates block diagram for face recognition-based attendance system.

A GUI is also developed for the proposed system for monitoring purpose. GUI is simple user interface where user can access the data remotely better understanding, and it is as user friendly as anyone can collect the data with the help of given buttons.

4 Result and Discussion

The proposed model of IoT-based smart campus includes five facilities. The results observed from each of the facility and their interpretation are mentioned in this section. Figure 5 is the image of actual prototype implemented. Figure 6 illustrates GUI after accessing the ENTC department and selecting air quality as parameter GUI displays the value of MQ-135 and MQ-7 sensor.

Smart energy management and alert system. The management of use of electricity in campus is essential to avoid its overuse and wastage. Time also plays significant role in work of both working and non-working staff. The proposed system aims to save electricity as well as time. The results illustrated in Fig. 7 are observed on the serial monitor and the status of occupancy is displayed. The attendance record is saved as .csv format on cloud and faculty can access the date wise data as illustrated in Fig. 8. The dataset consisting of student's photos is provided to system, and

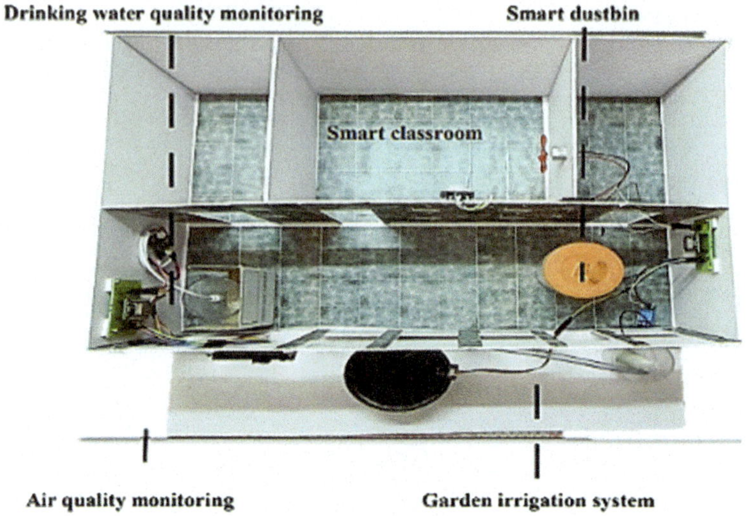

Fig. 5 Prototype of proposed IoT-based smart campus

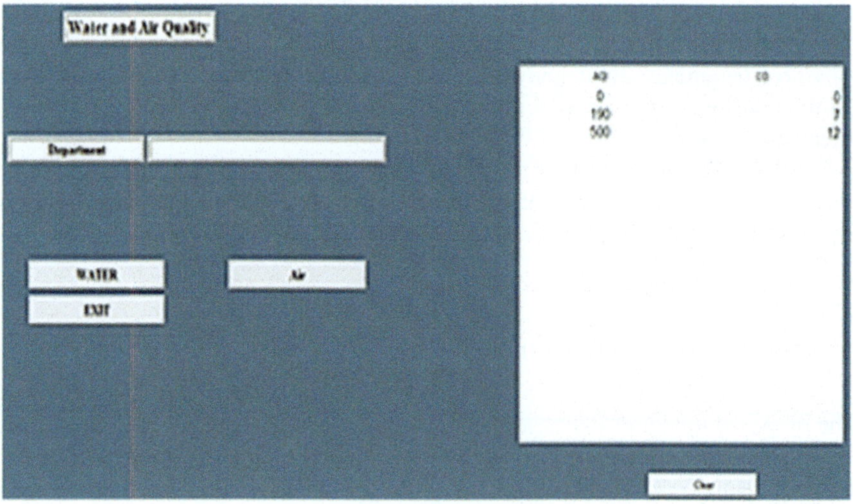

Fig. 6 GUI—after accessing air data

comparison of image from camera and dataset is performed. The time of entrance is marked along with the name of student (Fig. 9).

Traditional ceiling fan uses 0.023 kWh electricity and one incandescent light uses 0.43 kWh, so approx. per unit electricity charges is 7rs. As college has break of one hour, then there we can save ($0.023 \times 7 \times 6 + 0.43 \times 7 \times 6$) 19.5rs per hour. Considering three rooms and duration of 1 month total saving of 1170Rs will be

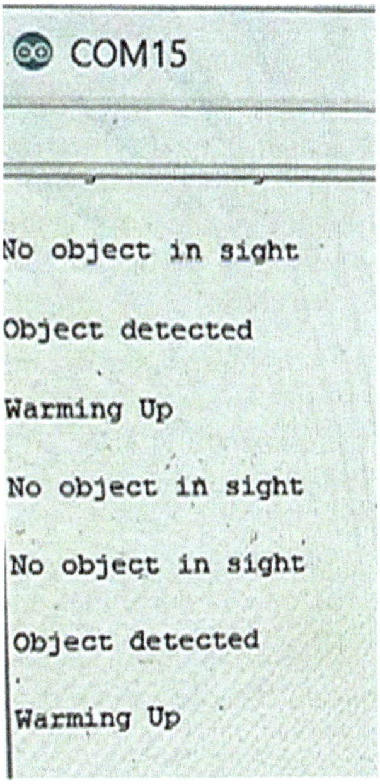

Fig. 7 Classroom occupancy detection

2023-03-04	04-03-2023 19:28	Microsoft Excel Co...
2023-03-05	05-03-2023 01:56	Microsoft Excel Co...
2023-03-19	19-03-2023 17:50	Microsoft Excel Co...
2023-03-21	21-03-2023 12:28	Microsoft Excel Co...
2023-04-18	18-04-2023 16:07	Microsoft Excel Co...
2023-04-19	19-04-2023 03:25	Microsoft Excel Co...
2023-05-04	04-05-2023 22:55	Microsoft Excel Co...

Fig. 8 Date wise maintaining attendance

Fig. 9 Attendance system

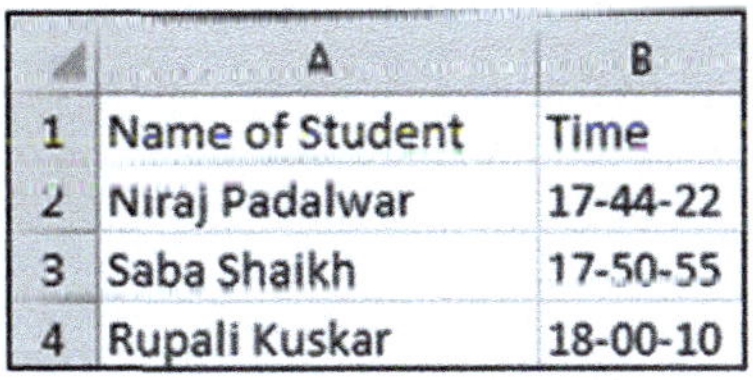

	A	B
1	Name of Student	Time
2	Niraj Padalwar	17-44-22
3	Saba Shaikh	17-50-55
4	Rupali Kuskar	18-00-10

achieved. The observations of this system are displayed on Blynk application. The status of light and fan in classroom, soil moisture level, and percentage of trash in dustbin are displayed. Blynk application is a freely available mobile application for IoT application. It allows user to add and remove devices for monitoring and control purpose. In Fig. 10, this result indicates occupant detected so lights and fans are ON. Figure 11 indicates that the bin is 40% full and the level of soil moisture is 9%, whereas Fig. 12 indicates bin is 40% full and soil moisture is 50%, so the pump is turned OFF.

Drinking water and air quality monitoring. Drinking water and air quality plays a significant role in health of person. It is necessary to monitor these parameters. The drinking water and air quality in smart campus is monitored and displayed on ThingSpeak server. Figure 13 illustrates the graph of TDS level of drinking water. The 30 TDS value indicates the water is of acceptable quality with some impurities. Figure 14 illustrates the graph of TDS level of drinking water. Acceptable value of drinking water is 50. The high level of TDS indicates the water is highly unsafe for

Fig. 10 Occupancy detected

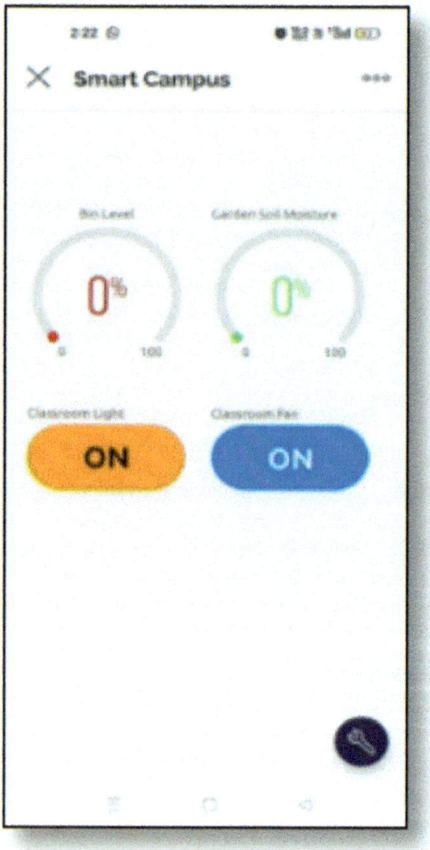

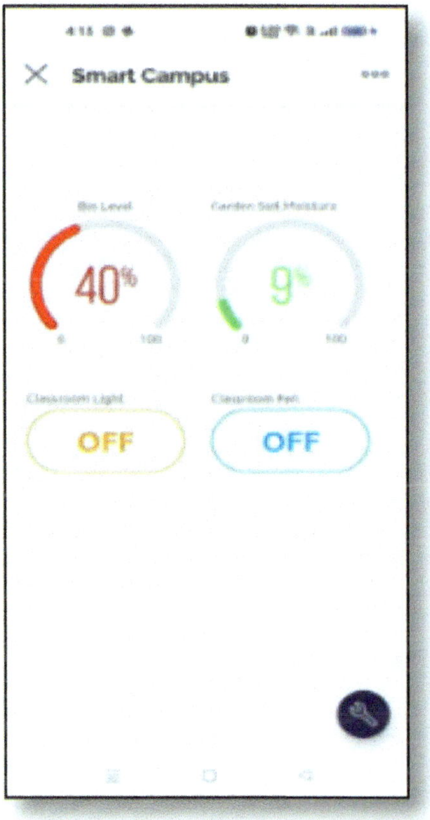

Fig. 11 Dustbin and soil moisture level

drinking purpose as it contains many impurities. Figure 15 illustrates graph of AQI. 190 PPM is value of AQI in Pune on the day of observation. This value indicates satisfactory level. The value of AQI increased when observation was carried in smoky environment and 500 PPM was observed and it is very poor air quality. Figure 16 illustrates the graph of CO concentration in campus at the time of observation. The CO concentration of value 7 is of satisfactory level. The value increased when observation is carried out in smoky environment. The smoke causes increase CO concentration from 7 to 12 PPM. This is poor value of CO as 15 is the maximum level of CO. Depending on results obtained, alerts can be generated as warnings.

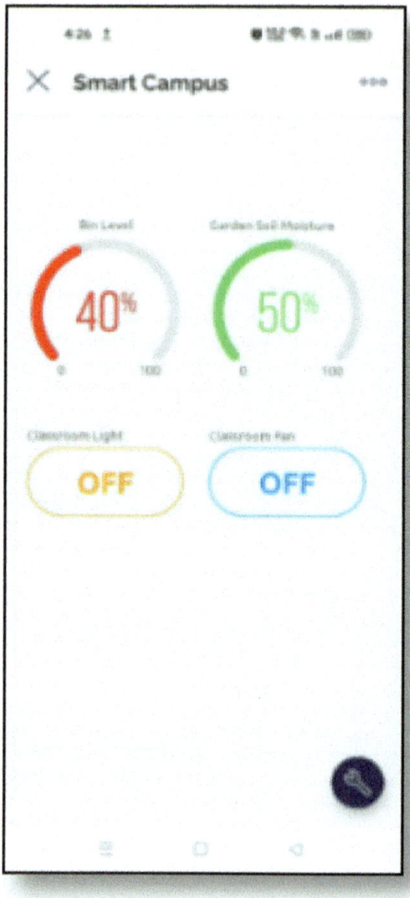

Fig. 12 Dustbin (40%) and soil moisture (50%) level

5 Conclusion and Future Scope

The development of smart campus plays an important role in digitization. This study is developed using IoT. It integrates the software with hardware components to ensure the efficient system which is connected to cloud. Cloud is a technology which enables store and access of data remotely from any place in its network. This paper gives an idea of IoT-based smart campus having facilities like smart classroom, dustbin, and garden irrigation system which are part of smart energy management and alert system. Drinking water and air quality monitoring is also included to alert users toward health.

Smart classroom consists of two applications first is automatic light and fan control. It saves 25% electricity. Second is face recognition-based attendance. This system will approximately reduce 5–7 min time of each lecture. The garden irrigation

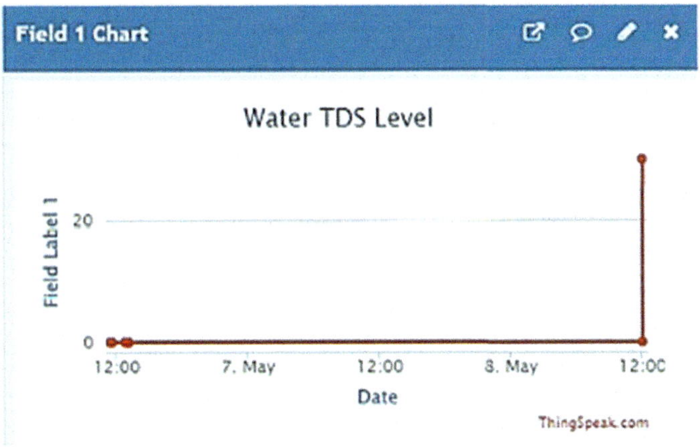

Fig. 13 ThingSpeak acceptable water quality graph

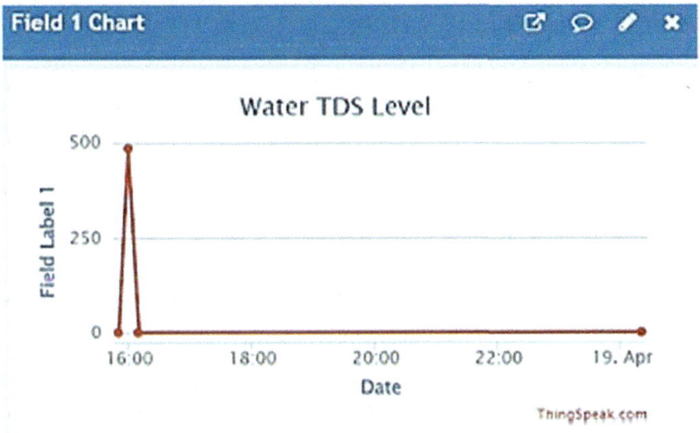

Fig. 14 ThingSpeak poor water quality graph

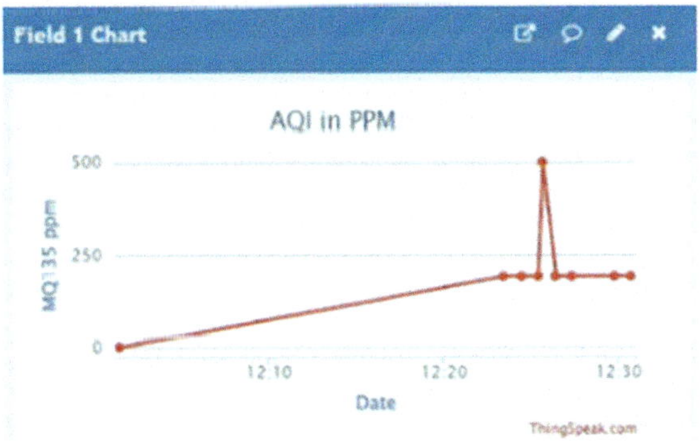

Fig. 15 ThingSpeak–AQI graph

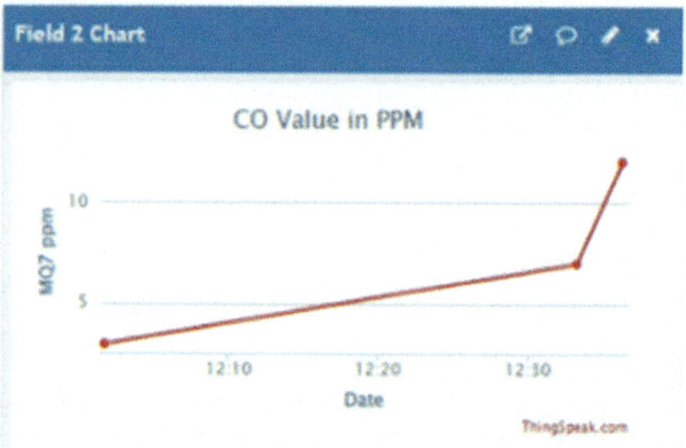

Fig. 16 ThingSpeak–CO in air graph

system maintains plant health by maintaining soil moisture. The smart dustbin application helps to manage the garbage in dustbin. Drinking water quality monitoring system monitors the TDS of water. Air quality monitoring monitors the quality of air in the campus and see if it is good or bad.

The applications related to security in smart campus can be implemented in future. The facilities like smart street lighting, remote control of devices, intruder detection, etc., can also be included in the future. The use of renewable energy is also necessary for a sustainable smart campus.

References

1. Sari MW, Ciptadi PW, Hafid Hardyanto R (2017) Study of smart campus development using internet of things technology. IOP Conf Ser: Mater Sci Eng 190(1)
2. Shivaraj Kumar TH, Sriraksha TA, Saba Noor U (2017) An IoT based secured smart e-campus. Int J Hum Soc Sci Invent 6(3)
3. Revathi R, Suganya M, Gladiss Merlin NR (2020) IoT based cloud integrated smart classroom for smart and a sustainable campus. Procedia Comput Sci 172:77–81
4. Ani R et al (2018) An approach towards building an IoT based smart classroom. In: 2018 international conference on advances in computing, communications and informatics (ICACCI). IEEE
5. Praveen A et al (2020) IoT based smart bin: a Swachh-Bharat initiative. In: 2020 international conference on electronics and sustainable communication systems (ICESC). IEEE
6. Leh NAM et al (2019) Smart irrigation system using internet of things. In: 2019 IEEE 9th international conference on system engineering and technology (ICSET). IEEE
7. Ayele TW, Mehta R (2018) Air pollution monitoring and prediction using IoT. In: 2018 second international conference on inventive communication and computational technologies (ICICCT). IEEE
8. Warungase P, Worlikar A, Mhatre J, Saha D, Salunkhe G. IoT based water monitoring system
9. Ravi RV, Dutta PK, Lai WC (2022) An IoT based air pollution monitoring system for smart city applications, pp 407–411
10. Kantipudi MVVP, Velamuri S (2021) Internet of things—based smart campus—a review, pp 490–495
11. Enugala VPR, Vuppala S (2018) Internet of Things—based smart classroom environment. In: 2018 fifth international conference on parallel, distributed and grid computing (PDGC). IEEE

12. Jia S-C et al (2021) Smart classroom system based on internet of things wireless communication technology. In: 2021 international conference on intelligent transportation, big data & smart city (ICITBS). IEEE
13. Roy P et al (2018) Microcontroller based automated room light and fan controller. In: 2018 emerging trends in electronic devices and computational techniques (EDCT). IEEE
14. Kolhatkar C et al (2018) Smart E-dustbin. In: 2018 international conference on smart city and emerging technology (ICSCET). IEEE
15. Hadi MS et al (2020) IoT based smart garden irrigation system. In: 2020 4th international conference on vocational education and training (ICOVET). IEEE
16. Kumar SS et al (2022) IoT based novel portable water quality monitoring for micro industries. In: 2022 6th international conference on devices, circuits and systems (ICDCS). IEEE
17. Barot V, Kapadia V (2020) Air quality monitoring systems using IoT: a review. In: 2020 international conference on computational performance evaluation (ComPE). IEEE

Wheel Alignment Detection Using Embedded ML and Tyre Pressure Detection

Soumya Sajjan, Poorva Rudrapur, H. M. Khushi, Shriraksha, Supriya Katwe, and Nalini Iyer

Abstract With its capacity to automate processes and controls, smart technology finds application in areas ranging from any industry to health care, helping to improve efficiency and optimise performance. These devices are made brainy by advanced computing frameworks like machine learning and artificial intelligence and connection to the network to form the Internet of Things (IoT). These advanced computing technologies have also made autonomous driving possible. Hence, to aid independent driving, the present-day need is to develop intelligent tyres capable of reporting parameter changes, such as pressure, temperature and alignment. The journeys with innovative technology can be smarter with real-time tracking of the tyre status providing efficient preventive measures. Innovative technology in the tyre assists in maintaining its health by providing its real-time status or predicting its possible parameters. This study focuses on two tyre features: misalignment status and pressure monitoring. It aims to notify the user about the tyre pressure and alignment status to take preventive measures. The project is realised with embedded ML technology.

Keywords Embedded ML · Wheel alignment · Tyre pressure

1 Introduction

The two essential aspects of tyre health monitoring are wheel alignment and tyre pressure maintenance. These two parameters play an important role in safety measures to be considered to take care of your car's tyres. Wheel alignment is achieved by adjusting suspension parts and maintaining the suspension angles of the elements. Various wheel angles are appropriately adjusted to ensure the wheels are parallel and perpendicular to the ground surface. The proper wheel alignment reduces the wear and tear of tyres, improving the tyre's life and the quality of the vehicle. It also enhances durability and fuel mileage. The car's stability is improved, giving the user

S. Sajjan (✉) · P. Rudrapur · H. M. Khushi · Shriraksha · S. Katwe · N. Iyer
KLE Technological University, Hubballi, India
e-mail: soumya172002@gmail.com

V. K. Gunjan and J. M. Zurada (eds.), *Proceedings of 4th International Conference on Recent Trends in Machine Learning, IoT, Smart Cities and Applications*, Lecture Notes in Networks and Systems 873, https://doi.org/10.1007/978-981-99-9442-7_46

enhanced experiences of handling and braking quality. Smart tyres with contemporary sensors monitor tyre conditions and pressure and notify drivers via smartphones or other electronic gadgets when the vehicle requires routine maintenance. Automotive smart tyres improve tyre performance by detecting changes in road conditions, adjusting tyre pressure, and adapting to terrain to operate with the least wear and tear. Various signs indicate disorders in the car's wheel alignment. To avoid this, monitoring the tyre tread and ensuring the tyres are properly inflated and aligned are essential. If any significant difference in the wear pattern of the tyres is noticed, it's time to get them checked by a professional.

2 Methodology

This paper presents an approach to monitor the health of tyres by considering two main features: tyre pressure and misalignment status. The misalignment status is monitored with the help of an accelerometer, and a barometric pressure sensor measures the pressure in the tyre.

2.1 Wheel Misalignment Status Indicator Module: Programming Flow

The misalignment status indication module is implemented with the embedded ML technology, and its flow is shown in Fig. 1. The Arduino Nano 33 BLE Sense [1] controller trains and tests the ANN model and accelerometer sensor. The phases involved are detailed below.

Phase 1: Generation of the dataset. The data is collected from the controller connected to the accelerometer sensor (LSM9DS1). The data is collected in four classes: stable, unstable-1, unstable-2, and idle. Twenty samples of ten seconds of data are collected for each class. The collected data is saved on Edge Impulse [2] cloud.

Stable. Vibrations are sensed in desired range in all three axes of the accelerometer.
Unstable type 1. Vibrations are sensed only along the x-axis of the accelerometer.
Unstable type 2. Vibrations are sensed only along the z-axis of the accelerometer.
Idle. No vibrations are sensed, and the vehicle is in a static position.

The accelerometer data collected along each axis are combined to extract a feature.

Features being extracted here are the RMS value. Feature extraction is being done to reduce the number of inputs being fed to the model, which would otherwise be 375, considering a sampling frequency of 62.5 Hz for 2 s along three axes. In total, twelve features are extracted, which are as follows:

Fig. 1 Implementation flow of accelerometer

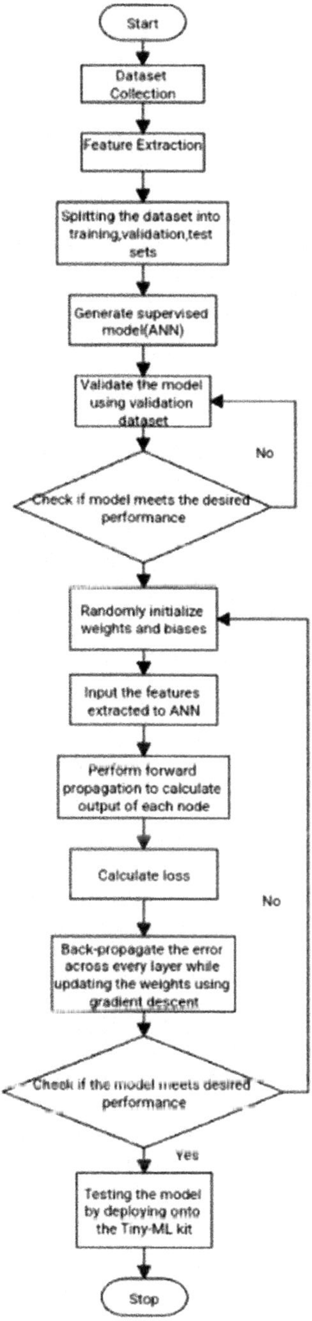

- accZ RMS
- accZ Skewness
- accX RMS
- accY RMS
- accY Spectral Power 1.95–5.86 Hz
- accY Skewness
- accY Kurtosis
- accZ Spectral Power 1.95–5.86 Hz
- accX Kurtosis
- accZ Kurtosis
- accX Spectral Power 1.95–5.86 Hz
- accX Skewness.

Phase 2: Training the Model. The system architecture used to train the model is an artificial neural network (ANN) with four layers. The twelve features extracted in the data collection phase using Root Mean Square (RMS) propagation are fed as inputs to the first layer of the model. The neural network is designed with the following default training settings (as is in Edge Impulse) with the number of training cycles (epochs) seeming to be 30, the learning rate set to 0.0005, and the minimum confidence rating taken as 0.60. The model executes a forward pass by performing mathematical operations on inputs. The rectified linear unit (ReLU) is a nonlinear function applied after summing to enable decision-making in each node. The output is passed to nodes in the second layer. This process is repeated with new nodes. Initial prediction is inaccurate due to random parameters. The cost is determined using the loss function, which uses the model for accurate predictions. Back-propagation computes gradients using the chain rule and adjusts model parameters through gradient descent. The four output nodes that represent the four classes of alignment of the tyre are totalled using the soft-max function. The soft-max function converts each network output into a set of probabilities. Each probability must be between 0 and 1. The output with the highest probability determines the output's class.

Phase 3: Validating and Testing the Model. The model is tested for accuracy using a live classification feature of Edge Impulse that enables validating the model with data captured directly from the Tiny-ML board. The dataset is divided into training and testing sets when collecting data. Only the training set is used to train the model, while the testing set assesses how well the model will perform on unseen data. This ensures that the model did not learn to overfit the training data. This is followed by a confusion matrix showing how well the modes perform in each class.

Phase 4: Deployment and Testing of the Model on the Controller. After training and validating the model, it is deployed to the controller. This allows the model to run without an Internet connection, with low latency and power consumption. A package is prepared that provides information about the deployment, such as the names of all classes, the frequency of the data, and the whole impulse configuration. The deployment options convert this package into a completely optimised C++ source code, which is customised and integrated with the application.

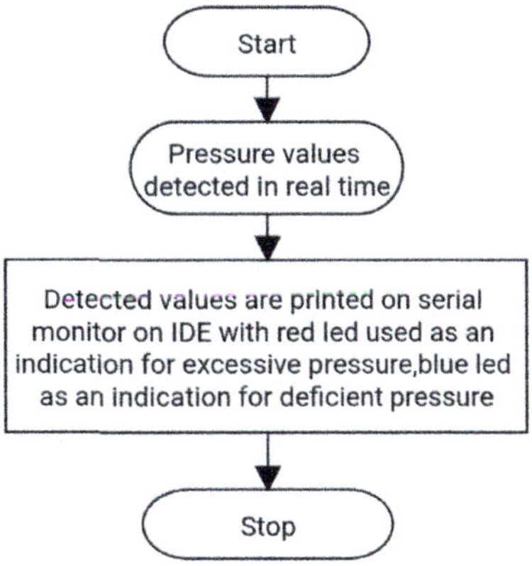

Fig. 2 Implementation of pressure module

2.2 *Implementation of Pressure Module*

The pressure monitoring uses the barometric pressure sensor (LPS22HB) embedded in the Arduino Nano BLE Sense 33 controllers. The **proposed** system aims to provide real-time tyre pressure monitoring to enhance safety and efficiency. Pressure thresholds are determined to trigger LED indications on the board based on low and high-pressure conditions. Red LED blinking indicates high pressure, while blue LED blinking indicates low pressure. The thresholds are adjustable to suit specific requirements. Figure 2 shows the implementation flow of the proposed method.

3 Results

The module is fixed to the wheel of a bicycle for study purposes. The model's output is deployed and tested in real time to determine the tyre's alignment. The output of the accelerometer is a probability function, where the probability of the stable class is close to 1 if the wheel is properly aligned; otherwise, if the tyre is out of alignment, the probabilities of unstable types 1 and 2 are high due to variations in the x and y axes, respectively, and the probability of the idle class is high because the tyre is in static motion. The results obtained at each phase of the project are listed below:

3.1 *Data Collection*

Edge Impulse development platform is used to collect the data under the four classes: stable, idle, unstable-type1, and unstable-type2. For every label, twenty samples lasting for ten seconds each are collected. The generated dataset is shown in Fig. 3.

Fig. 3 Data collection using Edge Impulse

3.2 *Feature Extraction*

To extract a feature, the accelerometer data collected along each axis are concatenated. RMS value is the feature being retrieved in this case. Feature extraction is done to cut down on the model's inputs, which would otherwise total 375 sample inputs when sampling at 62.5 Hz for 2 s along three axes. Twelve characteristics are retrieved during this phase. Figure 5 shows the features extracted (Fig. 4).

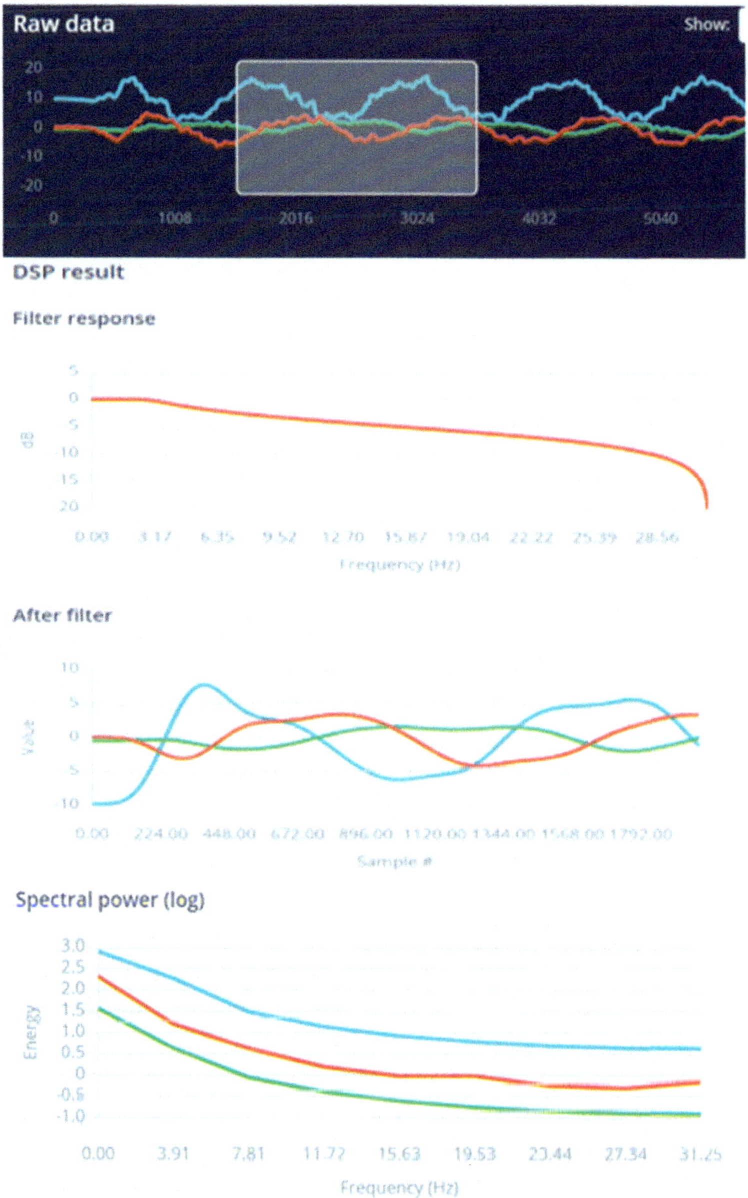

Fig. 4 Feature extraction process

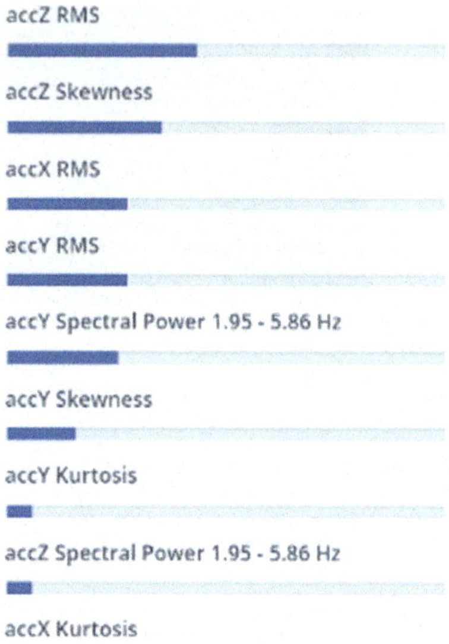

Fig. 5 Features extracted

3.3 *Data Collection Using Neural Network*

Confusion Matrix. Due to the vast amount of data acquired and the low learning rate value, the model achieves 100% accuracy (Fig. 6).

3.4 *Hardware Implementation Results*

Wheel alignment status indicator module. The real-time alignment status model is tested using an Arduino Nano 33 BLE Sense board. The board is mounted on the bicycle rim for real-time testing. The accelerometer determines the output class by detecting the tyre's alignment along the three axes. The class with the highest probability is the one to which output belongs (Fig. 7).

Pressure monitoring module. The pressure controller senses and compares the pressure against a set threshold. If the pressure sensed is above the threshold, a red led glows and displays the pressure value on the terminal, indicating excess pressure. If the tyre's pressure goes below the threshold value, the blue led glows, indicating deficient pressure, notifying the user to take preventive measures (Fig. 8).

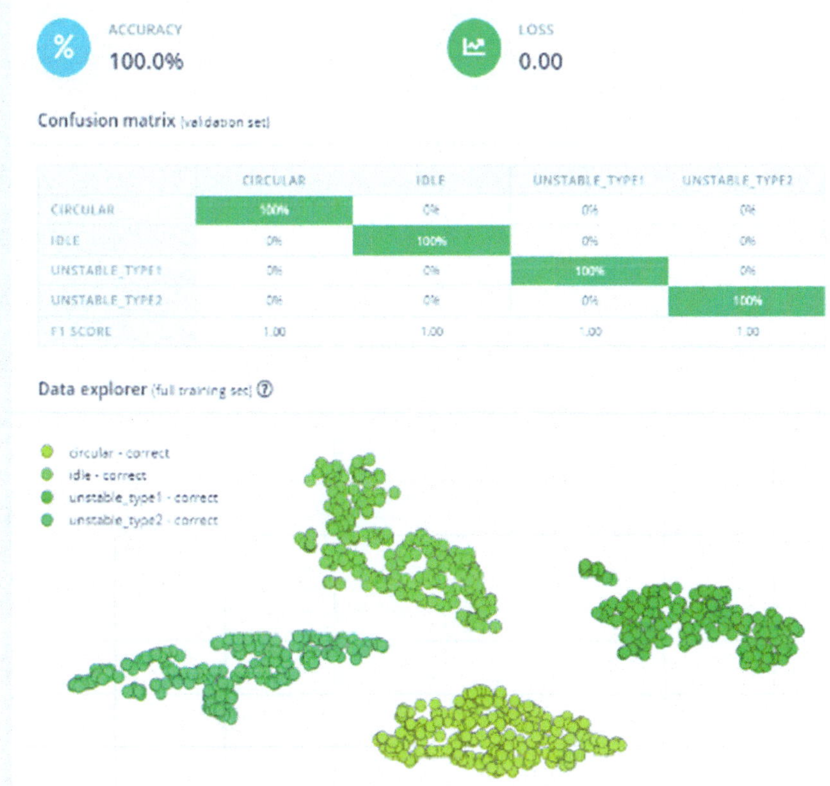

Fig. 6 Confusion matrix and classified data

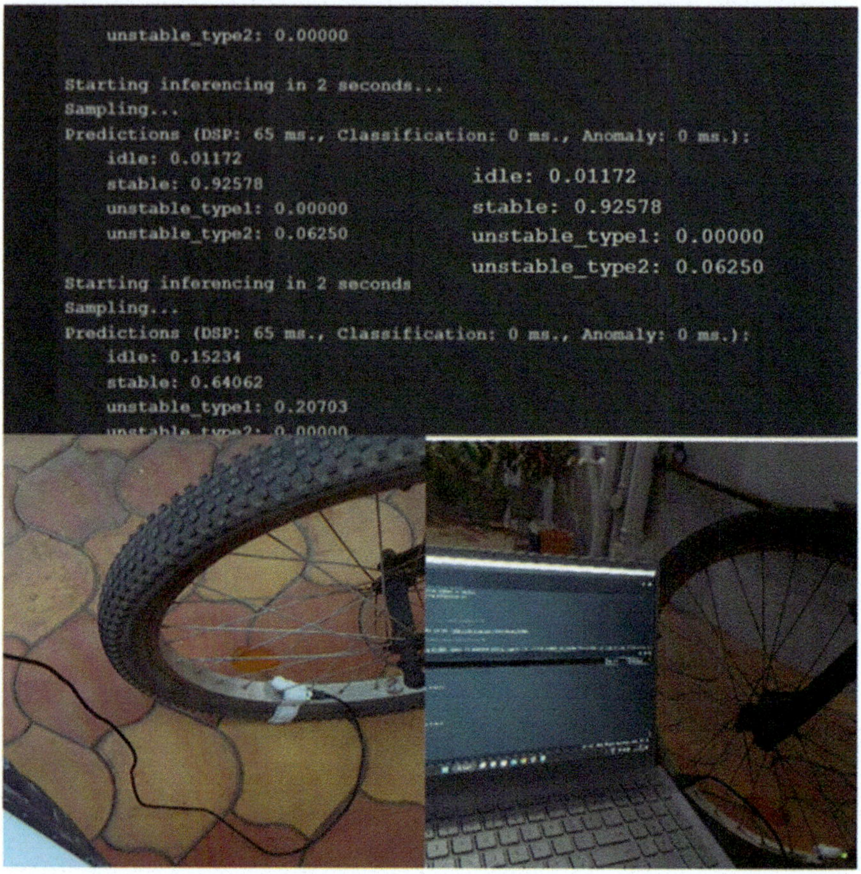

Fig. 7 Results of real-time accelerometer

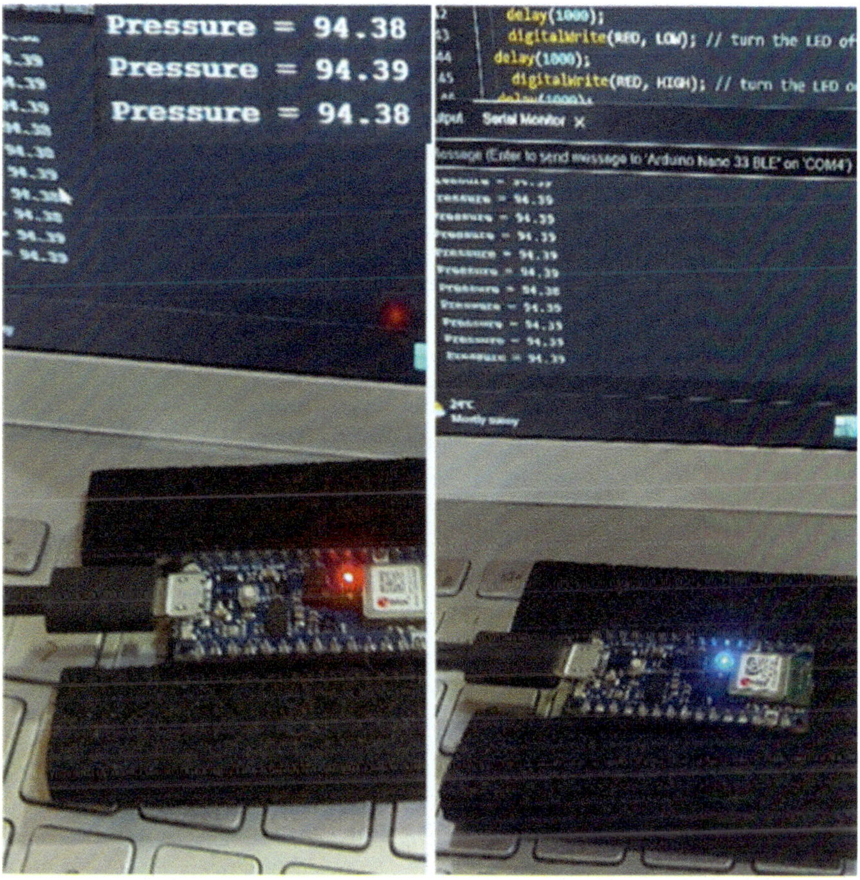

Fig. 8 Real-time pressure monitoring

4 Conclusion

The tyre pressure monitoring and wheel misalignment module proposed in this paper is implemented and tested using a bicycle for simplicity. The ANN model developed using the Edge Impulse platform is deployed and tested on the controller. The model gave accurate results up to the 99 percentile. The pressure monitor module displays and indicates the pressure. With two features of tyre health monitoring, the system can provide the tyre's status and aid in its preventive maintenance.

References

1. Arduino Nano 33 BLE Sense. https://docs.arduino.cc/hardware/nano-33-ble-sense
2. Edge Impulse. https://www.edgeimpulse.com/

Generalized Data Auditing Solution in Trustless Environments Using IoT and Hyperledger Fabric

Yerakaraju Rohit Kumar Varma, Kolli Chanakya Venkata Sai, Poornapragna Math, E. P. Vishnutheerth, K. N. Sreehari, and Anu Chalil

Abstract As the world progresses into the phase of Industrial IoT (IIoT) and Industry 4.0, IoT services are being rapidly deployed into Industrial, Environmental, and other critical domains. Prior to this, audits and quality checks in Industries were manual, presaging that fraud, corruption, and bribery were highly plausible events where sensitive data like Air Quality Index, Radiation Levels, and other mandatory quality assurance data from Industries can be effortlessly manipulated. This issue can create a false sense of security among all the stakeholders, the public, and a massive danger to the Environment. This paper proposes a Generalized Data Auditing Solution in trustless environments like Industries. This solution is built using IoT to automate data collection process and uses Hyperledger Fabric, an open-sourced permissioned blockchain, to store the data on a secure, tamper-proof blockchain data structure. This paper also demonstrates a webpage and services built on top of the data layer which can be used by audit and quality assurance authorities to query the historical data transmitted by the deployed IoT nodes. The proposed solution eliminates any manual data recording and data manipulation by combining the forces of IoT and blockchain.

Keywords Internet of Things · Blockchain · Hyperledger Fabric · Industrial IoT · Industry 4.0

Y. Rohit Kumar Varma · K. Chanakya Venkata Sai · P. Math · E. P. Vishnutheerth (✉) · K. N. Sreehari · A. Chalil
Department of Electronics and Communication Engineering, Amrita School of Engineering, Amrita Vishwa Vidyapeetham, Amritapuri, India

K. N. Sreehari
e-mail: sreeharikn@am.amrita.edu

A. Chalil
e-mail: anuchalil@am.amrita.edu

V. K. Gunjan and J. M. Zurada (eds.), *Proceedings of 4th International Conference on Recent Trends in Machine Learning, IoT, Smart Cities and Applications*, Lecture Notes in Networks and Systems 873, https://doi.org/10.1007/978-981-99-9442-7_47

1 Introduction

Bringing Industrial IoT (IIoT) in Industry 4.0 and aiding the problem of security in the system, i.e., IoT data, with a blockchain-based solution.

Industry 4.0 is the next phase of era-defining changes in the Industrial and Manufacturing sector by embracing smart automation technologies run by intelligent systems to enable data-driven events and processes. And one of the most important technologies shaping this revolution is Industrial IoT, also known as IIoT. The most essential need for IIoT in the future of Industry 4.0 is to solve the complex manual nature of modern manufacturing and Industrial operations. Previous methods of manufacturing relied on manual design and decision changes by the manual workforce. This results in the following major drawbacks:

1. Slower production process.
2. Increased cost of production.
3. Higher error rates.

However, using IIoT in the future of the Industrial revolution helps Industries connect manufacturing devices and equipment to the Internet via IoT devices and sensors, which can give the manufacturers a chance to enable massive leverage over:

1. Automation of data collection for important end parameters to maintain consistency in output.
2. Run analytics over the collected data to estimate future production values.
3. Real-time monitoring and set actions when an event trigger.

This being positive side of embracing IIoT into Industry 4.0, but there are some major problems in the current implementation of IIoT and this leads to our problem statement for the proposed solution in this paper.

The current implementation of IIoT largely focuses on automation and data collection using IoT edge nodes but lacks attention to security, transparency, and risks involving centralized data storage techniques. These raise the risk of fraud, data manipulation, corruption, and bribery among auditors, quality assurance officers, and Industries.

To get a better understanding of the problem statement, assume a shoe manufacturing Industry that uses leather and other materials which produce a large amount of Air and Water pollutants. Governments around the world have come together to restrict the number of untreated pollutants being released into the environment. But in many Industries, especially in developing countries like India, China, and other Southeast Asian nations, there are cases of data manipulation, bribery, and corruption to alter many critical quality parameters which quantify Air Pollution, Water Pollution, Cellular Radiation Levels, etc.

This manipulation results from centralized storage of data that can be changed even after the data is logged by the IoT edge nodes deployed in the Industrial complexes. This lack of tamper-proof data storage can cause a sense of false assurance to all stakeholders and the public over the quality of the product and its environmental by-products. This may lead to low-quality products that can leverage a high public health risk.

If this possibility of data manipulation is left untreated, this will create a sense of mistrust in IIoT infrastructure and automation in general. Here comes the concept called "Blockchain" [1] that can help IIoT solve the problem of data manipulation, lack of transparency, and centralized storage architecture. Blockchain is a distributed ledger technology that provides a secure and tamper-proof data storage solution to store and query IoT data. Some of the features of Blockchain that complement the use cases of IoT are:

- **Security**: Blockchains provide theoretically unbreakable cryptographic security, which is considered extremely difficult to break with the current hardware standards. This makes it extremely difficult for fraudulent manipulation of data by Industries.
- **Transparency**: Blockchains provide a transparent [2] and auditable record to store and query every transaction sent by a particular edge computing device in the IoT network. This makes it easy for quality assurance and audit authorities to check for critical parameters in Industries.
- **Tamper-Proof**: Blockchains leverage consensus algorithms like Proof-Of-Work (PoW), Proof-Of-Stake (PoS), Proof-Of-History (PoH), Proof-Of-Luck (PoL), etc. to make sure that the amount of effort required by a malicious actor to change the on-chain data is much greater than all the current computing power available.
- **Decentralization**: With blockchains, there is no need to depend on any central authority or intermediary to handle IoT data. This reduces the risk of a single point of failure and makes data resilient to a wide range of attacks.

In the next section of the paper, we will be exploring the proposed solution and the benefits it brings to solve the problem statement.

2 Proposed Solution and Its Benefits

2.1 Automated Data Collection

The first layer of benefits from the proposed solution is the "Automated Data Collection" layer, which consists of the IoT layer and the data transmission layer. The IoT layer consists of any microcontroller which can connect with the Internet, the proposed model uses the MQTT publish-subscribe protocol to connect the IoT microcontroller with the Node-RED data processing platform, which is followed by pushing the data to the data storage layer on the Hyperledger Fabric blockchain. Node-RED is a popular open-sourced platform used to build services on top of IoT hardware, to trigger actions based on the subscribed events set on the IoT data. This platform is used as a pre-processing solution before pushing the data into the blockchain, this also offloads the processing power from lightweight IoT microcontrollers like Arduino UNO or NodeMCU, etc.

2.2 Tamper-Proof Storage and Query Layer for IoT Data

The next layer of benefits from the proposed model is in the novel Hyperledger Fabric data storage layer, which can be generalized for any Industrial use case by configuring the required parameters and their data structures. This layer acts as an immutable data storage layer that is transparent, decentralized, and secure, thus eliminating any fraudulent attempt at data manipulation by Industries or Authorities. This acts as a read-only layer for external users, where the web platform built on top of the blockchain layer queries the transactions or logs from the Hyperledger fabric chain and displays it for authorities and quality assurance departments, ensuring complete transparency in real time.

2.3 A Simplified Web Platform for Auditors to Query Data

And the user layer of this proposed model provides the authorities with a simple webpage where they can monitor the deployed Hyperledger fabric chain and check the real-time logs or transactions being submitted by the edge IoT nodes. Also, when a value of a critical parameter crosses a previously set threshold, it raises an alert notification for the authorities to take action on the concerned Industries. This highly benefits the auditors to get real-time data from the deployed IoT nodes without the need to know any under the hood complexities of the proposed solution.

3 Hyperledger Fabric: Selection Criteria

In this section, we will explore the reasons and criteria behind selecting Hyperledger Fabric as the primary blockchain solution for the proposed model. Hyperledger Fabric is an open-sourced permissioned blockchain project developed by IBM [3]. This is aimed at enterprise-level permissioned projects, where all the nodes which send the transaction data are known in advance, analogous to the deployed IoT nodes in the proposed project.

Consensus is a key property that defines a blockchain [4]. It is an architecture that defines decentralization, i.e., the ways in which all the nodes in a blockchain reach an agreement, whether a new transaction or a new data input from an edge IoT node, is verified to be added to the blockchain. And the higher the chance of dishonest nodes which try to add data, the higher the needed computational power to achieve the consensus. This can be translated to the nature of public and permissioned blockchains.

This leads to high computational requirements demanding consensus algorithms [5] like Proof-of-Work and Proof-of-Stake, which are generally used in public blockchains like Bitcoin and Ethereum, respectively. These public blockchain networks can allow anyone on the open Internet to act as a miner, and verify the data. This open mining of data leads to a high chance of the presence of a dishonest or malicious node. Thus, public blockchains need Mining, Cryptocurrencies, and Computationally Intensive Consensus Algorithms, to compensate the miners and maintain the data verifiability on the blockchain.

But in the case of Hyperledger Fabric, it is a permissioned blockchain that only allows certified or permissioned nodes to send the data to the blockchain. This creates a trusted network of nodes, in our proposed model, this network of nodes are the IoT edge computing devices and sensors deployed in the Industrial regions. This allows Hyperledger Fabric to use a lightweight consensus algorithm that can operate with trusted nodes in the network, this also enables Hyperledger Fabric to run without mining or a cryptocurrency. The following are the key properties of Hyperledger Fabric and the primary selection criteria to choose a permissioned blockchain over other public blockchains:

- Lightweight Consensus Algorithm to run on low overhead microcontroller devices like Arduino, NodeMCU, etc.
- No need for Mining or Cryptocurrencies to run the network or compensate the nodes.

4 Working Architecture of the Proposed Solution

4.1 Overview of the Block Diagram

As shown in Fig. 1 the proposed solution can be described as a 4-layered model, where the base layer begins with the IoT layer, followed by a data processing layer to process and transmit the data from the hardware layer to the Hyperledger Fabric blockchain. The final layer in the block diagram would be the Audit Services, which consists of a webpage for real-time data querying of data from the blockchain.

4.2 IoT Layer

IoT layer of the proposed solution consists of a network of edge computing microcontroller devices that can be equipped with any sensor to record Industrial or Environmental data based on ground-level requirements. This enables the proposed model to be generalized to any use case. This hardware layer is connected to the internet

Fig. 1 Block diagram of the proposed solution

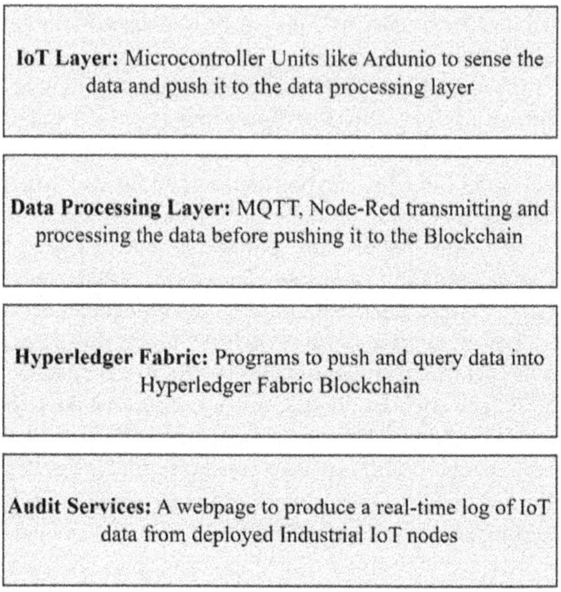

via other components like an ESP8266 Wi-Fi module. In the later section of demonstrating the proposed model, a NodeMCU with an integrated ESP8266 Wi-Fi module will be used to connect the model to the Internet.

4.3 Data Processing Layer

The data processing layer will be responsible for processing and transmitting the data from the IoT hardware to the deployed Hyperledger Fabric blockchain. This layer consists of two components:

- Message Queuing Telemetry Transport (MQTT), the standard open-sourced library to enable IoT communications with deployed software components.
- Node-RED is another standard open-sourced programming tool developed by IBM to wire together various layers of an IoT solution. This will be used to process the data and send the data to the deployed Hyperledger Fabric blockchain.

Node-RED is a powerful tool for building IoT applications that are connected via MQTT. Message Queuing Telemetry Transport (MQTT) is a lightweight messaging protocol that is commonly used in IoT applications to communicate between devices and applications with low data overhead.

Using Node-RED, the proposed solution can easily create a flow-based programming model that allows us to connect and process data from various IoT devices using

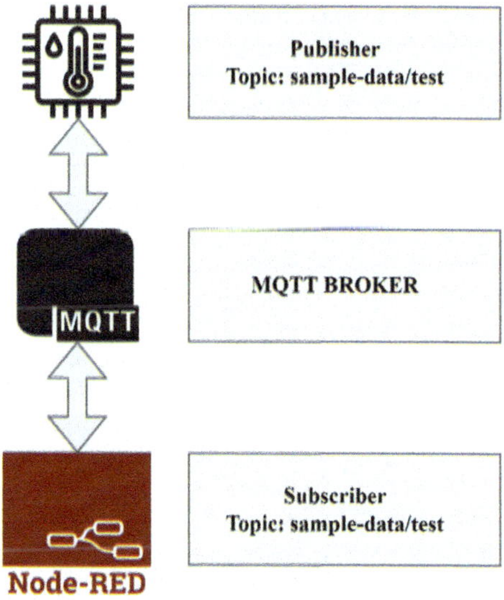

Fig. 2 MQTT publish-subscribe model

MQTT. This enables the solution to quickly and easily create complex workflows to integrate the IoT network and the Hyperledger Fabric.

One of the key advantages of using Node-RED is its ability to handle complex data transformations and integrations. For example, Node-RED can be used to collect data from an MQTT-integrated sensor and convert it into a format that is usable by other applications or systems down the line.

As shown in Fig. 2, MQTT will be utilizing an efficient publish-subscribe model to take data tagged to a topic as input from a publisher and deliver the same data to subscribers who are tagged to the same topic. In the case of our proposed solution the data, the publisher would be the IoT layer, which transmits any generalized Industrial or Environmental data to the MQTT broker, and the primary subscriber to our data topic would be the Node-RED platform which consumes [6] the data sent via the MQTT broker and process it into a format suitable for the Hyperledger Fabric blockchain.

4.4 *Hyperledger Fabric Layer*

The Hyperledger Fabric layer brings novelty to the proposed solution, as this layer acts as the secure, transparent, and tamper-proof data storage and querying medium for the IoT data. This layer intakes the data from the automated Node-RED platform which process and pushes the IoT data into the blockchain. This layer can be visualized as a 3-step process, as shown in Fig. 3.

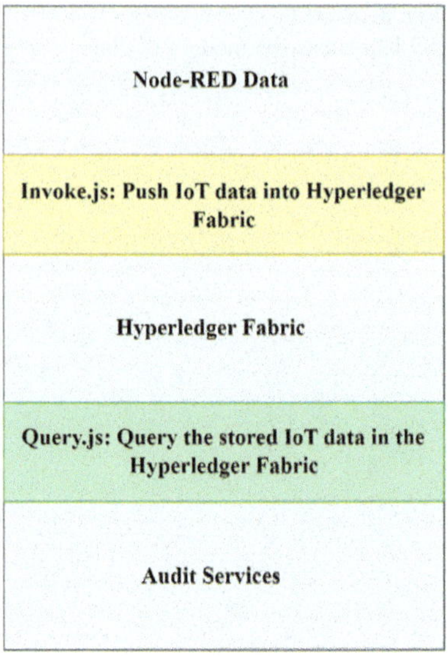

Fig. 3 Hyperledger Fabric working in the proposed solution

- The first step in this layer is to push the data to the Hyperledger Fabric blockchain, from the Node-RED platform after processing the data into the accepted format of the data structure defined in the Hyperledger Fabric blockchain.
- Next step involves a script file [7] developed as part of the proposed solution called Invoke.js. As the name suggests it is a javascript file written to automate the process of collecting the data sent by Node-RED and pushing the data into the blockchain after validation of certificates and data integrity.
- The final step in the working of this Hyperledger Fabric blockchain layer in the proposed solution is querying the stored data to display them in the Audit Services layer, which is a webpage that gets updated with real-time data logs of the IoT data, to give Auditors a real time and historical view of the logged data. To perform this querying operation, another script file called Query.js is developed to read the data on-chain and push the data onto the Audit Services webpage.

4.5 Audit Services and Webpage

Audit Services is the final piece in the proposed model which abstracts all of the under the hood operations involving IoT and Hyperledger Fabric blockchain, and gives the Auditors a real-time flow of ground-level sensor data taken from the deployed Industrial and Environmental regions. This webpage is built with the following features:

- Auditor and Quality Checker Login Authentication.
- Real-time log of data from multiple deployed IoT networks.

- Triggers a notification to Auditor emails once a threshold is crossed, based on the measured data from the IoT nodes.

This webpage acts as a dashboard for all the data being generated and stored on the blockchain, and this can be leveraged to perform data analytics and other critical operations on the stored data. As this is a dynamic read-only webpage, the data on the blockchain is secure, transparent, and tamper-proof.

5 Results: Proposed Model Demonstration

The proposed model can be demonstrated layer-wise as explained in the working architecture section. As the proposed solution is a generic solution to a wide range of potential use cases which can be explored in the next section, to demonstrate its working in this paper, we will be assuming a specific use case like "Monitoring Industrial Gas Emissions". In this demonstration the proposed solution would be used to securely store the data in the blockchain, to allow Auditors to verify critical LPG and Butane gas leaks and emission data from trustless Industrial regions to ensure the safety standards being practiced onsite; this data is sent via a network of IoT edge computing nodes. So in the IoT layer for this specific demonstrative use case, the proposed solution uses an MQ-6 gas sensor [8] coupled with a NodeMCU as shown in Fig. 4. Then the logged data as seen in Fig. 5, will be sent to the Node-RED platform in real time as shown in Fig. 6, via the MQTT library running on top of the NodeMCU, shown in Fig. 7, based on the code written in the Arduino IDE. As discussed earlier, MQTT enables a low overhead, lightweight data transfer from hardware to processing platforms like Node-RED.

Fig. 4 An MQ-6 gas sensor connected to NodeMCU

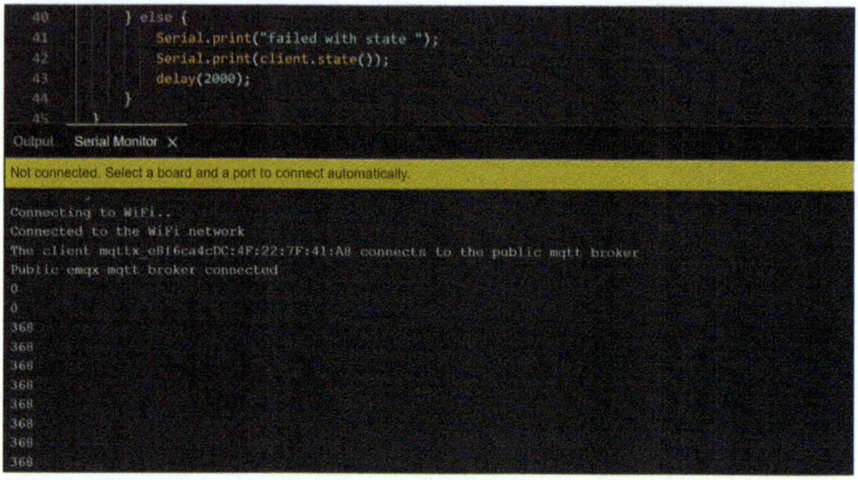

Fig. 5 Readings of MQ-6 sensor shown in Arduino IDE

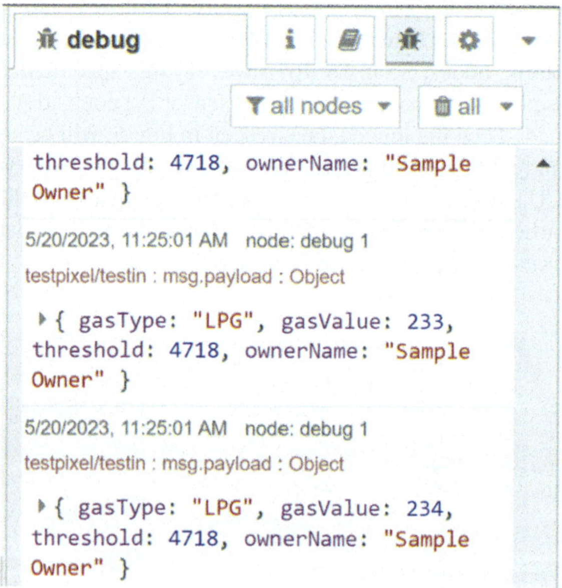

Fig. 6 Real-time gas values being updating in Node-RED as shown in debug menu

This data will be consumed by the Hyperledger Fabric blockchain, to store the data on-chain. A Hyperledger Fabric blockchain instance will verify the data using a chaincode, a file that will define the data structures of the stored data, and methods used to interact with the blockchain.

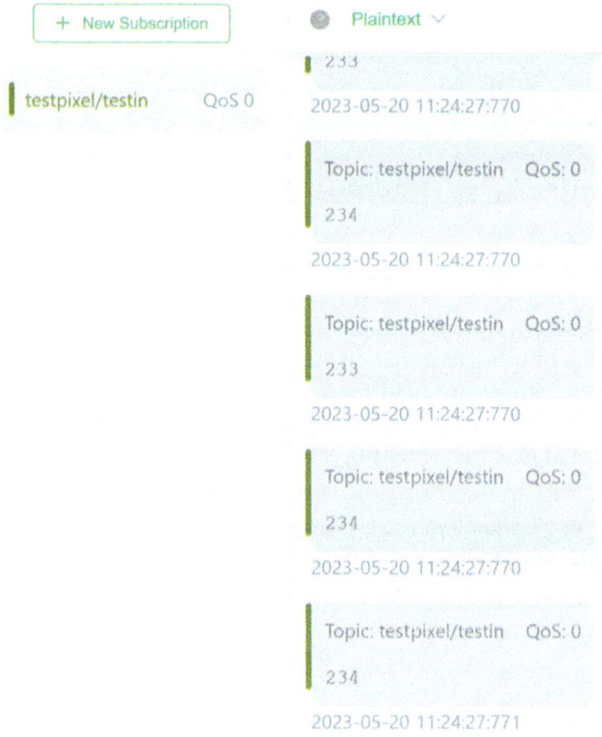

Fig. 7 Real-time gas values being sent to Node-RED from MQTT server

As shown in Figs. 8 and 9, after the chaincode is ready, the proposed model uses the programmed scripts Invoke.js and Query.js to interact with the blockchain, to push and query the data from the blockchain.

Querying the data stored on the blockchain is important to build Audit Services on top of the blockchain, to allow Auditors and Quality assurance authorities to verify the historical IoT logged data stored on the blockchain. To facilitate this and abstract

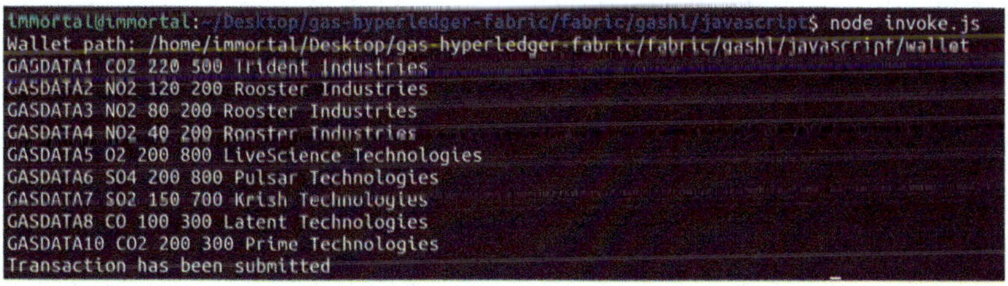

Fig. 8 Result of running Invoke.js

Fig. 9 Result of running Query.js

the complexity in the code, the proposed solution uses a webpage as shown in Fig. 10, to display the real-time log of the data stored on the blockchain.

The user can also set a certain threshold for a particular gas in the Node-RED description so that if the gas value ever exceeds the set threshold, an email is triggered to the user's registered email id as shown in Fig. 11.

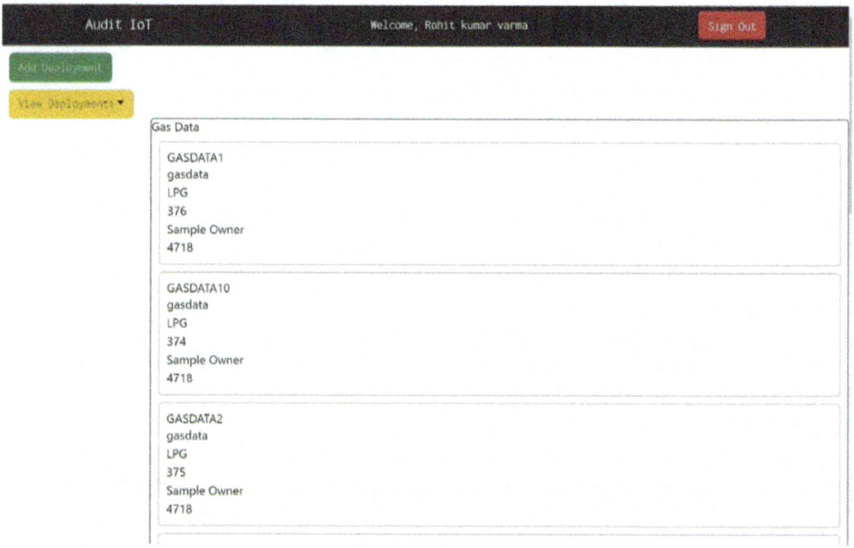

Fig. 10 Webpage built to serve auditors to verify data in real time

Fig. 11 An email sent to the user when the gas value exceeds a set threshold of 2000

6 Conclusion

The proposed solution presents a blockchain-based framework for a secure Industrial IoT (IIoT) experience in Industry 4.0 with a demonstrative example. This framework can be used across Industries, to automate data collection, and store and retrieve the data in a secure and transparent fashion. All the layers and components of the proposed solution are completely generalized and can be used for a wide range of applications just by changing the data structure and the required methods used in the Hyperledger Fabric's Chaincode. This proposed solution can tackle various Industrial use cases like:

- Preventing fraudulent manipulation of Pollution data in Industries.
- Preventing manipulation of cellular radiation levels before authorized audits.
- Tracking Food Supply chain [9] and historical food quality index of a manufacturer.
- Monitoring the storage temperatures of cold chain vaccines.

This proposed solution can be integrated into any Industrial or Environmental Audit and Quality checking lifecycle, to make the data collection automated using IoT, data storage secure, transparent, and tamper-proof using Hyperledger Fabric blockchain and organized with the help of a webpage.

References

1. Cui P, Guin U, Skjellum A, Umphress D (2019) Blockchain in IoT: current trends, challenges, and future roadmap. J Hardware Syst Secur 3
2. Samal D, Arul R (2020) A novel privacy preservation scheme for internet of things using blockchain strategy. In: International conference on communication, computing and electronics systems, 5 Mar 2020
3. Honar Pajooh H, Rashid M, Alam F, Demidenko S (2021) Hyperledger Fabric blockchain for securing the edge internet of things. Sensors 21
4. Raghav, Andola N, Venkatesan S, Verma S (2020) PoEWAL: a lightweight consensus mechanism for blockchain in IoT. Pervasive Mob Comput 69
5. Ambili KN, Sindhu M, Sethumadhavan M (2017) On federated and proof of validation based consensus algorithms in blockchain. In: IOP Conference Series: Materials Science and Engineering, vol 225. International conference on materials, alloys and experimental mechanics (ICMAEM-2017), 3–4 Jul 2017
6. Sanju S, Sankaran S, Achuthan K (2018) Energy comparison of blockchain platforms for internet of things. In: 2018 IEEE international symposium on smart electronic systems, 17–19 Dec 2018
7. Christidis K, Devetsikiotis M (2016) Blockchains and smart contracts for the internet of things. IEEE Access 4:2292–2303
8. Sparkfun.com (2013) Technical data MQ 6 gas sensor. Available from: https://www.sparkfun.com/datasheets/Sensors/Biometric/MQ-6.pdf. Accessed 20 Aug 2022
9. Muthuraj S, Babu S, Shijith Kumar PM (2022) Identify the key success factors and enabling strategies for successful blockchain implementation in supply chain. In: International conference on industrial engineering and operations management, Istanbul, Turkey, 7–10 Mar 2022

Detecting and Analyzing the Emotional Levels of a Person Through CBT Using MFCC and Lexicon-Based Approach

P. Yasaswini, K. Sri Devendra, G. Sai Sathwik, and T. Uday Kiran

Abstract Depression is a serious illness that usually goes ignored by friends, family, and society. Despite not being recognized as such mental illness, depression is a severe mental condition that requires careful management. The AI industry has been obliged to take matters under its control due to cost and availability issues. As part of this study, our chatbot makes use of the Librosa library to evaluate audio and music while also listening for the speaker's tone and voice. The lexicon-based approach is one approach for semantic and sentiment analysis from text. This technique establishes the sentiment orientations of the text. Both human and automated methods can be used to create the dictionary of lexicons. One way for extracting emotions from text is lexicon-based methods. Six distinct emotions may be recognized, and it contains certain predetermined phrases. A little amount of information from the voice signal is taken out using the feature extraction technique, which may then be utilized to identify each speaker individually. Using the Mel Frequency Cepstral Coefficient (MFCC) approach, an individual's mood may be inferred from their voice. The effectiveness of the MFCC system was found to be roughly 80% when the happy, sad, and angry emotions were tested. By asking open-ended inquiries like "How are you?" and "How was your day?" the chatbot carries on the conversation and gathers additional data to better understand the user. We will employ datasets for voice recognition like Ravdess and Tess in order to train the AI model using a variety of emotions and words.

K. Sri Devendra, G. Sai Sathwik and T. Uday Kiran contributed equally to this work.

P. Yasaswini · K. Sri Devendra (✉) · G. Sai Sathwik · T. Uday Kiran
Department of CSE, V R Siddhartha Engineering College, Vijayawada, Andhra Pradesh 520007, India
e-mail: devendrakonduru@gmail.com

P. Yasaswini
e-mail: yasaswini@vrsiddhartha.ac.in

G. Sai Sathwik
e-mail: sathwik221@gmail.com

T. Uday Kiran
e-mail: udaykiran.t2114@gmail.com

V. K. Gunjan and J. M. Zurada (eds.), *Proceedings of 4th International Conference on Recent Trends in Machine Learning, IoT, Smart Cities and Applications*, Lecture Notes in Networks and Systems 873, https://doi.org/10.1007/978-981-99-9442-7_48

Keywords Depression · Mental illness · Librosa library · Semantic analysis · Sentiment analysis · Lexicon-based approach · Feature extraction · MFCC · Emotion detection · Training datasets · Speech recognition

1 Introduction

Emotions are an essential aspect of human communication and behavior, playing a pivotal role in shaping our thoughts, actions, and well-being understanding and accurately detecting emotional states in individuals can be crucial for various applications, such as mental health assessment, human–computer interaction, sentiment analysis, and personalized therapy [1]. Cognitive Behavioral Therapy (CBT) is a widely recognized and effective approach in psychotherapy, focusing on identifying and modifying negative thought patterns to promote emotional well-being. The development of mental health chatbots, such as "Psykh," has emerged as a valuable tool for enhancing mental health therapy [2].

Studies on "multimodal emotion recognition" have demonstrated the effectiveness of using a variety of data modalities to improve emotion identification accuracy [3]. It's vital to identify the larger range of emotion detection methodologies, including ECG-based systems in healthcare [4], even though our main focus is on speech-based emotional analysis using MFCC and CNN based approaches [5].

A comprehensive review of emotion recognition methodologies highlighted the importance of MFCC and lexicon-based methods, reinforcing their significance in our project [6]. Text-based emotion analysis has also seen notable advancements in sentiment analysis and emotion recognition [7]. Multimodal emotion recognition, involving the integration of various data sources, shows great promise for improving emotion identification accuracy [8]. Our study aims to leverage this potential for more extensive emotional assessments. Significant advancements in emotional analysis, ranging from emotional speech recognition to multimodal emotion recognition and text-based emotion analysis, have paved the way for our project's integration of MFCC and lexicon-based methods with CBT principles and deep learning techniques [9]. By creating a comprehensive framework for emotional level detection, our research aims to contribute to improved mental health interventions and emotional well-being.

2 Methodology

The proposed methodology is described in the following steps:

Step 1: Data Collection

Data is collected for the analysis.

Step 2: Classification of Data

The collected data is classified into text or audio for further preprocessing.

Step 3: Text Data Preprocessing

For text data, the following preprocessing steps are performed:

Step 3.1: Text Tokenization

The text is tokenized by breaking it down into smaller units known as tokens.

Step 3.2: Stop Word Removal

Commonly used words, known as stop words, are removed from the text data as they are considered insignificant for analysis or modeling. Examples of stop words include "a," "an," "the," "is," "and," "in," "of," etc.

Step 3.3: Lemmatization

Words are reduced to their base or canonical form, known as the lemma, through morphological and contextual analysis. This step helps in standardizing word variations. For example, "run," "running," and "ran" would be reduced to "run," and words like "cats," "cat's," and "cats'" would be reduced to "cat."

Step 4: Audio Data Preprocessing

For audio data, the following preprocessing steps are performed:

Step 5: Frame Segmentation

The audio signal is divided into short overlapping frames, typically with a duration of 20–40 ms. This step captures the temporal dynamics of the audio signal.

Step 6: Windowing

Each frame is multiplied by a window function, such as the Hamming window, to reduce spectral leakage caused by abrupt transitions at the frame boundaries. The windowed frames emphasize the stationary segments of the audio signal.

Step 7: Fast Fourier Transform (FFT)

The Fourier transform is applied to each windowed frame, converting the audio signal from the time domain to the frequency domain. This transformation reveals the spectral content of the audio signal.

Step 8: Mel Filter Bank

A bank of triangular filters is applied to the power spectrum obtained from the FFT. This compresses the spectrum by emphasizing certain frequency regions while attenuating others.

Step 9: Logarithm

The logarithm of the filter bank energies is taken to account for the human perception of loudness, which is roughly logarithmic in nature. This step converts the filter bank energies into a more linear representation.

Step 10: Discrete Cosine Transform (DCT)

The DCT is applied to the logarithmic filter bank energies to obtain the Mel Frequency Cepstral Coefficients (MFCC). The DCT decorrelates the filter bank energies and produces a compact representation of the spectral features.

Step 11: Feature Selection

In this step, a subset of MFCC coefficients is selected as the final set of features. Typically, the lower-order coefficients, representing the overall spectral shape, are more informative and used for further analysis.

Step 12: Feature Normalization

The MFCC coefficients are often mean-normalized across frames or standardized to have zero mean and unit variance. This step helps in removing biases and scaling differences among different audio signals.

The above steps outline the methodology for preprocessing both text and audio data.

3 Results

Our project successfully developed an emotion detection system by combining MFCC (Mel Frequency Cepstral Coefficients) with a lexicon-based approach. Through extensive research and implementation, our system showcased effective emotion detection capabilities. By extracting MFCC features from audio signals and utilizing the NRC Emotion Lexicon, we achieved accurate classification and detection of various emotions expressed in speech. This signifies the system's potential to accurately identify and analyze emotional states based on audio inputs.

Moreover, the real-time analysis feature of our system is notable as it provides immediate feedback on emotional states, enabling individuals to regulate their emotions in the moment. This feature has far-reaching implications for mental health monitoring, stress management, and interactive virtual environments. Additionally, the integration of the lexicon-based approach using the NRC Emotion Lexicon enhanced the system's accuracy and robustness.

Figures 1, 2, and 3 show the output screens of our model.

4 Conclusion

The proposed application detects and analyses human emotions using MFCC and a lexicon-based approach. Through extensive research and implementation, we achieved significant outcomes in the field of emotion detection. While our project has achieved promising results, there are several avenues for future work and improvement in the field of emotion detection using MFCC and lexicon-based approaches. Some potential areas of future research and development include: **Multi-modal**

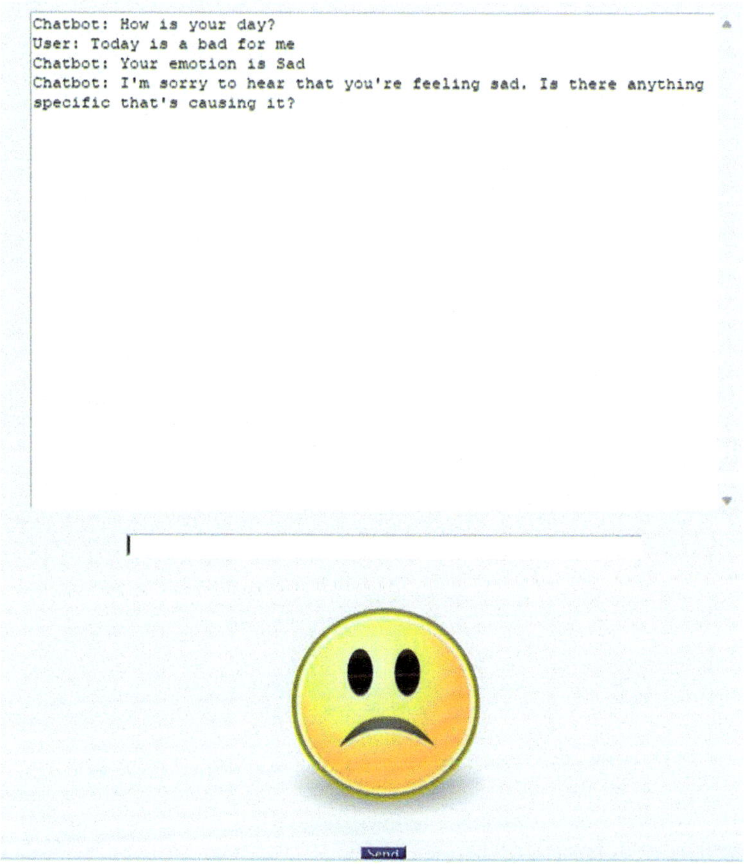

Fig. 1 Chatbot demonstrates the emotion of the user. The chatbot interface showcases a user-friendly and intuitive design, making it accessible to a wide range of users. As the user interacts with the chatbot, their speech input is transcribed and analyzed using the MFCC feature extraction technique, which captures the acoustic characteristics of their voice. Simultaneously, the chatbot processes the text input to identify emotional cues using the lexicon-based approach, mapping words and phrases to their corresponding emotional intensities

Analysis: Expanding the analysis to incorporate multiple modalities such as facial expressions, physiological signals, and linguistic cues can lead to a more comprehensive understanding of human emotions. Integrating MFCC analysis with other modalities can enhance the accuracy and richness of emotion detection.

Dataset Expansion: While the NRC Emotion Lexicon serves as a valuable resource, expanding the dataset with more diverse and extensive emotion annotations can enhance the system's generalization and adaptability. Collecting and curating larger emotion datasets specific to different languages, cultures, and contexts can lead to more robust emotion detection models.

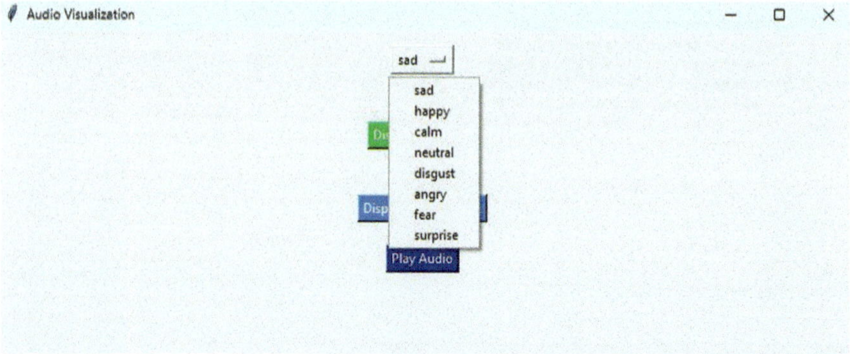

Fig. 2 Categorization of audio files in the RAVDESS dataset. Each emotion category is labeled with its corresponding emotional state, such as "happy," "sad," "angry," "fearful," "surprised," or "neutral," enabling users to gain insights into the emotional diversity captured within the dataset

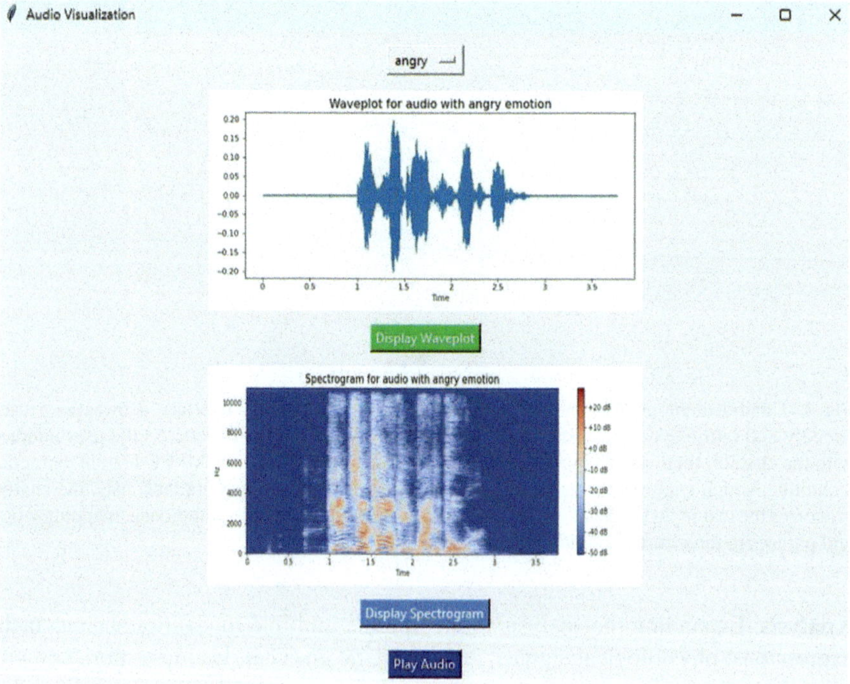

Fig. 3 Wave plot for the relevant audio file is displayed

References

1. Van Trinh L, Le Dao Thi T, Le Xuan T, Castelli E (2022) Emotional speech recognition using deep neural networks. Sensors 22:1414. https://doi.org/10.3390/s22041414
2. Patole A et al (2021) Mental Health Chatbot (Psykh). Int Res J Eng Technol 8(3):1084–1087
3. Tang H et al (2017) Multimodal emotion recognition using deep neural networks. In: Neural information processing: 24th international conference, ICONIP 2017, Guangzhou, China, 14–18 Nov 2017, proceedings, Part IV 24. Springer
4. Hasnul MA et al (2021) Electrocardiogram-based emotion recognition systems and their applications in healthcare—a review. Sensors 21(15):5015
5. Puri T, Soni M, Dhiman G, Khalaf O, Alazzam M, Khan E (2022) Detection of emotion of speech for RAVDESS audio using hybrid convolution neural network. J Healthc Eng 2022:1–9. https://doi.org/10.1155/2022/8472947
6. Dzedzickis A, Kaklauskas A, Bucinskas V (2020) Human emotion recognition: review of sensors and methods. Sensors 20(3):592
7. Nandwani P, Verma R (2021) A review on sentiment analysis and emotion detection from text. Soc Netw Anal Min 11:81. https://doi.org/10.1007/s13278-021-00776-6
8. Ahmed N, Al Aghbari Z, Girija S (2023) A systematic survey on multimodal emotion recognition using learning algorithms. Intell Syst Appl 17:200171. ISSN: 2667-3053
9. Shinohara S, Nakamura M, Omiya Y, Higuchi M, Hagiwara N, Mitsuyoshi S, Toda H, Saito T, Tanichi M, Yoshino A, Tokuno S (2021) Depressive mood assessment method based on emotion level derived from voice: comparison of voice features of individuals with major depressive disorders and healthy controls. Int J Environ Res Public Health 18(10):5435. https://doi.org/10.3390/ijerph18105435

AI-Enabled Fault Detection for Predictive Maintenance of Ball Bearings

Likhith Reddy Kuruvalli, Divyajyoti Morabad, H. M. Vijay, and P. Ratheesh

Abstract Manufacturing Industry 4.0 refers to the integration of advanced digital technologies and automation into the manufacturing sector. In the manufacturing industry 4.0, the Predictive Maintenance (PdM) plays a crucial role in optimizing equipment reliability, improving productivity, and reducing maintenance costs. The fault detection under predictive maintenance in ball bearing uses the sensor data and advanced analytics techniques to proactively identify faults, enabling timely maintenance actions and preventing equipment failures. The AI algorithms, such as Autoencoders, Artificial Neural Network (ANN), and Random Forest (RF) classification, enable the analysis of sensor data to detect and classify various types of bearing faults accurately making use of the Case Western Reserve University (CWRU) bench-marked dataset. The fault detection method is applied to ball bearings with fault diameters 0.007″, 0.014″, 0.021″ units. The Autoencoder model gets the greatest accuracy, outperforming the ANN and RF classification algorithms.

Keywords Fault detection · Predictive maintenance · Ball bearing

1 Introduction

The manufacturing sector is going through a major upheaval according to the industry 4.0 model [1] that aims to automate the majority of industrial processes. Predictive maintenance (PdM) [2] falls under the banner of Industry 4.0, and this technology aims to offer useful data about the overall health of the device and permit regular

L. Reddy Kuruvalli (✉) · D. Morabad · H. M. Vijay
KLE Technological University, Hubli, Karnataka, India
e-mail: rlikhith22@gmail.com

H. M. Vijay
e-mail: vijay.hm@kletech.ac.in

P. Ratheesh
Carriage Repair Workshop—South Western Railway, Hubli, Karnataka, India

V. K. Gunjan and J. M. Zurada (eds.), *Proceedings of 4th International Conference on Recent Trends in Machine Learning, IoT, Smart Cities and Applications*, Lecture Notes in Networks and Systems 873, https://doi.org/10.1007/978-981-99-9442-7_49

upkeep that identifies problems before they arise. Device condition monitoring is used to improve efficiency in production using automated machines. To avoid downtime, factory managers have to repair machine parts on a regular basis and perform scheduled maintenance. The majority of predictive maintenance actions are unproductive, resulting in productivity losses and the waste of resources. Therefore, machine learning-based predictive maintenance has rapidly developed to maximize uptime and reduce service costs. The induction motors, in particular, play a significant part in today's smart manufacturing sectors. These motors consists of a number of parts, including the stator, rotor, shaft, and bearings [3].

Electric machine defects can be categorized into a number of distinct subgroups, such as failure of the stator winding insulation, bearing problems, drive inverter problems, and eccentric air leakage [4]. According to research, bearing issues are most frequent kind of errors, 30–40% of all mechanical breakdowns are caused by bearing issues that are used in rotating machinery to support and guide the shafts. As a result, any defect in the bearings might result in losses in production and equipment, leading to the waste of time, money, and resources. The following parts of ball bearings are frequently employed in rotating machinery: the outer race (OR), inner race (IR), ball, and ball cage.

Even though the roller bearing is operating in a conventional workplace context, it has to endure several conditions like frequent lateral stress causes poor installation, irregular lubricant, improper speed, and operating temperature.

Four misalignment conditions which are most probable to occur in bearing failures [5] are (a) misalignment (out-of-line), (b) shaft deflection, (c) crooked or tilted OR, and (d) crooked or tilted IR [6]. This research focuses on how artificial intelligence supports the sector of manufacturing in producing predictive modeling for bearing health.

2 Literature Survey

The topic of this study is the diagnosis and prognosis of intelligent errors. They suggested CRNN methodology, which recognizes bearing flaws in following datasets. IMS and CWRU bearing datasets showed that the total classification accuracy was 97.13% and 99.77%, respectively. The report presents the tests for selecting each suggested network component as well as the network architecture that is used to attain high accuracy [7].

This study examines all articles on fault diagnosis in manufacturing that were written between 2007 and 2017 and used various machine learning techniques. They summarized four crucial techniques (Bayesian network, HMM, ANN, and SVM). They came to the conclusion that modeling benefits from both BN and HMM methodologies. The study also demonstrates that ANN produced flawless outcomes, but being a blackbox method, the model makes it challenging to understand what occurs inside of it [8].

Research study regarding intelligent misalignment of ball bearings was conducted by excluding time zone-related data from the shaking data; in this work, an ANN-based ball bearing fault identification technique is included [9].

In a study on the detecting and diagnosing of faults, it was suggested that mechanical induction motor flaws be selected and categorized using RF classification. There, it is discussed how RF classifications affect misdiagnosis. Outcomes were assessed against contemporary neural networks and artificial intelligence methods. The analysis of the data showed that, in comparison with previous technologies. The proposed method outperformed others with a higher level of accuracy [10].

A study was conducted on the CWRU dataset utilizing various deep learning methods. According to the findings, Stacked Denoising Autoencoder (SDAE) and Sparse Autoencoder (SAE) have an average accuracy of 99.83%. Additionally, they provide thorough advice on how to pick the best deep learning algorithm category for various application cases [6].

3 Architectures

3.1 Autoencoder

Autoencoder is a kind of network that is used for unsupervised learning and dimensionality reduction. An Autoencoder network includes of three parts: an encoder, a code, and a decoder. The encoder network takes three input features and compresses to lower dimensional representation which is fed to next layer. The decoder part takes the compressed representation and reconstructs original input data in the outer layer. The architecture is configured by ReLU activation function, Adam optimizer and loss mean square error [11]. The encoder is interpreted by π and the decoder by ψ, which makes up Autoencoders.

$$\pi = Y \rightarrow C, \tag{1}$$

$$\psi = C \rightarrow Y, \tag{2}$$

3.2 Random Forest Classification

RF classification is an entity learning methodology which combines multiple decision trees to construct a predictive model that is more resilient and accurate. Unlike a single decision tree, which can be prone to over-fitting, a RF aggregates the predictions of multiple trees to make more accurate and reliable predictions.

The exact equations used for splitting and prediction within each decision tree depend on the specific impurity measure and splitting criterion chosen, such as Gini impurity, entropy, or mean squared error. These equations involve comparing feature values to thresholds and evaluating the impurity or error reduction achieved by different splits. The power of RF lies in the ensemble approach, which reduces over-fitting and improves generalization performance by aggregating the predictions of multiple individual models [12].

Entropy can be employed to determine the branching of nodes within a decision tree. Entropy is used by the node to select how to branch based on the probability of a given result. Due to the usage of the logarithmic function in the computation, it is more mathematical than the Gini index.

3.3 Artificial Neural Network

They are computational models which take inspiration from the form and functionality of the neural networks found human brain. The learning process involves feeding the network with labeled training data, comparing its predictions to the desired outputs, and updating the weights using algorithms like gradient descent and backpropagation.

The neural network architecture consists of three layers: the input layer, the hidden layer, and the output layer. The initial data is received by the input layer which consists of two input features and transfers them to the subsequent layers. Acting as intermediaries, the hidden layers perform computations on the inputs and gradually extract more complex features [13]. The ReLU activation function is being used in both the layers. This equation shows the computation of each neuron in the network of an ANN. Here is the general equation for a neuron in a feed forward neural network, where z represents the weighted sum of the inputs and a bias term, w is the weight of input, x is input value, b bias term, and n is number of inputs to the neuron.

$$Z = \sum_{i=1}^{n}(w_i x_i) + b. \tag{3}$$

Finally, the output layer generates the final results or predictions by utilizing (Sigmoid activation) the computations performed by the preceding layers.

The neural network model is configured with the Adam optimizer, binary cross entropy loss function, and accuracy metrics for evaluating the model's performance.

4 Methodology

4.1 Description of Dataset

Dataset utilized here is sourced from CWRU bearing data center. MATLAB data chunks are converted into .csv files.

In this study, both Normal Baseline (NB) data and 12k Normal Baseline (DE) bearing fault data were utilized. The fault data consisted of both IR and OR faults with diameters of **0.007″, 0.014″, and 0.021″**. The OR faults were further categorized based on their load zone, including 'centered' (6 o'clock), 'orthogonal' (3 o'clock), and 'opposite' (12 o'clock). It is important to note that the data for the 12k DE bearing issue exhibited anomalies. A total of 121,155 data chunks were selected. The data was gathered with a motor operating at an average speed of 1750 revolutions per minute (rpm). NB data was combined with every dataset of 12k DE fault bearings for fault diameters of **0.007″, 0.014″, and 0.021″** simultaneously.

4.2 Block Diagram

System design employed in this study is illustrated in Fig. 1. Following data collection and preparation, feature extraction was conducted to eliminate irrelevant features and data from the database, while preserving essential features for model utilization. Moreover, only the unique features were standardized, ensuring that all features adhere to a standardized scale. Various algorithms are then used to model the datasets. To enhance the results, the datasets are scrutinized and parameters are fine-tuned. Three models, namely the ANN, Autoencoder, and RF classification are employed to get results related to precision, recall, $F1$-score, loss, accuracy, and anomalies.

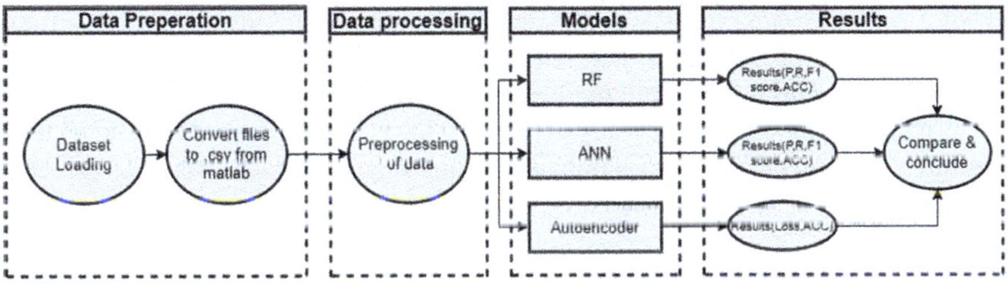

Fig. 1 Design and structure of the system

5 Results and Discussion

ANN and RF classification are used to display the precision, recall, accuracy, $F1$-score, and anomalies. They are used to evaluate the effectiveness of a model or system in terms of its ability to correctly classify or retrieve data. The loss (mean absolute error) and anomalies are computed using Autoencoders.

5.1 Random Forest Classification

Table 1 presents the comparative results obtained using RF classification with both NB and 12k DE bearing fault data. Various algorithms were employed, considering **0.007″, 0.014″, and 0.021″** fault diameters, and it also presents the precision, recall, $F1$-score, anomalies, and accuracy detected using RF classification.

5.2 Artificial Neural Network

Table 2 presents the comparison of results obtained using ANN with both NB and 12k DE bearing fault data. Various algorithms are employed, considering **0.007″, 0.014″ and 0.021″** fault diameters, and it also presents the precision, recall, $F1$-score, anomalies, and accuracy detected using ANN. The end product is determined through the utilization of Adam optimizer.

5.3 Autoencoder

Table 3 presents the comparison of results obtained using Autoencoders with both NB and 12k DE bearing fault data. Various models were employed, considering of **0.007″, 0.014″ and 0.021″** fault diameters, and it also presents the loss (mean absolute error), anomalies, and accuracy detected using Autoencoders. The OR—0.007″, exhibited the highest count of anomalies, while others had the least occurrence of anomalies.

5.4 Comparative Analysis on Accuracy Using Different Algorithms

Table 4 illustrates the accuracy results for all three models. The Autoencoder model demonstrated the highest accuracy rate compared to ANN and RF classification.

Table 1 RF classification results

RF classification—0.007″ fault diameter					
	Precision	Recall	$F1$-score	Anomalies	Accuracy
NB (%)	82	90	86	7093	85.36
IR (%)	89	81	85		
NB (%)	92	97	94	2827	94.16
OR @ 3 (%)	97	91	94		
NB (%)	81	89	85	7823	83.86
OR @ 6 (%)	88	78	83		
NB (%)	72	80	76	12,404	74.41
OR @ 12 (%)	77	69	73		
RF classification—0.021″ fault diameter					
	Precision	Recall	$F1$-Score	Anomalies	Accuracy
NB (%)	81	89	85	7801	83.9
IR (%)	88	79	83		
NB (%)	77	86	81	9517	80.36
OR @ 3 (%)	84	75	79		
NB (%)	72	80	76	12,443	80.36
OR @ 6 (%)	72	80	76		
NB (%)	66	71	68	16,071	66.83
OR @ 12 (%)	68	63	65		
RF classification—0.014″ fault diameter					
	Precision	Recall	$F1$-score	Anomalies	Accuracy
NB (%)	62	69	65	17,093	63.68
IR (%)	65	59	62		
NB (%)	61	67	64	18,368	62.09
OR @ 3 (%)	63	57	64		

These findings suggest that the Autoencoder model is the most suitable choice for this study, indicating its superior performance.

Table 2 ANN results

ANN—0.007″ fault diameter					
	Precision	Recall	$F1$-score	Anomalies	Accuracy
NB (%)	75	92	83	9846	79.68
IR (%)	89	68	76		
NB (%)	89	98	83	3378	93.02
OR @ 3 (%)	98	88	93		
NB (%)	75	92	83	9396	80.61
OR @ 6 (%)	90	69	78		
NB (%)	68	88	77	12,849	66.05
OR @ 12 (%)	83	59	69		
ANN—0.021″ fault diameter					
	Precision	Recall	$F1$-score	Anomalies	Accuracy
NB (%)	76	90	83	9088	81.24
IR (%)	90	70	79		
NB (%)	74	88	81	10,293	78.76
OR @ 3 (%)	85	75	77		
NB (%)	67	90	77	13,105	72.95
OR @ 6 (%)	85	56	67		
NB (%)	61	87	72	16,449	66.05
OR @ 12 (%)	78	45	57		
ANN—0.014″ fault diameter					
	Precision	Recall	$F1$-score	Anomalies	Accuracy
NB (%)	59	84	69	17,956	62.95
IR (%)	72	42	53		
NB (%)	58	86	68	18,482	61.86
OR @ 3 (%)	73	38	50		

Table 3 Autoencoder results

Autoencoder results			
	Loss (MAE)	Anomalies	Accuracy
IR—0.007″ (%)	0.002460065	0	97.6
IR—0.014″ (%)	0.002744782	0	96.17
IR—0.021″ (%)	0.001431925	0	97.88
OR—0.007″ @ 3 (%)	0.001619012	0	97.02
OR—0.007″ @ 6 (%)	0.014085592	967	96.72
OR—0.007″ @ 12 (%)	0.011286878	182	94.61
OR—0.014″ @ 6 (%)	0.000175512	0	99.94
OR—0.021″ @ 3 (%)	0.000437233	0	99.98
OR—0.021″ @ 6 (%)	0.000638956	0	98.8
OR—0.021″ @ 12 (%)	0.000657117	0	98.72

Table 4 Comparative evaluation of accuracy

Comparison of the models based on accuracy			
	RF (%)	ANN (%)	Autoencoder (%)
IR—0.007″ (%)	85.36	79.68	97.6
IR—0.014″ (%)	83.9	81.24	96.17
IR—0.021″ (%)	63.68	62.95	97.88
OR—0.007″ @ 3 (%)	94.16	93.02	97.02
OR—0.007″ @ 6 (%)	83.86	80.61	96.72
OR—0.007″ @ 12 (%)	74.41	73.48	94.61
OR—0.014″ @ 6 (%)	62.07	61.86	99.94
OR—0.021″ @ 3 (%)	80.36	78.76	**99.98**
OR—0.021″ @ 6 (%)	83.86	72.95	98.8
OR—0.021″ @ 12 (%)	66.83	66.05	98.72

6 Conclusion

This study focused on investigating and exploring various ML and DL algorithms for anomaly prediction in bearing machinery. The three algorithms examined were RF classification, ANN, and Autoencoder, implemented on the CWRU dataset.

Our findings revealed that Autoencoders achieved higher accuracy compared to RF classification and ANN. Additionally, Autoencoders required less feature preprocessing compared to RF classification and ANN.

References

1. Zonta T, Da Costa CA, da Rosa Righi R, de Lima MJ, da Trindade ES, Li GP (2020) Predictive maintenance in the industry 4.0: a systematic literature review. Comput Ind Eng 150:106889
2. Schwendemann S, Amjad Z, Sikora A (2021) A survey of machine-learning techniques for condition monitoring and predictive maintenance of bearings in grinding machines. Comput Ind 125:103380
3. Boglietti A, Cavagnino A, Lazzari M, Pastorelli A (2002) A simplified thermal model for variable speed self cooled industrial induction motor. In: Conference record of the 2002 IEEE industry applications conference. 37th IAS annual meeting (Cat. No. 02CH37344), vol 2. IEEE, pp 723–730
4. Han Y, Song YH (2003) Condition monitoring techniques for electrical equipment—a literature survey. IEEE Trans Power Delivery 18(1):4–13
5. Wang P, Xu H, Ma H, Han H, Yang Y (2022) Effects of three types of bearing misalignments on dynamic characteristics of planetary gear set-rotor system. Mech Syst Signal Process 169:108736
6. Zhang S, Zhang S, Wang B, Habetler TG (2020) Deep learning algorithms for bearing fault diagnostics—a comprehensive review. IEEE Access 8:29857–29881
7. Khorram A, Khalooei M (2019) Intelligent bearing fault diagnosis with convolutional long-short-term-memory recurrent neural network. Appl Intell 51:736–751
8. Ademujimi TT, Brundage MP, Prabhu VV (2017) A review of current machine learning techniques used in manufacturing diagnosis. In: Advances in production management systems. The path to intelligent, collaborative and sustainable manufacturing: IFIP WG 5.7 international conference, APMS 2017, Hamburg, Germany, 3–7 Sept 2017, proceedings, Part I. Springer, pp 407–415
9. Seryasat OR, Haddadnia J, Arabnia Y, Zeinali M, Abooalizadeh Z, Taherkhani A, Tabrizy S, Maleki F (2012) Intelligent fault detection of ball-bearings using artificial neural networks and support-vector machine. Life Sci J-Acta Zhengzhou Univ Overseas Ed 9(4):4186–4189
10. Patel RK, Giri VK (2016) Feature selection and classification of mechanical fault of an induction motor using random forest classifier. Perspect Sci 8:334–337
11. Kamat P, Marni P, Cardoz L, Irani A, Gajula A, Saha A, Kumar S, Sugandhi R (2021) Bearing fault detection using comparative analysis of random forest, ANN, and autoencoder methods. In: Communication and intelligent systems: proceedings of ICCIS 2020. Springer, pp 157–171
12. Singha Roy S, Dey S, Chatterjee S (2020) Autocorrelation aided random forest classifier-based bearing fault detection framework. IEEE Sens J 20(18):10792–10800
13. Samanta BI, Al-Balushi KR, Al-Araimi SA (2003) Artificial neural networks and support vector machines with genetic algorithm for bearing fault detection. Eng Appl Artif Intell 16(7–8):657–665

Designing of GUI-Based Home Automation System Using Arduino and MATLAB

Priyanshu Baran, Rahul Guleria, Mothukuru Phaneendra, and Harpreet Kaur Channi

Abstract The home automation industry has attracted the researcher's favorable focus for over ten years. The plan's primary goal is to manage and monitor electronic and electrical household appliances mechanically. The home automation system has launched and developed numerous industrial and research versions. Many technologies have been incorporated into an effective home system. This paper's main objective is to suggest a method for monitoring and managing household equipment that interfaces MATLAB and a microcontroller. The solution suggested uses serial communication to connect a microcontroller to MATLAB to control home appliances. The status of household appliances will be displayed constantly on a GUI developed with MATLAB, thanks to an image acquisition device that is interfaced with the program. Appropriate instructing is carried out from the MATLAB GUI, and home appliances connected to the Arduino through relay boards can be turned ON or OFF.

Keywords Circuit board · ESP8266 · Four-channel relay · Motor driver (L293D) · Temperature sensor LM35 · DC toy motor · Bulbs · Bulb holders · Fan · Connecting wires

P. Baran · R. Guleria · M. Phaneendra · H. Kaur Channi (✉)
Department of Electrical Engineering, Chandigarh University, Mohali, India
e-mail: harpreetchanni@yahoo.in

P. Baran
e-mail: 21bel1068@cuchd.in

R. Guleria
e-mail: 21bel1018@cuchd.in

M. Phaneendra
e-mail: 21bel1048@cuchd.in

V. K. Gunjan and J. M. Zurada (eds.), *Proceedings of 4th International Conference on Recent Trends in Machine Learning, IoT, Smart Cities and Applications*, Lecture Notes in Networks and Systems 873, https://doi.org/10.1007/978-981-99-9442-7_50

1 Introduction

In recent years, the field of home automation has grown rapidly. Home automation technology involves centralizing and automating control of houses, buildings, and industries, including safety measures against sudden, unanticipated events. Home automation generally refers to using electronic equipment to control household activities, such as electrical appliances, lighting, heating, air conditioning, and security systems. The concept of home automation allows people to maintain or automate things around their homes remotely. Appliances are devices or instruments designed for a specific purpose, especially electrical devices for home use, such as fans. Today's world is one where automation rules supreme. Appliances can be controlled without human intervention, handling the most basic tasks, such as powering essential tools ON/OFF. This can all be done in the vicinity of these devices or remotely. Home automation has expanded quickly. Home automation centralizes and automates control of residences, buildings, and industries, including safety measures against unexpected incidents [1].

Electronic devices control home appliances, lighting, heating, air conditioning, and security systems in home automation. Control systems are now used in homes and businesses worldwide. Security systems protect assets and live today. A single control unit integrates many subordinate subsystems into the primary security system. Some subsystems involve Recon, imposter control, admission control, and fire detection. In a smart home, a "smart device" is electrical equipment that a smartphone, Bluetooth, or Internet can remotely control. Home automation lets you remotely control appliances and other objects, as shown in Fig. 1. Electrical appliances like fans and lighting are purpose-built. Devices can turn crucial tools on/off without human participation. These gadgets may be controlled remotely or locally. MATLAB GUI and esp8266 microcontroller control motors in this smart home automation system. This power electronics lab teaching tool is outstanding. The Arduino algorithm programs driving components during construction. After connecting the motors, the relay cards and driver need an external power source. esp8266 powers this drive. MATLAB GUIs control the motor panel. An esp8266 microcontroller serially interfaces with a MATLAB-designed board [2].

1.1 Need for Home Automation System

A large and growing market for technology has developed around home automation. With various products to make life easier and more convenient, automation services are now more affordable, allowing them to reach a wider audience. Home automation can provide safety, convenience, control, comfort, and energy savings. The following are some of the benefits that home automation can offer:

- Home automation increases safety and security.
- It is convenient.

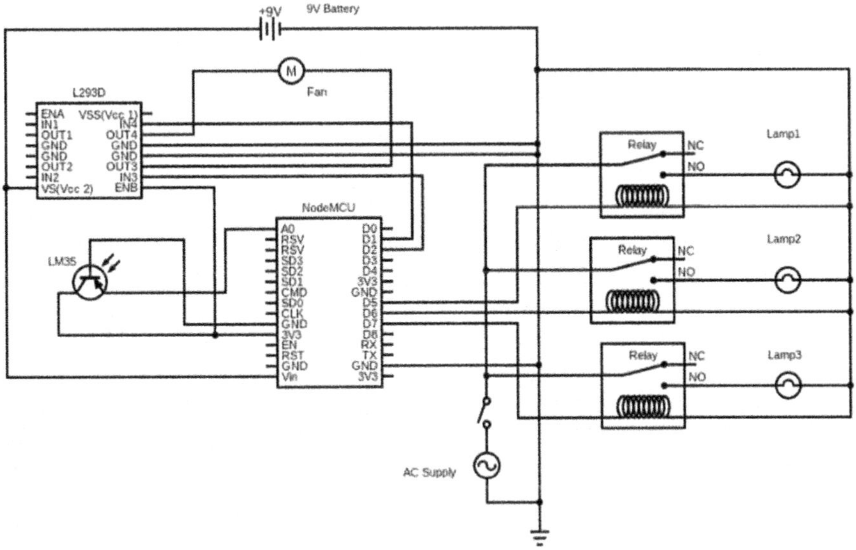

Fig. 1 Circuit diagram of home automation system

- It can offer data and control.
- Home automation creates comfortable routines.
- Significantly improves peace of mind.
- Home automation saves energy.

2 Literature Review

The Wi-Fi-based Home Management System is the first system discussed. It is made up of a server that hosts the website, an electronic interface module (which includes an Arduino Printed Circuit Board (PCB), a Wi-Fi shield PCB, input alarms PCB, and output actuators PCB), and actuators and sensors that govern and monitor the user's home. This approach is scalable and open to change, allowing remote users to connect to the server-based application through the Internet [1]. The second system under review is an Android-based Home Automation System. A gateway card was developed to ensure connectivity between the remote user, server host, Raspberry Pi card, and household appliances. To regulate the shutter of windows and another opening, the Android application on a smartphone sends signals to the Raspberry Pi card. This structure is efficient because it uses an interface card to update and modify signals between actuators, sensor devices, and the Raspberry Pi card [2]. The Cloud established Monitoring and Controlling System for Home Appliances is the third system discussed. It includes designing and setting up a gateway to collect metadata from household appliances and transfer it to a cloud-based data server host for storing and processing using MapReduce. The system allows remote users to monitor the system [3]. The Raspberry Pi-based Home Automation System is the fourth system discussed. The Raspberry Pi is utilized for reading the subject of emails and the algorithm. This system exhibits that it is a potent, economical, and efficient platform for executing smart home mechanization. It surpasses other home automation approaches, such as DTMF and web server-based home automation [4]. The fifth system discussed is the Smart House Monitor & Manager (SHMM). It is a ZigBee-based automation system, and all sensor devices and actuators are linked via the ZigBee communication channel. A modest smart socket is built for ZigBee remote operation, and a PC host serves as a collection device for data. The sensor data is uploaded to the cloud VM, which the end-user can monitor and control over the Internet [5]. The Arduino-based Home Automation System is the sixth system discussed. It collects user commands via an Ethernet shield through an Arduino microcontroller. The system uses smart scheduling of tasks with an exploratory for the Resource Constrained Scheduling Problem (RCPSP) and uses wired X10 and wireless ZigBee technologies. The system is highly adaptive, scalable, and expandable, and the portable device can be tethered to the chief controller or interface wirelessly with it. The Arduino IDE has a web-based server program that talks with web-built Android applications over the HTTP protocol [6]. The proposed system includes a home network that observes sensors and appliances, transmitting information to a cloud-based host server for user management through a mobile application [7]. It is cost-efficient with low power consumption and offers good modularity and configurability [8]. Various systems are being utilized more frequently to automatically control household appliances and devices, leading to a rising trend in home automation. As a result of these systems, household functions can be controlled with greater efficiency, safety, and convenience. Over time, the home automation system has been used effectively with different technologies and techniques. There are other home automation systems, such as voice-controlled home automation systems, where the system uses voice assistants, namely Alexa, Google Assistant, or Siri, to give commands to appliances. Smart home automation system uses a Bluetooth module, connecting the apparatus to Arduino and controlling it through Bluetooth. Home automation system using GSM, Bluetooth, IR, Android, wireless smart home automation system using IoT [9]. Similarly, with technological advancement, the automation system is developing by introducing new features. Home automation systems provide several benefits to users, such as efficient use of energy, time-saving, convenience, safety, and control over household appliances. These systems are being used more frequently and are easier to access, allowing more households to benefit from their advantages [10].

2.1 Problem Formulation

Nowadays, technology has surpassed all other factors in the globe. Everything we see around us is somehow linked to technology, significantly influencing it. We cannot imagine living without technology. However, as technology has advanced, its abuse has also increased. People's privacy and confidential data are vulnerable to theft. Our homes, where we secure our personal belongings, are no haven. However, we can protect and redefine our houses using the same technology. The problem is developing an effective and user-friendly MATLAB and Arduino home automation system. The home automation system includes features such as Arduino, GUI through MATLAB. The goal is to deliver a home automation system to help people control appliances from a particular position. In this project, we will research the most effective features for better designing GUI using MATLAB. The project's prime objective will be to develop GUI-established home automation through MATLAB and Arduino to control home appliances remotely. To achieve such a goal, a systematic step will be followed. All the libraries and tools in MATLAB required to link with Arduino will be downloaded. Then designing GUI using app design in MATLAB with various buttons to operate home appliances like lights, fans, etc. will be done. By coding, we have to link MATLAB with Arduino and the devices, and then our model will be complete. The unnecessary creation of complex features in the design will be avoided. Testing before implementation of the model is mandatory.

2.2 Objectives of the Work

- Designing of feasible and robust GUI-based Home Automation System.
- Monitoring and regulating of household appliances from distant point.
- Using MATLAB app designer to make an application.
- Using a cloud-based service to link the Automation system remotely.

3 Methodology

As home automation technology grows, various systems are being used to automate the control of household appliances and devices.

- ***Voice-controlled home automation***

Home automation based on voice control has a unique feature that allows the user to control home appliances using voice commands like google assistant, Alexa, or Siri. It is quite useful as we use voice assistants for various purposes and can be efficiently used while doing any work.

- ***MATLAB-based industrial computerization automation system***

The MATLAB-based industrial computerization automation system is mostly used in industries. This system uses MATLAB's GUI tool to create the interface, programming, and various sensors and signals generated by communication technologies like GSM, RF Transmitter, RFID, and DTMF for controlling multiple loads.

- ***Home automation using Arduino and Bluetooth***

The users of this system can operate equipment with the help of a smartphone application by connecting to a Bluetooth module connected to appliances through Arduino. As Bluetooth is used for many purposes and for controlling, we do not require any application to be downloaded additionally, as we can easily access it through Bluetooth, which is available on all mobiles and laptops. The features that are required in this project are:

- A friendly user graphical user interface using App design in MATLAB so that everyone can easily control without any problem.
- Through a Wi-Fi connection, a wireless connection between GUI and the Esp8266 (microcontroller).
- A feature that senses the room temperature.
- A feature that also regulates the speed of the fan.

3.1 Analysis of Features and Subject Finalization to Constraints

The MATLAB-based Home Automation System we designed works on active internet/ wireless hotspots. It requires continuous, uninterrupted connectivity; otherwise, the connection between Adafruit cloud service will be broken with Esp8266. As the system works on a good Internet connection, thus a 4G or 5G connection will result in a faster response time. The Esp8266 requires a 9 V DC voltage source for operation. Incessant use will rapidly discharge the battery, ceasing the Esp8266 to work. We have used commonly procurable electronic components that are very economical. They are very robust and can withstand minor difficulties easily. As most of the essential part of the project is covered in computerized software, the odds of the system failure can be easily fixed by going through the software and not dealing with the hardware part.

3.2 Design Selection

The three processes used in the design flow are the different ways to proceed with home automation. By comparing the three processes, process-3 is the best and most efficient than the other two processes. The reasons are

- In the first process, we do not use any IoT platforms, so it always requires connecting the microcontroller cable to the laptop to give the inputs, as each time we cannot join, the microcontroller may get damaged.
- In process-3, we use the Adafruit IO IOT platform, which has an MQTT protocol that is best suited for microcontrollers and gives us less delay time for each input.

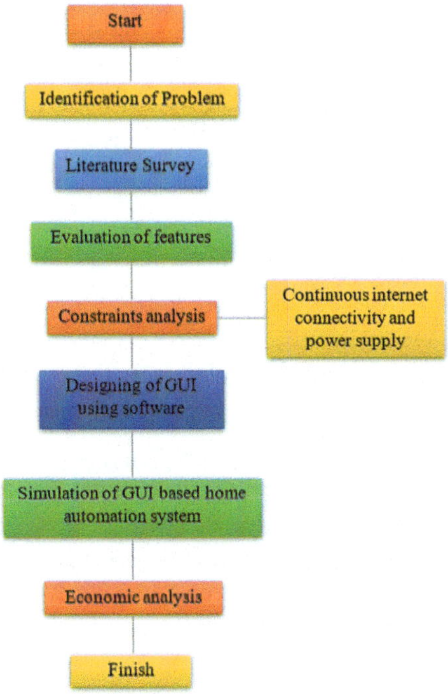

Fig. 2 Methodology flow chart

- This MQTT through Adafruit IO can be easily linked to MATLAB.
- In process-2, we use the Thingspeak IOT platform, which also has MQTT protocol, but it isn't easy to link MQTT to MATLAB through Thingspeak.
- Thingspeak with http protocol has a fixed delay time of approximately 10 s for whenever we give the input. Figure 2 shows the methodology to achieve the objectives based on the above reasons.

4 Working of the System

MATLAB stands for MATRIX LABORATORY. It is a programming platform that MathWorks develops. It is the most popular software for analyzing and designing systems. MATLAB provides many features, as it can also interact with other systems, applications, or IOT platforms. It also provides tools to create a friendly graphical user interface. Arduino IDE stands for Arduino Integrated Development Environment. It is open-source software that any operating system can use. The Arduino IDE provides a friendly user interface to write and upload the code to any Arduino board. This software also contains many libraries that can be installed within the software, allowing one to write and upload code to any other boards other than Arduino boards like ESP-32 boards and Raspberry Pi. Adafruit IO is an IOT platform created and developed by Adafruit Industries. Adafruit is a cloud-based service allowing users to

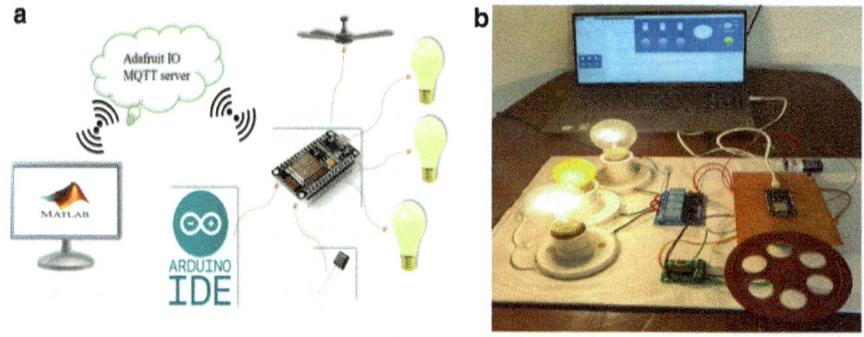

Fig. 3 **a** Design flow of the system. **b** Design flow of the system

connect their devices to the Internet. Data can be stored, retrieved, and can control the devices from anywhere if the Internet is available, as shown in Fig. 3a. The platform supports multiple connectivity protocols, such as MQTT, HTTP, and WebSocket, making connecting various sensors and devices convenient, as shown in the prototype in Fig. 3b.

5 Conclusion

The graphical user interface is created using the app design tool in MATLAB. The appliances are turned according to the input given through GUI. The temperature is shown in the GUI, which the temperature sensor senses. The project "Designing of GUI-based Home Automation System using Arduino and MATLAB" is designed to produce a system that allows the user to operate electrical appliances like lights and fans at home. In this paper, we presented designing a GUI-based home automation system that uses MATLAB, Arduino IDE, ESP8266, and Adafruit IO to offer a user-friendly interface. The proposed system would enable people who cannot move to improve their quality of life by increasing mobility and independence. This system is greatly helpful for those people who are unable to move as this system provides an interface to operate appliances remotely. With this system, millions of people with mobility impairments worldwide could have an affordable option to improve their mobility.

References

1. Gebhardt J, Massoth M, Weber S, Wiens T (2014) Ubiquitous smart home controlling Raspberry embedded system. In: UBICOMM: the eighth international conference on mobile ubiquitous computing, systems, services and technologies
2. Zanella A, Bui N, Castellani A, Vangelista L, Zorzi M (2014) Internet of Things for smart cities. IEEE Internet Things J 1(1)
3. ElShafee A, Hamed KA (2012) Design and implementation of a WiFi based home automation system. Int J Comput Electr Autom Control Inf Eng 6(8)
4. Lamine H, Abid H (2014) Remote control of a domestic equipment from an Android application based on Raspberry pi card. In: 15th international conference on sciences and techniques of automatic control & computer engineering—STA'2014, Hammamet, Tunisia, 21–23 Dec 2014
5. Cui Y, Kim M, Gu Y, Jung J-j, Lee H (2014) Home appliance management system for monitoring digitized devices using cloud computing technology in ubiquitous sensor network environment. Int J Distrib Sens Netw 2014, Article ID 174097
6. Sarthak J, Anant V, Lovely G (2014) Raspberry Pi based interactive home automation system through email. In: 2014 international conference on reliability, optimization and information technology (ICROIT 2014), India, 6–8 Feb 2014
7. Tseng S-P, Li B-R, Pan J-L, Lin CJ (2014) An application of Internet of Things with motion sensing on smart house. IEEE. 978-1-4799-6284-6
8. Baraka K, Ghobril M, Malek S, Kanj R, Kayssi A (2015) Low-cost Arduino/Android-based energy-efficient home automation system with smart task scheduling. In: 2013 fifth international conference on computational intelligence, communication systems and networks
9. Baraka K, Ghobril M, Malek S, Kanj R, Kayssi A (2015) Smart power management system for home appliances and wellness based on wireless sensors network and mobile technology. In: 2015 XVIII AISEM annual conference. IEEE. 978-1-4799-8591-3
10. Kumar S (2014) Ubiquitous smart home system using android application. Int J Comput Netw Commun (IJCNC) 6(1)

Recent Trends of Information Retrieval System: Review Based on IR Models and Applications

Vaishali Patel, Dilendra Hiran, and Kruti Dangarwala

Abstract Access to data on the web is incredibly increasing now a day with the expansion of web-based applications. Many traditional and proposed information retrieval models and systems are available for information retrieval from the extensive collection of documents on the web. These models and procedures don't consider exact information about the actual user and search content. This paper discusses the importance of information on IR on the web. Also, discuss the importance of various traditional information retrieval models with their pros and cons to enhance the research area for future research work. This paper also briefs on information retrieval, types of data used in the IRS, IRS process, and preprocessing of IR. The article also discusses the literature survey of information retrieval based on the system, traditional models, recent trends, and applications of IR. Current IR models suffer from significant problems of not having accurate information retrieval. This paper focuses on the need for various machine learning techniques, including advanced ML (deep learning), to overcome the exact and actual retrieval of information from a vast heterogeneous collection of databases, including various web-based applications, by improving performance.

Keywords Information retrieval system · IR models · Deep learning · Machine learning

V. Patel (✉) · D. Hiran
PAHER University, Udaipur, Rajasthan, India
e-mail: vaishalirajpatel@gmail.com

K. Dangarwala
SVM Institute of Technology, Bharuch, Gujarat, India
e-mail: kruti.dangarwala@svmit.ac.in

V. K. Gunjan and J. M. Zurada (eds.), *Proceedings of 4th International Conference on Recent Trends in Machine Learning, IoT, Smart Cities and Applications*, Lecture Notes in Networks and Systems 873, https://doi.org/10.1007/978-981-99-9442-7_51

1 Introduction

Recently, usage and access to the Internet have been incredibly increasing, including vast amounts of data. Information retrieval retrieves the relevant information of a user query submitted to the web. Internet is facing a problem of information overloaded, including retrieval. The IR scans all the documents from the database and then assigns scores and ranks each document according to the relevance degree to the user query to display results to the user. But it requires a long run time. Information retrieval (IR) stores, retrieves, and represents the data as knowledge discovery against the user query. IR plays an essential role in developing computer applications for various fields, generating effective and accurate information to satisfy user requests. IR is differentiated from the DB structure as it uses documents or other formats to store the data. Information retrieval is the leading research domain that searches and retrieves documents with indexing, ranking, and classification of the retrieved documents, which is used by the web search engines like Google. People use many web-based applications in their daily lives, which generate lots of data on the web. Information retrieved based on user query using brute force and linear search, which match with the ranked documents [1].

IRS must convert the retrieved information from unstructured, structured, or semi-structured documents into meaningful and valuable information to fill the research gap in information retrieval. Advanced mathematical techniques known as deep learning can overcome the said problem of IR [2]. The retrieval process retrieves relevant information to the user query by applying various steps, including filter, search, match, and rank of retrieved documents. To fulfill the agenda, IRS uses the process order including index, filter, and search, as shown in Fig. 1.

The following points discuss the types, preprocessing, and procedure of IR.

- Various techniques are available in IR depending on the different data types like textual, image, audio, and video [3]. Unstructured data can be in various formats such as textual, image, audio, video, and more and retrieved from different data sources like web search engines, social media, geographical, and more [4, 5].
- Preprocessing is the first, most important, and critical step of data analysis or mining. This section discusses phases of preprocessing of documents for information retrieval such as document parsing, tokenizing, stemming, and feature extraction [6].
- Web document is pass to the various steps of preprocessing as discussed above and prepare the document to generate a pattern or information which match with the user query as shown in Fig. 2 [7].

Now a day, IR systems are in the most demand in various areas, including industries, academics, and public and government sectors, due to the increased volume of data on the web [8]. Some examples of IRS applications are web search engines and media searches [8, 9].

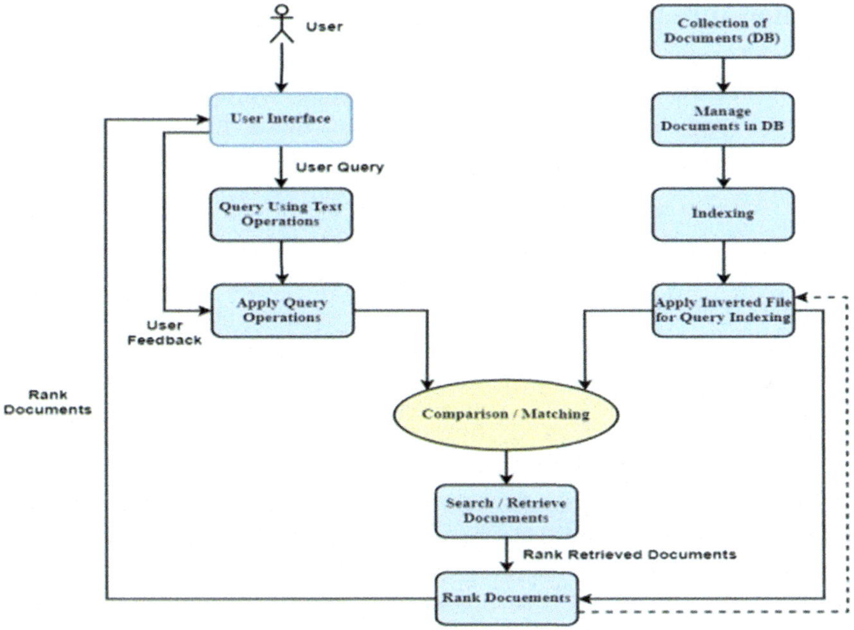

Fig. 1 Process diagram of information retrieval system

Fig. 2 Pro-processing steps of information retrieval

2 Literature Review of IR

Information retrieval process is to retrieve data which meets the user requirements in the form of query form huge collection of database. Researchers define information retrieval differently. Literature survey discusses the different meaning of information retrieval.

2.1 Review-Based IRS

As per the authors, users use information retrieval techniques to translate user needs or queries into the exact and actual information or documents from the large collection. It is a search and retrieval process for the stored information in collections [10]. According to Meenakshi Sinha, the information retrieval process is applied to search, retrieve, recover, and interpret the information based on the user request from the large collection of databases [11]. As per the authors, IR retrieves the data based on the user query or need from the extensive data collection. The researcher may generate useful and previously unknown knowledge using appropriate data analysis techniques against the user query, overcome the search limitations, and retrieve exact information [12]. As per McGill and Salton, information retrieval is defined as a system that stores information to be searched, processed, retrieved, and represented to the users based on their request [13]. Bouramoul's proposed work results show that considering the context of information retrieval increases the relevancy of the generated or retrieved information and reduces the noise or outliers [14]. Originally, information retrieval systems and fields were innovated in 1950 [15].

2.2 Review-Based IR Models

As per the authors, a Boolean model is based on the set of Boolean theory and algebra. It is a collection of objects, and each object is a part of this collection. It is a collection of logical and binary operators [16]. As per Dong and Chang's survey, a Boolean model is simple with less complexity, but fuzzy and semantic matches are not considered due to binary criteria. Also, it is difficult for complex user queries without ranking [17]. Authors have analyzed and evaluated the effectiveness of the Boolean method based on the number of documents retrieved, time taken for the retrieval, and relevancy of the retrieved documents based on the user query. Also, a Boolean search of string generates more relevant documents than the text-based query, which generates a huge set of documents with some irrelevancy [18]. Waller and Kraft presented a mathematical model with a weighted Boolean retrieval system that evaluates the document's relevancy against a user query [19]. Salton et al. have proposed and developed a Boolean search retrieval model with precision calculation [20].

Salton proposed a model where each document is considered a vector, and each dimension of the vector is assigned with respect to the indexed term [21, 22]. Once the term exists in the document, then the weight is assigned to the corresponding dimension of the vector. The query can also be considered as a vector with respect to indexed terms. Hence, the relevancy between a query and a document is considered as the cosine of the angle between the two vectors. As per Ribeiro and Muntz, a probabilistic model is applied to improve the probabilistic description of the accurate retrieval from the set of documents based on the user query [23]. Authors survey the vector space model where tf-idf term weighting with partial matching is used to improve the recall of information retrieval. But indexing of dynamic documents is difficult to maintain and also affects the performance of the information retrieval. As per Zhai, the similarity-based model assumes the relevancy status of the document based on the user query is related to the similarity between the query and the document. Any similarity measure can be used to correlate documents as well as to rank the document [24]. As per Salton, document and user query are considered as two different term vectors in multi-dimensional term space in vector space model, where weight is assigned to each term respectively a query. As per the authors' survey, the vector space model is based on some principles which represent the document or text in many areas [25]. This model splits, filters, and classifies the document, which is more abstract and statistics [26].

2.3 Review Based on Current Trends and Applications of IR

Authors clarify the confusion of synonymous concepts of IR like vagueness, uncertainty, inconsistency, and imprecision by suggesting an electronic library where researchers and librarians retrieve relevant information from the collection of documents. Authors noted in spite of the tremendous need for the web for information retrieval, a major issue of the semantic web which explores the IRS to model and manage the semantics of user queries and retrieved documents [27]. As per Belkin and Croft, Instead of partial matches of information, only documents that match exactly with the user query are retrieved. So, it is necessary to have good domain knowledge or information to write effective user queries to generate an effective or exact match of the retrieved information. The authors noted that documents that are retrieved against of the user query are not ranked. Also, the Boolean model retrieves so many or fewer documents based on a user query [28]. Gao stated that the IR model does not use the term weight. It is treated similarly if it occurs once or many times in the retrieved document [29].

IR fills the gap between the generated knowledge and retrieved information by managing various operations, including organization, storage, representation, and maintenance of the information. The performance of this process may be improved by applying machine learning techniques and algorithms [30]. Machine learning techniques and algorithms can be classified as classification and clustering [31]. Use of the Internet is increasing day by day, so data is also increasing, which is not handled efficiently. Classification algorithms rank the relevant documents better than IR models. Still, it is difficult to retrieve appropriate information from the large collection due to the highly increasing growth of data on the web. Authors have proposed a method for Boolean query expansion by applying labeled and ranking of retrieved documents to achieve effective information retrieval [32]. Diaz et al. proposed a query expansion model which an unconstrained amount of data [33]. Authors have proposed a medical diagnostic-based model which processes the medical data, including various clinical activities which retrieve the same nature of data [34]. This model classifies clinical data by applying a decision tree in conjunction with a random forest algorithm for better results. Song, He et al. have also proposed a model based on two re-ranking for clinical system [35]. Kumar et al. proposed a method where classification techniques are applied on image data where a label is assigned to each image, and labeled images are recursively divided into vertical and horizontal partitions with great accuracy [36]. Authors proposed a method where information retrieval was in the form of the image where a random forest algorithm is used [37].

Sathya et al. proposed a search engine based on intelligent clusters where a K-means clustering algorithm is applied [38]. But this model did not generate accurate results due to improper clustering of the retrieved web pages. The authors used proposed a method where ranking techniques are improved for cross-lingual information retrieval [39]. Authors applied a re-ranking for automation of answers to complicated user queries [40]. Hore and Bhattacharyya applied a machine learning algorithm to classify the suggestions from the user across India for the topic "save the girl child" under the slogan "Beti Bachao Beti Padhao" [41]. Authors have suggested Naive Bayesian and KNN, respectively, to improve the clinical analysis process of information retrieval [42, 43]. An unsupervised learning technique, including statistics, was applied to musical data of various categories to retrieve relevant features as per authors [15], deep learning is suggested in conjunction with IR to overcome the limitations of traditional machine learning algorithms, which may not retrieve efficient and accurate information in case of huge and complex data. As per the authors, deep learning techniques RNN with LSTM cell extracts the information from each word of the sentence [44]. Blaschke et al. applied deep learning to overcome the limitations of machine learning techniques to design and retrieve clinical data [45]. Authors show that classification algorithms generate better results of IR applications, including music data retrieval, clinical data retrieval, image data retrieval, and more, including various areas like medical, digital library, social media, and music systems [15]. Deep learning is also an enhancement of the machine learning technique, which is more suitable for retrieval from high and complex databases.

3 Performance Measures for IR System

Effective information retrieval means a measure of the actual retrieved documents against of the user query in the system. The measure to compute the effectiveness of the information retrieve against of the user query in considered as effectiveness evaluation of IR. Most widely popular effectiveness measure of IT system is precision and recall where prominent precision is obtained through subordinate of recall. High recall with low number of prominent precision of relevant documents is returned against of the user query by effective IRS. So, many effective IRS uses factor (F1) as a measure of precision and recall combination.

4 Traditional IR Models

Historically, the term information retrieval was invented by Calvin Mooers in 1950. Since then various approaches of information retrieval have been proposed and discussed by many researchers. Some of these approaches are popular as classical or traditional information retrieval models. Later on many extended and alternative IT models based on these classical models have been proposed and discussed to overcome the issues of the existing traditional models of IR with more scope and improvement in the existing models. Objective of the IR model is to search and retrieves relevant and exact information that matches user query from large collection of documents. Types of classical IR models are: Boolean, Vector, and Probabilistic.

4.1 Boolean Model

Boolean model is the first traditional model and novel approach for information retrieval from large collections of documents. This model works using logical functions (operations) for the user query submitted to retrieve appropriated information from the collections. In Boolean model, documents are represented by a set of terms which is known as index terms [20]. User queries submitted are converted into Boolean expressions to retrieve the information from the database the collection. Boolean model is used view and display each document as just of set of words from collection of documents. Boolean retrieval model is simply used to retrieve information from large collection of documents. Boolean model is one of the simplest and oldest model for IR which use classical set of theory and Boolean algebra operations [23]. This model represents the query and retrieved documents as a set of indexed word or term. This model has some cons including: only actual matching possible, partial matching is not possible, representation of query language is more complex and retrieved documents are not properly ranked.

4.2 Vector Space Model

Document is the collection of database is represented by the vector traditional model where each word is associated with the document and value of each word represents the importance of the word of the document. Retrieved ranked documents focus on the inability to retrieve complete match and any importance information. Existing vector space traditional approach contains rules and regulations to represent the document in the field of latest research era including pattern recognition. This approach divides, identify and partitions the submitted documents which represent the documents based on various statistics in abstract form.

4.3 Probabilistic Model

This model represents documents from the collection are represented based on the assumptions which may be relevant or non-relevant against of the specific user query. Probabilistic classical model is based on the theory and concepts probabilistic ranking principal. This model retrieves the documents based on the probabilistic operations and Bayes theorem which match the specific documents with weight values of the multi values words against of the user request.

5 Issues, Challenges, and Current Trends of IR

The primary issue of traditional IR models is high-quality information retrieval due to the selection of search vocabulary, search strategy formulations, and information overload on the web. The critical issues identified with the traditional IR model are summarized as follows: Information on the documents which may have a partial or sub-string match with the query is not retrieved. So, good and proper knowledge of the domain is necessary to form appropriate and effective user queries for a massive set of documents that need not be ranked, which results in the effective retrieval of information [29]. Also, for a large set of documents, traditional models retrieve either too many documents or few documents. An extended Boolean retrieval model was also proposed to overcome the main drawback of the basic Boolean conventional model, which is based on the Boolean expression over terms.

Vector classical model addresses the significant issues of Boolean models: partial fetch matching and absence of procedure to rank the retrieved documents. But in the vector space model, index terms of documents for information retrieval are independent of each other. Also, this model does not retrieve the semantics in the retrieved documents based on the user's request. And computation cost is comparatively high due to extensive collections of documents on the web. Retrieval and latent semantic index are based on algebraic methods which reduce the dimensions to overcome the limitation of high computing resource requirements. The probabilistic model provides the right technique to manage the improbability which occurs in elements retrieved in the set of documents collected comparatively Boolean and vector IR models. But this method is based on many assumptions which may have many side effects, and it is difficult to apply and implement real-world applications based on a large scale. Users also not getting accurate and exact information due to inaccurate and uncertainties in languages. The issue of IR, which is not retrieving precise and detailed information, can be overcome using various machine learning techniques, including classification. These days, usage of the Internet is increasing. So, data is growing tremendously on the web in multiple forms, including unstructured, structured, and semi-structured. And it becomes difficult to retrieve exact and accurate information from this massive collection of databases. Many researchers have started to use advanced computational techniques of machine learning: deep learning, which handles this enormous collection of unstructured, structured, and semi-structured databases and is applied to many web-based applications to solve real-world issue [30].

6 Conclusion

The primary goal of information retrieval is to retrieve the actual or exact matching of the documents from a set of extensive collections from the database. As we know, the increasing use of data on the Internet requires high-quality and accurate information retrieval. After studying and analyzing the different IR models of information retrieval systems, the vector space classical model is a frequently used model, and its results strictly depend on the word or term weight. But this method has the following two significant issues: term independence and term weighting. Based on the problems discussed, we need to propose and develop a more robust and exact information retrieval system that can produce outcomes that are superior to those of Boolean models without involving ourselves in the intricate computations of the probabilistic model. This article discusses and compares traditional IR models with their objectives, pros, and cons. It also identifies various issues and challenges in this particular sphere. With the increasing need and retrieval of data, including high-quality, robust, and accurate retrieval results, more than classical information retrieval methods are needed. Researchers have started to apply various machine learning techniques to retrieve precise and exact matches of retrieved documents from the massive collection of databases. But due to the incredible growth of the heterogeneous data, ML techniques may not recover the desired information. So, researchers have started to apply deep learning techniques to retrieve desired and exact information from the enormous volume of data, which improves the performance of many web-based applications.

References

1. Greengrass E (2000) Information retrieval: a survey. Technical Report
2. Kanimozhi S, Devi B (2018) A novel approach for deep learning techniques using information retrieval from big data. Int J Pure Appl Math 118:601–606
3. Sharma T, Bajaj A, Sangwan O (2022) Deep learning approaches for textual sentiment analysis. Research anthology on implementing sentiment analysis across multiple disciplines, pp 256–267
4. Adnan K, Akbar R (2019) An analytical study of information extraction from unstructured and multidimensional big data. J Big Data 6:1–38
5. Wang Y, Kung L, Byrd T (2018) Big data analytics: understanding its capabilities and potential benefits for healthcare organizations. Technol Forecast Soc Chang 126:3–13
6. Dzisevič R, Šešok D (2019) Text classification using different feature extraction approaches. In: 2019 open conference of electrical, electronic and information sciences (eStream), pp 1–4
7. Lomotey R, Jamal S, Deters R (2012) SOPHRA: a mobile web services hosting infrastructure in mHealth. In: 2012 IEEE first international conference on mobile services, pp 88–95
8. Roshdi A, Roohparvar A (2015) Information retrieval techniques and applications. Int J Comput Netw Commun Secur 3:373–377
9. Ibrihich S, Oussous A, Ibrihich O, Esghir M (2022) A review on recent research in information retrieval. Procedia Comput Sci 201:777–782
10. Hjørland B (2021) Information retrieval and knowledge organization: a perspective from the philosophy of science. Information 12:135
11. Sinha M (2018) Information retrieval and its legal impact on the society. Hidayatullah National Law University Raipur. http://www.legalserviceindia.com/articles/inret2.htm
12. Marcos-Pablos S, Garcıa-Peñalvo F (2020) Information retrieval methodology for aiding scientific database search. Soft Comput 24:5551–5560
13. Gondaliya T, Joshi H (2017) Journey of information retrieval to information retrieval tools—IRIRT a review. INFLIBNET Centre
14. Bouramoul A (2016) Contextualisation of information retrieval process and document ranking task in web search tools. Int J Space-Based Situat Comput 6:74–89
15. Singhal A (2001) Others modern information retrieval: a brief overview. IEEE Data Eng Bull 24:35–43
16. Baeza-Yates R, Ribeiro-Neto B et al (1999) Modern information retrieval. ACM Press, New York
17. Dong H, Hussain F, Chang E (2008) A survey in traditional information retrieval models. In: 2008 2nd IEEE international conference on digital ecosystems and technologies, pp 397–402
18. Aliyu M (2017) Efficiency of Boolean search strings for information retrieval. Am J Eng Res 6:216–222
19. Waller W, Kraft D (1979) A mathematical model of a weighted Boolean retrieval system. Inf Process Manag 15:235–245
20. Salton G, Fox E, Wu H (1983) Extended Boolean information retrieval. Commun ACM 26:1022–1036
21. Salton G (1971) The SMART retrieval system experiments in automatic document processing. Prentice-Hall, Inc.
22. Salton G, Lesk M (1968) Computer evaluation of indexing and text processing. J ACM (JACM) 15:8–36
23. Ribeiro B, Muntz R (1996) A belief network model for IR. In: Proceedings of the 19th annual international ACM SIGIR conference on research and development in information retrieval, pp 253–260
24. Zhai C (2007) A brief review of information retrieval models. Urbana-Champaign: UIUC
25. Tang M, Bian Y, Tao F (2010) The research of document retrieval system based on the semantic vector space model. J Intell 5:167–177
26. Yu B (2019) Research on information retrieval model based on ontology. EURASIP J Wirel Commun Netw 2019:1–8

27. Kraft D, Bordogna G, Pasi G (2018) Fuzzy set theory. Encyclopedia of library and information sciences, pp 1618–1635
28. Faridi A, Zafar A (2012) Trends and issues in modern information retrieval. J Adv Sci Technol 3
29. Gao K, Wang Y, Wang Z (2004) An efficient relevant evaluation model in information retrieval and its application. In: The fourth international conference on computer and information technology, 2004. CIT'04, pp 845–850
30. Bajaj A, Sharma T, Sangwan O (2020) Information retrieval in conjunction with deep learning. Handbook of research on emerging trends and applications of machine learning, pp 300–311
31. Alzubi J, Nayyar A, Kumar A (2018) Machine learning from theory to algorithms: an overview. J Phys: Conf Ser 1142:012012
32. Kim Y, Seo J, Croft W (2011) Automatic Boolean query suggestion for professional search. In: Proceedings of the 34th international ACM SIGIR conference on research and development in information retrieval, pp 825–834
33. Diaz F, Mitra B, Craswell N (2016) Query expansion with locally-trained word embeddings. ArXiv Preprint arXiv:1605.07891
34. Pollettini J, Pessotti H, Pazin Filho A, Ruiz E, Junior M (2015) Applying natural language processing, information retrieval and machine learning to decision support in medical coordination in an emergency medicine context. In: 2015 IEEE 28th international symposium on computer-based medical systems, pp 316–319
35. Song Y, He Y, Hu Q, He L (2015) ECNU at 2015 CDS track: two re-ranking methods in medical information retrieval. In: Proceedings of the 2015 text retrieval conference
36. Kumar J, Ye P, Doermann D (2012) Learning document structure for retrieval and classification. In: Proceedings of the 21st international conference on pattern recognition (ICPR2012), pp 1558–1561
37. Fu H, Qiu G (2012) Fast semantic image retrieval based on random forest. In: Proceedings of the 20th ACM international conference on multimedia, pp 909–912
38. Sathya M, Jayanthi J, Basker N (2011) Link based K-Means clustering algorithm for information retrieval. In: 2011 international conference on recent trends in information technology (ICRTIT), pp 1111–1115
39. Azarbonyad H, Shakery A, Faili H (2019) A learning to rank approach for cross-language information retrieval exploiting multiple translation resources. Nat Lang Eng 25:363–384
40. Herrera J, Poblete B, Parra D (2018) Learning to leverage microblog information for QA retrieval. In: Advances in information retrieval: 40th European conference on IR research, ECIR 2018, Grenoble, France, 26–29 Mar 2018, proceedings, vol 40, pp 507–520
41. Hore S, Bhattacharya T (2019) Analyzing social trend towards girl child in India: a machine intelligence-based approach. In: Recent developments in machine learning and data analytics: IC3 2018, pp 43–50
42. Lo Y, Rensi S, Torng W, Altman R (2018) Machine learning in chemoinformatics and drug discovery. Drug Discovery Today 23:1538–1546
43. Lo Y, Senese S, Li C, Hu Q, Huang Y, Damoiseaux R, Torres J (2015) Large-scale chemical similarity networks for target profiling of compounds identified in cell-based chemical screens. PLoS Comput Biol 11:e1004153
44. Lu Y, Wu C, Lu C, Lerch A (2016) An unsupervised approach to anomaly detection in music datasets. In: Proceedings of the 39th international ACM SIGIR conference on research and development in information retrieval, pp 749–752
45. Palangi H, Deng L, Shen Y, Gao J, He X, Chen J, Song X, Ward R (2016) Deep sentence embedding using long short-term memory networks: analysis and application to information retrieval. IEEE/ACM Trans Audio Speech Lang Process 24:694–707

S Band Tuned Graphene-Based Reconfigurable Antenna

G. Sivarama Subramanium, Suresh Kumar Muthuvel, A. Deepak, and G. Anitha

Abstract In this paper, we fabricated a reconfigurable rectangular patch antenna which is graphene PVDF reinforced composite microstrip antenna. The diodes used to shift the frequencies continuously using varactor diodes, mainly to shift frequency to another frequencies. By changing the value of the capacitance ranging tuning the antenna frequencies from 1.95 and 4.25 GHz frequency. The frequency shifting take place between different capacitance values of higher (1.95 GHz), mid (3.60 GHz), and lower (4.2 GHz).

Keywords Bakelite graphene substrate material · Rectangular patch · Varactor diodes · Frequency tuning · Antenna refection coefficient · Reconfigurable antenna's frequency

1 Introduction

Reconfigurable antennas are essential in wireless, mobile, and s band communications. The light weight patch antennas, less expensive, and both the planer surfaces have uniformity. The efficiency can be improved by increasing the antenna [1].

- The antenna extracts electrical signal from radio waves and vice-versa.
- For all particular submissions, the radiation pattern are most important factor for any antenna.
- There are different types of antennas, which it includes microstrip, loop, Horn, dipole, and patch antennas.
- The Bakelite substrate forms a ground plane and a feeding part of the antenna.

G. Sivarama Subramanium (✉) · S. Kumar Muthuvel · A. Deepak · G. Anitha
Saveetha School of Engineering, Chennai, India
e-mail: gsivaram47@gmail.com

A. Deepak
e-mail: deepakarun@saveetha.com

V. K. Gunjan and J. M. Zurada (eds.), *Proceedings of 4th International Conference on Recent Trends in Machine Learning, IoT, Smart Cities and Applications*, Lecture Notes in Networks and Systems 873, https://doi.org/10.1007/978-981-99-9442-7_52

- Patch antennas are lightweight, inexpensive, and simple to manufacture, and they are used in a wide range of wireless communication applications. Patch antennas are constructed with a ground plane, microstrip patch element, and a Bakelite substrate.
- Graphene which is sp-hybridized, is a carbon allotrope with extreme conductivity [2–4].
- On exposing to current over radio range of frequency, the graphene absorbs, and radiates them more effectively.
- The reconfigurable antennas are prevent to emit very less co2 and more advantage than other common types of antennas.
- The reduction of device's cost and hardware complexity is reduced by the interchangeable in the antenna. The varactor diode can be used to control the reversible antennas [5].
- The dual switches with two-fold set of feed and four double throw switches, regulates the variable impedance bandwidth of the antenna, which ranges from 1.95 to 4.2 GHz.
- The feed of the antenna [6] in which the patch and impedance bandwidth complement each other well.

The value of capacitance of varactor diode are adjusted over 7.40–0.74 pF to get the frequency tuning over the frequency range of 1.95–4.2 GHz.

- The antenna of rectangular microstrip have adjustable frequency of about (1.95–4.2 GHz) [7]. A varactor diode and a ground plane were used to tune the frequency indefinitely [8].

Using a BAR smv2022 varactor diode [9]. The reflection coefficients of the rectangular patch are 1.95–4.2 GHz with a lower value of capacitance of 0.73 pF, an inductance of 1.5 nH and resistance value of 2.2 Ω, as well as a higher capacitance of 7.41 pF. The capacitance of the varactor diodes is changed to get the reconfigurable state of the antenna.

The proposed work is modeled in HFSS-V15 (Ansoft/Ansys) [10, 11] and wave analysis can also be done [12].

The varactor diode equivalent circuit with biasing having the voltage source in series to inductance $L1$, $L2$, resistance $R1$, and capacitance $C1$, $C2$ in which biasing take place in the diode and the equivalent circuit is as follows (Fig. 1).

A reflection coefficient (RC) of − 20 dB or higher is desirable, but − 10 dB and lower is undesirable. An antenna's normal operation needs a reflection coefficient of around − 8 dB and it is find out by dividing the amplitude of the reflected wave by the amplitude of the incident wave.

$$\mathrm{RC} = (Z_r - Z_\ell)/(Z_r + Z_\ell)$$

where

$Z_r - Z_\ell$, amplitude of reflected wave,
$Z_r + Z_\ell$, amplitude of incident wave.

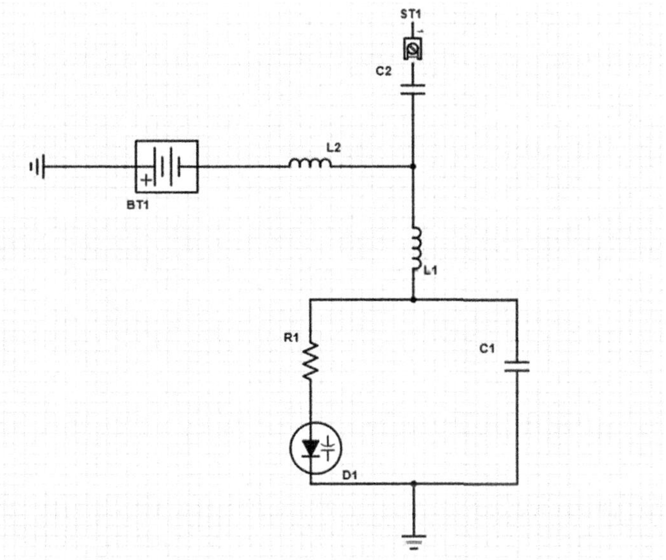

Fig. 1 Equivalent circuit diagram for the varactor diode biasing

The antenna gain G is measured by multiplying the efficiency and directivity of those antennas. It is denoted in decibels (dB), for isotropic antenna it is denoted by decibels relative (dBi), and for degree (dBd) (decibels relative to a dipole antenna) are used to measure gain. The antenna gain, G is given by

$$\text{Gain} = E * D.$$

where

E, antenna efficiency,

D, antenna directivity.

The eigen value of the TM_n mode for single index transmission mode it is denoted by "n", which is derived from the analysis of mathematical [9].

The following description of model current and voltage determines the TM_n admittance at all point

According to the network theorem, we cannot use the transmitting lines as passive. For constructing the circular, the current and voltage of the model is defined in terms of transmission by power, either inside or outside.

Where

$E_{zn}(\rho) = Z$ direction's radial function,

$H\varphi_n(\rho) = \phi$ TM_n mode's directed magnetic field.

Antenna used to convert the electrical signal into electromagnetic waves over a longer distance. The different frequencies of microwave [12–14].

2 Principle of Operation

By changing the value of the resistance of diode, the distribution factor of the current value for antenna can be changed. The feed of the line should be fixed for excitation of the antenna in mode of orthogonal near the diagonal area of the patch. When the capacitance value reaches 7.41 pF, the antenna behaves like a rectangular patch, with frequency shifting occurring over the frequency range of operation of 2.45 GHz. When the exciting frequency rises to 3.6 GHz, the antenna frequency of the antenna shifted with the value of capacitance 0.729 pF and 1.49 nH will be the inductance value. The pattern radiates as monopole for a range of frequency.

2.1 Antenna Radiation Pattern

Representing the radiation pattern in a graphical manner refers to radiation pattern of antenna in space coordinates. Designed for use in the distant field. In terms of angular coordinates.

The thickness of the Bakelite graphene substrate is 1.55 mm, with a 4.65 relative permittivity (loss tangent). The antenna's feed point has $y = 3.2$ mm, $x = 32$ mm coordinates which is fed through connector—51 co-axial. The antennas has $x = 29.6$ mm, $y = 37$ mm rectangular patch location and it is constructed as portrayed as [15].

Substrate (1.6.14.75.-2.3) positioned with the range of value. The varactor diode is inserted into the gap in patch area. The antenna patch is connected with three varactor diodes. The biasing line of the antenna and inductance are placed only on one sideways of the antenna patch to improve the efficiency of antenna. The antennas' resonant frequency is determined by changing the value of capacitor of the varactor diode.

The patch area is coupled via three varactor diodes. The resonance frequency in the antenna is achieved with variation in the capacitor value for the varactor diode. A biasing pad is placed on the patch's shorting pin to connect the ground for DC signal with the rectangular patch.

The DC supply power is coupled with the ground plane terminal, which provides DC zero voltage in anode terminal in varactor diode. An inductance value with 47 nH which divert the varactor from the RF signal.

3 Results and Discussion

By tuning the frequency from 1.96 to 4.25 GHz by changing the capacitor value of the diode from 0.73 to 7.41 pF. Figures 2 and 3 show the proposed antennas reflection coefficient (S11) with different capacitor value of the varactor diode and the variation of frequency is over 1.96 and 4.58 GHz. The antenna reflection coefficient and the gain (7.1 dB) of a rectangular patch with higher capacitance 1.95 GHz, mid capacitance 3.6 GHz, and lower capacitance 4.2 GHz are obtained. The efficiency and antenna gain are increased, by increasing the size of the antenna (Table 1).

Then the frequency drops to 4.2 GHz as the length of radiation increases. And the capacitance value varies in rectangular patch c = 07.42 pF, c = 01.51 pF, c =

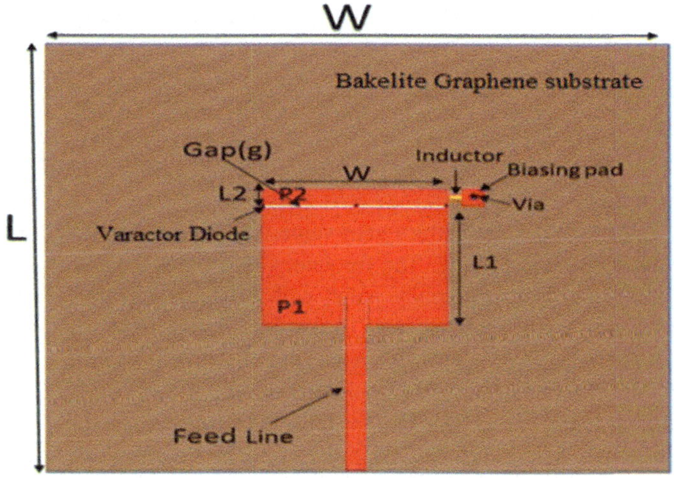

Fig. 2 Top view of the antenna geometry: $L = 102$ mm, $W = 105$ mm, $L1 = 32$ mm, $L2 = 5.6$ mm, $W = 32$ mm

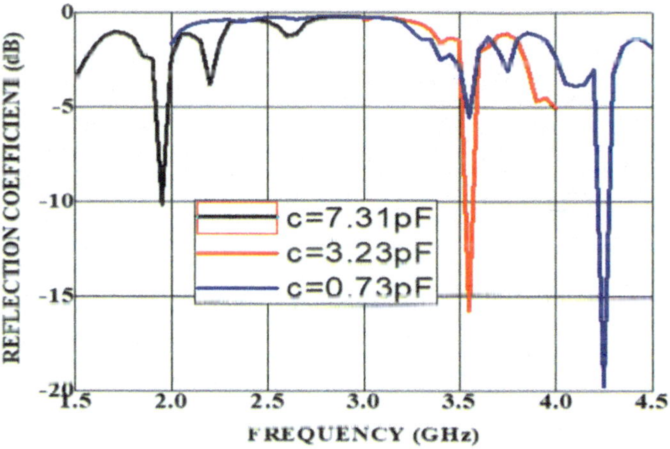

Fig. 3 The different capacitance value of the rectangular patch of 7.41, 1.50, 0.82, and 0.75 pF with different tuning frequencies of 1.95 and 3.60 GHz. 4.2 GHz at reflection coefficient

Table 1 Portraits RC—reflection coefficient of resonance in RLC (low and high capacitance)

References	Freq. (GHz)	BW	Reconfiguration		Pattern	Gain (dBi)
			Freq.	Pol		
[8]	2.10–2.9	34.52	No	Yes	0°, ± 50°	4.31
[7]	2.05–3.13	14	No	Yes	0°	3.04
[6]	2.0–2.96	34.5	No	Yes	0°, 90°	4.31
[5]	1.5–3.5	10	No	Yes	0°, ± 90°	3.31
Graphene Bakelite substrate	1.95–4.2	10.5, 7	Yes	No	0°	8.55

00.82 pF, $c = 00.74$ pF and the range of frequencies as 1.95, 3.60 GHz, 4.2 GHz at reflection coefficient.

Figure 4 portrays the 3D pattern gain value for the diode in RLC resonated for high and low capacitance value. The frequency drops to 4.2 GHz as the length of radiation increases. The rectangular patch with capacitances of 7.41, 1.50, 0.791, and 0.693 pF and a frequency of 1.95, 3.60, 4.2 GHz at reflection coefficient (− 11.46 dB), (− 16.28 dB), (− 19.7 dB) over gain of 05.88, 08.55 dBi.

Figure 5 shows the surface current distribution length of the antenna increases as the radiating length $R = 1.45\ \Omega$, $L = 0.45$ nH increases; as the surface current distribution length increase, to 1.95 GHz. The antenna's current on surface radiating length $R1 = 2.05\ \text{k}\Omega$, $L1 = 0.45$ nH, $C1 = 0.65$ pF decreases, and these frequency rises to 4.2 GHz as the surface current distribution length decreases.

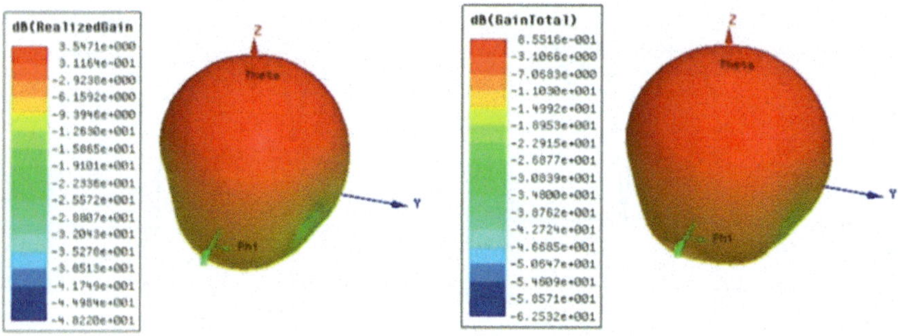

Realized Gain 3.54 dBi at C= 0.73pF **Realized Gain 8.55 dBi at C= 7.41pF**

Fig. 4 The different values of the capacitance of the rectangular patch with frequency 1.95, 3.60, 4.2 GHz at reflection coefficient

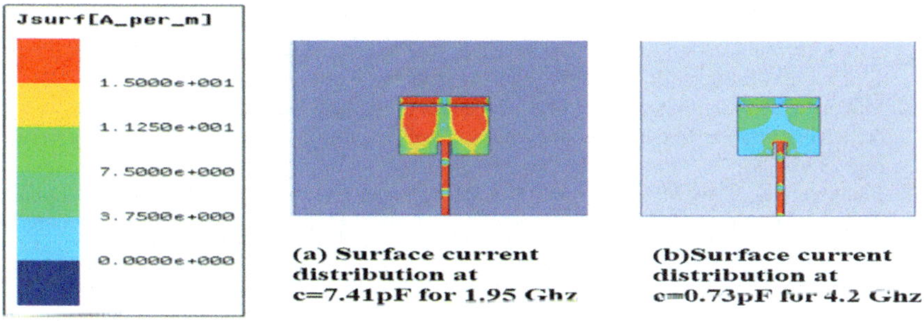

Fig. 5 Current distribution (i) at 1.95 GHz, $R1 = 1.46\ \Omega$, $L1 = 0.45$ nH; over 2.75 GHz, $R = 02.05\ \mathrm{k}\Omega$, $L1 = 0.46$ nH, $C1 = 0.65$ pF

4 Conclusions

The antenna with different varying tunable varactor diode frequency are designed and simulated in this paper. The outer and inner patch of the antenna. In the Bakelite substrate, biasing occurs by shorting the diodes. The frequency of resonance is varied with the varying capacitor value over 1.96–4.19 GHz.

At 4.2 GHz, the getting the very high gain radiation pattern of 8.56 dBi and attenuated to 3.55 dBi at 3.6 GHz. The far field and the near field characteristics for different orientations will be calculated at the antenna fabrication design in future work. With the bandwidth of roughly 15% of impedance matching of better than −27 dB, gives a significant frequency in the output of the proposed antenna.

References

1. Muthuvel SK, Choukiker YK (2021) Wideband frequency agile and polarization reconfigurable antenna for wireless applications. IETE J Res 69:1529–1538
2. Deepak A, Ramya S, Ganesan V, Shankar P (2014) Compos Lett 23:27
3. Deepak, Ganesan V, Karthik V, Shankar P (2014) Nanostruct Polym Nanocompos 10:5
4. Deepak, Srinivasan N, Ganesan V, Karthik V, Ramya S, Ganesan V, Shankar P (2015) Adv Mater Res 1101:314
5. Indrasen S, Tripathi VS (2011) Int J Comput Technol Appl 2:1595
6. Rop KV, Konditi DBO (2012) Innov Syst Des Eng 3:2222
7. Anzar K, Rajesh N (2012) Int J Comput Appl 55:40
8. Sajal SZ, Braaten BD, Marinov VR (2015) Microstrip patch antenna manufactured with flexible graphene-based conducting material. In: Proceedings of the 2015 IEEE international symposium on antennas and propagation & USNC/URSI national radio science meeting, pp 14–19
9. Sapna C, Luthra PK, Kumar A (2013) Int J Emerg Technol Comput Appl Sci 3:272
10. High Frequency Structure Simulator Software. https://www.ansys.com/products/electronics/ansys-hfss
11. Tawk Y, Costantine J, Christodoulou CG (2012) A varactor-based reconfigurable filtenna. IEEE Antennas Wirel Propag Lett 11:716–719. https://doi.org/10.1109/LAWP.2012.2204850

12. Kumar MS, Kumar Choukiker Y (2018) Tunable wideband frequency and switching polarisation reconfiguration antenna for wireless applications. IET Microwave Antennas Propag 12:2364–2371. https://doi.org/10.1049/iet-map.2018.5400
13. Muthuvel SK, Choukiker YK (2019) Frequency tunable circularly polarized antenna with branch line coupler feed network for wireless applications. Int J RF Microwave Comput Aided Eng 29:e21784
14. Muthuvel SK, Shaw M, Choukiker YK (2021) Concentric circular compound reconfigurable microstrip patch antenna for wireless applications. Int J Electron Commun. https://doi.org/10.1016/j.aeue.2021.153960
15. Yogesh B, Sangeeta S, Ashok K (2015) J Int Assoc Technol Sci 1:9

Capsule Network Approach for Image Classification

Moirangthem Marjit Singh and Nayan Kumar Sarkar

Abstract Numerous deep learning (DL) models based on convolutional neural network (CNN) have been developed for image classification. Though the CNN models attained outstanding performance, its architecture contains certain limitations such as loss of information in the pooling layer and no spatial relationship among the features of an object. The development of capsule network (CapsNet) has overcome the drawbacks of CNN. The CapsNet works on a set of capsules, and each capsule comprises a set of features. The CapsNet consists of a convolution layer, a primary capsule layer, image caps and a fully connected (FC) layer. In this paper, a variant of CapsNet approach is proposed by tuning the parameters of the original CapsNet and replacing ReLU with AReLU activation function in each FC layer. The classification performance of the proposed approach is evaluated on the Plant Village (PV) and Tomato datasets. 80% of the datasets were trained and 20% were tested. The proposed approach considering filter size of 5 × 5 in the feature extraction layer with AReLU activation function in each FC layer delivers 94.02% and 94.06% classification accuracies on PV and Tomato datasets respectively. Compared to other modern approaches, the proposed approach delivered better results.

Keywords DL · Capsule network (CapsNet) · Convolutional layer · Capsule layer · Decoder

1 Introduction

Image classification is an important application for various modern scientific advancements. It helps to make efficient decisions and is widely used in various areas such as disease detection in healthcare and agriculture, robotics, self-driving cars, etc. A number of image classification approaches that have been emerged over time have

M. Marjit Singh · N. Kumar Sarkar (✉)
Department of Computer Science and Engineering, North Eastern Regional Institute of Science and Technology, Itanagar, Arunachal Pradesh, India
e-mail: nayankrsarkar@gmail.com

V. K. Gunjan and J. M. Zurada (eds.), *Proceedings of 4th International Conference on Recent Trends in Machine Learning, IoT, Smart Cities and Applications*, Lecture Notes in Networks and Systems 873, https://doi.org/10.1007/978-981-99-9442-7_53

considered several datasets such as ImageNet, MNIST for evaluating performance. In this paper, the agriculture-related datasets, i.e., PV and Tomato datasets are being used for evaluating the performance. Image classification plays a key role in detecting various crop diseases in agriculture. Crop disease affects the productivity of food and crops. Even after adopting modern farming techniques, a good percentage of plants are affected by crop diseases [1]. Crop disease results in reduced crop productivity and quality [2]. Periodic monitoring of agricultural activities and ecosystem will minimize such threats [3, 4]. DL techniques can be suitably used in classifying crop diseases [5]. The DL-based automated image classification techniques have been applied in several domains. A survey on DL based image classification can be seen in [6].

Some limitations encountered in CNN are: (1) the requirement of large training data; (2) the pooling layer of CNN loses spatial and low-level features; and (3) the models do not consider the orientation and spatial relationship among various components of an image, i.e., the CNN cannot recognize the pose, deformation and texture of an image.

To overcome these limitations, capsule network (CapsNet) [7] approach can be considered. The CapsNet consists of a set of capsules, and each capsule represents a specific instance of an image at a particular position [8]. It has the capability to learn the pose, orientation and texture of an image. The CapsNet also consists of a convolution layer, a primary capsule layer, image caps and a FC layer.

The paper is structured as follows: Sect. 2 highlights the related works, Sect. 3 consists the proposed approach and datasets considered for result analysis. Implementation and result analysis are given in Sect. 4. Conclusion is given in Sect. 5 and the references at the end.

2 Related Works

A number of CNN-based image classification approaches have been developed for various applications. Since its emergence, the CapsNet has been applied to diverse domains, such as brain tumor categorization, lung cancer screening, cervical image classification, prediction of traffic flow, emotion recognition [9].

Dong et al. [10] proposed a CapsNet-based approach for detecting peanut diseases from its leaf images. They considered image with dimension 64×64. 80% and 20% of the were used for training and testing, respectively. The original CapsNet approach had only one convolution layer; they added three convolution layers in their approach for exploring more feature. As a result, the proposed approach with three convolution layers performed superior than the original model. Verma et al. [11] implemented the CapsNet for classifying the potato leaf diseases and compared the performance of their proposed approach with the performances of ResNet18, VGG16 and GoogleNet approaches. The potato dataset they considered was employed from the PV dataset, and their approach delivered 91.83% classification accuracy on it.

Kurup et al. [12] analyzed the performance of CapsNet on two distinct datasets. They analyzed the performance of their approach for both classifying crop diseases considering the PV dataset and classifying crop leaves considering the Leafsnap [13] dataset. The result analysis was done with different numbers of epochs and learning rate. But, the performance was found better with the setting of the number of epochs = 50 and learning rate = 0.01. Oladejo et al. [14] suggested an optimized capsule network approach for the categorization of banana leaf diseases. Their approach could categorize the diseased leaf images with 95% accuracy. Recently, Diana Andrushia et al. [15] also considered the CapsNet approach for detecting the diseased grape leaf images. For detecting effective features, the suggested approach comprises of five convolution layers with a variety of kernel sizes and strides. The result analysis of the approach was done considering both non-augmented and augmented datasets. The approach delivered 99.12% classification accuracy.

3 Proposed Approach

The proposed approach is a variant of the original capsule network (CapsNet) [7] approach. The approach is constructed by tuning the parameters of the original CapsNet approach, considering a filter of size 5×5 in the feature extraction layer and by replacing ReLU with the AReLU activation function in each fully connected layer. The CapsNet can be considered as opposite to CNN because it maintains a spatial relationship among components of an input image where CNN cannot maintain. It has deep learning architecture. In CapsNet, the capsule consists of a set of neurons (features) that learn the entities such as size, pose and orientation of an image. Every capsule has an activity vector that captures a variety of parameters (position, orientation and size) for detecting a specific type of object. The output of each capsule (activity vector) represents the probability of the existence of that object and its orientation. As CapsNet comprises multiple layers of capsules; data is routed from lower layer capsules to higher layer. Layer-wise coupling coefficient exists between the capsules with different values. The coefficient is determined by a softmax function:

$$O^{\wedge}_{c|b} = W_{bc} O_{b^c}, \tag{1}$$

where $O^{\wedge}_{c|b}$ is the output vector of cth capsule, W_{bc} is the weight matrix and O_{b^c} is the output of capsule b in the previous layer. The current capsule outputs a higher probability when it finds a strong cluster of previous predictions indicating the existence of an object. During training, the CapsNet learns the spatial relationship among the various parts of the object using dynamic routing.

3.1 Architecture of the Proposed Approach

The architecture of CapsNet comprised of two main parts: a convolutional encoder and a decoder. Each of these parts is briefly explained below:

Convolutional encoder: The convolutional encoder consists of the following layers:

- Convolutional layer: The convolutional layer extracts features from the input image. The layer of the proposed approach used a kernel of size = 5, stride = 1 and padding = 0.
- Primary capsules layer: This layer generates the first layer of capsules. The primary capsule layer takes the features of the convolutional layer as input and generates a set of output vectors. The proposed approach comprised 16 such primary capsules. The output of the primary capsule vector is a value between 0 and 1, which indicates the existence of visual parts of an image. The output is a probability between 0 and 1, which is determined by a squash function as shown in Eq. (2).

$$m_c = \frac{||K_c^2||K_c}{1 + ||K_c^2||K_c}, \tag{2}$$

where m_c is the normalized vector value for capsule c, K_c is the weighted sum of all the output vectors of the previous layer capsules.

- Image capsules (image caps) layer: It is the final layer of the convolutional encoder section. Image caps is comprised of the number of capsules and the total number of class labels in the dataset. For example, for 38 categories of samples, the number of capsules set is 38. Each of these capsules is to produce a 32 dimensional output vector space in the proposed approach.

Decoder: This is the last layer of the approach. It is also called the fully connected (FC) layer and is followed by the softmax function. The FC layer connects the neurons of one layer to another layer. The loss of the approach is computed as per equation (3).

$$N_c = L_c(\max(0, n^+ - m_c)) + \lambda(1 - L_c)(\max(0, m_c - n^-)), \tag{3}$$

where n^+ and n^- represent the value 0.9 and 0.1 respectively. L_c represent the labels, i.e., correct (1) and incorrect (0) of capsule vector, and λ is a variable.

The architecture of the approach comprised of convolutional, primary capsules, image capsules and the fully connected layer. The proposed CapsNet approach is shown in Fig. 1.

In the convolution layer, feature extraction is done. In the original approach, the considered kernel size was 9. We extracted features with both settings of kernel size = 5×5 and 9×9, stride = 1 and padding = 0. The second layer, i.e., the primary capsule layer, takes the extracted features as input and generates 16 numbers of capsules. The capsules of this layer are connected to the capsules of the image capsules layer. For

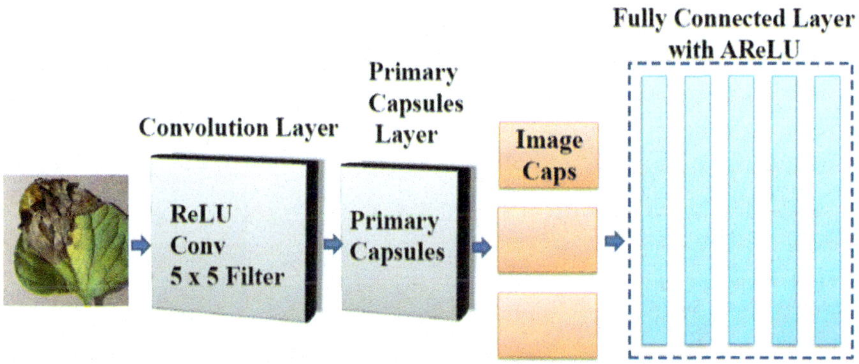

Fig. 1 Proposed CapsNet approach

checking layer-wise learning, the dynamic routing by agreement algorithm is applied in this layer. In CapsNet, the dynamic-routing algorithm is considered an alternative solution to the max-pooling operation of CNN, as the max-pooling operation loses information during layer-wise routing of features. The routing iterations in the dynamic routing define the number of times coupling coefficients are updated. In the proposed CapsNet approach, the routing iteration is set to 20. The last layer consists of five fully connected layers, with 512, 1024, 1024, 2048 and 2352 number of neurons in each layer. Though in the original approach, the ReLU activation function was considered in the FC layer; in the proposed approach, the ReLU is replaced with the AReLU in each FC layer.

3.2 *Datasets Considered*

Performance assessment of the proposed approach is done on the Plant Village (PV) [16] and the Tomato dataset. Tomato dataset is part of PV dataset. The dataset comprised of 54,305 number of RGB images of 256 × 256 dimensions. It consists of 38 types of diseased and healthy images of leaves from 14 categories of crops. An image of each category of the PV dataset is shown in Fig. 2.

Data samples are preprocessed before being passed to the approach. In preprocessing, images are augmented, resized and normalized before being sent to the approach. The input images are resized from a dimension of 256 × 256 to 28 × 28 so that the model gets samples of similar size. The dimensions changed to 28 × 28 also reduced training time and the number of parameters. Finally, the pixel values were normalized in a range of 0 and 1.

Fig. 2 Image of each category of the PV dataset

4 Implementation and Result Analysis

The proposed approach was implemented with the Pytorch framework on the Google Colab platform, as Colab provides GPU access with high RAM. The performance of the proposed approach was assessed on the PV and Tomato datasets. Classification accuracy is considered the performance metric of the proposed approach. The classification accuracy (CA) is computed as per the equation [4].

$$\text{CA} = (\text{Image predicted correctly}/\text{Total no. image}) * 100. \tag{4}$$

80% of the datasets were trained and 20% were tested. The approach was run for 100th epochs to train and delivered 94.02% classification accuracy on test set of PV dataset and 94.06% classification accuracy on the test set of Tomato dataset. The result comparison of the proposed approach with other approaches on the PV dataset is shown in Table 1. From Table 1, it is observed that the our approach delivered better accuracy considering the filter of size 5 × 5 and AReLU in each FC layer. We also assessed the result of our approach with a 10:90 train–test split. However, the approach delivered 78.56% and 79.20% classification accuracies on the Tomato and PV datasets, respectively.

The graphical plot of the training loss of the proposed approach on the PV dataset is shown in Fig. 3. The class-wise test accuracy delivered by the proposed approach on PV dataset is shown in Table 2.

Efficacy of the approach is also analyzed on the Tomato dataset. The result comparison of the approaches on the Tomato dataset is kept in Table 3. From Table 3, it is also observed that the suggested approach delivered better accuracy with the Tomato dataset, considering the filter size = 5 × 5 and AReLU in each FC layer. The graphical plot of training loss of the approach on the Tomato dataset is shown in Fig. 4.

Table 1 Performance comparison of CapsNet approaches on PV dataset

Approach	Year	Kernel size	Activation function	Epoch	Accuracy (%)
CapsNet [12]	2020	9 × 9	ReLU	100	94
CapsNet [3]	2021	9 × 9	ReLU	100	93.01
Proposed		9 × 9	ReLU	100	90.78
Proposed		5 × 5	ReLU	100	93.73
Proposed		9 × 9	AReLU	100	91.28
Proposed		5 × 5	AReLU	100	94.02

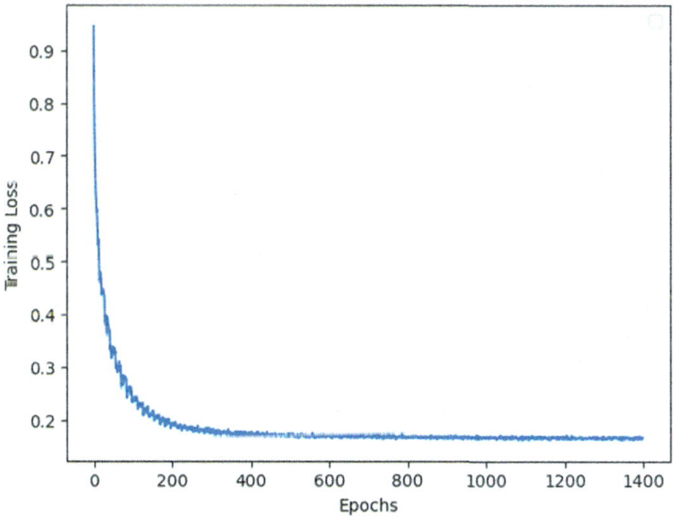

Fig. 3 Graphical plot of training loss of the proposed approach on PV dataset

Table 2 Class-wise classification accuracy delivered by the approach on PV dataset

Class	Obtained accuracy	Sample classified correctly	Test sample
C-1	83.33	105	126
C-2	92.80	116	125
C-3	87.27	48	55
C-4	94.22	310	329
C-5	96.67	291	301
C-6	95.26	201	211
C-7	95.90	164	171
C-8	67.96	70	103
C-9	99.58	238	239
C-10	95.93	189	197
C-11	100.00	233	233
C-12	80.50	190	236
C-13	98.91	274	277
C-14	96.29	208	216
C-15	95.29	81	85
C-16	99.45	1096	1102
C-17	94.34	434	460
C-18	94.44	68	72
C-19	85.50	171	200
C-20	91.55	271	296
C-21	95.50	191	200
C-22	79.50	159	200
C-23	70.96	22	31
C-24	89.33	67	75
C-25	98.82	1006	1018
C-26	97.54	358	367
C-27	89.63	189	222
C-28	95.65	88	92
C-29	96.71	412	426
C-30	62.50	125	200
C-31	84.03	321	382
C-32	85.86	164	191
C-33	89.85	319	355
C-34	94.34	317	336
C-35	84.69	238	281
C-36	99.06	1062	1072
C-37	97.33	73	75
C-38	97.80	312	319

The per-class test accuracy delivered by the approach on Tomato dataset is shown in Table 4.

Initially, result analysis was done considering images with dimensions of 224 × 224 and multiple convolution layers with different filter sizes and strides for feature extraction. However, the performance was not satisfactory and took longer training time due to large dimensions of samples. With a goal to reduce the training time, the dimensions of the input samples were reduced to 28 × 28, which resulted in an improvement in accuracy. Further, in the convolution layer of the original paper, the filter size was set to 9, the filter number was 256 and the number of capsules in the initial layer was 8. We changed the filter size to 5, the filter number to 400 and the capsules to 16. These changes improved the performance. We also analyzed the performance of the approach, considering both the 9 × 9 and 5 × 5 filters with ReLU and AReLU activation functions in the FC layer. The approach delivered better results with a 5 × 5 filter in comparison to a filter size of 9 × 9. The replacement of

Table 3 Performance comparison of the CapsNet approaches on Tomato dataset

Approach	Year	Kernel size	Activation function	Epoch	Accuracy (%)
CapsNet [11]	2020	9 × 9	ReLU	100	91.83
CapsNet [17]	2022	7 × 7	ReLU	100	93.03
Proposed		9 × 9	ReLU	100	90.95
Proposed		5 × 5	ReLU	100	92.96
Proposed		9 × 9	AReLU	100	90.98
Proposed		5 × 5	AReLU	100	94.06

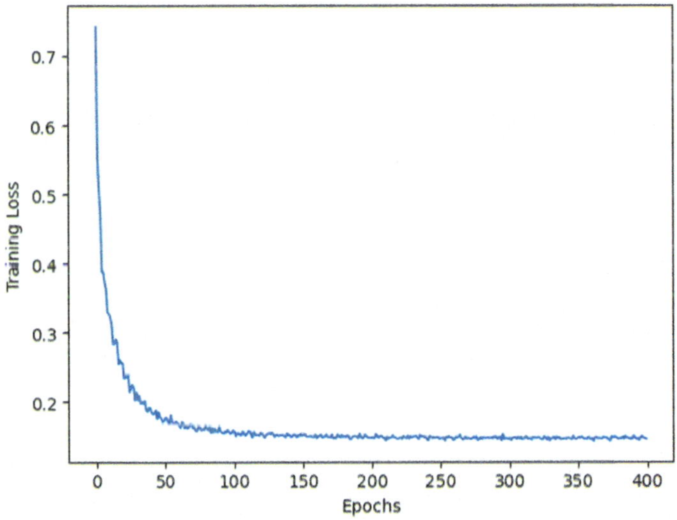

Fig. 4 Graphical plot of training loss of the proposed approach on Tomato dataset

Table 4 Class-wise classification accuracy delivered by the approach on Tomato dataset

Class	Obtained accuracy	Sample classified correctly	Test sample
C-1	99.26	410	426
C-2	77.00	154	200
C-3	87.17	333	383
C-4	92.14	176	191
C-5	96.33	342	355
C-6	95.83	322	326
C-7	84.34	237	281
C-8	99.44	1066	1072
C-9	97.33	73	75
C-10	96.55	308	319

ReLU with the AReLU activation function in the fully connected layer also improved performance.

5 Conclusion

The CNN-based DL approaches delivered better performance in various image classification domains. The paper has attempted to improve the image classification accuracy of the existing CapsNet approach by replacing ReLU with AReLU activation function and considering filter size 5×5. With dataset splits of 80% for training and 20% for testing, the proposed approach obtained 94.02% classification accuracy on the PV dataset and 94.06% classification accuracy on the Tomato dataset, respectively. The approach considering the filter size 5×5 and AReLU in the FC layer delivered better classification accuracy in comparison to other existing approaches. Other related datasets can also be considered to evaluate the performance.

References

1. Harvey CA, Rakotobe ZL, Rao NS, Dave R, Razafimahatratra H, Rabarijohn RH, Rajaofara H, Mackinnon JL (2014) Extreme vulnerability of smallholder farmers to agricultural risks and climate change in Madagascar. Philos Trans R Soc Lond B Biol Sci 369(1639):20130089. https://doi.org/10.1098/rstb.2013.0089. PMID: 24535397; PMCID: PMC3928894
2. Yin K, Qiu JL (2019) Genome editing for plant disease resistance: applications and perspectives. Philos Trans R Soc Lond B Biol Sci 374(1767):20180322. https://doi.org/10.1098/rstb.2018.0322
3. Samin OB, Omar M, Mansoor M (2021) CapPlant: a capsule network based framework for plant disease classification. PeerJ Comput Sci 7:e752. https://doi.org/10.7717/peerj-cs.752

4. Savary S, Ficke A, Aubertot JN et al (2012) Crop losses due to diseases and their implications for global food production losses and food security. Food Secur 4:519–537. https://doi.org/10.1007/s12571-012-0200-5
5. Sarkar NK, Singh MM, Nandi U (2022) Learning based image classification techniques. In: Mukhopadhyay S, Sarkar S, Dutta P, Mandal JK, Roy S (eds) Computational intelligence in communications and business analytics. CICBA 2022. Communications in computer and information science, vol 1579. Springer, Cham. https://doi.org/10.1007/978-3-031-10766-5_3
6. Sarkar NK, Singh MM, Nandi U (2022) Recent researches on image classification using deep learning approach. Int J Comput Digit Syst 12(1):1357–1374. https://doi.org/10.12785/ijcds/1201109
7. Sabour S, Frosst N, Hinton GE (2017) Dynamic routing between capsules. Advances in neural information processing systems, vol 30, pp 1–11. https://doi.org/10.5555/3294996.3295142
8. Afshar P, Heidarian S, Naderkhani F, Oikonomou A, Plataniotis KN, Mohammadi A (2020) COVID-CAPS: a capsule network-based framework for identification of COVID-19 cases from X-ray images. Pattern Recogn Lett 138:638–643. ISSN: 0167-8655. https://doi.org/10.1016/j.patrec.2020.09.010
9. Patrick MK, Adekoya AF, Mighty AA, Edward BY (2022) Capsule networks—a survey. J King Saud Univ Comput Inf Sci 34(1):1295–1310. ISSN: 1319-1578. https://doi.org/10.1016/j.jksuci.2019.09.014
10. Dong M, Mu S, Su T, Sun W (2019) Image recognition of peanut leaf diseases based on capsule networks. In: Knight K, Zhang C, Holmes G, Zhang ML (eds) Artificial intelligence. ICAI 2019. Communications in computer and information science, vol 1001. Springer, Singapore. https://doi.org/10.1007/978-981-32-9298-7_4
11. Verma S, Chug A, Singh AP (2020) Exploring capsule networks for disease classification in plants. J Stat Manag Syst 23(2):307–315. https://doi.org/10.1080/09720510.2020.1724628
12. Kurup RV, Anupama MA, Vinayakumar R, Sowmya V, Soman KP (2020) Capsule network for plant disease and plant species classification. In: Smys S, Tavares J, Balas V, Iliyasu A (eds) Computational vision and bio-inspired computing. ICCVBIC 2019. Advances in intelligent systems and computing, vol 1108. Springer, Cham. https://doi.org/10.1007/978-3-030-37218-7_47
13. Kumar N et al (2012) Leafsnap: a computer vision system for automatic plant species identification. In: Fitzgibbon A, Lazebnik S, Perona P, Sato Y, Schmid C (eds) Computer vision—ECCV 2012. ECCV 2012. Lecture notes in computer science, vol 7573. Springer, Berlin, Heidelberg. https://doi.org/10.1007/978-3-642-33709-3_36
14. Oladejo BF, Ademola OO (2020) Automated classification of banana leaf diseases using an optimized capsule network model. In: CS & IT conference proceedings, vol 10, no 9. https://doi.org/10.5121/csit.2020.100910
15. Diana Andrushia A, Mary Neebha T, Trephena Patricia A et al (2023) Image-based disease classification in grape leaves using convolutional capsule network. Soft Comput 27:1457–1470. https://doi.org/10.1007/s00500-022-07446-5
16. Mohanty SP. Colored. https://github.com/spMohanty/PlantVillage-Dataset
17. Mensah PK, Ayidzoe MA (2022) Overview of CapsNet performance evaluation methods for image classification using a dual input capsule network as a case study. Int J Comput Digit Syst 11(1):29–43. https://doi.org/10.12785/ijcds/120104

An Intelli BPS: An Intelligent Biopsychosocial Parameters-Based Machine Learning Model to Predict Hypertension

Satyanarayana Nimmala, Rella Usha Rani, Preeti Nutipalli, Usikela Naresh, and R. Ravinder Reddy

Abstract The biopsychosocial parameters of an individual are major influencing factors behind the good health or illness of a person. In recent times, these factors are extensively under study by doctors in the medical field to diagnose and treat different types of diseases. An individual's blood pressure (BP) might be elevated as a result of physical, psychological, or emotional trauma. If the systolic blood pressure (BP) consistently exceeds 140 mmHg and the diastolic BP consistently exceeds 90 mmHg in repeated measurements, it is classified as high blood pressure (HBP) or hypertension. The biological parameters such as aging, cholesterol, and obesity, psychological parameters such as anger, stress, and depression, and emotional parameters such as ego, jealousy, the family background, and the living environment play a major role in one's life and impacting the quality of life. In this paper, we have proposed an Intelligent Biopsychosocial Parameters-based Machine Learning (ML) Model to Predict Hypertension. The proposed model considers biological parameters obesity level, cholesterol level, and age (OCA), psychological parameters anger level, anxiety level (AA) of a person, and social parameter of individual socioeconomic status (SES). The proposed model works in three stages: at stage 1, the model calculates the absolute impact of the biopsychosocial input parameter on BP using Pearson Correlation Coefficient (PCC); at stage 2, the model is trained using linear

S. Nimmala (✉)
Department of CSE (Data Science), CVR College of Engineering, Hyderabad, India
e-mail: satyauce234@gmail.com

R. Usha Rani · U. Naresh
Department of CSE (AI & ML), CVR College of Engineering, Hyderabad, India

P. Nutipalli
Department of CSE, Anil Neerukonda Institute of Technology and Sciences, Vishakhapatnam, India
e-mail: npreethi.cse@anits.edu.in

R. R. Reddy
Department of CSE, Chaitanya Bharathi Institute of Technology, Hyderabad, India
e-mail: rravinderreddy_cse@cbit.ac.in

V. K. Gunjan and J. M. Zurada (eds.), *Proceedings of 4th International Conference on Recent Trends in Machine Learning, IoT, Smart Cities and Applications*, Lecture Notes in Networks and Systems 873, https://doi.org/10.1007/978-981-99-9442-7_54

regression analysis to design a threshold function Hypertension Identification Function (HIF); and at stage 3, the HIF is utilized to predict an individual's likelihood of being afflicted with hypertension.

Keywords Age · Obesity · Anger · Anxiety · Smoking · Alcohol consumption · Sodium intake

1 Introduction

The heart performs the vital function of circulating blood throughout the human body, supplying oxygen and nutrients to sustain the tissues. This process involves the heart pumping blood with a specific force during the systole and diastole cycle. This force exerted by the blood within the arteries and veins is commonly referred to as blood pressure [1]. Individuals who exhibit a blood pressure reading exceeding 140/90 mmHg in multiple measurements are diagnosed with high blood pressure, often known as hypertension [2]. Currently, hypertension (HBP) is exerting a profound impact on the lives of a substantial number of individuals worldwide [3]. HBP is a state developed inside the human body knowingly or unknowingly; if it is not noticed and treated properly in advance, it may lead to abnormal functioning of different organs in the human body and subsequently loss of life. To address this issue, this paper looked at biological factors like age, obesity, and cholesterol levels, as well as psychological ones like anger and anxiety levels, SES parameters, highest educational degree, current salary, job category, and marital status to predict hypertension using machine learning techniques (MLT) [4]. The BP of a person may fluctuate every second because of different reasons. It may go high when you run, and go low when you sleep. A person's blood pressure is naturally managed and controlled by the various BP control mechanisms that are operated by the circulatory system of the human body.

2 Background Work

HBP is a significant global health concern characterized by its insidious nature and increasing incidence. The elevation of blood pressure (BP) constitutes a significant risk factor for the development of cardiovascular diseases (CVDs), including brain stroke and heart attack, which account for approximately one-third of all mortality cases in India [5]. Based on data from the World Health Organization (WHO), it has been observed that just 12% of individuals in India possess blood pressure (BP) levels that fall within the healthy range. In the research work authored by Gavaghan et al., it was asserted that the development of models and simulations for physiological systems holds significant value, as they contribute to the diagnostic and therapeutic capabilities of medical practitioners. According to renowned medical scientist Arthur Guyton, the utilization of mathematical models and experimentation on medical data contribute to the creation of a knowledge database, hence facilitating a more comprehensive comprehension of the system being investigated (Guyton) [6]. According to the findings of Coleman and Randall, as well as Kofranek et al., it was determined that these models provide valuable assistance to medical professionals in the process of making decisions related to diseases. According to Guyton and Hall, there are multiple systems in the human body, such as activation of flight or fight mode, over or under-functioning of hormones like Angiotensin II, abnormal functioning of kidneys. Their research concluded that they are short-term, medium-term, and long-term blood pressure management systems in built-in humans, are the reason for the maintenance of the blood pressure. If the functionality of any of these mechanisms deviates from normal functioning, one may a victim of blood pressure on different scales.

As a person's age increases, heart muscle cells degenerate and thin, and the heart pacemaker system begins to accumulate fat. This reduces the heart's ability to pump blood. The research carried out by different research scholars to find the impact of age on BP like Mikko Peltokangas et al., Elisabete Pinto, and Julie K. K. Vishram et al. [7] concluded that an increase in BP in particular increase in SDP is most common in aging people. The researchers Meenakshi Kalyan et al., L. Landsberg et al., L. Mertens and Luc F. Van Gaal, and Richard N. Re (MD) [8], all revealed that obesity may lead to activation of the SNS, activation of the renin-angiotensin system, and sodium retention, thus increasing the BP. If the cholesterol level is high that of Low-Density Lipoproteins (LDL) in the bloodstream leads gathering of plaque inside the blood vessels, which reduces the diameter of arteries [9, 10]. The arteries reduced in diameter resist the normal blood flow, thus increasing the BP of a person [11]. The research article "High Cholesterol: Prevention, Treatment, and Research" reviewed by Michael Joseph Blaha, M. D., Roger Scott Blumenthal, M. D., revealed that higher LDL level in blood is the root cause of hypertension as well as heart attacks [12]. The INTERMAP Study, published in the Journal of Hypertension in 2011 by Masaru Sakuraia, Jeremiah, et al., also concluded that high LDL is strongly related to an increase in SBP of a person [13].

3 Proposed Methodology

This section presents a sophisticated system that utilizes the HIF function. The HIF function is a regression function that utilizes multiple variables to forecast the class label of a test record. The aforementioned prediction is derived by the comparison of the threshold value obtained during the training phase of the HIF function. During the training phase of the classifier under consideration, the HIF function integrates multiple attributes, including anxiety levels, rage levels, total blood cholesterol level (referred to as cholesterol level), age, obesity level, and socioeconomic status (SES) factors. The proposed model underwent evaluation using a real-time dataset, and its performance was compared to that of simple logistic regression, Naive Bayes, J48, and REPTree [14]. The proposed model exhibited exceptional performance across multiple parameters, encompassing accuracy, true positive (TP) rate, false positive (FP) rate, precision, recall, and *F*-measure [15]. The data about anger, anxiety, and socioeconomic position was obtained manually from the same cohort of individuals by administering a set of predetermined questionnaires [16], as provided by Doctor C. We have examined a total of ten standardized questionnaires for measuring anger and twenty questionnaires for measuring anxiety, all of which are accessible online [17]. Each item within these questions requires respondents to select one of many predetermined response alternatives. No, the term "never" implies that its value is regarded as zero. (b) Indeed, the term "rarely" typically denotes a number of one. (c) Similarly, the term "often" generally signifies a value of two. If the answer is yes, most often regarded as three. Experimental analysis involves considering the mean value of all the responses. The socioeconomic status (SES) characteristics, highest level of educational attainment, current wage, job type, and marital status are also gathered, and their influence is determined using a heuristic technique. In the training phase of the proposed model, the average value is employed as the bias value for the HIF function (Table 1).

Table 1 SES parameters considered and their impact on BP

SES attribute name	Value	Impact
Highest educational qualification	SSC or below SSC	2
	SSC to degree	1.5
	Post graduation	1
	Ph.D.	0.5
Job category	Driver	2
	Labor	1.5
	Non-teaching staff	1
	Teaching staff	0.5
Marital status	Unmarried	0.5
	Married	1
	Divorced	1.5
Salary	$\leq$ 30,000	2
	> 30,000 and $\leq$ 60,000	1.5
	> 60,000 and $\leq$ 120,000	1
	Above 120,000	0.5

Table 2 PCC value and absolute weight of risk factor

Attribute or RF_i	PCC value	Absolute weight of RF_i
Anxiety level	0.7014	30.94
Anger level	0.5546	24.44
Age	0.3976	17.54
Cholesterol level	0.2866	12.64
Obesity level	0.3266	14.40
Total		100.00

3.1 Data Preprocessing

Mean arterial pressure (MAP) is calculated in the data preprocessing phase. Later, PCC value between each attribute considered and MAP is calculated using Eq. 1 [18]. The rounded value of MAP is considered for experimental purposes. MAP represents the average blood pressure in blood vessels during a single cardiac cycle. The BP of every person involved in the study is monitored for six months to obtain the most consistent value. The most consistent SDP and DBP values on repeated measurements are taken into consideration before experiments are conducted. The PCC value represents whether the considered parameters and MAP are positively correlated or not. If they are positively correlated, the value represents the degree of influence of the selected attribute on the MAP (Table 2).

$$\text{PCC} = \frac{N\sum XY - \sum X \sum Y}{\left[N\sum X^2 - \left(\sum X\right)^2\right]\left[N\sum Y^2 - \left(\sum Y\right)^2\right]}. \tag{1}$$

3.2 Absolute Impact of Risk Factor

The suggested methodology initially partitions the complete domain space of each attribute into distinct domains, which are determined by the attribute's value. For instance, in cases when an individual's age is 30 or less, the influence on the increase in blood pressure is diminished due to their youthful age and the presence of elastic and thin blood vessels. Individuals between the age range of 30–50 may exhibit a certain level of resistance in their blood vessels, which might potentially result in a mild influence on blood circulation. Upon reaching the age of 50, an individual has a decline in the flexibility of their blood vessels, resulting in increased resistance to the flow of blood [19]. According to this idea, the risk factor domain space is categorized into low, medium, and high levels, as illustrated in Table 3.

Table 3 Effect based on the attribute value

Attribute or RF_i	Impact based on the value		
Anxiety level	≤ 1 Impact = low L1(1)	$> 1\&\& \leq 2$ Impact = medium M1(2)	> 2 Impact = high H1(3)
Anger level	≤ 1 Impact = low L2(1)	$> 1\&\& \leq 2$ Impact = medium M2(2)	> 2 Impact = high H2(3)
Age	≤ 30 Impact = low L3(1)	$> 30\&\& \leq 50$ Impact = medium M3(2)	> 50 Impact = high H3(3)
Cholesterol level	≤ 150 Impact = low L4(1)	$> 150\&\& \leq 200$ Impact = medium M4(2)	> 200 Impact = high H4(3)
Obesity level	≤ 25 Impact = low L5(1)	$> 25\&\& \leq 30$ Impact = medium M5(2)	> 30 Impact = high H5(3)

3.3 *Real Impact of Risk Factor*

The absolute weight of each risk factor is determined and presented in Table 2. This absolute weight signifies the entire contribution of each risk factor to the increase in blood pressure. The quantification of a characteristic is employed to compute the tangible and measurable consequences of a risk factor. Equations 2–4 are employed for the computation of the genuine impact of an attribute, taking into consideration its absolute impact. The distribution of the absolute weight of each attribute follows a ratio of 1:2:3, which is determined by the impact of the attribute value, as illustrated in Table 4.

$$L_i = \frac{L_i * (\text{Absolute Weight}(\text{RF}_i))}{L_i + M_i + H_i} \tag{2}$$

$$M_i = \frac{M_i * (\text{Absolute Weight}(\text{RF}_i))}{L_i + M_i + H_i} \tag{3}$$

$$H_i = \frac{H_i * (\text{Absolute Weight}(\text{RF}_i))}{L_i + M_i + H_i}. \tag{4}$$

The variables L_i, M_i, and H_i are used to denote the quantitative representation of the attribute's impact, as determined by the values specified in Table 3.

3.4 *The Establishment of a Threshold for the Proposed HIF Function*

The threshold value for the HIF is determined by the utilization of Eq. 5. The training dataset is utilized for both model training and determining the threshold value. The training procedure commences by setting a threshold value of 31 (Fig. 1).

Table 4 Real effect of attributes based on the value

Attribute	Real impact (RI) of risk factors based on the attribute value			Absolute weight (RF_i)
Anxiety level	≤ 1 Then RI = 5.1566	$> 1 \&\& \leq 2$ Then RI = 10.3133	> 2 Then RI = 15.47	30.94
Anger level	≤ 1 Then RI = 4.073	$> 1 \&\& < 2$ Then RI = 8.1466	> 2 Then RI = 12.22	24.44
Age	≤ 30 Then RI = 2.9233	$> 30 \&\& \leq 50$ Then RI = 5.8466	> 50 Then RI = 8.77	17.54
Cholesterol level	≤ 150 Then RI = 2.1066	$> 150 \&\& \leq 200$ Then RI = 4.2133	> 200 Then RI = 6.32	12.64
Obesity level	≤ 25 Then RI = 2.4	$> 25 \&\& \leq 30$ Then RI = 4.8	> 30 Then RI = 7.2	14.40

This threshold value determines the classification outcome for a specific training record based on the HIF function value. If the HIF function value is equal to or above 31, the predicted class label is "yes"; otherwise, it is "no." During each repetition, the threshold value was incremented by either 2 or 3 units, and the corresponding accuracy measurements were recorded and documented in Table 5. In this context, r_i represents the quantifiable measure of the true impact, while x_i denotes the attribute's normalized value. When the threshold value is configured as 55, the classifier attains an accuracy of 87.78%. Nevertheless, the accuracy of the classifier exhibits a decline in successive iterations, commencing from an initial threshold value of 58. Consequently, the threshold value for the suggested HIF function is established at 55.

$$\text{HIF}(Y) = \sum_{i=1}^{n} ri_i * x_i + \text{SES}_i. \tag{5}$$

4 Experimental Results and Discussion

The proposed HIF model has exhibited a prediction accuracy of 87.78%, outperforming commonly utilized classifiers such as basic logistic regression, Naïve Bayes, J48, and REP tree implemented in the WEKA.

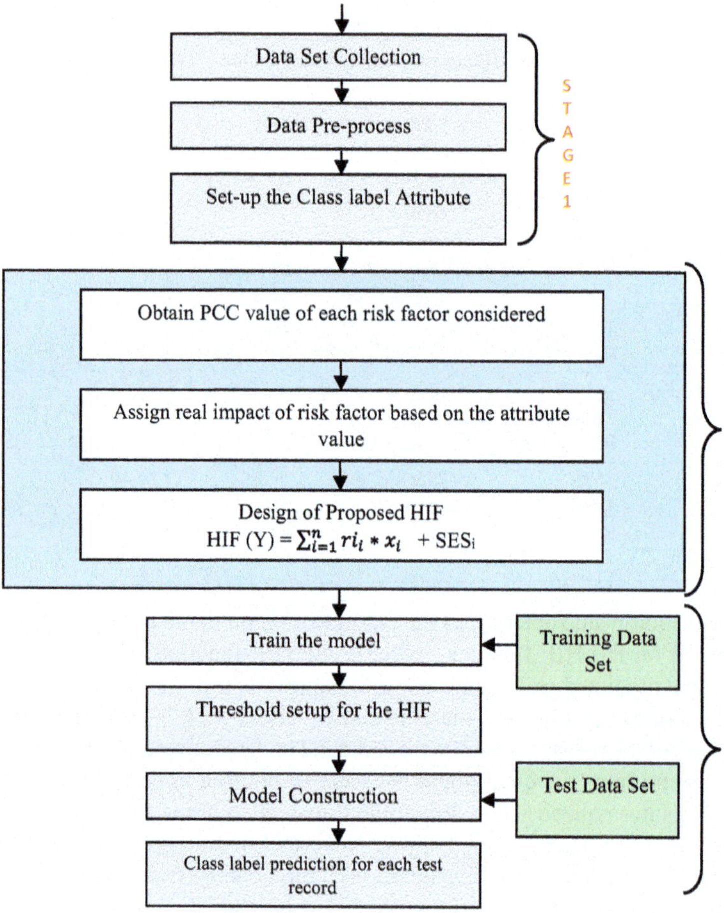

Fig. 1 Architecture of the proposed model

Table 6 and Fig. 2 present the performance metrics, including accuracy, of the proposed yes class classifier and other classifiers. While J48 and REPTree demonstrate proficiency in predicting negative classes, they significantly lag behind the proposed HIF in terms of positive class predictions. The TPR also called recall represents how accurate the classifier is in predicting the positive class records as positive and FPR represents the false positive rate of the classifier. The precision of the positive class represents how exact the classifier is at positive class predictions. The *F*-measure represents the performance of the classifier concerning the right and wrong predictions of each decision class. Although the FPR of simple logistic regression, J48, and REPTree is smaller than the proposed HIF, the *F*-measure of the proposed is more than the simple logistic regression, J48, and REPTree.

Table 5 Threshold set up for HIF

S. No.	HIF values set	Accuracy	TP rate/recall	FP rate	Precision
1	31	63.0	0.963	0.435	0.423
2	33	66.7	0.951	0.440	0.406
3	36	69.4	0.952	0.396	0.439
4	39	73.4	0.945	0.343	0.512
5	42	78.2	0.947	0.247	0.596
6	46	83.0	0.918	0.179	0.637
7	49	84.0	0.913	0.156	0.653
8	52	85.2	0.878	0.138	0.658
9	**55**	**87.78**	**0.849**	**0.136**	**0.706**
10	58	87.4	0.749	0.115	0.619
11	60	84.5	0.778	0.105	0.711
12	63	82.8	0.698	0.085	0.723
13	66	81.9	0.548	0.074	0.708

Bold represents the threshold value of the classifier, that is set during the trianing phase of classifier

Table 6 Performance analysis of different classifiers

Algorithm	Accuracy %	TPR	FPR	Precision	Recall	*F*-measure
Simple logistic regression	78.56	0.454	0.092	0.645	0.454	0.533
Naïve Baye's	81.76	0.769	0.164	0.634	0.769	0.695
J48 algorithm	85.58	0.657	0.072	0.772	0.657	0.710
REPTree	85.79	0.704	0.086	0.752	0.704	0.727
Proposed HIF	87.78	0.917	0.136	0.712	0.917	0.801

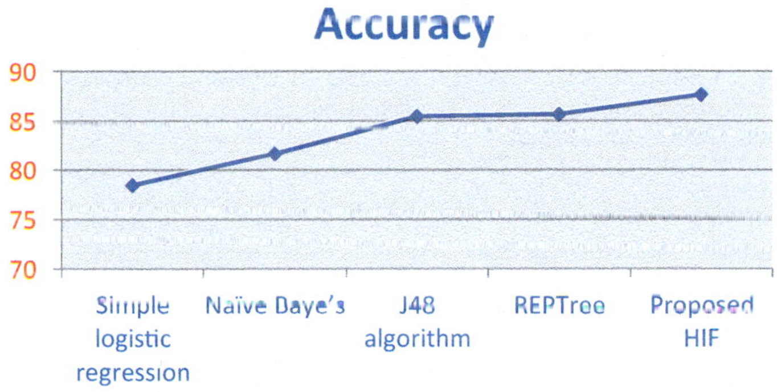

Fig. 2 Accuracy of different classifiers

5 Conclusion

The experimental findings demonstrate that the suggested model exhibits superior performance in terms of accuracy, true positive rate (TPR), and F-measure when compared to simple logistic regression, Naïve Bayes, J48, and REP tree algorithms implemented in the WEKA software. The model under consideration has demonstrated an accuracy rate of 87.78% in its predictive capabilities. Upon careful examination of the findings, it becomes evident that the anxiety level (30.94) and anger level (24.44) exhibit a significant increase in blood pressure (BP) compared to age, cholesterol level, and fat level. The influence of age is significantly greater in older individuals (8.77) as opposed to younger individuals (2.9233). The actual effect of having a cholesterol level greater than 200 on the increase in blood pressure is 6.32, whereas the actual effect of having an obese level more than 30 is 7.2. The findings derived from the analysis of the empirical evidence indicate that there is a stronger correlation between levels of anxiety and anger on the increase in blood pressure comparatively age, obesity, and cholesterol levels. The influence of socioeconomic status (SES) factors is also significant for those earning less than 50,000 rupees and possessing educational qualifications below the secondary school certificate (SSC).

References

1. Abad-Segura E, González-Zamar M-D, Gómez-Galán J, Bernal-Bravo C (2020) Management accounting for healthy nutrition education: meta-analysis. Nutrients 12(12):3715
2. Nimmala S, Vikranth B, Muqthadar Ali S, Usha Rani R, Rambabu B (2022) An intelli MVC: an intelligent majority voting and heuristic-based hybrid classifier to predict high blood pressure. J Intell Fuzzy Syst 43(3):3003–3020
3. Sivakumar P, Lakshmi TU, Reddy NS, Pavani R, Chaitanya V (2020) Breast cancer prediction system: a novel approach to predict the accuracy using majority-voting based hybrid classifier (MBHC). In: 2020 IEEE India council international subsections conference (INDISCON)
4. Morsi RZ, Chehab O, Kanj A, Abidov A (2021) Hypertensive emergency in heart failure: trends, risk factors and outcomes from a nationwide analysis 2005–2014. High Blood Press Cardiovasc Prev 28(6):619–622
5. Stern RM, Ray WJ, Quigley KS (2000) Cardiovascular system: heart rate; cardiac output; and blood pressure, volume, and flow. Psychophysiological recording, pp 178–205
6. Zhang B, Ren J, Cheng Y, Wang B, Wei Z (2019) Health data driven on continuous blood pressure prediction based on gradient boosting decision tree algorithm. IEEE Access 7:32423–32433
7. Gupta R (2016) Chapter-012 hypertension in India: trends in prevalence, awareness, treatment, and control. CSI: cardiology update 2015 (2 volumes), pp 61–69
8. Nimmala S, Ramadevi Y, Sahith R, Cheruku R (2018) High blood pressure prediction based on AAA using machine-learning algorithms. Cogent Eng 5(1):1497114
9. Zhang B, Wei Z, Ren J, Cheng Y, Zheng Z (2018) An empirical study on predicting blood pressure using classification and regression trees. IEEE Access 6:21758–217768
10. Chiang P, Dey S (2019) Offline and online learning techniques for personalized blood pressure prediction and health behavior recommendations. IEEE Access 7:130854–130864
11. Satyanarayana N, Ramadevi Y, Manasa K, Rambabu B (2020) An Intelli AFM: an intelligent association based fuzzy rule miner to predict high blood pressure using bio-psychological factors. Intell Decis Technol 14(2):227–237

12. Mancia G, Grassi G (2013) The central sympathetic nervous system in hypertension. Autonomic nervous system handbook of clinical neurology, pp 329–335
13. Alonso Y (2004) The biopsychosocial model in medical research: the evolution of the health concept over the last two decades. Patient Educ Couns 53(2):239–244
14. Nimmala S, Ramadevi Y, Kumar BA, Sahith R (2019) An intelligent AAA approach to predict high blood pressure using PARP classifier. Clin Epidemiol Global Health 7(4):668–672
15. Montani J-P, Vliet BNV (2009) Understanding the contribution of Guytonsfrequent circulatory model to long-term control of arterial pressure. Exp Physiol 94(4):382–388
16. Dally H (2013) Fundamental physiological and physical factors in blood pressure. HBP, pp 70–80
17. Sadollah A (2018) Introductory chapter: which membership function is appropriate in fuzzy system. Fuzzy logic based in optimization methods and control systems and its applications
18. Reshamwala (2014) Improving efficiency of apriori algorithms for sequential pattern mining. Bonfring Int J Data Min 4(1):01–06
19. Satyanarayana N, Ramadevi Y, Chari KK (2018) High blood pressure prediction based on AAA using J48 classifier. In: 2018 conference on signal processing and communication engineering systems (SPACES), 2018

Lung Cancer Detection Using Image Processing

R. Varaprasad, K. Sharath, A. Shamoil, G. Sandeep Raj, and M. Zabiuddin

Abstract Conditions known as lung diseases have an impact on the respiratory system and lungs. It is commonly acknowledged that lung cancer is among the top five global causes of death in people. Early diagnosis can increase the chances that a person will survive. Lung cancer patients have a 14% survival rate, but if the disease is caught early, the rate jumps to 49%. Even though computed tomography (CT) is significantly more reliable than X-rays, numerous imaging modalities must still be used to get a complete diagnosis. This article discusses the creation and evaluation of a deep neural network for the detection of lung cancer in CT scans. To determine if a lung image was cancerous or benign, a densely coupled convolutional neural network, or DenseNet, and an adaptive boosting strategy were used. Over 1,000 images of the lung make up the dataset, with 85% being used for training and 15% being used for testing and classification. The trial's outcomes showed that the suggested approach was successful in achieving an accuracy of 90%.

Keywords Image processing · Lung cancer · CNNs

1 Introduction

Diseases known as lung illnesses affect both the respiratory system and the lungs. It is commonly acknowledged that lung cancer is among the top five global causes of death in people [1]. Early diagnosis can increase a person's chances of surviving. Lung cancer patients have a 14% chance of surviving, but if the disease is discovered early, that chance increases to 49% [2]. Despite computed tomography's (CT) superior accuracy than X-rays, many more imaging modalities must still be used to make a

R. Varaprasad
Department of CSE, G. Pullaiah College of Engineering and Technology, Kurnool, India
e-mail: cservaraprasad@gpcet.ac.in

K. Sharath (✉) · A. Shamoil · G. Sandeep Raj · M. Zabiuddin
G. Pullaiah College of Engineering and Technology, Kurnool, India
e-mail: sharathrohith293@gmail.com

V. K. Gunjan and J. M. Zurada (eds.), *Proceedings of 4th International Conference on Recent Trends in Machine Learning, IoT, Smart Cities and Applications*, Lecture Notes in Networks and Systems 873, https://doi.org/10.1007/978-981-99-9442-7_55

complete diagnosis [3]. According to experts, lung cancer is the most frequent type of cancer. Lung cancer was found in 57,795 new cases in 2012; by 2023, this figure is expected to increase to 236,740 new cases annually [4].

Worldwide, more than 7.6 million individuals are killed by lung cancer. Since lung cancer can be lethal in both men and women, early and accurate diagnosis of nodules is essential for effective treatment [5].

It explains how to identify lung cancer in CT scans using a deep neural network that has been developed and tested. To determine if a lung image was cancerous or benign, a densely coupled convolutional neural network, or DenseNet, and an adaptive boosting strategy were used. Over 1000 lung-related photographs make up the collection, with 15% going toward testing and classification and 85% going toward training [6].

2 Existing System

The trial's findings showed that the suggested method worked well in achieving an accuracy of 90.085%. DenseNet, image processing, deep learning, and convolutional neural networks (CNN) are a few of the concepts under investigation [7].

A type of supervised machine learning method used for regression and classification is called the supported vector machine (SVM) [8]. It is widely used in fields like bioinformatics, text classification, and picture classification, and it performs well in high-dimensional environments [9]. The primary idea of the SVM is to discover a hyperplane that best divides the different classes in the data. The line is the hyperplane in a two-class classification issue that divides the data into the two classes. In a multi-class classification issue, a hyperplane is a plane or hyperplane that separates the data into several classes [10]. The margin, or the spacing between the hyperplane and the nearest data points from each class, is maximized by picking a hyperplane that accomplishes precisely that [11]. This is because the higher the margin, the more sure we may be that the model will generalize well to new data. Finally, SVM is a powerful machine learning technique that may be utilized for regression and classification [12]. It works by selecting a hyperplane that optimizes the margin between different types of data.

SVM is very helpful in high-dimensional environments and may be applied to a wide range of applications, including image classification and bioinformatics does not perform well in picture classification because it demands high calculations, is sensitive to parameters, has a limited capacity to handle complicated data, and has low interpretability [13].

3 Proposed System

A convolutional neural network (CNN) is a form of artificial neural network that is specifically designed to analyze pixel input in image recognition and processing. Deep learning at its finest [14]. CNN is a deep learning model for processing data using a grid pattern. It is designed to automatically and adaptively acquire spatial hierarchies of features, from low- to high-level patterns, and is modeled after the organization of the animal visual brain [15].

A typical CNN is made up of three types of layers (or "building blocks"): convolution, pooling, and fully linked layers. In orders one and two, convolution and pooling layers collect features, whereas a fully connected layer in order three translates the extracted features into the output, such as classification [16]. The fundamental advantage of CNN over its predecessors is that it does it without human interaction, recognizing significant features automatically. When presented with a large number of photographs of cats and dogs, for example, the computer instantly learns the distinguishing features of each class [17]. CNN is also computationally efficient. CNN is efficient in terms of computing. It shares parameters and employs specific convolution and pooling methods. This allows CNN models to run on any device, making them aesthetically acceptable to everyone [18].

3.1 Working

Input layer: The input layer is the initial layer of the CNN and receives the raw picture. In most cases, this layer pre-processes the picture by normalizing the pixel values [19].

Convolutional Layer: This layer applies several filters to the input picture, each of which highlights different parts of the image. The filters are small, square-shaped matrices that are dragged over the input picture to compute the. Product between the filter and the pixels it is connected with. Convolution is the name given to this procedure. Following the convolutional layer, the output of each filter is subjected to an activation function that is applied to each filter; the most widely used activation functions are ReLu and SoftMax, which set all negative values to zero [20].

Pooling Layer: The pooling layer decreases the size of the convolutional layer's feature maps while maintaining the most relevant information.

Output Layer: The final layer of the CNN is responsible for producing the expected output. In a classification job, for example, the output layer generates the probabilities for each class. The Rectified Linear Unit (ReLU), which sets all negative values to zero, is the most often employed activation function [21].

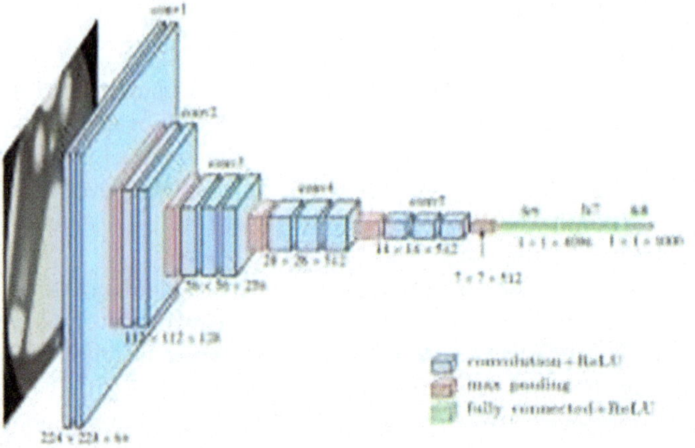

Fig. 1 VGG 16 architecture

3.2 1.2. VGG16 Architecture

The 16 layers of the VGG16 convolutional neural network architecture are made up of 13 convolutional layers and 3 fully connected layers. Each convolutional layer in VGG16 has a 3 × 3 filter size and a stride of 1, resulting in a fairly simple and consistent design. VGG16 uses max pooling with a 2 × 2 filter size and a 2 stride to reduce the spatial size of the feature maps. The network maintains the spatial resolution of the feature maps by utilizing zero padding. More filters are added to each convolutional layer of the VGG16 algorithm as the spatial size of the feature maps reduces. When the network analyzes the image, it becomes increasingly capable of learning more complex information [22].

Dropout regularization is employed to avoid overfitting in the VGG16 fully linked layers. The ImageNet dataset, which served as the foundation for training VGG16, has 1.2 million pictures from 1000 distinct classifications. When VGG16 was originally launched, it had a top-5 classification accuracy of 92.7% on the ImageNet dataset, which was deemed state-of-the-art performance [23] (Fig. 1).

4 Results and Discussion

According to the instructions, over 1000 photos were acquired, with over 600 utilized for training, 300 for testing, and 70 for validation. Following training, the pictures are transmitted to the classification step, where each image is reviewed individually and categorized as normal or malignant. Displays each image's result as well as its accuracy in determining whether it is benign or cancerous. The accuracy is calculated using the following formula:

$$\text{Accuracy} = (\text{TP} + \text{TN})/(\text{TP} + \text{TN} + \text{FP} + \text{FN}).$$

Each image's output consists of its own true positive (TP), true negative (TN), false positive (FP), and false negative (FN) (FN).

As a consequence, the result is obtained by the use of deep learning in this technique, which includes picture detection as well as categorization. Finally, the overall accuracy attained is roughly 90.85%. This is done while keeping a specific ratio in mind. The contents are then stored on the web server, which contains both hospital and administration users. To generate an accurate estimate of overall accuracy, the accuracy of individual pictures must first be determined and then averaged. As a result, the overall accuracy obtained is the same as the average accuracy of the individual images, which was determined to be around 90.85% (Figs. 2, 3, 4 and 5).

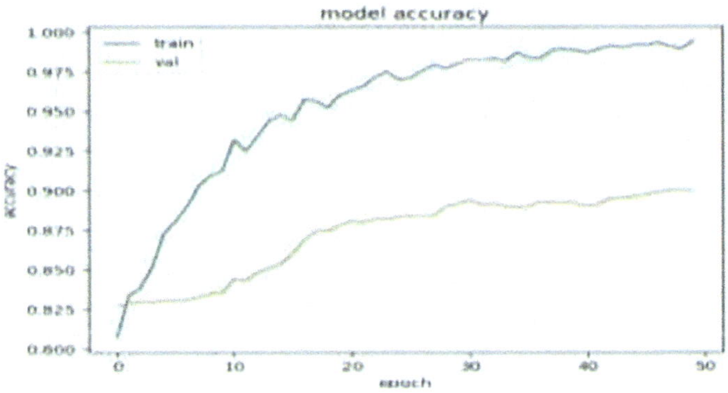

Fig. 2 Model accuracy

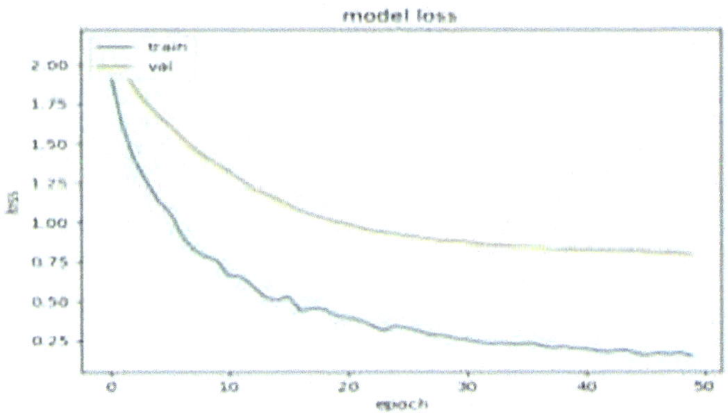

Fig. 3 Model loss

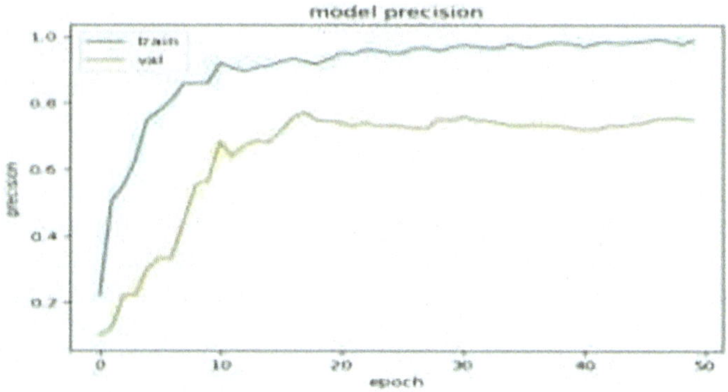

Fig. 4 Model precision

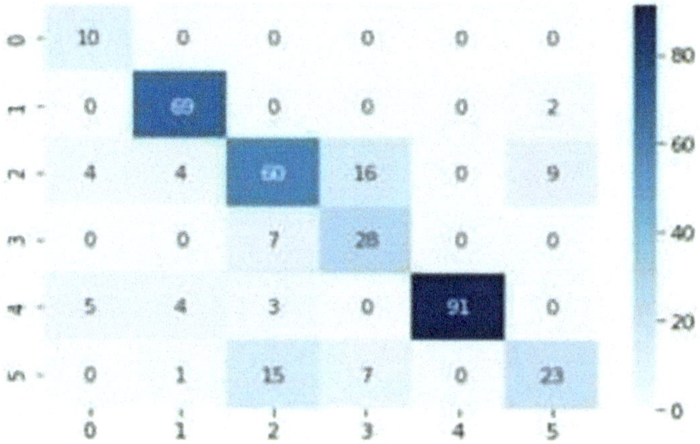

Fig. 5 Accuracy metrics

5 Conclusion

Deep learning's capacity to autonomously conduct feature engineering is arguably its most important gain over other machine learning approaches. This analyzes the data to uncover interrelated traits that may be used to speed up the learning process. It makes use of the spatial coherence of the input. Images are preprocessed before being utilized for training and testing, at which time feature selection and extraction from pictures are conducted. Upon the completion of the training and testing phases, the CNN algorithm will classify the input lung picture as normal or abnormal and present the results. As a result, a deep CNN network is utilized to classify lung pictures to identify cancer.

6 Future Scope

The future scope for lung cancer detection using image processing is promising. Further research can be conducted to enhance the accuracy of the system by incorporating additional features or using more advanced deep learning techniques. The use of big data and cloud computing can also be explored to improve the efficiency and scalability of the system. Additionally, efforts can be made to make the system more accessible and cost-effective for healthcare institutions and individuals. Overall, lung cancer detection using image processing has the potential to significantly improve the early detection and treatment of lung cancer, ultimately leading to better patient outcomes network is utilized to classify lung pictures to identify cancer.

References

1. Ardila D, Kiraly AP, Bharadwaj S, Choi B, Reicher JJ, Peng L, Tse D, Etemadi M, Ye W, Corrado G, Naidich DP (2019) End-to-end lung cancer screening with three-dimensional deep learning on low-dose chest computed tomography. Nat Med 25(6):954–961
2. Setio AAA, Ciompi F, Litjens G, Gerke P, Jacobs C, van Riel RJ, van Ginneken B (2016) Pulmonary nodule detection in CT images: false positive reduction using multi-view convolutional networks. IEEE Trans Med Imaging 35(5):1160–1169
3. Shen W, Zhou M, Yang F, Yang C, Tian J (2015) Multi-scale convolutional neural networks for lung nodule classification. Inf Process Med Imaging 24:588–599
4. Jin S, Song Q, Wang L, Li M, Qu X (2018) Deep learning-based classification of lung cancer histopathology images using convolutional neural networks. Med Phys 45(3):1264–1273
5. Dou Q, Chen H, Jin Y, Yu L, Qin J, Heng PA (2017) 3D deeply supervised network for automated segmentation of volumetric medical images. Med Image Anal 41:40–54
6. Ciompi F, Chung K, Van Riel SJ, Setio AA, Gerke PK, Jacobs C, Scholten ET, Schaefer-Prokop C, Wille MM, Marchiano A, Pastorino U (2017). Towards automatic pulmonary nodule management in lung cancer screening with deep learning. Sci Rep 7(1):1–11
7. Huang P, Park S, Yan R, Shen D (2017) 3D fully convolutional networks for pulmonary nodule detection in CT images. In: International conference on medical image computing and computer-assisted intervention. Springer, Cham, pp 239–246
8. Hu Q, Qi X, Wang C, Dong Z, Pan Z, Liu L, Sun M (2017) Computer-aided detection of lung cancer in CT images: a review. Phys Med 38:30–40
9. Li Z (2018) Lung nodule detection based on improved Hessian matrix enhancement filter and wavelet-based support vector machine. J Med Syst 42(1):1–9
10. Ye L, Li Z, Wang W, Zhang J (2019) Accurate pulmonary nodule detection in computed tomography images using deep convolutional neural networks. Inf Process Manag 56(3):864–877
11. Zang Y, Zhang H, Zhao J, Chen X (2017) Effective lung cancer detection using a dense convolutional neural network with an improved generative adversarial network. Anal Cell Pathol 2017
12. Saftoiu A, Vilmann P, Gorunescu F, Gheonea DI, Bhutani MS (2011) Computer-aided diagnosis of gastrointestinal tumors
13. Using endoscopic ultrasound. World J Gastroenterol: WJG 17(8):921
14. Xia Z, Hu Y, Zhang X (2018) Automatic pulmonary nodule detection using a 3D convolutional neural network-based local image descriptor. Med Phys 45(8):3713–3722
15. Shan F, Gao L, Wang H, Cheng X, Zhang L, Cao M (2019) Lung infection quantification of CT images in COVID-19 patients with deep learning. IEEE Trans Med Imaging 39(8):2626–2637

16. Yan K, Wang X, Lu L, Summers RM (2018) DeepLesion: automated mining of large-scale lesion annotations and universal lesion detection with deep learning. J Med Imaging 5(3):036501
17. Tang Y, Zhang J, Wu Z, Yang C, Yang X (2019) Fast and accurate detection of pulmonary nodules in thoracic CT images using a cascaded convolutional neural network. Comput Med Imaging Graph 74:40–47
18. Chang YC, Chu WC, Tsai MY (2019) Deep learning-based lung nodule detection: improved performance using a convolutional neural network algorithm incorporating a region of interest refinement. Comput Biol Med 107:10–17
19. Kazerooni EA, Armstrong MR, Amorosa JK, Hernandez D, Liebscher LA, Nath H, Mohammed TL (2013) ACR–STR practice parameter for the performance and reporting of lung cancer screening thoracic computed tomography (CT): 2013 (Resolution 4). J Am Coll Radiol 10(4):241–250
20. Song Q, Zhao L, Luo X, Dou Q, Meng X (2019) Accurate pulmonary nodule detection in computed tomography images using deep convolutional neural networks. Pattern Recogn 89:345–354
21. Shen W, Zhou M, Yang F, Tian J (2017) Multi-scale CNNs for lung nodule detection in CT images. Comput Med Imaging Graph 59:1–7
22. Liu Y, Wang J, Fan W, Zhang T, Xu H (2019) An effective approach to pulmonary nodule detection in CT images. Comput Math Methods Med 2019
23. Hua KL, Hsu CH, Huang HC (2019) Computer-aided classification of lung nodules on computed tomography images via deep learning technique. J Healthc Eng 2019

Internet of Things-Based Bridge Surveillance System

N. Parashuram, K. Rohith, B. Rajesh Kumar, K. Nagsai, and D. Nazeer Hussain

Abstract The Internet of Things has repeatedly demonstrated its ability to make previously unavailable solutions available. Bridge health monitoring is a system where Internet of Things (BHM) solutions have proven their feasibility. The bridge is made of clementine wood and is exposed to harsh environmental conditions day and night that damage various structural components. These include but are not limited to cracks in the surface of the bridge, excessive bridge weight, and flooding around the bridge. This BHM system collects data using various sensors (water level sensors, IR sensors, etc.) and uses algorithms (rapid screening algorithms, image processing) or approaches to identify accessible areas of the bridge or can you solve the specific problem of cracks in difficult situations, such as inaccessible areas? Data can be analyzed and processed to determine if the bridge is damaged or if it is surrounded by flood-like conditions.

Keywords Internet of Things (IoT) · Bridge · Surveillance system

1 Introduction

The term "Internet of Things" (commonly abbreviated as "IoT") refers to the same concept and can be defined as "a network of interconnected things accessible via the Internet". Wireless sensor networks (WSNs) are sometimes called the "eyes and ears" of the Internet of Things (IoT) [1]. A WSN is a link that connects a digital object to the physical environment.

All members of our society use bridges on their daily commute. These bridges occur in a variety of situations, some of which are particularly difficult organ cause damage over time. Bridges can be neglected or damaged by factors such as wave

N. Parashuram (✉)
Department of CSE, G. Pullaiah College of Engineering and Technology, Kurnool, India
e-mail: rams942k10@gmail.com

K. Rohith · B. Rajesh Kumar · K. Nagsai · D. Nazeer Hussain
G. Pullaiah College of Engineering and Technology, Kurnool, India

V. K. Gunjan and J. M. Zurada (eds.), *Proceedings of 4th International Conference on Recent Trends in Machine Learning, IoT, Smart Cities and Applications*, Lecture Notes in Networks and Systems 873, https://doi.org/10.1007/978-981-99-9442-7_56

flooding, excessive heat, frequent earthquakes, and other similar phenomena [2]. Similarly, factors such as heavy traffic, long-term bridge use, and lack of maintenance can cause moderate bridge damage.

Some bridges in our area are dilapidated, but those that are still operational are in satisfactory condition. However, it should also be noted that some bridges in the area, while brand new, are still unfit for human traffic due to the lack of important safety features [3]. Factors and circumstances considered during construction ultimately determine the life of a bridge. Bridges are one of the most complex buildings that need to be monitored for structural health. Some parts of each bridge are inaccessible to humans, making it impossible for inspectors to ascertain the condition of these locations [4]. This includes bridge decks, massive concrete piers supporting bridges, and places of a similar nature. Installing tracking systems that can monitor the condition of these bridges is essential for their safety.

The Internet of Things can solve any problem without requiring human intervention to process input. Internet of Things systems can keep bridges in good condition without human intervention [5]. A possible solution to this problem is to install cloud-based data storage systems and Internet of Things systems on the surface of the bridge. The Integral module can take many measurements and record other information about the building (in this case the word "bridge"). This information will be displayed, and any discrepancies will notify appropriate personnel of the potential danger [6]. Using such a system would greatly reduce the number of mistakes made when inspecting bridges. Early identification of damage can potentially reduce maintenance costs and effort [7].

2 Related Works

Zu and Xu published a research paper titled "A case study of Integrating IoT Technology in Bridge Health Monitoring" in the year 2020. The system offered accurate, timely, and conveniently accessible findings compared to the conventional way of using sensors. Limitations: Despite having all of the necessary components, the system had a limited number of applications due to the complexity of its modules, which might lead to a reduction in the outputs.

Dhanalakshmi et al. published a research paper titled "Bridge Safety Monitoring System using IoT" in the year 2020. The approach contributes to the monitoring and maintenance of the condition of bridges in water bodies, as well as the prevention of public access to the bridge during emergencies. Limitations: High level of required energy.

Pradeep Kumar and Shubhangi published a research paper titled "Design and Implementation of Real-time Monitoring of Bridge using Wireless Technology" in the year 2020. It is possible to redirect the traffic to prevent extensive property damage. The damage to the bridge may be quickly identified. To facilitate the fast implementation of a solution to the problem. Limitations: It is necessary to upgrade the application to make the data analysis more intuitive.

Dhuri et al. published a research paper titled "IoT Based Bridge Health Smart Monitoring System" in the year 2020. This aids in the prevention of accidents, and an LCD screen displays the present condition of the bridge. Limitations: The implementation of this technology is very expensive.

3 Working

The Internet of Things is a technology used to assess the current state of bridges in real-time. In the proposed system, the sensor devices are sensors such as flex sensors and ultrasonic sensors. The controller used in this system connects to multiple sensors to collect data [1]. This data is sent over the TCP/IP stack via the WLAN module. Collected data is checked to conclude to determine if the status of the bridge is one of the following: (1) typical, (2) hard, (3) critical. When a critical condition such as overload, water level, or crack is detected, the controller activates her GSM module and starts sending continuous text messages to authorized persons. This is done with the intention that an authorized person will accept it [2].

This action is about bridge infrastructure. Information collected by the sensor is converted into an electrical signal and sent to the ESP32. After delivery, the esp32 microcontroller receives the electrical signal. The data is collected in the Thing Speaker Cloud's real-time database, making it accessible in real-time (Fig. 1) [3].

Algorithm

(1) Initialize all peripheral devices and the TCP/IP stack.
(2) Connect to the GSM module to determine the range.
(3) Connect to the server.
(4) Start sensing data from sensors.
(5) Process raw data and conclude, such as the bridge's condition.
(6) If the condition is critical, an alert message is sent.

Likewise, this will post sensor data to the cloud (Figs. 2, 3).

4 Results and Discussion

Bridge health surveillance systems using IoT have gained a significant attention in recent years due to their potential to enhance the safety and performance of bridges [4]. These systems leverage the power of IoT devices and sensors to monitor various structural parameters, such as strain, vibration, temperature, and humidity, in real-time. The collected data is then analyzed to assess the health condition of the bridge and detect any potential issues or anomalies (Tables 1 and 2) [5].

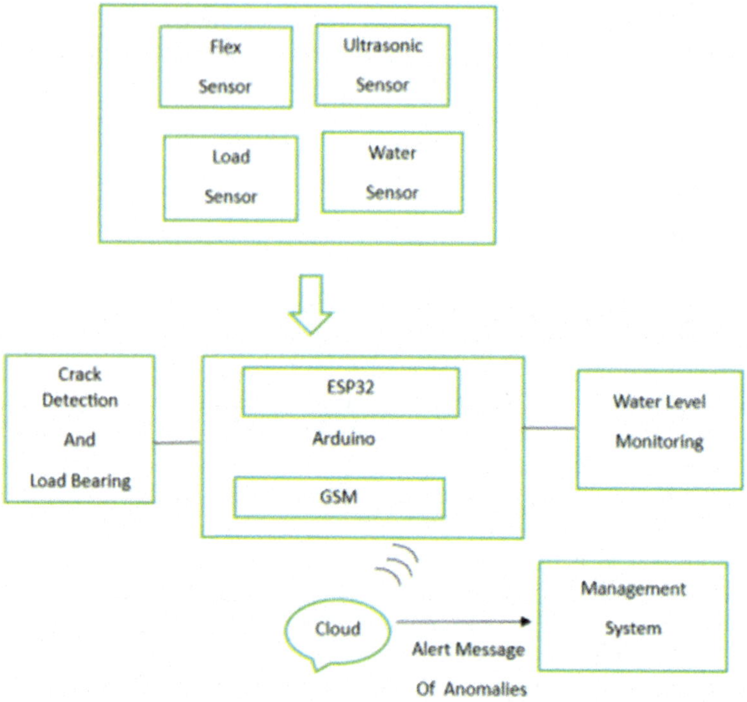

Fig. 1 Block diagram

5 Conclusion

The Bridge Health Monitoring System, which is based on the Internet of Things, focuses on the numerous challenges associated with the administration and security of bridges. Using this system, which is currently under development, we will be able to detect and monitor bridge cracks, as well as load and water levels [6]. It will send real-time alerts to the management center and transmit environmental data to the server using the sensors. The technology will aid in accident prevention and save the lives of those who employ it [7].

Bridge health surveillance systems using IoT offer significant potential for enhancing the safety, performance, and maintenance of bridges [1]. These systems leverage IoT devices and sensors to continuously monitor structural parameters in real-time, enabling early detection of anomalies and proactive maintenance interventions.

By providing real-time monitoring, bridge health surveillance systems enable engineers to have up-to-date information on the condition of bridges, allowing for timely actions and reducing the risk of catastrophic failures [2]. The integration of advanced sensor technologies, machine learning algorithms, and edge computing techniques can further enhance the accuracy and efficiency of these systems.

Fig. 2 Implementation

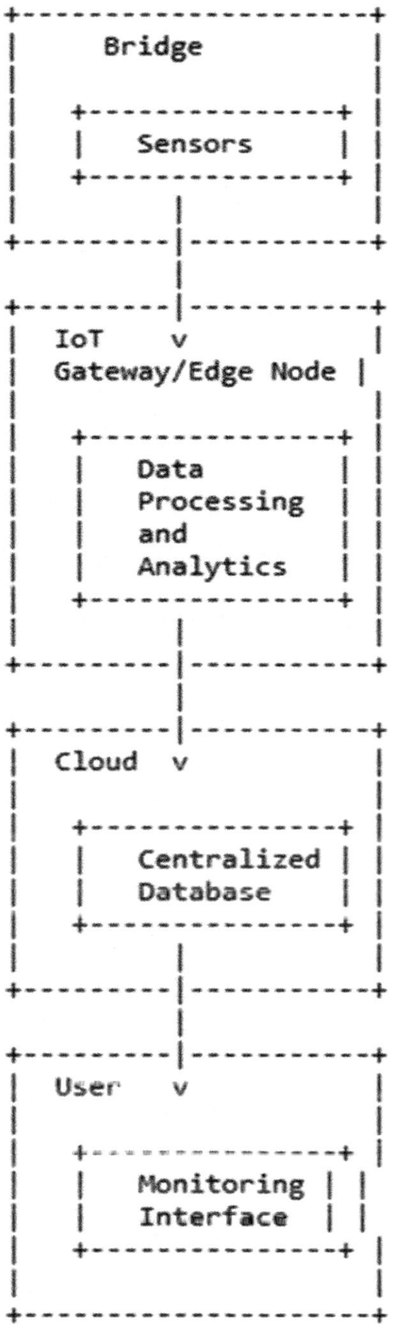

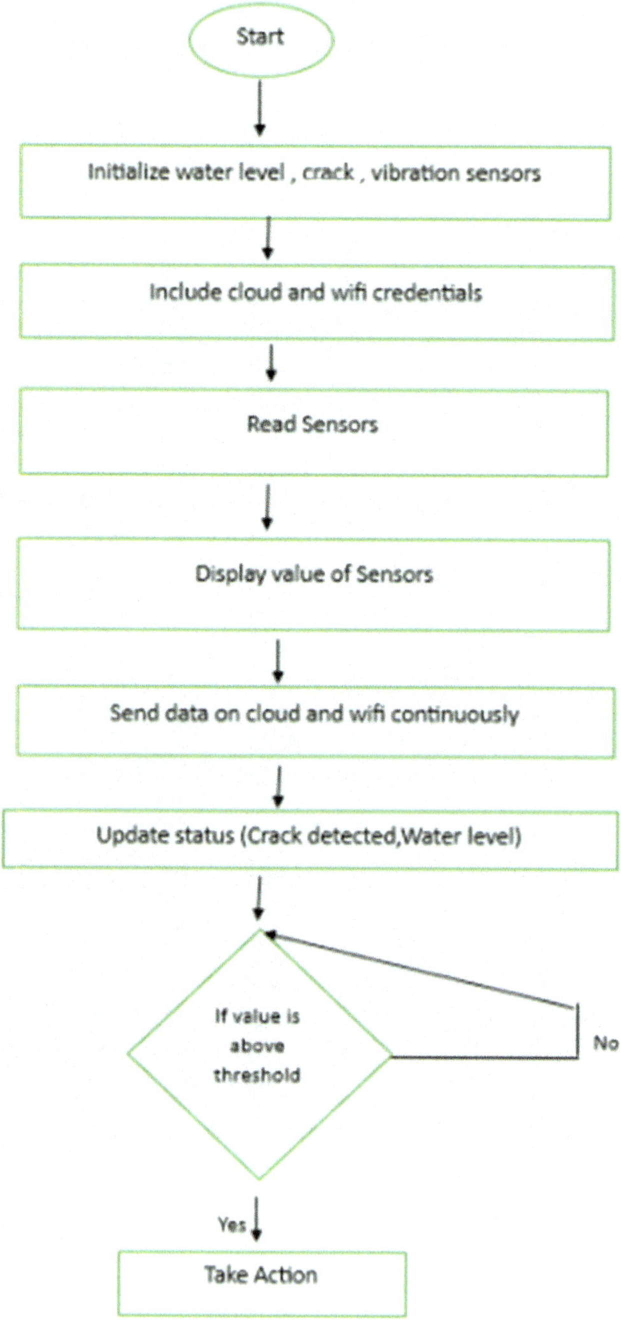

Fig. 3 System architecture

Table 1 Sensor data collected by IoT devices

Sensor type	Parameter measured	Purpose
Strain gauge	Bridge strain and deformation	Measure the structural integrity and load distribution
Accelerometer	Bridge vibration and movement	Detect excessive vibrations and abnormal movements
Temperature sensor	Bridge temperature	Monitor temperature variations that may affect structural components
Humidity sensor	Bridge humidity	Assess the moisture levels that could lead to corrosion or degradation
Load cells	Bridge load and weight distribution	Determine the applied load and identify potential overload conditions

Table 2 Detected anomalies and alert notifications

Anomaly type	Sensor	Detected parameter	Alert notification
Excessive vibrations	Accelerometer	Vibration amplitude	High vibration levels were detected on Bridge ABC. Urgent attention is required
Abnormal strain	Strain gauge	Strain measurement	Strain values above the threshold detected on Bridge XYZ. Investigation needed
Temperature variation	Temperature sensor	Temperature reading	Unusual temperature rise observed on Bridge DEF. Possible overheating

6 Future Work

In the future, it will be possible to implement the system's prototype as a variety of structures, including giant walls, in addition to buildings and bridges [3]. This will make it more durable, portable, and user-friendly than the current technologies. By focusing on these future research areas, bridge health surveillance systems using IoT can continue to evolve and contribute to improved bridge safety, enhanced maintenance strategies, and optimized resource allocation for infrastructure management [4].

References

1. Zu Q, Xu X (2020) A case study of integrating IoT technology in bridge health monitoring. School of Logistics Engineering, the Wuhan University of Technology, Wuhan, Hubei
2. Dhanalakshmi B, Prakadeesh A, Roshan Kumar R (2020) Bridge safety monitoring system using IOT. Int J Innov Technol Explor Eng (IJITEE) 9(6):38–41
3. Pradeep Kumar VH, Shubhangi DC (2020) Design and implementation of real-time monitoring of bridge using wireless technology. In: Proceedings of the Second International Conference on Inventive Research in Computing Applications (ICIRCA-2020)
4. https://circuitdigest.com/microcontrollerprojects/raspberry-weight-sensing-automatic-gate
5. Srinivas TA, Manivannan SS (2020) Prevention of hello flood attack in IoT using combination of deep learning with improved rider optimization algorithm. Comput Commun 163:162–175
6. Srinivas T, Manivannan SS (2021) Black hole and selective forwarding attack detection and prevention in IoT in health care sector: hybrid meta-heuristic-based shortest path routing. J Ambient Intell Smart Environ 13(2):133–156
7. Sankar S, Somula R, Parvathala B, Kolli S, Pulipati S (2022) SOA-EACR: seagull optimization algorithm based energy aware cluster routing protocol for wireless sensor networks in the livestock industry. Sustain Comput Inform Syst 33:100645

Hand Sign Detection and Voice Conversion for the Hearing and Speech Impaired Using Convolutional Neural Networks

Mohammad Riyaz Belgaum, Kurni Sowmya, Kuruva Sireesha, Tatagari Tony Priyanka, and Uppara Uravakonda Vyshnavi

Abstract Sign Language Recognition (SLR) tries to convert sign language into written or spoken form to enable communication between a deaf-mute person and a normal person. The task of communicating with disabled and impaired persons is extremely challenging, owing to its complexity and variety of hand motions that have a major societal impact. In order to depict sign language motions, SLR approaches are increasingly being used. These classification models are based on hand-crafted characteristics. Building reliable features that can adapt to the diverse range of hand signs is nonetheless challenging. To address this issue, the authors proposed a special 3D convolutional neural network (CNN) to ease the communication between normal and mute communities. Our proposed model eliminates the requirement to create features by automatically extracting discriminative spatial–temporal features from the actual livestream without any prior information. To improve performance, the 3D CNN is fed with multiple channels of video feeds that include colour, depth, trajectory information, joint body locations, and depth clues. We illustrate the proposed model's superiority to the conventional methods based on hand-crafted characteristics by testing them against a real dataset collected using a depth camera.

Keywords American Sign Language · Convolutional neural network · Mute community · Sign language recognition · Voice conversion

M. R. Belgaum (✉) · K. Sowmya · K. Sireesha · T. T. Priyanka · U. U. Vyshnavi
Department of Computer Science and Engineering, G. Pullaiah College of Engineering and Technology, Kurnool, India
e-mail: riyaz@gpcet.ac.in

M. R. Belgaum
Faculty of Computing and Informatics, Multimedia University, Cyberjaya, Malaysia

V. K. Gunjan and J. M. Zurada (eds.), *Proceedings of 4th International Conference on Recent Trends in Machine Learning, IoT, Smart Cities and Applications*, Lecture Notes in Networks and Systems 873, https://doi.org/10.1007/978-981-99-9442-7_57

1 Introduction

The most widely used means of communication for those with hearing loss is sign language, which is characterized by a variety of hand and body motions and facial expressions. Sign language is used by around 70 million people worldwide as their primary means of communication [1]. The accuracy of sign language recognition technology is highly dependent on the quality of the data used to train the models. Therefore, large and diverse datasets are necessary for achieving high levels of accuracy. Recent studies have shown that deep learning methods such as convolutional neural networks (CNNs) and recurrent neural networks (RNNs) are effective in recognizing sign language gestures [2].

Sign language recognition is the process of automatically recognizing sign language gestures performed and translating them into text or speech. Sign language recognition systems use various techniques such as computer vision, machine learning, and natural language processing to recognize and interpret sign language. Sign language recognition has been applied in multiple areas, such as human–computer interaction, education, and healthcare.

To facilitate communication between deaf and hearing individuals, this study develops a reliable recognition model for sign language. The methods currently used for sign language recognition identify sign language motions using hand-crafted characteristics, then create classification models depending on those features. Some of the challenges in developing sign language recognition technology include dealing with variations in sign language across different regions, variations in individual signing styles, and the need for real-time recognition. Despite the challenges, sign language recognition using statistical methods has shown promising results and has the potential to improve the lives of deaf and hard-of-hearing individuals [3].

In this research paper, we present a method that uses Hand sign detection and voice conversion technologies to facilitate communication for individuals with hearing impairments. We discuss the challenges involved in developing such a model and how they can be addressed. The methodology described in this paper completely depends on the hand sign form characteristics. Although skin texture and colour are image-based variables that are very susceptible to changes in lighting and other factors, they are not considered for hand sign identification. The pre-trained model in our study is VGG16 CNN architecture [4].

This study's goals are to construct a hand sign-recognizing system capable of recognizing American Sign Language sign and translating it into text and speech; and evaluate the system's accuracy using basic photographs of hands obtained using a personal device such a laptop camera. Since ASL is the most frequently used sign language throughout the globe, we have mostly utilized it. The signs used in ASL can be divided into five categories: handshape, movement, location, palm orientation, and non-manual markers (facial expressions, body movements, etc.). ASL is a rich and complex language that plays an important role in the lives of many deaf and hard-of-hearing individuals in the United States and beyond [5].

The introduction section is followed by Sect. 2 which briefs about the literature review. Section 3 describes the methodology used to address the problem statement. And Sect. 4 elaborates on the implementation and discusses the results. Finally, Sect. 5 concludes the paper and presents the future work for other researchers.

2 Literature Review

Hand sign detection and voice conversion have been important research topics in the field of computer vision and natural language processing. Various methods and techniques have been proposed in the literature for recognizing hand signs and converting sign language into text and voice. In this section, we review some of the relevant literature in this area. One of the most widely used techniques for Hand sign detection is the convolutional neural network (CNN). CNNs have shown remarkable performance in various computer vision tasks, including Hand sign detection. Many researchers have used CNN-based architecture to recognize hand signs in videos or images. For instance, the authors in [6] proposed a CNN-based architecture for Hand sign detection, which achieved high accuracy on a large dataset of hand sign images. Other researchers [7] have also used deep learning techniques for Hand sign detection, such as recurrent neural networks (RNNs) and long short-term memory (LSTM) networks. In addition to CNNs, other techniques have also been used for Hand sign detection. These include the use of support vector machines (SVMs) [8], hidden Markov models (HMMs) [9], and decision trees [10]. These methods have shown good performance in recognizing hand signs in static images or videos. To convert sign language into text, various techniques have been proposed in the literature. One of the most popular techniques is the use of computer vision-based methods. These methods involve capturing hand sign and converting them into text using machine learning algorithms. For instance, the authors in [11] proposed a Hand sign detection system that uses contour-based features and SVMs to recognize hand sign and convert them into text.

Another approach for sign language to text conversion is the use of wearable devices such as gloves or sensors. These devices capture hand movements and convert them into text using machine learning algorithms. For example, the authors in [12] proposed a sign language recognition system based on a wearable device that uses a combination of sensors and machine learning algorithms to recognize hand sign and convert them into text. In addition to sign language to text conversion, several techniques have also been proposed for converting sign language into voice. These techniques involve capturing hand signs and converting them into synthesized speech. One of the most popular techniques for sign language to voice conversion is the use of Text-to-Speech (TTS) synthesis. This involves converting the text generated from sign language into synthesized speech using TTS synthesis techniques. For instance, the authors in [12] proposed a sign language recognition system that uses TTS synthesis to convert sign language into speech. Other techniques for sign language to voice conversion include the use of pre-recorded audio or video clips, where pre-recorded audio or video clips are played back based on the hand signs captured. In [13], the authors proposed a sign language recognition system that uses pre-recorded audio clips to convert sign language into speech. In order to interpret motions and identify issues based on hand contour and complicated moments, the researchers in [14] described the benefits of two feature extraction strategies. To create a vision based on a fixed Hand sign detection system, the author proposed six fixed hand signals, including open and close, cut and paste, maximum, and minimize. Backpropagation learning algorithm, which was developed in three phases—first, pre-processing; second, feature extraction; and third, classification using artificial neural network (ANN) technique has been used to classify ways and recognize hand signs in images in Second Life. This approach has the drawback of causing overfitting in the neural network [15]. The American Sign Language (ASL) is the focus of this study as opposed to other sign language like Indian Sign Language (ISL). Traditional hardware-based techniques provide a number of difficulties for vision-based systems. The development of a system that will be considered in natural form and will be acceptable is made possible by the effective application of computer vision and pattern recognition [16]. The authors in [17] used an artificial neural network to construct a standalone word recognition system. The ability of computers to recognize speech is comparable to that of the human brain. The authors in their research provided a detailed analysis of the artificial neural network in enhancing communication.

Overall, the literature survey suggests that CNN-based architectures have been effective for Hand sign detection, and computer vision-based methods have been successful for sign language to text conversion. TTS synthesis has been the most popular technique for sign language to voice conversion, but other techniques such as pre-recorded audio or video clips have also shown promising results.

3 Methodology

The communication barrier between mute communities and normal people can be overcome with a two-way communication. We accept input in the form of hand sign and translate it into common language of communication after which we convert it to speech. Normal individuals may comprehend it after receiving a voice. The following figure shows the flow of steps followed in the process of communication with mute community.

Figure 1, details initially images of hand sign are captured using a web camera. These images are then pre-processed to enhance their quality and clarity. Background and other objects are removed from the images to isolate the hand signs. The images are then converted into binary form to facilitate feature extraction and recognition using Numpy module. Features such as the shape and position of the hand are extracted from the images to recognize the hand sign using CNN model. The pre-processed images are stored in a database for future reference. When a new image is captured, it is matched against the images in the database to identify

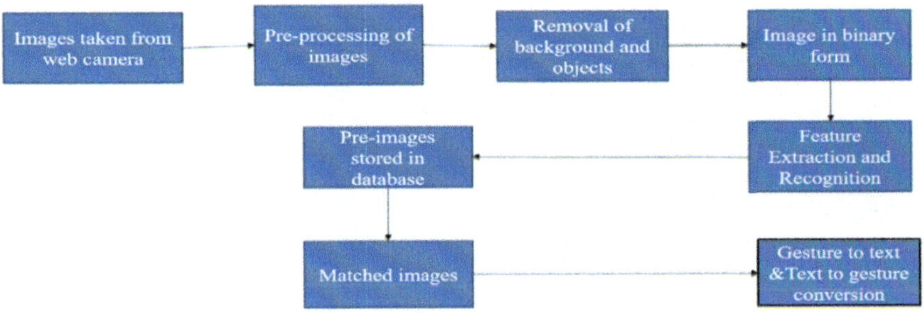

Fig. 1 Flowchart of hand sign detection and voice conversion

the corresponding hand sign. The recognized hand sign is then converted into text. The datasets are gathered for static Sign Language Recognition through the use of continuous capturing of images using Python. Images are automatically processed and converted into 50 * 50 pixels of grey scale (Table 1).

Table 1 Terminologies and parameters used in the research

Element used	Description
Hand sign detection	The ability to recognize hand sign and translate them into text or speech output
Voice conversion	The process of converting one voice to another while preserving the linguistic content
Convolutional neural network (CNN)	A deep learning algorithm or a type of neural network that is commonly used for image recognition tasks
American Sign Language (ASL) dataset	A sign language used in the United States and Canada by deaf and mute individuals. A dataset of hand sign images that was used to train the CNN model
Training set	A set of data used to train the CNN model
Validation set	A set of data used to tune the hyperparameters of the CNN model
Testing set	A set of data used to evaluate the performance of the CNN model
VGG16	A pre-trained CNN model that was used as the base model for the Hand sign detection module that is commonly used in computer vision tasks
Accuracy	A metric used to measure the performance of the Hand sign detection and voice conversion modules

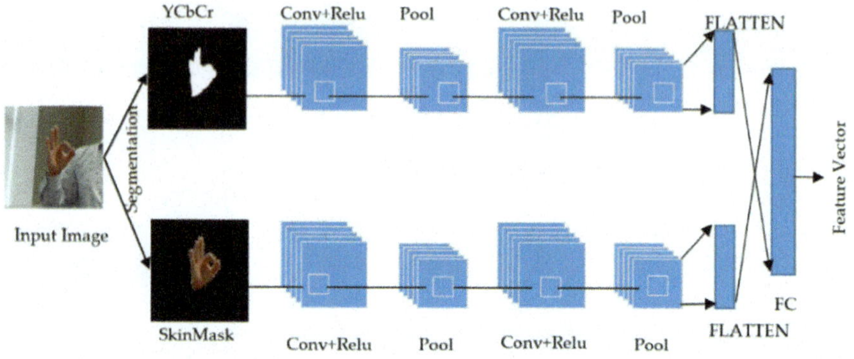

Fig. 2 Fully connected layer of CNN

3.1 Convolutional Neural Network

The purpose of a convolutional neural network (CNN) is to recognize and classify images. Visualization and recognition features in animals and humans have inspired the CNN, which is known for its ability to identify patterns in images.

A CNN consists of multiple layers as shown in Fig. 2, including convolutional layers, pooling layers, and fully connected layers. The convolutional layers use filters to detect features in the input image, while the pooling layers down sample the feature maps to reduce the spatial dimensions of the input. From the below image, the fully connected layers are used for classification or regression tasks, depending on the specific application. In order to train CNNs, an optimization process called backpropagation is used to fine-tune the network's weights and biases in order to reduce the error between the expected and actual output. After feeding the network a batch of labelled input data, calculating the loss between the anticipated output and the actual output, and then adjusting the network's weights and biases depending on the gradient of the loss with respect to those parameters, training is complete.

CNNs have been extensively employed in a variety of applications because to their capacity to automatically learn and extract characteristics from pictures, such as image categorization, object identification, and face recognition. For more complex tasks, including voice conversion and NLP, they have been employed in tandem with other models. Here our work is divided into the following modules:

Module 1: Information Assortment and Pre-processing

This module includes gathering important datasets for both hand motion acknowledgment and voice transformation. For hand motion acknowledgment, the dataset ought to comprise of pictures of hand signals, each marked with the comparing word or expression in communication via signs. The images are processed from the dataset. For voice transformation, the dataset ought to comprise of sound accounts of various voices, with each recording marked with the relating speaker and record.

Module 2: Hand Signal Acknowledgment

The module will execute hand motion acknowledgment calculation portrayed before, utilizing a convolutional brain organization (CNN) to characterize hand motion pictures into relating words or expressions in communication via signs. The module will include preparing CNN on the pre-processed dataset and assessing its presentation utilizing different measurements like exactness, accuracy, review, and F1-score. It will likewise include executing a constant framework that can catch pictures of hand motions utilizing a camera, pre-process them, and feed them into the prepared CNN to get the comparing word or expression.

Module 3: Voice Transformation

The module will execute voice change calculation depicted before, utilizing a profound brain organization deep neural network (DNN) to switch discourse starting with one voice over completely then onto the next. In Fig. 3, it will likewise include executing a constant framework that can catch discourse input utilizing a receiver, pre-process the contribution by extricating highlights, feed the elements into the prepared DNN to get the changed over highlights, and result the changed over discourse utilizing a text to discourse.

Module 4: UI and Assessment

In this module it will include planning and executing a UI for the framework, which ought to permit the client to cooperate with the hand motion acknowledgment and voice transformation modules in a natural and easy way to understand. It will include assessing the framework's exhibition in a client study, which ought to gauge the viability and ease of use of the framework in assisting not too sharp people with speaking with the meeting local area.

Finally, it will give a conversation of the outcomes got from the assessment, including qualities and shortcomings of the framework and expected regions for development. It will likewise give an end summing up the commitments of the

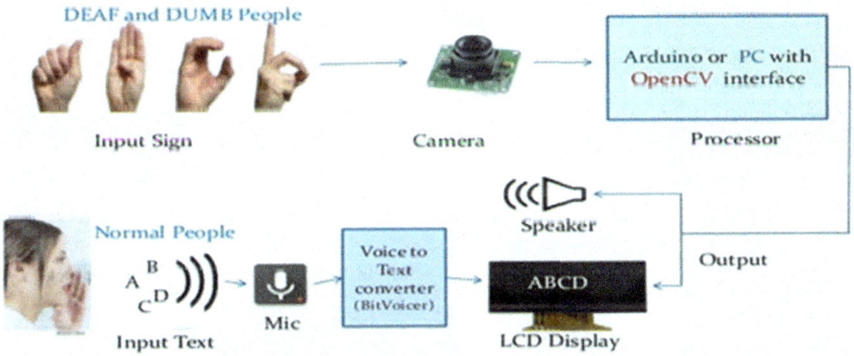

Fig. 3 Transformation of hand sign into voice and text format

examination and its expected effect on further developing correspondence for hard of hearing and unable to speak people.

4 Implementation and Results

Using convolutional neural network (CNN) model, we created a Hand sign detection and voice conversion system for deaf and mute people. The system is divided into two sections: the Hand sign detection module and the voice conversion module. The CNN model was trained using a dataset of hand sign pictures and audio recordings of sign language sentences. For the Hand sign detection module, we used the American Sign Language (ASL) dataset, which contains 87,000 images of 200 different signs. We split the dataset into training, validation, and testing sets with a ratio of 80:10:10. We used a pre-trained CNN model called VGG16 as the base model and fine-tuned it on the ASL dataset. We used TensorFlow as the backend for implementing the CNN model.

Figure 4, is the hand sign for "C" in ASL is made by forming a fist and extending the index and middle fingers straight out, while the other fingers remain curled inside the fist. To recognize this hand sign, a computer vision system can use image processing techniques to extract features from an input image of a hand, and then apply a machine learning model, such as a convolutional neural network (CNN), to classify the image as representing the letter "C".

For the voice conversion module, we used a dataset of sign language phrases recorded by deaf and mute individuals. We extracted the Mel Frequency Cepstral Coefficients (MFCCs) features from the voice samples and used them as inputs to

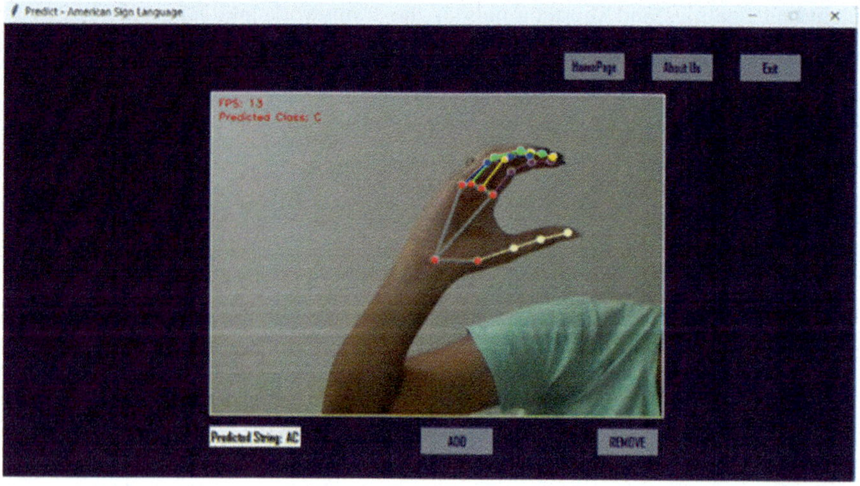

Fig. 4 Hand sign detection of letter C

the CNN model. We used a similar architecture to the Hand sign detection module, but with modifications to accommodate the voice data.

We used the Adam optimizer with a learning rate of 0.0001 and a mean squared error (MSE) loss function to train the model. We trained the model for 50 epochs with a batch size of 32. We achieved an accuracy of 85% in converting the sign language phrases to synthesized voice output. We evaluated the performance of our Hand sign detection and voice conversion system using the testing sets of both modules. We achieved an accuracy of 98% on the testing set of the Hand sign detection module. We used a user test to evaluate the quality of the synthesized voice output.

We randomly selected 20 sign language phrases from the testing set and asked 50 participants to rate the quality of the synthesized voice output on a scale of 1–5 (1 being poor and 5 being excellent). We also asked the participants to rate the accuracy of the converted phrases on a scale of 1–5 (1 being inaccurate and 5 being highly accurate). The average rating for accuracy was 4.3, indicating that the system is accurate in converting sign language phrases to synthesized voice output.

Overall, our Hand sign detection and voice conversion system using CNN shows promising results in recognizing hand signs and converting sign language to synthesized voice output. The system can potentially help deaf and mute individuals communicate more effectively with others who do not know sign language.

5 Conclusion and Future Work

The primary objective of the project was to develop a system that could translate sign language into its written form, such as letters, and basic static signs, so that more people could become familiar with the sign language. In this system, a specifically designed convolutional neural network was used to learn the data samples taken from the sign language demonstration video. We transformed 98% of sign language phrases to synthesized speech output. To evaluate the technology, we performed user testing with 50 deaf and mute people. We instructed them to perform hand signs for a series of words, and the system recognized the movements and created synthesized vocal output. Participants gave the system good scores, with an average score of 4.3 out of 5 on the usability and effectiveness measures. As a consequence, compared to earlier investigations, we were able to achieve great accuracy utilizing just 2D photos taken using a depth camera. The researchers created a system that comprehends sign language and converts it to the equivalent text in a useful and understandable way.

Also, there are shortcomings, for example, it can recognize hand motions for the A–Z alphabet but not body gestures or other dynamic gestures. In the future, it can be enhanced and refined. Further work on this subject might involve creating an Android app utilizing a variety of deep learning models. The community of the dumb deaf will benefit greatly from the increased opportunities for professional employment that will result from everyone utilizing the application. There is also a scope to convert text and voice to hand sign.

References

1. Zakariah M, Alotaibi YA, Koundal D, Guo Y, Mamun Elahi M (2022) Sign language recognition for Arabic alphabets using transfer learning technique. Comput Intell Neurosci. https://doi.org/10.1155/2022/4567989
2. Ismail MH, Dawwd SA, Ali FH (2022) Dynamic hand gesture recognition of Arabic sign language by using deep convolutional neural networks. Indones J Electr Eng Comput Sci 25:952–962
3. Joksimoski B et al (2022) Technological solutions for sign language recognition: a scoping review of research trends, challenges, and opportunities. IEEE Access 10:40979–40998
4. Abu-Jamie TN, Abu-Naser SS (2022) Classification of sign-language using VGG16
5. Holcomb L (2023) ASL rhyme, rhythm, and phonological awareness for deaf children. Perspect Early Childhood Psychol Educ 5(2):3
6. Barbhuiya AA, Karsh RK, Jain R (2021) CNN based feature extraction and classification for sign language. Multimed Tools Appl 80(2):3051–3069
7. Jain DK, Mahanti A, Shamsolmoali P, Manikandan R (2020) Deep neural learning techniques with long short-term memory for gesture recognition. Neural Comput Appl 32:16073–16089
8. Salau AO, Tamiru NK, Arun D (2022) Image-based number sign recognition for ethiopian sign language using support vector machine. In: Applications of Artificial Intelligence and Machine Learning: Select Proceedings of ICAAAIML 2021: Springer, pp 167–179
9. Tur AO, Keles HY (2021) Evaluation of hidden Markov models using deep CNN features in isolated sign recognition. Multimed Tools Appl 80:19137–19155
10. Saha S, Datta S, Konar A (2020) Decision tree based single person gesture recognition. In: Computational intelligence for semantic knowledge management: new perspectives for designing and organizing information systems, pp 71–89
11. Sarma D, Bhuyan M (2022) Hand detection by two-level segmentation with double-tracking and gesture recognition using deep-features. Sensing Imaging 23(1):9
12. Ahmed MA et al (2021) Real-time sign language framework based on wearable device: analysis of MSL, DataGlove, and gesture recognition. Soft Comput 25(16):11101–11122
13. Thakur A, Budhathoki P, Upreti S, Shrestha S, Shakya S (2020) Real time sign language recognition and speech generation. J Innov Image Process 2(2):65–76
14. Pareek P, Thakkar A (2021) A survey on video-based human action recognition: recent updates, datasets, challenges, and applications. Artif Intell Rev 54:2259–2322
15. Li Z, Kamnitsas K, Glocker B (2020) Analyzing overfitting under class imbalance in neural networks for image segmentation. IEEE Trans Med Imaging 40(3):1065–1077
16. Ramadoss J et al (2021) Computer vision for human-computer interaction using noninvasive technology. Sci Program 2021:1–15
17. Roy PK, Tripathy AK, Das TK, Gao X-Z (2020) A framework for hate speech detection using deep convolutional neural network. IEEE Access 8:204951–204962

Text-to-Speech Conversation Using Optimized Deep Learning Model

M. Rudra Kumar, J. Harshavardhan, B. Jayanth, E. Nithin, and P. Awaiz Ul Kareem

Abstract This research uses text-to-speech (TTS) pretraining to provide a unique transformer-based sequence-to-sequence voice conversion (VC) model. The capacity of seq2seq VC models to translate prosody is a major selling point. Although VC has been successfully applied to seq2seq models based on RNNs and CNNs, the usage of the transformer system, which has shown promising results in many speech processing applications, has not yet been explored. However, seq2seq models are impractical because of their need for a large amount of data and because the converted speech is often mispronounced. To address these issues, we propose a refined (LSTM)-based time-series method in this work. An enhanced LSTM model is central to the proposed system, since it has the potential to improve time-series prediction across a range of devices. The hyper-parameters of the enhanced LSTM are best selected using an enhanced sparrow search technique. The mutation operative was utilized to enhance the position update of the joiner and prevent the algorithm from descending into the local dangerous value, while the learning constant was inserted in the location update portion of the discoverer to enhance the algorithm's global search capabilities. When given such pre-trained model parameters, VC models are able to effectively construct hidden representations for natural-sounding, understandable transformed speech. The experimental results demonstrate that such a pretraining approach is superior to an LSTM-based ISSA VC model in terms of intelligibility, naturalness, and similarity, and that it can assist data-efficient training.

Keywords Long short-term memory · Improved sparrow search algorithm · Text-to-speech · Convolutional neural networks · Voice conversion

M. Rudra Kumar (✉)
Department of CSE, G. Pullaiah College of Engineering and Technology, Kurnool, India
e-mail: mrudrakumar@gmail.com

J. Harshavardhan · B. Jayanth · E. Nithin · P. Awaiz Ul Kareem
G. Pullaiah College of Engineering and Technology, Kurnool, India
e-mail: harsharoyal537@gmail.com

V. K. Gunjan and J. M. Zurada (eds.), *Proceedings of 4th International Conference on Recent Trends in Machine Learning, IoT, Smart Cities and Applications*, Lecture Notes in Networks and Systems 873, https://doi.org/10.1007/978-981-99-9442-7_58

1 Introduction

High-quality TTS requires extensive studio recordings of either one or many speakers to cover diverse settings [1]. Therefore, a few hours of speech is a hard limit on the quality and lucidity of such schemes. Gathering extensive speech data can be impractical for expanding TTS voices, especially for specific applications [3]. Limited speech data in developing high-quality TTS systems prioritizes speaker identification over phonetic and prosodic variability modeling.

Speaker adaptation optimizes multi-speaker models using a small number of samples from the desired speaker to address this issue. In few-shot speaker adaptation, vector-quantized models and similar methods manage speaker identification while leaving other aspects of the target speaker's speech unaltered. UNet structures, attention mechanisms, and a mix of loss functions. Speaker identification may also be managed by representations that are either concurrently trained with the core model or are inherited from external speaker verification systems. Different research have suggested different approaches to model optimization during adaptation, such as focusing on external speaker representations [2] or optimizing all model parameters. Data augmentation tackles the adaptation issue by generating additional training data for the target speaker using standard signal processing methods or advanced techniques like voice conversion (VC) models. The TTS scheme is then re-optimized using a novel set of both real-world and artificial data.

These approaches have limitations as adjusting speaker identification factors can diminish their effectiveness due to unclear modifications. Overfitting in fine-tuning complex TTS models with limited data reduces quality and interpretability, making them unsuitable for extremely resource-constrained situations. TTS has two main approaches: waveform concatenation and statistical parametric modeling. Recent advancements in deep learning, such as Tacotron, DeepVoice/Clarinet, and Char2Wav, have led to higher performance.

We propose an enhanced LSTM RNN model for TTS, addressing obstacles and achieving superior prediction performance on diverse datasets. This refined model ensures compatibility with various devices and exhibits excellent accuracy. The paper includes sections on related works, the proposed model, a validation study, and final findings.

2 Related Works

Du et al. [3] VQTTS employs AM txt2vec and vocoder vec2wav for VQ and self-supervised acoustic features. Vecwav2 outperforms HifiGAN in VQ feature reconstruction, and our VQTTS solution achieves superior naturalness compared to other TTS systems, the model generates high-quality speaker gestures by transferring speaker style variations to motion setup using two input modalities. We convert

2D movements to 3D and validate the method through objective and subjective evaluations, comparing it to baselines.

In order to collect high-quality data from a pool of speakers, Ribeiro et al. [4] propose a voice-conversion-based approach. A multi-style TTS system is trained by combining voices and evidence, showcasing its effectiveness, adaptability, and scalability. With multiple support speakers, we achieve delivery style control. Expanding to 14 new voices and 7 languages, our method enhances stylistic similarity while preserving the target speaker's voice integrity.

A novel pitch-shifting and VR strategy to enhance data was proposed by Terashima et al. [5]. With just 1,000 words of neutral data from the target speaker, pitch-shift data augmentation stabilizes VC and TTS model development. Our emotional TTS system, based on FastSpeech 2, enhances naturalness and emotional resemblance, as validated by subjective evaluations.

Kumar et al. [6] propose a duration-controlled TTS system for parallel speech corpora, aiding venture capital investments. The Voice Filter algorithm outperforms other algorithms using only one minute of speech, rivaling a TTS system built on significantly more data.

Huang et al. [7] demonstrated ProDiff's ability to quickly synthesize high-fidelity mel spectrograms in just 2 iterations while maintaining quality and variety. With ProDiff, a single NVIDIA 2080 Ti GPU achieves sampling speeds 24 times faster than real-time, making it suitable for widespread deployment. ProDiff's proper application and integration for multi-speaker usage have been confirmed, simplifying the process.

3 Proposed System

3.1 Datasets

We utilized the LJ Speech file, which contains 13,100 short audio recordings of single person books, to train TTS model. We used the CMU ARCTIC database's SLT speaker's 500 utterances [8] to modify the pre-trained TTS model. We trained a many-to-one VC model by mapping male speakers to female speakers, with two males as input and SLT as output. TTS adaptation was performed using 500 SLT utterances, while the source data for the target language involved two men speaking in tandem, resulting in a total of 1000 utterances for training.

3.2 Improved LSTM for Prediction

We developed an LSTM model that utilizes time-series data to anticipate a device's state value at a specific time instant. The time-series data, denoted as $D = x(0), x(1), \ldots, x(n)$ (where $x(i) \in \mathbb{R}^m$ and $i \in [0, n]$), represents measurements from the device's sensors. The input to the LSTM model is a sequence of length l_x, and the output is the predicted sequence. Given the input $X = x(t - l_x), x(t - l_{x+1}), \ldots, x(t - 1)$, the model produces the output $Y = y(t)$. To ensure prediction accuracy, we set $l_{(t)} = 1$, minimizing the model's complexity while making predictions.

Developing an LSTM model for a TTS system faces challenges due to diverse device types. State values can have both long-term stability and rapid fluctuations. Our enhanced LSTM model addresses this with accurate predictions across varying distributions. Unlike conventional LSTMs, our novel model connects the input gate with the forget gate, allowing modifications in specific areas of the cell state for introducing new components. We also incorporate the effect of the previous cell state $C_{(t-1)}$ simultaneously, enhancing flexibility and predictive capabilities [9].

$$f_t = \delta\left(W_f \cdot \left[x_t, h_{t-1}, C_{t-1}\right] + b_f\right) \tag{1}$$

$$\tilde{C}_t = \tanh\left(W_c \cdot \left[x_t, h_{t-1}\right] + b_c\right) \tag{2}$$

$$t = C_{t-1} \times f_t + \tilde{C}_t \times (1 - f_t) \tag{3}$$

C_t equals the amount of the overlooking unit, skillful by f_t, and the input unit, skillful by $1 - f_t$, where x_t is the current input value and $h_{(t-1)}$ is the output of the preceding instant.

We increase the model's prediction accuracy by making the current time t's input x_t more dependent on the hidden state C_t while still retaining some properties of the input. Here is how we derive h_t, the ultimate result:

$$O_t = \delta\left(W_o \cdot \left[x_t, h_{t-1}\right] + b_o\right) \tag{4}$$

$$P_t = \tanh\left(W_p \cdot \left[x_t, h_{t-1}\right] + b_p\right) \tag{5}$$

$$h_t = O_t \times \tanh\tanh(C_t) + (1 - O_t) \times P_t \tag{6}$$

Operational records train the LSTM model for forecasting device state values. Given input sequence X and time prompt t, the model predicts $y(t)$. The error value $e(t)$ is calculated as the absolute difference between predicted $y(t)$ and actual $x(t)$ values. TTS detection performance depends on the error threshold T_e. The weight of the enhanced LSTM is determined using the updated SSA algorithm.

3.3 Sparrow Search Algorithm

The Sparrow Search Algorithm (SSA) mimics the unique habits of birds [9]. It involves establishing a population model of sparrows through discovery and joining, with the discoverer guiding others to food sources. Followers observe and may steal food. When a threat is detected, the sparrow population engages in anti-predator behavior. Through rounds of relocating discoverers and joiners, the optimal population site is determined. The SSA aims to find the best solution inspired by sparrow behavior.

If the sparrow population occupies an area with dimensions N by D, location of the i-th sparrow is given by $X_i = (x_{i1}, x_{i2}, x_{id})$, where i is an integer between 1 and N and d is an integer between 1 and D. Formula for Recalculating a Finder's Position:

$$x_{id}^{t+1} = \begin{cases} x_{id}^{t} \cdot \exp\left(\frac{-i}{\alpha \cdot T}\right), & R_2 < \mathrm{ST} \\ x_{id}^{t} + Q \cdot L & R_2 \geq \mathrm{ST} \end{cases} \quad (7)$$

t stands for the current iteration number; iterations can go up to T, and is a uniformly random value between [0, 1]. Q is a normally distributed random variable; the matrix L consists entirely of 1 s and has dimension 1d; $R_2 \in [0, 1]$ value that serves as a caution; the range of the safety value, ST, is $[-1, 1]$.

The population is safe and the explorer can keep looking if R_2 is less than ST. When R_2 is smaller than ST, it indicates that the watchful bird has seen the predator and has sounded the alarm. People acted quickly to avoid being eaten by predators, fleeing to a secure spot.

Update formula for joiner position:

$$x_{id}^{t+1} = \begin{cases} Q \cdot \exp\left(\frac{x_{\mathrm{worst}d}^{t} - x_{id}^{t}}{i^2}\right) & i > \frac{N}{2} \\ x_{\mathrm{best}d}^{t+1} + \frac{1}{D} \sum_{d=1}^{D} \left(\mathrm{rand}(-1, 1) \cdot \left|x_{id}^{t} - x_{\mathrm{best}d}^{t+1}\right|\right), & i \leq \frac{N}{2} \end{cases} \quad (8)$$

In particular, $x^t{}_{(\mathrm{worst}\, d)}$ signifies the worst possible global site at iteration t, while $x^{(t+1)}_{(\mathrm{best}\, d)}$ represents the greatest possible global position at repetition $t + 1$.

If i is more than N2/3, then the i-th joiner has not yet found sufficient sustenance and will have to make a flight somewhere. If i is less than N2/3, then the i-th joiner is searching for food in a region around the global ideal location. The watchful stance's formula for status updates:

$$x_{id}^{t+1} = \begin{cases} x_{\mathrm{worst}\, d}^{t} + \beta\left(x_{id}^{t} - x_{\mathrm{worst}\, d}^{t}\right), & f_i \neq f_g \\ x_{id}^{t} + K\left(\frac{x_{id}^{t} - x_{\mathrm{worst}\, d}^{t}}{|f_i - f_w| + e}\right), & f_i = f_g \end{cases} \quad (9)$$

The step parameter, denoted by, is a random variable with a normal distribution having a mean of 0 and a alteration of 1; the sparrow's movement direction, denoted by K, is also a random variable with a value in the interval $[-1, 1]$; e is a very small constant; f_i denotes the fitness of the i-th sparrow; f_g denotes the population. Hence f_i, f_g, the i-th sparrow is on the periphery of the populace and is vulnerable to predators; when $f_i = f_g$, the i-th bird is in the center needs to be near to other dunnocks to decrease the capture risk; and is in the middle of the populace and is unaware of the danger.

3.3.1 Position Inform Based on Learning Coefficient and Mutation Worker

In this study, we address the issue of local extrema trapping in the SSA by incorporating learning coefficients to enhance the discoverer's global search capability. The discoverer, an expert forager, can become stuck in local maxima, leading the algorithm as a whole to converge to a local minimum. To assist the joiners in escaping local extrema, a mutation operator is integrated. The relocation procedure for the pioneer is executed after an advancement has been made.

$$x_{id}^{t+1} = \begin{cases} \{v(t)x_{id}^{t} \cdot \exp\left(\frac{-i}{a \cdot T}\right), & R_2 < \mathrm{ST} \\ v(t)x_{id}^{t} + Q \cdot L, & R_2 \geq ST \end{cases} \tag{10}$$

Among them, $v(t)$ is the learning coefficient of the discoverer.

The appearance of $v(t)$ is:

$$v(t) = v_{\min} + (v_{\max} - v_{\min}) \times \sin\left(\frac{t}{T}\pi\right) \tag{11}$$

Among them, $v_{\max}$ and $v_{\min}$ are the extreme and least learning constants correspondingly.

The formula to use when repositioning the joiner after an upgrade:

Cauchy(t) is a random mutable with a Cauchy distribution, and represents the degree to which variation is under control.

4 Results and Discussion

4.1 Experimental Setup

Operational records train the LSTM model for forecasting device state values. Given input sequence X and time prompt t, the model predicts $y(t)$. The error value $e(t)$ is calculated as the absolute difference between predicted $y(t)$ and actual $x(t)$ values. TTS detection performance depends on the error threshold T_e. The weight of the enhanced LSTM is determined using the updated SSA algorithm.

Table 1 Confusion matrix

	Predicted negative	Predicted positive
Actual negative	(TN)	(FN)
Actual positive	−(FP)	(TP)

4.2 *Parameter Evaluation*

Regular measurement of results is crucial to assess the effectiveness of the proposed system, known as the "parameter performance factor." Performance is evaluated by collecting, storing, and analyzing data. Table 1 displays confusion metrics used to assess binary data categorization:

In binary categorization, "true positive" (TP) refers to correctly identified positives, "false positive" (FP) represents misidentified negatives, "false negative" (FN) indicates mislabeled negatives, and "true negative" (TN) represents correctly identified negatives.

Accuracy, F-measure, Precision, and Recall are Eqs. (12), (13), (14), and (15), respectively.

$$\text{Accuracy} = \frac{\text{TN} + \text{TP}}{\text{TP} + \text{TN} + \text{FN} + \text{FP}} \times 100 \tag{12}$$

$$F\text{-measure} = \frac{2\text{TP}}{(2\text{TP} + \text{FP} + \text{FN})} \times 100 \tag{13}$$

$$\text{Precision} = \frac{\text{TP}}{(\text{FP} + \text{TP})} \times 100 \tag{14}$$

$$\text{Recall} = \frac{\text{TP}}{(\text{FN} + \text{TP})} \times 100 \tag{15}$$

where TP is signified as true positive, TN is indicated as true negative, FP is specified as false positive, and FN is indicated as false negative (Table 2).

Existing models achieved 90–94% accuracy in the PR analysis, while the proposed model achieved 96.05%. Poor performance in existing models is due to suboptimal hyperparameter selection and high computation time. In terms of accuracy, RNN

Table 2 Analysis of various DL classifiers without ISSA

Techniques	PR	*R*	F-M	ACC
DBN	90.21	90.28	90.75	90.65
AE	90.45	90.43	90.90	91.23
RNN	93.05	93.23	94.07	94.09
CNN	94.24	94.78	94.90	94.90
LSTM	96.05	96.25	95.23	96.30

Table 3 Analysis of DL models with ISSA

Techniques	PR	*R*	F-M	ACC
DBN	97.32	97.54	97.42	97.55
AE	97.86	97.64	97.35	97.60
RNN	97.90	97.89	97.87	98.01
CNN	97.12	97.64	98.01	98.05
LSTM	98.05	98.25	98.23	98.30

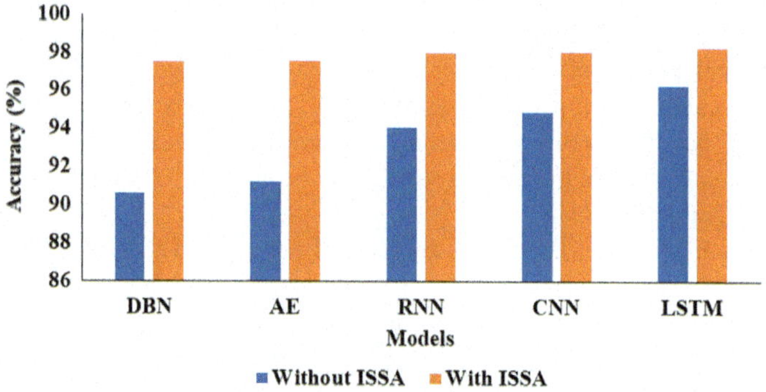

Fig. 1 Accuracy analysis

and CNN achieved 94%, DBN achieved 90.65%, AE achieved 91.23%, and the proposed LSTM model achieved 96.30%. The proposed model also outperformed existing models in *F*-measure (95.23%) and recall (96.25%), where existing models achieved 90–94% (Table 3).

Existing models achieved 97–98% accuracy, while the proposed model achieved 98.30%. After optimization, all models showed improved performance, with precision and *F*-measure reaching 97–98%. However, the proposed LSTM model outperformed others due to its ability to retain sequence information in its memory. Figures 1 and 2 provide graphical analysis comparing the proposed model with other techniques.

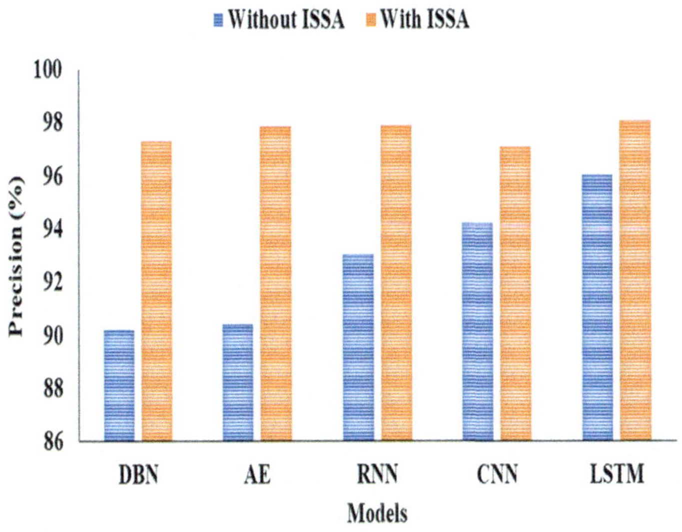

Fig. 2 Precision comparison

5 Conclusion

This study introduces a combined paradigm for TTS and VC applications. To address data efficiency and mispronunciation issues, we propose initializing VC models using pre-trained TTS models. Information transfer from publicly available large-scale TTS corpora enables a complete voice synthesis process from input characters. Our deep learning-based LSTM-ISSA model for voice conversion outperforms a baseline technique in both objective and subjective evaluations of speech intelligibility improvement. Our system demonstrates effective training even with limited data, without significant performance loss. In the future, we aim to rigorously evaluate and compare RNN-based models with transformer-based models. Additionally, we plan to develop more effective training algorithms and explore methods for masking input encoders.

References

1. Zhou K, Sisman B, Rana R, Schuller BW, Li H (2023) Emotion intensity and its control for emotional voice conversion. IEEE Trans Affective Comput 14(1):31–48. https://doi.org/10.1109/TAFFC.2022.3175578
2. Sang DV, Thu LX (2021) FastTacotron: a fast, robust and controllable method for speech synthesis. In: 2021 International Conference on Multimedia Analysis and Pattern Recognition (MAPR), Hanoi, Vietnam: IEEE, pp 1–5. https://doi.org/10.1109/MAPR53640.2021.9585267

3. Du C, Guo Y, Chen X, Yu K (2023) Speaker adaptive text-to-speech with timbre-normalized vector-quantized feature. IEEE/ACM Trans Audio Speech Lang Process 31:3446–3456. https://doi.org/10.1109/TASLP.2023.3308374
4. Sam Ribeiro M, Roth J, Comini G, Huybrechts G, Gabrys A, Lorenzo-Trueba J (2022) Cross-speaker style transfer for text-to-speech using data augmentation. In: ICASSP 2022—2022 IEEE International Conference on Acoustics, Speech and Signal Processing (ICASSP), Singapore. IEEE, Singapore, pp 6797–6801. https://doi.org/10.1109/ICASSP43922.2022.9746179
5. Terashima R et al (2023) Cross-speaker emotion transfer for low-resource text-to-speech using non-parallel voice conversion with pitch-shift data augmentation. Preprint at arXiv, http://arxiv.org/abs/2204.10020
6. Kumar VKA, Kumar MR, Shribala N, Singh N, Gunjan VK, Siddiquee KN, Arif M (2022) Dynamic wavelength scheduling by multiobjectives in OBS networks. J Mathemat 2022:3806018. https://doi.org/10.1155/2022/3806018
7. Huang R, Zhao Z, Liu H, Liu J, Cui C, Ren Y (2022) ProDiff: progressive fast diffusion model for high-quality text-to-speech. In: Proceedings of the 30th ACM International Conference on Multimedia. ACM, Lisboa, pp 2595–2605. https://doi.org/10.1145/3503161.3547855
8. Rudra Kumar M, Gunjan VK (2022) Peer level credit rating: an extended plugin for credit scoring framework. In: Kumar A, Mozar S (eds) ICCCE 2021. Lecture Notes in Electrical Engineering, vol 828. Springer, Singapore. https://doi.org/10.1007/978-981-16-7985-8_128
9. Xue J, Shen B (2020) A novel swarm intelligence optimization approach: sparrow search algorithm. Syst Sci Control Eng 8(1):22–34. https://doi.org/10.1080/21642583.2019.1708830

IoT-Based Automated Dustbin Using Arduino and Global System for Mobile Communication

T. Durga Devi, M. Shanmuga Sundari, P. Swarnalatha, K. Aishwarya, and S. Nikhitha

Abstract A smart dustbin using IoT technology is a revolutionary solution for waste management that uses sensors to detect the level of garbage and send notifications when it is full. The system comprises a sensor placed inside the dustbin that measures the level of waste and forwards data to a cloud server for analysis. The cloud server then sends alerts to the user or concerned authorities via a text message. The smart dustbin helps in reducing the overflow of garbage, promoting cleanliness, and maintaining hygiene in public areas. It can be used in households, offices, public places, and hospitals to optimize the waste collection process and reduce the environmental impact caused by waste. The smart dustbin is an excellent addition to the smart home and smart city concept, providing a sustainable and efficient solution for waste management. Its real-time monitoring, notification alerts, and data analytic features can help in improving waste management operations and reducing the associated costs. In summary, the smart dustbin using IoT technology is an innovative solution that has the potential to revolutionize the waste management industry. Its features make it an efficient and sustainable solution for waste collection and can help in creating a cleaner, healthier, and more sustainable environment. The main objective of the model is to receive notifications when the bin is 80% full.

Keywords Arduino UNO · Alarm · Dustbin · GSM · Servomotor · Ultrasonic sensor

T. Durga Devi (✉) · M. Shanmuga Sundari · P. Swarnalatha · K. Aishwarya · S. Nikhitha
BVRIT Hyderabad College of Engineering for Women, Hyderabad, India
e-mail: durgadevi.t@bvrithyderabad.edu.in

M. Shanmuga Sundari
e-mail: sundari.m@bvrithyderabad.edu.in

P. Swarnalatha
e-mail: 19wh1a0516@bvrithyderabad.edu.in

K. Aishwarya
e-mail: 19wh1a0518@bvrithyderabad.edu.in

S. Nikhitha
e-mail: 20wh5a0501@bvrithyderabad.edu.in

V. K. Gunjan and J. M. Zurada (eds.), *Proceedings of 4th International Conference on Recent Trends in Machine Learning, IoT, Smart Cities and Applications*, Lecture Notes in Networks and Systems 873, https://doi.org/10.1007/978-981-99-9442-7_59

1 Introduction

IoT consists of things that are connected to the internet and have unique identities. IoT devices [1] are made of sensors and actuators that possess sensing, actuating, and processing ability. The IoT devices contain various interfaces to establish a wired or wireless connection among devices. Interfaces include audio, video, storage, memory, input/output interfaces for sensors, interfaces for connection to the internet and also helps in data analytics. These devices communicate and share data with each other over a network. Sensors can collect data on rainfall [2], humidity, temperature, and soil content and other factors that help to automate the farming techniques. Google Home voice controller [3] provides voice-enabled services like volume control, alarms, lights, thermostats. Self-Configuring [4] provides particular functionality by allowing no devices to work with coordination. Unique Identity: Every IoT device is identified by a unique identifier (IP address) and unique identity.

This project is developed for real-time issues that we will face in our day-to-day life. It is developed based on IoT. We generally observe that roadside GHMC bins overflow [5], they will take a longer time to clean up the bins which causes bacteria to spread, and people can get infected. And some people will throw cigarettes into dustbins without smashing it properly which can cause a fire accident. To overcome these problems, this project is developed. It requires some hardware components.

An ultrasonic sensor [6] is placed at the front side of the dustbin; whenever it detects a human being, it will get open; then a person can throw wastage in it; after the person moved from there, it will get close; and even if the person is still not moving, it will get close after 15 s. This part is developed using Arduino [7], ultrasonic sensor, and servomotor. Arduino requires a code which contains the instructions for operating this dustbin. Ultrasonic sensor is used to detect the persons when they come in front of the dustbin to throw it. Power supply can be supplied using a 12 V adapter or we can also use a 9 V battery.

To measure the level of the bin, one more ultrasonic sensor is used which is placed at the top of the dustbin where it can have the entire look of the dustbin from inside. It measures based on the distance, if the distance is less than 5 cm it will consider the dustbin is 80% full, and then a text message will be sent notifying that the bin is about too full please take action. A flame sensor is placed inside this smart dustbin; whenever it detects fire, it gives an alarm. This part is developed using a flame sensor [8] which detects fire, a buzzer which gives alarms, and a 9 V battery for power supply.

The next step of this project is sending the text messages directly to the GHMC admin/toll free number, so that they can know that a particular bin from a particular location is filled, and they will send their crew members to that lactation to clean up the bin and the bin status will get updated.

2 Literature Survey

Nukavarupu and Nadeem [10] proposed a system that identifies devices with accuracy of 96% and training data of 3%. This system has 90% accuracy while giving the device type as IoT, non-IoT, or any other. They showed us that the model will support features like continual learning ability, zero-day malware attacks, noise robustness, and novelty detection. To start this smart IoT network services, they integrated a software-defined network with a classifier.

So, the proposed system in the paper developed a multistage classifier [11] based on semi-supervised generative adversarial networks that can perform automatic feature extraction with minimal labeled data. Pasolini et al. [11] are saying that (LPWAN) because an essential enabler for IoT connectivity.

They contributed Aoe-driven architecture [12], Qoe metrics, and Qoe-aware resource management. The Quality of Experience (Qoe) model has become a bright approach to give actual users experience of services. So, we are depending on the threefold contribution system in this paper. Velez-Estevez et al. [13] presented IOT-TEG 4.0.

Here IOT-TEG 4.0 contributed the real-time testing of test cases which includes the types of data that the 14.0 devices which are connected under study are managed. Industry 4.0 is the most important and major part of the ally company's digital transformation which is also defined as the rise of digital industrial technology.

Camps-Mur et al. [14] gave us a solution of using satellite-based non-terrestrial networks which can provide connectivity with global coverage of all areas from offshore and rural areas, which are principal use cases for future networks. The paper discusses contemporary 3GPP Architectural Approach and limitations of 3GPP NTN IOT standardization. Khamayseh et al. [15], this paper talks about challenges faced due to IoT growing exponentially in the past decade, which are central server overload, illegal use of private data, resource constraints.

3 Proposed System

The proposed system for a smart dustbin using IoT notification using a GSM module involves incorporating a GSM module (Global System for Mobile communication) with the already active system.

3.1 *Workflow*

The workflow for a smart dustbin using IoT notification involves the following steps:

- The sensor placed inside the dustbin estimates the level of trash inside the bin and updates data to the microcontroller.

- The server analyzes the data and sends notifications to the user or concerned authorities via SMS through the GSM module.
- The GSM module is connected to the microcontroller, and it uses the SIM card to send SMS alerts to the user or authorities.
- When the dustbin reaches its capacity, the sensor detects it and sends the data to the microcontroller, which triggers the server to send notifications through SMS via the GSM module.
- The user or concerned authorities receive the SMS alert and can take necessary actions to empty the dustbin. The system continues to monitor the waste level, and the process repeats.

Figure 1 depicts the workflow for the smart dustbin using IoT notification using a GSM module aims to optimize the waste collection process.

Fig. 1 Flowchart for proposed dustbin

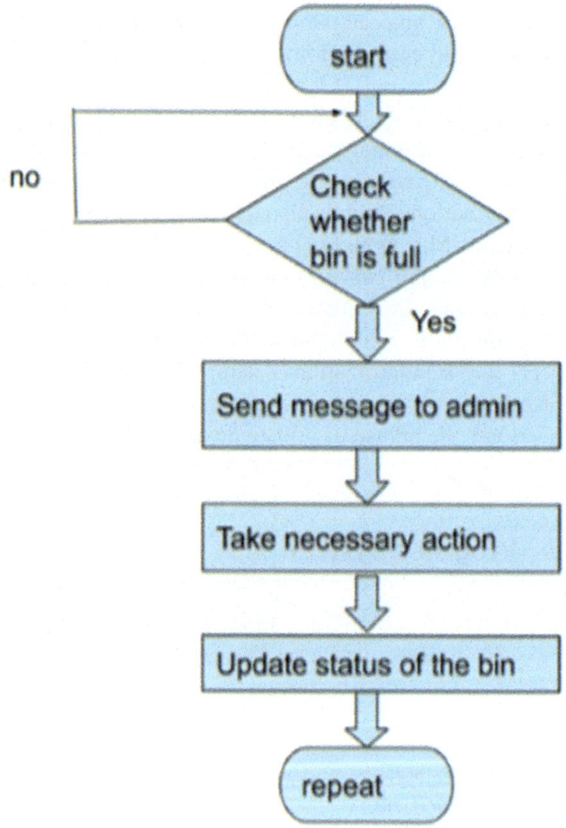

4 Sensors Used for Smart Dustbin

Sensors used for smart dustbin for various requirements in circuits to make the system work. In this system, different sensors and hardware components are used. Those components are specified below.

Figure 2a is a standard microcontroller. Arduino contains analog and digital pins. Pins include 6 analog pins and 14 other digital pins: power jack, USB connector, ICSP header. Figure 2b is a mobile communication module. It is used to transmit the data. It has a capacity to carry 65 kbps–120 mbps rates of data. It requires a sim card to send the data to the registered number. It sends data through SMS. Figure 2c is used to measure the distance of the objects irrespective of size and shape. Figure 2d is designed for detecting and responding to the flame or fire. Photosensitivity is high and detection angle is 600. Flame sensors operate at a voltage of 3.3–5 V.

Figure 3a shows a motor that can rotate with great precision. It is made up of a simple motor which runs by servomechanisms. They are two types of servos, AC servos and DC servos. Figure 3b shows the wires that are used to connect two ends. These wires are thin and lightweight. Figure 3c shows an audio signaling device like a beeper or buzzer made of piezoelectric material. It converts audio signals to sound signals. Figure 3d shows a 9v battery voltage which provides power to the circuit.

(a) (b) (c) (d)

Fig. 2 **a** Arduino UNO, **b** GSM module, **c** ultrasonic sensor, and **d** flame sensor

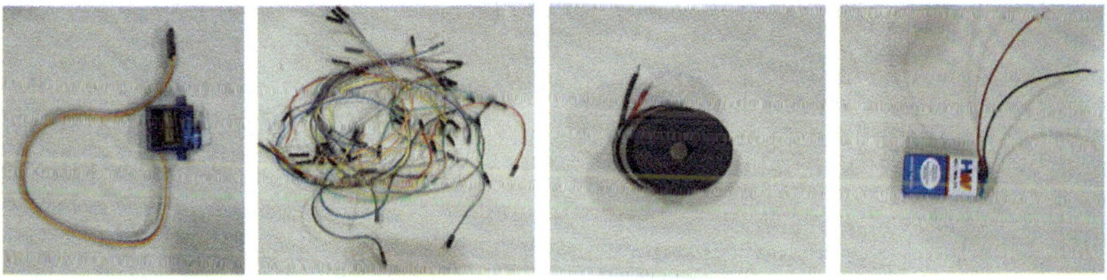

(a) (b) (c) (d)

Fig. 3 **a** Servo motor, **b** jumper wires, **c** buzzer, **d** battery

(a) (b) (c)

Fig. 4 **a** Fire detection, **b** notifications

5 Analyses

This is the fire alarm setup in the projects which gives us a buzzer warning when the fire is detected. It keeps on giving a buzzer sound until the fire is completely gone, so someone can take action to avoid such fire accidents. In this project when the bin is 80% full, it gives an alert message saying that the bin is almost full, please take action. Next time when it gets 100% full, it gives a text message alerting that the bin is 100% full, to the admin or the person whose number is registered.

In Fig. 4a whenever a fire accident occurs, this system detects the fire and gives the buzzer until someone takes action to stop the fire. Figure 4b shows the received notifications that when the bin is 80% full, then someone should clean the garbage as soon as possible to prevent dustbin overflow. Shows alert messages given to the admin or the concerned authorities via SMS through the GSM module. Whenever the bin is full about 80%, this system will give the alert SMS to the admin.

6 Conclusions

In this project, IoT plays an important role, where it can be able to track the status of the dustbin and send alert notifications and text messages when it is almost 80% full.

And the buzzer blares when the fire is detected in the initial stages so the accident minimizes. It diminishes the need for manual monitoring and presents real-time updates on the waste level inside the dustbin, enabling timely actions to be taken to prevent overflowing of garbage. The Arduino UNO and GSM module are the major parts which acts as a minicomputer to operate the whole product with the help of servomotor, ultrasonic sensor, flame sensor, buzzer, jumper wires, and power supply. In future, the main aim is to contact the GHMC organization to send the text messages directly to the GHMC administration, then they can send their staff members to clean up the bin. Another thing that we would like to develop is sending the text message to the fire department whenever the bin detects fire, so the fire accident can be stopped as soon as possible.

References

1. https://en.wikipedia.org/wiki/Internet_of_things
2. Syarifuddin SD, Khurniawan A, Aulia S, Ramadan DN, Hadiyoso S (2021) Rainfall Information System Using Geometry Algorithm on IoT Platform. In: 2021 IEEE.
3. Wang K, Liu X, Chen CM, Kumari S, Shojafar M, Hossain MS (2020) Voice-transfer attacking on industrial voice control systems in 5G-aided IIoT domain. IEEE Trans Industr Inf 17(10):7085–7092
4. Mittal EA, Yadav EP, Yadav H (2018) IoT: challenges and issues in Indian perspective
5. Lincy FA, Sasikala T (2021) Smart dustbin management using IOT and Blynk application
6. Vanitha CN, Sharma B, Praveen Raja S, Sathyamoorthy M, Sharma AK, Chowdhury S (2023) Smart city waste management system using IOT
7. Sri VB, Veerendra ODM, Lakshmi PM, Chand RP, Chakravarthy SS, Rao CV (2021) Arduino based smart dustbin for waste management during Covid-19
8. Ashok A, Savarinathan P, Karuppasamy T, Ram RS, Jayapalan A (2022) Garbage monitoring system using IOT
9. Joshi B, Kolhatkar C, Choudhari P, Bhuva D (2018) Smart E-dustbin
10. Nukavarapu S, Nadeem T (2022) iKnight—guarding IoT infrastructure using generative adversarial networks. IEEE
11. Pasolini G, Marini R, Mikhaylov K, Buratti C (2022) Low-power wide-area networks: comparison of LoRaWAN and NB-IoT performance. IEEE Internet Things J 9(21):21051–21063
12. Charnsripinyo C, Saovapakhiran B, Naruephiphat W, Baydere S, Özdemir S (2022) QoE-driven IoT architecture: a comprehensive review on system and resource management. IEEE Access 8(10):84579–84621
13. Velez-Estevez A, Gutiérrez-Madroñal L, Medina-Bulo I (2021) IoT-TEG 4.0: a new approach 4.0 for test event generation. IEEE Trans Reliab 71(3):1368–1380
14. Camps-Mur D, Kellermann T, Centelles RP, Ferrús R, Guadalupi M, Augé AC (2022) Novel architecture for cellular IoT in future non-terrestrial networks: store and forward adaptations for enabling discontinuous feeder link operation. IEEE Access 10:68922–68936
15. Khamayseh YM, Bataineh MR, Mardini W, Yassein MMB (2022) Novel and secure blockchain framework for health applications in IoT. IEEE Access 10:14914–14926

An Enhanced WOA and MI-Based Feature Selection Method for Attack Detection in Smart Meter Communication

R. Vijayanand, N. Naveen Kumar, M. Ulaganathan, D. Devaraj, and B. Kannapiran

Abstract Smart meter is a customer locality component of modern electrical grid. It is an easy target for the cyber-attacks. Learning algorithm-based intrusion detection system has good performance in various applications. It is a notable mechanism that provides security by early detection of intrusions from the meter traffic communication. The noisy data collected from the sources diminish the functioning of the attack finding algorithms. Thus, the feature selection algorithms are used to enhance the detection ratio of machine learning classifier. In this work, a collaborative method using wrapper-based whale optimization algorithm and filter-based mutual information is used to recognize the informational features. The identified features are feed as input to support vector machine classifier. The standard dataset ADFA-LD is employed to inspect the effectiveness of recommended method. The outputs certified the suitability of selected hybrid method for providing security to smart meter communication network.

Keywords Whale optimization algorithm · Mutual information · Hybrid feature selection method · IDS for meter communication · Smart meter security

R. Vijayanand (✉)
Nagarjuna College of Engineering and Technology, Bengaluru, India
e-mail: rkvijaynand@gmail.com

N. Naveen Kumar
Muthayammal Engineering College, Rasipuram, India

M. Ulaganathan
PSR Engineering College, Sivakasi, India

D. Devaraj
Kalasalingam Academy of Research and Education, Srivilliputhur, India

B. Kannapiran
Dr. Mahalingam College of Engineering and Technology, Pollachi, India

V. K. Gunjan and J. M. Zurada (eds.), *Proceedings of 4th International Conference on Recent Trends in Machine Learning, IoT, Smart Cities and Applications*, Lecture Notes in Networks and Systems 873, https://doi.org/10.1007/978-981-99-9442-7_60

1 Introduction

Smart grid revolutionizes the electrical industry by replacing the working model of old-fashioned electric grid with latest technologies. The conventional grid is not able to satisfy the necessity of modern industries and generates more pollution. Thus, smart grid is emerged with the collaboration of computational devices and communication network. It offers solutions to various difficulties of power industries like blackouts, real-time observation, transmission loss, etc. [1]. It also has advantages like bi-directional power flow, demand response, etc. Advanced components like smart meter, power line communication, etc. helps the effective functioning of smart grid.

Smart meter is an end device transmits the customer power utilization details at real time using duplex data network. A large number of researchers recommend mesh fashioned wireless network as an ideal choice to ensure availability of network in the meter data communication to concentrator [2]. The proposed setup model provides various good things but brings cyber-attacks to power grid. The accessible security mechanisms are not suitable to safeguard the meter data because of poor configurations of smart meter. Thus the specialized security mechanism is vital for the reliable transmission of meter data to the control server.

Intrusion detection system (IDS) is a noteworthy method that proves its reliability in a lot of applications. It silently monitors the data movement of the network and report to the authority if any suspicious events happened in the system [3]. Learning algorithms like artificial neural network (ANN), deep neural network, etc. are employed to finding the attacks. It exhibits better performance against the trending attacks [4]. In [5], an IDS is designed in distributed approach for smart grid using ANN, support vector machine (SVM), and random forest.

The machine learning algorithms suffer from replicate and unrelated features that affect the learning ratio of detectors. Thus, the feature selection processes are hired to upgrade the detector's finding rate. The renowned methods are graded under filter or wrapper technique. In first technique, the informational features are recognized on the support of the input feature relation with output class [6]. Alternatively, the randomly picked features are iteratively analyzed by the evaluation algorithms in latter methods. In [7], the genetic algorithm elects the informative features for hierarchical structured SVM. The issues of many conventional methods are irrelevant feature selection, cost in-effective, and poor detection [8]. In some cases, both filter and wrapper methods are integrated to raise the finding rate of detectors.

The notable intention of this research is to design a collaborative feature selection models to enhance the working nature of SVM algorithm in smart meter communication. The developed method has the property of whale optimization algorithm (WOA) and mutual information (MI) technique that increases the effectiveness and the attack finding ratio of detection algorithm. The ADFA-LD dataset is recruited in this research to validate the effectiveness of the suggested method.

2 WOA + MI-Based Feature Selection Method

Figure 1 represents the block diagram of proposed collaborative feature selection method for IDS. The normal and attacks data of smart meter communication are more similar and are highly complex to categorize into particular group. The selected informational features boost the attack finding ratio of IDS. In this method, the MI technique and WOA are hybridized to select the optimal features set. The random positioning of whales is generated in the early stage is marked as initial population. The best feature set of the population is considered as semi informative features. Then the MI technique [9] helps to choose the most informational features. MI algorithm utilizes entropy and conditional entropy formula given in Eqs. (1) and (2) respectively to calculate the correlation value.

$$h_y = -\sum_{j=1}^{N_y} P(Y_j) \cdot \log(P(Y_i)) \tag{1}$$

$$h_{(y/x)} = -\sum_{i=1}^{N_x} P(X_i) \sum_{j=1}^{N_y} P_{YX} \tag{2}$$

Then, the MI value of each individual feature is determined using,

$$I(Y; X) = \mathrm{h}_y - \mathrm{h}_{(y/x)} \tag{3}$$

where h_y—entropy, $h_{(y/x)}$—conditional entropy, N_y—total classes, $P(y_j)$—probability of individual class, P_{YX}—$P(Y_j/X_i)\cdot\log_2 P(Y_j/X_i)$, N_x—Total input variable, $P(Y_j/X_i)$—probability of specific value in input to the particular output class, $P(X_i)$—probability of presented values in the network input features.

The features having maximum value is selected as most informational features and are feed as input to classifier. Based on the fitness value of each position, i.e., the position for the next generation is created by circle hunting, bubble-net attacking, and prey hunting methods [10]. This process replicates till maximum iteration or

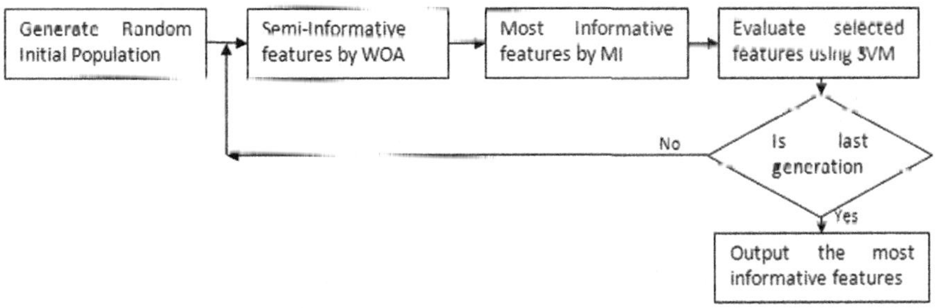

Fig. 1 Proposed hybrid feature selection method

threshold limit reached. The top features of concluding round are considered as most informational features and are used to train the detection algorithm.

The multi-faceted meter communication faces various attacks and the use of multi-attack detection classifiers is recommended. In this paper, SVM classifier is used that has proven record in multi-attack detection and is suitable to implement as classifier for IDS in WMN [11]. The classifier is initially trained with train data and validated with test data before applied to real environment.

3 Review of SVM Classifier

SVM is a popular algorithm, gets its importance after the research paper published by Vapnik et al. [12]. Basically, it is a binary classifier that has better results in 2-class problems. Mathematically, it is written as,

$$W^T X + B = 0 \tag{4}$$

where X—input data, W—weight vector, and B—bias. In SVM, a linear line or hyper-plane is constructed between the classes to categorize it. But the real applications need curve instead of line. Thus, the kernel functions are designed to map the given data in multi-dimensional format and then converged by separable line. In multi-class problems, multiple SVMs are employed with the consideration of specific class as '1' and all others as '0'. The final output is decided by the highest confidence value among the classifiers and is represented as,

$$\text{Output} = \arg \max_{i=1-n} \tag{5}$$

where n is total number of classifiers used in the particular problem. It shows better result in the classification of non-linear data and is implemented in numerous forms like hierarchical, distributed, etc. to get maximum accuracy.

4 Simulation Results

The effectiveness of the developed method is experimented using the software 'Matlab 2014' in a system having 'I3' processor system. In the applications like WMN, the normal and attack data features are much similar and are difficult to separate.

The ADFA-LD dataset is used to inspect the attack finding ratio of the developed system [13]. The dataset has 49 features and are categorized into 10 classes. In this experiment, the attacks related to WMN only considered for evaluation as 6 attacks and a normal data class. The total data considered for analyzing for this work is 8004. From that, half of the network traffic is used to train the model and the remaining as testing data. Table 1 shows the informational features selected by the proposed method. The selected features are introduced to the classifier as input and the finding rate is verified. Table 2 shows the performance analysis of proposed method using ADFA-LD dataset. The less true positive rate and positive predictive value of 'Backdoor' and 'Worms' are due to the presence of more false positive data. The inclusion of more data in training will raise the detection ratio of the attacks. The comparative results of various feature selection algorithms are given in Table 3.

Table 4 shows the comparison of execution time of SVM classifier with MI, WOA, and WOA + MI feature selection methods. It indicates the training of classifier in real environment requires large time that degrade the detection ratio of IDS. Thus, the offline training for classifier is suggested. The simulation results clearly demonstrate that the proposed method has better detection rate compared to the conventional methods.

Table 1 Informational features chosen by the proposed method

Label number	Informational feature
8	Destination bytes
11	Source ttl
28	Source Mean
29	Destination mean

Table 2 Performance analysis of proposed method using ADFA–LD dataset

Class	Test data used	Detection ratio	True positive rate	True negative rate	Positive predictive value
Normal	1762	0.9892	0.9958	0.9899	0.9866
Exploits	438	0.9805	0.8922	0.9902	0.9104
DOS	96	0.9542	0.4179	0.9728	0.3478
Reconnaissance	538	0.9775	0.9494	0.9811	0.8657
Backdoor	11	0.9862	0.2857	0.9911	0.1860
Generic	1395	0.9702	0.9482	0.9803	0.9566
Worms	2	0.9307	0.0137	0.9652	0.0147

Table 3 Comparative analysis of the proposed method

Technique	Detection rate (%)
MI	79.45
WOA	85.65
Proposed method	96.80

Table 4 Execution time analysis of proposed method

Techniques	Execution time (in seconds)
MI	12.91
WOA	92.65
Proposed method	93.80

5 Conclusion

The finding of informational features is a major task in the designing of good IDS for the applications like smart meter communication. This paper contains a collaborative method using WOA and MI technique to improve the effectiveness of IDS. The selected informational features boost the detection ratio of SVM classifier. The effectiveness of the proposed method is proved by comparing with the features selected by the mutual information and whale optimization algorithms. It clearly indicates that the recommended method has better attack finding ratio and suitable for giving protection to meter communication of smart grid.

References

1. Shi H, Xie L, Li P (2021) Detection of false data injection attacks in smart grid based on a new dimensionality-reduction method. Comput Electric Eng 91:107058
2. Scott S (2014) Selfish insider attacks in IEEE 802.11s wireless mesh networks. IEEE Commun Mag 52(6):227–233
3. Phanindra K, Noorullah S, Rajkumar L (2022) An anomaly based intrusion detection system using recursive feature elimination technique for improved attack detection. Theoret Comput Sci 931:56–64
4. Kevric J, Jukic S, Subasi A (2017) An effective combining classifier approach using tree algorithms for network intrusion detection. Neural Comput Appl 28:1051–1058
5. Zheng Y, Wang L, Sun W, Green RC II, Alam M (2011) Distributed intrusion detection system in a multi-layer network architecture of smart grids. IEEE Trans Smart Grid 2(4):796–808
6. Ganesh Kumar P, Devaraj D (2010) Intrusion detection using artificial neural network with reduced input features. ICTACT J Soft Comput 1:30–36
7. Vijayanand R, Devaraj D, Kannapiran B (2018) Intrusion detection system for wireless mesh network using multiple support vector machine classifiers with genetic-algorithm-based feature selection. Comput Secur 77:304–314
8. Anurag T, Amrita C (2022) A hybrid feature selection approach based on information theory and dynamic butterfly optimization algorithm for data classification. Expert Syst Appl 196:116621
9. Vijayanand R, Devaraj D, Kannapiran B (2017) Support vector machine based intrusion detection system with reduced input features for advanced metering infrastructure of smart grid. In: 2017 International Conference on Advanced Computing and Communication Systems, Coimbatore, 2017, IEEE, NY
10. Mirjalili S, Lewis A (2016) The whale optimization algorithm. Adv Eng Softw 95:51–67
11. Biswas SK, Mia MMM (2015) Image reconstruction using multi layer perceptron and support vector machine classifier and study of classification accuracy. Int J Sci Technol Res 04:226–231
12. Cortes C, Vapnik V (1995) Support-vector networks. Mach Learn 20(3):273–297

13. Creech G, Hu J (2013) Generation of a new IDS test dataset: time to retire the KDD collection. In: Proceedings of IEEE Wireless Communications and Networking Conference, 2013, pp 4487–4492. IEEE, New York

Review on Depression Detection Using Machine Learning Techniques

S. Vidya, G. T. Raju, and M. Vinayaka Murthy

Abstract Machine learning has achieved notable advancements in the medical field by improving the accuracy, precision, and analysis of diagnostics while also reducing labour-intensive tasks. With strong evidence, machine learning has exhibited the capability of detecting mental distress including depression. It is fact that a large proportion of the population is currently experiencing depression, which is considered to be a global burden of disease. Hence there is a critical need to develop an early depression detection model. This article aims to improve our understanding the present state of depression prediction models by exploring issues and identifying difficulties. This review relies on the examination of audio, video, images, PHQ questions, Social media (Twitter and Facebook), EEG signal depression detection models using different machine learning approaches. This paper reviews indicators of depression and provides a summary evaluation of diverse research studies encompassing their performance parameters.

Keywords Depression detection · Machine learning/ML · Anxiety · Architecture · CNN · ANN · LSTM

S. Vidya (✉)
REVA University, Bengaluru 560064, India
e-mail: R21PCA10@reva.edu.in

Department of MCA, Bangalore Institute of Technology, Bengaluru 560004, India

G. T. Raju
Department of Computer Science and Engineering, Sri Jagadguru Chandrashekaranatha Swamiji Institute of Technology, Chickballapur, Karnataka 562101, India
e-mail: gtraju@sjcit.ac.in

M. Vinayaka Murthy
School of Computer Science and Applications, REVA University, Kattigenahalli, Yelahanka, Bengaluru 560064, India
e-mail: mvinayaka.murthy@reva.edu.in

V. K. Gunjan and J. M. Zurada (eds.), *Proceedings of 4th International Conference on Recent Trends in Machine Learning, IoT, Smart Cities and Applications*, Lecture Notes in Networks and Systems 873, https://doi.org/10.1007/978-981-99-9442-7_61

1 Introduction

During 2021 World Health Organization (WHO) has conducted the study states that the depression is the second most common source of global burden of physical and mental illness. Now a days, across the world depression stands as a significant contributor to disability. It has become the most common and harmful mental disorder affecting the mankind all over the world [1]. Globally, lacking sufficient medical facilities mental illness is one of the highest prevalence's in India to help for needy people. The World Health Organization states that there is a critical shortage of trained specialists in India's mental health field. There exists a considerable disparity between the number of individuals afflicted with severe mental health conditions and the available psychiatrists to treat them. According to statistics from international organizations, hardly three psychiatrists and psychologists are there for every lakh individuals. It is projected that approximately 20% of individuals in the country are experiencing mental health conditions due to the pandemic and various contributing factors. Focusing on India specifically, there are around 38 million individuals having an anxiety problem and 56 million who have a mental ailment. According to a study, there would be 56,675,969 depression cases in India in 2022, with a prevalence rate of 4.5% [2].

The study of adaptive computer systems that become better over time is known as machine learning. Machine learning has become tremendously important in almost all well-known fields, and it has been used in medical diagnostics to diagnose and classify a wide range of illnesses, including neurodegenerative diseases. As compared to other current conventional techniques, the use of machine learning in the medical field has improved prediction accuracy [3]. The study of variation in speech, facial expressions, and emotions may also be investigated utilizing machine learning (ML) techniques.

Currently, depression affecting a large number of population. Majority of teenagers suffer from depression due to several reasons. Depressed people feel exhaust, unpleasant, energy lost, it could lead to fatigue, loss of concentration. Unable to handle on responsibility, lost interest in life, and many more reasons [4]. If this goes on for a while, it damages relationships and weakens the human mind. Consequently, it is recommended to get mental health therapy at the earliest. The following symptoms exists daily for at least 2 weeks:

a. Tendency of depression.
b. No interest in activities.
c. Conception of suicide.
d. Feeling of inefficiency.
e. Intensified potential to think and concentrate.

Figure 1 gives a brief statistics about the suicide throughout the India. A significant proportion of suicides in India are linked to mental health problems, such as depression, anxiety, and substance abuse. Unfortunately, there is still lack of knowledge concerning mental health problems in the country, and many people do not seek help or have access to services for mental health.

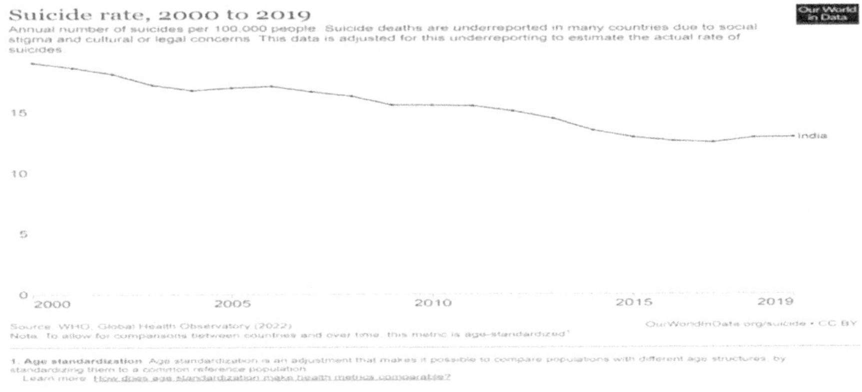

Fig. 1 Suicide statistics in India Courtesy: https://ourworldindata.org/suicide

This study employed a brief interactive human–computer evaluation utilizing deep learning AI to accurately predict depression status. The evaluation process is adhered to HIPAA compliance standards for the storage and access of participant data from demographics, mental health questionnaires, and video responses. The limitation of this methodology is that the self-report assessment is administered at specific points in time, which may lead to the behavioural outcomes reflecting an individual's state-dependent affect, rather than the long-term emotional characteristics associated with depression. Another limitation which is a common caveat of supervised learning methods [5].

Orabi et al. [6] uses a social networking site like Twitter text data to determine if any Twitter users are depressed. It was challenging due to unstructured text posted by the users. The Twitter posts consists of new words, syntactic errors, limitations in characters and misspelled words while composing a message. Authors used a novel optimized word embedding for classification task, they used comparative evaluation on deep learning depression detection models from tweets on the user level. They used datasets like CLPsych2015 and Bell Lets Talk in there experiment. In order to evaluate optimized embedding's they used 3 word embedding's like skip-gram, CBOW, and random trainable. The outcome of the study showed that CNN-based models performs better than RNN-based model.

The major goal of this study [7] is to use face traits to identify sadness in college students, specifically the frontal face features. To accomplish this, the researchers have developed a system that utilizes a combination of positive and negative facial emotions to train a video-based model. This model captures a video of the student and extracts their facial features using Gabor filters. The extracted features are then classified using an SVM classifier, which is used to determine the level of depression present in the video by quantifying the amount of negative emotions detected. If the level of depression is deemed significant, a notification is sent to the appropriate counsellor, such as the class advisor, department counsellor, or university counsellor, to alert them of the student's mental state. This paper takes a brief look at the architecture of the neural network-based machine learning model. Under the spectre of depressive illness signs, many visual and video indicators, text data, audio samples are examined to detect the depression. The system has been shown to achieve an accuracy of 64.38%.

2 Related Work on Depression Detection Technique

Deshpande et al. [8] in this study, the objective is to use methods of natural language processing on Twitter feeds to perform emotion analysis specifically targeting depression. Each individual tweet is categorized as either neutral or negative, utilizing a carefully curated word-list designed to identify indicators of depression tendencies. To predict the classification, support vector machine and Naive Bayes classifiers are employed. The outcomes are reported through key classification metrics, such as F1-score, accuracy, and a confusion matrix. The findings indicate that Multinomial Naive Bayes yielded the highest performance with an F1-score of 83.29, surpassing SVM which obtained a lower F1-score of 79.73. This trend is consistent when considering precision and recall, with Multinomial Naive Bayes demonstrating superior results compared to SVM.

Supratak et al. [9], this research introduces Deep Sleep Net, a deep learning model, designed for the automatic scoring of sleep stages using raw single-channel EEG data. In this work, CNN is used to extract time-invariant characteristics from EEG epochs, and LSTM is used to automatically learn bidirectional transition rules between various phases of sleep. A two-step training algorithm is implemented to efficiently train the proposed model. The evaluation is performed on diverse single-channel EEG datasets [F4-EOG (Left), Fpz-Cz, and Pz-Oz] obtained from two publicly available sleep datasets that exhibit distinct properties and adhere to different scoring standards (AASM and R&K). In comparison to state-of-the-art approaches (MASS: 85.9%–80.5, Sleep-EDF: 78.9%–73.7), the findings show that the proposed model achieves overall accuracy and macro F1-score (MASS: 86.2%–81.7, Sleep-EDF: 82.0%–76.9) on both datasets. It suggests that the proposed model could maintain consistent performance across different datasets while learning features for sleep stage scoring automatically from a variety of raw single-channel EEGs.

Hou et al. [10], this research provides two remarkable contributions. Firstly, it introduces an audio depression regression model named DR AudioNet, which combines a convolutional neural network and a long short-term memory network (LSTM). The model utilizes a multi-scale audio differential normalization (MADN) feature extraction algorithm to characterize non-personalized speech. Two network models are designed based on MADN features extracted from adjacent audio segments. Comparative experiments demonstrate the effectiveness of this method in accurately identifying depression patients. Secondly, a human–computer interaction system is designed. Users voice has given as input into the system, and the final recognition result is generated using the network model. This system offers a visually intuitive interface, facilitating convenient user interaction. The practical application value of this system is evident, as it enhances the user experience and aids in depression identification.

Asare et al. [11] used observations of human behaviour to forecast depression. Authors used datasets from smart phones for prediction. This research used an exploratory contingent analysis, they gathered data from 629 participants regarding their smart phone usage and self-reported depression assessments using the eight-item Patient Health Questionnaire (PHQ-8). The study lasted for an average of 22.1 days (SD 17.90; range 8–86). From the smart phone data, they analysed 22 behavioural markers related to regularity, entropy, and standard deviation. To analyse the behavioural features and depression relationship, this study employed correlation analysis and bivariate linear mixed models (LMMs). Additionally, it utilizes five supervised ML algorithms, incorporating hyper parameter optimization, nested cross-validation, and imbalanced data handling techniques to predict depression. Finally, they identified influential behavioural markers for predicting depression through the permutation importance method. 96.44–98.14% accuracy is achieved in this model.

Betul et al. [12], to determine depression using EEG signals they proposed a deep hybrid model using LSTM and CNN. The deep model utilized in this study incorporates CNN layers to learn the temporal properties of signals and LSTM layers for sequence learning. Specifically, both the left and right hemispheres of the brain's EEG signals were used. The results reveals that the developed CNN-LSTM model achieved classification accuracies of 99.12% and 97.66% for the right and left hemisphere EEG signals, respectively. Based on these findings, suggest that the CNN-LSTM model is both accurate and efficient in detecting depression using EEG signals.

Wu et al. [13] in this study researchers collected the data from Facebook. They enrolled 1453 Facebook users and collected their Facebook data. Study introduces an automated approach called Deep Learning-based Depression Detection with Heterogeneous Data Sources (D3 HDS) to predict an individual's depression status by analysing various factors such as their living environment, behaviour, and social media posting contents. To compute the representation of each individual's posts, authors employed Recurrent Neural Networks. These representations were combined with different features related to content, behaviour, and living environment. They used deep neural networks to predict the individual's depression label based on this combined information.

Zhao et al. [14] researchers proposed a multimodal fusion algorithm for depression diagnosis which is based on facial image and speech signal sequence. In this research, a technique called spectrum subtraction was introduced to improve the quality of depressed speech signals. Additionally, the cepstrum method was utilized to extract pitch frequency features, which exhibit a high variation rate, as well as formant features that display significant differences. Analysis in the temporal and spectral domains was done on the short-time energy and Mel-frequency cepstral coefficients to characterize different emotional speeches. Based on these results, a model for training and identification was created. In addition, the orthogonal match pursuit method was used to a sparse linear combination of test samples for facial expression.

This method was then proportionally combined with voice and facial expressions of emotion. The study's findings revealed the proposed depression detection algorithm, which combines speech and facial emotions, achieved a recognition rate of 81.14%.

Monaro et al. [15] to receive compensation from an insurance provider, major depression symptoms may be made up or exaggerated. This study comes with a novel approach for detecting the depression automatically. Patients must react to simple and difficult questions concerning depressive symptoms using mouse movements based on an electronic, double-choice task. Mouse dynamics characteristics gathered during subject responses and how many symptoms have been reported by participants were used to develop artificial intelligence models. Statistical study revealed that people who were malingering depression reported a higher number of both depressive and non-depressive symptoms compared to depressed participants. Additionally, individuals with depression took longer to complete mouse-based tasks compared to both truth-tellers and liars. The machine learning models demonstrated impressive performance, achieving a classification accuracy of up to 96% in effectively distinguishing between liars, truth-tellers, and depressed patients. Since the algorithm's accuracy does not have compared to that of clinicians, it is crucial to understand that these results are preliminary.

Freeman and Ambady [16] used the Mouse Tracker application to conduct their investigation. 76 randomly selected questions were shown for every participant in the top area of the computer screen. The buttons with the labels for the NO and YES responses were situated in the top left and top right corners of the screen. Participants has to click the appropriate label to reply to the questions that appeared when they pressed the START. On the basis of response time of the participant it will be decided whether participant is depressed or not.

Keshavarz et al. [17] to identify depression from photographs or recordings the depressed face must be defined precisely and clearly. It is crucial to differentiate the facial expression of depressed and sad state of a person, since the emotions are similar. They conducted screening procedures to assess the relevance of the data collected and data extraction procedures to reduce errors and biases. They used extensive analysis procedures to examine the risk of bias along with duplicate quality assessments. They conducted a meta-analysis using the general inverse variance approach. The results were evaluated using Grading of Recommendations Assessment, Development, and Evaluation (GRADE) system.

Researchers conducted their studies in identifying depression using social markers along with visual and audio analyses. It has been noticed that a depressed person exhibits weak body language and negative social behaviour. Many of these markers can be used to identify depression. Substance abuse and usage, social and family alienation, and poor performance are all social effects of grief [18].

Soundariya et al. [19] used 4 common prototypes classification, including Bayes Network, Support Vector Machine (SVM), C 4.5 Decision Tree, and Artificial Neural Network (ANN). ANN found out the best result to determine the prior depression symptoms in ageing patients.

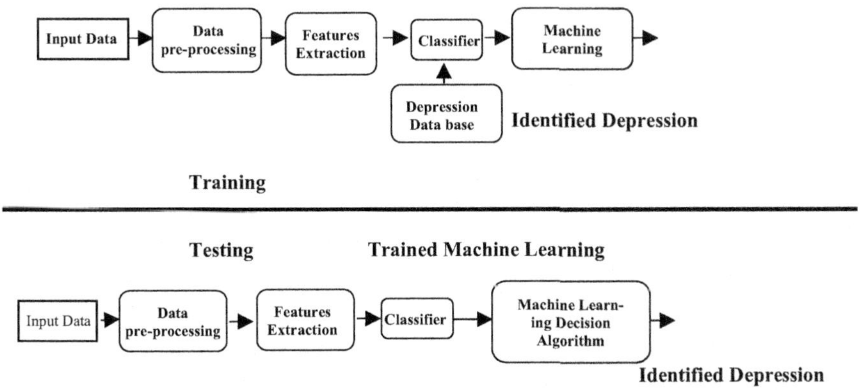

Fig. 2 Depression detection architecture

Haque et al. [20] used an embedded convolutional neural network (CNN) model to examine 3D face attributes and vocalized words to determine the severity of depression. This model indicated 83.3% sensitivity and 82.6% specificity.

Zulfiker et al. [21] around 0.8 million suicide instances occur each year in the world. In the light of [22] forecast the start of a future downturn. Using the Random Forest classifier, the predictive model was created. SMOTE was utilized this study also showed that the problems of class inequalities. The accuracy of this research was 86.20%. Additionally, this study found that the most important elements influencing the onset of depression are health and happiness with social and familial relationships.

Figure 2 shows the general depression detection architecture to identify the depression using various input parameters. Input parameter can be audio, video, questioners, EEG signals, Twitter/Facebook data.

2.1 Data Pre-processing

Pre-processing serves as the initial step in working with datasets allowing us to extract valuable and meaningful information essential for both testing and training phases. Any type of input data, including audio, video, images, text data can be used. In the case of image/video input, pre-processing encompasses various tasks such as normalization, illumination adjustment, alignment, segmentation, and face detection. Once the datasets have undergone pre-processing they are often divided into training sets used to train machine learning models and testing sets used to assess the model's performance during the testing phase.

2.1.1 Feature Extraction

It is a vital and foundational step in both training and testing processes as it lays the groundwork for subsequent stages. When it comes to identifying depression relevant features can be extracted from facial expressions, head position, audio, and text. Principal Component Analysis (PCA) is widely recognized as a prominent technique in this domain. However, the selection of these features requires careful consideration to create a well-suited system. Insufficient feature selection may lead to under fitting while an excessive number of features during the training phase can result in over fitting. The selection of features varies among researchers, influenced by their individual approaches and specific requirements.

2.1.2 Performance Analysis

Research has been conducted using several algorithms with various parameters. Puthankattil et al. [13] has used Wavelet packet and non-linear features/PNN algorithm and achieved accuracy of 99.5%, sensitivity of 99.2%, and specificity of 99.7% with a less data set, Zulfiker et al. [21] used K-Best Features (Select K-Best), Boruta feature selection algorithm with an accuracy of 92.56% with a huge dataset. In Table 1 gives a brief review of various techniques to detect depression using different ML algorithms. After the review of several article it shows that different algorithms can be used to identify the depression with an accurate result. In the light of the analysis, with less dataset, the accuracy is more when compared to huge number of dataset; it is analysed that the accuracy can be increased by considering huge dataset with different algorithm. Different techniques via video, audio, social media, images, EEG signals, and sleeping patterns are used as a parameters to identify the depression using various machine learning algorithms.

3 Analysis of Machıne Learning Methods

The methods of machine learning have completely transformed the area of prediction and classification. Several machine learning techniques, such as SVM, deep learning, and neural networks are used by researchers. For problems involving classification and regression, support vector machines may be used. However, classification problems are most frequently used. In an N-dimensional space, the support vector machine looks for a hyper plane that can categorize the data points in a certain manner. In order to identify emotions, the SVM classifier is trained using a range of emotional characteristics during the training phase. The SVM classifier can evaluate the features retrieved from each frame to determine whether these emotional traits are exist in the frame. It solely categorizes based on the features that are present in the frame; as

Table 1 Comparative study of literature review on depression detection

Ref./year	Feature extraction methods/ techniques used	Data set	Parameters	Results	Limitations
Mali et al. [4] 2021	XGBoost Tree, Random Trees, Neural Network, SVM, Random Forest, C5.0, and Bay Net	Kaggle web site	Attribute: Age, Married, Number of children, Education, total members...	Accuracy: Random Trees—76.61% Neural Network—82.34% SVM—76.38% Random Forest—80.28% C5.0—83.94 Bay Net—79.13 XGBoost Tree—83.49%	Filters are not applied in results
Zulfiker et al. [21] 2021	K-Best Features (Select K-Best), Boruta feature selection algorithm	604 individual questionnaire	Attribute in questionnaires: Gender Profession Marital Status Residing Place...	Accuracy: 92.56%	Dataset doesn't includes biological marker to identify depression
Betul et l. [12] 2019	End-to-end/ CNN-LSTM	15 normal 15 depressed	EEG signal (Left) EEG signal (Right)	Accuracy: 97.66% (Left) 99.12% (Right)	Model is expensive, less data sets are used
Hague et al. [20] 2018	CNN, SVM, LSTM	DAIC-WOZ dataset	3D facial features, language vocalized, Patient Health Questionnaire (PHQ) scale	Sensitivity: 83.3% and specificity:82.6%	The 3D facial scan has a low resolution
Ashraf et al. [18] 2018	Logistic Model Tree (LMT) Naïve Bayes Sequential Minimal Optimization (SMO) and Random Forest	60 participants	Mouse Tracker software using Beck Depression Inventory (BDI)	Accuracy up to 96%	The algorithm's accuracy hasn't been contrasted with clinician accuracy

(continued)

Table 1 (continued)

Ref./year	Feature extraction methods/ techniques used	Data set	Parameters	Results	Limitations
Victor et al. [5] 2018	Data Augmentation /LSTM-RNN	671 Participants	Video questions PHQ-9	Accuracy: 87.77%	The use of self-report exams, short-term emotional influences, and supervised learning's limitations need categorization of the data
Deshpande and Rao [8] 2017	Naive Bayes Classifier, Support Vector Machines	10,000 Tweets collected using the Twitter API	App Authentication, Keyword list, Extracting text from JSON/ Multinomial	Accuracy: Multinomial Naive Bayes: 83% Support Vector Machine: 79%	The accuracy of supervised learning classification in predicting depression from text data is limited and cannot reach human levels
Ref. [27] 2018	End-to-end/ CNN	15 normal 15 depressed	EEG Signal (Left) EEG signal (Right)	Accuracy: 93.54% (Left) 95.96% (Right)	Limited EEG data available
Wu et al. [13] 2014	Wavelet packet and non-linear features/PNN	15 normal 15 depressed	Bipolar EEG recordings from locations FP1–T3 (left half) and FP2–T4 (right half) of the brain in resting state	Classification accuracy of 99.5%, sensitivity of 99.2%, and specificity of 99.7%	System has only been evaluated in a modelling context with well-behaved data

a result, if a frame contains an emotion trait, the frequency of that feeling is correspondingly reduced. According to the hyperplane, an emotion is also deemed to be high if the frequency of the qualities is high in a given frame [7].

Other machine learning techniques, deep learning algorithms, such as neural networks, are being researched in the field of depression prediction from images and videos. Modular neural network, convolutional neural network, residual depth wise separable CNN [23], neural network with radial basis functions [24, 25], some examples of neural networks include the Kohonen Self-Organizing Neural Network, the Recurrent Neural Network (RNN)—Long Short-Term Memory and the Feed Forward Neural Network—Artificial Neuron. From the facial images, the depression can be identified using some unsupervised ML algorithms, such as K-means.

Ihsanto et al. [25] discusses the possibility of using facial images to identify whether a person who is not depressed would likely experience depression two years from now. The prediction method makes use of a standard preparation and testing organization strategy. Images from teenagers who were either "at risk" for depression or "not at risk" were used to create the class models. Two feature extraction techniques were examined at the Eigenface (PCA features) and the Fisherface (PCA + LDA features). This paper implement the closest neighbour (NN) arrangement using both individual dependent and individual independent methods. The Fisherface (PCA + LDA) approach led to the better result, with expected precision of 51% when employing the unique independent methodology and 61% when using the individual dependent technique.

Table 1 describes about the different set of parameter for the different ML algorithms. The results gives the sensitivity, specificity, and accuracy of the corresponding algorithms. The last column provides the research limitation.

The three criteria that are generally employed in a binary classification process—accuracy, sensitivity, and specificity—were utilized to evaluate the classification results. Following are the definitions of these parameters:

$$\text{Sensitivity} = \frac{\text{TP}}{\text{TP} + \text{FN}} \times 100\% \quad (1)$$

$$\text{Specificity} = \frac{\text{TN}}{\text{TN} + \text{FP}} \times 100\% \quad (2)$$

$$\text{Accuracy} = \frac{\text{TP} + \text{TN}}{\text{TP} + \text{TN} + \text{FP} + \text{FN}} \times 100\% \quad (3)$$

AR class represents positive class label and NAR class represents the negative label. True positive represented as TP (the number of AR adolescents classified as AR), the false positive outcomes number is represented by FP (the number of NAR adolescents classified as AR), the true negative outcomes number is represented by TN (the number of NAR adolescents classified as NAR), and the false negative outcomes number is represented by FN (the number of AR adolescents classified as NAR).

4 Results

This section covers a summary of the study findings, and some gaps in the literature have been identified. Research work uses different methods of identifying depression through social media like Facebook, tweeter Data, and these are supervised learning's constraints. Only through the text data there is no way to provide the human level accuracy. Identifying depression through facial expressions will be having a short-term emotional influence need to address this issue. Predicting depression through the EEG signals requires efforts towards the data set collection. Considering all these facts, a unique hybrid depression detection model needs to be developed using an appropriate machine learning algorithm. Most of the included articles highlight limitations in datasets and a lack of comparison with clinical accuracy and absence of biological marker. The application of various machine learning techniques could potentially boost the efficiency. Further research is needed to address research gaps, such as the need for qualitative and quantitative datasets. Also need to compare the result by considering the clinical parameters, so that more accurate level of depression can be identified.

5 Conclusions

This research examines various machine learning methods for detecting depression. For this intent, many models have been created and it has various advantages and disadvantages. But a more accurate and tested detection model is needed. This review identifies numerous open-ended questions and offers several new insights. In order to enhance results and improve living standards, it is crucial to mention that machine learning-based algorithms with image/video-based, social media, EEG signals, and PHQ/BDI questions are used as a parameters in early depression detection systems which is valuable for both clinical practise and research. To date, only a minimal amount of research has been conducted on early detection of depression within the farming community. In Indian economy agriculture plays an important role and this domine provides proximately 52% of jobs available in India and contributes around 18.1% to the GDP [26]. GDP from agriculture in India increased to 6934.75 INR Billion in the fourth quarter of 2022 from 4297.55 INR Billion in the third quarter of 2022 [27]. The future works can be concentrated on finding the severity of depression among the farmer community. Analysing how negative emotions affects the mental state of the depressed farmer by considering the biological factors too. So, directly or indirectly this work will be helpful for social cause like, saving the life of the farmers and to motivate more youths towards agriculture and we can create more jobs. At the same time we can increase the GDP of our country. At most importantly we are going to save the life of the farmers.

References

1. World Health Organization. Depression. https://www.who.int/news-room/fact-sheets/detail/depression. Accessed 2 Dec 2022
2. Mindvoyage. https://mindvoyage.in/depression-in-india-latest-statistics. Accessed on 2 Feb 2023
3. Salvatore C et al (2014) Machine learning on brain MRI data for differential diagnosis of Parkinson's disease and progressive supranuclear palsy. J Neurosci Methods 222:230–237. https://doi.org/10.1016/j.jneumeth.2013.11.016
4. Mali D, Kumawat K, Kumawat G, Chakrabarti P, Poddar S, Chakrabarti T, Hussaine J, Kamali AM, Bolsev V, Kateb B, Nami M (2021) A machine learning technique to analyze depressive disorders. ResearchSquare 1, 1–11. https://doi.org/10.21203/rs.3.rs-322564
5. Victor E et al (2019) Detecting depression using a framework combining deep multimodal neural networks with a purpose-built automated evaluation. Psychol Assess 31(8):1019–1027. https://doi.org/10.1037/pas0000724
6. Orabi AH, Buddhitha P, Orabi MH, Inkpen D (2018) Deep learning for depression detection of Twitter users. In: Proceedings 5th Workshop on computational linguistics and clinical psychology: from keyboard to clinic, pp 88–97
7. Parameswaran NS, Venkataraman D (2019) A computer vision based image processing system for depression detection among students for counseling. Indones J Electr Eng Comput Sci (IJEECS) 14(1):503–512
8. Deshpande M, Rao V (2017) Depression detection using emotion artificial intelligence. In: 2017 International Conference on Intelligent Sustainable Systems (ICISS), Palladam, pp 858–862. https://doi.org/10.1109/ISS1.2017.8389299
9. Supratak A, Dong H, Wu C, Guo Y (2017) Deep sleep net: a model for automatic sleep stage scoring based on raw single channel EEG. In: IEEE transactions on neural systems and rehabilitation engineering, p 99
10. Hou J, Gao J (2021) A novel smart depression recognition method using human-computer interaction system. Wireless Commun Mobile Comput 2021:1–8
11. Opoku Asare K, Terhorst Y, Vega J, Peltonen E, Lagerspetz E, Ferreira D (2021) Predicting depression from smartphone behavioral markers using machine learning methods, hyperparameter optimization, and feature importance analysis: exploratory study. JMIR mHealth uHealth 9(7):e26540. https://doi.org/10.2196/26540
12. Betul et al (2019) Automated depression detection using deep representation and sequence learning with EEG signals. J Med Syst 43(7):205. https://doi.org/10.1007/s10916-019-1345-y
13. Wu M, Shen C-Y, Wang ET, Chen ALP (2020) A deep architecture for depression detection using posting, behavior, and living environment data. J Intell Inform Syst 54(2):225–244
14. Zhao J, Su W, Jia J, Zhang C, Lv T (2017) Research on depression detection algorithm combine acoustic rhythm with sparse face recognition. Clust Comput 2017:1–12
15. Monaro M et al (2018) The detection of malingering: a new tool to identify made-up depression. Front Psychiatry 9:249. https://doi.org/10.3389/fpsyt.2018.00249
16. Freeman JB, Ambady N (2010) MouseTracker: software for studying real-time mental processing using a computer mouse-tracking method. Behav Res Methods 42(1):226–241. https://doi.org/10.3758/BRM.42.1.226
17. Keshavarz H et al (2013) Screening for depression: a systematic review and meta-analysis. CMAJ Open 1(4):E159–E67. https://doi.org/10.9778/cmajo.20130030
18. Ashraf A, Gunawan TS, Riza BS, Haryanto EV, Janin E (2020) On the review of image and video based depression detection using machine learning. Indones J Electr Eng Comput Sci 19(3):1677–1684
19. Soundariya RS, Nivaashini M, Tharsanee RM, Thangaraj P (2019) Application of various machine learning techniques in sentiment analysis for depression detection. Int J Innov Technol Explor Eng (IJITEE) 8(10S):292–296
20. Haque A, Guo M, Miner AS, Li F-F (2018) Measuring depression symptom severity from spoken language and 3D facial expressions. arXiv preprint arXiv:1811.08592. Version 2

21. Zulfiker MS, Kabir N, Biswas AA, Nazneen T, Uddin MS (2021) An in-depth analysis of machine learning approaches to predict depression. Curr Res Behav Sci 2(100044):2021. https://doi.org/10.1016/j.crbeha.2021.100044
22. Na K-S et al (2020) Predicting future onset of depression among community dwelling adults in the Republic of Korea using a machine learning algorithm. Neurosci Lett 721:134804. https://doi.org/10.1016/j.neulet.2020.134804
23. Gunawan TS, Alghifari MF, Morshidi MA, Kartiwi M (2018) A review on emotion recognition algorithms using speech analysis. Indones J Electr Eng Inform (IJEEI) 6(1):12–20
24. Ihsanto E, Ramli K, Sudiana D, Gunawan TS (2020) An efficient algorithm for cardiac arrhythmia classification using ensemble of depthwise separable convolutional neural networks. Appl Sci 10(2):483. https://doi.org/10.3390/app10020483
25. Ihsanto E, Ramli K, Sudiana D, Gunawan TS (2020) Fast and accurate algorithm for ECG authentication using residual depthwise separable convolutional neural networks. Appl Sci 10(9):3304. https://doi.org/10.3390/app10093304
26. Arjun KM (2013) Indian agriculture—status, ımportance and role in Indian economy. Int J Agric Food Sci Technol 4(4):343–346
27. https://tradingeconomics.com/india/gdp-from-agriculture. Accessed 24 Apr 2023

Empowering Fast-Moving Consumer Goods Supply Chain Leveraging Blockchain

Karan Singh Thakur and Rohit Ahuja

Abstract This blockchain project aims to develop a decentralized supply chain management system leveraging blockchain technology. The primary objectives are to improve transparency, traceability, and security within the supply chain process. To achieve this, a permissioned blockchain will be utilized, granting access exclusively to authorized participants. By implementing this solution, all stakeholders involved in the supply chain will have access to an immutable and comprehensive record of all transactions and activities. Consequently, the risk of fraudulent activities will be minimized, efficiency will be increased, and costs associated with conventional supply chain management systems will be reduced. Smart contracts will be employed to automate various aspects of the supply chain process, including order processing, inventory management, and payment processing. This automation will streamline the overall process, resulting in reduced time and effort required for supply chain management. Moreover, the system will feature a user-friendly interface enabling participants to conveniently access and monitor their transactions and activities. Advanced analytics capabilities will also be integrated into the system, empowering participants to analyze their data and make well-informed decisions. In summary, the proposed blockchain-based supply chain management system holds immense potential to transform conventional supply chain practices. It will enhance transparency, traceability, and security while simultaneously driving down costs and improving overall efficiency.

K. S. Thakur (✉) · R. Ahuja
Thapar Institute of Engineering and Technology, Patiala, India
e-mail: kthakur1_be20@thapar.edu

R. Ahuja
e-mail: rohit.ahuja@thapar.edu

V. K. Gunjan and J. M. Zurada (eds.), *Proceedings of 4th International Conference on Recent Trends in Machine Learning, IoT, Smart Cities and Applications*, Lecture Notes in Networks and Systems 873, https://doi.org/10.1007/978-981-99-9442-7_62

1 Introduction

Fast-Moving Consumer Goods (FMCG) is a significant industry and is essential to the nation's economy. It encompasses a wide range of consumer products that are in high demand, have a short shelf life, and are sold at a relatively low cost [1, 2]. Consumers nowadays have the option to purchase FMCG products online, providing convenience and access to a wide range of products. FMCG companies have also embraced digital marketing and advertising strategies to engage with consumers and promote their brands [3, 4]. The FMCG supply chain is a complex network that involves the movement of goods from manufacturers to end consumers. It comprises various stages and stakeholders, each playing a crucial role in ensuring the timely availability of FMCG products in the market. Here are the key components of the FMCG supply chain:

1. Procurement: The supply chain begins with the procurement of raw materials and components needed for manufacturing FMCG products. This involves identifying reliable suppliers, negotiating contracts, and ensuring the availability of raw materials to meet production demands.
2. Manufacturing: Once the raw materials are procured, FMCG products are manufactured in large quantities. The manufacturing process may involve multiple production facilities, each specializing in specific product categories. Quality control measures are implemented to maintain product standards.
3. Warehousing and Distribution: After manufacturing, the FMCG products are stored in warehouses before distribution. Warehouses act as an intermediate storage facilities that hold inventory until it is needed for delivery. These warehouses are strategically located to ensure proximity to the target markets and optimize transportation costs.
4. Logistics and Transportation: Logistics companies are responsible for managing the transportation process, including selecting appropriate modes of transport (trucks, trains, ships) and ensuring timely delivery. Efficient transportation is essential to meet the demands of time-sensitive FMCG products.
5. Distributors and Wholesalers: Distributors and wholesalers act as an intermediaries between manufacturers and retailers, buying products in bulk and distributing them to retailers across different regions. Distributors and wholesalers often have established networks and relationships with retailers, enabling them to reach a wide customer base.
6. Retailers: Retailers are at the another end in the FMCG supply chain, selling the products directly to consumers. They include supermarkets, convenience stores, department stores, and online retailers. Retailers stock and display FMCG products, manage inventory levels, and provide a point of sale for consumers to purchase the goods.
7. Demand Forecasting and Planning: FMCG companies utilize demand forecasting techniques to estimate consumer demand accurately. This helps in production planning, inventory management, and ensuring adequate supply to meet

consumer needs. Advanced forecasting tools and data analytics are employed to analyze historical sales data, market trends, and other factors influencing demand.
8. Reverse Logistics: In the FMCG supply chain, reverse logistics handles the return of products from retailers or consumers due to factors like expiration, damage, or product recalls. FMCG companies establish processes to manage these returns, including product collection, sorting, disposal, or refurbishment, depending on the nature of the products.
9. Efficiency and agility are crucial in the FMCG supply chain to meet the high demands and short product lifecycles. Companies invest in technologies, such as supply chain management software, inventory tracking systems and real-time analytics to streamline processes, reduce costs as well as to improve customer experience. Continuous monitoring, collaboration, and coordination among all stakeholders are essential to ensure seamless product flow and maximize the efficiency of the FMCG supply chain.

Our research aims to develop a decentralized supply chain management system using blockchain technology [5, 6] to enhance transparency, traceability, and security in the FMCG supply chain process. The proposed solution will enable all participants in the supply chain to have a complete and immutable record of all transactions and activities. This will help to minimize the risk of fraud, increase efficiency and reduce costs associated with traditional supply chain management systems. The system will use smart contracts to automate several facets of the supply chain process including order processing, inventory management, and payment processing. The procedure will be streamlined as a result, and supply chain management will take less time and effort.

2 Existing System

The diagram in Fig. 1 represents the existing FMCG supply chain, encompassing production processes, transportation, and distribution of products [7, 8]. Transparency is crucial to address cost and customer engagement concerns, which can harm a brand's reputation. Access to information fosters trust among participants, including suppliers, manufacturers, couriers, and end users. Risk management is vital due to fraud and code of conduct violations. Complex supply chains pose variability risks [9, 10]. Multiple stakeholders, such as suppliers, distributors, and customers, amplify these risks. Blockchain technology automates operations and adapts to a changing business environment [11–13].

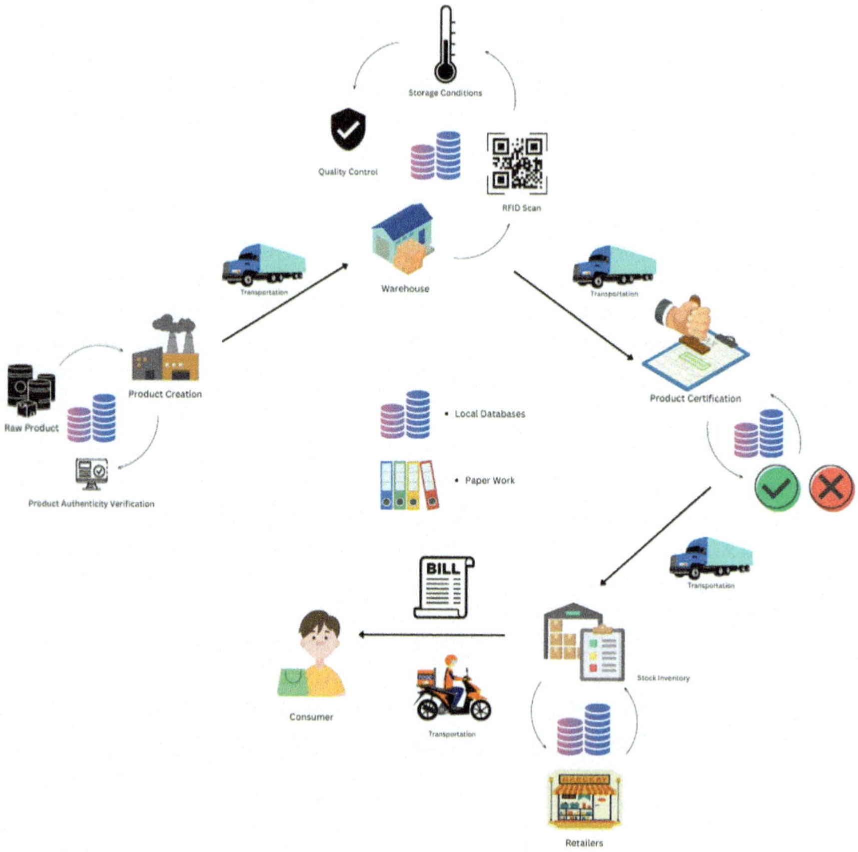

Fig. 1 Existing system

2.1 Proposed Framework

The management of the mentioned supply chain on blockchain can be achieved using the actors primarily a consumer, manufacturer, transporter, distributor, and an inspecting agent. The actors can be directed to their respective portals using a SQL-Database containing the login records. Once the login is achieved now we start updating the nodes on our blockchain.

New Item Creation: To create a new item, the Algorithm 1 is called with the parameters `_text` and `_date`.

Algorithm 1 CreateItem: Create a New Item.

Require: Takes as input the string `_text` representing the product name and the string `_date` representing the date.

Ensure: Returns `true` indicating the successful creation of a new item.

```
function NEWITEM(string text, string date) returns bool
    Product newItem ← Product({creator: msg.sender, totalStates: 0, productName: _text, productId: items, date: _date})
    allProducts[items] ← newItem
    items ← items + 1
    emit Added(items − 1)
    return true
end function
```

Add State: To add a state to a product, Algorithm 2 is called using the parameters `_productId` and `info`.

Algorithm 2 AddState: Add State to a Product.

Require: Takes as input the uint `_productId` representing the product ID and the string `info` representing the state description.

Ensure: Returns the string `info`.

```
function ADDSTATE(uint_productId, string info) returns string
    require(_productId ≤ items)
    State newState ← State({person: msg.sender, description: info})
    allProducts[_productId].positions[_allProducts[_productId].totalStates] ← newState
    allProducts[_productId].totalStates ← allProducts[_productId].totalStates + 1
return info
end function
```

Find Product: To search for a product, Algorithm 3 is called using the parameter `_productId`.

Algorithm 3 Query: To find a Product.

Require: Takes as input the uint `_productId` representing the product ID

Ensure: Returns a string `output` containing the product information

```
function SEARCHPRODUCT(uint_productId) returns string
    require(_productId ≤ items).
    string output ← "Product Name:"
    output ← concat(output, allProducts[_productId].productName)
    output ← concat(output, "!br.Manufacture Date:")
    output ← concat(output, allProducts[_productId].date)
    for uint256 j ← 0; j < allProducts[_productId].totalStates; j ← j + 1
      do
```

```
            output ← concat(output, allProducts[_productId].positions[j].description)
    end for
    return output
end function.
```

Concatenate Strings: To concatenate two strings Algo 4 `ConcatStr` is called with the parameters `_a` and `_b`.

Algorithm 4 ConcatStr: Concatenate Strings.

Require: Takes as input two strings `_a` and `_b`.
Ensure: Returns a string result obtained by concatenating `_a` and `_b`.

```
function CONCAT(string_a, string_b) returns string.
    bytes bytes_a ← bytes(_a).
    bytes bytes_b ← bytes(_b)
    string length_ab ← new string(bytes_a.length + bytes_b.length)
    bytes bytes_c ← bytes(length_ab)
    uint k ← 0
    for uint i ← 0; i < bytes_a.length; i ← i + 1 do
        bytes_c[k++] ←bytes_a[i]
    end for
    for uint i ← 0; i < bytes_b.length; i ← i + 1 do
        bytes_c[k++] ← bytes_b[i]
    end for
    return string(bytes_c)
end function
```

Tag Geo Location: To tag the geolocation of a product, Algorithm 5 is called and the parameters `_productId` and `_location` are taken as input.

Algorithm 5 TGeoLoc: Tag Geolocation of a Product.

Require: Takes as input the uint `_productId` representing the product ID and the string `_location` representing the geolocation.
Ensure: Returns true indicating the successful tagging of the geolocation.

```
function TAGLOCATION(uint_productId, string_location) returns bool
    require(_productId ≤ items)
    allProducts[_productId].location ← _location
    return true
end function
```

The proposed framework as depicted in Fig. 2.

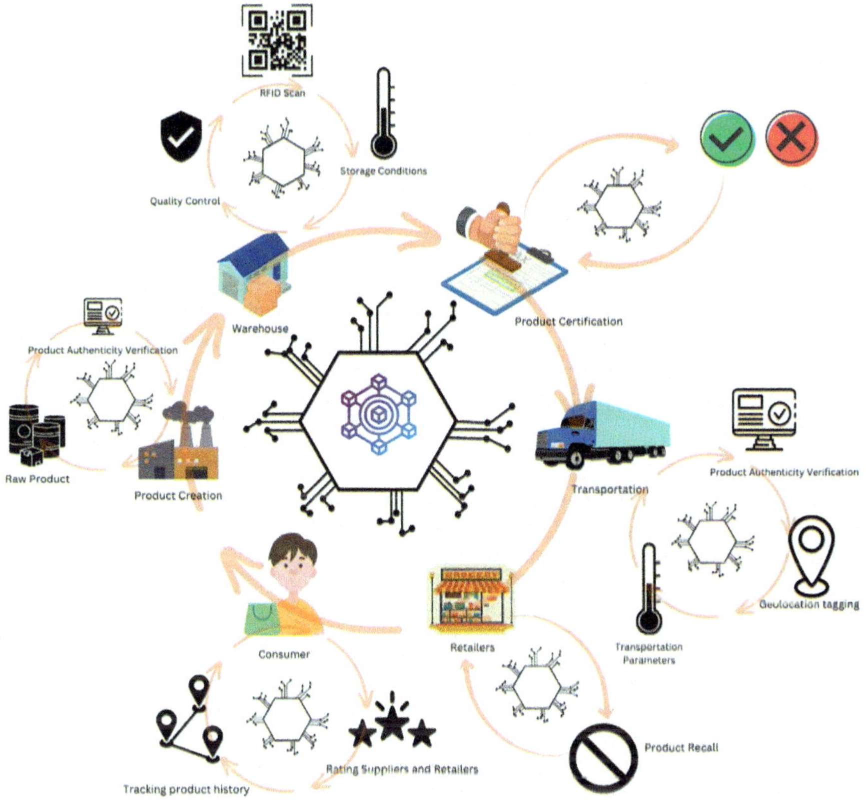

Fig. 2 Proposed framework

3 Implementation

3.1 Technology Stack Used

The technology stack discusses below allow us to develop, test, and deploy smart contracts on the Sepolia test network using the Hardhat development environment.

1. **Sepolia Test Network:** Sepolia is a test network based on Ethereum that allows developers to deploy and test their smart contracts without using real Ether. It provides a sandbox environment for development and testing purposes. In our solution, we utilize the Sepolia test network to deploy and test our smart contracts.
2. **Hardhat:** Hardhat is an Ethereum development environment and testing framework that streamlines smart contract compilation, deployment, and testing. Hardhat supports various Ethereum networks, including Sepolia. We use Hardhat as our development environment to write, compile, deploy, and test our smart contracts on the Sepolia test network.

3. **Solidity:** Solidity is an object-oriented programming language for creating smart contracts on blockchains. We use Solidity to write the smart contracts that govern the FMCG supply chain processes in our solution.
4. **Web3.js:** Web3.js is a library collection that facilitates the interaction with an Ethereum node via HTTP, IPC, or WebSocket. In order to communicate with smart contracts it offers an API to perform various operations on the blockchain. We use Web3.js to interact with the Sepolia test network and execute transactions on our deployed smart contracts.
5. **Remix IDE:** Remix is an IDE for writing, compiling, and deploying smart contracts. It provides a user-friendly interface and features such as code highlighting, syntax checking, and debugging. We can connect Remix to the Sepolia test network and use it to deploy and interact with our smart contracts during development and testing.

3.2 Architecture of Proposed System

Figure 3 depicts the architecture of the proposed system. A blockchain-based supply chain management solution can be designed using a layered architecture that incorporates various components to ensure transparency, security, and efficiency. The following is an overview of the architecture for a blockchain-based supply chain management solution:

1. Data Layer: This layer is responsible for storing data related to the supply chain transactions. The data could include information about product details, supplier details, manufacturing details, shipping details, and customer details. The data layer could use distributed ledger technology (DLT) to store data in a decentralized manner, ensuring transparency and security.
2. Smart Contract Layer: This layer is responsible for executing smart contracts that facilitate the various transactions in the supply chain. Smart contracts are self-executing contracts that automate the terms and conditions of an agreement between two parties. For instance, a smart contract could be used to automatically release payment to a supplier once the goods are received by the customer.
3. Consensus Layer: This layer is responsible for ensuring consensus among the nodes in the network. The consensus algorithm ensures that all the nodes in the network agree on the state of the blockchain. This layer could use a consensus algorithm such as proof-of-work or proof-of-stake.
4. Application Layer: This layer is responsible for developing the user interface and applications that interact with the blockchain. The applications could include a supplier portal, a customer portal, a logistics portal, and a financial portal. The application layer could use technologies such as RESTful APIs to interact with the blockchain.
5. Integration Layer: This layer is responsible for integrating the blockchain-based supply chain management solution with existing enterprise systems such as ERP,

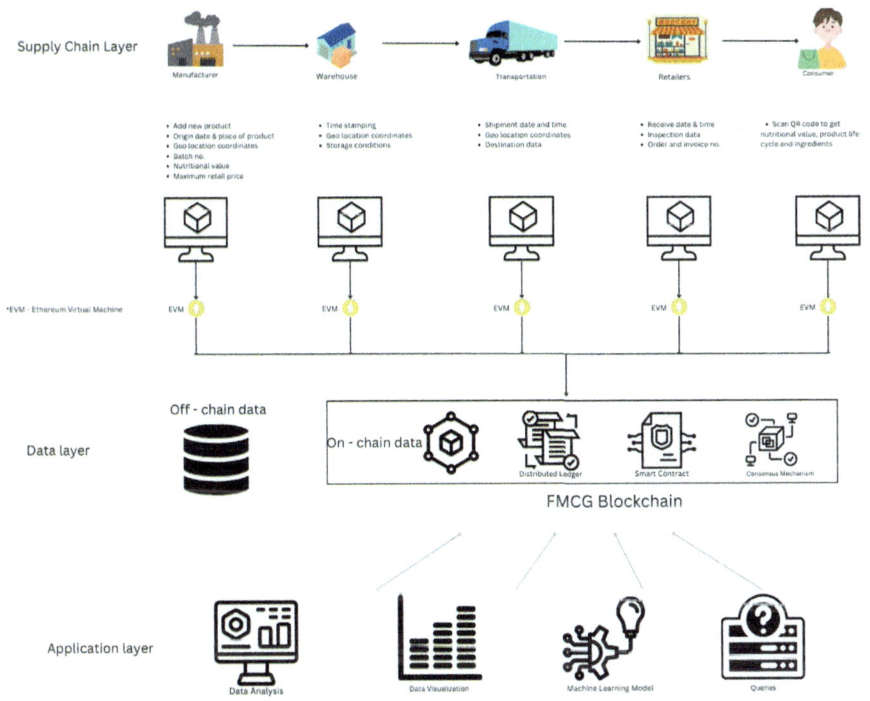

Fig. 3 Architecture of proposed framework

CRM, and SCM. The integration layer could use middleware technologies such as ESB to facilitate the integration.

In conclusion, a blockchain-based supply chain management solution can be designed using a layered architecture that incorporates data storage, smart contract execution, consensus algorithms, user interface development, and system integration. The architecture can ensure transparency, security, and efficiency in the supply chain transactions.

3.3 *Designing Smart Contracts*

Smart contracts need to be designed keeping in consideration of all the actors we have defined previously in the use cases, mainly the manufacturer, consumer, transporter, distributor, and the inspecting agent. We need to incorporate all these services in the form of functions that would communicate with the blockchain. The smart contract is written in Solidity. We can also extend the proposed framework into a marketplace using the standard ERC-20 token functionalities. Designing smart contracts for a blockchain-based supply chain management solution involves creating a set of self-executing codes that automate the terms and conditions of an agreement between two parties. Here are some examples of smart contracts that could be used in the above architecture:

1. Supplier Contract: The supplier contract could include details such as the quantity of goods to be supplied, the delivery schedule, and the price of the goods. The smart contract could automatically execute payment to the supplier once the goods are delivered to the customer.
2. Logistics Contract: The logistics contract could include details such as the shipping details, the delivery schedule, and the mode of transportation. The smart contract could automatically update the status of the shipment on the blockchain and trigger payment once the goods are delivered to the customer.
3. Customer Contract: The customer contract could include details such as the product specifications, the delivery schedule, and the payment terms. The smart contract could automatically release payment to the supplier once the customer confirms the delivery and quality of the goods.
4. Financial Contract: The financial contract could include details such as the payment terms, the payment schedule, and the payment methods. The smart contract could automatically execute payment to the supplier based on the agreed payment terms.

The above smart contracts could be developed using Solidity, a programming language used for writing smart contracts on the Ethereum blockchain. The smart contracts could be tested using a test blockchain network, such as Ganache, before deploying them on the main network. It's essential to ensure that the smart contracts are secure, audited, and free from vulnerabilities to avoid any potential breaches (Fig. 4).

Fig. 4 Transaction costs

Table 1 Comparison of existing and proposed technique, where –: Absence of feature, ✓: presence of feature

Features	Existing	Proposed
Transparency	Low	High
Trust	Low	High
Authenticity verification	–	✓
Product recall	Manual	Automatic
Ratings	–	✓
Certification management	–	✓
Data Integrity	Low	High
Efficiency	Low	High
Cost	–	✓
Geo location	–	✓
Tagging	–	✓
Consumer role	Limited	Flexible

4 Experimental Evaluation and Results

4.1 Feature Analysis

The features of the proposed framework are contrasted with the existing scheme depicted in Table 1.

5 Conclusion

The current research work proposes a blockchain-based FMCG supply chain distribution to make things better than the traditionally adopted practices. The proposed system achieves transparency, immutability, security, and reduces fraud. In summary, a blockchain-based supply chain management solution to achieve significant benefits for businesses, including increased transparency, improved efficiency, and reduced costs. The proposed architecture and smart contract design can be customized based on the specific requirements of the business. The integration with existing enterprise systems can be achieved using middleware technologies such as ESB. Overall, the implementation of a blockchain-based supply chain management solution can help businesses stay competitive in a rapidly changing market.

References

1. Manders JHM, Caniëls MCJ, Paul W (2016) Exploring supply chain flexibility in a FMCG food supply chain. J Purchas Supply Manag 22(3):181–195
2. Bottani E, Montanari R, Volpi A (2010) The impact of RFID and EPC network on the bullwhip effect in the Italian FMCG supply chain. Int J Prod Econ 124(2):426–432
3. Ahuja R, Mohanty SK, Sakurai K (2016) An identity preserving access control scheme with flexible system privilege revocation in cloud computing. In: 2016 11th Asia Joint Conference on Information Security (AsiaJCIS). IEEE, pp 39–47
4. Devrani S, Ahuja R, Goel A, Kharbanda SS (2023) A blockchain-driven framework for issuance of NFT-based warranty to customers on E-commerce. In International Conference on Multi-disciplinary Trends in Artificial Intelligence. Springer Nature Switzerland, Cham, pp 265–276
5. Shrestha AK, Joshi S, Vassileva J (2020) Customer data sharing platform: a blockchain-based shopping cart. In: 2020 IEEE International Conference on Blockchain and Cryptocurrency (ICBC). IEEE, pp 1–3
6. Shakila UK, Sultana S (2021) A decentralized marketplace application based on ethereum smart contract. In: 2021 24th International Conference on Computer and Information Technology (ICCIT). IEEE, pp 1–5
7. Končar J, Grubor A, Marić R, Vučenović S, Vukmirović G (2020) Setbacks to IoT implementation in the function of FMCG supply chain sustainability during COVID-19 pandemic. Sustainability 12(18):7391
8. Bottani E, Montanari R (2010) Supply chain design and cost analysis through simulation. Int J Prod Res 48(10):2859–2886
9. Bottani E, Rizzi A (2008) Economical assessment of the impact of RFID technology and EPC system on the fast-moving consumer goods supply chain. Int J Prod Econ 112(2):548–569
10. Manders JH, Caniëls MC, Paul W (2016) Exploring supply chain flexibility in a FMCG food supply chain. J Purchas Supply Manage 22(3):181–195
11. Bala M, Kumar D (2011) Supply chain performance attributes for the fast moving consumer goods industry. J Transp Supply Chain Manage 5(1):23–38
12. Alicke K, Rexhausen D, Seyfert A (2017) Supply chain 4.0 in consumer goods, vol 1(11). Mckinsey and Company
13. Ahuja R, Mohanty SK, Sakurai K (2016) A traceable signcryption scheme for secure sharing of data in cloud storage. In: 2016 IEEE International Conference on Computer and Information Technology (CIT). IEEE, pp 524–531

DeEffNet: A CNN Model for Optimizing the Image Classification of Comorbid Patients

H. S. Suresh Kumar, N. Preethi, C. N. Pushpa, J. Thriveni, and K. R. Venugopal

Abstract Amidst the increasing global mortality caused by the COVID-19 virus, researchers are committed to finding technological solutions that can support healthcare professionals in their daily responsibilities. Artificial Intelligence (AI) techniques are being employed to provide fast and accurate predictions of disease severity in patients with comorbidities, aiding doctors in their assessment. Currently, X-ray images are used as initial indicators to detect comorbid patients. This research focuses on developing classification models, specifically DenseNet121 and EfficientNetB0. The performance of these models is compared iteratively against a threshold value. The proposed models utilize DenseNet121 and EfficientNetB0 with ReLU activation function and softmax pooling, achieving accuracies of 94% and 77%, respectively. Based on the results, DenseNet121 is considered an efficient classification model.

Keywords AI · Comorbid · DeEffNet · DenseNet121 · EfficientNetB0

1 Introduction

COVID-19 was declared a global pandemic by WHO on March 11, 2020 [1]. The utilization of machine learning techniques in COVID-19 severity prediction has gained substantial attention due to its potential to provide valuable insights into patient prognosis and guide resource allocation. By leveraging diverse ML algorithms, researchers have been able to extract meaningful patterns and correlations from an extensive range of clinical and demographic data. These machine learning models aim to identify risk factors and predict disease progression, enabling early intervention and appropriate allocation of healthcare resources.

H. S. S. Kumar (✉) · N. Preethi · C. N. Pushpa · J. Thriveni
Department of CSE, University of Visvesvaraya College of Engineering, Bangalore University, Bengaluru 560001, India
e-mail: kumar10harobande@gmail.com

K. R. Venugopal
Bangalore University, Bangalore, India

V. K. Gunjan and J. M. Zurada (eds.), *Proceedings of 4th International Conference on Recent Trends in Machine Learning, IoT, Smart Cities and Applications*, Lecture Notes in Networks and Systems 873, https://doi.org/10.1007/978-981-99-9442-7_63

One of the main aims in utilizing machine learning techniques for COVID-19 severity prediction is to improve patient triage and allocation of resources. By accurately identifying patients at higher risk of severe disease outcomes, healthcare providers can prioritize critical care resources, such as ventilators and ICU beds, for those who need them the most. Moreover, ML models can aid in identifying potential comorbidities or underlying conditions that contribute to disease severity, allowing healthcare professionals to tailor treatment plans accordingly.

In this research study, we purpose to explore the applications of different machine learning techniques for COVID-19 severity prediction. By examining various ML algorithms and their performance on diverse datasets, we seek to identify the strengths, limitations, and potential areas of improvement in existing approaches. Additionally, we aim to shed light on the ethical considerations, interpretability, and generalizability of ML models when applied to COVID-19 severity prediction.

Overall, the integration of ML techniques for COVID-19 severity prediction holds great promise in improving patient care and healthcare resource allocation during this unprecedented pandemic. By leveraging the power of ML algorithms and combining them with comprehensive datasets, we can enhance our understanding of COVID-19 severity and make informed decisions to mitigate the impact of this disease on individuals and communities.

Motivation: COVID-19, a pandemic disease first identified in 2019, has had a global impact. Throughout history, humanity has faced other pandemic diseases like the Black Death, Flu, and Plague. The lack of historical knowledge and delayed medical treatment during the early stages of COVID-19 resulted in significant loss of lives. It is crucial to implement effective screening measures and provide timely medical attention to individuals infected with COVID-19 in instruction to contain the banquet of the disease. Unfortunately, the situation is further complicated by the utilization of multiple screening tools with delayed test results. To address this challenge, our proposed method serves as a foundation for developing machine learning models. These models are then refined using the limited number of available datasets based on results.

Contribution: We have designed and developed a Fusion Model DeEffNet using DenseNet121 and EfficientNetB0 models for image classification. In this work, the experimental results lead to the conclusion that deep features mined from DenseNet121 and EfficientnetB0 significantly improve the accuracy of COVID-19 disease classification. Furthermore, there is potential for further exploration of augmentation techniques and experimentation with different combinations of CNNs as feature extractors. Through comparative analysis, we determined that the approach employed in this research achieves a classification accuracy of 94%, surpassing the accuracy scores reported in the existing literature and outperforming EfficientnetB0.

Organization: The paper is organized into several sections. Section 2 offers an extensive review of relevant articles, while Sect. 3 presents the problem definition. The methodology employed and algorithm is explained in Sect. 4, and the results of the system are presented in Sect. 5. Finally, Sect. 6 encompasses the conclusions drawn from the study as well as directions for future work.

2 Related Work

The increasing interest in machine learning stems from its potential in diagnosing COVID-19 disease. Various ML algorithms have been proposed as diagnostic tools, aiding doctors in making informed treatment decisions. This section reviews several relevant studies that contribute to the understanding of this topic.

In this study conducted by Alaiad et al. [2], the objective is to assess the severity of COVID-19 by examining lung CT scans of people who have been detected with the virus. The dataset used for this study comprises a total of 2205 CT images. A radiologist categorized the images into four severity levels: normal, mild, moderate, and severe. Here to predict lung disease severity, the Resnet 101 algorithm, yields the highest performance with accuracy of 99.5%.

To create a novel ensemble DL model that integrates multiple neural networks to enhance prediction accuracy got Sidra et al. [3]. To tackle the matter of data imbalance, employed slicing functions and data augmentation techniques. Furthermore, we implemented test time data augmentation to further enhance the performance. By leveraging a straightforward yet powerful ensemble approach with robust test time augmentations, the proposed model delivered results on par with more intricate techniques.

The researchers Suresh et al. [4] have been studying the correlation between COVID-19 and several coexisting health conditions, including respiratory problems, diabetes, hypertension, cardiovascular diseases, and mental health disorders. This study provides a wide-ranging analysis of the impacts of COVID-19 and its related comorbidities. Furthermore, it explores the advancements in AI techniques for finding the severity of COVID-19 and comorbidities, drawing upon existing literature.

Jibril et al. [5] present an automated framework for detecting COVID-19 by utilizing a modified VGG-19 CNN, along with traditional data augmentation methods and GAN. This framework is capable of categorizing chest radiographs into three classes: Normal, COVID-19, and pneumonia. The proposed model demonstrates impressive performance, achieving a remarkable accuracy rate of 91.3%, with precision and recall both at 91%, and an F1-score of 90.4% [6]. These results highlight the model's effectiveness in accurately identifying COVID-19 cases from chest X-rays (Table 1).

Table 1 Performance of the existing work on VGG16/19 and other models

References	Model / Architecture	Datasets	Accuracy (%)
Ding et al. [7]	VGG 16, DenseNet121, MobileNetV2, Exception, Inception V3	X-ray image and chest CT image	98.6
Paul et al. [8]	ResNet-50, DenseNet-121	CXR image	99.07
Kulkarni et al. [9]	DenseNet121	X-ray images	90.06
Suresh et al. [10]	VGG19, MobileNetV2	X-ray, Ultrasound, and CT images	98, 98, and 100
Rafi [11]	Base Convolutional Neural Network, DenseNet121, ResNEt50, and EfficientNetB4	X-ray images	98.86

3 Problem Definition

Traditional methods for processing the COVID-19, Lung-Opacity, Normal, and Viral-Pneumonia images are challenging in terms of time consumption and false positive rates.

To address this issue, we aim to design and develop a fusion model using Densenet121 and EfficientNetB0 models for image classification. Our objectives are:

i. Implement DenseNet121 and EfficientNetB0 models for training and testing.
ii. Validate the outcomes of these models.
iii. Retrain, retest, and revalidate the models to get improved accuracy.

4 Methodology

The primary goal of this study is to create a robust and efficient model that can accurately classify comorbidities based on X-ray images. To achieve this goal, the study utilized a dataset comprising images from various classes, including COVID-19, Lung-Opacity, Normal, and Viral-Pneumonia cases. Data augmentation is a technique for initializing the required parameters, such as rotation range, zoom range, and horizontal and vertical shifts shown in Fig. 1.

4.1 X-Ray Images

The X-ray images utilized in this study were sourced from the Kaggle repository and encompassed diverse cases. The dataset consisted of 105,825 images of varying sizes. Of these, 74,077 images were allocated for model training, 21,166 images

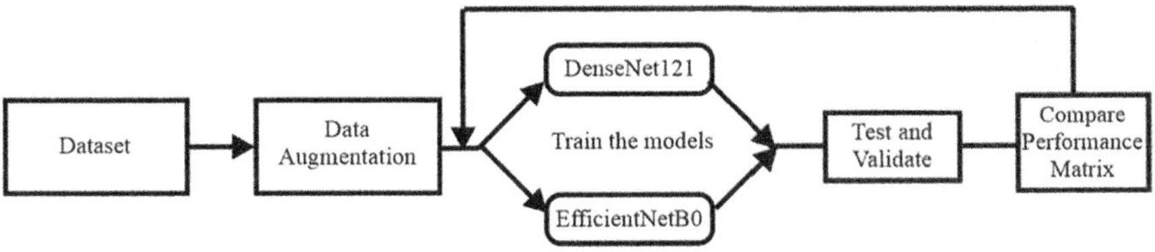

Fig. 1 Block diagram of the DeEffNet model

Table 2 Sample dataset for experiments

Images	Condition	Source images	Train	Test	Validation
X-ray	COVID-19	18,080	12,656	3616	1808
	Lung-opacity	30,060	21,042	6012	3006
	Normal	50,960	35,672	10,192	5096
	Viral-pneumonia	6725	4707	1345	673

were used for model testing, and validation was performed using 10,582 images, as detailed in Table 2.

5 DeEffNet Algorithm

Begin

Step 1: Data Collection:

Collect the following types of data: {COVID-19, Lung-Opacity, No Findings, Viral-Pneumonia.}

Step 2: Data Preprocessing and Augmentation:

Prepare and enhance the collected data for further analysis

Step 3: Classification of Pre-Processed Data:

Split the pre-processed data into training, testing, and validation sets

Step 4: Train, Test, and Validate Models:

4.1 Train DenseNet121 model:

- Add average pooling and fully connected layers with ReLU activation
- Include a SoftMax activation function
- Test the model with 40% dropout
- Repeat the training process until the model's performance is optimized with respect to a threshold

4.2 Train EfficientNetB0 model:

- Add average pooling and fully connected layers with ReLU activation
- Include a SoftMax activation function

- Test the model with 40% dropout
- Repeat the training process until the model's performance is optimized with respect to a threshold

Step 5: Compare the output matrices of both models to capture diverse and robust representations

Step 6: Analyze the Evaluation Matrix to derive final implications and draw concluding insights

End

5.1 *Pre-trained Models for Feature Extraction*

DenseNet121: The traditional feed-forward CNN architecture faces challenges as it becomes deeper, leading to the "vanishing gradient" problem where information can get lost during training. DenseNets address this issue by simplifying the connectivity pattern between layers. In a DenseNet, each layer is directly connected to every other layer, allowing for better feature reuse and improving classification accuracy [12].

As shown in Fig. 2. DenseNet121 consists of several components, including connectivity, Dense-Blocks, growth rate, and bottleneck layers. The connectivity pattern involves concatenating the feature maps from previous layers as inputs for each layer. DenseNets utilize Dense-Blocks, where the dimensions of feature maps remain constant within a block while the number of filters changes. Transition layers are inserted between Dense-Blocks to reduce the number of channels. The growth rate parameter controls the number of features added in each layer, and bottleneck layers improve computational efficiency.

The DenseNet121 architecture, for example, consists of multiple convolutional and pooling layers, Dense-Blocks with repeated convolutions, transition layers, and a global average pooling layer for classification. In total, DenseNet121 has 120 convolutions and 4 average pooling layers. The weights of all layers are spread over multiple inputs, allowing deeper layers to utilize features from earlier layers. The concentration of high-level features tends to increase toward the final layers.

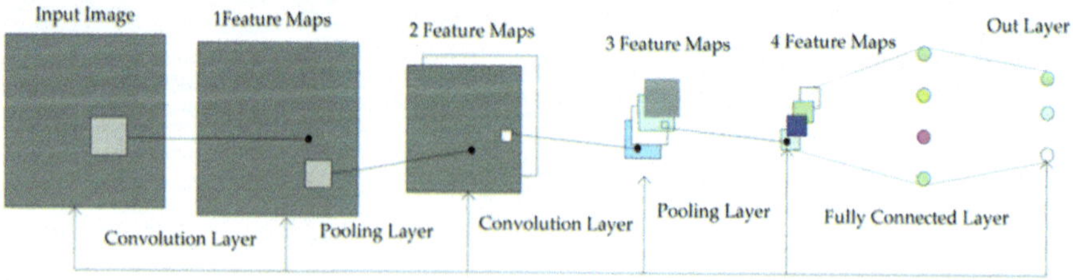

Fig. 2 DenseNet121 layered architecture for classification of X-ray images

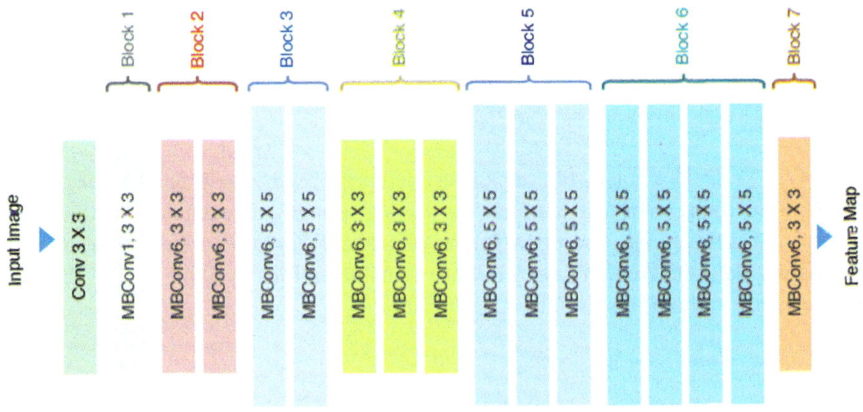

Fig. 3 EfficientNetB0 layered architecture for feature mapping of X-ray images

Through experiments, DenseNet has shown promising results in capturing complex features and achieving high accuracy in image classification tasks.

EfficientNetB0: Compound scaling is a technique that expands a neural network's width, depth, and resolution without changing the operations within each layer. In the case of EfficientNet, the authors designed a mobile-size baseline network called EfficientNet-B0, which optimizes accuracy and FLOPS using a multi-objective neural architecture. The building block of this architecture is the mobile inverted bottleneck MBConv, which includes an additional Squeeze and Excitation (SE) [13] block.

To apply compound scaling, the authors set certain parameters (α, β, γ) based on grid search and the constraint $\alpha \cdot \beta^2 \cdot \gamma^2 \approx 2$, resulting in values of $\alpha = 1.2$, $\beta = 1.1$, and $\gamma = 1.15$. Using these parameters, they derived a family of neural networks called EfficientNet-B1 to B7 are shown in Fig. 3.

The MBConv block, used in EfficientNet, follows an inverted pattern compared to traditional residual blocks. It starts with a narrow channel, widens through a point-wise convolution, applies a depth-wise convolution to reduce parameters, and finally uses another point-wise convolution to narrow the channels before the addition operation. This design makes EfficientNet more efficient and adaptable for mobile platforms.

The Squeeze and Excitation (SE) block is another component used in EfficientNet to improve interdependencies between channels. It performs dynamic feature channel-wise recalibration, assigning higher weights to important channels. This helps optimize the flow of information within the network.

Overall, EfficientNet combines compound scaling, MBConv blocks, and SE blocks to create efficient and effective neural network architectures for various tasks.

6 Results

6.1 Dataset

X-Ray COVID-19 dataset for our experiments by removing a single image with an incorrect label. We also curated the normal images to exclude any that were mislabeled projections. The open-source platform Kaggle repository was used for the X-Ray dataset. We summarized these transformations in Table 2, and the train and test, and validation split into a 70:20:10 ratio.

6.2 Experimental Setup

The evaluation of the proposed model involved assessing its performance using four essential metrics, including accuracy, etc. To implement the proposed model and other pre-trained models, Python and TensorFlow were utilized. All experiments were conducted on a system featuring an Intel (R) Core (TM) i7-7700 CPU @ 3.60 GHz processor and 8 GB RAM. For training purposes, both the proposed and pre-trained models employed the ReLu activation function and softmax pooling to process images. For a comprehensive overview of the image setup utilized in the experiments.

6.3 Outcomes

The experimental outcomes utilizing DenseNet121 and EfficientNetB0 models are presented in Table 3. Among these models, the proposed approach demonstrated superior performance with the DenseNet121 model. Specifically, the DenseNet121 model achieved impressive results across various metrics, including accuracy (94%), precision (98%), recall (97%), and F1-score (98%). In contrast, the EfficientNetB0 model exhibited lower performance with accuracy (77%), precision (85%), recall (92%), and F1-score (88%). The corresponding confusion matrices for DenseNet121 can be found in Fig. 4a, while the accuracy results are displayed in Fig. 4b. Similarly, the confusion matrices for EfficientNetB0 are illustrated in Fig. 5a, and the accuracy results are presented in Fig. 5b.

Table 3 Results of the experiment for the proposed model

Images	DenseNet121	EfficientNetB0
Accuracy	94	77
Precision	98	85
Recall	97	92
F1-score	98	88

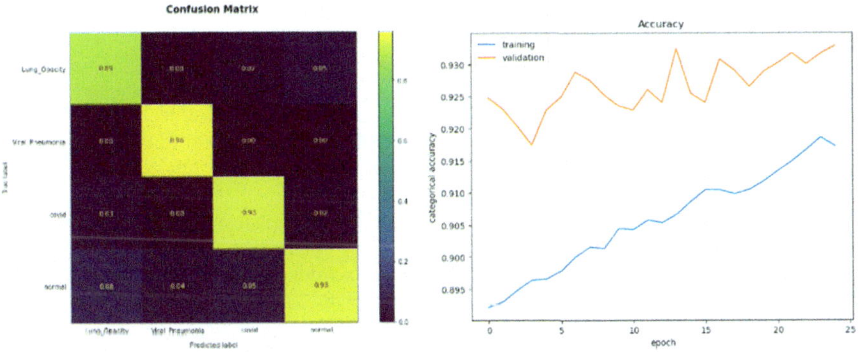

(a) (b)

Fig. 4 **a** Confusion Matrix and **b** accuracy for DenseNet121 model

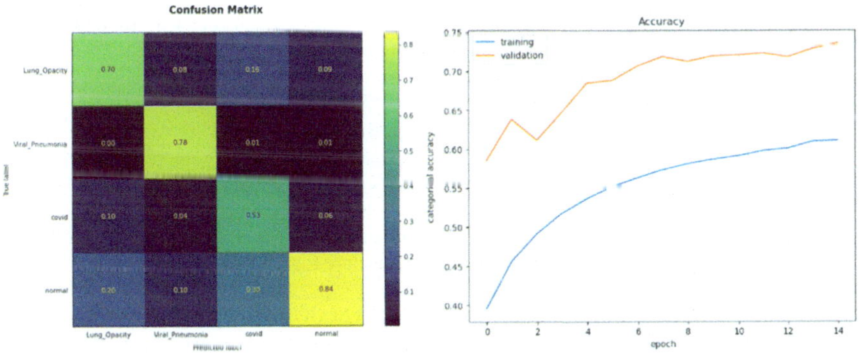

(a) (b)

Fig. 5 **a** Confusion Matrix and **b** accuracy for EfficientNetB0 model

7 Conclusions and Future Work

The study's findings reveal that X-ray images exhibit the highest accuracy in detecting comorbid patients. In our experiments, we compared the performance of DenseNet121, and EfficientNetB0, which were meticulously fine-tuned with appropriate parameters to achieve superior accuracy in classifying images of comorbid patients. However, the limited availability of data presented a challenge in training more complex models. Conspicuously, our experimental results highlight the effectiveness of the fusion model, which combines the outputs of individual models, leading to improved image classification. To further enhance the overall conclusions of the study, future research endeavors could incorporate laboratory images, as they have the potential to provide more accurate validation data.

References

1. BBC News (2020) Coronavirus Confirmed as a Pandemic by World Health Organization. BBC
2. Alaiad AI, Mugdadi EA, Hmeidi II et al (2023) Predicting the severity of COVID-19 from lung CT images using novel deep learning. J Med Biol Eng 43:135–146. https://doi.org/10.1007/s40846-023-00783-2
3. Aleem S, Maniparambil M, Little S, O'Connor N, McGuinness K (2023) An ensemble deep learning approach for COVID-19 severity prediction using chest CT scans (arXiv:2305.10115v1 [eess.IV] for this version). https://doi.org/10.48550/arXiv.2305.10115
4. Suresh Kumar HS, Pushpa CN, Thriveni J, Venugopal KR (2023) Comorbidities and COVID-19 pandemic dynamics: a review. Gis Science Journal 10(3):1852–1876
5. Jibril F, Adeshina SA, Nnanna AA (2021) A deep learning based COVID-19 detection framework. In: 2021 1st International Conference on Multidisciplinary Engineering and Applied Science (ICMEAS) IEEE, pp 1–6
6. Pushpa CN, Thriveni J, Venugopal KR, Patnaik LM (2013) Web search engine based semantic similarity measure between wards using pattern retrieval algorithm. J CS IT-CSCP, pp 1–11
7. Ding W, Nayak J, Swapnarekha H, Abraham A, Naik B, Pelusi D (2021) Fusion of intelligent learning for COVID-19: a state-of-the-art review and analysis on real medical data. Neurocomputing 457:40–66. https://doi.org/10.1016/j.neucom.2021.06.024
8. Paul S, Das R, Khanal B (2022) Covid-19 detection using chest X-Ray images along with corresponding metadata of the chest X-Ray. https://doi.org/10.21203/rs.3.rs-2102953/v1
9. Kulkarni AR, Athavale AM, Sahni A, Sukhal S, Saini A, Itteera M, Zhukovsky S, Vernik J, Abraham M, Joshi A, Amarah A, Ruiz J, Hart PD, Kulkarni H (2021) Deep learning model to predict the need for mechanical ventilation using chest X-ray images in hospitalized patients with COVID-19. BMJ Innov 24(6):100419
10. Suresh Kumar HS, Boomika S, Pushpa CN, Thriveni J, Venugopal KR (2023) MobNetCov19: detection of COVID-19 using MobileNetV2 architecture for multi-mode images. In: Presented ICCSST 2023 in Christ University, Bengaluru
11. Rafi TH (2020) A holistic approach to identification of COVID-19 patients from chest X-ray images utilizing transfer-based learning. medRxiv 2020–07
12. Yu H, Yang LT, Zhang Q, Armstrong D, Deen MJ (2021) Convolutional neural networks for medical image analysis: state-of-the-art, comparisons, improvement, and perspectives. Neurocomputing 444:92–110
13. Thanh NC, Long TQ (2021) CRF-EfficientUNet: an improved UNet framework for polyp segmentation in colonoscopy images with combined asymmetric loss function and CRF-RNN layer. IEEE Access. 9:156987–157001

A Comparative Study of Approaches to Short Text Document Clustering

N. Meena, D. S. Jayalakshmi, and J. Geetha

Abstract With the increasing volume of information and data, both structured and unstructured, clustering has found applications in various domains such as browsing and search engines. Text document clustering focuses on organizing documents into clusters based on their content similarity. Feature extraction techniques are employed to enhance the results, and the similarity between documents is measured to evaluate the quality of the clusters. This paper presents a literature review that concentrates on the utilization of preprocessing techniques, feature extraction, clustering algorithms, and evaluation measures in document clustering. Through an extensive survey, the paper investigates which algorithms yield superior performance in feature extraction, clustering, and evaluation. According to the survey, nature-inspired algorithms like particle swarm optimization (PSO) offer optimal solutions and are particularly effective for feature extraction. Furthermore, the paper suggests employing a hybrid approach to address the clustering problem, utilizing principal component analysis (PCA) for dimensionality reduction.

Keywords Text documents · Clustering · Feature extraction

1 Introduction

Due to the rapid advancements in technology, we now have the capability to gather vast amounts of diverse data, both structured and unstructured. However, analyzing such data quickly and efficiently has become a challenge. Data mining refers to the

N. Meena (✉) · D. S. Jayalakshmi · J. Geetha
Department of CSE, Ramaiah Institute of Technology, Bangalore, India
e-mail: meenagowda78@gmail.com

D. S. Jayalakshmi
e-mail: jayalakshmids@msrit.edu

J. Geetha
e-mail: geetha@msrit.edu

V. K. Gunjan and J. M. Zurada (eds.), *Proceedings of 4th International Conference on Recent Trends in Machine Learning, IoT, Smart Cities and Applications*, Lecture Notes in Networks and Systems 873, https://doi.org/10.1007/978-981-99-9442-7_64

process of discovering patterns, relationships, and insights from large volumes of data, including both structured and unstructured data. Text document clustering is a specific data mining technique that aims to identify underlying patterns and relationships within the documents by analyzing their content, word frequencies, and other linguistic features. Clustering algorithms, such as K-means, hierarchical clustering, or density-based methods, are applied to assign documents to clusters based on similarity measures, such as cosine similarity or distance metrics. The objective is to maximize the intra-cluster similarity while minimizing the inter-cluster similarity, resulting in cohesive and well-separated clusters. The outcome of text document clustering allows for efficient data organization, information retrieval, and data exploration, enabling tasks such as topic modeling, sentiment analysis, text categorization, and document recommendation systems. This process also allows search engines to provide more accurate and focused results. Moreover, text document clustering aids in knowledge discovery, enabling researchers and analysts to explore patterns and gain insights from large textual datasets.

Cluster analysis offers users a means to gain insights into a dataset's main properties without prior information. However, due to the numerous input parameters required by clustering algorithms, determining the optimal configuration can be challenging, often involving a time-consuming trial and error process. Popular clustering techniques like K-means can be sensitive to user-defined parameters, such as the initial selection of points or the choice of distance metric to measure similarity between data points. Measuring similarity and evaluating the quality of clusters are crucial steps in the process.

This paper presents a survey on text document clustering, discussing the processes of preprocessing, feature extraction, vectorization, and clustering of documents, along with evaluation methods. The research literature survey examines the results, summarizes findings, and outlines potential future work.

2 Text Document Clustering Techniques

Text document clustering employs various techniques to group similar documents together based on their content. Figure 1 depicts the categorization of text document clustering techniques.

There are two main types of clustering techniques, namely hierarchical clustering and partitional clustering, which can be classified into different subcategories. Hierarchical clustering is a widely used method that groups objects and visually represents them in a hierarchical tree called a dendrogram. It can be further divided into two methods: agglomerative clustering and divisive clustering. Agglomerative clustering can be classified based on different approaches such as single link, complete link, and average link. On the other hand, partitional clustering is a method that classifies observations within a dataset into multiple groups based on their similarities. Partitional clustering can be further categorized into four types: centroid/distance-based,

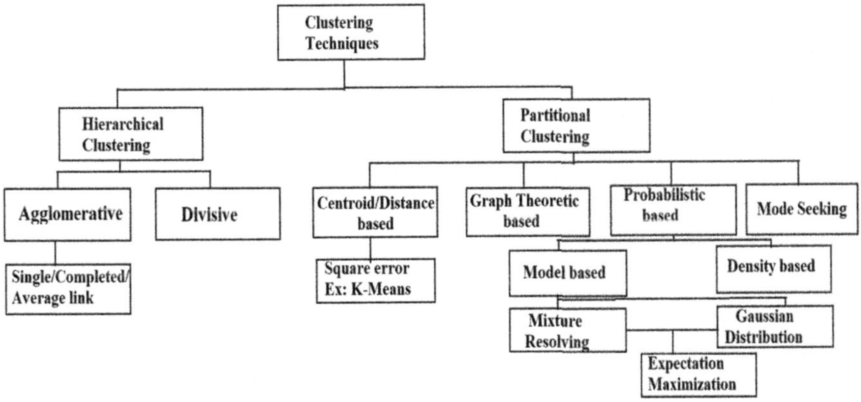

Fig. 1 Taxonomy of clustering techniques in text document clustering

graph theoretic-based, probabilistic-based, and mode seeking. Distance-based clustering, for example, uses squared error methods like K-means. Probabilistic-based clustering includes model based approaches such as Gaussian distribution, which can be further divided into mixture resolving and distribution methods. An example of mixture resolving is the Expectation Maximization algorithm. In the next section, a review of related work under several of the above mentioned categories is presented.

3 Literature Survey

3.1 Distance-Based Clustering Model

The paper [1] explores the document-to-vector paradigm and investigates the accuracy of clustering techniques, namely K-Means, Spherical K-Means, LDA, and DBSCAN. To preserve semantic connections between words, word embedding-based approaches are used to reduce the dimensionality of document representations. Notably, the most significant improvement in Adjusted Mutual Information is observed in DBSCAN when combined with the Doc2Vec embedding. In [2], clustering algorithms are effectively utilized within a cloud computing infrastructure to address the problem of extracting textual information from vast collections of documents. Three alternative algorithms, namely K-Means, Bisecting K-Means, and Gaussian Mixture Model (GMM), were employed for clustering. To evaluate the effectiveness of Map Reduce in clustering documents, a parallel K-means approach was employed. Although this approach requires more computation time, it can discover intricate patterns and generate superior clustering results with distinct boundaries. In [3], researchers have developed Patent Trend Analysis (PTA) software that leverages Artificial Intelligence (AI) and Big Data (BD) techniques to mine patent data. Experimental results demonstrate that the system provides valuable information on patent trends, aiding in patent trend analysis and comprehension.

Traditional text document clustering approaches that employ the vector space model and word embedding fail to consider polysemy (multiple meanings) or the semantic relationships between words. However, in [4], the authors utilize a pre-trained language representation model called BERT to create contextualized sentence embeddings as an initial step. Document dimensions are a crucial concern in [5]. The primary focus of this research has been dimension reduction using the fundamental similarity principle. Stack Random Projection, a greedy layer for dimensional reduction, is recommended. Rapid search and Density Peak Clustering (DPC) approaches for clustering are followed by the K-means clustering strategy. In [6], document clustering method is applied to find interesting features in text. Dimensionality reduction (DR) approach is used to minimize dimensions of data matrices. The system calculates K-means, singular vector decomposition, and nonnegative matrix factorization scores, including homogeneity, completeness, adjusted Rand, and adjusted mutual information scores. In comparison to NMF, it has been seen that K-means and SVD produce lower scores. The correct clusters can be obtained more accurately with NMF. In [7], it is suggested to use limited clustering to learn the background information and improve the document clustering process. This paper examined effects of constraint extraction on document clustering and created an autonomous constraint learning approach from document gathering in order to effectively include constraints into the document clustering process. In [8], representing document models for explaining document level semantics is investigated using the document representation topic-aware approach. It suggests several features that would be desirable for an effective document representation, including (i) word-to-word relationships can be captured, (ii) topic units–which are larger compositions of words-can be derived from fundamental words and their associations, (iii) a feature-weighting method based on frequent topic units and document correlation can be used. The authors in [9] introduce several novel concepts and semantics that focus on improving the "bag of words" method. A measurement is needed for document clustering to determine how unexpected two pieces of information can be. This dissimilarity is frequently calculated using distance measures like Euclidean distance and Jaccard Coefficient as well as similarity measures like Pearson's correlation coefficient. The authors assessed and looked into the effectiveness of these measures in partitioned clustering of text document datasets. While Pearson and Jaccard approach is more suited to finding logical clusters with high clarity values, Euclidean distance analysis is more effective at grouping documents. In [10], document clustering as applied to tasks like search engine optimization and the extraction of secret data from IoT sensor data is the focus. Swarm intelligence (SI) algorithms leverage stochastic and heuristic principles that adhere to a few basic rules to execute incredibly difficult jobs. SI algorithms can produce answers in a flexible, reliable, decentralized, and self-organized manner by associating problem-specific properties with algorithmic parameters. In [10], PSO, BA, and GWO SI formulae are applied to the document clustering problem.

3.2 Hybrid Approach

In [11], the Hybrid Jaya Optimization method (HJO), also known as HJO-DC, is applied for text document clustering. In order to gauge a solution's quality, the Silhouette index is employed as a statistic. Thorough examination the numerical outcomes proved beyond a reasonable doubt that the suggested algorithm produces the greatest quality clustering, outperforming previous clustering techniques, and is superior to them. In [12], a brand-new hybrid strategy hybrid black hole (BH) to document clustering is presented. This paper proposes three distinct model combinations in particular for expressing text documents: (1) K-Mean+ + with fix k, BH, and TF-IDF, (2) BH + W2V with fix k in K-Mean++, (3) Automated *k* hybrid black hole. In [13], The H-MVO approach, a novel hybrid of the MVO and K-means clustering algorithms, is used to improve the quality of the initial candidate solutions as well as the best solution that MVO develops at each iteration. The study in [14], suggests two methods for effective document clustering that can be using intelligent hybrid approach PSO algorithm and soft computing. The suggested method combines K-means and particle swarm optimization, fuzzy C-means algorithm and particle swarm optimization. The study in [15] recommends that an improved version of Arithmetic Optimization Algorithm (AOA) be used to solve the text document clustering problem by utilizing several LABIC text clustering datasets. In [16], the K-means and hybridization Cuckoo Search is used for clustering of high-dimensional documents.

3.3 Mode Seeking (KNN) Approach

In [17], text databases are clustered using a hybrid strategy (eSQ + KNN). The initial clusters are formed by the set of several single word search queries and all unassigned documents are then assigned to the cluster that is closest to them using the KNN algorithm. The genetic algorithm (GA) is used by evolved Search Queries (eSQ) to evolve a collection of search queries (requests). In [18], in order to increase effectiveness of clustering similarities, K-means and KNN algorithms is used. According to test findings, the hybrid K-means and KNN algorithm performs well when grouping text.

3.4 Graph-Theoretic-Based Model

In [19], a graph-based method for representing documents with constraints was presented for document clustering. Documents frequently include a variety of feature terms that can be related nonlinearly, making it simple for graph data structure to express this information. The resulting dependency network is changed into a weighted dependency graph by utilizing the TF-IDF measure to determine the weight of each vertex. Constraints can be divided into three categories: instance level, corpus-level, and constraint-level. In graph-based method, VSM built on ontology is used to represent documents in novel algorithm. Constrained restricted hierarchical agglomerative clustering (HAC) is also suggested since it will improve results by employing instance level limits as prior knowledge to direct the clustering process. In paper [20], it is evident that choice of the text representation technique has a significant impact on clustering accuracy. Semantic and sparsity issues are common in textual documents that cannot fixed. This work suggests a graph-based text representation method called dependency graph to address the challenges of sparsity and semantic to improve the accuracy of document clustering.

3.5 Probabilistic-Based Model

In this paper [21], topic modeling method, which divides documents into themes and subjects into words, is used a text mining technique. Finding hidden patterns in the text is done using a technique called topic modeling. To identify the buried topics in the data, Latent Dirichlet Allocation (LDA), a probabilistic modeling technique, being applied a group of documents.

3.6 Distance-Based and Hierarchical (Agglomerative)-Based Model

In [22], a technique for classifying documents n the semantic similarity basis using the NLTK vocabulary is demonstrated. The K-mean algorithm and the Ward's approach are used for clustering, and TF-IDF is used to create a vector space. IMDB and Wikipedia dataset synopses are defined, tokenized, and stemmed as part of the process. To thematically group papers, WordNet is also employed. Three implementation scenarios are studied for each algorithm: (1) no preprocessing; (2) without stemming preprocessing; and (3) with stemming preprocessing.

4 Conclusion

This survey explores document clustering, which involves grouping documents into different clusters based on their similarity in content. Collecting unstructured data is a key step in the clustering process. It is important to select appropriate algorithms to achieve improved clustering results with text data. Based on our survey, we conclude that a hybrid approach that combines different techniques is beneficial to achieve optimal clustering solutions. Nature-inspired optimization techniques can be utilized to enhance the clustering results by optimizing the problem-solving process.

References

1. Radu R-G, Rădulescu I-M, Truică C-O, Apostol E-S, Mocanu M (2020) Clustering Documents using the document to vector model for dimensionality reduction. In: 2020 IEEE International Conference on Automation, Quality and Testing, Robotics (AQTR), pp 1–6. https://doi.org/10.1109/AQTR49680.2020.9129967
2. Dritsas E, Trigka M, Vonitsanos G, Kanavos A, Mylonas P (2022) An apache spark implementation for text document clustering. In: 2022 17th International Workshop on Semantic and Social Media Adaptation and Personalization (SMAP), 2022, pp 1–6. https://doi.org/10.1109/SMAP56125.2022.9941983
3. Liang R, Li Y, Chen X, Chen J (2020) Patent trend analysis through text clustering based on K-means algorithm. ISCEIC 2020:115–118. https://doi.org/10.1109/ISCEIC51027.2020.00032
4. Li Y, Cai J, Wang J (2020) A text document clustering method based on weighted BERT model. In: 2020 IEEE 4th Information Technology, Networking, Electronic and Automation Control Conference (ITNEC), pp 1426–1430. https://doi.org/10.1109/ITNEC48623.2020.9085059
5. Kotte VK, Vuppu S, Thadishetti R (2021) High dimensional text document clustering and classification using machine learning methods. In: 2021 5th International Conference on Intelligent Computing and Control Systems (ICICCS), pp 1612–1617. https://doi.org/10.1109/ICICCS51141.2021.9432128
6. Kumbhar R, Mhamane S, Patil H, Patil S, Kale S (2020) Text document clustering using K-means algorithm with dimension reduction techniques. In: 2020 5th International Conference on Communication and Electronics Systems (ICCES), pp 1222–1228. https://doi.org/10.1109/ICCES48766.2020.9137928
7. Rafi M, Abid F, Khan HM, Mirza A (2021) Towards the constraint learning and optimization approach to document clustering. In: 2021 6th International Multi-Topic ICT Conference (IMTIC), pp 1–7. https://doi.org/10.1109/IMTIC53841.2021.9719790
8. Rafi M, Khan H, Nadeem H, Shakeel H (2021) Unsupervised topic aware document-level semantic representation for document clustering. In: 2021 22nd International Arab Conference on Information Technology (ACIT), 1–10. https://doi.org/10.1109/ACIT53391.2021.9677217
9. Kumar N, Yadav SK, Yadav DS (2020) Similarity measure approaches applied in text document clustering for information retrieval. In: 2020 Sixth International Conference on Parallel, Distributed and Grid Computing (PDGC), pp 88–92. https://doi.org/10.1109/PDGC50313.2020.9315851
10. Selvaraj S, Choi E (2021) Swarm intelligence algorithms in text document clustering with various benchmarks. Sensors 21:3196
11. Thirumoorthy K, Muneeswaran K (2021) A hybrid approach for text document clustering using Jaya optimization algorithm. Exp Syst Appl 178:115040. https://doi.org/10.1016/j.eswa.2021.115040
12. Malik F, Khan S, Rizwan A, Atteia G, Samee NA (2022) A novel hybrid clustering approach based on black hole algorithm for document clustering. IEEE Access 10:97310–97326. https://doi.org/10.1109/ACCESS.2022.3202017
13. Abasi AK, Khader AT, Al-Betar MA, Naim S, Alyasseri ZAA, Makhadmeh SN (2020) A novel hybrid multi-verse optimizer with K-means for text documents clustering. Neural Comput Appl
14. Bezdan T, Stoean C, Namany AA, Bacanin N, Rashid AT, Zivkovic M, Venkatachalam K (2021) Hybrid fruit-fly optimization algorithm with k-means for text document clustering. Mathematics 9:1929

15. Abualigah L, Almotairi KH, Al-qaness MAA et al (2022) Efficient text document clustering approach using multi-search Arithmetic Optimization Algorithm. Knowledge-Based Syst 248:108833
16. Boushaki SI, Bendjeghaba O, Brakta N (2021) Document clustering analysis based on hybrid cuckoo search and K-means algorithm. In: 2021 IEEE 12th Annual Information Technology, Electronics and Mobile Communication Conference (IEMCON), pp 58–62. https://doi.org/10.1109/IEMCON53756.2021.9623204
17. Hirsch L, Nuovo AD, Haddela P (2021) Document clustering with evolved single word search queries. IEEE Cong Evol Comput (CEC) 2021:280–287. https://doi.org/10.1109/CEC45853.2021.9504770
18. Habibpour R, Khalilpour K (2014) A new hybrid k-means and k-nearest-neighbor algorithms for text document clustering. Int J Acad Res. https://doi.org/10.7813/2075-4124.2014/6-3/A.12
19. Amin F, Rafi M, Shaikh MS (2014) Document clustering using graph based document representation with constraints
20. Abdulsahib AK, Kamaruddin SS (2015) Graph based text representation for document clustering. J Theor Appl Inform Technol 76(1)
21. Gupta M, Wasi A, Verma A, Awasthi S (2021) Document clustering and topic classification using latent Dirichlet allocation. In: 2021 International Conference on Innovative Computing, Intelligent Communication and Smart Electrical Systems (ICSES), pp 1–6. https://doi.org/10.1109/ICSES52305.2021.9633830
22. Salih NM, Jacksi K (2020) Semantic document clustering using K-means algorithm and ward's method. In: 2020 International Conference on Advanced Science and Engineering (ICOASE), pp 1–6. https://doi.org/10.1109/ICOASE51841.2020.9436588

A Review of Satellite Image Analysis Tools

Bhushan Pawar, Vijay Prakash, Lalit Garg, Charles Galdies, Sandra Buttigieg, and Neville Calleja

Abstract The rapid growth of urban areas, population, and pollution presents a significant environmental challenge. Addressing these challenges requires innovative analytical approaches and data sources. This study focuses on the three-scale phenomena of the environmental system, physical environment, and regional surroundings, considering their implications on a global network scale. Remote sensing data, with its easy accessibility, offers new toolkits and methods to improve the global environment. It provides self-consistent measurements of vital physical attributes and surpasses traditional data sources regarding accuracy, efficiency, and coverage of larger geographical areas. This study aims to identify and analyze appropriate satellite image analysis tools for healthcare. The study objectives include studying available tools, analyzing their features, categorizing tools into open-source and commercial categories, and providing recommendations. This study emphasizes

B. Pawar · V. Prakash (✉) · L. Garg
Department of Computer Information Systems, Faculty of Information and Communication Technology, University of Malta, Msida 2080, MSD, Malta
e-mail: vijaysoni200@gmail.com

B. Pawar
e-mail: bhushan.d.pawar.21@um.edu.mt

L. Garg
e-mail: lalit.garg@um.edu.mt

V. Prakash
Department of Computer Science and Engineering, Graphics Era Hill University, Dehradun, India

C. Galdies
Environmental Management and Plan Division, Institute of Earth Systems, University of Malta, Msida, Malta
e-mail: charles.galdies@um.edu.mt

S. Buttigieg
Faculty of Health Sciences, University of Malta, Msida, Malta
e-mail: sandra.buttigieg@um.edu.mt

N. Calleja
Faculty of Medicine and Surgery, University of Malta, Msida, Malta
e-mail: neville.calleja@um.edu.mt

V. K. Gunjan and J. M. Zurada (eds.), *Proceedings of 4th International Conference on Recent Trends in Machine Learning, IoT, Smart Cities and Applications*, Lecture Notes in Networks and Systems 873, https://doi.org/10.1007/978-981-99-9442-7_65

feature extraction and image enhancement parameters, which are crucial for image analysis. Further, this study provides a functional overview of various commercial and open-source satellite image analysis tools and suggests suitable tools for healthcare applications.

Keywords Satellite image analysis · Geospatial data · ArcGIS · ENVI · 1. ERDAS IMAGINE · SAGA GIS

1 Introduction

1955 was the historical year of human history. The space race was started between two superpowers (the Soviet Union and the United States) to acquire spaceflight and space launching capability [1]. This event extended the ability of human beings and attracted researchers to apply this new technology for the welfare of Earth residents. This event also brought revolutionary robotic space pro and artificial satellite launches to Venus, Mars, and our natural satellite "Moon." After some years, scientists could also rotate the artificial satellite in low earth orbit for detection and surveillance [2]. This leads to another challenge for researchers: finding the appropriate methodology to process the enormous amount of data gathered by the satellites, simplifying the gathering process, and increasing the availability of satellite data [3].

Many governments (ESA, NASA, ISRO, and many more) and private business organizations (Apple, Google, SpaceX, Etc.) released their space program by launching artificial satellites and developing a platform for accessing the collected data [4]. All the data is available for further enhancement or development for the research community. These efforts encourage researchers and scholars to add innovative approaches to solve real-world problems. Remote sensing is a promising technology that allows us to observe the changes happening to an object without modifying its circumstances [5, 6]. Nowadays, remote sensing expands its applicability into many fields, such as weather, forestry, agriculture, surface changes, and biodiversity. As mentioned, each area requires remote sensing characteristics to accommodate the prerequisites; researchers must focus on the authorized dataset platforms and tools [7, 8]. Selecting appropriate datasets and tools has a significant role in the research study. This research study will,

(1) Describe the taxonomy for understanding the suitable satellite image analysis tools.
(2) Inspect the limitations and strengths of the described satellite image analysis tools.
(3) Recommend a list of satellite image analysis tools for future researchers or research studies.

2 eCognition

Cognition Network Technology (CNT) was introduced, "eCognition," well known as Definiens Cognition Network Technology. This tool consists of an innovative approach based on object-oriented image analysis, extracting the image information with the help of image objects hierarchy (Collection of Pixel) [9]. The eCognition suite consists of three modules typically utilized independently or collectively to address various image analysis problems. One of the notable features of eCognition is its server software, which offers a batch-processing environment for the execution of image analysis tasks [8]. The software suite offers a wide range of data types to support the implementation of various remote sensing and geospatial applications in different scenarios and environments. These data types are outlined below:

- **Generic:** Change Detection, Rapid Mapping, Object Recognition.
- **By environment:** Urban Analysis, Diverse Land Cover Mapping, Agriculture, Forestry, Riparian, and Marine [10].
- **Other:** Security, Defense, Climate, and Atmosphere.

3 ArcGIS

Eris developed a client–server GIS program, "ArcGIS." In 2015, Eris released "ArcGIS Pro" for Web Maps to handle 2D and 3D visualization, cartography, and AI [11]. Eris created "ArcGIS Server." It simplifies workflows like consuming and publishing feature layers with ArcGIS online [12]. Below are some amazing ArcGIS Pro 2.6 features [13].

- **Voxel Layers:** These layers are designed to visualize composite layers that often occur in oceanic and atmospheric space–time cubes of data, represent 3D data for space and time, and are stored under the nerCDF file extension. These voxel layers are applied for analyzing unique patterns (spatial) of data in required circumstances [14].
- **Trace Networks:** This facility checks the connectivity models such as highways, roads, railways. The concurrence of geometric features determines the network's connectedness. Trace networks are utilized with network topology to offer supplementary functionalities, including validation and traceability.
- **Interactive suitability analysis using the new suitability:** This feature searches for the ideal location for a residential building or other construction work. It requires unique attributes (criteria) input to the model for predicting optimal location. The modeler allows every user to observe the changes per the modification in criteria, update itself based on changes before decision-making for the project, and generate feedback to know the model.

- **Graphics layers:** It play a vital role in storing geometric properties; it overlaps on other map layers for better demonstration. It also adds information to the map, like significant features or text. It is possible to superimpose numerous graphics layers on a map or assemble them for better visualization.
- **Project recovery:** A useful feature is provided by ArcGIS Pro 2.8, which automatically saves the project in the case of emergency power loss or interruption. When software loads, it prompts the end-user to select the backup option that keeps every unsaved modification in the backup folder and later used without losing any data.
- **Link analysis:** Determines the relevance of network patterns and uncovers existing ones. Link analysis helps understand the network by identifying the shortest path between nodes, visualizing nodes with the strongest connections, and identifying the nearest nodes.

4 ENVI

ENVI is the most advanced software program used to analyze, process, and visualize geospatial imagery, developed by Harris Geospatial in 1994, well known for creating various applications for environmental, defense and intelligence, engineering, medical imaging, and aerospace. ENVI supports recently launched airborne and satellite sensors. These include hyperspectral, multispectral, LiDAR, panchromatic, thermal, infrared, radar, full-motion video, HDF5, ASCII, NET-CDF-4, and LAZ [15]. This remote sensing software suite allows users to access established algorithms for easy, accurate, and fast image analysis [16, 17]. Some of the features of ENVI are listed below:

- **ENVI functionality for science applications** includes Image Orthorectification, Image Registration, Atmospheric Correction, Image Calibration, Feature Identification, Vector Overlay Creation, Region of Interest Identification, Pan Sharpening, Masking, Mosaicking and Digital Elevation Models, etc.
- **ENVI features for GIS applications** include Image Classification, Image Auto Registration, Feature Extraction, Post Classification Methods Change Detection, etc.
- **More ENVI products** include ENVI, ENVI EX, ENVI for ArcGIS Server-SARScape and ENVI Modules.

5 ERDAS Imagine

ERDAS, Inc. produces remote sensing and raster graphics editing software ERDAS IMAGINE. Geospatial raster data processing is the main focus of this software. It helps users visualize, improve, and prepare digital images for CAD or GIS [18]. The provided toolbox facilitates the execution of various picture operations, enabling users to generate solutions for diverse real-time geographical challenges. The software suite offers a range of functionalities for processing satellite images obtained from Landsat MSS, AVHRR, SPOT imaging, and Landsat TM. These images are utilized in creating land use/land cover maps, deforestation maps, and in identifying oil reserves. ERDAS IMAGINE is a comprehensive software package that offers a range of functionalities inside a single product. These functionalities include LiDAR analysis, Radar processing, remote sensing, fundamental vector analysis, and photogrammetry [19]. Some features of the product are as follows:

- User-friendly ribbon, GIS, remote sensing, and image analysis.
- The ability to effectively process and analyze both hyperspectral and multispectral imagery, optical panchromatic, LiDAR, and radar data, and spatial modeling with vector, raster, and point cloud operators with real-time results preview.
- Multi-core and distributed processing with a variety of change detection tools.
- The software can transform more than 190 image formats into many prominent file formats, such as NITF, GeoTIFF, JPEG, CADRG, MrSID, JPEG2000, and ECW.
- Conversion of data, efficient terrain mosaicking, and preparation.
- Complete OGC web services, including WPS, WCS, WMS Catalogue Services, and Web (CS-W).

ERDAS IMAGINE is widely recognized as a very influential software package that provides comprehensive support for various workflows involved in the generation of derived information, particularly in the realm of data production including:

- Color balancing, compression, mosaicking, and orthorectification.
- LiDAR classification and editing.
- Terrain categorization and land cover mapping.
- Feature capture and update.
- Spatial modeling and analysis.
- Report and map generation and printing with the help of Word or PowerPoint map composer.

6 SAGA GIS (System for Automated Geoscientific Analyses)

The software is freely available and open-source, catering specifically to the research community. It offers spatial data manipulation capabilities [20]. SAGA GIS has experienced significant growth in the range of geoscientific methodologies accessible through its module libraries. SAGA GIS necessitates a little 10 MB of disc space and does not support direct installation from a USB drive [21]. SAGA GIS encompasses a set of standard modules, which are as follows:

- **File access**: Vector, table, grid, and image file formats are supported by this software. These formats include Esri binary and ASCII grids, shapefiles, and GDAL grid file formats. Additionally, it supports SAGA GIS' native format, SGRD.
- **Filter for grids**: Multi-directional Lee, Laplacian, and Gaussian filter.
- **Gridding**: The process of interpolating vector data can be facilitated by utilizing many techniques, including triangulation, inverse distance, and nearest neighbor.
- **Geostatistics**: Single and multiple regression, residual, variance analysis, and universal and ordinary kriging.
- **Grid calculator**: Use user-defined functions to combine grids.
- **Grid discretization**: Segmentation and skeletonization.
- **Grid tools**: Gap filling, merging, and resampling.
- **Image classification**: Box classification, pattern recognition, cluster analysis, region growing, and maximum likelihood.
- **Projections**: (Proj4 and GeoTrans libraries) Coordinate transforms for vector and grid data, and grid georeferencing.
- **Simulation of dynamic processes**: Nitrogen distributions, TOPMODEL, landscape development, and erosion.
- **Terrain analysis**: Geomorphometric computations like curvatures, aspect, slope, analytical hill shading, curvature classification, flow path analysis, sink elimination, solar radiation, catchment delineation, relative altitudes, and channel lines.

7 Grass GIS

The "Geographic Resources Analysis Support System" software suite is commonly utilized for geospatial data analysis, temporal and spatial modeling, image processing, map and graphic generation, and visualization [22]. It can be run on multiple operating system platforms such as Windows, Linux, and OS X. End-users can access features provided by Grass GIS through Graphical User Interface (GIS) or using plugins in QGIS Software [23]. It also provides an interface to the module with the help of a custom-made shell, called an individual shell from the standard shell [24]. The recently released version of Grass GIS is GRASS GIS 7.8.6 in 2021. Some important features of GRASS GIS 7.8.6 are as follows:

- By using the GDAL-OGR library, it supports almost all GIS file formats.
- **Raster analysis:** Interpolation, map algebra, land space analysis, masking.
- **3D raster (Voxel) analysis:** 3D interpolation, 3D visualization, 3D map algebra.
- **SQL database:** Interface to SQLite, PostgreSQL, etc.
- Geo-coding of vector and raster maps.
- **Terrain analysis**: Surface and contour generation, hydrological tools, slope-aspect, and cost-path analysis.
- **Vector analysis:** Topology generalization and corrections, network analysis, overlays, buffer.

- **Temporal framework:** Big spatiotemporal environmental data offers vector and raster analysis and time series management.
- **Image processing:** Satellite, aerial, and UAV (Landsat, MODIS, Sentinel, etc.), classification (supervised and unsupervised), whole Object-Based Image Analysis (OBIA) chain.
- **Spatial statistics**: Covariance and correlation analysis, zonal statistics, and regression.

GRASS GIS can be used through the following interfaces,

- The most accessible and user-friendly Graphical User Interface allows users to access the software tool easily.
- It also provides a text-based Command-line Interface and
- C API, Object-oriented Python API, and Scripting library.
- Provides Jupyter Notebook programming editor.
- QGIS offers two ways to execute GRASS GIS modules: Plugin and processing toolbox.
- R language provides an interactive interface to access GRASS GIS through rgrass7.
- It also provides an online version of GRASS Online.

Table 1 depicts the characteristics of each tool, whereas Table 2 compares the different features of all tools in a summarized way.

8 Conclusion and Future Scope

One of the findings of our research study is that the open-source availability of analysis tools is simple, rare, and could provide the scope of improvement and scalability for future development. Orfeo toolbox is one of the best satellite analysis tools available in open-source software series, which can be operated on multiple operating systems and take input as raster and vector data. It supports various output formats such as text, binary, and image. This tool has numerous features applicable in multiple applications like defense, healthcare, town planning, agriculture, etc.

In commercial software suits, ENVI has a user-friendly interface, which is simple and efficient. The software suite can operate on multiple input formats (flat binary format, Landsat files, Spot files etc.) and generates output in all image processing, vectors, and generic image formats. We strongly recommend that the above software suits researchers using satellite image processing and GIS for various domains. Those also apply to other research domains under GIS and satellite image processing. We want to convey that the applicability of suggested GIS software suits/tools is subject to the research study's requirement and goal; opinion may vary per the project needs.

Table 1 Tools attributes/characteristics

Tool	Developer	Purpose	User-friendliness	Data handling	Analysis capabilities	Integration Capabilities
eCognition	Trimble	Object-based image analysis and classification	Moderate	Excellent	Advanced	Limited
ArcGIS	Esri	Geographic Information System (GIS)	User-friendly	Excellent	Extensive	Excellent
ENVI	Harris Geospatial	Remote sensing data analysis and image processing	Moderate	Excellent	Advanced	Good
ERDAS IMAGINE	Hexagon	Remote sensing data processing and analysis	Moderate	Excellent	Extensive	Good
SAGA GIS	SAGA GIS	Geoscientific analysis and data processing	Moderate	Good	Advanced	Good
Grass GIS	Open-Source	Geospatial analysis and data management	Moderate	Good	Advanced	Excellent

Table 2 Individual features of each tool

Features/tools	eCognition	ArcGIS	ENVI	ERDAS IMAGINE	SAGA GIS	Grass GIS
Operating System	Win-64	Win/Linux	MacOS/Win/Linux	Win/Linux	Win/Linux	Win/Mac OS/Linux
Platform	–	.Net/Java	Java	C and CPP	CPP	CPP
Input format	JPEG, Raster	All image formats	Lansat files, flat binary files, SPOT files. Etc.	Raster data and all Image formats	Raster data	XCF, JPEG and PNG
Output format	Raster and image data	Raster and image data	Image processing formats, generic image formats, vector formats	Image and raster	TIFF and JPEG	Vector and image
Extensibility	–	Yes	Yes	Yes	Yes	Yes
Feature extraction	Yes	Yes	Yes	Yes	Yes	Yes
Editing	Yes	Yes	–	Yes	Yes	Yes
Availability	Paid	Paid	Free	Paid	Free	Free
Geometric transformation	–	Yes	Yes	Yes	Yes	–
Classification	Yes	Yes	Yes	Yes	Yes	Yes
Segmentation	Yes	Yes	Yes	Yes	Yes	Yes
Color processing	Yes	Yes	Yes	Yes	Yes	Yes
Image enhancement	Yes	Yes	Yes	Yes	–	Yes
Change detection	Yes	Yes	Yes	Yes	Yes	–
Visualization	Yes	Yes	Yes	Yes	Yes	Yes
Image analysis	Yes	Yes	Yes	Yes	Yes	Yes

Acknowledgements This work is funded by the Malta Council for Science and Technology (MCST) Space Research Fund 2021 (Grant number SRF-2021-3S1, Project Satellite4Health).

References

1. Neufeld MJ (2007) Wernher von Braun's ultimate weapon. Bull Atom Sci 63(4):50–78. https://doi.org/10.1080/00963402.2007.11461094
2. Hegazy IR, Kaloop MR (2015) Monitoring urban growth and land use change detection with GIS and remote sensing techniques in Daqahlia governorate Egypt. Int J Sustain Built Environ 4(1):117–124. https://doi.org/10.1016/j.ijsbe.2015.02.005
3. Pawar B, Humbe VT, Kundnani L (2017) Morphology based moving vehicle detection. In: 2017 International Conference on Big Data Analytics and Computational Intelligence (ICBDAC), IEEE, pp 217–223. https://doi.org/10.1109/ICBDACI.2017.8070837
4. Jat MK, Garg PK, Khare D (2008) Monitoring and modelling of urban sprawl using remote sensing and GIS techniques. Int J Appl Earth Obs Geoinf 10(1):26–43. https://doi.org/10.1016/j.jag.2007.04.002
5. Wästfelt A, Tegenu T, Nielsen MM, Malmberg B (2012) Qualitative satellite image analysis: mapping spatial distribution of farming types in Ethiopia. Appl Geogr 32(2):465–476. https://doi.org/10.1016/j.apgeog.2011.04.001
6. Rembold F, Carnicelli S, Nori M, Ferrari GA (2000) Use of aerial photographs, Landsat TM imagery and multidisciplinary field survey for land-cover change analysis in the lakes region (Ethiopia). Int J Appl Earth Obs Geoinf 2(3–4):181–189. https://doi.org/10.1016/S0303-2434(00)85012-6
7. Rapp L, Santos C (2019) Satellite imagery, very high-resolution and processing-intensive image analysis: potential risks under the GDPR. Air Space Law 44(3):275–295. https://doi.org/10.54648/AILA2019018
8. Buczak AL, Koshute PT, Babin SM, Feighner BH, Lewis SH (2012) A data-driven epidemiological prediction method for dengue outbreaks using local and remote sensing data. BMC Med Inform Decis Mak 12(1):124. https://doi.org/10.1186/1472-6947-12-124
9. Nussbaum S, Menz G (2008) eCognition image analysis software. In: Object-Based Image Analysis and Treaty Verification. Springer Netherlands, Dordrecht, pp 29–39. https://doi.org/10.1007/978-1-4020-6961-1_3
10. Trimble Inc (2019) The Power of eCognition. https://docs.ecognition.com/v9.5.0/eCognition_documentation/Modules/1%20eCognition%20at%20a%20glance/01%20Introduction%20to%20the%20Power%20of%20eCognition.htm. Accessed 21 Jul 2023
11. ESRI (2020) ArcGIS 9. https://downloads.esri.com/support/documentation/ao_/698What_is_ArcGis.pdf. Accessed 21 Jul 2023
12. Crosier S, Booth B, Dalton K, Mitchell A, Clark K (2020) ArcGIS® 9 Getting Started With ArcGIS®. http://downloads.esri.com/support/documentation/ao_/1003Getting_Started_with_ArcGIS.pdf. Accessed 21 Jul 2023
13. esri (2022) ArcGIS Developers. *esri*. https://developers.arcgis.com/features/. Accessed 21 July 2023
14. ArcGIS Pro (2022) What is a voxel layer? *esri*. https://pro.arcgis.com/en/pro-app/latest/help/mapping/layer-properties/what-is-a-voxel-layer-.htm. Accessed 21 Jul 2023
15. NV5 Geospatial (2023) Getting Started with ENVI. https://www.l3harrisgeospatial.com/docs/GettingStartedWithENVI.html#Visualiz. Accessed 21 July 2023
16. ITT Visual Information Solutions (2023) Using ENVI and geographic information systems (GIS). https://vip.arizona.edu/documents/software/Using_ENVI_and_GIS.pdf. Accessed 21 Jul 2023

17. Research Systems (2001) ENVI Tutorials. http://www.glyfac.buffalo.edu/courses/gly560/Lessons/OLD/ENVI35_Tutorial/TUTORIAL.PDF. Accessed 21 July 2023
18. HEXAGON (2022) ERDAS IMAGINE 2022. https://bynder.hexagon.com/m/163c59089b0cf77/original/Hexagon_GSP_ERDAS_IMAGINE_2022_Update2_Release_guide.pdf. Accessed 21 Jul 2023
19. HEXAGON (2020) ERDAS IMAGINE product description. https://bynder.hexagon.com/m/3e734d4e8fcfdfd7/original/Hexagon_GSP_ERDAS_IMAGINE_Product_Description.pdf. Accessed 21 Jul 2023
20. G. Dept. of Physical Geography (2017) What is SAGA ? https://saga-gis.sourceforge.io/en/about/about.html. Accessed 21 Jul 2023
21. Conrad O et al (2015) System for automated geoscientific analyses (SAGA) v. 2.1.4. Geosci Model Dev 8(7):1991–2007. https://doi.org/10.5194/gmd-8-1991-2015
22. Lemenkova P (2023) A GRASS GIS scripting framework for monitoring changes in the ephemeral salt lakes of Chotts Melrhir and Merouane, Algeria. Appl Syst Innov 6(4):61. https://doi.org/10.3390/asi6040061
23. Neteler M, Mitasova H (2008) Open source GIS: a GRASS GIS approach, Third, vol 773. Springer
24. OSGeo (2023) Grass, G. I. S. 2023. https://grasswiki.osgeo.org/w/images/Flyer_0.pdf. Accessed 21 Jul 2023

Secure Image Retrieval in an Untrusted Cloud Environment

Vijay Prakash, Kavya Khandelwal, Disha Gupta, and Lalit Garg

Abstract Image processing in the scrambled area has received increasing attention for its many anticipated uses, such as providing efficient and safe solutions for protection-saving applications in untrustworthy environments. The high degree of ciphertext development caused by the existing homomorphic encryptions is one barrier to the widespread use of these algorithms. This paper considers the problem of managing images in scrambled space. We develop an image encryption scheme that limits ciphertext expansion while preserving the homomorphic characteristic using image design elements. The proposed encryption scheme first uses a probabilistic homomorphic cryptosystem to jumble the image's pixels, then packs the complete scrambled image to conserve space. Our approach offers a significantly lower ciphertext development factor while maintaining the homomorphic characteristic than the component-wise encryption scheme. It is unnecessary to demand additional intuitive norms when using secure sign-handling tools on the compressed encoded image. We provide a rapid calculation for the encryption and pressure of the proposed image encryption plot, which expedites the computation and significantly increases the effectiveness of our strategy. The security, ciphertext extension percentage, and computing complexity are also being examined. Our validation and testing approach shows that the proposed computations are valid. The proposed algorithms for image encryption make sense as an image encryption scheme for safe picture management applications.

V. Prakash (✉) · L. Garg
University of Malta, Msida, Malta
e-mail: vijaysoni200@gmail.com

L. Garg
e-mail: lalit.garg@um.edu.mt

V. Prakash · K. Khandelwal · D. Gupta
School of Computer Science, University of Petroleum and Energy Studies, Dehradun, India

V. Prakash
Department of Computer Science & Engineering, Graphics Era Hill University, Dehradun, India

V. K. Gunjan and J. M. Zurada (eds.), *Proceedings of 4th International Conference on Recent Trends in Machine Learning, IoT, Smart Cities and Applications*, Lecture Notes in Networks and Systems 873, https://doi.org/10.1007/978-981-99-9442-7_66

Keywords Image retrieval · Homomorphic encryption · Paillier cryptosystem and untrusted cloud

1 Introduction and Background

Due to the increase in the web's data transfer limit, people are now used to accessing and challenging audio and visual material stored on a server [1]. Furthermore, clients collect their information on the other side and utilize the computational power the other side provides to carry out their tasks, presuming the distant server has strong calculation/stockpiling capability and a wealth of resources. In this case, the net offers passive query management and is connected with sophisticated tools. This condition is identical to cloud processing and helpful for sight and sound information that necessitates extensive calculation and communication. Although encryption is a standard method for securing transmitted data, the encrypted data will hinder the functions typically performed on plaintexts [2]. Various studies [3–6] have focused on encrypted domain operations to process ciphertexts further and retrieve the respective results in the plaintext domain. Rather than multimedia retrieval of information in the realm of encryption, text document retrieval has received considerable attention in the field of literature for protecting privacy.

Image retrieval is required in today's world for various reasons, including large amounts of visual data; with the rise of digital cameras, smartphones, and social media, we now have more visual data than ever. Image retrieval allows us to search and organize this data efficiently in the following domains:

- **E-commerce**: Many e-commerce websites use image retrieval to help customers find the products they are looking for. By allowing users to search for products based on images rather than text, e-commerce sites can improve their shopping experience, which can help increase sales.
- **Medical diagnosis**: Image retrieval can help doctors and researchers quickly search through large amounts of medical images to identify patterns and diagnose diseases.
- **Law enforcement**: Image retrieval can be used to search through large databases of images to identify suspects and solve crimes.

Overall, image retrieval is essential in managing and analyzing large amounts of visual data in various fields, making it a vital tool today. In this paper, we are securely retrieving an image in an untrusted cloud environment, which refers to retrieving images from a cloud server while ensuring the security and privacy of the images [7]. In an untrusted cloud environment, the server is not under the user's control and can potentially be compromised. Therefore, it is crucial to employ security measures to protect the privacy of the images during retrieval [7]. This can be achieved through homomorphic encryption, secure multi-party computation, and fast indexing. Secure image retrieval in an untrusted cloud environment aims to provide users convenient and efficient access to their images while ensuring data confidentiality, privacy, and integrity [8, 9].

With the increasing adoption of cloud storage and services, there has been a growing concern about the security and confidentiality of information stored in the cloud [10–12]. In particular, images containing sensitive information such as medical records, financial information, or personal identification can be vulnerable to unauthorized access, manipulation, or theft. Such methods should provide end-to-end encryption and authentication to ensure that only authorized users can access the images and that the images have not been tampered with during transmission or storage. The need for secure image retrieval from untrusted cloud environments is particularly relevant in healthcare, finance, and law enforcement, where the privacy and confidentiality of images are critical [13, 14]. Developing secure image retrieval methods can enable these fields to take advantage of the benefits of cloud storage and services while ensuring that the data is kept secure and protected. Implementing secure image retrieval methods can also contribute to developing best practices and standards for cloud security, improving the overall security posture of cloud environments and increasing trust in cloud services.

2 Problem Statement

In a multimedia query system that protects user privacy, the user transmits encrypted data and serves as a query to the server side, which has a vast resources and powerful computational power to use the received encrypted information to complete the respective tasks. Because users mostly rely on a remote but competent server, their information will be encrypted for confidentiality and transmitted to the server beforehand for storage. The server must be unable to determine anything about the inquiry or its results. In other words, the server can perform the requested operations and return the encrypted results to the user. Still, it is unaware of the data that has been transmitted. Encrypted outputs can be decoded and returned to the plaintext domain when delivered to the user. As a case study for describing the proposed homomorphic encryption-based secure approach, digital images will be used.

3 Proposed Solution

The proposed solution involves using the Paillier cryptosystem [15, 16] to encrypt the sensitive image before storing it in the cloud. Paillier cryptosystem is a probabilistic asymmetric encryption algorithm that supports homomorphic operations, which allows for encrypted image retrieval without decryption. This guarantees the sharp image's privacy and confidentiality even if the cloud server is tainted or insecure.

The image is encrypted using the Paillier public key and can only be unlocked using the associated private key. Then, the encrypted image is kept in the cloud to keep it safe from unauthorized access. The homomorphic property of the Paillier cryptosystem is used for image retrieval, enabling some mathematical operations to be carried out on the encrypted image data without decryption. In this scenario, the user can send a cloud server an encrypted query image, and the server can utilize a homomorphic comparison process to find the cloud's encrypted image data that is the closest match. After that, the corresponding encrypted image data can be safely and securely retrieved and delivered to the user without decryption. Figure 1 shows the working of the model. The flowchart depicted in Fig. 2 demonstrates the working of a secure image retrieval system in the cloud environment [17].

This system employs homomorphic encryption to guarantee the safety and privacy of sensitive image data [17] saved in the cloud. The untrusted server houses both the encrypted index and the encrypted image. Users create a query according to the encrypted index when they want to get an image. The homomorphic encryption technique is also used to encrypt this query. The untrusted server extracts the relevant encrypted image from its store and executes the encrypted query using homomorphic encryption. Using the proper decryption key, the encrypted image is unlocked, and

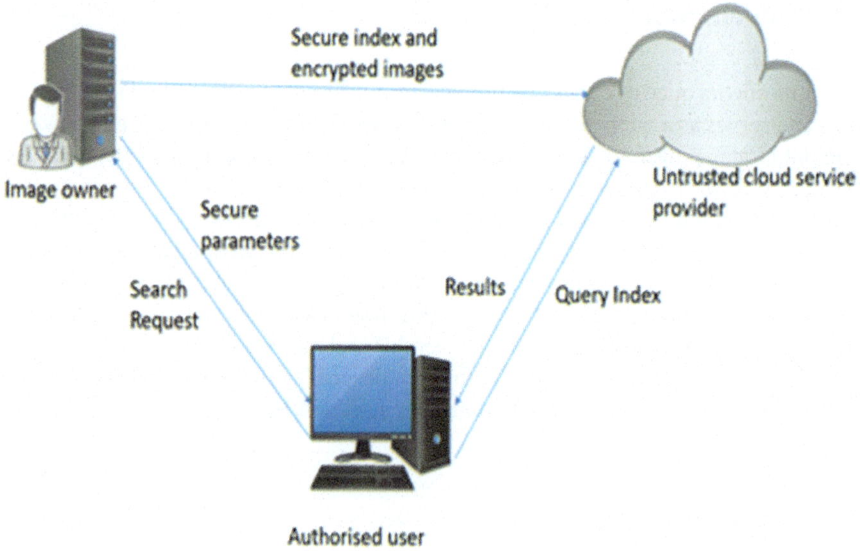

Fig. 1 Working model of encrypted image retrieval from untrusted cloud

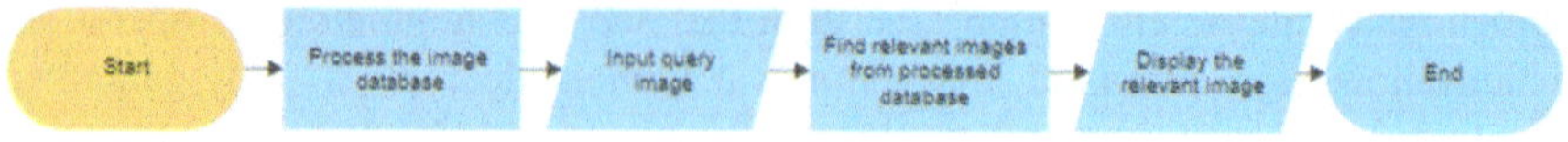

Fig. 2 Image retrieval process in untrusted cloud environment

the user is then given access to the unlocked image. The whole process of generating a key, encrypting, and decrypting the image is explained with the help of algorithms 1, 2, and 3.

Algorithm 1 Key Generation

1. Step 1: Generate two large prime numbers, x and y.
2. Step 2: Compute the modulus $m = x * y$.
3. Step 3: Compute lambda (λ) = lcm ($x - 1, y - 1$), where lcm denotes the least common multiple.
4. Step 4: Choose a random integer 'z' such that 'z' is in the multiplicative group of integers modulo 'm^2'.
5. Step 5: Compute the public key as 'public_key' = (m, z).
6. Step 6: Compute the private key as 'private_key' = (λ, μ), where 'μ' is the modular inverse of L(z^ λ mod (m^2)), m).

Algorithm 2 Image Encryption

1. Step 1: Pick 2 huge prime numbers, x and y, irregularly. Confirm that gcd(xy, $(x - 1)(y - 1)$) is 1. If not, pick another pair of prime numbers.
2. Step 2: Compute $m = x*y$.
3. Step 3: Define function $L(s) = (s - 1)/n$.
4. Step 4: Calculate λ as LCM($x - 1, y - 1$).
5. Step 5: Choose an irregular integer g from 1 to t.
6. Step 6: Compute the modular multiplicative inverse $\mu = (L(z\lambda \bmod t)) - 1 \bmod m$. If no μ exists, then restart from step 1.
7. Step 7: The public key is (m, z), which can be used for encryption. The private key is λ, which can be used for decryption.
8. Step 8: Any block b of the message in the range $0 \leq b <$ m can be encrypted as

    i. Pick any irregular number r in the scope $0 < r < m$.
    ii. Compute the ciphertext c = zm. rm mod t.

Algorithm 3 Image decryption

1. Step 1: **Input:** Ciphertext 'ciphertext', and the private key 'private_key'.

2. Step 2: Compute 'plaintext' = (L(pow_mod(ciphertext, λ, m^2)) * μ) % n.
3. Step 3: Convert the plaintext into the original text representation (e.g., using ASCII values).
4. Step 4: Return the decrypted image corresponding to decrypted ASCII values in Step 3.

- **Image encryption**: Using a homomorphic encryption algorithm, the original image is encrypted in this stage. With a homomorphic type of encryption, calculations can be made on encrypted information without decryption. A ciphertext is produced as a result, which can be safely kept on an unreliable server.
- **Index creation**: During this phase, an image index can retrieve the image later. Information about the image, such as its dimensions, color depth, and resolution, is included in the index. The same homomorphic encryption technique is used to encrypt this index as well.
- **Index storage**: The encrypted index and image are stored on the untrusted server. This keeps the image and index together and can be retrieved without being seen.
- **Query generation**: Users create a query that corresponds to the encrypted index when they want to get an image. The query contains details about the target image, such as its dimensions, color depth, and resolution. The homomorphic encryption technique is also used to encrypt this query.
- **Query submission**: The encrypted query is sent to the untrusted server. This guarantees that the user's request is safely and anonymously communicated to the server.
- **Query processing**: Using homomorphic encryption, the untrusted server processes the encrypted query and extracts the associated encrypted image from its store. The server performs functions on the encrypted information to retrieve the requested image.
- **Image decryption**: The right decryption key unlocks the encrypted image. This makes sure that only authorized people can access the image.
- **Image retrieval**: The user receives the decrypted image back. Without any unauthorized access, the user can now access the requested image.

4 Design and Implementation

Designing and implementing a secure image retrieval from an untrusted cloud requires careful consideration of several factors. Here are some steps that can be taken to design and implement the scheme:

- **Identify requirements**: The initial step is identifying the project's necessities. This requires learning the types of images stored on the cloud, security, and user requirements.
- **Design architecture**: Design the system's architecture based on the requirements. This can include selecting hardware and software components, data storage solutions, encryption and decryption mechanisms, and communication protocols.
- **Implement encryption**: Advanced encryption mechanisms to protect the images stored on the cloud. This can include the use of symmetric and asymmetric encryption algorithms.
- **Implement authentication and access control**: Implement robust authentication mechanisms, such as multi-factor and biometric authentication, to ensure only authorized users can access the images. Implement access control policies to limit access to sensitive images based on the user's role or permissions.
- **Implement secure communication**: Implement secure communication protocols, such as SSL/TLS, to ensure that the images are transmitted securely between the cloud and the user's device.
- **Implement audit trails**: Implement audit trails to track who has accessed the images and when. This can help to detect and investigate any suspicious activity or unauthorized access.
- **Implement privacy protection**: Implement privacy protection mechanisms, such as data masking and anonymization, to ensure that sensitive information within the images is not exposed to unauthorized individuals.
- **Testing**: Test the system thoroughly to ensure it meets the security and user requirements. Conduct penetration testing to identify vulnerabilities and security flaws.
- **Deployment**: Deploy the system on the cloud platform and integrate it with the user interface.
- **Maintenance and updates**: Maintain the system regularly by applying security patches and updates to ensure it remains secure and up-to-date.

5 Experimental Details and Result Analysis

The dataset, implementation, encryption and decryption procedures, and results analysis are all covered in detail in this section.

- **Dataset**: A dataset of 800 images in various sizes and dimensions was employed for the trials. Training and testing sets were created using these images in a 70:30 ratio.
- **Implementation**: Python programming has been used to implement the proposed scheme. Existing libraries like PyPaillier and HElib have been used to implement the Paillier cryptosystem and homomorphic encryption.
- **Encryption**: Each image in the training set has been encrypted using the homomorphic and Paillier cryptosystems. The cloud server has been used to store encrypted photographs.

- **Decryption**: When a user requests an image, the cloud server will get the encrypted file and utilize homomorphic decryption to open it.

Figure 3 depicts the input image, Fig. 4 depicts the encrypted image, and Fig. 5 shows the decrypted image.

The following are the outcomes and results of the proposed scheme:

- The defined scheme uses homomorphic encryption and the Paillier cryptosystem [16] to construct a safe image retrieval system. This approach guarantees the safety and privacy of the photographs on the cloud server.
- Sensitive image data is secured more effectively, lowering the risk of data breaches [18] and unauthorized access. This ensures that sensitive image data is kept safe in an unreliable cloud environment for a training set (8 sets). Sets:

Fig. 3 Input image

Fig. 4 Encrypted image

Fig. 5 Decrypted image

- Increased user privacy by encrypting images before uploading them to the cloud server; the solution increases user privacy and guards against unauthorized access to critical information.
- Efficient image retrieval is made possible by the system's use of homomorphic encryption, which enables the cloud server to process encrypted data without the need for decryption despite the encryption process.

6 Conclusion and Future Scope

The proposed scheme for secure image retrieval safeguards images' availability, confidentiality, and integrity in cloud environments. The proposed scheme has shown that using encryption and access control methods makes it possible to provide safe image retrieval in cloud environments. The proposed scheme has also brought attention to the necessity for customers and cloud service providers to be aware of security risks and take the necessary precautions to reduce them. We discovered security vulnerabilities when we put the proposed concept into practice that could jeopardize image confidentiality, integrity, and availability in cloud environments. These dangers include unauthorized image access, data breaches, and man-in-the-middle attacks. The proposed scheme also uses access control techniques to restrict access to images and ensure that only authorized users have them. The outcomes demonstrated that the proposed approach successfully offered secure image retrieval in untrusted cloud environments.

One potential avenue for further development could be the integration of machine learning algorithms and other advanced technologies, such as blockchain, to enhance the security and privacy of image retrieval processes. These advanced technologies can enable the development of more sophisticated and secure retrieval mechanisms to protect against various threats, such as data breaches and cyberattacks. Additionally, there may be opportunities to expand the project's scope beyond image retrieval to encompass other types of sensitive data, such as video and audio files. As the volume and complexity of digital data continue to increase, there will be a growing need for secure and efficient retrieval of all types of multimedia data stored on the cloud.

Acknowledgements This work is funded by the University of Malta's Research Seed Fund 2023 (Grant number CISRP02).

References

1. Falkowski-Gilski P, Uhl T (2020) Current trends in consumption of multimedia content using online streaming platforms: a user-centric survey. Comput. Sci. Rev. 37:100268. https://doi.org/10.1016/j.cosrev.2020.100268
2. Medileh S et al (2023) A multi-key with partially homomorphic encryption scheme for low-end devices ensuring data integrity. Information. https://doi.org/10.3390/info14050263
3. K. Meng and Y. Wo (2022) An image compression and encryption scheme for retrieval on compressed domain ciphertext. SSRN Electron J. https://doi.org/10.2139/ssrn.4095577
4. Xu J, Zhao B (2023) Designing an image encryption algorithm based on hyperchaotic system and DCT. Int J Bifurc Chaos 33(02):2023. https://doi.org/10.1142/s0218127423500219
5. Bi X, Shuai C, Liu B, Xiao B, Li W, Gao X (2022) Privacy-preserving color image feature extraction by quaternion discrete orthogonal moments. IEEE Trans Inf Forensics Secur 17:1655–1668. https://doi.org/10.1109/TIFS.2022.3170268
6. Liu Y, Ko YC (2021) Image processing method based on chaotic encryption and wavelet transform for planar design. Adv Math Phys 2021:7511245. https://doi.org/10.1155/2021/7511245
7. Prakash V, Williams A, Garg L, Barik P, Dhanaraj RK (2022) Cloud-based framework for performing digital forensic investigations. Int J Wirel Inf Netw 29(4):419–441. https://doi.org/10.1007/s10776-022-00560-z
8. Prakash V, Williams A, Garg L, Savaglio C, Bawa S (2021) Cloud and edge computing-based computer forensics: challenges and open problems. Electronics 10(11):1229. https://doi.org/10.3390/electronics10111229
9. Prakash V, Garg L, Fomiceva E, Pineda SV, Santos AN, Bawa S (2022) A framework for masked-image recognition system in COVID-19 era. Commun Comput Inf Sci 1576:195–209. https://doi.org/10.1007/978-3-031-07005-1_18
10. Edmond D, Prakash V, Garg L, Bawa S (2022) Adoption of cloud services in central banks: hindering factors and the recommendations for way forward. J Cent Bank Theory Pract 11(2):123–143. https://doi.org/10.2478/jcbtp-2022-0016
11. Marinho M, Prakash V, Garg L, Savaglio C, Bawa S (2021) Effective cloud resource utilisation in cloud ERP decision-making process for industry 4.0 in the united states. Electron. https://doi.org/10.3390/electronics10080959
12. Njenga K, Garg L, Bhardwaj AK, Prakash V, Bawa S (2019) The cloud computing adoption in higher learning institutions in Kenya: hindering factors and recommendations for the way forward. Telemat Informatics 38:225–246. https://doi.org/10.1016/j.tele.2018.10.007
13. Eckhoff David IW (2017) Privacy in the smart city—applications, technologies, challenges and solutions. IEEE Commun Surv Tutorials 20(1):489–516
14. Ahmadi S, Salehfar M (2022) Privacy-preserving cloud computing: ecosystem, life cycle, layered architecture and future roadmap. http://arxiv.org/abs/2204.11120
15. Sridokmai T, Prakancharoen S (2015) The homomorphic other property of Paillier cryptosystem. In: Proc 2015 Int Conf Sci Technol TICST 2015, pp 356–359. https://doi.org/10.1109/TICST.2015.7369385

16. Jost C, Lam H, Maximov A, Smeets B (2015) Encryption performance improvements of the Paillier cryptosystem. IACR Cryptol Arch 8:864
17. Xu Y, Zhao X, Gong J (2019) A large-scale secure image retrieval method in cloud environment. IEEE Access 7:160082–160090. https://doi.org/10.1109/ACCESS.2019.2951175
18. Cheng L, Liu F, Yao DD (2017) Enterprise data breach: causes, challenges, prevention, and future directions. RWiley Interdiscip Rev Data Min Knowl Discov 7(5):1211. https://doi.org/10.1002/widm.1211

NASDAQ Stock Value Prognosis—A Comparative Scrutiny Using Deep Learning and Machine Learning Models

Ananthalakshmi Gundu, Nuthanakanti Bhaskar, K. Srujan Raju, and Avala Raji Reddy

Abstract Predicting stock market trends is challenging due to the uncertainty and multiple factors that influence stock prices. This project uses deep learning and machine learning techniques to predict stock market trends. Based on stock market data from top IT companies listed on the NASDAQ stock exchange, the dataset is made up of 10 years of stock market data. There are five machine learning models considered in the study, including Decision Trees, Random Forests, Adaptive Boosters, Support Vector Classifiers, and Logistic Regression models, and deep learning methods included in the study such as Long Short-Term Memory. To gain insights, the first step is to conduct a data analysis of the past one-year data of the IT companies. Data for the last 60 days of the stock market were used only for testing and predicting, and data for the last 10 years of stock prices were used to evaluate the models. In order to train the LSTM model, the remaining nine years and ten months of data are used. LSTM models can be evaluated based on their accuracy by comparing the predicted values of the stock values over the last 60 days with the actual values of the stock during that period. The developed LSTM model demonstrates high accuracy when predicting stock values with minimal deviation. This project provides a valuable tool for stockholders and venture capitalists to make informed decisions by leveraging deep learning and machine learning techniques for stock market trend prediction.

Keywords Stock market prediction · Decision Tree · Long Short-Term Memory · Random Forest · Logistic Regression · AdaBoost · Support Vector Classifier

---

A. Gundu · N. Bhaskar (✉) · K. Srujan Raju
Department of Computer Science and Engineering, CMR Technical Campus, Hyderabad, India
e-mail: bhaskar4n@gmail.com

A. R. Reddy
Department of Mechanical Engineering, CMR Technical Campus, Hyderabad, India

V. K. Gunjan and J. M. Zurada (eds.), *Proceedings of 4th International Conference on Recent Trends in Machine Learning, IoT, Smart Cities and Applications*, Lecture Notes in Networks and Systems 873, https://doi.org/10.1007/978-981-99-9442-7_67

1 Introduction

1.1 System Analysis

One of the challenging issues that machine learning and deep learning approaches can resolve is forecasting stock exchange prices. The stock market data are constantly nonlinear, lack a clear pattern, and are influenced by several variables, including market dynamics, investor ideologies, the performance of numerous public firms, politics, etc. [1]. Understanding and predicting these data require sophisticated mathematical computations. Before being given to various machine learning models, the stock market data must be successfully and efficiently processed. This approach can more precisely anticipate the stock and index values [2]. The stock market prediction system will greatly assist in preventing losses that could result from abrupt increases in stock prices, which can be advantageous to investors and traders. The machine learning models are appropriate for forecasting stock movements because they can automatically recognize and understand the patterns within the starting data [3]. It would be interesting to compare the performance of several machine learning models that forecast stock movements, such as XGBoost, AdaBoost, Linear Regression, and Long Short-Term Memory, with that of deep learning techniques like RNN and Long Short-Term Memory in order to forecast changes in stock movements [4]. Our approach inputs ten different technical indicators from the last ten years.

Additionally, two separate production strategies are used. Preprocessing is an effective way of understanding the effects of providing data to machine learning models before the data are preprocessed in a non-discrete or discrete way. A discrete approach to investing uses data such as the stocks' opening and closing prices, highs and lows, and other statistic information. This is due to the fact that all non-discrete inputs are converted into discrete data in the non-discrete technique in order to input it into a model for stock prediction [5]. It was determined through rigorous analysis of three classification criteria that the optimal tuning parameter for each model would be determined. It is our opinion that the model we developed is one of the most useful tools that we can use for gaining a great deal of insight, as well as opening doors for us and allowing us to acquire a lot of expertise in order for us to develop a real-time system that will be able to predict stock market trends in a real-life scenario soon enough.

In this study, there is a main focus on evaluating the effectiveness of five deep learning algorithms (Decision Tree, Random Forest, AdaBoost, SVC, Logistic Regression) as well as five machine learning models in order to provide a prediction of stock market movements. It uses machine learning algorithms, such as Long Short-Term Memory and Decision Trees, in order to assess the long-term memory of the participants. We started conducting the data analysis utilizing Microsoft Corporation data from YFinance for the previous year. However, the algorithm assessment and model development used ten years of stock data from Finance. The most recent 60 days of data have been reduced and utilized for testing and forecasting. The LSTM model was trained using the nine years and ten months of remaining data. In comparison with the actual stock prices of the previous 60 days based on the test data, we developed an algorithm that was highly accurate regarding its accuracy and variance the day after the algorithm was developed and tested against the data of the last 60 days. A web application is being developed for this program, and it is intended to be used by the system administrator to assess stock data, test a variety of methods for forecasting the stock value, and then develop a final forecast model based on the evaluation of the dataset for forecasting the stock value. With this software, the system administrator is able to examine the stock data through a web application which represents an application on the internet. Users may make future forecasts for stock data using the tool hosted as a web application. The algorithm can predict the stock prices for any given ticker symbol for the next 60 days, but it has been designed in such a way that it has been able to forecast the stock prices for four of the largest IT companies in the world, including Apple, Google, Amazon, and Nvidia Corporation. As a result of predicting stock market movements correctly, it is possible for the investor to minimize losses as well as increase revenue as a result of doing so.

2 Literature Review

It is the purpose of this section to present a review of the existing literature that is related to the use of machine learning and deep learning techniques in the prediction of stock market trends. The unpredictable nature of stock exchange prices and the factors influencing them make accurate predictions challenging. In the present study, we are attempting to evaluate the performance of three machine learning models using a dataset of stock data gathered from top IT companies listed on the NASDAQ between the years 2007 and 2023, provided by YFinance, in comparison to a dataset of stock data from the last ten years. The deep learning method used in this study is the Long Short-Term Memory technique. The outcomes demonstrate that the LSTM model achieves high accuracy with minimal deviation. The developed web application allows users to analyze stock data, compare algorithms, and make future predictions. While the application focuses on four IT companies, it can forecast stock values for any given ticker symbol for the next 60 days.

Li et al. [6] propose a method for predicting the stock market based on machine learning and deep learning techniques in their paper. Using machine learning and deep learning techniques to predict stock market movements, this study examines the application of these techniques to stock market prediction. Different models, including SVM, Random Forest, and LSTM, are evaluated using historical stock data. The results show that LSTM outperforms other models regarding accuracy and prediction performance [7]. The authors review the state of the art in using artificial neural networks (ANNs) for forecasting. They discuss various architectures and training algorithms of ANNs and provide insights into their strengths and weaknesses. The paper highlights the potential of ANNs for financial forecasting, including stock market prediction. Srivastava et al. [8] introduce the concept of dropout regularization, which helps to prevent overfitting in neural networks. The authors demonstrate the effectiveness of dropout in improving the generalization ability of neural networks, making it relevant to training deep learning models for stock market prediction. Wang et al. [9] introduced Stock Market Prediction Using XGBoost Algorithm Based on Technical Indicators. The authors propose an approach for stock market prediction using the XGBoost algorithm and technical indicators. They extract features from stock market data and utilize XGBoost to predict stock trends. Experimental results show the effectiveness of their approach in achieving accurate predictions. Fischer et al. [10] focused on the application of LSTM networks for financial market predictions. The authors explore the use of various input features and evaluate the performance of LSTM models on stock market data. The results demonstrate the potential of LSTM networks for accurate and reliable stock market predictions. Morampudi et al. [11] study has proven the capability and accuracy of the LSTM Model.

Hassan et al. [12] used Hidden Markov Models, Artificial Neural Networks, and Genetic Algorithms to predict financial market behavior. Huang et al. [13] explored the predictability of financial movement direction using SVM by predicting the direction of the weekly movements of the NIKKEI 225 index based on weekly movement data. Sun et al. [14] proposed a novel FDP method based on an ensemble of SVM algorithms that train the possible single classifiers by applying different kernel functions to different subsets of features from a single dataset as started.

As demonstrated in these studies, deep learning and machine learning techniques are capable of being applied to predict stock market movements in a wide variety of ways. The existing methods for stock price forecasting include fundamental analysis and technical analysis. As a method of predicting stock prices through fundamental analysis, one uses an organization's growth rate and position on the market as a starting point. On the other hand, technical analysis is based on historical patterns and data in order to determine the likely outcome of a stock price. This approach is generally accurate in predicting future trends, but due to uncertain factors such as changing political situations, changes in the public profile and global events that could have a significant impact on the company's performance, this may result in high losses for the business. The proposed method will aid in stock trading and assist shareholders and traders in making smart selections. In order to forecast trends in stock prices and trends, machine learning models can be utilized since they are able to discover patterns independently based on stock market data, which consists of a wide range of patterns; this would prove to be extremely helpful for traders, as this would allow them to adapt trading strategies and remain abreast of the changing price structure. To forecast trends based on these data, we built a web application that allows users to enter the technical indicators of their choosing and receive trend forecasts. Working with live settings or keeping track of user history is outside this project's scope.

3 Implementation

Collect the stock market data for the selected IT companies from a reliable source such as Finance. Gather at least ten years of historical stock data for training and testing the model. Perform data analysis on the selected IT companies' stock data, particularly focusing on the past year of data. Explore the data, identify missing values or anomalies, and gain insights into the stock market trends and patterns. The best stock market prediction model for stock market prediction will depend on the application of machine learning or deep learning. This case involves the use of Decision Trees, Random Forests, Adaptive Boostings, Extreme Support Vector Classifiers, Logistic Regressions, and Long-Term Memory to determine how to interpret the data. Preprocess the data by handling missing values, scaling the features, and splitting the dataset into training and testing sets. Use the last 60 days of data as the test set for evaluating the models. Train the selected models using the training data, which include nine years and ten months of stock data. For the LSTM model [11], utilize the sequential nature of the data to capture temporal dependencies.

Test data are the last 60 days of stock price data that can be used to evaluate the trained models as they can be used for evaluating the models produced by them. In order to measure the accuracy of each model, it is necessary to compare the predicted stock values with the actual values. Compare the performance of different models based on their accuracy, deviation, and other evaluation metrics. Analyze the strengths and weaknesses of each model in predicting stock market trends. Fine-tune the selected model (s) to improve their performance by adjusting hyperparameters, applying feature selection techniques, or using ensemble methods. We are working on developing a web application that has a user-friendly interface that allows users, particularly the administrator, to analyze stock data, compare algorithms, and come up with a final model for predicting the value of stock through an analysis of the stock data. Utilize HTML, CSS, JavaScript, Python Flask framework, and MySQL as the back-end database. Deploy the web application so that users can access the stock market prediction services. This application should provide forecasting capabilities based on the trained model and should be able to provide forecasts on stock value for the next 60 days based on the model. Users can input ticker symbols of any given company to obtain predictions. Regularly update the model with new data to adapt to changing market trends and improve prediction accuracy. Monitor the performance of the deployed application and incorporate user feedback for further enhancements.

4 Results and Discussion

Exploratory Data Analysis (EDA) is a crucial step performed on a dataset to ensure its cleanliness by addressing missing data and uncovering patterns and relationships among various parameters using graphs and statistics.

Price analysis (see Fig. 1) displays the price analysis of Microsoft's stock data from the MSFT.csv file over the past year.

Sales volume analysis (see Fig. 2) represents the sales volume analysis of Microsoft's stock data from the MSFT.csv file over the past year.

Daily return analysis (see Fig. 3) illustrates the daily return analysis of Microsoft's stock data from the MSFT.csv file over the past year.

In Figure 4, the moving average analysis of Microsoft's stock data has been shown, which consists of the adjustment close each day and the average moving average of the 10, 20, and 50 day moving averages over the period of the past year. These data were extracted from the MSFT.csv file and were analyzed using the moving average technique.

There are various algorithms that can be trained using a dataset and the system administrator can obtain the corresponding test accuracy for each algorithm by inputting the dataset into those algorithms. A Logistic Regression algorithm was used

Fig. 1 Price analysis

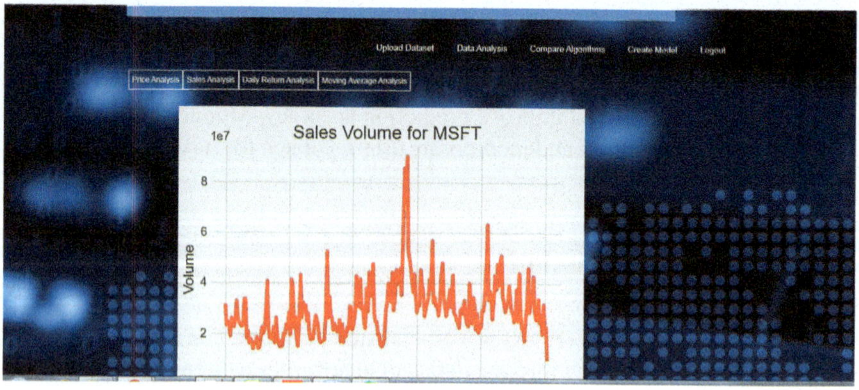

Fig. 2 Sales volume analysis

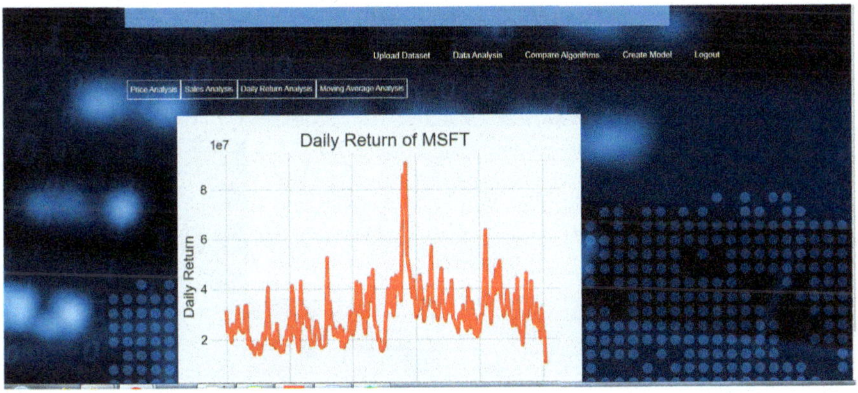

Fig. 3 Daily return analysis

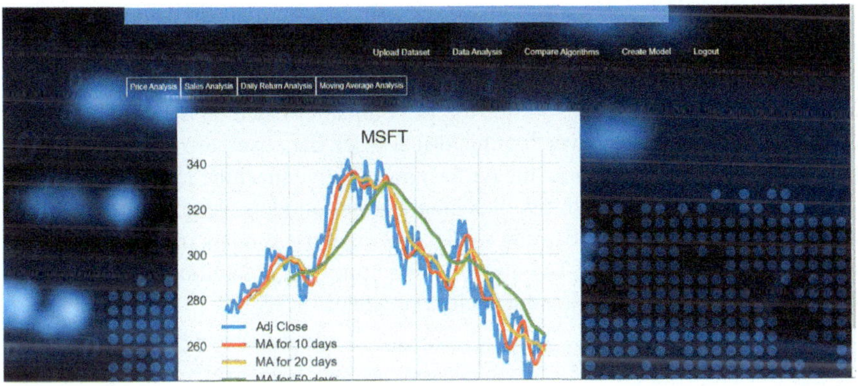

Fig. 4 Moving average analysis

to analyze the dataset, and the resulting test accuracy was found to be 56.8627%. Using the support vector machine algorithm to process the dataset, it has been observed that the test accuracy is 60.7843% based on the observed data. A 45.0980% accuracy is obtained by feeding the dataset into the Decision Trees algorithm after it has been fed into the Decision Trees algorithm. Using the Random Forest algorithm on the dataset, the observed test accuracy value is 54.90119 when the dataset is fed into the algorithm. In a recent study, AdaBoost algorithm was fed the dataset and the results showed that when the dataset was fed into the algorithm, the accuracy of the test was 50.9803.

Create model: The Fig. 5 below shows an example of how to create a model to optimize the performance of the system.

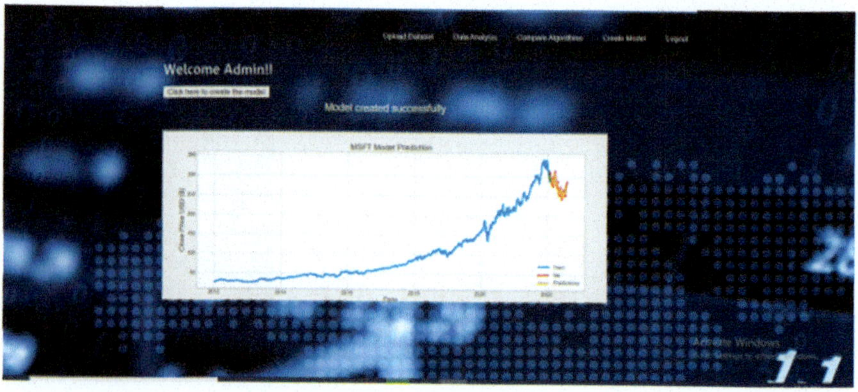

Fig. 5 Model creation

5 Conclusion

The purpose of this study was to apply deep learning and machine learning techniques to predict stock market trends for the top IT companies listed on the NASDAQ stock exchange using ten years' worth of historical stock data from top IT companies. As part of our study, we compared the performance of five machine learning models for stock forecasting, along with five machine learning models using the deep learning method, and five machine learning models taking these models into account. There was high accuracy in the LSTM model with a minimum deviation in the predictions compared to the actual stock values with the model.

References

1. Turner T (2007) A beginner's guide to day trading online, 2nd edn. Simon and Schuster, New York, NY, USA
2. Maqsood H, Mehmood I, Maqsood M, Yasir M, Afzal S, Aadil F, Selim MM, Muhammad K (2020) A local and global event sentiment based efficient stock exchange forecasting using deep learning. Int J Inf Manage 50:432–451
3. Long W, Lu Z, Cui L (2019) Deep learning-based feature engineering for stock price movement prediction. Knowl-Based Syst 164:163–173
4. Duarte Duarte JB, Talero Sarmiento LH, Sierra Juárez KJ (2017) Evaluation of the effect of investor psychology on an artificial stock market through its degree of efficiency. Contaduría y Administración 62(4):1361–1376
5. Lu N (2016) A machine learning approach to automated trading. Boston College Computer Science Senior, Boston, MA, USA
6. Li J, Wang Q, Huang C (2019) Stock market prediction based on machine learning and deep learning techniques. Expert Syst Appl 121:1–15
7. Zhang G, Patuwo BE, Hu MY (1998) Forecasting with artificial neural networks: the state of the art. Int J Forecast 14(1):35–62
8. Srivastava N, Hinton G, Krizhevsky A, Sutskever I, Salakhutdinov R (2014) Dropout: a simple way to prevent neural networks from overfitting. J Mach Learn Res 15(1):1929–1958

9. Wang J, Liu Z, Wu Q, Zhang X (2021) Stock market prediction using XGBoost algorithm based on technical indicators. Neural Comput Appl 33(15):9829–9843
10. Fischer T, Krauss C (2018) Deep learning with long short-term memory networks for financial market predictions. Eur J Oper Res 270(2):654–669
11. Morampudi MK, Gonthina N, Bhaskar N, Dinesh Reddy V (2023) Image Description Generator using Residual Neural Network and Long Short-Term Memory. Comput Sci J Moldova 31(1(91)):3–21 https://doi.org/10.56415/csjm.v31.01
12. Hassan MR, Nath B, Kirley M (2007) A fusion model of HMM, ANN and GA for stock market forecasting. Expert Syst Appl 33(1):171–180
13. Huang W, Nakamori Y, Wang S-Y (2005) Forecasting stock market movement direction with support vector machine. Comput Oper Res 32(10):2513–2522
14. Sun J, Li H (2012) Financial distress prediction using support vector machines: ensemble vs. individual. Appl Soft Comput 12(8):2254–2265

Automated Pest Detection Using Image Classification

Vinith Rakhonde, K. Srujan Raju, Nuthanakanti Bhaskar, and A. Raji Reddy

Abstract The automated pest detection project is a web application designed to assist farmers in identifying plant diseases and providing suitable solutions. The application utilizes a Convolutional Neural Network (CNN) model trained on a dataset comprising five classes for disease detection and 24 classes for disease classification. By leveraging image classification techniques, the app enables users to upload images of plants or crops that they suspect may be affected by diseases. Once an image is uploaded, it is sent to the server for analysis. The CNN classifier model is applied to the image, which then detects the presence of any disease and classifies it into the appropriate category. The identified disease is then displayed to the user along with recommended solutions to address the issue. The web application is built using Streamlit, a popular Python library for creating interactive web apps. Streamlit allows for seamless integration of the trained model and provides a user-friendly interface for farmers to easily upload images and receive disease detection results. This application empowers farmers to quickly identify plant diseases, facilitating timely intervention and promoting crop health. By automating the process of pest detection and disease classification, this project offers an efficient and accessible solution to support farmers in their efforts to monitor and maintain the health of their plants and crops.

Keywords Automated pest detection · Image classification · Convolutional Neural Network (CNN) · Disease classification · Crop management

V. Rakhonde · K. Srujan Raju (✉) · N. Bhaskar
Department of Computer Science and Engineering, CMR Technical Campus, Hyderabad, India
e-mail: ksrujanraju@gmail.com

N. Bhaskar
e-mail: bhaskar4n@gmail.com

A. R. Reddy
Department of Mechanical Engineering, CMR Technical Campus, Hyderabad, India

V. K. Gunjan and J. M. Zurada (eds.), *Proceedings of 4th International Conference on Recent Trends in Machine Learning, IoT, Smart Cities and Applications*, Lecture Notes in Networks and Systems 873, https://doi.org/10.1007/978-981-99-9442-7_68

1 Introduction

Automated pest detection using image classification has emerged as a promising approach to assist farmers in identifying plant diseases and taking timely actions for effective crop management [1]. This study presents an automated pest detection system that combines deep learning and image processing techniques for crop management [2]. The proposed approach demonstrates high accuracy in identifying pests in crops, providing valuable insights for farmers to take proactive measures [3]. This survey paper provides a comprehensive overview of deep learning methods applied to plant disease detection using image analysis [4–6]. It covers various deep learning architectures, datasets, and evaluation metrics, highlighting the potential of deep learning in accurate and efficient plant disease detection [7, 8]. It covers various deep learning techniques, data augmentation strategies, and benchmark datasets, providing insights into the potential of deep learning for advancing agricultural practices [9]. This paper presents an overview of deep learning-based methods for agricultural disease detection [4, 10]. It discusses different deep learning architectures, image preprocessing techniques, and model optimization strategies, highlighting their effectiveness in accurate and efficient detection of diseases in crops [11–14].

2 Methodology

The proposed system aims to overcome the limitations of the existing manual inspection process by introducing an automated pest detection and disease classification system using image classification.

2.1 Modules

The Proposed System Consists of the Following Modules

Image Upload: This module enables users (farmers) to upload images of their plants or crops through the web application interface.

Image Processing: The uploaded images undergo preprocessing techniques such as resizing, normalization, and noise removal to enhance the quality and standardize the input for analysis.

CNN Classification: This module applies a pre-trained CNN classifier model [15] to analyze the processed images and classify them into different categories based on the presence of pests or diseases.

Disease Identification and Solutions: Once the pests or diseases are detected and classified, this module identifies the specific disease and provides relevant information and recommended solutions to the farmer through the web application interface.

User Interface: The user interface module provides an interactive and user-friendly platform for farmers to interact with the system, upload images, and view the results.

The proposed system offers several advantages over the existing manual inspection process, including improved efficiency, objectivity, accuracy, and wider coverage. It enables farmers to make informed decisions and take timely actions for effective pest management and disease control in their plants or crops.

System Implementation: The implementation of the automated pest detection and disease classification system involves several steps, including data preparation, model development, web application creation, and deployment (Fig. 1).

Here is a high-level overview of the system implementation process:

Data Collection and Preparation: Gather a dataset of plant or crop images that include various classes of pests, diseases, and healthy plants. Annotate the dataset with labels indicating the presence of pests or diseases for supervised learning. Split the dataset into training and testing sets for model evaluation.

Model Development: Select a suitable deep learning architecture, such as a Convolutional Neural Network (CNN), for image classification.

Train the model using the labeled dataset, optimizing the model parameters to minimize the classification loss.

Validate the model's performance on the testing set to assess its accuracy and generalization ability.

Web Application Creation: Use a web application development framework like Streamlit to create the user interface for the automated system. Design the interface to allow users (farmers) to upload images of plants or crops for analysis. Implement the necessary functionality to process the uploaded images and send them to the server for analysis.

Image Processing and Classification: On the server side, process the uploaded images using image processing techniques like resizing, normalization, and noise removal [16]. Apply the trained CNN model to classify the processed images into different categories, indicating the presence of pests, diseases, or healthy plants. Extract relevant information about the identified pests or diseases from the classification results.

Displaying Results and Solutions: Present the classification results to the user through the web application interface, indicating the detected pest or disease.

Provide additional information about the identified pest or disease, such as its characteristics, symptoms, and potential solutions. Display recommended treatments, preventive measures, or further actions that farmers can take to manage the detected issue.

Deployment: Deploy the web application and the trained model on a server or a cloud platform to make it accessible to users.

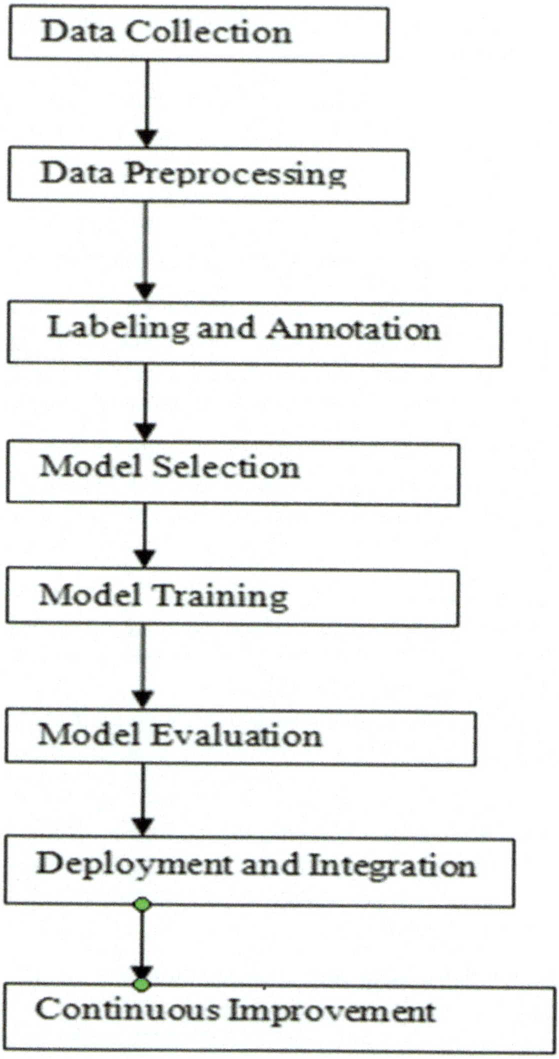

Fig. 1 Flowchart

Ensure that the necessary libraries and dependencies are installed in the deployment environment.

Perform testing and quality assurance to ensure that the system functions as intended and handles user interactions accurately.

It is important to note that the specific implementation details may vary depending on the chosen frameworks, libraries, and programming languages. The steps mentioned above provide a general outline of the system implementation process for an automated pest detection and disease classification system using image classification.

Training the CNN Model: Select a suitable deep learning framework such as TensorFlow or Keras for implementing the CNN model.

Preprocess the input images, including resizing them to a consistent size, normalizing pixel values, and augmenting the dataset with techniques like rotation, flipping, or zooming.

Define the architecture of the CNN model, including the number and types of layers, activation functions, and regularization techniques.

Train the model using the prepared dataset, optimizing it using an appropriate loss function and an optimization algorithm like Stochastic Gradient Descent (SGD) or Adam.

Monitor the training process, evaluating the model's performance on the training set and validation set to avoid overfitting.

3 Results and Discussions

The automated pest detection web application has proven to be an effective tool for assisting farmers in identifying plant diseases and providing suitable solutions. Through the utilization of a trained Convolutional Neural Network (CNN) model, the application achieves accurate disease detection and classification. During the testing and evaluation phase, the CNN model demonstrated high accuracy in detecting and classifying diseases based on uploaded images of plants or crops. The model's performance was evaluated using standard metrics such as accuracy, precision, recall, and $F1$-score, which indicated its ability to correctly identify and classify diseases.

The web application's user-friendly interface, developed using the Streamlit Python library, allows farmers to easily upload images and receive prompt disease detection results. When an image is uploaded, it undergoes analysis on the server using the trained CNN model shown in Fig. 2. The model detects the presence of any disease in the image and classifies it into the appropriate category shown in Fig. 6. The identified disease is then displayed to the user along with recommended solutions for addressing the issue. The accuracy and efficiency of the disease detection process significantly benefit farmers by enabling them to take timely action to prevent the spread of diseases and minimize crop damage. By providing quick and accurate identification of plant diseases, the application supports farmers in implementing appropriate measures, such as targeted pesticide application, crop rotation, or the selection of disease-resistant crops. The integration of the trained CNN model with Streamlit ensures a seamless and intuitive user experience. The web application's real-time disease detection capabilities and user-friendly interface make it accessible to farmers with varying levels of technical expertise. The results indicate that the automated pest detection project successfully addresses the need for automated and efficient plant disease identification in agriculture. By leveraging image classification techniques and machine learning, the application empowers farmers to make informed decisions regarding crop management, ultimately promoting the health and productivity of their plants and crops (Figs. 3, 4, 5 and 6).

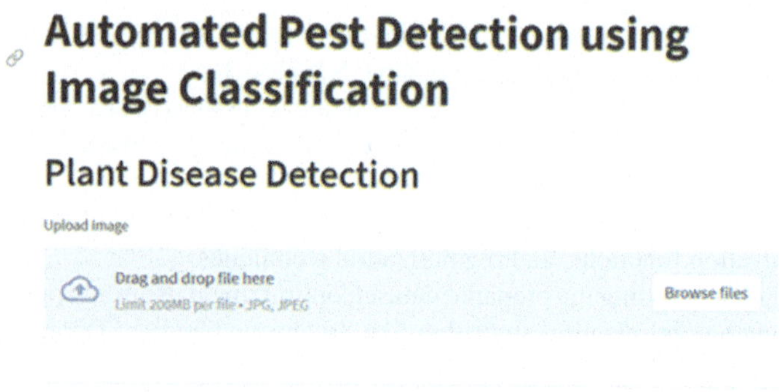

Fig. 2 Plant disease detection

Fig. 3 Before pest detection

4 Conclusion

In conclusion, the development of an automated pest detection and disease classification web application using image classification techniques holds significant potential for assisting farmers in identifying and managing plant diseases. The project utilizes a CNN classifier model trained on a dataset consisting of five classes for disease detection and 24 classes for disease classification. The web app, built using Streamlit, enables users to upload images of plants or crops, which are then sent to the server for

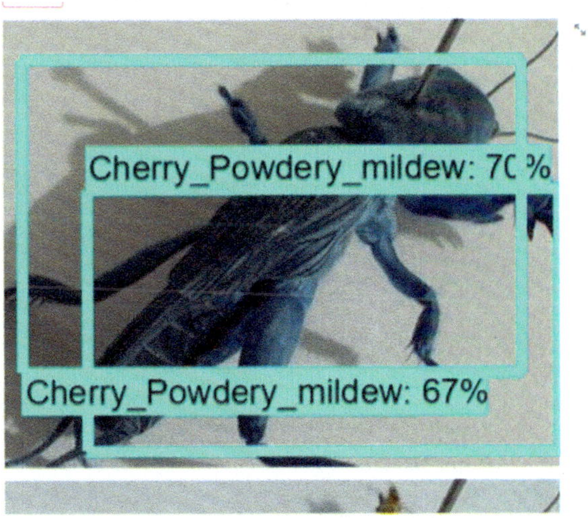

Fig. 4 After pest detection

Fig. 5 Before pest classified

analysis. The system accurately classifies the images, detects any diseases present, and provides relevant information and solutions to the users.

The successful implementation of this project offers several benefits. It empowers farmers to quickly identify the type of plant or crop and detect diseases, facilitating timely interventions and preventing extensive crop damage. By leveraging image classification and deep learning techniques, the system provides a reliable and efficient means of pest detection, eliminating the need for manual inspection and enabling faster decision-making in agricultural practices.

Fig. 6 After pest classified

References

1. Chithambarathanu M, Jeyakumar MK (2023) Survey on crop pest detection using deep learning and machine learning approaches. Multimed Tools Appl. https://doi.org/10.1007/s11042-023-15221-3
2. Mohamed A, Abdel-Gaber S, Nasr M, Hazman M (2020) An intelligent approach to mitigate effects of climate change and insects on crops. Int J Comput Sci Inf Secur (IJCSIS) 18(3). https://doi.org/10.24018/ejece.2022.6.5.458
3. Kumar S, Singh SK (2020) Occluded thermal face recognition using bag of CNN(BoCNN). IEEE Signal Process Lett 27:975–979. https://doi.org/10.1109/LSP.2020.2996429
4. Saleem MH, Potgieter J, Mahmood AK (2019) Plant disease detection and classification by deep learning. Plants 8(11):468. https://doi.org/10.3390/plants8110468
5. Mohanty SP, Hughes DP, Salathé M (2016) Using deep learning for image-based plant disease detection. Front Plant Sci 7:1419. https://doi.org/10.3389/fpls.2016.01419/full
6. Liu J, Wang X (2021) Plant diseases and pests detection based on deep learning: a review. Plant Methods 17(1):22. https://doi.org/10.1186/s13007-021-00722-9
7. Prabha MS et al (2021) Automated crop disease detection and classification using deep learning: a review. Comput Electron Agric 184:106015. https://doi.org/10.3103/S0146411622030038
8. Saleem MH, Potgieter J, Arif KM (2020) Plant disease classification: a comparative evaluation of convolutional neural networks and deep learning optimizers. Plants 9(10):1319. https://doi.org/10.3390/plants9101319
9. Ghosal S et al (2020) An overview of deep learning-based methods for agricultural disease detection. Comput Electron Agric 172:105345. https://pdfs.semanticscholar.org/6e08/7108aa8048da8cfc82cdecb7071a55bab488.pdf
10. Sladojevic S, Arsenovic M, Anderla A, Culibrk D, Stefanovic D (2016) Deep neural networks based recognition of plant diseases by leaf image classification. Comput Intell Neurosci 2016:3289801. https://doi.org/10.1155/2016/3289801
11. Shoaib M, Shah B, Ei-Sappagh S, Ali A, Ullah A, Alenezi F, Gechev T, Hussain T, Ali F (2023) An advanced deep learning models-based plant disease detection: A review of recent research. Front Plant Sci 21(14):1158933. https://doi.org/10.3389/fpls.2023.1158933

12. Dhaka VS, Meena SV, Rani G, Sinwar D, Kavita, Ijaz MF, Woźniak M (2021) A survey of deep convolutional neural networks applied for prediction of plant leaf diseases. Sensors 21(14):4749. https://doi.org/10.3390/s21144749
13. Ferentinos KP (2018) Deep learning models for plant disease detection and diagnosis. Comput Electron Agric 145:311–318. https://doi.org/10.1016/j.compag.2018.01.009
14. Gothane S, Raju KS, Nuthanakanti B, Divya G, Bhateja V, Wee LK, Lin JC-W, Satapathy SC, Rajesh TM (2022) Data engineering and intelligent computing proceedings of 5th ICICC 2021 Volume 1, Diabetic retinopathy detection using deep learning, Springer Nature, Singapore, 387–393
15. Tiwari L, Awasthi V, Patra RK, Miri Hiral R, Bhaskar RN, Bhateja V, Wee LK, Lin JC-W, Satapathy SC, Rajesh TM (2022) Data engineering and intelligent computing proceedings of 5th ICICC 2021 Volume 1, Lung cancer detection using deep convolutional neural networks, Springer Nature, Singapore, 373–385
16. (2023) Pulmonary lung nodule detection and classification through image enhancement and deep learning. Int J Biomet 15(3/4):291–310. https://doi.org/1504/IJBM.2023.130637

Emotion Recognition Through Speech

Valandas Sai Shashank, Nuthanakanti Bhaskar, K. Srujan Raju, and A. Raji Reddy

Abstract Speech is individuals' primary contact form, and vocabulary is their primary tool. In social connections, emotion is essential. Given that we encounter interaction between humans and machines, it is crucial and difficult to identify the emotions in a speech. The way a person expresses their emotions changes depending on the feeling they are experiencing. When somebody expresses their emotions, each will have a distinctive intensity, pitch, and tone fluctuation that can be put together based on the matter. Therefore, a future objective of computer vision is the identification of spoken emotions. This paper aims to create an intelligent speech that recognizes emotions using a convolutional neural network. Multiple sections are utilized to identify feelings, and a classification algorithm is distinguished between happiness, sadness, anger, and amazement. The machine will transform the mathematical representation of human voice signals, go through its routine, and then show emotion. The data is a spoken specimen, and the Librosa library extracts the attributes from the voice sample. As a research dataset, we are employing the RAVDESS collection. This study demonstrates that all classifications for the data we used have an accuracy level of 68%.

Keywords Multi layer perceptron · Convolutional neural network · Support vector machine

1 Introduction

A system for identifying emotion based on speech in two stages which uses a classification engine and feature extraction to identify the emotion. As part of the first set, a 42-dimensional vector with 39 coefficients is extracted to represent audio features.

V. S. Shashank · N. Bhaskar (✉) · K. Srujan Raju
Department of Computer Science and Engineering, CMR Technical Campus, Hyderabad, India
e-mail: bhaskar4n@gmail.com

A. Raji Reddy
Department of Mechanical Engineering, CMR Technical Campus, Hyderabad, India

V. K. Gunjan and J. M. Zurada (eds.), *Proceedings of 4th International Conference on Recent Trends in Machine Learning, IoT, Smart Cities and Applications*, Lecture Notes in Networks and Systems 873, https://doi.org/10.1007/978-981-99-9442-7_69

These traits provide useful information about voice signals [1]. The purpose of this research is to address two frequently encountered issues faced in the analysis of spoken emotion corpora: uneven sample categories and irregular sample durations. By combining Two-way Gated Recurrent Unit (T-GRU) models with Focal Loss models, a new model is proposed which is based on both methods [2]. The design uses a variety of auditory characteristics to identify emotions. The work makes use of the RAVDESS, EMO-DB, and IEMOCAP datasets. An incremental method is utilized to improve classification accuracy. The proposed models do not convert acoustic data into a visual representation but operate directly on it. According to testing results, the best-performing model obtains favorable performance for EMO-DB [3]. In order to achieve speech emotion recognition (SER), it is important to consider feature representation. The researchers propose a supervised NMF model called DSNMF that integrates discriminative and similarity information to enhance NMF's recognition capability. The model divides the data into representations with smaller dimensions that are simpler to distinguish [4]. In order to develop an Automatic Voice Emotion Recognition (SER) system that can accurately identify different emotional classes in voice signals, we must first develop an automated voice emotion recognition system. The researchers propose Adaptive Time–Frequency Features, a novel method for feature extraction that combines Cepstral features and Fractional Fourier Transform. The three databases utilized to assess the system are EMO-DB, SAVEE, and PDREC [5].

2 Literature Survey

Using a merged convolutional neural network (CNN) that extracted deep features from different data sources to recognize the emotion of speech, Zhao et al. developed a speech emotion recognition system [6]. There is a study by Banu et al. which uses Random Forest, SVM, and Naive Bayes algorithms in order to identify negative and positive tweets on Twitter. These algorithms had $F1$-scores of 0.224, 0.410, and 0.702, respectively, and accuracy values of 50%, 52%, and 73% on Twitter tweets, respectively [7]. According to Palo et al., speech emotions can be divided into high and low arousal classes that can represent a range of emotions, such as anger and surprise. In order to recognize emotions using multi-layer perception, there have been a number of features that have been incorporated, which are expressed in the form of linear prediction coefficients, linear prediction cepstral coefficients, Mel frequency cepstral coefficients, and perceptual linear prediction coefficients [8]. In analyzing 2442 sentences containing emotional prosody from Emotional Prosody Speech and Transcripts, Dai et al. have identified 62 features from the sentences in the corpus, and 62 of them have been evaluated subjectively. Using a neural network classifier, it was found that utterances could be classified into different emotional states by recognizing in them different types of emotions, and that a neural network classifier was able to distinguish 90% of the utterances correctly [9]. A novel system has been proposed by New et al., for detecting emotional and stress-related speech signals. It uses both linear (FFT-based) low-frequency power coefficients (LFPC) based on short-time log frequency power and nonlinear (TEO-based) low-frequency power coefficients in the time and frequency domains simultaneously [10]. Using convolutional neural networks (CNNs), Mao et al. have proposed a model that can learn affect-salient features for SER. The model outperformed a number of well-established features for SER [11]. A study conducted by Sainath et al. compared CNNs to DNNs and GMMs on various long-vowel task speech retrieval backend tasks, comparing CNNs to features extracted from neural networks in large vocabulary speech retrieval backend tasks [12]. Using DNN-based recognizers, Yu et al. found that they were able to produce better recognition accuracy than systems based on GMMs or shallow networks without explicitly adapting the models or normalizing features [13].

3 Methodology

MFCC was successfully employed in the research endeavor as an identifier for categorizing voice recordings into various mental types using artificially unaffected networking, as shown in Fig. 1. Thanks to deep neural networks, we benefit from classifying various emotional states in a variable-length audio stream in a real-time environment. This technique achieves a good balance between the computational amount and the instantaneous processes' efficiency and reliability. It may be used with any machine that supports the Python programming language. In terms of handling music, it is very user-friendly and very quick to use. The device can handle audio files of different lengths.

The purpose of this is to ensure that the listeners will be able to understand it completely. Since the volume of the audio will vary, it will be difficult to extract the features of the audio from it. A 32-bit numeric expression will be used to represent the audio's duration. This expression has higher relevance values. The MLP classifier is given several hyperparameters. A differential function is a logistic activation.

We utilized sections because each has processes like data collection, preprocessing, and emotion detection as represented in Fig. 2.

3.1 Data Collection

A collaboration between artists and consumers has resulted in the creation of this work as a result of the RAVDESS series. The number of male performers is odd, whereas the number of female performers is even. There are several emotions that are recorded in the data collection process, such as enjoyment, independence, antagonism, dissatisfaction, unexpected fear, and peace of mind. This dataset contains all the statements in each voice-video, voice-only, and no-video form. Our primary objective is to identify emotion in communication; hence, our model was trained to

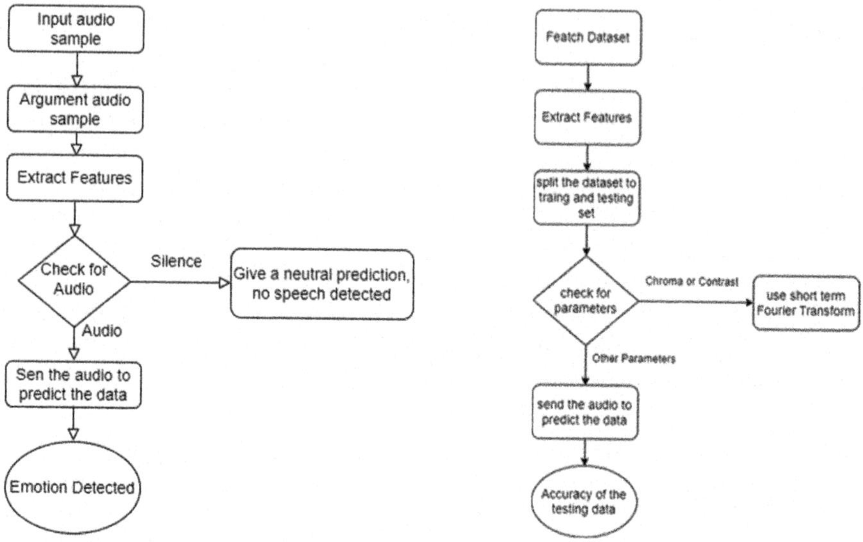

Fig. 1 Training process

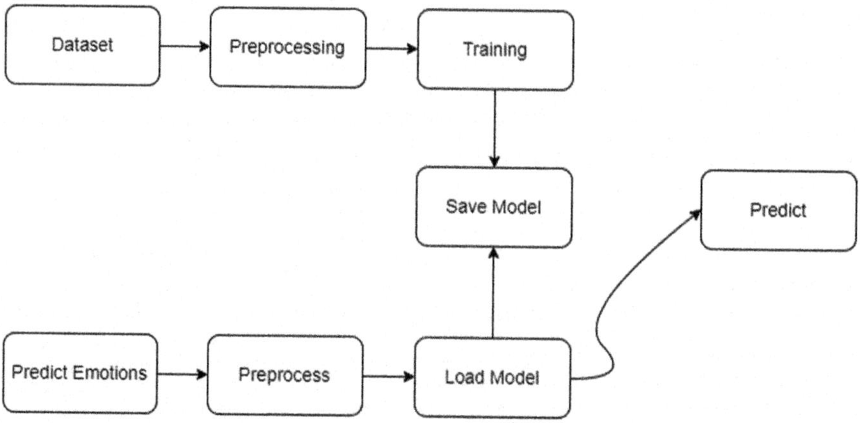

Fig. 2 Workflow diagram

employ audio-only data. All 24 actors speak two predetermined words twice to represent each of these eight sensations. All sentiments are conveyed at two distinct levels of focus. She accepts a factor emotion, which can only be produced at regular intensity: common and potent. The data group is given a name based on the hexadecimal encoded data. Each file is given a unique filename. The filename's third numerical digit designates an identifier for the corresponding feelings. The seven-part numerical identifier that makes up the document's name is composed. The words used for the emotions are neutral, happy, and sad.

3.2 Preprocessing

In most cases, unspoken emotions are revealed through the quality and tone of the voice. As a result of this process, the relevant characteristics of conversational patterns, such as feelings, are extracted. Based on the indications of the informative discussion, there are five features that can be derived from the discussion. The five main characteristics of the MFCC are the contrast, Mel's Spectrograph, the frequency color, and the tonettes.

3.3 Emotion Detection

Unspoken emotions are typically revealed through voice quality and accent. The process extracts pertinent characteristics from conversational structures, such as feelings. Five aspects are obtained from the signals of the informative conversation [14]. Melanic's Spectrograph, MFCC, and competing for the five primary qualities are Tonettes, number, and color. Use the supplied input to forecast emotion using an MLP class. We get outcomes using the five extracted attributes. We give the framework five different properties. Since only one particular factor is necessary to generate a reliable forecast, using the differences between passing each quality alone and collectively is substantial. Raves knowledge set, split into the testing and lesson subsets at a proportion of 75:25, is provided to the Multi-layer Learner to train the system. We explore these eight behavioral subcategories. The trainee uses timing series-based statistics throughout our circumstance and profit from it.

4 Results

The training process was initiated by executing command prompts and running specific commands in the command-line interface (CLI) to obtain the results. The model was trained using the established methodology outlined in the research paper. Subsequently, a web page was created to present the output to the users as shown in Figs. 3 and 4. The web page was linked to the trained data by configuring the Flask framework and setting the appropriate paths as represented in Fig. 5. Upon successful configuration, the web page was accessible through a generated IP address. Users could access the web page using a browser and select a specific WAV file to retrieve the corresponding output.

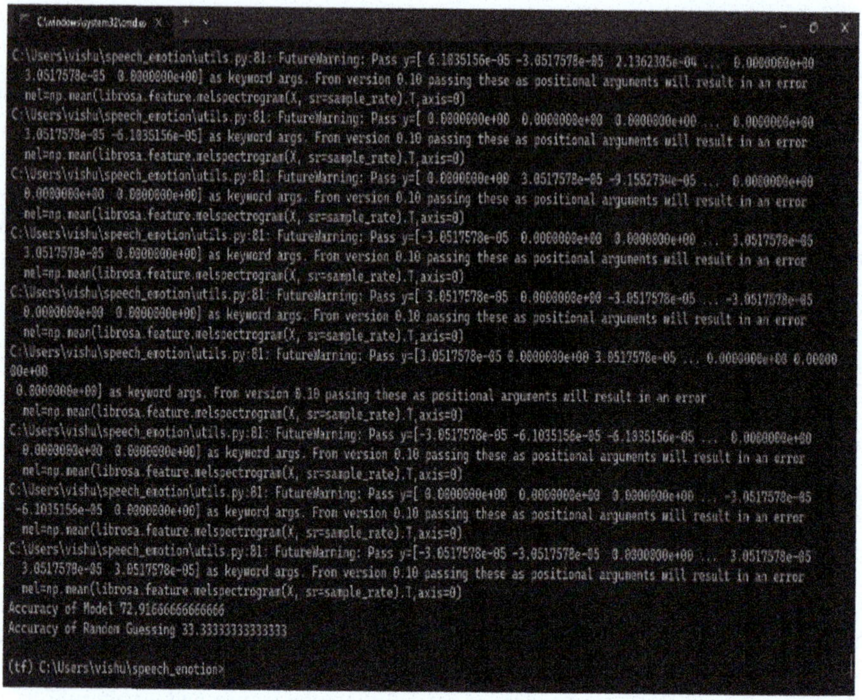

Fig. 3 Dataset screen

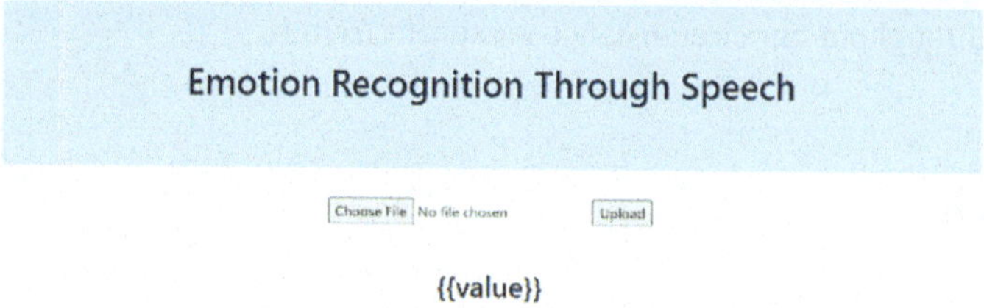

Fig. 4 Home screen

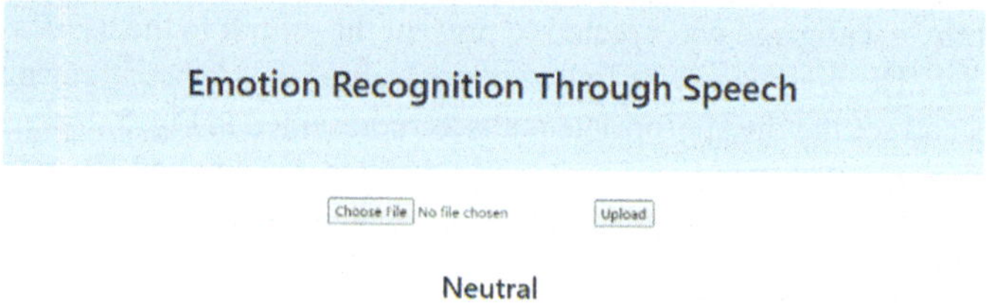

Fig. 5 Emotion detected

5 Conclusion

It demonstrates the success that multi-level partnerships are in categorizing voice signals. Even with reduced models, it is simple to recognize a small number of letters. Compared to previous methodologies, we are achieving greater accuracy about specific sentiments. The effectiveness of the preparation has a significant impact on a module's speed. The reliability of Mel Frequency Cestrum parameters is excellent. Each mood had been carefully examined and dissected, and its correctness is verified. This study is one of the first on the market to demonstrate that speech recognition can be handled by MLPs, regardless of which type of recognition problem it is referring to, and that it can be used in the detection of all emotions expressed in language.

References

1. Aouani H, Ayed YB (2020) Speech emotion recognition with deep learning. Procedia Comput Sci 176:251–260. https://doi.org/10.1016/j.procs.2020.08.027
2. Zhu Z, Dai W, Hu Y, Li J (2020) Speech emotion recognition model based on Bi-GRU and focal loss. Pattern Recogn Lett 140:358–365. https://doi.org/10.1016/j.patrec.2020.11.009
3. Issa D, Fatih Demirci M, Yazici A (2020) Speech emotion recognition with deep convolutional neural networks. Biomed Signal Process Control 59:101894. https://doi.org/10.1016/j.bspc.2020.101894
4. Hou M, Li J, Lu G (2020) A supervised non-negative matrix factorization model for speech emotion recognition. Speech Commun 124:13–20. https://doi.org/10.1016/j.specom.2020.08.002
5. Langari S, Marvi H, Zahedi M (2020) Efficient speech emotion recognition using modified feature extraction. Inform Med Unlock 20:100424. https://doi.org/10.1016/j.imu.2020.100424
6. Zhao J, Mao X, Chen L (2018) Learning deep features to recognise speech emotion using merged deep CNN. IET Signal Process. https://doi.org/10.1049/iet-spr.2017.0320
7. Banu R, Ahammed FGA, Divya G, Reddy VD, Bhaskar N, Kanthi M (2023) Sentiment analysis for real-time micro blogs using twitter data. In: 2023 2nd international conference for innovation in technology (INOCON), Bangalore, India, pp 1–6. https://doi.org/10.1109/INOCON57975.2023.10101366
8. Palo HK, Mohanty MN, Chandra M (2015) Use of different features for emotion recognition using MLP network. In: Computational vision and robotics, advances in intelligent systems and computing. Springer, India. https://doi.org/10.1007/978-81-322-2196-8_2
9. Dai K, Fell HJ, MacAuslan J (2013) Recognizing emotion in speech using neural networks In: IEEE conference on "neural networks and emotion recognition". https://doi.org/10.1109/ICONIP.1999.845644
10. New TL, Foo SW, De Silva LC (2003) Detection of stress and emotion in speech using traditional and FFT based log energy features. IEEE. 0-7803-8185-8/03 2003. https://doi.org/10.1109/ICICS.2003.1292741
11. Mao Q, Dong M, Huang Z, Zhan Y (2014) Learning salient features for speech emotion recognition using convolutional neural networks. IEEE Trans Multimedia 16:2203–2213. https://doi.org/10.1109/TMM.2014.2360798

12. Sainath TN, Mohamed A-R, Kingsbury B, Ramabhadran B (2013) Deep convolutional neural networks for LVCSR. In: 2013 IEEE international conference on acoustics, speech and signal processing, pp 8614–8618. https://doi.org/10.1109/ICASSP.2013.6639347
13. Yu D, Seltzer ML, Li J, Huang J-T, Seide F (2013) Feature learning in deep neural networks-studies on speech recognition tasks. arXiv preprint arXiv:1301.3605. https://doi.org/10.48550/arXiv.1301.3605
14. Tejaashwini Goud J, Bhaskar N, Kumar VN, Mubeen S, Narasimharao J, Unnisa R, Bhateja V, Yang X-S, Ferreira MC, Sengar SS, Travieso-Gonzalez CM (2023) Evolution in computational intelligence proceedings of the 11th international conference on frontiers of intelligent computing: theory and applications (FICTA 2023) a survey: classifying and predicting features based on facial analysis springer nature singapore singapore 297–307

Author Index

V. K. Gunjan and J. M. Zurada (eds.), *Proceedings of 4th International Conference on Recent Trends in Machine Learning, IoT, Smart Cities and Applications*, Lecture Notes in Networks and Systems 873, https://doi.org/10.1007/978-981-99-9442-7

SPRINGER NATURE

GPSR Compliance

The European Union's (EU) General Product Safety Regulation (GPSR) is a set of rules that requires consumer products to be safe and our obligations to ensure this.

If you have any concerns about our products, you can contact us on ProductSafety@springernature.com

In case Publisher is established outside the EU, the EU authorized representative is:

Springer Nature Customer Service Center GmbH
Europaplatz 3
69115 Heidelberg, Germany

The manufacturer's authorised representative in the EU is Springer Nature Customer Service Centre GmbH, Europaplatz 3, 69115 Heidelberg, Germany. If you have any concerns regarding our products, please contact ProductSafety@springernature.com

Printed and bound by CPI Group (UK) Ltd, Croydon, CR0 4YY
01/10/2026
02225901-0015